SINGLE-VARIABLE
Calculus

Donald Trim
University of Manitoba

Prentice-Hall Canada Inc.
Scarborough, Ontario

Canadian Cataloguing in Publication Data
Trim, Donald W.
Single-variable calculus

ISBN 0-13-118142-4

1. Calculus.   I. Title.
QA303.T75 1992   515   C91-094762-7

© 1992 Prentice-Hall Canada Inc., Scarborough, Ontario
ALL RIGHTS RESERVED

No part of this book may be reproduced in any form without permission in writing from the publisher.

Prentice Hall, Inc., Englewood Cliffs, New Jersey
Prentice-Hall International, Inc., London
Prentice-Hall of Australia, Pty., Ltd., Sydney
Prentice-Hall of India Pvt., Ltd., New Delhi
Prentice-Hall of Japan, Inc., Tokyo
Prentice-Hall of Southeast Asia (Pte.) Ltd., Singapore
Editora Prentice-Hall do Brasil Ltda., Rio de Janeiro
Prentice-Hall Hispanoamericana, S.A., Mexico

ISBN 0-13-118142-4

Acquisitions Editor: Jacqueline Wood
Developmental Editor: David Jolliffe
Copy Editor: Cecilia Chan
Production Editor: Amy Lui-Ma
Production Coordinator: Florence Rousseau
Interior and Cover Design: Full Spectrum Art
Cover Image Credit: Mark Tomalty/Masterfile
Page Layout: Pronk&Associates

1 2 3 4 5 AGC 96 95 94 93 92

Printed and bound in the U.S.A. by Arcata Graphics

Contents

Preface   vii

Chapter 1 ■ Plane Analytic Geometry and Functions   1
Section 1.1 Real Numbers and Coordinates   1
Section 1.2 Rectangular Coordinates in the Plane   8
Section 1.3 Equations and Curves   13
Section 1.4 Conic Sections   23
Section 1.5 Functions and Their Graphs   36
Section 1.6 Further Graphing Techniques   46
Summary   54
Review Exercises   55

Chapter 2 ■ Limits and Continuity   57
Section 2.1 Limits   57
Section 2.2 Infinite Limits   65
Section 2.3 Limits at Infinity   68
Section 2.4 Continuity   73
Section 2.5 A Mathematical Definition of Limits   82
Summary   87
Review Exercises   88

Chapter 3 ■ Differentiation   89
Section 3.1 The Derivative   89
Section 3.2 Rules for Differentiation   97
Section 3.3 Differentiability and Continuity   103
Section 3.4 Product and Quotient Rules   106
Section 3.5 Higher-Order Derivatives   113
Section 3.6 The Chain Rule and the Extended Power Rule   116
Section 3.7 Implicit Differentiation   122
Section 3.8 Derivatives of the Trigonometric Functions   130
Section 3.9 Derivatives of the Exponential and Logarithm Functions   136
Section 3.10 Logarithmic Differentiation   142
Section 3.11 Rolle's Theorem and the Mean Value Theorems   145
Summary   149
Review Exercises   150

Chapter 4 ■ Applications of Differentiation   153
Section 4.1 Newton's Iterative Procedure for Solving Equations   153
Section 4.2 Increasing and Decreasing Functions   160
Section 4.3 Relative Maxima and Minima   164
Section 4.4 Concavity and Points of Inflection   171
Section 4.5 Curve Sketching   178
Section 4.6 Absolute Maxima and Minima   184

Section 4.7 Velocity and Acceleration 198
Section 4.8 Related Rates 203
Section 4.9 Indeterminate Forms and L'Hôpital's Rule 211
Section 4.10 Differentials 220
Section 4.11 Problem Solving 225
Summary 230
Review Exercises 231

Chapter 5 ■ The Indefinite Integral or Antiderivative 233
Section 5.1 The Reverse Operation of Differentiation 233
Section 5.2 Velocity and Acceleration Revisited 241
Section 5.3 Change of Variable in the Indefinite Integral 248
Section 5.4 An Introduction to Differential Equations 251
Summary 259
Review Exercises 260

Chapter 6 ■ The Definite Integral 261
Section 6.1 Sigma Notation 261
Section 6.2 The Need for the Definite Integral 267
Section 6.3 The Definite Integral 271
Section 6.4 The Fundamental Theorem of Integral Calculus 276
Section 6.5 An Alternative Version of the Fundamental Theorem 280
Section 6.6 Average Values 284
Section 6.7 Change of Variable in the Definite Integral 288
Summary 291
Review Exercises 292

Chapter 7 ■ Applications of the Definite Integral 293
Section 7.1 Area 293
Section 7.2 Volumes of Solids of Revolution 302
Section 7.3 Lengths of Curves 314
Section 7.4 Work 319
Section 7.5 Fluid Pressure 326
Section 7.6 Centres of Mass and Centroids 331
Section 7.7 Moments of Inertia 343
Section 7.8 Additional Applications 348
Section 7.9 Improper Integrals 356
Summary 365
Review Exercises 367

Chapter 8 ■ Further Transcendental Functions and Their Derivatives 369
Section 8.1 Inverse Functions 369
Section 8.2 The Inverse Trigonometric Functions 374
Section 8.3 Derivatives of the Inverse Trigonometric Functions 385
Section 8.4 Hyperbolic Functions and Their Derivatives 390
Summary 397
Review Exercises 398

Contents

Preface   vii

Chapter 1 ■ Plane Analytic Geometry and Functions   1
Section 1.1 Real Numbers and Coordinates   1
Section 1.2 Rectangular Coordinates in the Plane   8
Section 1.3 Equations and Curves   13
Section 1.4 Conic Sections   23
Section 1.5 Functions and Their Graphs   36
Section 1.6 Further Graphing Techniques   46
Summary   54
Review Exercises   55

Chapter 2 ■ Limits and Continuity   57
Section 2.1 Limits   57
Section 2.2 Infinite Limits   65
Section 2.3 Limits at Infinity   68
Section 2.4 Continuity   73
Section 2.5 A Mathematical Definition of Limits   82
Summary   87
Review Exercises   88

Chapter 3 ■ Differentiation   89
Section 3.1 The Derivative   89
Section 3.2 Rules for Differentiation   97
Section 3.3 Differentiability and Continuity   103
Section 3.4 Product and Quotient Rules   106
Section 3.5 Higher-Order Derivatives   113
Section 3.6 The Chain Rule and the Extended Power Rule   116
Section 3.7 Implicit Differentiation   122
Section 3.8 Derivatives of the Trigonometric Functions   130
Section 3.9 Derivatives of the Exponential and Logarithm Functions   136
Section 3.10 Logarithmic Differentiation   142
Section 3.11 Rolle's Theorem and the Mean Value Theorems   145
Summary   149
Review Exercises   150

Chapter 4 ■ Applications of Differentiation   153
Section 4.1 Newton's Iterative Procedure for Solving Equations   153
Section 4.2 Increasing and Decreasing Functions   160
Section 4.3 Relative Maxima and Minima   164
Section 4.4 Concavity and Points of Inflection   171
Section 4.5 Curve Sketching   178
Section 4.6 Absolute Maxima and Minima   184

Section 4.7   Velocity and Acceleration   198
Section 4.8   Related Rates   203
Section 4.9   Indeterminate Forms and L'Hôpital's Rule   211
Section 4.10 Differentials   220
Section 4.11 Problem Solving   225
Summary   230
Review Exercises   231

Chapter 5 ■ The Indefinite Integral or Antiderivative   233
Section 5.1   The Reverse Operation of Differentiation   233
Section 5.2   Velocity and Acceleration Revisited   241
Section 5.3   Change of Variable in the Indefinite Integral   248
Section 5.4   An Introduction to Differential Equations   251
Summary   259
Review Exercises   260

Chapter 6 ■ The Definite Integral   261
Section 6.1   Sigma Notation   261
Section 6.2   The Need for the Definite Integral   267
Section 6.3   The Definite Integral   271
Section 6.4   The Fundamental Theorem of Integral Calculus   276
Section 6.5   An Alternative Version of the Fundamental Theorem   280
Section 6.6   Average Values   284
Section 6.7   Change of Variable in the Definite Integral   288
Summary   291
Review Exercises   292

Chapter 7 ■ Applications of the Definite Integral   293
Section 7.1   Area   293
Section 7.2   Volumes of Solids of Revolution   302
Section 7.3   Lengths of Curves   314
Section 7.4   Work   319
Section 7.5   Fluid Pressure   326
Section 7.6   Centres of Mass and Centroids   331
Section 7.7   Moments of Inertia   343
Section 7.8   Additional Applications   348
Section 7.9   Improper Integrals   356
Summary   365
Review Exercises   367

Chapter 8 ■ Further Transcendental Functions and Their Derivatives   369
Section 8.1   Inverse Functions   369
Section 8.2   The Inverse Trigonometric Functions   374
Section 8.3   Derivatives of the Inverse Trigonometric Functions   385
Section 8.4   Hyperbolic Functions and Their Derivatives   390
Summary   397
Review Exercises   398

Chapter 9 ■ Techniques for Integration   399
Section 9.1. Integration Formulas and Substitutions   399
Section 9.2 Exponential Growth and Decay   403
Section 9.3 Integration by Parts   411
Section 9.4 Trigonometric Integrals   415
Section 9.5 Trigonometric Substitutions   422
Section 9.6 Completing the Square and Trigonometric Substitutions   429
Section 9.7 Partial Fractions   434
Section 9.8 Using Tables and Reduction Formulas   442
Section 9.9 Numerical Integration   448
Summary   460
Review Exercises   460

Chapter 10 ■ Parametric Equations and Polar Coordinates   463
Section 10.1 Parametric Equations   463
Section 10.2 Polar Coordinates   472
Section 10.3 Curves in Polar Coordinates   475
Section 10.4 Areas in Polar Coordinates   484
Section 10.5 Conic Sections   488
Section 10.6 Conic Sections in Polar Coordinates   499
Summary   505
Review Exercises   506

Chapter 11 ■ Infinite Sequences and Series   507
Section 11.1 Infinite Sequences of Constants   508
Section 11.2 Applications of Sequences   518
Section 11.3 Infinite Series of Constants   524
Section 11.4 Nonnegative Series   531
Section 11.5 Further Convergence Tests for Nonnegative Series   537
Section 11.6 Series With Positive and Negative Terms   542
Section 11.7 Approximating the Sum of a Series   547
Section 11.8 Power Series and Intervals of Convergence   551
Section 11.9 Taylor's Remainder Formula and Taylor Series   560
Section 11.10 Taylor Series Expansions of Functions   569
Section 11.11 Sums of Power Series and Series of Constants   580
Section 11.12 Applications of Taylor Series and Taylor's Remainder Formula   583
Summary   591
Review Exercises   593

Appendix A ■ Trigonometry Review   595

Appendix B ■ The Exponential and Logarithm Functions   605

Appendix C ■ Mathematical Induction   615

Appendix D ■ Answers to Even-numbered Exercises   619

Appendix E ■ Additional Exercises   651

Index   671

Donald W. Trim received his honours degree in mathematics and physics in 1965. Masters and doctorate degrees followed with a dissertation in general relativity. As a graduate student at the University of Waterloo, he discovered that teaching was his life's ambition. He quickly became known for his teaching, and in 1971 he was invited to become the first member of the Department of Applied Mathematics in the Faculty of Science at the University of Manitoba. In 1976, after only five years at the university, he received one of the university's highest awards for teaching excellence, based on submissions by faculty and graduating students. At the University of Manitoba no individual can receive the other award within ten years of receiving the first. In 1988, Professor Trim received the second award. In addition, the Faculty of Engineering presented him with a gold replica of the engineer's iron ring in appreciation for his service.

Professor Trim's skills in teaching are reflected in the notes and books that he has written. These include an Introduction to Applied Mathematics, an Introduction to the Theory and Applications of Complex Functions, and Applied Partial Differential Equations, and this calculus text. All have received most favourable reviews from students.

Professor Trim is a member of the Canadian Applied Mathematics Society and the Mathematical Association of America. His present interests are in partial differential equations and mathematical education.

PREFACE

To the Instructor

Approach

This calculus book evolved from a set of notes developed by the author over a number of years of teaching first- and second-year calculus, and an earlier version of the text. For the most part our approach is intuitive, making free use of geometry and familiar physical settings to motivate and illustrate concepts. For instance, limits are introduced intuitively, but an optional section is included to introduce more advanced students to the mathematical definition of a limit. Derivatives are motivated and defined algebraically as instantaneous rates of change, and interpreted geometrically as slopes of tangent lines. The definite integral is motivated through two geometric and two physical problems, but its definition is algebraic.

Order of Topics

The order of topics in the book is fairly standard with a few exceptions:

1. Conic sections are introduced in Chapter 1 as illustrations of how the form of the equation of a curve dictates its shape and conversely how the shape of a curve influences its equation. Detailed discussions of properties of conic sections are delayed until Chapter 10.

2. Limits at infinity and infinite limits are presented in Chapter 2 along with general discussions on limits. Students have little difficulty mastering the three situations (at least on an intuitive level), and completion of the topic allows for an early introduction to the use of limits in curve sketching.

3. It is assumed that students are familiar with trigonometric, exponential, and logarithm functions, but for those students whose background is suspect, Appendices A and B review this material. Derivatives of these functions are developed in Chapter 3 making them available for the applications of differentiation in Chapter 4.

4. In-depth discussions on inverse trigonometric functions and hyperbolic functions, and their derivatives are delayed until Chapter 8. To discuss these functions in Chapter 3 would certainly bring the topic of differentiation to an early conclusion, but this argument is overshadowed by what we call the "one-at-a-time" principle. *Do not do too many new things at one time; two or more new ideas presented simultaneously leads to confusion.* By restricting the domain of functions that can be differentiated to relatively simple ones in Chapter 3, students can concentrate on differentiation itself. With difficult functions, such as inverse trigonometric and hyperbolic, differentiation principles could be lost in the complexity of the functions.

5. Newton's iterative procedure for solving equations is the first application of differentiation. Numbers in subsequent applications need not then be contrived in order to lead to equations with elementary solutions. We use the intermediate value theorem to discuss accuracy of approximations from two viewpoints — number of decimal places and maximum possible error, and demand that the accuracy of every approximation be clearly identified.

6. Velocity and acceleration are treated as applications of differentiation in Chapter 4, and from an integration point of view in Chapter 5. We do not encourage the use of standard physics formulas; instead we expect students to state time conventions, choose coordinate systems, and develop whatever equations are necessary to solve a problem.

7. Differential equations are introduced early in the book in order that students become familiar with the ideas surrounding this important branch of calculus. Only the simplest of equations are solved, however; we use them more as an illustration of mathematical modelling.

8. We place applications of the definite integral before techniques of integration. In this way the student sees the diversity of applications of integration and can appreciate why it is necessary to antidifferentiate more difficult functions. At the same time, we can review the applications in the examples and exercises of the techniques chapter.

Pedagogy

We have, as does every instructor, certain views on the best pedagogical approach to calculus. Some of these are iterated below:

1. "Curve sketching" is developed one step at a time as calculus unfolds. We begin with basic ideas of symmetries, translations, and addition and multiplication of ordinates in Chapter 1. Limits in Chapter 2 lead to discontinuities, and vertical and horizontal asymptotes. The geometric interpretation of the derivative as the slope of the tangent line in Chapter 3 hints at its value as a sketching tool. With critical points, relative extrema, and points of inflection, we bring curve sketching techniques together for final discussion in Section 4.5. It is not the purpose of these techniques to develop an extremely accurate graph of a function; a computer can do this more easily. We use calculus to obtain as detailed a graph of a function as is required for a particular application. We see how the graph becomes more and more refined as additional techniques are brought into play, and we discover how the techniques complement one another. In addition, we see how calculus can verify what a calculator or computer may suggest — the existence of vertical, horizontal, or slanted asymptotes, or vertical or horizontal points of inflection.

2. Relative extrema and absolute extrema are discussed in separate sections, the former in Section 4.3 and the latter in Section 4.6. Most applied extrema problems involve absolute extrema — curve sketching uses relative extrema. They are different and they are used differently. Let the student see them "one-at-a-time."

3. We have included a section with hints on how to manage word problems. We stress how to think about a problem, how to key on certain words, and how a series of questions and answers leads from problem to solution.

4. Antiderivatives or indefinite integrals are treated in a separate chapter. Relegating them to a section in the chapter on differentiation does not do them justice. Conceptually integration is simple — backwards differentiation — but this operation is so important, and it is so much more difficult to perform, only a separate chapter will bring students to this realization. In Chapter 5 we introduce the three most basic integration techniques — straightforward recognition, "guess-and-fix-up", and simple

substitutions. Other techniques (integration by parts, trigonometric substitutions, partial fractions, etc.) are discussed in Chapter 9.

5. We treat antidifferentiation from an organizational viewpoint, rather than from formulas. The student is taught to organize an integrand into a form in which integration is obvious; he is encouraged not to memorize formulas unnecessarily.

6. Definite integrals are motivated with four problems — area, volume, work, and fluid flow — but they are defined algebraically as limits of Riemann sums. This permits us to use the definition of a definite integral in a wide variety of applications without fear that it has been associated with any one in particular. Definitions of double, triple, line, and surface integrals in multivariable calculus as limit-summations follow quite naturally.

7. Antiderivatives are used to evaluate definite integrals; this is the fundamental theorem of integral calculus. The fact that a definite integral with a variable upper limit can be regarded as an antiderivative is given secondary importance. Students do not, at the earliest stage, appreciate or understand why we would want to make the upper limit of a definite integral a variable quantity. Do not detract from the importance of the antiderivative as a calculational tool for definite integrals (one-thing- at-a-time).

8. Applications of definite integrals (area, volume, lengths of curves, etc.) are approached from a "representative element" viewpoint. Instead of providing formulas for such problems, we stress the need for understanding both the application and the definite integral. We establish representative elements to approximate the required quantity and use definite integrals to add over all elements. Students will then be able to introduce definite integrals into other applications, and the transition to double, triple, line, and surface integrals will be reasonably smooth.

9. We choose to keep the polar coordinate r nonnegative. Nothing is lost by this convention; curves may have slightly different equations depending on which convention is used, but by choosing $r > 0$, considerable simplification is achieved when equations of curves in polar coordinates are rewritten in Cartesian coordinates.

10. The limit ratio test for convergence of infinite series is applied only to series with positive terms. It is not used directly as a test for convergence of series with positive and negative terms. Many students have difficulty with the concept of convergence and divergence of series. Let us take it one step at a time. First discuss convergence of series of positive terms using the comparison, limit comparison, integral, limit ratio, and limit root tests. Now introduce series with positive and negative terms, and use the above tests for absolute convergence.

11. The basis for power series expansions is Taylor's remainder formula; when remainders approach zero, the Taylor series converges to the function. This is discussed first. It then becomes a matter of illustrating that various techniques exist to find power series (such as integrating and differentiating known series), and these techniques conveniently avoid remainders.

12. Appendix E contains 370 review problems for the entire book. Students should use them in preparation for their final examination. We have provided answers to only the first 150 even-numbered problems. The last 70 or so become more and more challenging. We usually throw some of them out to the class, one at a time, for a bonus of 3, 4, or 5 marks. Perhaps the first correct solution, or the best solution, earns the bonus. This often sparks the interest of the more advanced student.

Supplements

Two supplements have been prepared to aid instructors and students. Available to students is a manual containing detailed solutions to all even-number exercises. For instructors there is a manual with answers to odd-numbered exercises. Any errors in the text or supplements are the responsibility of the author, and we would appreciate having them brought to our attention.

Acknowledgements

I wish to thank the following individuals who reviewed the manuscript and offered many helpful suggestions:

    R. Biggs, The University of Western Ontario
    K. Dunn, Dalhousie University
    P. Ponzo, University of Waterloo
    R. Ross, University of Toronto
    C. Roth, McGill University
    G.P. Wright, University of Ottawa

Thanks also to D. Greenwood and K. Khan, University of Guelph, who proofed the calculations.

I also wish to express my appreciation to the staff of Prentice-Hall Canada Inc. for a superb effort in editing, design, and illustration. Special thanks to J. Wood, D. Jolliffe, C. Chan, and A. Lui-Ma for their suggestions and encouragement.

To the Student

This book was written for you, the student of calculus. It begins with an introduction to the fundamentals of plane analytic geometry in order to give you a way to visualize problems geometrically. You will be surprised at the number and variety of problems that you will be able to solve using calculus, but as it so often happens in mathematics, the most difficult part of the solution to a problem is the initial step. Once started, the solution frequently unfolds smoothly and easily, but that first step sometimes seems impossible. One of the best ways to start a problem is with a diagram. A picture, no matter how rough, is invaluable in giving you a "feeling" for what is going on. It displays the known facts surrounding the problem; it permits you to see what the problem really is, and how it relates to the known facts; and it often suggests that all-important first step. Do not underestimate the value of a diagram; get into the habit of making one at every opportunity — not just to solve problems, but to understand what calculus is all about.

Our study of calculus really begins in Chapter 2 with a discussion of limits. The idea of limits is fundamental to all of calculus, and it is surprising that the founders of calculus, Gottfried Leibniz (1646–1716) and Isaac Newton (1645–1727), had not formulated the idea of a limit. Their development of the subject was based on the concept of "infinitesimals" — quantities that are extremely close to zero, but not equal to zero. It was the work of such mathematicians as L. Euler (1707–1783), J. D'Alembert (1717–1783), and later, C. Cauchy (1789–1857) that led to the rigorous formulation of limits and a firm foundation on which to develop the rest of calculus.

All calculus originates with two basic ideas — the derivative and the definite integral. If one quantity depends on another, and the first quantity is in a state of change, so does the second. For example, the speed of a car depends on how far the accelerator is depressed — more depression, more speed. As soon as the accelerator depression is changed, so is the speed of the car, and derivatives describe how these changes are related. When a furnace is turned on, the amount of heat it produces affects room temperature — more heat, higher temperature. Derivatives relate the rate at which temperature rises to the rate at which heat is produced.

The definite integral is a completely different kind of entity. Suppose that we use a pair of scissors to cut some strange shape from a piece of paper and ask for its area. This could be very difficult since there are very few shapes for which we have an area formula. We could, however, estimate its area by drawing rectangles on the piece of paper and adding together their areas. By drawing smaller and smaller rectangles near the curved edge, we could get a better and better estimate of the true area. In essence, this is the idea of the definite integral; it is a sum of terms where the number of terms is increased indefinitely.

Surprising as it may seem, these two concepts which appear so different are intimately related, and it is this connection that makes calculus so fascinating. In fact, calculus is a subject wherein every topic depends heavily on those that went before it, and at the same times leads naturally into the next. Once we identify these interdependences, we begin to understand the inner beauty of calculus.

You will find that we place considerable emphasis on problem solving, and we do this because you will never learn calculus by going to class and/or reading the textbook. You must involve yourself in the subject, and the best way to do this is with problems. Only when you solve problems do you begin to understand. The exercises in each section begin with routine, drill-type problems designed to reinforce fundamentals. It follows, therefore, that if you have difficulties with early exercises, then you have not understood the fundamentals, and you should therefore review material in that section. In later exercises we begin to make you think. Some of the exercises are of the same type as were found earlier, but with more involved calculations; some require you to think about a concept from a different point of view, others may draw on ideas from previous sections; still others may do all of these. It is in these exercises that you find most of the applications, for only when you have mastered fundamentals can you begin to apply calculus to problems from other fields. Certain exercises are marked with a calculator icon to indicate the need for an electronic calculator. There are many other exercises for which a calculator could eliminate laborious hand calculations and use of trigonometry and logarithm tables, but exercises marked with the icon definitely require a calculator. In addition, a calculator is essential for those sections that deal with approximations and iterations (Sections 4.1, 9.9, 11.2, 11.7, and 11.12).

At the end of each chapter there is a brief summary of the results in that chapter and a set of review exercises. Do not underestimate the value of these exercises; they test your understanding of the material in the entire chapter. Section exercises develop expertise in small areas; chapter exercises test knowledge of major blocks of subject matter. In addition, Appendix E contains a set of exercises on the entire book. Use these exercises as a review when you have completed your calculus studies, or in preparation for your final examination.

Answers to even-numbered problems can be found in Appendix D. In addition, a student supplement containing detailed solutions to all even-numbered problems is available.

Throughout the book you will encounter an icon ◉ to warn you about pitfalls which frequently entrap students. These pitfalls may be inappropriate calculations or misinterpretations of ideas. Pay close attention to each one.

They key word in our approach to calculus is "think." Above all else, we hope that you will train your mind to think and to think logically. We want you to learn how to organize facts and interpret them mathematically. If there is a problem, we want you to be able to decide exactly what that problem is, and finally, to produce a step-by-step procedure by which to solve it. Do not, therefore, read this book expecting formulas that can be used to solve all problems; we will not give them to you. We will give you a few formulas and, we hope, a great deal of insight into the ideas surrounding these formulas. Then you begin to think for yourself. You use what you know to solve the first problem. In this way, your calculus studies will be a true learning experience — not an exercise in memorization and regurgitation of facts and formulas, but a developmental period during which you both learn calculus and increase your mental capacity. When you reach the end of the road and successfully complete your calculus studies, think back and compare how much you knew at the beginning to how much you know at the end. Congratulate yourself on accomplishing a great deal, and mostly through your own efforts.

Donald Trim
Winnipeg, Manitoba
February, 1992

CHAPTER 1

Plane Analytic Geometry and Functions

Analytic geometry is a union of algebra and geometry. On the one hand, it provides a way to describe geometric objects algebraically; on the other hand, it permits a geometric visualization of algebraic statements. In the first four sections of this chapter we introduce the principles of plane analytic geometry, proceeding from coordinates along a line in Section 1.1 to coordinates in a plane in Section 1.2. In Sections 1.3 and 1.4 we discuss curves and their representation by use of equations, paying particular attention to straight lines, circles, parabolas, ellipses, and hyperbolas. Students who have previously studied analytic geometry could skip these sections and begin with functions in Section 1.5, but we strongly recommend against this. Even the best-prepared student can benefit by a cursory reading of these sections. Not only is it helpful to refresh one's memory on concepts learned some time ago, it is also wise to become familiar with the terminology, notation, and conventions set forth in these sections.

SECTION 1.1

Real Numbers and Coordinates

At an early stage in our mathematical education we learned how the real number line serves as a geometric representation of the real-number system. In this section we reverse these roles and use the real-number system to *coordinatize* or identify points on a line. Essentially the two approaches are one and the same; they set up a correspondence between real numbers and points on a line. To coordinatize a line, which we call the x-axis, we begin by choosing some point of reference O on the line. This point, called the **origin**, is identified or coordinatized by the real number zero. We now start at O and mark off equal distances along the line in both directions (Figure 1.1). These marks could, for example, be one centimetre apart or one metre apart. The first point to the right of O is one unit of distance from O and is coordinatized by the real number 1; the second point is two units from O and is identified by the number 2, and so on. Unit distances to the left of O are identified by the negative integers $-1, -2, -3$, and so on. Fractional numbers such as $3/4$ and $-5/3$ identify points that are $3/4$ of our unit distance to the right of O, and $5/3$ of our unit distance to the left of O, respectively. Continuation of this process leads to a coordinatization of the entire x-axis. Every point on the axis is identified by a real number that we denote by x, and this number is the **directed distance** from the origin O to the point. For example, point Q in Figure 1.1 is two and one-half units of distance to the right of O, and therefore its x-coordinate is $5/2$. Point R is three and one-quarter units to the left of O; its x-coordinate is -3.25.

FIGURE 1.1

To describe Q we should say "the point whose x-coordinate is $5/2$," but this is somewhat cumbersome. We usually say "the point $x = 5/2$." To refer to a general point P on the x-axis, we say "the point x" meaning, of course, the point with coordinate x. In this book we use the convention that directed distances from one point to another are represented by line segments which have an arrowhead on the "to" end but not on the "from" end. In Figure 1.1 we have the directed distance from O to P with an arrowhead at P. Were we interested in the length of the line segment joining O and P, rather than the directed distance from O to P, we would place arrowheads on both ends of the line segment (Figure 1.2). We denote this length by $\|OP\|$. When P is to the right of O, $\|OP\|$ and x are equal, but when P is to left of O, $\|OP\|$ is the negative of x. Coordinates are always directed distances and should not be confused with lengths.

FIGURE 1.2

Figure 1.3 shows a second coordinatization of the x-axis using a unit of distance smaller than that in Figure 1.1. Since the unit of length can be chosen arbitrarily, as can the point of reference O, the number of ways to coordinatize a line is infinite.

FIGURE 1.3

To represent sets of points on the x-axis we use open and closed intervals as defined below.

Definition 1.1

> The set of points on the x-axis between two given points $x = a$ and $x = b$, where $a < b$, but not including these points, is called an **open interval**. It is denoted by (a, b).

The open interval (a, b) is described algebraically as all points on the x-axis whose x-coordinates satisfy the inequality $a < x < b$. For example, $1 < x < 2$ describes the open interval $(1, 2)$. We have shown this interval in Figure 1.4, where the open circles at $x = 1$ and $x = 2$ indicate that these points are not included in the set.

FIGURE 1.4

The interval (1, 2)

```
—+———+———+———○———○———+———
 -2   -1    0    1    2    3
```

Definition 1.2

> The set of points on the x-axis between two given points $x = a$ and $x = b$, where $a < b$, including these points, is called a **closed interval**. It is denoted by $[a, b]$.

Algebraically, $[a, b]$ is the set of points whose x-coordinates satisfy $a \leq x \leq b$. The closed interval $[1, 2]$ in Figure 1.5 is therefore described by $1 \leq x \leq 2$. The solid dots at $x = 1$ and $x = 2$ indicate that these points are included in the interval. A closed interval contains its end points, whereas an open interval does not.

FIGURE 1.5

Intervals need not be open or closed. For example, the set of points whose coordinates satisfy $1 < x \leq 2$ is denoted by $(1,2]$. It is a half-open and half-closed interval; it contains one end but not the other (Figure 1.6). Intervals need not be of finite length. The set of points whose coordinates satisfy $x > 2$ (or $2 < x < \infty$) has only one end, namely, $x = 2$, which is not contained in the set; the set is also denoted by $(2,\infty)$ (Figure 1.7). The notations $2 < x < \infty$ and $(2,\infty)$ have introduced the symbol ∞, called *infinity*. Infinity is *not* a number; it is a symbol that is notationally convenient in many parts of calculus. In $2 < x < \infty$, it means that there is no largest value for x; x can be chosen arbitrarily large. The interval described by $x \leq 2$ is also denoted by $(-\infty, 2]$; it contains its only end point $x = 2$ (Figure 1.8). A square bracket is never placed on the end of an interval corresponding to ∞ or $-\infty$; that is, we never write $(2,\infty]$ or $[-\infty, 2)$. A square bracket indicates that the point next to it is included in the interval. Since neither $-\infty$ or ∞ is a point, they cannot be included.

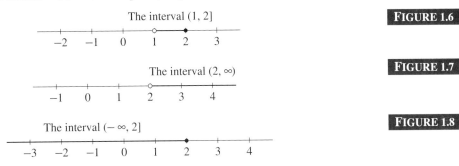

FIGURE 1.6

FIGURE 1.7

FIGURE 1.8

Intervals are used to describe more complicated sets of points on the x-axis. For example, the set of points between $x = 1$ and $x = 3$ but not including $x = 1$, $x = 2$, or $x = 3$ consists of all points in the open intervals $1 < x < 2$ and $2 < x < 3$ (Figure 1.9). The set of points satisfying the inequality $x^2 \geq 9$ consists of all points in the intervals $x \geq 3$ and $x \leq -3$ (Figure 1.10). Inequalities that are even more complicated can often be simplified to one or more of the four basic inequalities $a < x < b$, $a \leq x < b$, $a < x \leq b$, $a \leq x \leq b$ using the following rules:

FIGURE 1.9

FIGURE 1.10

If $a < b$, then

$$a + c < b + c \quad \text{for any real } c. \tag{1.1}$$

If $a < b$, then

$$ac < bc \quad \text{when } c > 0, \tag{1.2a}$$
$$ac > bc \quad \text{when } c < 0. \tag{1.2b}$$

If $0 < a < b$, then

$$\frac{1}{a} > \frac{1}{b}. \tag{1.3}$$

Each inequality $<$ in 1.1 and 1.2 may be replaced by $\leq$ provided that $>$ is at the same time replaced by $\geq$.

EXAMPLE 1.1

Simplify the following inequalities:

$$\text{(a) } 4 - 2x \leq 5 \qquad \text{(b) } 1 + 2x^2 < 5 \qquad \text{(c) } \frac{2x-1}{x+2} > 1$$

SOLUTION

(a) If we add -4 to both sides of the inequality (using 1.1), we obtain

$$-2x \leq 1.$$

Multiplication by $-1/2$ then gives the interval

$$x \geq -1/2.$$

Thus, equivalent to the inequality $4 - 2x \leq 5$ is the simpler statement $x \geq -1/2$.

(b) If we subtract 1 from both sides of the inequality, which is equivalent to adding -1,

$$2x^2 < 4,$$

and divide by 2 (this is equivalent to multiplying by $1/2$),

$$x^2 < 2.$$

Points must therefore lie in the open interval $-\sqrt{2} < x < \sqrt{2}$.

(c) For fractional inequalities it is often useful to bring all terms to one side of the inequality. To do this we add -1 to both sides,

$$0 < \frac{2x-1}{x+2} - 1 = \frac{(2x-1)-(x+2)}{x+2} = \frac{x-3}{x+2}.$$

Figure 1.11 contains what we call a "sign diagram". It is a convenient way to analyze the sign of an expression which is the product and/or quotient of a number of factors. The first line indicates that $x - 3$ is negative for $x < 3$ and positive for $x > 3$. The second line shows that $x + 2$ is negative for $x < -2$ and positive for $x > -2$. The last line combines these two. The vertical bars at $x = 3$ and $x = -2$ in the first two lines are brought down, thus dividing the x-axis into three intervals $x < -2$, $-2 < x < 3$, and $x > 3$. The sign of $(x-3)/(x+2)$ in each of these intervals is determined by counting the corresponding number of negative signs in lines one and two. It follows that $(x - 3)/(x + 2) > 0$ [or $(2x - 1)/(x + 2) > 1$] when $x < -2$ or $x > 3$.

FIGURE 1.11

In many applications of mathematics we find it necessary to discuss the "sign" of a quantity, and sign diagrams like that in Figure 1.11 are very helpful. We shall use them in many places in succeeding chapters.

When we say that the coordinate of a point P on the x-axis is x, we mean that the directed distance from the origin O to P is x. The quantity $x - a$ for any real number a has a similar geometric interpretation. It is the directed distance from the point $x = a$ to P. For instance, $x - 5$ is the directed distance from the point Q at $x = 5$ in Figure 1.12 to any point P. If P is at $x = -4$, the directed distance from Q to P is $-4 - 5 = -9$. Similarly, in Figure 1.13, $x + 2$ is the directed distance from the point Q at $x = -2$ to any point P.

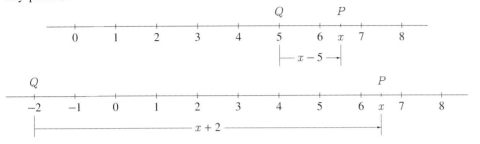

FIGURE 1.12

FIGURE 1.13

One of the most useful notations for specifying sets of points on the x-axis is provided by the next definition.

Definition 1.3

The **absolute value** of a real number x is

$$|x| = \begin{cases} x & x \geq 0 \\ -x & x < 0. \end{cases} \quad (1.4)$$

For example, $|10| = 10$, $|-9| = 9$, and $|2(4)^2 - 9(5)| = |32 - 45| = |-13| = 13$. The absolute value then is an operation that defines the size or "magnitude" of a number without regard for sign.

To see how the absolute value can be used to specify sets of points on the x-axis, consider first the set described by the inequality $|x| < 2$. This inequality states that the magnitude of x must be less than 2. It means that x must lie between -2 and 2; hence $|x| < 2$ describes all points in the open interval

$$-2 < x < 2.$$

Similarly, the inequality $|x - 5| < 2$ states that the magnitude of $x - 5$ must be less than 2, and therefore

$$-2 < x - 5 < 2.$$

Addition of 5 to each of these terms gives the interval

$$3 < x < 7.$$

Alternatively, we could have reasoned that $x - 5$ is the directed distance from 5 to x on the x-axis (Figure 1.12). The inequality $|x - 5| < 2$ requires this distance to be less in magnitude than 2; in other words, the distance from 5 to x must be less than 2. Hence x must lie between 3 and 7.

The inequality

$$|x^2 + 1| \leq 2$$

requires that $x^2 + 1$ be between -2 and 2, including ± 2;

$$-2 \leq x^2 + 1 \leq 2.$$

When 1 is subtracted from each of these terms,

$$-3 \leq x^2 \leq 1.$$

Now the left-hand inequality $-3 \leq x^2$ is really not a restriction on x at all — every real x satisfies this inequality. Consequently, x must satisfy only $x^2 \leq 1$, or $-1 \leq x \leq 1$. The original inequality $|x^2 + 1| \leq 2$ is therefore equivalent to $-1 \leq x \leq 1$ or $|x| \leq 1$.

EXAMPLE 1.2 Find all points satisfying each of the following inequalities:

(a) $|3 - 5x| \leq 2$ (b) $|4x^2 - 7| > 2$

SOLUTION (a) The absolute value may be replaced by the double inequality

$$-2 \leq 3 - 5x \leq 2,$$

and when we subtract 3, we find

$$-5 \leq -5x \leq -1.$$

Multiplication of each term in these inequalities by $-1/5$ gives

$$1 \geq x \geq \frac{1}{5} \quad \text{or} \quad \frac{1}{5} \leq x \leq 1.$$

(b) The inequality $|4x^2 - 7| > 2$ is satisfied if x satisfies either $4x^2 - 7 > 2$ or $4x^2 - 7 < -2$. From the first, $4x^2 > 9$ or $x^2 > 9/4$. This implies that either $x > 3/2$ or $x < -3/2$. From the inequality $4x^2 - 7 < -2$, we find that $x^2 < 5/4$ or $-\sqrt{5}/2 < x < \sqrt{5}/2$. Consequently, in order to satisfy $|4x^2 - 7| > 2$, x must lie in one of the three intervals $x < -3/2$, $x > 3/2$, or $-\sqrt{5}/2 < x < \sqrt{5}/2$ (Figure 1.14). These may be stated more compactly as $|x| < \sqrt{5}/2$ and $|x| > 3/2$.

FIGURE 1.14

   This example uses a convention adopted by all mathematicians. The square root symbol "$\sqrt{}$" means positive square root. For instance $\sqrt{4} = 2$; it is not equal to -2. This convention is often forgotten in situations like $\sqrt{x^2}$. This is equal to x, if x is positive. But if x is negative, it is equal to $-x$. For example, when $x = -3$, $\sqrt{x^2} = \sqrt{(-3)^2} = 3 = -x$. Thus, we can write that

$$\sqrt{x^2} = \begin{cases} x & x \geq 0 \\ -x & x < 0. \end{cases}$$

Absolute values combine these conveniently into

$$\sqrt{x^2} = |x|, \tag{1.5}$$

valid for all x.

EXAMPLE 1.3

Describe the set of points between 1 and 3, but not including 1, 2 or 3, in terms of absolute values.

SOLUTION To specify that x must be within a distance of 1 on either side of 2, we write
$$|x - 2| < 1.$$
The condition that $x \neq 2$ is now easily incorporated
$$0 < |x - 2| < 1.$$
∎

The absolute value of the product of two real numbers is always equal to the product of their absolute values:
$$|ab| = |a||b|. \tag{1.6a}$$

The absolute value of a quotient is the quotient of the absolute values:
$$\left|\frac{a}{b}\right| = \frac{|a|}{|b|}. \tag{1.6b}$$

These results are not valid for sums and differences. The absolute value of the sum of a and b is equal to the sum of their absolute values when a and b have the same sign; that is,
$$|a + b| = |a| + |b|$$
when a and b are either both positive or both negative. When a and b have opposite signs, the correct statement is
$$|a + b| < |a| + |b|.$$
In general, then, we may state that for any two real numbers a and b,
$$|a + b| \leq |a| + |b|. \tag{1.6c}$$

This inequality finds its way into many applications of mathematics. It can also be used to prove another inequality that is quite useful,
$$|a + b| \geq ||a| - |b||, \tag{1.6d}$$

(see Exercise 45).

EXERCISES 1.1

1. Coordinatize the x-axis with unit distance equal to (a) 2 cm; (b) 3/4 cm. Mark on the axis points with coordinates -3, -2.5, -2, -1.5, -1, -0.5, 0, 0.5, 1, 1.5, 2, 2.5, and 3.

In Exercises 2–9 express the interval as an inequality.

2. $(-3, 4)$
3. $[5, 8]$
4. $(-1, 6]$
5. $[-12, 4)$
6. $[3, \infty)$
7. $(-\infty, -4)$
8. $(-\infty, \infty)$
9. $(-10, \infty)$
10. Is $(3, 2)$ an acceptable interval?
11. Is $(2, \infty]$ an acceptable interval?

In Exercises 12–21 express the inequality in interval notation.

12. $3 \leq x \leq 10$
13. $-2 \leq x < 5$
14. $4 < x \leq 22$
15. $x > 1$
16. $x \leq -2$
17. $1 < x < \infty$

18. $-\infty < x \leq -2$
19. $x \geq -3$
20. $-\infty < x < \infty$
21. $-0.2 < x \leq -0.05$

In Exercises 22–39 find all points that satisfy the inequality.

22. $3x - 2 < 7$
23. $2x^2 + 12 \geq 19$
24. $3/x > 5$
25. $3/x > -5$
26. $x^2 - 4 \leq 5$
27. $15 + x^2 < 14$
28. $|3x - 2| < 7$
29. $|2 - 3x| \geq 7$
30. $|2x^2 + 5| \leq 6$
31. $|2x^2 + 6| \leq 5$
32. $|2x + 6| \leq 5$
33. $|x^2 - 5| > 4$
34. $|x^2 - 4| \geq 5$
35. $|x^2 + 5| > 4$
36. $|x^2 + 4| \geq 5$
37. $\dfrac{2x - 6}{x - 2} > 3$
38. $\dfrac{3x + 5}{4x - 2} < 3$
39. $\dfrac{9x + 5}{3x - 2} < 2$

40. Verify that the inequality

$$x^2 + 4x < 6$$

can be written in the form

$$(x + 2)^2 < 10,$$

and thereby find all points satisfying the inequality.

In Exercises 41–44 use the technique of Exercise 40 to find all points that satisfy the inequality.

41. $x^2 + 6x + 1 \leq 0$
42. $2x^2 + 4x - 5 > 0$
43. $-x^2 - 2x \geq 1$
44. $|x^2 + 2x + 5| \geq 3$

45. Use inequality 1.6c to verify 1.6d.

46. Is

$$|a - b| \geq ||a| - |b||$$

a correct variation of 1.6d?

In Exercises 47–60 find all points that satisfy the inequality.

47. $x + \dfrac{1}{x} \geq 2$
48. $\dfrac{1}{x + 2} - \dfrac{1}{x - 2} < 5$
49. $11 < x^2 + 6x + 4 < 20$
50. $|x^2 + 2x - 3| \leq 1$
51. $\dfrac{3x^2 + 5}{x^2 - 4} > 1$
52. $\dfrac{4x^2 - 64}{x^2 - 9} > 3$
53. $\left|\dfrac{3x + 4}{x - 2}\right| < 4$
54. $\dfrac{x + 5}{x - 3} < \dfrac{x + 2}{x + 1}$
55. $|x^2 + 2x + 3| \geq 5$
56. $|3x + 5| \leq 4x$
57. $|3x + 5| \leq |4x|$
58. $|x^2 + 5| + |x - 2| \leq 8$
59. $\dfrac{x - 3}{x + 1} > \dfrac{2x + 4}{x}$
60. $\dfrac{2x + 3}{x - 5} > x + 4$

In Exercises 61–64 express the condition algebraically and find all points that satisfy the condition.

61. The sum of the distances from a point x to $x = 1$ and $x = 5$ must be less than 7.

62. The sum of the distances from a point x to $x = -1$ and $x = 2$ must be greater than 4.

63. The absolute value of the difference of the distances from a point x to $x = 5$ and $x = 12$ must be less than 14.

64. The absolute value of the difference of the distances from a point x to $x = 2$ and $x = -6$ must be less than 3.

65. The **arithmetic mean** (average) of any two real numbers a and b is defined as $(a + b)/2$. The **geometric mean** of any two positive numbers a and b is $\sqrt{ab}$. Show that the arithmetic mean of any two distinct positive numbers is always greater than their geometric mean.

66. Find all points that satisfy the following inequalities:
   (a) $x^2 + 12x + 11 < 0$
   (b) $\dfrac{2x^2 + 11x + 5}{x + 2} < x - 3$
   (c) $2x + 1 < \dfrac{x^2 - x - 6}{x + 5}$
   (d) $\dfrac{2x + 1}{x + 2} < \dfrac{x - 3}{x + 5}$

67. Find all points that satisfy

$$\dfrac{|3x + 1|}{x + 5} < x.$$

SECTION 1.2

Rectangular Coordinates in the Plane

The technique of coordinatizing a line by means of real numbers can be extended to define rectangular coordinates in a plane and rectangular coordinates in space. This turns out to be indispensable to our approach to calculus. At every stage in our development of the subject we shall have pictures — pictures to introduce ideas, pictures to illustrate concepts, and pictures to reinforce principles. These pictures can only be described adequately with coordinate systems.

To coordinatize a plane, we begin by choosing any point O in the plane, called the origin, and any line through O, called the x-axis. This line, which we coordinatize as in Section 1.1 is drawn horizontally with positive direction to the right (denoted by an arrowhead in Figure 1.15). Next we draw through O a line perpendicular to the x-axis called the y-axis. This line is coordinatized with positive direction usually upward. In Figure 1.15, we coordinatized the x- and y-axes with exactly the same unit of distance, but we shall see in Section 1.3 that it is sometimes more convenient to have different scales along the two axes. Unless we indicate otherwise, unit distances along the axes are chosen equal.

FIGURE 1.15

The two coordinatizations of the axes can be used to identify every point in the plane. Through the point P in Figure 1.16 we draw lines parallel to the x- and y-axes to intersect the axes at points A and B. The x-coordinate of A is $5/2$, and the y-coordinate of B is $3/2$. We call these the x- and y-coordinates of P, and write them together in the form $(5/2, 3/2)$. The x-coordinate of P, also called the **abscissa** of P, is therefore the distance cut off on the x-axis by the line through P parallel to the y-axis (Figure 1.17). Since the x- and y-axes are perpendicular, we can state alternatively that the x-coordinate of P is the directed (perpendicular) distance from the y-axis to P. Similarly, the y-coordinate of P, also called the **ordinate** of P, is the directed distance from the x-axis to P, or the distance cut off on the y-axis by the line through P parallel to the x-axis.

FIGURE 1.16

FIGURE 1.17

By this construction, every point in a plane can be identified or coordinatized by an ordered pair of real numbers (x, y) called its x- and y-coordinates. Although the ordered pair notation for the coordinates of a point is identical to that for an open interval, context always makes it clear which interpretation is intended. Conversely, given any pair of real numbers x and y, there is one and only one point in the plane with these as x- and y-coordinates. Since coordinates (x, y) are defined in terms of a pair of perpendicular axes, we call them **rectangular** or Cartesian **coordinates**. Figure 1.18 shows four points together with their Cartesian coordinates. They are called Cartesian coordinates after the French mathematician René Descartes (1596-1650), one of the founders of analytic geometry.

The axes divide the plane into four parts which, beginning at the upper right and proceeding counterclockwise, are called respectively the first, second, third, and fourth **quadrants**. The axes themselves are not considered part of any quadrant. Points in the first quadrant have x- and y-coordinates that are both positive, points in the second quadrant have a negative x-coordinate and a positive y-coordinate, and so on (Figure 1.19).

FIGURE 1.18

FIGURE 1.19

If P and Q are any two points with coordinates (x_1, y_1) and (x_2, y_2) respectively (Figure 1.20), then by the Pythagorean relation for the right-angled triangle PQS,

$$\|PQ\|^2 = \|PS\|^2 + \|QS\|^2,$$

where $\|PQ\|$ denotes the length of the line segment joining P and Q, and similarly for $\|PS\|$ and $\|QS\|$. But

$$\|PS\| = |x_2 - x_1| \quad \text{and} \quad \|QS\| = |y_2 - y_1|,$$

and therefore

$$\|PQ\|^2 = |x_2 - x_1|^2 + |y_2 - y_1|^2 \quad \text{or}$$

$$\|PQ\| = \sqrt{(x_2 - x_1)^2 + (y_2 - y_1)^2}. \tag{1.7}$$

This very important result expresses the length of a line segment joining two points P and Q in terms of their coordinates (x_1, y_1) and (x_2, y_2). For example, if the coordinates of P and Q are $(-3, 4)$ and $(2, -1)$, then

$$\|PQ\| = \sqrt{(2+3)^2 + (-1-4)^2} = 5\sqrt{2}$$

(Figure 1.21).

FIGURE 1.20

FIGURE 1.21

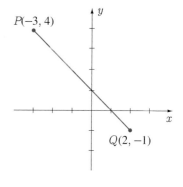

EXAMPLE 1.4

Points with coordinates $(0,0)$, $(7,0)$, and $(3,5)$ form three vertices of a parallelogram. What are possible coordinates for the fourth vertex?

SOLUTION Figures 1.22(a,b,c) show that there are three possibilities. The coordinates of R in Figures (a) and (b) should be fairly clear. Since line QR in (a) is horizontal and of length 7, the coordinates of R are $(10,5)$. Similarly, in Figure (b), R is the point $(-4,5)$. To find the coordinates (x,y) of R in Figure (c), we express the facts that $||OR|| = ||QP||$ and $||OQ|| = ||PR||$ algebraically,

$$\sqrt{(x-0)^2 + (y-0)^2} = \sqrt{(7-3)^2 + (0-5)^2},$$
$$\sqrt{(3-0)^2 + (5-0)^2} = \sqrt{(x-7)^2 + (y-0)^2}.$$

When these equations are squared, and simplified, the result is

$$x^2 + y^2 = 41,$$
$$x^2 - 14x + y^2 = -15.$$

If we subtract the second of these from the first, we obtain $14x = 56$, or $x = 4$. Substitution of this into either of the above equations gives $y = \pm 5$. But only $x = 4$ and $y = -5$ yield a parallelogram. The coordinates of R are therefore $(4, -5)$.

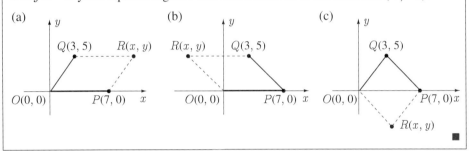

FIGURE 1.22

Suppose point $R(x,y)$ in Figure 1.23 is the midpoint of the line segment joining $P(x_1, y_1)$ and $Q(x_2, y_2)$.

FIGURE 1.23

Since triangles PTR and RUQ are congruent, it follows that $||RU|| = ||PT||$, and for P and Q as shown in Figure 1.23, this condition becomes $x_2 - x = x - x_1$. Consequently,

$$x = \frac{x_1 + x_2}{2}. \quad (1.8\text{a})$$

Similarly, equality of $||RT||$ and $||QU||$ gives

$$y = \frac{y_1 + y_2}{2}. \quad (1.8\text{b})$$

Derivation of these results assumed that $x_2 > x_1$ and $y_2 > y_1$, but the same formulas are valid for any x_1, x_2, y_1, and y_2 whatsoever. Coordinates of the midpoint of a line segment are therefore averages of coordinates of the ends of the line segment.

EXERCISES 1.2

1. Show a plane coordinatized with Cartesian coordinates in each of the following ways:
   (a) equal unit distances of 1 cm along the x- and y-axes;
   (b) unit distances along the x- and y-axes of $1/2$ and 1 cm respectively;
   (c) unit distances along the x- and y-axes of 1 cm and 1 mm respectively.

2. On a plane with rectangular coordinates show the points $(2,5)$, $(-3,6)$, $(7,-2)$, $(6,-3)$, $(-4,1)$, $(1,3)$, $(-1/3,1)$, $(2,3/2)$, $(-1/2,-3/4)$.

In Exercises 3–6 find the distance between the points.

3. $(1,3)$, $(3,4)$ 4. $(-2,1)$, $(4,-2)$
5. $(-1,-2)$, $(-3,-8)$ 6. $(3,2)$, $(-4,-1)$

7. –10. Find the midpoint of the line segment joining the points in Exercises 3–6.

11. A square has sides of length 4 units. Find the coordinates of its vertices if:
    (a) one vertex is at the origin, two of its sides lie along the coordinate axes, and one vertex lies in the third quadrant;
    (b) its diagonals lie along the axes.

12. Show all points in a plane with Cartesian coordinates x and y that have each of the following:
    (a) an x-coordinate equal to 4;
    (b) a y-coordinate equal to -1;
    (c) equal x- and y-coordinates;
    (d) a y-coordinate equal to the negative of the x-coordinate.

13. Determine whether each of the triangles with the following vertices is right-angled:
    (a) $A(2,9)$, $B(-1,3)$, $C(6,7)$
    (b) $A(-1,4)$, $B(3,6)$, $C(1,1)$

14. A ship sails N30°E from a pier at 5 km/h. If coordinates are set up with the pier as origin and the positive x- and y-axes in the east and north directions, respectively, what are the coordinates of the ship after 3 hours?

15. A ladder 10 m long rests against a wall with its feet 4 m from the base of the wall (Figure 1.24). If a man stands on the middle rung of the ladder, what are the coordinates of his feet?

FIGURE 1.24

16. What are the coordinates of points A, B, and C in Figure 1.25 if BC and BA form a right angle and the side lengths are as shown?

FIGURE 1.25

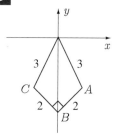

17. Lines that intersect at angles other than right angles (Figure 1.26) can also be used to coordinatize the plane. If l_1 and l_2 are coordinatized as shown, and P is any point in the plane, we draw through P lines parallel to l_2 and l_1. The lengths x and y cut off on l_1 and l_2 are then called the coordinates (x,y) of P.

FIGURE 1.26

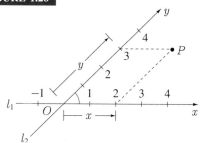

(a) Show the plane coordinatized with equal unit distances of 1 cm along two lines that intersect at $\pi/6$ radians.
(b) On your diagram mark the points $(1,3)$, $(-2,3)$, $(4,-5)$, $(-1,-2)$, $(1/2,-1/2)$, $(-2,1/2)$.

18. If the radius of the circle in Figure 1.27 is 10 cm, find the length of line segment AB.

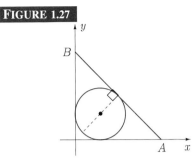

FIGURE 1.27

19. If the minute and hour hands of a clock are, respectively, 10 cm and 7.5 cm long (Figure 1.28), find coordinates for the tips of the hands when the time is:
(a) 9:00   (b) 4:00   (c) 7:30   (d) 1:40

FIGURE 1.28

20. If a circle passes through three points $(0,0)$, $(4,0)$, and $(0,3)$, what are the coordinates of its centre?

21. Plot points $P(2,1)$ and $Q(5,5)$ on a Cartesian plane coordinatized with unit distances of 1 cm and 0.5 cm along the x- and y-axes, respectively. According to distance formula 1.7, $\|PQ\| = \sqrt{(5-2)^2 + (5-1)^2} = 5$. Measure $\|PQ\|$ with a ruler. Is it 5 cm? Explain.

22. A man 2 m tall walks along the edge of a straight road 10 m wide (Figure 1.29). On the other edge of the road stands a streetlight 8 m high. If rectangular coordinates (with unit distances equal to 1 m) are set up as shown, find coordinates for the tip of the man's shadow when he is 10 m from the point directly opposite the light.

FIGURE 1.29

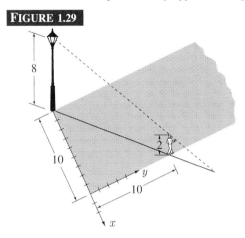

23. Generalize the result of equations 1.8 to prove that if a point R divides the length PQ so that

$$\frac{\text{length } PR}{\text{length } RQ} = \frac{r_1}{r_2},$$

where r_1 and r_2 are positive integers, then the coordinates of R are

$$x = \frac{r_1 x_2 + r_2 x_1}{r_1 + r_2} \quad \text{and} \quad y = \frac{r_1 y_2 + r_2 y_1}{r_1 + r_2}.$$

24. Prove that in any triangle the sum of the squares of the lengths of the medians is equal to three-quarters of the sum of the squares of the lengths of the sides. (A median of a triangle is a line segment drawn from one vertex to the midpoint of the opposite side.)

SECTION 1.3

Equations and Curves

In Section 1.2 we began the study of plane analytic geometry by using ordered pairs of real numbers (x, y) to describe the positions of points in a plane, an algebraic description of a geometric object. In this section and Section 1.4 we bring out the other aspect of analytic geometry: that algebraic statements can be visualized geometrically.

An equation such as $y = x^2$ specifies a relationship between numbers represented by the letter x and those represented by the letter y — y must always be the square of x. Algebraically, we speak of pairs of values x and y that satisfy this equation. To illustrate this we have constructed Table 1.1, called a **table of values**.

TABLE 1.1

x	−6	−5	−4	−3	−2	−1	0	1	2	3	4	5	6
y	36	25	16	9	4	1	0	1	4	9	16	25	36

Each pair of entries in this table represents a pair of values of x and y that satisfies the equation $y = x^2$. If we write these pairs in the form (x, y), and interpret each as coordinates of a point, we obtain the thirteen points in Figure 1.30(a). Note that since the values of y range from 0 to 36 and those for x range only from −6 to 6, we have used different scales along the x- and y-axes. We could add more points to this illustration by choosing noninteger values of x between −6 and 6, as well as values of x beyond −6 and 6. We would find that all such points lie on the curve in Figure 1.30(b). This curve then is a geometric visualization of solution pairs of the equation $y = x^2$. Every pair of values x and y that satisfies $y = x^2$ is represented by a point on the curve. Conversely, the coordinates (x, y) of every point on the curve provide a pair of numbers that satisfies $y = x^2$. Algebraic solutions then are represented geometrically as points on a curve; points on the curve provide algebraic solutions.

FIGURE 1.30

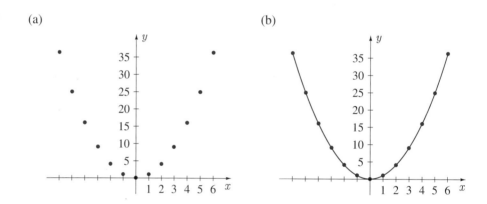

If (x, y) are the coordinates of a point in the plane, then the point is on the curve in Figure 1.30(b) if and only if x and y satisfy the equation $y = x^2$. In other words, this equation completely characterizes all points on the curve. We therefore call $y = x^2$ the **equation of the curve**. The equation of a curve then is an equation that the coordinates of every point on the curve satisfy and at the same time is an equation that no point off the curve satisfies. In the remainder of this section and Section 4 we discuss five classes of curves — straight lines, parabolas, circles, ellipses, and hyperbolas — and show that each can be described by a particular type of equation. Once again note the algebraic-geometric interplay; the form of an equation dictates the shape of the curve, and conversely, the shape of a curve determines the form of its equation.

The Straight Line

Every point on the vertical line in Figure 1.31 has an x-coordinate equal to 2; the equation of the line is therefore $x = 2$. The equation of the horizontal line in the same diagram is $y = -5/2$.

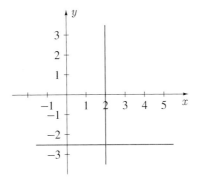

FIGURE 1.31

To find the equation of a straight line that is not parallel to either axis, we introduce the idea of *slope* in Figure 1.32(a).

> **Definition 1.4**
>
> The **slope** of a line through points (x_1, y_1) and (x_2, y_2) is
> $$m = \frac{y_2 - y_1}{x_2 - x_1}. \qquad (1.9)$$

It is customary to call the difference $y_2 - y_1$ the **rise** because it represents the vertical distance between the points, and $x_2 - x_1$ the **run**, the horizontal distance between the points. It is easy to show using similar triangles that m is independent of the two points chosen on the line; that is, no matter what two points we choose on the line to evaluate m, the result is always the same. The four numbers x_1, x_2, y_1, and y_2 vary, but the ratio $(y_2 - y_1)/(x_2 - x_1)$ remains unchanged. Note in Figure 1.32(b) that a horizontal line has slope zero (since $y_2 - y_1 = 0$), whereas the slope of a vertical line is undefined (since $x_2 - x_1 = 0$). In Figure 1.32(c), line l_1, which leans to the right, has a positive slope, and line l_2, which leans to the left, has a negative slope.

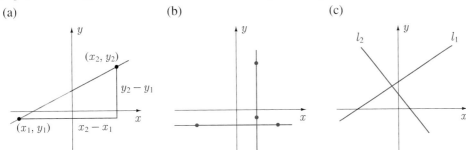

FIGURE 1.32

In Figure 1.33 we have shown the line through the points $(1, 2)$ and $(3, 5)$. Its slope is $(5 - 2)/(3 - 1) = 3/2$.

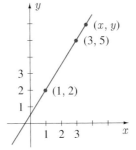

FIGURE 1.33

To find the equation for this line, we let (x, y) be the coordinates of any other point on the line. Since the slope of the line must also be given by $(y - 2)/(x - 1)$, it follows that
$$\frac{y - 2}{x - 1} = \frac{3}{2},$$
and when this equation is simplified
$$2y = 3x + 1.$$

If (x, y) are the coordinates of any point not on this line, then they do not satisfy this equation because the slope $(y - 2)/(x - 1)$ joining (x, y) to $(1, 2)$ is not equal to $3/2$. Thus, $2y = 3x + 1$ is the equation for the straight line through $(1, 2)$ and $(3, 5)$.

We can use this procedure to find the equation for the straight line through any point (x_1, y_1) with any slope m (Figure 1.34). If (x, y) are the coordinates of any other point on the line, then the slope of the line is $(y - y_1)/(x - x_1)$; therefore

$$\frac{y - y_1}{x - x_1} = m \quad \text{or}$$
$$y - y_1 = m(x - x_1). \tag{1.10}$$

This is called the **point-slope** formula for the equation of a straight line; it uses the slope m of the line and a point (x_1, y_1) on the line to determine the equation of the line.

FIGURE 1.34

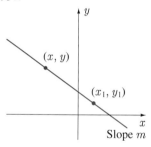

Slope m

Other formulas for the equation of a line are used when different characteristics of the line are given. They are listed below:

Form of Equation	Name of Equation	Description of Characteristics Determining the Line
$\dfrac{y - y_1}{y_2 - y_1} = \dfrac{x - x_1}{x_2 - x_1}$	Two-point formula	Two points (x_1, y_1) and (x_2, y_2) on the line
$y = mx + b$	Slope y-intercept formula	m is the slope; b is the y-intercept (point where line crosses y-axis)
$y = m(x - a)$	Slope x-intercept formula	m is the slope; a is the x-intercept (point where line crosses x-axis)
$\dfrac{x}{a} + \dfrac{y}{b} = 1$	Two-intercept formula	a and b are x- and y-intercepts

Each of these formulas immediately determines the equation of the line when the specified characteristics are known. Alternatively, with minimal calculations, point-slope formula 1.10 can be used in each situation. For example, if we know that the y-intercept of a line is b, then a point on the line is $(0, b)$. Hence, 1.10 gives $y - b = m(x - 0)$, or, $y = mx + b$, the slope y-intercept formula.

There is one other equation that is worth mentioning. Vertical lines, which do not have slopes, cannot be represented in form 1.10; they have the form $x = k$, for some constant k. On the other hand, all lines can, for various values of the constants A, B, and C, be represented in the form

$$Ax + By + C = 0. \qquad (1.11)$$

This is often called the **general equation** of a line.

EXAMPLE 1.5

Find the equation of the line through the points $(-1, 1)$ and $(2, 3)$.

SOLUTION Since the slope of the line is $(3 - 1)/(2 + 1) = 2/3$, we can use the point $(-1, 1)$ and the slope $2/3$ in point-slope formula 1.10 to give

$$y - 1 = \frac{2}{3}(x + 1) \qquad \text{or} \qquad 3y = 2x + 5.$$

The same result is obtained if the point $(2, 3)$ is used in place of $(-1, 1)$. Furthermore, the two-point formula could have been used.

EXAMPLE 1.6

Find equations for the lines in Figure 1.35.

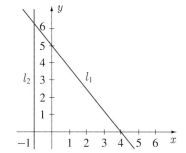

FIGURE 1.35

SOLUTION Since the slope of l_1 is $-5/4$ and its y-intercept is 5, we use the slope y-intercept formula,

$$y = -\frac{5}{4}x + 5 \qquad \text{or} \qquad 4y = 20 - 5x.$$

The equation of l_2 is clearly $x = -1$.

EXAMPLE 1.7

What is the slope of the line $3x + 4y = 2$?

SOLUTION By solving this equation for y in terms of x,

$$y = -\frac{3}{4}x + \frac{1}{2},$$

we express the equation in slope y-intercept form. Thus the slope of the line must be $-3/4$.

To find the point of intersection of two straight lines — say, $3y = 2x + 5$ and $4y + x = 14$ (Figure 1.36) — we must find the point whose coordinates (x, y) satisfy both equations. If we solve each equation for x and equate the expressions, we obtain $(3y - 5)/2 = 14 - 4y$, which immediately yields $y = 3$. Either of the original equations then gives $x = 2$, and the point of intersection of the lines is $(2, 3)$. Note once again the algebraic-geometric interplay. Geometrically, $(2, 3)$ is the point of intersection of two straight lines. Algebraically, the two numbers constitute the solution of the equations $3y = 2x + 5$ and $4y + x = 14$.

FIGURE 1.36

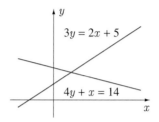

Of particular importance to the study of lines are the concepts of parallelism and perpendicularity.

Definition 1.5 Two lines are said to be **parallel** if they have no point of intersection.

EXAMPLE 1.8 Verify that the lines with equations $2x - y = 4$ and $4x - 2y = 7$ are parallel.

SOLUTION If we solve each of these equations for y and equate the resulting expressions, we obtain
$$2x - 4 = 2x - \frac{7}{2},$$
an obvious impossibility. Consequently, the lines do not intersect.

Geometrically, the following is clear.

Theorem 1.1 *Two lines are parallel if and only if they have the same slope.*

For example, if we write the equations of the lines in Example 1.8 in slope y-intercept form $y = 2x - 4$ and $y = 2x - 7/2$, coefficients of the x-terms identify slopes of the lines. Since each has slope 2, the lines are parallel.

Definition 1.6 Two lines are said to be **perpendicular** if they intersect at right angles.

The following theorem gives a test for perpendicularity of straight lines in terms of slopes.

Theorem 1.2

Two lines with nonzero slopes m_1 and m_2 are perpendicular if and only if

$$m_1 = -\frac{1}{m_2} \quad \text{or} \quad m_1 m_2 = -1. \tag{1.12}$$

Proof Suppose nonparallel lines l_1 and l_2 with nonzero slopes m_1 and m_2 intersect at a point Q (Figure 1.37).

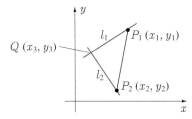

FIGURE 1.37

Suppose the coordinates of Q are (x_3, y_3), and let $P_1(x_1, y_1)$ and $P_2(x_2, y_2)$ be any two points on l_1 and l_2 respectively. Lines l_1 and l_2 are perpendicular if and only if triangle QP_1P_2 is right-angled at Q; that is, if and only if,

$$\|P_1P_2\|^2 = \|QP_1\|^2 + \|QP_2\|^2.$$

Using distance formula 1.7, this condition can be expressed as

$$(x_1 - x_2)^2 + (y_1 - y_2)^2 = (x_1 - x_3)^2 + (y_1 - y_3)^2 + (x_2 - x_3)^2 + (y_2 - y_3)^2.$$

When terms are expanded, and the equation is simplified, the result is

$$x_3^2 - x_1 x_3 + x_1 x_2 - x_2 x_3 = -y_3^2 + y_1 y_3 - y_1 y_2 + y_2 y_3.$$

After factoring each side, this equation can be rearranged into the form

$$\frac{y_3 - y_2}{x_3 - x_2} = -\frac{x_3 - x_1}{y_3 - y_1} = -\frac{1}{\dfrac{y_3 - y_1}{x_3 - x_1}}.$$

Since $m_1 = (y_3 - y_1)/(x_3 - x_1)$ and $m_2 = (y_3 - y_2)/(x_3 - x_2)$, we have shown that lines l_1 and l_2 are perpendicular if and only if

$$m_2 = -\frac{1}{m_1}.$$

EXAMPLE 1.9

Find the equation of the straight line that passes through the point $(2, 4)$ and is perpendicular to the line $3x + y = 5$.

SOLUTION Since the slope of the given line $y = -3x + 5$ is -3, the required line has slope $1/3$. Using point-slope formula 1.10, we find that the equation of the required line is

$$y - 4 = \frac{1}{3}(x - 2) \quad \text{or} \quad x - 3y = -10.$$

Perpendicularity and parallelism deal with lines that make a right angle at their point of intersection, or make no angle, as parallel lines do not intersect. Lines that intersect usually do so at angles other that $\pi/2$ radians. In this book all angles are measured in radians. We shall see why in Chapter 3. In order to determine the angle at which two lines intersect, we first define the inclination of a line.

Definition 1.7

The **inclination** of a line l is the angle of rotation ϕ ($0 \leq \phi < \pi$) from the positive x-direction to the line.

The line $y = x - 3$ in Figure 1.38(a) has an inclination of $\pi/4$ radians, and the line $y = -x - 4$ in Figure 1.38(b) has an inclination of $3\pi/4$ radians.

FIGURE 1.38

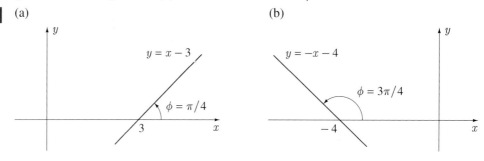

When the slope m of a line is positive, as shown in Figure 1.39(a), it is clear that ϕ must be in the interval $0 < \phi < \pi/2$, and m and ϕ are related by $\tan \phi = m$. When the slope of l is negative as in Figure 1.39(b), we use identity A.12b in Appendix A to write

$$m = -\tan(\pi - \phi) = -\frac{\tan \pi - \tan \phi}{1 + \tan \pi \tan \phi} = \tan \phi.$$

FIGURE 1.39

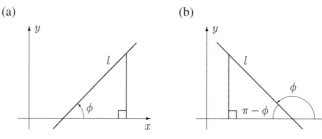

Thus, whenever the slope m of a line is defined, the angle of inclination is related to m by the equation

$$\tan \phi = m. \qquad (1.13)$$

In some sense this equation is true even when m is not defined. This occurs for vertical lines, which have no slope. For a vertical line, $\phi = \pi/2$, and $\tan \phi$ is undefined. Thus, equation 1.13 is also valid for vertical lines from the point of view that neither side of the equation is defined.

EXAMPLE 1.10

What is the angle of inclination of the line $2x - 3y = 4$?

SOLUTION From $y = 2x/3 - 4/3$, the slope of the line is $2/3$. The angle whose tangent is equal to $2/3$ is $\phi = 0.588$ radians. ∎

When two lines l_1 and l_2 with nonzero slopes m_1 and m_2 intersect (Figure 1.40), the angle θ $(0 < \theta < \pi)$ between the lines is given by the equation

$$\phi_1 = \theta + \phi_2 \quad \text{or} \quad \theta = \phi_1 - \phi_2.$$

By applying the tangent function to both sides of this equation and using identity A.12b from Appendix A once again, we can express θ in terms of m_1 and m_2,

$$\tan \theta = \tan(\phi_1 - \phi_2) = \frac{\tan \phi_1 - \tan \phi_2}{1 + \tan \phi_1 \tan \phi_2} = \frac{m_1 - m_2}{1 + m_1 m_2}.$$

This equation determines θ when $\phi_1 > \phi_2$. When $\phi_1 < \phi_2$, the equation is replaced by

$$\tan \theta = \frac{m_2 - m_1}{1 + m_1 m_2}.$$

In both cases we may write

$$\tan \theta = \left| \frac{m_1 - m_2}{1 + m_1 m_2} \right|, \qquad (1.14)$$

for the acute angle between the lines.

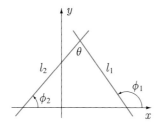

FIGURE 1.40

EXAMPLE 1.11

Find the angle between the lines $2x - 3y = 4$ and $x + 4y = 6$.

SOLUTION Since slopes of these lines are $2/3$ and $-1/4$, it follows that

$$\tan \theta = \left| \frac{2/3 - (-1/4)}{1 + (2/3)(-1/4)} \right| = \frac{11}{10}.$$

The acute angle whose tangent is equal to $11/10$ is $\theta = 0.833$ radians. ∎

EXERCISES 1.3

In Exercises 1–10 draw the line described and find its equation.

1. Through the points $(1, 2)$ and $(-3, 4)$

2. Through the points $(3, -6)$ and $(5, -6)$

3. Through the point $(-2, -3)$ with slope 3

4. Through the point $(1, -3/2)$ with slope $-1/2$

5. The y-axis   **6.** The x-axis

7. Through the point $(4, 3)$ and crossing the x-axis at -2

8. Through the point $(-1, -2)$ and crossing the y-axis at 4

9. Crossing the x- and y-axes at 1 and -3, respectively

10. Through the origin and the midpoint of the line segment joining $(3,4)$ and $(-7,8)$

In Exercises 11–16 find the angle of inclination of the line.

11. $x - y + 1 = 0$
12. $x + 2y = 3$
13. $3x - 2y = 1$
14. $y - 3x = 4$
15. $x = 4$
16. $y = 2$

In Exercises 17–24 determine whether the lines are perpendicular, parallel, or neither perpendicular nor parallel. In the last case, find the angle between the lines.

17. $y = -x + 4$, $y = x + 6$
18. $x + 3y = 4$, $2x + 6y = 7$
19. $x = 3y + 4$, $y = x/3 - 2$
20. $2x + 3y = 1$, $3x - 2y = 5$
21. $y = 3x + 2$, $y = -x/2 + 1$
22. $x - y = 5$, $2x + 3y = 4$
23. $x = 0$, $y = 5$
24. $x + y + 2 = 0$, $3x - y = 4$

In Exercises 25–29 find the point of intersection of the lines.

25. $x + y = 0$, $x - 2y = -3$
26. $x = 1$, $y = 2$
27. $3x + 4y = 6$, $x - 6y = 3$
28. $y = 2x + 6$, $x = y + 4$
29. $x/2 + y/3 = 1$, $2x - y/4 = 15$
30. $14x - 2y = 5$, $3x + 10y = 12$

In Exercises 31–38 find the equation of the line described.

31. Parallel to $x + 2y = 15$, and through the point of intersection of $2x - y = 5$ and $x + y = 4$

32. Perpendicular to $x - y = 4$, and through the point of intersection of $2x + 3y = 3$ and $x - y = 4$

33. Parallel to the line through $(1,2)$ and $(-3,0)$, and through the point $(5,6)$

34. Perpendicular to the line through $(-3,4)$ and $(1,-2)$, and through the point $(-3,-2)$

35. Crossing through the first quadrant to form an isosceles triangle with area 8 square units

36. Through $(3,5)$ and crossing through the first quadrant to form a triangle with area 30 square units

37. Has slope 2, and that part in the second quadrant has length 3

38. Parallel to the x-axis, below the point of intersection of the lines $x = y$ and $x + y = 4$, and forms with these lines a triangle with area 9 square units

39. A median of a triangle is a line segment drawn from a vertex to the midpoint of the opposite side. Find equations for the three medians of the triangle with vertices $(1,1)$, $(3,5)$, and $(0,4)$. Show that all three medians intersect in a point called the centroid of the triangle.

40. If (x_1,y_1), (x_2,y_2), (x_3,y_3), and (x_4,y_4) are vertices of any quadrilateral, show that the line segments joining the midpoints of adjacent sides form a parallelogram.

41. Find the equation of the perpendicular bisector of the line segment joining $(-1,2)$ and $(3,-4)$. The perpendicular bisector is the line that cuts the line segment in half and is perpendicular to it.

42. Find coordinates of the point that is equidistant from the three points $(1,2)$, $(-1,4)$, and $(-3,1)$.

43. Prove Theorem 1.1.

44. Let P be any point inside an equilateral triangle (Figure 1.41). Show that the sum of the distances of P from the three sides is always equal to the height h of the triangle.

FIGURE 1.41

45. Prove that the perpendicular distance from a point (x_1,y_1) to a line $Ax + By + C = 0$ is given by

$$\frac{|Ax_1 + By_1 + C|}{\sqrt{A^2 + B^2}}.$$

SECTION 1.4

Conic Sections

In this section we discuss parabolas, circles, ellipses, and hyperbolas, the so-called **conic sections**. As we are only interested in illustrating the algebraic-geometric interplay of analytic geometry, we do not give a complete development of these curves together with their many properties. We show only how the form of the equation for each conic section dictates its shape.

The Parabola

When the y-coordinate of a point (x,y) on a curve is related to its x-coordinate by an equation of the form

$$y = ax^2 + bx + c, \qquad (1.15)$$

where a, b, and c are constants (with $a \neq 0$), the curve is called a **parabola**. To determine the shape of a parabola we consider a number of examples. In Section 1.3 we used Table 1.1 to draw the simplest of all parabolas $y = x^2$ (Figure 1.30(b)). For every point (x,y) to the right of the y-axis on this curve, there is a point equidistant to the left of the y-axis which has the same y-coordinate; that is, the point $(-x,y)$ is also on the curve. Putting it another way, that part of the parabola to the left of the y-axis is the image in the y-axis (thought of as a mirror) of that part to the right of the y-axis. Such a curve is said to be symmetric about the y-axis. It happens whenever the equation of a curve is unchanged when each x therein is replaced by $-x$. In other words, we have the test:

*A curve is **symmetric** about the y-axis if its equation remains unchanged when x is replaced by* $-x$.

The values in Table 1.2 for the parabola $y = 9 - x^2$ lead to the curve in Figure 1.42. This parabola is also symmetric about the y-axis. It is said to "open downward", whereas the parabola in Figure 1.30(b) "opens upward". The sign in front of the x^2-term dictates which way a parabola opens.

TABLE 1.2

x	−4	−3	−2	−1	0	1	2	3	4
y	−7	0	5	8	9	8	5	0	−7

A table of values for the parabola $y = 2x^2 + 4x - 6$ leads to the curve in Figure 1.43. It is not symmetric about the y-axis; notice that its equation does change if x is replaced by $-x$. The parabola appears to be symmetric about the line $x = -1$, and its lowest point is $(-1, -8)$. To prove that this is indeed the case, we rewrite the equation as

$$y = 2(x^2 + 2x) - 6 = 2(x+1)^2 - 8.$$

This form clearly indicates that the smallest value for y is -8, and it occurs when $x = -1$. In addition, whenever the values $x = -1 \pm k$, where k is some fixed number, are substituted into the equation of the parabola, the same value for y is obtained, namely $y = 2k^2 - 8$. This means that the points $(-1+k, 2k^2 - 8)$ and $(-1-k, 2k^2 - 8)$ are both on the parabola, and these points are equidistant from the line $x = -1$; that is, the parabola is symmetric about $x = -1$.

The technique used above to rewrite $y = 2x^2 + 4x - 6$ as $2(x+1)^2 - 8$ is called *completing the square*. If we apply the same technique to the general parabola $y = ax^2 + bx + c$, we obtain

$$y = a\left(x^2 + \frac{b}{a}x\right) + c = a\left(x + \frac{b}{2a}\right)^2 + \left(c - \frac{b^2}{4a}\right). \qquad (1.16)$$

This form for the equation of the parabola shows the following:

1. When $a > 0$, the parabola opens upward and has a minimum at the point

$$\left(-\frac{b}{2a}, c - \frac{b^2}{4a}\right).$$

2. When $a < 0$, the parabola opens downward and has a maximum at the point

$$\left(-\frac{b}{2a}, c - \frac{b^2}{4a}\right).$$

3. The parabola is symmetric about the line $x = -b/(2a)$ (Figure 1.44); that is, for every point P on one side of the line, there is a point Q on the other side that is the mirror image of P in the line.

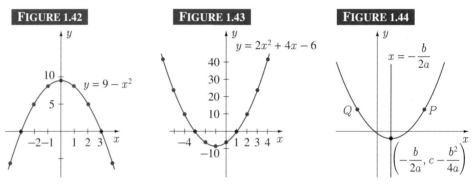

FIGURE 1.42

FIGURE 1.43

FIGURE 1.44

4. The parabola crosses the x-axis when

$$0 = y = a\left(x + \frac{b}{2a}\right)^2 + \left(c - \frac{b^2}{4a}\right).$$

To solve this equation for x, we first write

$$\left(x + \frac{b}{2a}\right)^2 = \frac{b^2}{4a^2} - \frac{c}{a},$$

and then take square roots of each side,

$$x + \frac{b}{2a} = \pm\sqrt{\frac{b^2}{4a^2} - \frac{c}{a}}.$$

Finally then,

$$x = -\frac{b}{2a} \pm \sqrt{\frac{b^2}{4a^2} - \frac{c}{a}} = \frac{-b \pm \sqrt{b^2 - 4ac}}{2a}. \qquad (1.17)$$

This is called the **quadratic formula**. It determines points where the parabola $y = ax^2 + bx + c$ crosses the x-axis. Equivalently, it defines roots of the quadratic equation $ax^2 + bx + c = 0$. Note that when $b^2 - 4ac > 0$, there are two solutions (Figure 1.45(a)); when $b^2 - 4ac = 0$, there is only one solution (Figure 1.45(b)); and when $b^2 - 4ac < 0$, there is no solution (Figure 1.45(c)); that is, no real solution.

(a)
(b)
(c)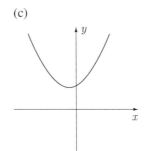

FIGURE 1.45

When x and y in equation 1.15 are interchanged, the resulting equation

$$x = ay^2 + by + c \qquad (1.18)$$

still defines a parabola. In this case we construct a table of values by listing various values of y and calculating corresponding values for x. For example, Tables 1.3 and 1.4 lead to the curves in Figures 1.46 and 1.47 for the parabolas $x = y^2 + 1$ and $x = -y^2 + 4y - 4$ respectively. The parabola $x = y^2 + 1$ is symmetric about the x-axis; for any point (x, y) on the parabola, so also is $(x, -y)$. In general, *a curve is symmetric about the x-axis if its equation remains unchanged when y is replaced by $-y$.*

TABLE 1.3

y	−4	−3	−2	−1	0	1	2	3	4
x	17	10	5	2	1	2	5	10	17

TABLE 1.4

y	−1	0	1	2	3	4	5
x	−9	−4	−1	0	−1	−4	−9

FIGURE 1.46

FIGURE 1.47

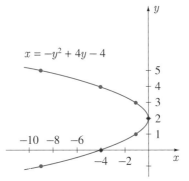

EXAMPLE 1.12

Find equations for the parabolas in Figures 1.48 and 1.49.

FIGURE 1.48

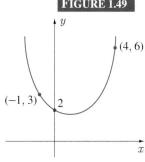

FIGURE 1.49

SOLUTION The fact that the parabola in Figure 1.48 is symmetric about the x-axis means that b in equation 1.18 must vanish; that is, its equation must be of the form $x = ay^2 + c$. Since the points $(2, 0)$ and $(0, 3)$ are on the parabola, their coordinates must satisfy the equation of the parabola,

$$2 = a(0)^2 + c,$$
$$0 = a(3)^2 + c.$$

These imply that $c = 2$ and $a = -c/9 = -2/9$. Thus, the equation of the parabola is $x = -2y^2/9 + 2$.

The parabola in Figure 1.49 has no special attributes that we can utilize (such as the position of the line of symmetry in Figure 1.48). We therefore use the facts that its equation must be of the form $y = ax^2 + bx + c$, and the three points $(-1, 3)$, $(0, 2)$, and $(4, 6)$ are on the parabola. Substitution of the coordinates of these points into the equation gives

$$3 = a(-1)^2 + b(-1) + c,$$
$$2 = a(0)^2 + b(0) + c,$$
$$6 = a(4)^2 + b(4) + c.$$

The second equation yields $c = 2$, and when this is substituted into the other two equations,

$$a - b = 1,$$
$$16a + 4b = 4.$$

These can be solved to obtain $a = 2/5$ and $b = -3/5$, and therefore the required equation is $y = 2x^2/5 - 3x/5 + 2$. ∎

EXERCISES 1.4A

In Exercises 1–12 draw the parabola.

1. $y = 2x^2 - 1$
2. $y = -x^2 + 4x - 3$
3. $y = x^2 - 2x + 1$
4. $3x = 4y^2 - 1$
5. $x = y^2 + 2y$
6. $2y = -x^2 + 3x + 4$
7. $x + y^2 = 1$
8. $2y^2 + x = 3y + 5$
9. $y = 4x^2 + 5x + 10$
10. $x = 10y^2$
11. $y = -x^2 + 6x - 9$
12. $x = -(4 + y)^2$

13. Find x- and y-intercepts for the following parabolas:
    (a) $y = x^2 - 2x - 5$
    (b) $x = 4y^2 - 8y + 4$

In Exercises 14–17 find the equation for the parabola shown.

14.

15.

16. **17.**

In Exercises 18–23 find all points of intersection for the curves. In each case draw the curves.

18. $y = 1 - x^2$, $y = x + 1$ **19.** $y + 2x = 0$, $y = 1 + x^2$

20. $y = 2x - x^2 - 6$, $25 + x = 5y$

21. $x = y(y - 1)$, $2y = 2x + 1$

22. $x = -y^2 + 1$, $x = y^2 + 2y - 3$

23. $y = 6x^2 - 2$, $y = x^2 + x + 1$

24. When a shell is fired from an artillery gun (Figure 1.50), it follows a parabolic path

$$y = -\frac{4.905}{v^2 \cos^2 \theta}x^2 + x \tan \theta,$$

where v is the muzzle velocity of the shell.

FIGURE 1.50

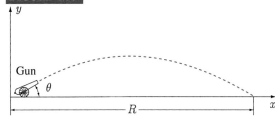

(a) Find the range R of the shell.

(b) What is the maximum height attained by the shell?

25. The cable of the suspension bridge in Figure 1.51 hangs in the shape of a parabola. The towers are 200 m apart and extend 50 m above the roadway. If the cable is 10 m above the roadway at its lowest point, find the length of the supporting rods 30 m from the towers.

FIGURE 1.51

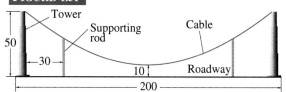

26. Find points of intersection for the parabolas

$$y = (x - 2)^2, \qquad 5x = y^2 + 4.$$

27. Find the height of the parabolic arch in Figure 1.52.

FIGURE 1.52

28. Find a parabola of type 1.15 passing through the points $(1, 2)$, $(-3, 10)$, and $(3, 4)$.

29. Find lines through the point $(3, 4)$ that intersect the parabola $y = x^2 - 1$ at only one point.

The Circle

When the x- and y-coordinates of points on a curve are related by an equation of the form

$$(x - h)^2 + (y - k)^2 = r^2, \qquad (1.19)$$

where h, k, and $r > 0$ are constant, the curve is called a **circle**. It takes but a quick recollection of distance formula 1.7 to convince ourselves that this definition of a circle coincides with our intuitive idea of a circle. If we write equation 1.19 in the form

$$\sqrt{(x - h)^2 + (y - k)^2} = r,$$

the left side is the distance from the point (x,y) to the point (h,k). Equation 1.19 therefore describes all points (x,y) at a fixed distance r from (h,k), a circle centred at (h,k) with radius r (Figure 1.53). For example, the radius of the circle in Figure 1.54 is equal to 2, and its equation is therefore

$$(x+1)^2 + (y-2)^2 = 4.$$

FIGURE 1.53

FIGURE 1.54

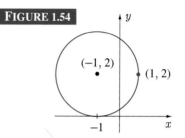

When equation 1.19 is expanded, we have

$$x^2 - 2hx + h^2 + y^2 - 2ky + k^2 = r^2 \quad \text{or}$$

$$x^2 + y^2 - 2hx - 2ky + h^2 + k^2 - r^2 = 0.$$

This shows that the equation of a circle may be given in another form, namely,

$$x^2 + y^2 + fx + gy + e = 0. \qquad (1.20)$$

Given this equation, the centre and radius can be identified by reversing the expansion and completing the squares of $x^2 + fx$ and $y^2 + gy$. For instance, if $x^2 + y^2 + 2x - 3y - 5 = 0$, then

$$0 = (x+1)^2 + (y-3/2)^2 - 5 - 1 - 9/4 = (x+1)^2 + (y-3/2)^2 - 33/4.$$

The centre of the circle is therefore $(-1, 3/2)$ and its radius is $\sqrt{33}/2$.

When the centre of a circle is the origin $(0,0)$, equation 1.19 simplifies to

$$x^2 + y^2 = r^2. \qquad (1.21)$$

 Be careful to use equation 1.19, not 1.21, when the centre of the circle is not the origin. It is a common error to use (1.21).

EXAMPLE 1.13 Figure 1.55 shows an arc of a circle. Find the equation for the circle.

FIGURE 1.55

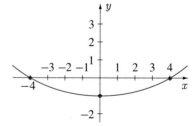

SOLUTION From the symmetry of the figure, we see that the centre of the circle is on the y-axis. Its equation must be of the form

$$x^2 + (y-k)^2 = r^2.$$

Because $(4,0)$ is a point on the circle, these coordinates must satisfy its equation; that is,

$$16 + (0-k)^2 = r^2 \quad \text{or} \quad 16 + k^2 = r^2.$$

Similarly, since $(0,-1)$ is on the circle,

$$(-1-k)^2 = r^2 \quad \text{or} \quad 1 + 2k + k^2 = r^2.$$

If we subtract these two equations, we obtain $2k - 15 = 0$, from which we see that $k = 15/2$. Consequently, $r^2 = 16 + k^2 = 16 + 225/4 = 289/4$, and the equation of the circle is

$$x^2 + (y - 15/2)^2 = 289/4.$$

■

EXERCISES 1.4B

In Exercises 1–10 draw the circle.

1. $x^2 + y^2 = 50$
2. $(x+5)^2 + (y-2)^2 = 6$
3. $x^2 + 2x + y^2 = 15$
4. $x^2 + y^2 - 4y + 1 = 0$
5. $x^2 - 2x + y^2 - 2y + 1 = 0$
6. $2x^2 + 2y^2 + 6x = 25$
7. $3x^2 + 3y^2 + 4x - 2y = 6$
8. $x^2 + 4x + y^2 - 2y = 5$
9. $x^2 + y^2 - 2x - 4y + 5 = 0$
10. $x^2 + y^2 + 6x + 3y + 20 = 0$

In Exercises 11–16 find an equation for the circle.

11.

12.

13.

14.

15.

16.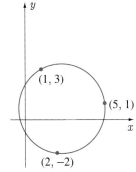

17. Find the equation of a circle that passes through the points $(3,4)$ and $(1,-10)$, and has its centre:
  (a) on the line $2x + 3y + 16 = 0$;
  (b) on the line $x + 7y + 19 = 0$.

In Exercises 18–21 find points of intersection for the curves.

18. $x^2 + 2x + y^2 = 4$, $y = 3x + 2$

19. $x^2 + y^2 - 4y + 1 = 0$, $2x + y = 1$

20. $x^2 + y^2 = 9$, $y = 3x^2 + 4$

21. $(x+3)^2 + y^2 = 25$, $y^2 = 16(x+1)$

22. Show that every equation of the form 1.20 represents a circle, a point, or nothing at all.

23. Prove that the perpendicular bisector of a chord of a circle always passes through the centre of the circle.

24. The *circumcircle* for a triangle is that circle which passes through all three of its vertices. Find the circumcircle for the triangle with vertices $A(1,1)$, $B(-3,3)$, and $C(2,4)$ by:
  (a) finding its centre and radius (see Exercise 23);
  (b) taking the equation of the circle in form 1.19 and requiring A, B, and C to be on the circle.

25. Prove that the three altitudes of the triangle in Exercise 24 intersect in a point called the *orthocentre* of the triangle.

26. Show that if a line $Ax + By + C = 0$ and a circle $(x-h)^2 + (y-k)^2 = r^2$ do not intersect, then the shortest distance between them is the smaller of the two numbers

$$\frac{|(Ah + Bk + C) \pm r\sqrt{A^2 + B^2}|}{\sqrt{A^2 + B^2}}.$$

Hint: Use the result of Exercise 45 in Section 1.3.

27. The *incircle* of a triangle is that circle which lies interior to the triangle but touches all three sides. The centre of the incircle is called the *incentre*. Use the result of Exercise 45 in Section 1.3 to show that the incentre (x,y) of the triangle with vertices $(0,0)$, $(2,0)$, and $(0,1)$ must satisfy the equations

$$|x| = |y| = \frac{|x + 2y - 2|}{\sqrt{5}}.$$

Solve these equations for the incentre, and explain why there are four points that satisfy these equations.

The Ellipse

The set of points whose coordinates (x, y) satisfy an equation of the form

$$\frac{x^2}{a^2} + \frac{y^2}{b^2} = 1, \qquad (1.22)$$

where a and b are positive constants, is said to constitute an **ellipse**. Since this equation is so similar to equation 1.21, and is exactly the same when $a = b = r$, it is not unreasonable to expect that the shape of this curve might be similar to a circle, especially when values of a and b are close together. To see that this is the case, we first note that x- and y-intercepts of the ellipse are $\pm a$ and $\pm b$, respectively. Next we write equation 1.22 in the form

$$\frac{x^2}{a^2} = 1 - \frac{y^2}{b^2} = \frac{b^2 - y^2}{b^2}.$$

Since the left side is nonnegative, so also must be the right side. But this implies that values of y must be confined to the interval $-b \leq y \leq b$. Similarly, by writing

$$\frac{y^2}{b^2} = \frac{a^2 - x^2}{a^2},$$

we conclude that x-coordinates of points on the ellipse must satisfy $-a \leq x \leq a$. What we have shown is that the ellipse must lie in the rectangle of Figure 1.56, and that the midpoints of the four sides of the rectangle are on the ellipse. Now the ellipse must be symmetric about both the x-axis and the y-axis because replacing x by $-x$ and y by $-y$ in 1.22 changes nothing. As a result, if we determine the shape of the ellipse in the first quadrant portion of the rectangle in Figure 1.56, mirror images lead to the remaining three parts of the curve.

When we take positive square roots of the last equation, the result is

$$y = \frac{b}{a}\sqrt{a^2 - x^2}.$$

What does this equation tell us? As x increases from 0 to a, y steadily decreases from b to 0; y values do not go up and down, they drop steadily from b to 0. Similarly as y decreases from b to 0, x steadily increases from 0 to a. We have drawn this in Figure 1.57(a) giving the curve an outward "bulge" to resemble a circle. Reflections in the axes give the full ellipse in Figure 1.57(b).

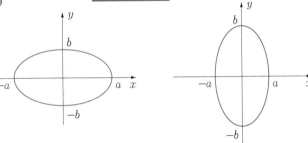

When $b > a$, the ellipse is elongated in the y-direction rather than the x-direction (Figure 1.58). The point of intersection of the lines of symmetry of an ellipse is called the **centre** of the ellipse. For equation 1.22 the centre is the origin since the x- and y-axes are the lines of symmetry.

EXAMPLE 1.14

Find the equation of the ellipse that has its centre at the origin, the x- and y-axes as axes of symmetry, and passes through the points $(4, 21/5)$ and $(-15\sqrt{5}/7, 2)$.

SOLUTION If we substitute the coordinates of the two given points into equation 1.22 (since they are both on the ellipse),

$$\frac{4^2}{a^2} + \frac{(21/5)^2}{b^2} = 1,$$

$$\frac{(-15\sqrt{5}/7)^2}{a^2} + \frac{2^2}{b^2} = 1.$$

When the first equation is divided by 16 and solved for $1/a^2$, the result is

$$\frac{1}{a^2} = \frac{1}{16} - \frac{441}{400 b^2}.$$

This is now substituted into the second equation,

$$\frac{1125}{49}\left(\frac{1}{16} - \frac{441}{400\,b^2}\right) + \frac{4}{b^2} = 1 \quad \text{or}$$

$$\frac{1}{b^2}\left(4 - \frac{1125 \cdot 441}{49 \cdot 400}\right) = 1 - \frac{1125}{49 \cdot 16}.$$

Thus,

$$b^2 = \left(\frac{4 \cdot 49 \cdot 400 - 1125 \cdot 441}{49 \cdot 400}\right)\left(\frac{49 \cdot 16}{49 \cdot 16 - 1125}\right) = 49 \quad \text{and}$$

$$\frac{1}{a^2} = \frac{1}{16} - \frac{441}{49 \cdot 400} = \frac{1}{25}.$$

The required equation is therefore $x^2/25 + y^2/49 = 1$.

When equation 1.22 is changed to

$$\frac{(x-h)^2}{a^2} + \frac{(y-k)^2}{b^2} = 1, \qquad (1.23)$$

where h and k are constants, the curve is still an ellipse; its shape remains the same. Just as a change from equation 1.19 to 1.21 for a circle moves the centre of the circle from $(0,0)$ to (h,k), equation 1.23 moves the centre of the ellipse to (h,k). Lines $x = h$ and $y = k$ are the new lines of symmetry (Figure 1.59), and a and b are the distances between the centre and where the ellipse crosses the lines of symmetry.

FIGURE 1.59

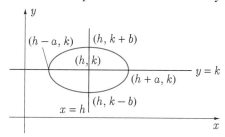

EXAMPLE 1.15

Sketch the ellipse $16x^2 + 25y^2 - 160x + 50y = 1175$.

SOLUTION When we complete squares on the x- and y-terms,

$$16(x-5)^2 + 25(y+1)^2 = 1600 \quad \text{or} \quad \frac{(x-5)^2}{100} + \frac{(y+1)^2}{64} = 1.$$

The centre of the ellipse is $(5, -1)$; it cuts the lines $y = -1$ and $x = 5$ at distances of 10 and 8 units from the centre respectively (Figure 1.60).

FIGURE 1.60

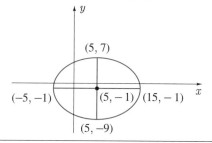

EXERCISES 1.4C

In Exercises 1–8 draw the ellipse.

1. $x^2/25 + y^2/36 = 1$
2. $7x^2 + 3y^2 = 16$
3. $9x^2 + 289y^2 = 2601$
4. $3x^2 + 6y^2 = 21$
5. $x^2 + 16y^2 = 2$
6. $x^2 + 2x + 4y^2 - 16y + 13 = 0$
7. $9x^2 + y^2 - 18x - 6y = 26$
8. $x^2 + 4x + 2y^2 + 16y + 32 = 0$
9. Find the equation of an ellipse through the points $(-2, 4)$ and $(3, 1)$.
10. Find the width of the elliptic arch in Figure 1.61.

FIGURE 1.61

In Exercises 11–16 find all points of intersection for the curves. In each case draw the curves.

11. $x^2 + 4y^2 = 4$, $y = x$
12. $16x^2 + 9y^2 = 144$, $y = x + 3$
13. $9x^2 - 18x + 4y^2 = 27$, $2y = \sqrt{3}x + \sqrt{3}$
14. $9x^2 - 18x + 4y^2 = 27$, $2y = -\sqrt{3}x + 5\sqrt{3}$
15. $x^2 + 4y^2 - 8y = 0$, $y = x^2$
16. $x^2 + 4y^2 = 4$, $y = x^2 - 4$

The Hyperbola

Changing one sign in the equation of an ellipse leads to a curve with totally different characteristics. The set of points whose coordinates (x, y) satisfy an equation of the form

$$\frac{x^2}{a^2} - \frac{y^2}{b^2} = 1 \quad \text{or} \quad (1.24\text{a})$$

$$\frac{y^2}{b^2} - \frac{x^2}{a^2} = 1, \quad (1.24\text{b})$$

where a and b are positive constants, is called a **hyperbola**. All hyperbolas have similar properties. Let us discover them with the specific example

$$\frac{x^2}{9} - \frac{y^2}{36} = 1.$$

This hyperbola crosses the x-axis at $x = \pm 3$, but does not cross the y-axis. Since the equation remains unchanged when x is replaced by $-x$ and y is replaced by $-y$, the hyperbola is symmetric about both the x-axis and the y-axis. This means that we can concentrate on drawing that part of the hyperbola in the first quadrant, and obtain the parts in the remaining three quadrants by reflections in the axes. To discover its shape in the first quadrant we begin by solving the equation for y in terms of x. By taking positive square roots of

$$\frac{y^2}{36} = \frac{x^2}{9} - 1,$$

we obtain

$$\frac{y}{6} = \sqrt{\frac{x^2}{9} - 1},$$

from which

$$y = 2\sqrt{x^2 - 9}.$$

What does this equation, for that part of the hyperbola in the first quadrant, tell us? First, it indicates that there can be no point on the curve with x-coordinate between 0 and 3. Secondly, as values of x get bigger and bigger, so also do corresponding values of y. In other words, the curve must proceed upwards and to the right from the point $(3,0)$. Finally, when large values of x are considered, say, 1000, 10 000, or 100 000, values of y are approximately equal to $2\sqrt{x^2} = 2x$ — the 9 becoming less and less significant the larger the value of x. This means that for large values of x, the hyperbola is very close to the line $y = 2x$. We have shown these facts in Figure 1.62(a). The complete hyperbola is shown in Figure 1.62(b). The lines $y = \pm 2x$ which the hyperbola approaches for large positive and negative values of x are called **asymptotes** of the hyperbola.

FIGURE 1.62

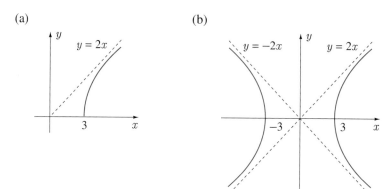

Let us now return to the more general equation 1.24a of a hyperbola. The x-intercepts of this hyperbola are $\pm a$, and it is symmetric about both axes. By writing the equation in the form

$$y = \pm \frac{b}{a}\sqrt{x^2 - a^2},$$

we see that for very large positive or negative x, y can be approximated by $\pm bx/a$; that is, the asymptotes of the hyperbola are the lines $y = \pm bx/a$. The hyperbola is shown in Figure 1.63. A similar analysis of equation 1.24b leads to the hyperbola in Figure 1.64.

FIGURE 1.63

FIGURE 1.64

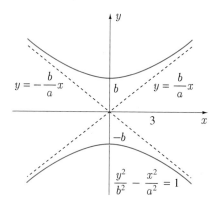

> **EXAMPLE 1.16**
>
> Find the equation of a hyperbola that cuts the y-axis at $y = \pm 5$ and has the lines $y = \pm x/\sqrt{3}$ as asymptotes.
>
> **SOLUTION** Since the hyperbola crosses the y-axis at $y = \pm 5$, we take its equation in form 1.24b with $b = 5$,
>
> $$\frac{y^2}{25} - \frac{x^2}{a^2} = 1.$$
>
> When we solve this equation for y in terms of x, the result is
>
> $$y = \pm \frac{5}{a}\sqrt{x^2 + a^2}.$$
>
> Since asymptotes of this hyperbola are $\pm 5x/a$, it follows that $5/a = 1/\sqrt{3}$, or, $a = 5\sqrt{3}$. The equation of the hyperbola is therefore
>
> $$\frac{y^2}{25} - \frac{x^2}{75} = 1.$$

When x and y in equations 1.24a,b are replaced by $x - h$ and $y - k$, the resulting equations

$$\frac{(x-h)^2}{a^2} - \frac{(y-k)^2}{b^2} = 1 \quad \text{and} \tag{1.25a}$$

$$\frac{(y-k)^2}{b^2} - \frac{(x-h)^2}{a^2} = 1, \tag{1.25b}$$

still describe hyperbolas. They are shown in Figures 1.65 and 1.66. These are the hyperbolas of Figures 1.63 and 1.64 shifted so that the asymptotes intersect at the point (h, k). Equations of the asymptotes are $y = k \pm b(x - h)/a$, and the lines $x = h$ and $y = k$ are now axes of symmetry.

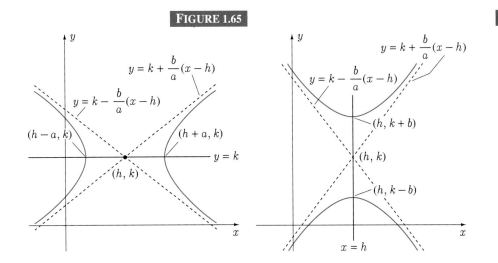

FIGURE 1.65   **FIGURE 1.66**

EXAMPLE 1.17

Sketch the hyperbola $x^2 - y^2 + 4x + 10y = 5$.

SOLUTION If we complete squares on the x- and y-terms, we obtain

$$(x+2)^2 - (y-5)^2 = -16 \quad \text{or} \quad \frac{(y-5)^2}{16} - \frac{(x+2)^2}{16} = 1.$$

The axes of symmetry of the hyperbola are $x = -2$ and $y = 5$, intersecting at the point $(-2, 5)$. When we solve for y in terms of x, the result is

$$y = 5 \pm \sqrt{(x+2)^2 + 16}.$$

Asymptotes are therefore $y = 5 \pm (x+2)$, and the hyperbola can be drawn as in Figure 1.67.

FIGURE 1.67

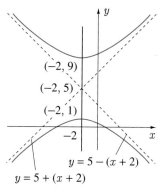

EXERCISES 1.4D

In Exercises 1–10 draw the hyperbola.

1. $y^2 - x^2 = 1$
2. $x^2 - y^2 = 1$
3. $x^2 - y^2/16 = 1$
4. $25y^2 - 4x^2 = 100$
5. $y^2 = 10(2 + x^2)$
6. $3x^2 - 4y^2 = 25$
7. $x^2 - 6x - 4y^2 - 24y = 11$
8. $9x^2 - 16y^2 - 18x - 64y = 91$
9. $4y^2 - 5x^2 + 8y - 10x = 21$
10. $x^2 + 2x - 16y^2 + 64y = 79$

11. Find the equation of a hyperbola that passes through the point $(1, 2)$ and has asymptotes $y = \pm 4x$.

In Exercises 12–18 find all points of intersection for the curves. In each case draw the curves.

12. $x^2 - 2y^2 = 1$, $x = 2y$
13. $9y^2 - 4x^2 = 36$, $y = x$
14. $9y^2 - 4x^2 = 36$, $x = 3y$
15. $3x^2 - y^2 = 3$, $2x + y = 1$
16. $x^2 - 2x - y^2 = 0$, $x = y^2$
17. $x^2 - 2x - y^2 = 0$, $x = -y^2$
18. $9(x-1)^2 - 4(y-1)^2 = 36$, $27x = 5(y-1)^2$

SECTION 1.5

Functions and Their Graphs

Most quantities that we encounter in everyday life are dependent on many, many other quantities. For example, think of what might be affecting room temperature as you read this sentence — thermostat setting, outside temperature, wind conditions, insulation of the walls, ceiling, and floors, and perhaps other factors that you can think of. Functional notation allows interdependences of such quantities to be represented in a very simple way.

When one quantity depends on a second quantity, we say that the first quantity is a function of the second. For example, the volume V of a sphere depends on its radius r; in particular, $V = 4\pi r^3/3$. We say that V is a function of r. When an object is dropped, the distance d (metres) that it falls in time t (seconds) is given by the formula $d = 4.905t^2$. We say that d is a function of t. Mathematically, we have the following definition.

> **Definition 1.8**
>
> A quantity y is said to be a function of a quantity x if there exists a rule by which we can associate exactly one value of y with each value of x. The rule that associates the value of y with each value of x is called the **function**.

If we denote the rule or function in this definition by the letter f, then the value that f assigns to x is denoted by $f(x)$, and we write

$$y = f(x). \qquad (1.26)$$

We should read this equation as "y is the value that f assigns to x", but usually we simply say "y equals f of x". In our first example above, we write $V = f(r) = 4\pi r^3/3$, and the function f is the operation of cubing a number and then multiplying the result by $4\pi/3$. For $d = f(t) = 4.905t^2$, the function f is the operation of squaring a number and multiplying the result by 4.905.

We call x in equation 1.26 the **independent variable** because values of x are substituted into the function, and y the **dependent variable** because its values depend on the assigned values of x. The **domain** of a function is the set of all specified (real) values for the independent variable. It is an essential part of a function and should always be specified. Whenever the domain of a function is not mentioned, we assume that it consists of all possible values for which $f(x)$ is a real number.

As the independent variable x takes on values in the domain, a set of values of the dependent variable is obtained. This set is called the **range** of the function. For the function $V = f(r) = 4\pi r^3/3$, which represents the volume of a sphere, the largest possible domain is $r > 0$, and the corresponding range is $V > 0$. Note that mathematically the function $f(r)$ is defined for negative as well as positive values of r, and $r = 0$; it is because of our interpretation of r as the radius of a sphere that we restrict $r > 0$. The function $d = f(t) = 4.905t^2$ represents the distance fallen in time t by an object that is dropped at time $t = 0$. If it is dropped from a height of 20 m, then it is clear that the range of this function is $0 \le d \le 20$. The domain that gives rise to this range is $0 \le t \le \sqrt{20/4.905}$.

> **EXAMPLE 1.18**
>
> If $f(x) = \sqrt{x^2 - 3}$, calculate each of the following:
>
> (a) $f(2)$ (b) $f(-4)$ (c) $f(t)$ (d) $f(h+4)$
>
> (e) $\dfrac{f(x)}{f(1-x)}$ (f) $[f(x)]^2 + 3f(x)f(-x)$
>
> **SOLUTION** (a) Setting $x = 2$ gives

$$f(2) = \sqrt{2^2 - 3} = 1.$$

(b) Setting $x = -4$ gives

$$f(-4) = \sqrt{(-4)^2 - 3} = \sqrt{13}.$$

(c) To find $f(t)$ we substitute t for each x in $f(x)$:

$$f(t) = \sqrt{t^2 - 3}.$$

(d) For $f(h+4)$, we replace each x in $f(x)$ with $h+4$,

$$f(h+4) = \sqrt{(h+4)^2 - 3} = \sqrt{h^2 + 8h + 13}.$$

(e)
$$\frac{f(x)}{f(1-x)} = \frac{\sqrt{x^2 - 3}}{\sqrt{(1-x)^2 - 3}} = \frac{\sqrt{x^2 - 3}}{\sqrt{x^2 - 2x - 2}}$$

(f)
$$[f(x)]^2 + 3f(x)f(-x) = \left[\sqrt{x^2 - 3}\right]^2 + 3\sqrt{x^2 - 3}\sqrt{(-x)^2 - 3}$$
$$= x^2 - 3 + 3(x^2 - 3) = 4(x^2 - 3)$$

EXAMPLE 1.19

Find the largest possible domain for the function

$$f(x) = \sqrt{\frac{x-4}{x+1}}.$$

SOLUTION The sign diagram in Figure 1.68 shows that the quotient $(x-4)/(x+1)$ is positive when $x > 4$ and when $x < -1$. Since the quotient is equal to zero when $x = 4$ and is undefined when $x = -1$, the largest possible domain for $f(x)$ consists of all values of x in the intervals $x < -1$ and $x \geq 4$.

FIGURE 1.68

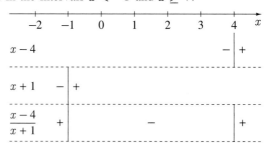

EXAMPLE 1.20

Does the equation $x - 4 - y^2 = 0$ define y as a function of x for $x \geq 4$?

SOLUTION For any $x > 4$, the equation has two solutions for y:

$$y = \pm\sqrt{x - 4}.$$

Since the equation does not define exactly one value of y for each value of x, it does not define y as a function of x. ∎

If we add to the equation in Example 1.20 an additional restriction such as $y \geq 0$, then y is defined as a function of x, namely,

$$y = \sqrt{x - 4}.$$

Note that $x - 4 - y^2 = 0$ does, however, define x as a function of y:

$$x = y^2 + 4.$$

In this case, y is the independent variable and x is the dependent variable. In other words, whenever an equation (such as $x - 4 - y^2 = 0$) is to be regarded as defining a function, it must be made clear which variable is to be considered as independent and which as dependent.

In the study of calculus and its applications we are interested in the behaviour of functions; that is, for certain values of the independent variable, what can we say about the dependent variable? The simplest and most revealing method for displaying characteristics of a function is a graph. To obtain the graph of a function $f(x)$ we use a plane coordinatized with Cartesian coordinates x and y (in short, the Cartesian xy-plane). The **graph** of the function $f(x)$ is defined to be the curve with equation $y = f(x)$, where x is limited to the domain of the function. We therefore take each value of x in the domain of the function, calculate $y = f(x)$, and plot the point (x, y) in the Cartesian plane. For example, if $f(x) = x^3 - 27x + 1$, the values in Table 1.5 lead to the curve in Figure 1.69.

TABLE 1.5

x	−6	−5	−4	−3	−2	−1	0	1	2	3	4	5	6
y	−53	11	45	55	47	27	1	−25	−45	−53	−43	−9	55

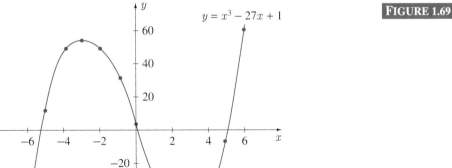

FIGURE 1.69

This curve is a pictorial or geometric representation of the function $f(x) = x^3 - 27x + 1$. The value of the function for given x is visually displayed as the y-coordinate of that point on the curve with x-coordinate equal to the given x. This graph clearly illustrates properties of the function that may not be so readily obvious from the algebraic definition $f(x) = x^3 - 27x + 1$. For instance,

1. For negative values of x, the largest value of $f(x)$ is $f(-3) = 55$; for positive values of x, the smallest value of $f(x)$ is $f(3) = -53$.

2. As x increases, values of $f(x)$ increase for $x < -3$ and $x > 3$, and values of $f(x)$ decrease for $-3 < x < 3$.

3. $f(x)$ is equal to zero for three values of x — one a little less than -5, one a little larger than 0, and one a little larger than 5.

Graphs of functions can be drawn very accurately or very roughly. An accurate graph of a function is usually obtained through an extensive table of values; this is more or less the method emphasized in Sections 1.3 and 1.4, and the above example. Although an accurate graph certainly has its advantages, and is indeed necessary in some instances, for many problems its construction is too time-consuming. Quite often we need only a rough sketch of the graph of a function. A sketch gives a general idea of the behaviour of the function; that is, a general indication of the shape of the graph. In the following examples we illustrate some useful techniques for obtaining sketches. Others will be introduced in Chapters 2, 3, and 4.

EXAMPLE 1.21

Sketch a graph of the function $f(x) = |x^3 + 5|$.

SOLUTION To obtain the curve $y = |x^3 + 5|$, we first sketch $y = x^3$, as in Figure 1.70(a). Next we sketch the curve $y = x^3 + 5$ by adding 5 to every ordinate in Figure 1.70(a). The result in Figure 1.70(b) is the curve in Figure 1.70(a) shifted upward 5 units. It crosses the x-axis at $-5^{1/3}$. The last step is to take the absolute value of every ordinate on the curve $y = x^3 + 5$. This changes no ordinate that is already positive, but changes the sign of any ordinate that is negative. The result and final sketch is shown in Figure 1.70(c).

FIGURE 1.70

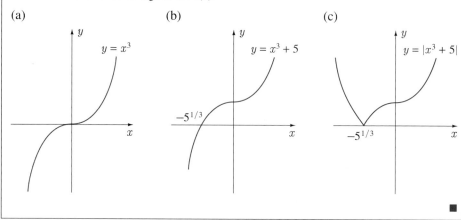

Note that in our "building" procedure we did not attempt to sketch $y = |x^3 + 5|$ in one step. By a series of steps we began with the curve $y = x^3$, proceeded to $y = x^3 + 5$, and finally obtained $y = |x^3 + 5|$. An essential feature of this graph is the sharp point where $x = -5^{1/3}$. For those readers with plotting calculators or computer access, ask the calculator or computer to draw this function. Does the curve touch the x-axis, and does there appear to be a sharp point on the curve at $x = -5^{1/3}$? The answer is probably no to both questions. This is a serious shortcoming of plotting routines in calculators and computers. They sometimes fail to divulge important features of functions. We shall see that one of the greatest accomplishments of calculus is its ability to recognize every characteristic of a function.

EXAMPLE 1.22

Sketch a graph of the function $f(x) = 2\sqrt{4 - x^2}$.

SOLUTION If we square $y = 2\sqrt{4 - x^2}$, we find that $y^2 = 4(4 - x^2)$, or, $4x^2 + y^2 = 16$. We recognize this equation as that for the ellipse in Figure 1.71(a). This is not the required graph, however, since it is not even the graph of a function, there being two y-values for each x in $-2 < x < 2$. The required graph is the upper half of the ellipse in Figure 1.71(b). We know this because the square root in $y = 2\sqrt{4 - x^2}$ demands that y be nonnegative. By squaring this equation we introduced the lower half of the ellipse.

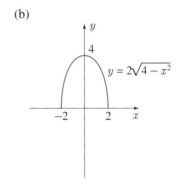

FIGURE 1.71

EXAMPLE 1.23

Sketch a graph of the function $f(x) = x + |x|$.

SOLUTION To sketch this function we first draw the straight line $y = x$ in Figure 1.72(a). The curve $y = |x|$ is obtained therefrom (shown dotted) by taking absolute values of all ordinates. The required curve $y = x + |x|$ has as its ordinates the sums of the ordinates x and $|x|$ just sketched. Consequently, to obtain $y = x + |x|$, we add corresponding ordinates of these two curves in Figure 1.72(b). The final result is shown in Figure 1.72(c).

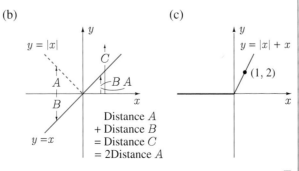

FIGURE 1.72

This building procedure, called **addition of ordinates**, is used again in the following example.

EXAMPLE 1.24 Sketch a graph of the function $f(x) = x + \sqrt{x^2 - 4}$.

SOLUTION If we square $y = \sqrt{x^2 - 4}$, then $y^2 = x^2 - 4$ or $x^2 - y^2 = 4$. This equation describes a hyperbola, the top of half of which is shown in Figure 1.73(a). On the same axes we draw the straight line $y = x$. Addition of ordinates now gives the required graph in Figure 1.73(b). Since the hyperbola is asymptotic to the line $y = x$ for $x > 0$, the graph of $y = f(x)$ is asymptotic to the line $y = 2x$ for $x > 0$. It is also asymptotic to the negative real axis.

FIGURE 1.73

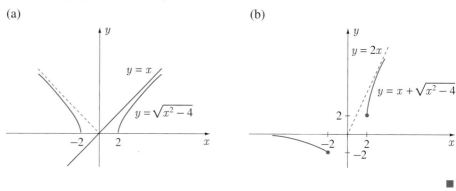

EXAMPLE 1.25 Sketch a graph of the function $f(x) = x^2\sqrt{1-x}$.

SOLUTION We first sketch the parabola $y = x^2$ in Figure 1.74(a). Next we consider a sketch of $y = \sqrt{1-x}$, which implies that $y^2 = 1 - x$ or $x = 1 - y^2$. The top half of this parabola is shown on the same axes. The required curve $y = x^2\sqrt{1-x}$ has ordinates which are products of the ordinates x^2 and $\sqrt{1-x}$ just sketched. Thus to obtain $y = x^2\sqrt{1-x}$ we multiply corresponding ordinates of the two curves in Figure 1.74(a). The result is shown in Figure 1.74(b).

FIGURE 1.74

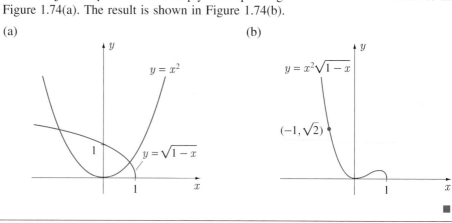

This building procedure is called **multiplication of ordinates**.

EXAMPLE 1.26

When a ball is dropped from the top of a building 20 m high at time $t = 0$, the distance d (in metres) that it falls in time t is given by the function $d = f(t) = 4.905 t^2$. Sketch its graph.

SOLUTION   In this example independent and dependent variables are denoted by letters t and d which suggest their physical meaning — time and distance — rather than the generic labels x and y. With the axes labeled correspondingly as the t-axis and d-axis, we sketch that part of the parabola $d = 4.905 t^2$ shown in Figure 1.75. The remainder of the parabola has no physical significance in the context of this problem.

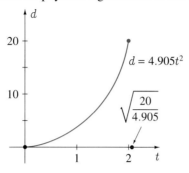

FIGURE 1.75

Polynomials and Rational Functions

Two very important classes of functions are polynomials and rational functions. A **polynomial of degree n** is a function of the form

$$f(x) = a_n x^n + a_{n-1} x^{n-1} + \cdots + a_2 x^2 + a_1 x + a_0 \qquad (1.27)$$

where n is a nonnegative integer, and $a_0, a_1, \ldots, a_n$ are real numbers ($a_n \neq 0$). Special names have become associated with some of the lower-degree polynomials because of the frequency with which they arise:

$f(x) = a_1 x + a_0$ polynomial of degree 1, linear polynomial, linear function

$f(x) = a_2 x^2 + a_1 x + a_0$ polynomial of degree 2, quadratic polynomial, quadratic function

$f(x) = a_3 x^3 + a_2 x^2 + a_1 x + a_0$ polynomial of degree 3, cubic polynomial, cubic function

Polynomials are defined for all values of x. Graphs of linear and quadratic polynomials are straight lines and parabolas. Cubic and higher degree polynomials are easy to graph when they can be factored into linear polynomials. For example, the cubic polynomial $f(x) = x^3 - 2x^2 - x + 2$ can be factored into $f(x) = (x-1)(x+1)(x-2)$. In this form we see that the x-intercepts of the graph of $f(x)$ are ± 1 and 2, and the sign diagram of Figure 1.76 leads to the curve in Figure 1.77. The factored form of the "quartic" polynomial $f(x) = x^4 - 13x^2 + 36 = (x-2)(x+2)(x-3)(x+3)$ leads to the graph in Figure 1.78.

FIGURE 1.76

FIGURE 1.77

FIGURE 1.78

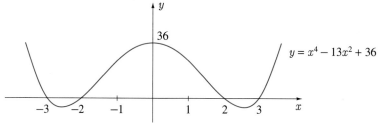

Graphing polynomials that cannot be factored completely into linear terms is more difficult. In Chapter 4 we shall see that calculus provides a complete set of tools for doing this. One might ask, however, whether addition of ordinates could be used to sketch polynomials. For example, could $f(x) = x^3 - 2x^2 - x + 1$ be sketched by adding ordinates of x^3 and $-2x^2 - x + 1$, or of x^3, $-2x^2$, and $1 - x$. In general the answer is no. (Try it.) Usually, so many questions arise in the process, that one is not at all confident in the result.

A **rational function** $R(x)$ is defined as the quotient of two polynomials $P(x)$ and $Q(x)$:

$$R(x) = \frac{P(x)}{Q(x)}. \tag{1.28}$$

Rational functions are undefined only at points where $Q(x) = 0$. To graph many rational functions we require "limits" from Chapter 2 and "derivatives" from Chapter 3.

Every function $f(x)$ can be represented pictorially by its graph, the curve with equation $y = f(x)$. But what about the reverse situation? Does every curve in the xy-plane represent a function $f(x)$? The curves in Figure 1.79, which both extend between $x = a$ and $x = b$, illustrate that the answer is no. The curve in Figure 1.79(a) represents a function, whereas the curve in Figure 1.79(b) does not, because for values of x between a and c there are two possible values of y. In other words, *a curve represents a function f(x) if every vertical line that intersects the curve does so at exactly one point.*

FIGURE 1.79 (a) (b)

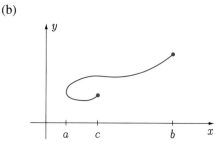

Basic to all mathematics is the concept of a function, a rule that assigns to each number x in a domain, a unique number y in the range. A function is simply another way of saying "a quantity y depends on x". The notation $y = f(x)$ for a function immediately suggests that a function can be represented geometrically by a curve — the curve with equation $y = f(x)$ — and we call this curve the graph of the function. Although graphs can always be drawn by plotting points from tables of values and joining these points smoothly, other methods such as addition and multiplication of ordinates, symmetry properties, and translations are frequently more expedient. In Chapters 2 and 4, we will discover further curve-sketching aids, and these will all be combined for a complete discussion in Section 4.5.

Key Terms and Formulas

In reviewing this chapter, you should be able to define or discuss the following key terms:

Open interval
Closed interval
Infinity
Sign diagram
Absolute value
Origin
Abscissa
Ordinate
Coordinates
Quadrant
Length of line segment
Midpoint of line segment
Equation of a curve
Slope of a line
Rise
Run
Point-slope formula
Parallel lines
Perpendicular lines
Inclination of a line

Parabola
Symmetry about x-axis
Symmetry about y-axis
Quadratic formula
Circle
Ellipse
Hyperbola
Asymptotes
Function
Domain
Range
Graph
Addition of ordinates
Multiplication of ordinates
Polynomial
Rational function
Even function
Odd function
Translation of a curve

REVIEW EXERCISES

In Exercises 1–6 find all values of x satisfying the inequality.

1. $2x - 6 < 3x + 5$
2. $|x^2 + 1| \leq 5$
3. $\dfrac{x-4}{2-x} \geq 5$
4. $\dfrac{x^2 - 14}{x - 3} < 2$
5. $|x^2 - 4| \geq 5$
6. $|x^2 - 5| > 4$

In Exercises 7 and 8 find the distance between the points and the midpoint of the line segment joining the points.

7. $(-1, 3), \quad (4, 2)$
8. $(2, 1), \quad (-3, -4)$

In Exercises 9–12 find the equation for the line described.

9. Parallel to the line $x - 2y = 4$ and through the point $(2, 3)$

10. Perpendicular to the line joining $(-2, 1)$ to the origin and through the midpoint of the line segment joining $(1, 3)$ and $(-1, 5)$

11. Perpendicular to the line $x = 4y - 11$ and through the point of intersection of this line and $x = \sqrt{y^2 + 9}$

12. Joining the points of intersection of the curves $y = x^2$ and $5x = 6 - y^2$

In Exercises 13–22 find the largest possible domain for the function.

13. $f(x) = \sqrt{x^2 + 5}$
14. $f(x) = \sqrt{x^2 - 5}$
15. $f(x) = \dfrac{1}{x^2 + 3x + 2}$
16. $f(x) = \dfrac{x+4}{x^3 + 2x^2 + x}$
17. $f(x) = (x^3 - 8)^{1/3}$
18. $f(x) = x^{3/2}$
19. $f(x) = \sqrt{x^2 + 4x - 6}$
20. $f(x) = \dfrac{1}{\sqrt{2x^2 + 4x - 5}}$

21. $f(x) = \sqrt{\dfrac{2x+1}{x-3} + 2}$ 22. $f(x) = \sqrt{x - \dfrac{1}{x}}$

In Exercises 23–34 identify the curve as a straight line, parabola, circle, ellipse, hyperbola, or none of these.

23. $x + 2y = 4$
24. $x = y^2 - 2y + 3$
25. $y = x^3 + 3$
26. $x^2 + 2y^2 = 4$
27. $y^2 - x^2 = x$
28. $x^2 + y^2 + 5 = 0$
29. $x^2 - 2x + y^2 = 16$
30. $x + y^2 = 3$
31. $x^2 + 2y^2 + y = 2x$
32. $x^2 - x + y^2 + y = 0$
33. $2x^2 + 20x + 38 = 3y^2 + 12y$
33. $2y^2 - x = 3x^2 - y$

In Exercises 35–60 sketch the curve.

35. $y = 2x^2 + 3$
36. $x^2 = 4 - y^2$
37. $y = x^3 - 1$
38. $|y| = |x|$
39. $y = x^3\sqrt{1 - x^2}$
40. $y = \sqrt{-x^2 + 4x + 4}$
41. $4x^2 + y^2 = 0$
42. $x^2 + 3y^2 = 6$
43. $2y^2 - x^2 = 3$
44. $x^2 - 2x - y^2 + 4y = 1$
45. $x = y\sqrt{y+3}$
46. $x^2 - 4y + 2 = 4x - 2y^2$
47. $y = \sin 3x$
48. $y = \cos(2x + \pi/2)$
49. $y = \cos(2x - \pi/4)$
50. $y = 2\sin(3x + \pi/2)$
51. $y = |x| + |x - 1|$
52. $y = \sqrt{|x - 1| - 1}$
53. $x = \tan y$
54. $y = 2\ln(3x + 4)$
55. $x = e^{-y}$
56. $y = \sec^2(x + \pi/4)$
57. $y = xe^{-x}$, $x \geq 0$
58. $y = \sin|x|$
59. $|y| = |\sin x|$
60. $x = \sqrt{\sin y}$

In Exercises 61-64 give an example of a function $y = f(x)$ with the indicated properties.

61. The range of the function consists of one number only.

62. The largest possible domain of the function is $-1 \leq x \leq 2$.

63. The domain of the function consists of all reals except $x = \pm 1$.

64. The domain of the function is $x \leq 0$ and the range is $y \geq 1$.

65. Is the triangle with vertices $(1, 3)$, $(2, -1)$, and $(-2, -2)$ right-angled?

66. Show that the lines that successively join the points $(0, 1)$, $(3, 4)$, $(3 - \sqrt{2}, 0)$, and $(-\sqrt{2}, -3)$ form the sides of a rhombus, which is a parallelogram with all sides of equal length.

67. Find the distance from the point $(2, 1)$ to the line $2x + 3y = 4$.

68. A lighthouse is 6 km off shore and a cabin on the straight shoreline is 9 km from the point on the shore nearest the lighthouse (Figure 1.96).

FIGURE 1.96

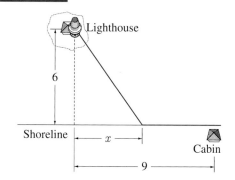

Show that if a man rows at 3 km/h and walks at 5 km/h, and he beaches the boat at x, then the total travel time from lighthouse to cabin is

$$t = f(x) = \dfrac{\sqrt{x^2 + 36}}{3} + \dfrac{9 - x}{5}, \quad 0 \leq x \leq 9.$$

Sketch a graph of this function.

69. Prove that the diagonals of a rhombus intersect at right angles.

CHAPTER 2

Limits and Continuity

The concept of a "limit" is crucial to calculus, for the two basic operations in calculus are differentiation and integration, each of which is defined in terms of a limit. For this reason we must have a clear understanding of limits from the beginning. In Sections 2.1–2.4 we give an intuitive discussion of limits of functions; in Section 2.5 we show how these ideas can be formalized mathematically.

SECTION 2.1

Limits

The value of the function $f(x) = x^2 - 4x + 5$ at $x = 2$ is $f(2) = 1$. A completely different consideration is contained in the question, "What number does $f(x) = x^2 - 4x + 5$ get closer and closer to as x gets closer and closer to 2?" Table 2.1 shows that *as x gets closer and closer to 2, $x^2 - 4x + 5$ gets close to 1 and stays close to 1.*

TABLE 2.1

x	$f(x) = x^2 - 4x + 5$	x
2.1	1.01	1.9
2.01	1.0001	1.99
2.001	1.000001	1.999
2.0001	1.00000001	1.9999

Likewise, the graph of the function (Figure 2.1) clearly shows that $f(x)$ approaches 1 as x approaches 2. This statement is not precise enough for our purposes. For instance, the graph also indicates that as x gets closer and closer to 2, $f(x)$ gets closer and closer to 0. It does not get very close to 0, but nonetheless, $f(x)$ does get closer and closer to 0 as x gets closer and closer to 2. In fact, we can make this statement for any number less than 1.

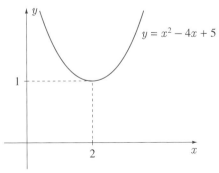

FIGURE 2.1

To distinguish 1 from all numbers less 1, we say that $x^2 - 4x + 5$ *can be made arbitrarily close to* 1 *by choosing* x *sufficiently close to* 2. Numbers less than 1 do not satisfy this statement. For instance, it is not true that $x^2 - 4x + 5$ can be made arbitrarily close to 0 by choosing x sufficiently close to 2. The closest the function gets to 0 is one unit when $x = 2$. In calculus we say that **the limit of** $x^2 - 4x + 5$ **as** x **approaches** 2 **is** 1 to represent the more lengthy statement "$x^2 - 4x + 5$ can be made arbitrarily close to 1 by choosing x sufficiently close to 2". In addition, we have a notation to represent both statements:

$$\lim_{x \to 2} (x^2 - 4x + 5) = 1. \tag{2.1}$$

This notation is read "the limit of (the function) $x^2 - 4x + 5$ as x approaches 2 is equal to 1", and this stands for the statement "$x^2 - 4x + 5$ can be made arbitrarily close to 1 by choosing x sufficiently close to 2".

We emphasize that the limit in 2.1 is not concerned with the value of $x^2 - 4x + 5$ at $x = 2$. It is concerned with the number that $x^2 - 4x + 5$ approaches as x approaches 2. These numbers are not always the same.

Generally we say that a function $f(x)$ has limit L as x approaches a, and write

$$\lim_{x \to a} f(x) = L, \tag{2.2}$$

if $f(x)$ can be made arbitrarily close to L by choosing x sufficiently close to a. Sometimes it is more convenient to write $f(x) \to L$ as $x \to a$ to mean that $f(x)$ approaches L as x approaches a. This is especially so in the middle of a paragraph, as opposed to a displayed equation such as 2.2.

EXAMPLE 2.1

Evaluate $\lim_{x \to 1} (x^2 + 2x + 5)$.

SOLUTION As x gets closer and closer to 1, $x^2 + 2x + 5$ gets closer and closer to 8, and we therefore write

$$\lim_{x \to 1} (x^2 + 2x + 5) = 8.$$

Once again this is corroborated by the graph of the function in Figure 2.2.

FIGURE 2.2

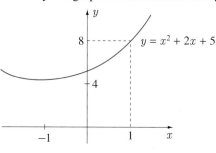

To calculate the limit of a function $f(x)$ as x approaches a, we evaluate $f(x)$ at values of x that get closer and closer to a. For limits of complicated functions such as

$$\lim_{x \to 5} \frac{x^2(3 - x)}{x^3 + x},$$

it would be tedious to evaluate $x^2(3 - x)/(x^3 + x)$ at many values of x approaching 5. The following theorem provides a much easier method.

> **Theorem 2.1**
>
> If $\lim_{x \to a} f(x) = F$ and $\lim_{x \to a} g(x) = G$, then
>
> (i) $\quad \lim_{x \to a} [f(x) + g(x)] = F + G;$ (2.3)
>
> (ii) $\quad \lim_{x \to a} [f(x) - g(x)] = F - G;$ (2.4)
>
> (iii) $\quad \lim_{x \to a} [cf(x)] = cF, \quad$ when c is a constant; (2.5)
>
> (iv) $\quad \lim_{x \to a} [f(x)g(x)] = FG;$ (2.6)
>
> (v) $\quad \lim_{x \to a} \dfrac{f(x)}{g(x)} = \dfrac{F}{G}, \quad$ provided $G \ne 0$. (2.7)

What this theorem says is that a limit such as $\lim_{x \to 5} [x^2(3-x)/(x^3+x)]$ can be broken down into smaller problems and reassembled later. For instance, since

$$\lim_{x \to 5} x^2 = 25, \quad \lim_{x \to 5} (3-x) = -2, \quad \lim_{x \to 5} x^3 = 125, \quad \lim_{x \to 5} x = 5,$$

we may write

$$\lim_{x \to 5} \frac{x^2(3-x)}{x^3+x} = \frac{25(-2)}{125+5} = \frac{-50}{130} = -\frac{5}{13}.$$

Although the results of Theorem 2.1 may seem evident, to prove them mathematically is not a simple task. In fact, because we have not yet given a precise definition for limits, a proof at this time is impossible. When we give definitions for limits in Section 2.5, it will then be possible to verify the theorem (see Exercises 31–35 in Section 2.5).

EXAMPLE 2.2

Evaluate $\lim_{x \to -2} [(x+2)/(x^2+9)]$.

SOLUTION Since $\lim_{x \to -2} (x+2) = 0$ and $\lim_{x \to -2} (x^2+9) = 13$, part (v) of Theorem 2.1 gives

$$\lim_{x \to -2} \frac{x+2}{x^2+9} = \frac{0}{13} = 0.$$

∎

EXAMPLE 2.3

Evaluate $\lim_{x \to -1} [x^2(1-x^3)/(2x^2+x+1)]$.

SOLUTION Using Theorem 2.1, we can write

$$\lim_{x \to -1} \frac{x^2(1-x^3)}{2x^2+x+1} = \frac{(1)(2)}{2+(-1)+1} = 1.$$

Be sure that you understand how we obtained the expression

$$\frac{(1)(2)}{2+(-1)+1}.$$

In particular, we *did not* set $x = -1$ in $x^2(1-x^3)/(2x^2+x+1)$. Indeed, this is not permitted because to evaluate a limit as x approaches -1, we are not to set $x = -1$; we are to let x get closer and closer to -1. What we did do is take limits of x^2, $1-x^3$, $2x^2$, and x as x approaches -1, and then use Theorem 2.1.

∎

The following example illustrates what can happen if we substitute $x = a$ into $f(x)$ in the evaluation of $\lim_{x \to a} f(x)$.

EXAMPLE 2.4 Evaluate $\lim_{x \to 3} [(x^2 - 9)/(x - 3)]$.

SOLUTION Because $\lim_{x \to 3} (x - 3) = 0$, we cannot use Theorem 2.1. Nor can we set $x = 3$ in $(x^2 - 9)/(x - 3)$ because it is inherent in the limiting procedure that we do not put $x = 3$. Besides, if we did, we would obtain the meaningless expression $0/0$. The correct procedure is to factor $x^2 - 9$ into $(x-3)(x+3)$ and cancel a factor of $x - 3$ from numerator and denominator:

$$\lim_{x \to 3} \frac{x^2 - 9}{x - 3} = \lim_{x \to 3} \frac{(x-3)(x+3)}{x - 3} = \lim_{x \to 3} (x + 3) = 6.$$

Cancellation of the factor $x-3$ would not be permissible if $x-3$ were equal to 0; that is, if x were equal to 3. But once again this cannot happen, because in the limiting operation we let x get closer and closer to 3, but do not set $x = 3$. ∎

Note in this example that although the limit is 6, there is no value of x for which the function $(x^2 - 9)/(x - 3)$ is ever equal to 6. The graph of the function in Figure 2.3 is a straight line with the point at $x = 3$ removed. The function is undefined at $x = 3$, but as x approaches 3, $f(x)$ clearly gets arbitrarily close to 6.

FIGURE 2.3

EXAMPLE 2.5 Evaluate $\lim_{x \to 0} [(-2x + 3x^2)/(4x - 5x^2)]$.

SOLUTION We have

$$\lim_{x \to 0} \frac{-2x + 3x^2}{4x - 5x^2} = \lim_{x \to 0} \frac{x(-2 + 3x)}{x(4 - 5x)} = \lim_{x \to 0} \frac{3x - 2}{4 - 5x} = \frac{-2}{4} = -\frac{1}{2}.$$

∎

EXAMPLE 2.6 Evaluate

$$\lim_{x \to 0} \frac{\sqrt{1 + x} - 1}{\sqrt{x}}.$$

SOLUTION Since the limit of the denominator as x approaches 0 is 0, we cannot immediately use Theorem 2.1. The best procedure is to rationalize the numerator, that is, rid the numerator of the square root by multiplying numerator and denominator by $\sqrt{1 + x} + 1$:

$$\lim_{x\to 0}\frac{\sqrt{1+x}-1}{\sqrt{x}} = \lim_{x\to 0}\left(\frac{\sqrt{1+x}-1}{\sqrt{x}}\frac{\sqrt{1+x}+1}{\sqrt{1+x}+1}\right)$$
$$= \lim_{x\to 0}\frac{x}{\sqrt{x}\left(\sqrt{1+x}+1\right)}$$
$$= \lim_{x\to 0}\frac{\sqrt{x}}{\sqrt{1+x}+1}$$
$$= 0.$$

EXAMPLE 2.7

Evaluate
$$\lim_{x\to 0}\frac{\sin 2x}{\sin x}.$$

SOLUTION Once again we cannot immediately use Theorem 2.1 since the limit of the denominator is zero. But using the double-angle formula $\sin 2x = 2\sin x\cos x$, we find that

$$\lim_{x\to 0}\frac{\sin 2x}{\sin x} = \lim_{x\to 0}\frac{2\sin x\cos x}{\sin x} = \lim_{x\to 0}(2\cos x) = 2.$$

EXAMPLE 2.8

Do the functions
$$f(x) = \sin\left(\frac{1}{x}\right) \quad\text{and}\quad g(x) = x\sin\left(\frac{1}{x}\right)$$
have limits as x approaches 0?

SOLUTION The best way to find out whether these particular functions have limits as x approaches 0 is to sketch their graphs. Suppose we divide the positive x-axis into two parts $0 < x \leq 1$ and $1 < x < \infty$. Now the function $\sin x$, with period 2π, has part of a cycle in the interval $0 < x \leq 1$, and an infinite number of oscillations in $1 < x < \infty$ (Figure 2.4).

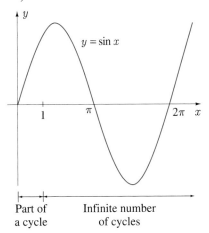

FIGURE 2.4

The function $\sin(1/x)$ performs two operations; it first inverts a value of x to give $1/x$ and then takes the sine of this number. Given a value of x in the interval $0 < x < 1$, $\sin(1/x)$ first evaluates $1/x$ which is in the interval $1 < 1/x < \infty$, and then takes the sine of $1/x$. Conversely, given an x in $1 < x < \infty$, $\sin(1/x)$ first finds $1/x$, which is in the interval $0 < 1/x < 1$, and then calculates $\sin(1/x)$. The infinite number of oscillations of $\sin x$ in the interval $1 < x < \infty$ take place in the interval $0 < x < 1$ for $\sin(1/x)$. Furthermore, the part cycle of $\sin x$ in $0 < x < 1$ is spread out over $1 < x < \infty$ by $\sin(1/x)$. This is shown in Figure 2.5(a). Notice in particular that $\sin(1/x)$ is equal to 0 when $x = 1/(n\pi)$ for every positive integer n. Since $\sin(1/x)$ is an odd function, its full graph is shown in Figure 2.5(b). The function oscillates more and more rapidly as x approaches 0, and clearly $f(x) = \sin(1/x)$ cannot have a limit there.

FIGURE 2.5

FIGURE 2.6

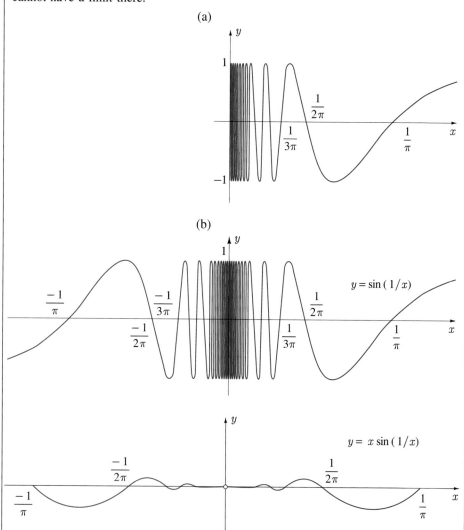

The graph of $g(x) = x\sin(1/x)$ is easily obtained by multiplication of ordinates of $y = x$ and $y = \sin(1/x)$. The result is shown in Figure 2.6 between $-1/\pi$ and $1/\pi$. Once again there is an infinite number of oscillations, but the factor x before $\sin(1/x)$ in $g(x)$ causes the "amplitude" to approach 0 as x approaches 0. In other words, $\lim_{x \to 0} g(x) = 0$. ∎

One-sided Limits

When we write $L = \lim_{x \to a} f(x)$, we mean that $f(x)$ gets arbitrarily close to L as x gets closer and closer to a. But how is x to approach a? Does x approach a through numbers larger than a, or does it approach a through numbers smaller than a? Or does x jump back and forth between numbers larger than a and numbers smaller than a, gradually getting closer and closer to a? We have not previously mentioned "mode" of approach simply because it would have made no difference to our discussion. In each of the preceding examples, all possible modes of approach lead to the same limit. In particular, Table 2.1 and Figure 2.1 illustrate that 1 is the limit of $f(x) = x^2 - 4x + 5$ as x approaches 2 whether x approaches 2 through numbers larger than 2 or through numbers smaller than 2.

Approaching a number a either through numbers larger than a or through numbers smaller than a are two modes of approach that will be very important, and we therefore give them special notations:

$$\lim_{x \to a^-} f(x)$$
indicates that x approaches a through numbers smaller than a (often called a **left-hand limit** since x approaches a along the x-axis from the left of a)

$$\lim_{x \to a^+} f(x)$$
indicates that x approaches a through numbers larger than a (often called a **right-hand limit** since x approaches a along the x-axis from the right of a)

Example 2.6 should, in fact, be designated a right-hand limit,

$$\lim_{x \to 0^+} \frac{\sqrt{1+x} - 1}{\sqrt{x}} = 0,$$

since the presence of $\sqrt{x}$ in the denominator demands that x be positive.

A natural question to ask is "What should we conclude if for a function $f(x)$

$$\lim_{x \to a^+} f(x) \neq \lim_{x \to a^-} f(x)?"$$

Our entire discussion has suggested (and indeed it can be proved; see Exercise 20 in Section 2.5) that if a function has a limit as x approaches a, then it has only one such limit; that is, the limit must be the same for every possible method of approach. Consequently, if we arrive at two different results depending on the mode of approach, then we conclude that the function does not have a limit. This situation is illustrated in the following example.

EXAMPLE 2.9

Evaluate, if possible, $\lim_{x \to 0} (|x|/x)$.

SOLUTION If $x < 0$, then $|x| = -x$, and

$$\lim_{x \to 0^-} \frac{|x|}{x} = \lim_{x \to 0^-} \frac{-x}{x} = -1.$$

If $x > 0$ then $|x| = x$, and

$$\lim_{x \to 0^+} \frac{|x|}{x} = \lim_{x \to 0^+} \frac{x}{x} = 1.$$

The function has a right-hand limit and a left-hand limit at $x = 0$, but because they are not the same, $\lim_{x \to 0} (|x|/x)$ does not exist. The graph of $f(x) = |x|/x$ in Figure 2.7 clearly illustrates the situation.

FIGURE 2.7

EXERCISES 2.1

In Exercises 1–40 find the indicated limit, if it exists.

1. $\lim_{x \to 7} \dfrac{x^2 - 5}{x + 2}$

2. $\lim_{x \to -2} \dfrac{x^3 + 8}{x + 5}$

3. $\lim_{x \to -5} \dfrac{x^2 + 3x + 2}{x^2 + 25}$

4. $\lim_{x \to 0} \dfrac{x^2 + 3x}{3x^2 - 2x}$

5. $\lim_{x \to 3^+} \dfrac{2x - 3}{x^2 - 5}$

6. $\lim_{x \to 2^-} \dfrac{2x - 4}{3x + 2}$

7. $\lim_{x \to 0^-} \dfrac{x^4 + 5x^3}{3x^4 - x^3}$

8. $\lim_{x \to 2^+} \dfrac{x^2 + 2x + 4}{x - 3}$

9. $\lim_{x \to 2} \dfrac{x^2 - 4}{x - 2}$

10. $\lim_{x \to 3^+} \dfrac{x^2 - 9}{x - 3}$

11. $\lim_{x \to 5^-} \dfrac{x^2 - 25}{x - 5}$

12. $\lim_{x \to 3} \dfrac{x^2 - 2x - 3}{3 - x}$

13. $\lim_{x \to 2} \dfrac{x^2 - 4x + 4}{x - 2}$

14. $\lim_{x \to 2} \dfrac{x^3 - 6x^2 + 12x - 8}{x^2 - 4x + 4}$

15. $\lim_{x \to 1} \dfrac{x^3 - 6x^2 + 11x - 6}{x^2 - 3x + 2}$

16. $\lim_{x \to 2} \dfrac{x^3 - 6x^2 + 11x - 6}{x^2 - 3x + 2}$

17. $\lim_{x \to 3^+} \dfrac{x^3 - 6x^2 + 11x - 6}{x^2 - 3x + 2}$

18. $\lim_{x \to 3^-} \dfrac{x^3 - 6x^2 + 11x - 6}{x^2 - 3x + 2}$

19. $\lim_{x \to 0} \dfrac{x^3 - 6x^2 + 11x - 6}{x^2 - 3x + 2}$

20. $\lim_{x \to -1} \dfrac{12x + 5}{x^2 - 2x + 1}$

21. $\lim_{x \to 1} \sqrt{\dfrac{2 - x}{2 + x}}$

22. $\lim_{x \to 5} \dfrac{\sqrt{1 - x^2}}{3x + 2}$

23. $\lim_{x \to 0} \dfrac{\sin 4x}{\sin 2x}$

24. $\lim_{x \to 0^+} \dfrac{\sin 6x}{\sin 3x}$

25. $\lim_{x \to 0} \dfrac{\tan x}{\sin x}$

26. $\lim_{x \to \pi/4} \dfrac{\sin x}{\tan x}$

27. $\lim_{x \to 0^+} \dfrac{\sin 2x}{\tan x}$

28. $\lim_{x \to 2} \dfrac{x - 2}{\sqrt{x} - \sqrt{2}}$

29. $\lim_{x \to 0} \dfrac{\sqrt{1 - x} - \sqrt{1 + x}}{x}$

30. $\lim_{x \to 5^+} \dfrac{|x^2 - 25|}{x^2 - 25}$

31. $\lim_{x \to 5^-} \dfrac{|x^2 - 25|}{x^2 - 25}$

32. $\lim_{x \to 5} \dfrac{|x^2 - 25|}{x^2 - 25}$

33. $\lim_{x \to 0^+} \dfrac{\sqrt{x + 2} - 2}{\sqrt{x}}$

34. $\lim_{x \to 0} \dfrac{1 - \sqrt{x^2 + 1}}{2x^2}$

35. $\lim_{x \to -2} \dfrac{\sqrt{-x} + \sqrt{2}}{x + 2}$

36. $\lim_{x \to 0} \dfrac{\sqrt{1 + x} - \sqrt{1 - x}}{x}$

37. $\lim_{x \to -2^+} \dfrac{\sqrt{x + 3} - \sqrt{-x - 1}}{\sqrt{x + 2}}$

38. $\lim_{x \to 0} \dfrac{x}{\sqrt{x + 4} - 2}$

39. $\lim_{x \to 0} \dfrac{\sqrt{1 + x} - \sqrt{1 - x}}{\sqrt{2 + x} - \sqrt{2 - x}}$

40. $\lim_{x \to 0} \dfrac{\sqrt{x + 1} - \sqrt{2x + 1}}{\sqrt{3x + 4} - \sqrt{2x + 4}}$

41. Does the greatest integer function $[[x]]$ of Exercise 36 in Section 1.5 have a limit, a right-hand limit, or a left-hand limit as x approaches integer values?

42. Prove or disprove the following statement: If

$f(x) < g(x)$ for all $x \neq a$, then

$$\lim_{x \to a} f(x) < \lim_{x \to a} g(x).$$

43. If n is a positive integer, evaluate

$$\lim_{h \to 0} \frac{(x+h)^n - x^n}{h}.$$

Hint: Use either the binomial theorem or the result that

$$a^n - b^n = (a-b)(a^{n-1} + a^{n-2}b + \cdots + ab^{n-2} + b^{n-1}).$$

44. Evaluate

$$\lim_{h \to 0} \frac{\sqrt{x+h} - \sqrt{x}}{h}.$$

45. The weight W of an object depends on its distance d from the centre of the earth. If d is less than the radius R of the earth, then W is directly proportional to d; and if d is greater than or equal to R, then W is inversely proportional to d^2. If the weight of the object on the earth's surface is W_0, find a definition for W as a function of d, and sketch its graph.

46. At the present time it is impossible for us to calculate algebraically

$$\lim_{x \to 0} \frac{\sin x - x}{x^3}.$$

What can we do? One suggestion might be to use a calculator to evaluate the function $(\sin x - x)/x^3$ for various values of x that approach 0. Try this with $x = 0.1, 0.001, 0.00001, 0.0000001, 0.000000001$, and make any conclusion you feel is justified. How would you feel about the use of the calculator if you knew that the value of the limit were $-1/6$?

47. If

$$\lim_{x \to a} \frac{f(x) - g(x)}{x - a}$$

exists, and $\lim_{x \to a} f(x) = L$, find $\lim_{x \to a} g(x)$.

48. If $f(x)$ is an even function and $\lim_{x \to a} f(x) = L$, find, if possible, $\lim_{x \to -a} f(x)$.

49. If $f(x)$ is an even function and $\lim_{x \to a^+} f(x) = L$, find, if possible, $\lim_{x \to -a^-} f(x)$.

50. If $f(x)$ is an even function and $\lim_{x \to a^+} f(x) = L$, find, if possible, $\lim_{x \to -a^+} f(x)$.

51.–54. Repeat Exercises 48.–50. for an odd function $f(x)$.

SECTION 2.2

Infinite Limits

Functions do not always have limits. Sometimes this is due to the fact that right-hand and left-hand limits are not identical; sometimes it is a result of erratic oscillations. Examples of both of these situations were discussed in Section 2.1 (see Examples 2.9 and 2.8). Nonexistence of a limit may also be due to excessively large values of the function. For instance, consider the function $f(x) = 1/(x-2)^2$ which is not defined at $x = 2$. Does it have a limit as x approaches 2? Table 2.2 indicates that as x approaches 2, values of $1/(x-2)^2$ become very large; in fact, values of the function can be made arbitrarily large by choosing x sufficiently close to 2. Thus, the function does not have a limit as x approaches 2. To express this fact we write

$$\lim_{x \to 2} \frac{1}{(x-2)^2} = \infty,$$

where ∞ once again represents infinity. This statement does not say that the limit exists, since ∞ is not a number. It states that the *limit does not exist*, and indicates that the reason it does not exist is that values of the function become arbitrarily large as x approaches 2. The graph of $f(x)$ in Figure 2.8 further illustrates this point. We say that the line $x = 2$ is a **vertical asymptote** for the graph.

TABLE 2.2

x	$f(x)$	x
1.9	10^2	2.1
1.99	10^4	2.01
1.999	10^6	2.001
1.9999	10^8	2.0001
1.99999	10^{10}	2.00001

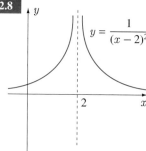

FIGURE 2.8

EXAMPLE 2.10 Evaluate $\lim_{x \to 1}[1/(x-1)]$, if it exists.

SOLUTION In this example we consider right-hand and left-hand limits as x approaches 1. We find that

$$\lim_{x \to 1^+} \frac{1}{x-1} = \infty \quad \text{and} \quad \lim_{x \to 1^-} \frac{1}{x-1} = -\infty,$$

the latter meaning that as x approaches 1 from the left, the function takes on arbitrarily "large" negative values. Either one of these expressions is sufficient to conclude that the function $1/(x-1)$ does not have a limit as x approaches 1. The function is shown in Figure 2.9; the line $x = 1$ is a vertical asymptote.

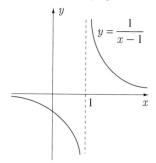

FIGURE 2.9

EXAMPLE 2.11 What are left-hand and right-hand limits of the function

$$f(x) = \frac{x^2 - 9}{x^2 + x - 2}$$

at $x = 1$ and $x = -2$?

SOLUTION The function is undefined at $x = 1$ and $x = -2$ (why?). To take limits as $x \to 1$ and $x \to -2$, it is advantageous to factor numerator and denominator of $f(x)$ as much as possible,

$$f(x) = \frac{(x+3)(x-3)}{(x+2)(x-1)}.$$

Consider now the (right-hand) limit as $x \to 1^+$. Each of the four factors has a limit as $x \to 1^+$:

$$x+3 \to 4, \qquad x-3 \to -2, \qquad x+2 \to 3, \qquad x-1 \to 0.$$

Were we to combine them according to Theorem 2.1, we would write

$$\frac{4(-2)}{3(0)}.$$

In spite of the fact that this is not correct — part (v) of the theorem does not allow a 0 in the denominator, this expression does tell us what is happening to $f(x)$ as $x \to 1^+$. It states that as $x \to 1^+$, the numerator of $f(x)$ approaches -8 and the denominator approaches 0. But this means that $f(x)$ must be taking on larger and larger values as $x \to 1^+$. Are these values positive or negative? The numerator is clearly negative, but the sign of the denominator depends on whether the 0 is approached through positive or negative numbers. Recalling that 0 arose from the fact that $x - 1 \to 0$ as $x \to 1^+$, we can be more specific; $x-1$ must approach 0 through positive numbers since $x > 1$ for $x \to 1^+$. We indicate this by writing $x - 1 \to 0^+$ as $x \to 1^+$. The above fraction is therefore replaced by

$$\frac{4(-2)}{3(0^+)},$$

and it is clearly negative. In other words,

$$\lim_{x \to 1^+} \frac{x^2 - 9}{x^2 + x - 2} = \lim_{x \to 1^+} \frac{(x+3)(x-3)}{(x+2)(x-1)} = -\infty.$$

Similarly, as $x \to 1^-$, the fraction

$$\frac{4(-2)}{3(0^-)},$$

which shows limits of the four factors of $f(x)$, indicates that

$$\lim_{x \to 1^-} \frac{(x+3)(x-3)}{(x+2)(x-1)} = \infty.$$

In the following, the expressions at the right of the page yield limits as $x \to -2^+$ and $x \to -2^-$:

$$\lim_{x \to -2^+} \frac{(x+3)(x-3)}{(x+2)(x-1)} = \infty, \qquad\qquad \frac{(1)(-5)}{(0^+)(-3)}$$

$$\lim_{x \to -2^-} \frac{(x+3)(x-3)}{(x+2)(x-1)} = -\infty. \qquad\qquad \frac{(1)(-5)}{(0^-)(-3)}$$

These expressions will be useful when limits are used to sketch a graph of this function in Section 2.4.

EXERCISES 2.2

In Exercises 1–24 evaluate the limit, if it exists.

1. $\lim\limits_{x \to 2^+} \dfrac{1}{x-2}$
2. $\lim\limits_{x \to 2^-} \dfrac{1}{x-2}$
3. $\lim\limits_{x \to 2} \dfrac{1}{x-2}$
4. $\lim\limits_{x \to 2^+} \dfrac{1}{(x-2)^2}$
5. $\lim\limits_{x \to 2^-} \dfrac{1}{(x-2)^2}$
6. $\lim\limits_{x \to 2} \dfrac{1}{(x-2)^2}$
7. $\lim\limits_{x \to 1} \dfrac{5x}{(x-1)^3}$
8. $\lim\limits_{x \to 1/2} \dfrac{6x^2 + 7x - 5}{2x - 1}$
9. $\lim\limits_{x \to 1} \dfrac{2x + 3}{x^2 - 2x + 1}$
10. $\lim\limits_{x \to 2} \dfrac{x - 2}{x^2 - 4x + 4}$
11. $\lim\limits_{x \to \pi/2^+} \tan x$
12. $\lim\limits_{x \to \pi/2^-} \tan x$
13. $\lim\limits_{x \to 0} \csc x$
14. $\lim\limits_{x \to \pi/4} \sec(x - \pi/4)$
15. $\lim\limits_{x \to 3\pi/4} \sec(x - \pi/4)$
16. $\lim\limits_{x \to 0^+} \cot x$
17. $\lim\limits_{x \to 1} \dfrac{x^2 - 2x + 1}{x^3 - 3x^2 + 3x - 1}$
18. $\lim\limits_{x \to 0} \dfrac{\sqrt{1+x} - 1}{x^2}$
19. $\lim\limits_{x \to 0} \dfrac{2x}{1 - \sqrt{x^2 + 1}}$
20. $\lim\limits_{x \to 4} \dfrac{|4 - x|}{x^2 - 8x + 16}$
21. $\lim\limits_{x \to 0^+} \ln(4x)$
22. $\lim\limits_{x \to 1} \dfrac{1}{\ln|x - 1|}$
23. $\lim\limits_{x \to 0} e^{1/x}$
24. $\lim\limits_{x \to 0} e^{1/|x|}$

25. It is not clear whether the limit

$$\lim_{x \to 0^+} x^2 \ln x$$

exists due to the fact that $\lim_{x \to 0^+} x^2 = 0$ and $\lim_{x \to 0^+} \ln x = -\infty$. It depends on which term is more dominant in the product, x^2 or $\ln x$. Calculate x^2, $\ln x$, and $x^2 \ln x$ for $x = 10^{-2}$, 10^{-4}, 10^{-6}, and 10^{-8}, and use this information to decide on a value for the limit.

26. Repeat Exercise 25 for the limit

$$\lim_{x \to 0^+} x^{10} e^{1/x},$$

but pick your own values of x at which to evaluate x^{10} and $e^{1/x}$.

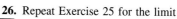

SECTION 2.3

Limits at Infinity

In many applications we are concerned with the behaviour of a function as its independent variable takes on very large values, positively or negatively. For instance, consider finding, if possible, a number that the function $f(x) = (2x^2 + 3)/(x^2 + 1)$, with domain $x \geq 0$, gets closer and closer to as x becomes larger and larger and larger. When x is very large, it is certainly not equal to zero, and we may therefore divide numerator and denominator by x^2:

$$f(x) = \frac{2x^2 + 3}{x^2 + 1} = \frac{2 + \dfrac{3}{x^2}}{1 + \dfrac{1}{x^2}}.$$

For very large x, the terms $3/x^2$ and $1/x^2$ are very close to zero, and therefore $f(x)$ is approximately equal to 2. Indeed $f(x)$ can be made arbitrarily close to 2 by choosing x sufficiently large. In calculus we express this fact by saying **the limit of** $(2x^2 + 3)/(x^2 + 1)$ **as x approaches infinity is** 2, and write

$$\lim_{x \to \infty} \frac{2x^2 + 3}{x^2 + 1} = 2.$$

Once again we stress that ∞ is not a number. The notation $x \to \infty$ simply means "as x gets larger and larger and larger". The graph of this function is shown in Figure 2.10. We say that the line $y = 2$ is a **horizontal asymptote**.

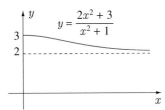

FIGURE 2.10

We may also find limits of functions as x takes on arbitrarily large negative numbers, denoted $x \to -\infty$.

EXAMPLE 2.12

Evaluate
$$\lim_{x \to -\infty} \frac{5x^4 - 3x + 5}{x^4 - 2x^2 + 5},$$
if it exists.

SOLUTION Division of numerator and denominator by x^4 leads to
$$\lim_{x \to -\infty} \frac{5 - \frac{3}{x^3} + \frac{5}{x^4}}{1 - \frac{2}{x^2} + \frac{5}{x^4}} = 5.$$

If the limit in the above example is taken as $x \to \infty$, the same result is obtained. The graph of this function is shown in Figure 2.11. The line $y = 5$ is a horizontal asymptote.

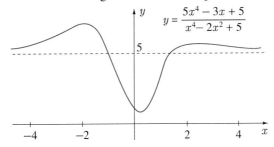

FIGURE 2.11

In general we say that a line $y = L$ is a horizontal asymptote for the graph of a function $f(x)$ if either of the following situations exists:

$$\lim_{x \to -\infty} f(x) = L \quad \text{or} \quad \lim_{x \to \infty} f(x) = L.$$

In Example 2.12, both of these conditions are satisfied for $L = 5$.

EXAMPLE 2.13

Evaluate the following limits:

(a) $\lim_{x \to \infty} \dfrac{2x^3 - 4}{x^3 + x^2 + 2}$ (b) $\lim_{x \to -\infty} \dfrac{2x^2 - 14}{3x^3 + 5x}$ (c) $\lim_{x \to \infty} \dfrac{2x^4 - 14}{3x^3 + 5x}$

SOLUTION (a) To obtain this limit we divide numerator and denominator of the fraction by x^3:

$$\lim_{x \to \infty} \frac{2x^3 - 4}{x^3 + x^2 + 2} = \lim_{x \to \infty} \frac{2 - \dfrac{4}{x^3}}{1 + \dfrac{1}{x} + \dfrac{2}{x^3}} = 2.$$

(b) Once again we divide numerator and denominator by x^3:

$$\lim_{x \to -\infty} \frac{2x^2 - 14}{3x^3 + 5x} = \lim_{x \to -\infty} \frac{\dfrac{2}{x} - \dfrac{14}{x^3}}{3 + \dfrac{5}{x^2}}.$$

Since the numerator approaches 0 and the denominator approaches 3, we conclude that

$$\lim_{x \to -\infty} \frac{2x^2 - 14}{3x^3 + 5x} = 0.$$

We could also have obtained this limit by dividing numerator and denominator by x^2 instead of x^3,

$$\lim_{x \to -\infty} \frac{2x^2 - 14}{3x^3 + 5x} = \lim_{x \to -\infty} \frac{2 - \dfrac{14}{x^2}}{3x + \dfrac{5}{x}}.$$

Now the numerator approaches 2, but since the denominator becomes very large, the fraction once again approaches zero.

(c) Division by x^3 in this case gives

$$\lim_{x \to \infty} \frac{2x^4 - 14}{3x^3 + 5x} = \lim_{x \to \infty} \frac{2x - \dfrac{14}{x^3}}{3 + \dfrac{5}{x^2}}.$$

Since the numerator becomes arbitrarily large as $x \to \infty$ and the denominator approaches 3, it follows that

$$\lim_{x \to \infty} \frac{2x^4 - 14}{3x^3 + 5x} = \infty.$$

∎

EXAMPLE 2.14

Sketch a graph of the function

$$f(x) = \frac{1}{x} \sin x, \quad x \geq \pi.$$

Does it have a horizontal asymptote?

SOLUTION Multiplication of ordinates of the graphs of $y = 1/x$ and $y = \sin x$ in Figure 2.12(a) gives the graph in Figure 2.12(b). The positive x-axis is a horizontal asymptote. The graph actually crosses the asymptote an infinite number of times.

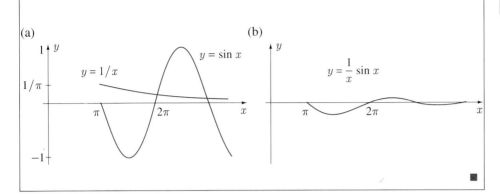

FIGURE 2.12

EXAMPLE 2.15

Evaluate

$$\lim_{x\to\infty} \frac{\sqrt{2x^2+4}}{x+5},$$

if it exists. What is the limit as $x \to -\infty$?

SOLUTION When we divide numerator and denominator by x, and take the x inside the square root as x^2,

$$\lim_{x\to\infty} \frac{\sqrt{2x^2+4}}{x+5} = \lim_{x\to\infty} \frac{\frac{1}{x}\sqrt{2x^2+4}}{\frac{1}{x}(x+5)} = \lim_{x\to\infty} \frac{\sqrt{\frac{2x^2+4}{x^2}}}{1+\frac{5}{x}}$$

$$= \lim_{x\to\infty} \frac{\sqrt{2+\frac{4}{x^2}}}{1+\frac{5}{x}} = \sqrt{2}.$$

In evaluating the limit as $x \to -\infty$, we must be more careful,

$$\lim_{x\to-\infty} \frac{\sqrt{2x^2+4}}{x+5} = \lim_{x\to-\infty} \frac{\frac{\sqrt{2x^2+4}}{x}}{1+\frac{5}{x}}.$$

It is not correct in this case to take the x inside the square root as

$$\frac{\sqrt{2x^2+4}}{x} = \sqrt{\frac{2x^2+4}{x^2}}$$

since for negative x, the expression on the left is negative and that on the right is positive. In this case, we should replace x by $-\sqrt{x^2}$, and write

$$\frac{\sqrt{2x^2+4}}{x} = \frac{\sqrt{2x^2+4}}{-\sqrt{x^2}} = -\sqrt{\frac{2x^2+4}{x^2}}.$$

Hence,

$$\lim_{x\to-\infty} \frac{\sqrt{2x^2+4}}{x+5} = \lim_{x\to-\infty} \frac{-\sqrt{\frac{2x^2+4}{x^2}}}{1+\frac{5}{x}} = \lim_{x\to-\infty} \frac{-\sqrt{2+\frac{4}{x^2}}}{1+\frac{5}{x}} = -\sqrt{2}.$$

The graph of this function (Figure 2.13) has two horizontal asymptotes, $y = \sqrt{2}$ and $y = -\sqrt{2}$.

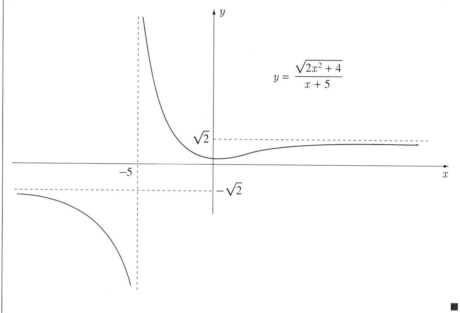

FIGURE 2.13

EXERCISES 2.3

In Exercises 1–38 evaluate the limit, if it exists.

1. $\lim\limits_{x\to\infty} \dfrac{x+1}{2x-1}$
2. $\lim\limits_{x\to\infty} \dfrac{1-x}{3+2x}$
3. $\lim\limits_{x\to\infty} \dfrac{x^2+1}{2x^3+5}$
4. $\lim\limits_{x\to\infty} \dfrac{1-4x^3}{3+2x-x^2}$
5. $\lim\limits_{x\to-\infty} \dfrac{2+x-x^2}{3+4x^2}$
6. $\lim\limits_{x\to-\infty} \dfrac{x^3-2x^2}{3x^3+4x^2}$
7. $\lim\limits_{x\to-\infty} \dfrac{x^3-2x^2+x+1}{x^4+3x}$
8. $\lim\limits_{x\to-\infty} \dfrac{x^3-2x^2+x+1}{x^2-x+1}$
9. $\lim\limits_{x\to\infty} \dfrac{\sqrt{x^2+1}}{2x+1}$
10. $\lim\limits_{x\to\infty} \dfrac{3x-1}{\sqrt{5+4x^2}}$
11. $\lim\limits_{x\to-\infty} \dfrac{\sqrt{1-2x^2}}{x+2}$
12. $\lim\limits_{x\to-\infty} \dfrac{\sqrt{1-2x}}{x+2}$
13. $\lim\limits_{x\to\infty} \sqrt{\dfrac{2+x}{x-2}}$
14. $\lim\limits_{x\to\infty} \dfrac{\sqrt{3+x}}{\sqrt{x}}$
15. $\lim\limits_{x\to\infty} (x^2-x^3)$
16. $\lim\limits_{x\to\infty} \left(x+\dfrac{1}{x}\right)$
17. $\lim\limits_{x\to\infty} \dfrac{x}{\sqrt{x+5}}$
18. $\lim\limits_{x\to-\infty} \dfrac{x^2}{\sqrt{3-x}}$
19. $\lim\limits_{x\to-\infty} \dfrac{x}{\sqrt[3]{4+x^3}}$
20. $\lim\limits_{x\to\infty} \dfrac{3x}{\sqrt[3]{2+4x^3}}$
21. $\lim\limits_{x\to\infty} \dfrac{1}{2x}\cos x$
22. $\lim\limits_{x\to-\infty} \dfrac{1}{2x}\cos x$
23. $\lim\limits_{x\to\infty} \dfrac{\sin 4x}{x^2}$
24. $\lim\limits_{x\to\infty} \dfrac{\sin^2 x}{x}$
25. $\lim\limits_{x\to-\infty} \tan x$
26. $\lim\limits_{x\to\infty} \dfrac{1}{x}\tan x$
27. $\lim\limits_{x\to\infty} \left(\sqrt{x^2+1}-x\right)$
28. $\lim\limits_{x\to\infty} \left(\sqrt{x^2+4}-x\right)$
29. $\lim\limits_{x\to\infty} \left(\sqrt{2x^2+1}-x\right)$
30. $\lim\limits_{x\to-\infty} \left(\sqrt{2x^2+1}-x\right)$

31. $\lim_{x \to \infty} \dfrac{\sqrt{3x^2 + 2}}{x + 4}$

32. $\lim_{x \to \infty} \dfrac{\sqrt{4x^2 + 7}}{2x + 3}$

33. $\lim_{x \to -\infty} \dfrac{\sqrt{3x^2 + 2}}{x + 4}$

34. $\lim_{x \to -\infty} \dfrac{\sqrt{4x^2 + 7}}{2x + 3}$

35. $\lim_{x \to \infty} \left(\sqrt{x^2 + 4} - \sqrt{x^2 - 1} \right)$

36. $\lim_{x \to \infty} \left(\sqrt[3]{1 + x} - \sqrt[3]{x} \right)$

37. $\lim_{x \to \infty} \left(\sqrt{x^2 + x} - x \right)$

38. $\lim_{x \to -\infty} \left(\sqrt{x^2 + x} - x \right)$

39. What is the value of

$$\lim_{x \to -\infty} \dfrac{\sqrt{ax^2 + bx + c}}{dx + e}$$

where $a > 0$, b, c, d, and e are constants?

40. What conditions on the constants a, b, c, d, e, and f will ensure that the

$$\lim_{x \to \infty} \left(\sqrt{ax^2 + bx + c} - \sqrt{dx^2 + ex + f} \right)$$

exists? What is the value of the limit in this case?

SECTION 2.4 Continuity

We have noticed that sometimes the limit of a function $f(x)$ as $x \to a$ is the same as the value $f(a)$ of the function at $x = a$. This property is described in the following definition.

Definition 2.1

A function $f(x)$ is said to be **continuous** at $x = a$ if

$$\lim_{x \to a} f(x) = f(a). \qquad (2.8)$$

There are three conditions imposed by this definition:

(1) $f(x)$ must be defined at $x = a$;

(2) $\lim_{x \to a} f(x)$ must exist;

(3) the numbers in (1) and (2) must be the same.

The graph in Figure 2.14 illustrates a function that is discontinuous (that is, not continuous) at $x = a$, b, c, d, and e. For instance, at $x = a$ conditions (1) and (2) are satisfied but (3) is violated; at $x = b$, (2) is satisfied, but (1) and (3) are not. Figure 2.14 suggests that discontinuities of a function are characterized geometrically by separations in its graph. This is indeed true, and it is often a very informative way to illustrate the nature of a discontinuity.

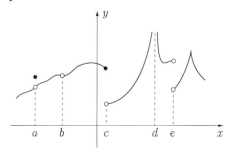

FIGURE 2.14

When a function is defined on a closed interval $a \leq x \leq b$, Definition 2.1 must be rephrased in terms of right- and left-hand limits for continuity at $x = a$ and $x = b$. Specifically, we say that $f(x)$ is **continuous from the right** at $x = a$ if $\lim_{x \to a^+} f(x) = f(a)$; and that $f(x)$ is **continuous from the left** at $x = b$ if $\lim_{x \to b^-} f(x) = f(b)$.

A function is said to be **continuous on an interval** if it is continuous at each point of that interval. In the event that the interval is closed, $a \leq x \leq b$, continuity at $x = a$ and $x = b$ is interpreted as continuity from the right and left, respectively. Geometrically, a function is continuous on an interval if a pencil can completely trace its graph without being lifted from the page.

EXAMPLE 2.16

Sketch graphs of the following functions indicating any discontinuities:

(a) $f(x) = \dfrac{x^2 - 16}{x - 4}$ (b) $f(x) = \dfrac{1}{(x-4)^2}$ (c) $f(x) = \dfrac{|x^2 - 25|}{x^2 - 25}$

SOLUTION (a) Since $f(x)$ is undefined at $x = 4$, the function is discontinuous there. For $x \neq 4$,

$$y = f(x) = \frac{(x+4)(x-4)}{x-4} = x + 4.$$

Consequently, the graph of $f(x)$ is a straight line with a hole at $x = 4$ (Figure 2.15). Note that $\lim_{x \to 4} [(x^2 - 16)/(x - 4)]$ exists and is equal to 8.

FIGURE 2.15

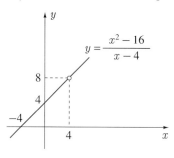

(b) Since $1/(x-4)^2$ is undefined at $x = 4$, the function is discontinuous there. The limits

$$\lim_{x \to 4^+} \frac{1}{(x-4)^2} = \infty, \qquad \lim_{x \to 4^-} \frac{1}{(x-4)^2} = \infty,$$

$$\lim_{x \to \infty} \frac{1}{(x-4)^2} = 0, \qquad \lim_{x \to -\infty} \frac{1}{(x-4)^2} = 0$$

and the fact that $f(x)$ is always positive lead to the graph in Figure 2.16.

FIGURE 2.16

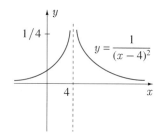

(c) The function $f(x) = |x^2 - 25|/(x^2 - 25)$ is undefined at $x = \pm 5$, and is therefore discontinuous at these values. When $-5 < x < 5$,

$$f(x) = \frac{25 - x^2}{x^2 - 25} = -1,$$

and when $|x| > 5$,

$$f(x) = \frac{x^2 - 25}{x^2 - 25} = 1.$$

The graph is shown in Figure 2.17. In this example, neither of the following limits exists:

$$\lim_{x \to 5} \frac{|x^2 - 25|}{x^2 - 25} \quad \text{or} \quad \lim_{x \to -5} \frac{|x^2 - 25|}{x^2 - 25}.$$

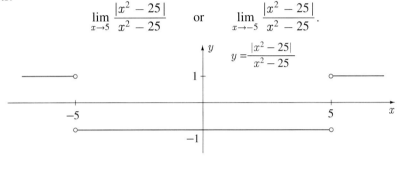

FIGURE 2.17

In Chapter 1 we introduced various techniques for sketching curves — addition and multiplication of ordinates, symmetries, and translations to name a few. Right-hand limits, left-hand limits, limits at infinity, and infinite limits can all be useful in sketching graphs of functions or curves. Part (b) of Example 2.16 illustrates this, and the following two examples contain further illustrations.

EXAMPLE 2.17

Sketch a graph of the function $f(x) = (3x - 6)/(x^2 + 5)$.

SOLUTION We first note that the graph has a horizontal asymptote given by

$$y = \lim_{x \to \infty} \frac{3x - 6}{x^2 + 5} = \lim_{x \to \infty} \frac{3 - \dfrac{6}{x}}{x + \dfrac{5}{x}} = 0.$$

Figure 2.18 shows two ways that $y = 0$ could be a horizontal asymptote as $x \to \infty$.

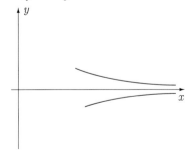

FIGURE 2.18

We need to decide whether $f(x) \to 0$ through positive or negative values. But this is easy to decide. For large positive x, $3 - 6/x$ is clearly positive, as is $x + 5/x$, and therefore $f(x)$ must approach 0 through positive numbers. We indicate this by writing

$$\lim_{x \to \infty} \frac{3x - 6}{x^2 + 5} = 0^+.$$

Similarly, for large negative values of x,

$$\lim_{x \to -\infty} \frac{3x - 6}{x^2 + 5} = \lim_{x \to -\infty} \frac{3 - \dfrac{6}{x}}{x + \dfrac{5}{x}} = 0^-$$

as the numerator is positive and the denominator is negative for large negative x. This information is shown in Figure 2.19(a), where we have also added the points $(-1, -3/2)$ and $(4, 2/7)$ and x- and y-intercepts of 2 and $-6/5$, respectively. There are many ways we could finish the graph; the simplest, and correct, way is shown in Figure 2.19(b). In Chapter 3, when we deal with differentiation, verification that this is indeed the correct graph will be straightforward. We will be able to show that the lowest point on the graph is at $x = -1$ and the highest point is at $x = 5$.

FIGURE 2.19

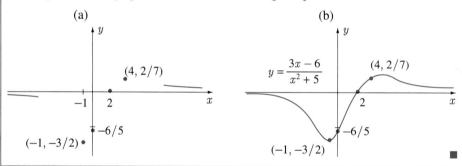

EXAMPLE 2.18

Sketch a graph of the function $f(x) = (x^2 - 9)/(x^2 + x - 2)$.

SOLUTION This function is discontinuous at $x = 1$ and $x = -2$. We should therefore calculate limits of $f(x)$ as $x \to -2^-$, $x \to -2^+$, $x \to 1^-$, and $x \to 1^+$. This was done in Example 2.11, the results being

$$\lim_{x \to 1^+} f(x) = -\infty, \quad \lim_{x \to 1^-} f(x) = \infty, \quad \lim_{x \to -2^+} f(x) = \infty, \quad \lim_{x \to -2^-} f(x) = -\infty.$$

We have shown these limits in Figure 2.20(a) together with a y-intercept of $9/2$, and x-intercepts of ± 3. We now calculate limits as $x \to \pm \infty$. First we note that

$$\lim_{x \to \infty} \frac{x^2 - 9}{x^2 + x - 2} = \lim_{x \to \infty} \frac{1 - \dfrac{9}{x^2}}{1 + \dfrac{1}{x} - \dfrac{2}{x^2}} = 1.$$

To commit this to a graph we must know whether 1 is approached through numbers larger than 1 or smaller than 1. This is most easily determined by noting that for very large x, the 9 and 2 in $f(x)$ are insignificant relative to x^2 and x, and therefore for such x, $f(x)$ is very closely approximated by

$$f(x) \approx \frac{x^2}{x^2 + x}.$$

But clearly the numerator is less than the denominator, and we conclude that

$$\lim_{x \to \infty} \frac{x^2 - 9}{x^2 + x - 2} = 1^-.$$

On the other hand, for large negative values of x, the numerator of $x^2/(x^2 + x)$ is greater than its denominator, and therefore

$$\lim_{x \to -\infty} \frac{x^2 - 9}{x^2 + x - 2} = 1^+.$$

These facts are all shown in Figure 2.20(b). The graph is completed in Figure 2.20(c).

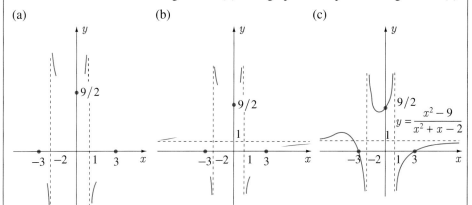

FIGURE 2.20

Once again determination of the exact positions of the high point to the left of $x = -2$ and the low point between $x = -2$ and $x = 1$ requires differentiation from Chapter 3. ∎

According to the following theorem, when continuous functions are added, subtracted, multiplied, and divided, the result is a continuous function.

> **Theorem 2.2**
>
> If functions $f(x)$ and $g(x)$ are continuous at $x = a$, then so also are the functions $f(x) \pm g(x)$, $f(x)g(x)$, and $f(x)/g(x)$ (provided $g(a) \neq 0$ in the case of division).

This is easily established with Theorem 2.1. For instance, to verify that $f(x) + g(x)$ is continuous, we note that since $\lim_{x \to a} f(x) = f(a)$ and $\lim_{x \to a} g(x) = g(a)$,

$$\lim_{x \to a} [f(x) + g(x)] = \lim_{x \to a} f(x) + \lim_{x \to a} g(x) \qquad \text{(by Theorem 2.1)}$$

$$= f(a) + g(a) \qquad \left(\text{by continuity of } f(x) \text{ and } g(x)\right).$$

Corollary *If functions $f(x)$ and $g(x)$ are continuous on an interval, then so also are the functions $f(x) \pm g(x)$, $f(x)g(x)$, and $f(x)/g(x)$ (provided $g(x)$ never vanishes in the interval in the case of division).*

It is an immediate consequence of this corollary that polynomials are continuous for all real numbers. Rational functions $P(x)/Q(x)$, where P and Q are polynomials, are continuous on intervals in which $Q(x) \neq 0$.

Besides adding, subtracting, multiplying, and dividing functions, functions can also be "composed", or substituted one into another. For instance, when $f(x) = \sqrt{x + 1/x}$ and $g(x) = x^2 + 3$, the function obtained by replacing each x in $f(x)$ by $g(x)$ is called the **composition** of f and g. It is often denoted by $(f \circ g)$, so that

$$(f \circ g)(x) = f(g(x)) = \sqrt{x^2 + 3 + \frac{1}{x^2 + 3}}.$$

Were we required to find the limit of this function as $x \to 1$, say, we would probably write nonchalantly that

$$\lim_{x \to 1} \sqrt{x^2 + 3 + \frac{1}{x^2 + 3}} = \sqrt{4 + \frac{1}{4}} = \frac{\sqrt{17}}{2}.$$

Effectively, we have interchanged the operation of taking square roots and that of taking limits; that is, without thinking we have written

$$\lim_{x \to 1} \sqrt{x^2 + 3 + \frac{1}{x^2 + 3}} = \sqrt{\lim_{x \to 1} \left(x^2 + 3 + \frac{1}{x^2 + 3} \right)} = \sqrt{4 + \frac{1}{4}} = \frac{\sqrt{17}}{2}.$$

This is correct according to the following theorem because the square root function is continuous.

Theorem 2.3

If $\lim_{x \to a} g(x) = L$, and $f(y)$ is a function that is continuous at $y = L$, then

$$\lim_{x \to a} f(g(x)) = f\left[\lim_{x \to a} g(x)\right] = f(L). \qquad (2.9)$$

An immediate consequence of this result is that the composition of a continuous function with a continuous function yields a continuous function. We state this as a corollary.

Corollary *If $g(x)$ is continuous at $x = a$, and $f(y)$ is continuous at $g(a)$, then $f(g(x))$ is continuous at $x = a$.*

EXAMPLE 2.19

Evaluate

$$\lim_{x \to 2} \sin\left(\frac{x^2 - 2}{2x - 3}\right).$$

SOLUTION Since the sine function is continuous for all values of its argument, we may take the limit operation inside the function and write

$$\lim_{x \to 2} \sin\left(\frac{x^2 - 2}{2x - 3}\right) = \sin\left[\lim_{x \to 2} \frac{x^2 - 2}{2x - 3}\right] = \sin 2.$$

Discontinuous functions are encountered in many applications of mathematics. To illustrate this, consider the phenomenon of radioactive disintegration, during which radioactive elements such as uranium-238 emit high-energy particles and in the process become new elements. If we analyze an ore sample that contains, along with various impurities, an amount A_0 of uranium, then disintegrations of uranium nuclei into thorium-234 nuclei gradually reduce the amount of uranium in the sample. Suppose we let $A = f(t)$ represent the amount of uranium in the sample at any given time, and choose time $t = 0$ when $A = A_0$. If we were to carefully measure the amount of uranium in the sample over a very short time interval (perhaps 10^{-10} second), a graph of $A = f(t)$ might appear as shown in Figure 2.21.

FIGURE 2.21

Each jump (discontinuity) in this graph signals the disintegration of a uranium nucleus into a thorium nucleus. Were we to complete this graph, and assume that no two disintegrations would occur at exactly the same instant, then the total number of discontinuities would be equal to the original number of uranium nuclei in the sample. In practice, we cannot hope to make such infinitesimal measurements. Instead, we must take measurements of the amount of uranium in the sample at much larger time intervals, and plot the data as a sequence of points, as in Figure 2.22(a). If these points are joined with a smooth curve as in Figure 2.22(b), we obtain a reasonable picture of A as a function of time. This macroscopic view (view in the large) of $A = f(t)$ suppresses the true nature of the graph, but it would be entirely adequate if we were interested in knowing, perhaps, when disintegrations would reduce the uranium to half its original amount. Questions such as these will be answered when we continue our discussion of radioactive disintegration in Section 9.2.

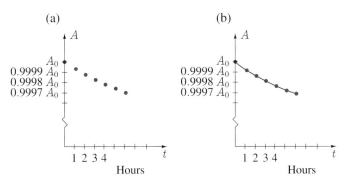

FIGURE 2.22

Our second example is taken from economics. A sugar manufacturer sells sugar in one-kilogram bags at the following prices:

40¢ per bag for the first 100 bags;

35¢ per bag for each bag after 100 up to 1000;

30¢ per bag for each bag after 1000.

If we denote the selling price of the xth bag by $f(x)$, then

$$f(x) = \begin{cases} 40 & 1 \leq x \leq 100 \\ 35 & 101 \leq x \leq 1000 \\ 30 & x \geq 1001. \end{cases}$$

This function is unlike any we have previously seen; its domain consists of only the positive integers. Its "graph", shown in Figure 2.23, therefore consists of dots above these integers. In economics, the domain of the function is extended to all positive values of x according to

$$f(x) = \begin{cases} 40 & 0 < x \leq 100 \\ 35 & 100 < x \leq 1000 \\ 30 & x > 1000, \end{cases}$$

with the resulting graph shown in Figure 2.24.

FIGURE 2.23

FIGURE 2.24

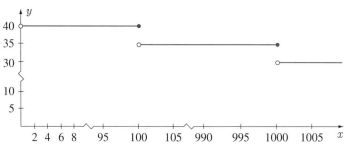

In spite of the fact that $f(x)$ is now defined for all $x > 0$, it still has significance only for x a positive integer. For instance, $f(2.1)$ would indicate the price of a tenth of a bag of sugar after two bags have already been purchased. To say that this fraction of a bag would cost 40¢ is pointless, since the manufacturer doesn't sell sugar in one-tenth kilogram bags. The advantage of extending the domain of $f(x)$ to noninteger values is therefore not one of physical significance. The real reason is that calculus can be applied to the extended function in order to analyze the economics of sugar sales. For the present, it is clear that the function $f(x)$ in Figure 2.24 is discontinuous at $x = 100$ and $x = 1000$.

We can derive from $f(x)$ a function that is far more useful in economic theory; the average price $r(x)$. It is the average price of a one-kilogram bag of sugar when x bags are purchased. Using the fact that average price is total selling price divided by total number of bags, we calculate that

$$r(x) = \begin{cases} 40 & 0 < x \leq 100 \\ \dfrac{40(100) + 35(x - 100)}{x} & 100 < x \leq 1000 \\ \dfrac{40(100) + 35(900) + 30(x - 1000)}{x} & x > 1000 \end{cases}$$

$$= \begin{cases} 40 & 0 < x \leq 100 \\ 35 + \dfrac{500}{x} & 100 < x \leq 1000 \\ 30 + \dfrac{5500}{x} & x > 1000. \end{cases}$$

Again, $r(x)$ is meaningful only for x an integer, but as shown in Figure 2.25 we extend its domain to all positive values of x. In addition, note that unlike $f(x)$, $r(x)$ has no discontinuities for $x > 0$.

FIGURE 2.25

EXERCISES 2.4

In Exercises 1–30 sketch a graph of the function indicating any discontinuities.

1. $f(x) = \dfrac{1}{x+2}$
2. $f(x) = \dfrac{16 - x^2}{x+4}$
3. $f(x) = |x^2 - 5|$
4. $f(x) = \dfrac{12}{x^2 + 2}$
5. $f(x) = \dfrac{12}{x^2 + 2x}$
6. $f(x) = \dfrac{12}{x^2 + 2x + 2}$
7. $f(x) = \dfrac{3 + 2x - x^2}{x+1}$
8. $f(x) = \dfrac{x^3 + x^2 - 2x}{x^2 - x}$
9. $f(x) = \dfrac{x^3 - 2x^2 + 5x - 10}{x - 2}$
10. $f(x) = \tan x$
11. $f(x) = \sec 2x$
12. $f(x) = \sin(1/x)$
13. $f(x) = x^2 \sin(1/x)$
14. $f(x) = \dfrac{x + 12}{x^2 - 9}$, $-5 \leq x \leq 4$
15. $f(x) = \dfrac{x^2 + 2x}{x^2 - 9}$, $-3 \leq x \leq 6$
16. $f(x) = \dfrac{x^3 - 27}{|x - 3|}$
17. $f(x) = \dfrac{x}{x^2 - 1}$
18. $f(x) = \dfrac{3x + 2}{x^2 - x - 2}$
19. $f(x) = \dfrac{x^2 - 3x + 2}{x^2 + 4x - 5}$
20. $f(x) = \dfrac{3x^2 - 6x}{x^2 - 6x - 7}$
21. $f(x) = \dfrac{x^2 + 3x + 2}{x + 4}$
22. $f(x) = \dfrac{x^2 - 2x + 4}{x - 1}$
23. $f(x) = \dfrac{1}{\sqrt{x - 1}}$
24. $f(x) = \dfrac{1}{\sqrt{5 + x}}$
25. $f(x) = \dfrac{1}{\sqrt{5 - x}}$
26. $f(x) = \sqrt{\dfrac{x - 3}{x + 2}}$
27. $f(x) = \dfrac{1}{x^3 - 4x}$
28. $f(x) = \dfrac{2x^3 - 2}{x^3 + 5x^2}$
29. $f(x) = \dfrac{1}{x^4 + 3x^2}$
30. $f(x) = \dfrac{|3x + 1|}{x + 5}$

31. Is the function
$$f(x) = \begin{cases} x \sin(1/x) & x \neq 0 \\ 0 & x = 0 \end{cases}$$
continuous at $x = 0$?

32. A function $f(x)$ is said to have a *removable discontinuity* at $x = a$ if $\lim_{x \to a} f(x)$ exists, but either $f(a)$ is not defined or $\lim_{x \to a} f(x) \neq f(a)$. Which of the discontinuities in Exercises 1–30 are removable?

33. Can rational functions have removable discontinuities?

34. Verify that the condition
$$\lim_{h \to 0} f(a + h) = f(a)$$
is equivalent to 2.8.

35. The number of bacteria in a culture is a function of time t, say, $N = f(t)$. Make two graphs of this function, one "in the small" to show birth and death situations, and a second "in the large" to get an overview of the rate of growth of the culture.

36. The *signum function* (or *sign function*), denoted by sgn, is defined by

$$\operatorname{sgn} x = \begin{cases} -1 & x < 0 \\ 0 & x = 0 \\ 1 & x > 0. \end{cases}$$

Sketch its graph, indicating any discontinuities.

37. Sketch a graph of the function $f(x) = \operatorname{sgn}(x+1) - \operatorname{sgn}(x-1)$, indicating any discontinuities.

38. Three of the most important functions in physics and engineering are defined below. Sketch a graph of each, indicating any discontinuities.
    (a) The Heaviside unit step function $H(x-a)$:
    $$H(x-a) = \begin{cases} 0 & x < a \\ 1 & x > a. \end{cases}$$
    (b) The unit pulse function $P_\epsilon(x-a)$:
    $$P_\epsilon(x-a) = \frac{1}{\epsilon}[H(x-a) - H(x-a-\epsilon)].$$
    (c) The Dirac delta function $\delta(x-a)$:
    $$\delta(x-a) = \lim_{\epsilon \to 0} P_\epsilon(x-a).$$

39. Where is the greatest integer function $[[x]]$ of Exercise 36 in Section 1.5 discontinuous?

40. (a) Sketch a graph of the function $f(x) = [[10x]]/10$. Where is the function discontinuous?
    (b) Prove that $f(x)$ truncates positive numbers after the first decimal.

41. What function truncates negative numbers after two decimals?

42. (a) Sketch a graph of the function $f(x) = [[x + 1/2]]$. Where is the function discontinuous?
    (b) Prove that $f(x)$ rounds positive numbers to the nearest integer.

43. What function rounds positive numbers to:
    (a) the nearest tenth;
    (b) the nearest hundredth;
    (c) 10^{-n} where n is a positive integer?

44. A toy manufacturer has two pricing policies for one of his dolls:
    Policy I: $15.00 each for the first 10 dolls,
                $12.50 for each doll after 10 up to 100,
                $10.00 for each doll after 100;
    Policy II: $15.00 each for the first 10 dolls,
                10% off the total price on orders of 11 or more,
                20% off the total price on orders of 101 or more.

    Sketch average price functions for each of these policies on the same graph. Which policy is better for a retailer?

45. A retailer of building supplies sells sand by the bag and in bulk. He quotes prices as follows:
    $3.00 per 50-kilogram bag up to 20 bags,
    $50.00 per tonne delivered only in integer tonnes.

    If, for example, a customer wants 2.2 tonness of sand, he receives 2 tonnes in bulk and 4 bags. Find an algebraic expression for the average price function and sketch its graph. Hint: The greatest integer function may be useful.

46. Determine points of continuity, if any, for the function
    $$f(x) = \begin{cases} 1 & x \text{ is a rational number} \\ 0 & x \text{ is an irrational number.} \end{cases}$$

47. Determine points of continuity, if any, for the function
    $$f(x) = \begin{cases} x & x \text{ is a rational number} \\ 0 & x \text{ is an irrational number.} \end{cases}$$

SECTION 2.5

A Mathematical Definition of Limits

Our work on limits has been intuitive, but some later topics in this book require a precise definition of limits. To obtain such a definition, we begin with our intuitive statement of a limit and make a succession of paraphrases, each of which is one step closer to the definition. We do this because the definition of a limit is at first sight quite overwhelming, and we wish to show that it can be obtained by a fairly straightforward sequence of steps. We hope that this will be a convincing argument that the definition of a limit does indeed describe in mathematical terms the concept of a limit. Our intuitive statement is:

- A function $f(x)$ has limit L as x approaches a if $f(x)$ can be made arbitrarily close to L by choosing x sufficiently close to a.

Next we paraphrase "$f(x)$ can be made arbitrarily close to L".

- A function $f(x)$ has limit L as x approaches a if the difference $|f(x) - L|$ can be made arbitrarily close to zero by choosing x sufficiently close to a.

Now we take the important step — make "$|f(x) - L|$ can be made arbitrarily close to zero" mathematical.

- A function $f(x)$ has limit L as x approaches a if given any real number $\epsilon > 0$, no matter how small, we can make the difference $|f(x) - L|$ less than ϵ by choosing x sufficiently close to a.

Penultimately, we paraphrase "by choosing x sufficiently close to a".

- A function $f(x)$ has limit L as x approaches a if given any $\epsilon > 0$, we can make $|f(x) - L| < \epsilon$ by choosing $|x - a|$ sufficiently close to zero.

Finally by making "choosing $|x - a|$ sufficiently close to zero" precise, we arrive at the definition of a limit.

> **Definition 2.2**
>
> A function $f(x)$ has limit L as x approaches a if given any $\epsilon > 0$, we can find a $\delta > 0$ such that
> $$|f(x) - L| < \epsilon$$
> whenever $0 < |x - a| < \delta$.

Notice that by requiring $0 < |x - a|$, this definition states explicitly that as far as limits are concerned, the value of $f(x)$ at $x = a$ is irrelevant. In taking limits we consider values of x closer and closer to a, but we do not consider the value of $f(x)$ at $x = a$. This definition states in precise terms our intuitive idea of a limit; that $f(x)$ can be made arbitrarily close to L by choosing x sufficiently close to a. A graphical interpretation may help to consolidate the definition. Figure 2.26(a) indicates a function that has limit L as x approaches a. In practice, we will not have a graph of the function, and verification that $\lim_{x \to a} f(x) = L$ will have to be done algebraically using Definition 2.2. Let us illustrate what must be done. We are given a value $\epsilon > 0$, which we should envisage as being very small, but we are never told exactly what it is. We are to restrict x so that $|f(x) - L| < \epsilon$. This inequality is equivalent to $-\epsilon < f(x) - L < \epsilon$ or $L - \epsilon < f(x) < L + \epsilon$, and the latter describes a horizontal band of width 2ϵ around the line $y = L$ [shaded in Figure 2.26(b)]. What Definition 2.2 requires is that we find an interval of width 2δ around $x = a$, as $|x - a| < \delta$ is equivalent to $a - \delta < x < a + \delta$, such that whenever x is in this interval, the values of $f(x)$ are all within the shaded horizontal band around $y = L$. Such an interval is shown in Figure 2.26(c) for the given ϵ. Now Definition 2.2 requires us to verify that the δ-interval can always be found no matter how small ϵ is chosen. This is always possible for the function illustrated in Figure 2.26, and it is clear that the smaller the given value of ϵ, the smaller δ will have to be chosen. For instance, for the value of ϵ in Figure 2.26(d), δ is smaller than that in Figure 2.26(c). In other words, the value of δ depends on the value of ϵ. Herein lies the difficulty in using Definition 2.2. We must determine precisely how δ depends on ϵ in order to ensure that δ can always be found. We illustrate with two examples.

FIGURE 2.26

(a)

(b)

(c)

(d)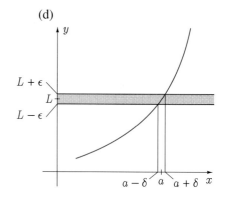

Consider proving first that

$$\lim_{x \to 3} (2x + 4) = 10.$$

Now, it is true that based on Theorem 2.1, this is quite obvious, but remember, this theorem has not yet been verified. For the next two examples we want to verify the limits, assuming that Theorem 2.1 is not at our disposal; verification must be through Definition 2.2. We must show that given any $\epsilon > 0$, there is a $\delta > 0$ such that

$$|(2x + 4) - 10| < \epsilon$$

whenever $0 < |x - 3| < \delta$; i.e., we must find a δ. Since

$$|2x + 4 - 10| = |2x - 6| = 2|x - 3|,$$

we must find a δ such that

$$2|x - 3| < \epsilon$$

whenever $0 < |x - 3| < \delta$. But this will be true if $\delta = \epsilon/2$; in other words, if we choose x to satisfy $0 < |x - 3| < \delta = \epsilon/2$, then

$$|(2x + 4) - 10| = 2|x - 3| < 2(\epsilon/2) = \epsilon.$$

The verification is now complete. We have shown that we can make $2x + 4$ as close to 10 as we want (within ϵ) by choosing x sufficiently close to 2 (within $\delta = \epsilon/2$).

Verifying that

$$\lim_{x \to 2} (x^2 + 5) = 9$$

is somewhat more complicated, but the idea is the same. We must show that given any $\epsilon > 0$, there is a $\delta > 0$ such that

$$|(x^2 + 5) - 9| < \epsilon \quad \text{or} \quad |x^2 - 4| < \epsilon$$

whenever $0 < |x - 2| < \delta$; i.e., we must find a δ. By working backward from the ϵ-inequality we can see how to choose δ. In addition, because the δ-inequality involves x in the combination $x - 2$, we rework the ϵ-inequality into the same combination. Since

$$|x^2 - 4| = |(x-2)^2 + 4x - 8| = |(x-2)^2 + 4(x-2)|,$$

we must choose δ such that

$$|(x-2)^2 + 4(x-2)| < \epsilon$$

whenever $0 < |x - 2| < \delta$. Recalling inequality 1.6c from Section 1.1, we can state that

$$|(x-2)^2 + 4(x-2)| \leq |x-2|^2 + 4|x-2|.$$

As a result, if we can find a $\delta > 0$ such that

$$|x-2|^2 + 4|x-2| < \epsilon$$

whenever $0 < |x - 2| < \delta$, then we will have a satisfactory δ. But now we have the key. Suppose we set $z = |x - 2|$ and consider the parabola $Q(z) = z^2 + 4z - \epsilon$ shown in Figure 2.27.

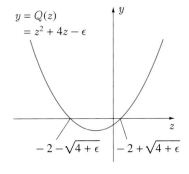

FIGURE 2.27

It crosses the z-axis when

$$z^2 + 4z - \epsilon = 0,$$

a quadratic with solutions

$$z = \frac{-4 \pm \sqrt{16 + 4\epsilon}}{2} = -2 \pm \sqrt{4 + \epsilon}.$$

The graph shows that whenever $0 < z < -2 + \sqrt{4 + \epsilon}$, then

$$z^2 + 4z - \epsilon < 0.$$

In other words, if $0 < z < \delta = \sqrt{4 + \epsilon} - 2$, then

$$z^2 + 4z < \epsilon.$$

Since $z = |x - 2|$, we can say that if $0 < |x - 2| < \delta = \sqrt{4 + \epsilon} - 2$, then

$$|x - 2|^2 + 4|x - 2| < \epsilon,$$

and therefore

$$|(x - 2)^2 + 4(x - 2)| < \epsilon.$$

The verification is now complete, We have shown that we can make $x^2 + 5$ as close to 9 as we want (within ϵ) by choosing x sufficiently close to 2 (within $\delta = \sqrt{4 + \epsilon} - 2$).

These examples have indicated the difference between "evaluation of" and "verification of" a limit. If you have learned only an appreciation for this difference, then our discussion has been worthwhile. If your comprehension goes beyond this and includes a working understanding of the definition of a limit, we could hope for no more. Results in succeeding chapters rely heavily on our intuitive understanding of limits and the ability to calculate limits, but use of Definition 2.2 is kept to a minimum.

EXERCISES 2.5

In Exercises 1–9 use Definition 2.2 to verify the limit.

1. $\lim\limits_{x \to 1} (x + 5) = 6$
2. $\lim\limits_{x \to 2} (2x - 3) = 1$
3. $\lim\limits_{x \to 0} (x^2 + 3) = 3$
4. $\lim\limits_{x \to 1} (x^2 + 4) = 5$
5. $\lim\limits_{x \to -2} (3 - x^2) = 1$
6. $\lim\limits_{x \to 3} (x^2 - 7x) = -12$
7. $\lim\limits_{x \to -1} (x^2 - 3x + 4) = 7$
8. $\lim\limits_{x \to 1} (x^2 + 3x + 5) = 9$
9. $\lim\limits_{x \to 2} \dfrac{x + 2}{x - 1} = 4$

In Exercises 10–19 give a mathematical definition for each statement.

10. $\lim\limits_{x \to a^+} f(x) = L$
11. $\lim\limits_{x \to a^-} f(x) = L$
12. $\lim\limits_{x \to \infty} f(x) = L$
13. $\lim\limits_{x \to -\infty} f(x) = L$
14. $\lim\limits_{x \to a} f(x) = \infty$
15. $\lim\limits_{x \to a} f(x) = -\infty$
16. $\lim\limits_{x \to \infty} f(x) = \infty$
17. $\lim\limits_{x \to \infty} f(x) = -\infty$
18. $\lim\limits_{x \to -\infty} f(x) = \infty$
19. $\lim\limits_{x \to -\infty} f(x) = -\infty$

20. Use Definition 2.2 to prove that a function $f(x)$ cannot have two limits as x approaches a.

In Exercises 21–28 use the appropriate definition from Exercises 10–19 to verify the limit.

21. $\lim\limits_{x \to 1} \dfrac{1}{(x - 1)^2} = \infty$
22. $\lim\limits_{x \to -2} \dfrac{-1}{(x + 2)^2} = -\infty$
23. $\lim\limits_{x \to \infty} (x + 5) = \infty$
24. $\lim\limits_{x \to \infty} (5 - x^2) = -\infty$
25. $\lim\limits_{x \to \infty} \dfrac{x + 2}{x - 1} = 1$
26. $\lim\limits_{x \to -\infty} \dfrac{x + 2}{x - 1} = 1$
27. $\lim\limits_{x \to -\infty} (5 - x) = \infty$
28. $\lim\limits_{x \to -\infty} (3 + x - x^2) = -\infty$

29. Use the definitions in Exercises 10 and 11 to verify that the Heaviside function of Exercise 38 in Section 2.4 has a right- and left-hand limit at $x = a$.

30. Prove that if $\lim_{x \to a} f(x) = L > 0$, then there exists an open interval I containing a in which $f(x) > 0$ except possibly at $x = a$.

In Exercises 31–35 we use Definition 2.2 to prove Theorem 2.1. In each exercise assume that $\lim_{x \to a} f(x) = F$ and that $\lim_{x \to a} g(x) = G$.

31. Show that given any $\epsilon > 0$, there exists numbers $\delta_1 > 0$ and $\delta_2 > 0$ such that

$$|f(x) - F| < \epsilon/2 \quad \text{whenever } 0 < |x - a| < \delta_1 \text{ and}$$
$$|g(x) - G| < \epsilon/2 \quad \text{whenever } 0 < |x - a| < \delta_2.$$

Use these results along with identity 1.6c to prove part (i) of Theorem 2.1.

32. Use a proof similar to that in Exercise 31 to verify part (ii) of Theorem 2.1.

33. Verify part (iii) of Theorem 2.1.

34. (a) Verify that

$$|f(x)g(x) - FG| \leq |f(x)||g(x) - G| + |G||f(x) - F|.$$

(b) Show that given any $\epsilon > 0$, there exists numbers $\delta_1 > 0$, $\delta_2 > 0$, and $\delta_3 > 0$ such that

$$|f(x)| < |F| + 1 \quad \text{whenever } 0 < |x - a| < \delta_1,$$
$$|g(x) - G| < \dfrac{\epsilon}{2(|F| + 1)} \quad \text{whenever } 0 < |x - a| < \delta_2,$$
$$|f(x) - F| < \dfrac{\epsilon}{2|G| + 1} \quad \text{whenever } 0 < |x - a| < \delta_3.$$

(c) Use these results to prove part (iv) of Theorem 2.1.

35. (a) Verify that when $G \neq 0$,

$$\left|\frac{f(x)}{g(x)} - \frac{F}{G}\right| \leq \frac{|f(x) - F|}{|g(x)|} + \frac{|F||G - g(x)|}{|G||g(x)|}.$$

(b) Show that given any $\epsilon > 0$, there exist numbers $\delta_1 > 0$, $\delta_2 > 0$, and $\delta_3 > 0$ such that

$$|g(x)| > \frac{|G|}{2} \quad \text{whenever } 0 < |x - a| < \delta_1,$$

$$|f(x) - F| < \frac{\epsilon |G|}{4} \quad \text{whenever } 0 < |x - a| < \delta_2,$$

$$|g(x) - G| < \frac{\epsilon |G|^2}{4|F| + 1} \quad \text{whenever } 0 < |x - a| < \delta_3.$$

(c) Now prove part (v) of Theorem 2.1.

36. Does the function

$$f(x) = \begin{cases} x \sin(1/x) & x \neq 1/(n\pi) \\ 1 & x = 1/(n\pi), \end{cases}$$

where n is an integer, have a limit as $x \to 0$?

SUMMARY

In Section 2.1 we introduced limits of functions. For the most part our discussion was intuitive, beginning with the statement "$\lim_{x \to a} f(x) = L$ if $f(x)$ can be made arbitrarily close to L by choosing x sufficiently close to a." This idea was then extended to include the following:

Right-hand limits: $\lim_{x \to a^+} f(x)$;

Left-hand limits: $\lim_{x \to a^-} f(x)$;

Limits at infinity: $\lim_{x \to \infty} f(x)$, $\lim_{x \to -\infty} f(x)$;

Infinite limits: $\lim_{x \to a} f(x) = \infty$, $\lim_{x \to a} f(x) = -\infty$.

Keep in mind that the term "infinite limits" is somewhat of a misnomer since in both situations the limit does not exist.

A function $f(x)$ is continuous at a point $x = a$ if

$$\lim_{x \to a} f(x) = f(a).$$

Note that the conditions that $f(x)$ is defined at $x = a$ and $\lim_{x \to a} f(x)$ exists are inherent in this definition. The function is continuous on an interval if it is continuous at each point of that interval. Geometrically, this means that one must be able to completely trace its graph without lifting pencil from page.

We illustrated that limits can be used to sketch the graph of a function near a point of discontinuity, as well as to indicate its behaviour for very large positive and negative values of x. To our list of techniques for sketching curves which already contains such methods as addition and multiplication of ordinates, symmetries, and translations, we therefore add the use of limits.

In Section 2.5 we developed the mathematical definition of limits.

Key Terms and Formulas

In reviewing this chapter, you should be able to define or discuss the following key terms:

Limit
Right-hand limit
Left-hand limit
Infinite limit
Vertical asymptote

Limit at infinity
Horizontal asymptote
Continuity at a point
Continuity on an interval
Composite function

REVIEW EXERCISES

In Exercises 1–20 evaluate the limit, if it exists.

1. $\lim\limits_{x \to 1} \dfrac{x^2 - 2x}{x + 5}$

2. $\lim\limits_{x \to -1} \dfrac{x^2 - 1}{x + 1}$

3. $\lim\limits_{x \to -2} \dfrac{x^2 + 4x + 4}{x + 3}$

4. $\lim\limits_{x \to \infty} \dfrac{x + 5}{x - 3}$

5. $\lim\limits_{x \to -\infty} \dfrac{x^2 + 3x + 2}{2x^2 - 5}$

6. $\lim\limits_{x \to -\infty} \dfrac{5 - x^3}{3 + 4x^3}$

7. $\lim\limits_{x \to \infty} \dfrac{3x^3 + 2x - 5}{x^2 + 5x}$

8. $\lim\limits_{x \to \infty} \dfrac{4 - 3x + x^2}{3 + 5x^3}$

9. $\lim\limits_{x \to 2^+} \dfrac{x^2 - 2x}{x^2 + 2x}$

10. $\lim\limits_{x \to 2^-} \dfrac{x^2 - 4x + 4}{x - 2}$

11. $\lim\limits_{x \to 0} \dfrac{x^2 + 2x}{3x - 2x^2}$

12. $\lim\limits_{x \to 1} \dfrac{x^2 + 5x}{(x - 1)^2}$

13. $\lim\limits_{x \to 1} \dfrac{\sqrt{x} - 1}{x}$

14. $\lim\limits_{x \to 1} \dfrac{\sqrt{x} - 1}{x - 1}$

15. $\lim\limits_{x \to 1/2} \dfrac{(2 - 4x)^3}{x(2x - 1)^2}$

16. $\lim\limits_{x \to \infty} \dfrac{\cos 5x}{x}$

17. $\lim\limits_{x \to -\infty} x \sin x$

18. $\lim\limits_{x \to -\infty} \dfrac{\sqrt{3x^2 + 4}}{2x + 5}$

19. $\lim\limits_{x \to \infty} \dfrac{\sqrt{3x^2 + 4}}{2x + 5}$

20. $\lim\limits_{x \to \infty} \left(\sqrt{2x + 1} - \sqrt{3x - 1} \right)$

In Exercises 21–32 sketch a graph of the function, indicating any discontinuities.

21. $f(x) = \dfrac{1}{x - 2}$

22. $f(x) = \dfrac{x}{x - 2}$

23. $f(x) = \dfrac{x^2}{x - 2}$

24. $f(x) = \dfrac{x^2 - 36}{x - 6}$

25. $f(x) = \dfrac{x + 1}{x - 1}$

26. $f(x) = \left| \dfrac{x + 1}{x - 1} \right|$

27. $f(x) = \dfrac{|x + 1|}{x - 1}$

28. $f(x) = \dfrac{x + 1}{|x - 1|}$

29. $f(x) = \dfrac{2x}{x^2 - 3x - 4}$

30. $f(x) = \dfrac{2x^2}{x^2 - 3x - 4}$

31. $f(x) = \dfrac{x^3 - 3x^2 + 3x - 1}{x - 1}$

32. $f(x) = \dfrac{x^3 - 3x^2 + 3x - 1}{x^2 - 2x + 1}$

33. For what values of x is the function $f(x) = [\![x^2]\!]$ discontinuous? Sketch its graph.

CHAPTER 3

Differentiation

Chapters 1 and 2 have prepared the way for calculus. The functions and curves in these chapters yield a wealth of examples for our discussions, and limits from Chapter 2 provide the tool by which calculus is developed. Calculus has two major components, *differentiation* and *integration*. In this chapter we study differentiation. We learn what a *derivative* is algebraically and geometrically and develop some of its properties; we learn how to differentiate polynomials, rational functions, trigonometric functions, and exponential and logarithm functions; and we see glimpses of the applications that are to follow in Chapter 4.

SECTION 3.1

The Derivative

Very few quantities in real life remain constant; most are in a state of change. For example, room temperature, the speed of a car, and the angle of elevation of the sun are three commonplace quantities which are constantly changing. A few more technical ones are the current in a transmission line, barometric pressure, moisture content of the soil, and stress in vertical members of tall buildings during hurricanes. Rates at which these quantities change are called derivatives. To study the concept of a rate of change more thoroughly, we consider two commonplace situations. First, suppose a car moves along a straight street between the two stop lights in Figure 3.1.

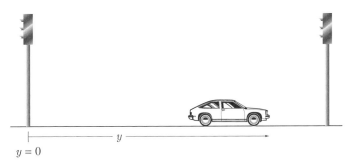

FIGURE 3.1

If its distance from the first light is denoted by y, then y is a function of time t, say, $y = f(t)$. When $y_1 = f(t_1)$ and $y_2 = f(t_2)$ are positions of the car at times t_1 and t_2, where $t_2 > t_1$, the quotient

$$\frac{y_2 - y_1}{t_2 - t_1} = \frac{f(t_2) - f(t_1)}{t_2 - t_1}$$

is called the *average velocity* of the car between y_1 and y_2. It is the average rate of change of y with respect to t between t_1 and t_2. Suppose time t_1 is kept fixed, and we take a number of times t_2 that are closer and closer to t_1 so that the time interval $t_2 - t_1$ gets smaller and smaller. If we demand that t_2 get arbitrarily close to t_1, then we are really asking for the limit of the average velocity as t_2 approaches t_1, written

$$\lim_{t_2 \to t_1} \frac{y_2 - y_1}{t_2 - t_1} = \lim_{t_2 \to t_1} \frac{f(t_2) - f(t_1)}{t_2 - t_1}.$$

This quantity is called the *instantaneous velocity* of the car at time t_1. It is the instantaneous rate of change of y with respect to t at time t_1.

As a second example, consider an open cylindrical container placed outside a house. The depth D of water in the container collected during a rainstorm (Figure 3.2) is a function of time t, say, $D = f(t)$.

FIGURE 3.2

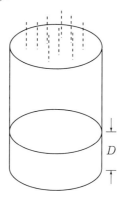

The quotient

$$\frac{D_2 - D_1}{t_2 - t_1} = \frac{f(t_2) - f(t_1)}{t_2 - t_1}$$

represents the *average rate of rainfall* during the time interval from t_1 to t_2. If we take the limit of this quotient as t_2 approaches t_1,

$$\lim_{t_2 \to t_1} \frac{D_2 - D_1}{t_2 - t_1} = \lim_{t_2 \to t_1} \frac{f(t_2) - f(t_1)}{t_2 - t_1},$$

we obtain the *instantaneous rate of rainfall* at time t_1. It is the instantaneous rate of change of D with respect to t at time t_1.

These rates of change are examples of rates of change that occur in a multitude of applications — applications from such diverse fields as physics, economics, psychology, and medicine, to name a few. Let us establish a mathematical framework within which to examine rates of change more closely, at the same time introducing notation which eliminates subscripts. Suppose the function $f(x)$ is defined for all x, and $x = a$ and $x = a + h$ are two values of x. The quotient

$$\frac{f(a + h) - f(a)}{h}$$

is the **average rate of change** of $f(x)$ with respect to x in the interval $a \leq x \leq a + h$. The limit of the quotient as h approaches zero,

$$\lim_{h \to 0} \frac{f(a + h) - f(a)}{h}$$

is called the **instantaneous rate of change** of $f(x)$ with respect to x at $x = a$. It is also called the derivative of $f(x)$ at $x = a$.

> **Definition 3.1**
>
> The **derivative** of a function $f(x)$ with respect to x at a point $x = a$, denoted by $f'(a)$, is defined as
>
> $$f'(a) = \lim_{h \to 0} \frac{f(a+h) - f(a)}{h}, \qquad (3.1)$$
>
> provided the limit exists.

The operation of taking the derivative of a function is called **differentiation**. We say that we differentiate the function when we find its derivative.

Tangent Lines and the Geometric Interpretation of the Derivative

We have defined the derivative of a function algebraically as its instantaneous rate of change. Its geometric interpretation is also important. Derivatives are intimately connected to tangent lines to curves, and most students of calculus have an intuitive idea of what it means for a line to be tangent to a curve. Often it is the idea of "touching". A line is tangent to a curve if it touches the curve at exactly one point. For curves such as circles, ellipses and parabolas, this notion is adequate, but in general it is unsatisfactory. For instance, in Figure 3.3 we have drawn what would look like the tangent line to the curve $y = x^3$ at the point $P(1, 1)$. But this line intersects the curve again at the point $Q(-2, -8)$. The tangent line at P does not touch the curve at precisely one point; it intersects the curve at a second point.

The idea that a tangent line touches a curve is good, but it needs to be phrased properly. Consider defining what is meant by the tangent line at the point P on the curve $y = f(x)$ in Figure 3.4. When P is joined to another point Q_1 on the curve by a straight line l_1, certainly l_1 is not the tangent line to $y = f(x)$ at P. If we join P to a point Q_2 on $y = f(x)$ closer to P than Q_1, then l_2 is not the tangent line at P either, but it is closer to it than l_1. A point Q_3 even closer to P yields a line l_3 that is even closer to the tangent line than l_2. Repeating this process over and over again, leads to a set of lines $l_1, l_2, l_3, \ldots$ which get closer and closer to what we feel is the tangent line to $y = f(x)$ at P. We therefore define the **tangent line** to $y = f(x)$ at P as the limiting position of these lines as points $Q_1, Q_2, Q_3, \ldots$ get arbitrarily close to P.

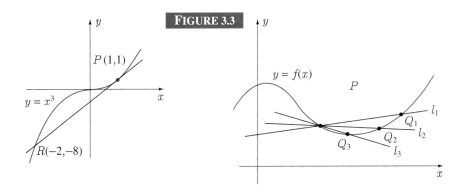

FIGURE 3.3

FIGURE 3.4

The line in Figure 3.3 satisfies this definition; it is the limiting position of lines joining P to other points on the curve which approach P. It is irrelevant whether this line intersects the curve again at some distance from P. What is the tangent line to the curve $y = x^3$ in Figure 3.3 at the origin $(0,0)$? According to the above definition the limiting position of lines joining $(0,0)$ to other points on the curve is the x-axis; that is, the x-axis is tangent to $y = x^3$ at $(0,0)$ (Figure 3.5). Notice that the tangent line actually crosses from one side of the curve to the other at $(0,0)$. To the left of $x = 0$, the tangent line is above the curve whereas to the right of $x = 0$, it is below the curve.

FIGURE 3.5

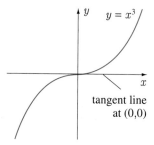

Now that we understand what it means for a line to be tangent to a curve, let us make the connection with derivatives. Suppose $P(a, f(a))$ is a point on the curve $y = f(x)$ in Figure 3.6.

If $Q(a + h, f(a + h))$ is another point on the curve, then the quotient

$$\frac{f(a+h) - f(a)}{h}$$

in Definition 3.1 is the slope of the line joining P and Q. As $h \to 0$, point Q moves along the curve toward P, and the line joining P and Q moves toward the tangent line at P. It follows that limit 3.1, the derivative $f'(a)$, is the slope of the tangent line to the curve $y = f(x)$ at $x = a$ (Figure 3.7).

FIGURE 3.6   **FIGURE 3.7**

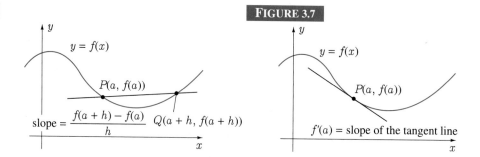

To summarize, algebraically the derivative $f'(a)$ of a function $f(x)$ at $x = a$ is its instantaneous rate of change; geometrically it is the slope of the tangent line to the graph of $f(x)$ at the point $(a, f(a))$.

Let us use Definition 3.1 to calculate derivatives of some simple functions.

EXAMPLE 3.1

Find $f'(1)$ if $f(x) = x^2$.

SOLUTION According to equation 3.1,

$$f'(1) = \lim_{h \to 0} \frac{f(1+h) - f(1)}{h} = \lim_{h \to 0} \frac{(1+h)^2 - 1}{h}$$
$$= \lim_{h \to 0} \frac{(1 + 2h + h^2) - 1}{h} = \lim_{h \to 0} \frac{2h + h^2}{h} = \lim_{h \to 0} (2 + h) = 2.$$

Algebraically, the instantaneous rate of change of $f(x) = x^2$ when $x = 1$ is equal to 2. Geometrically, the slope of the tangent line to the curve $y = x^2$ at the point $(1, 1)$ is 2 (see Figure 3.8).

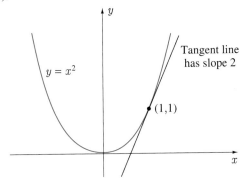

FIGURE 3.8

EXAMPLE 3.2

Find $f'(2)$ if $f(x) = x^3 - 12x$.

SOLUTION Using equation 3.1,

$$f'(2) = \lim_{h \to 0} \frac{f(2+h) - f(2)}{h} = \lim_{h \to 0} \frac{[(2+h)^3 - 12(2+h)] - (8 - 24)}{h}$$
$$= \lim_{h \to 0} \frac{6h^2 + h^3}{h} = \lim_{h \to 0} (6h + h^2) = 0.$$

This result is substantiated in Figure 3.9 where we see that the tangent line to $y = x^3 - 12x$ is horizontal (has zero slope) at $x = 2$.

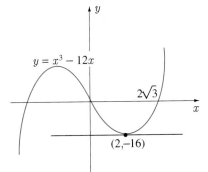

FIGURE 3.9

In Examples 3.1 and 3.2 we required the derivative for the function at only one value of a, and therefore set a in equation 3.1 equal to this value. An alternative, which would be far more advantageous, especially in an example where the derivative were required at a number of points, would be to evaluate $f'(a)$ and then set a to its desired value later. For instance, in Example 1 we could calculate that

$$f'(a) = \lim_{h \to 0} \frac{f(a+h) - f(a)}{h} = \lim_{h \to 0} \frac{(a+h)^2 - a^2}{h}$$
$$= \lim_{h \to 0} \frac{2ah + h^2}{h} = \lim_{h \to 0} (2a + h) = 2a.$$

This is the derivative of $f(x) = x^2$ at any value $x = a$. For $a = 1$, we obtain $f'(1) = 2(1) = 2$. But it is also easy to calculate $f'(a)$ at other values of a. For example, $f'(0) = 0$, $f'(-1) = -2$, and $f'(4) = 8$.

We now carry this idea to its logical conclusion. The derivative of $f(x)$ at a is denoted by $f'(a)$. But what is a? It is a specific value of x at which to calculate the derivative, and it can be any value of x. Why not simply drop references to a, and talk about the derivative of $f(x)$ at values of x? Following this suggestion, we denote by $f'(x)$ the derivative of the function $f(x)$ at any value of x. With this notation, equation 3.1 is replaced by

$$f'(x) = \lim_{h \to 0} \frac{f(x+h) - f(x)}{h}, \qquad (3.2)$$

for the derivative of $f(x)$ at x. We call $f'(x)$ the **derivative function**.

When a function is represented by the letter y, as in $y = f(x)$, another common notation for the derivative is $\frac{dy}{dx}$. Be careful when using this notation. Do not interpret it as a quotient; it is one symbol representing an accumulation of all the operations in equation 3.2. It is not therefore to be read as "dy divided by dx". Typographically, it is easier to print dy/dx rather than $\frac{dy}{dx}$, and we will take this liberty whenever it is convenient to do so. But remember, dy/dx is not a quotient, it is one symbol representing the limit operation in 3.2.

Sometimes it is more convenient to use parts of each of these notations and write

$$\frac{d}{dx} f(x).$$

In this form we understand that d/dx means to differentiate with respect to x whatever follows it, in this case $f(x)$. Let us use these new notations in calculating two more derivatives.

EXAMPLE 3.3

Find dy/dx if $y = f(x) = (x-1)/(x+2)$.

SOLUTION Using 3.2,

$$\frac{dy}{dx} = \lim_{h \to 0} \frac{f(x+h) - f(x)}{h}$$
$$= \lim_{h \to 0} \frac{1}{h} \left(\frac{x+h-1}{x+h+2} - \frac{x-1}{x+2} \right).$$

If we bring the terms in parentheses to a common denominator, the result is

$$\frac{dy}{dx} = \lim_{h \to 0} \frac{1}{h} \left(\frac{(x+h-1)(x+2) - (x+h+2)(x-1)}{(x+h+2)(x+2)} \right).$$

When we simplify the numerator, we find

$$\frac{dy}{dx} = \lim_{h \to 0} \frac{1}{h} \left(\frac{3h}{(x+h+2)(x+2)} \right)$$

$$= \lim_{h \to 0} \frac{3}{(x+h+2)(x+2)}$$

$$= \frac{3}{(x+2)^2}.$$

■

EXAMPLE 3.4

Find dv/dt if $v = f(t) = 1/t$.

SOLUTION In terms of variables v and t, equation 3.2 takes the form

$$\frac{dv}{dt} = \lim_{h \to 0} \frac{f(t+h) - f(t)}{h}$$

$$= \lim_{h \to 0} \frac{1}{h} \left(\frac{1}{t+h} - \frac{1}{t} \right)$$

$$= \lim_{h \to 0} \frac{t - (t+h)}{t(t+h)h}$$

$$= \lim_{h \to 0} \frac{-1}{t(t+h)}$$

$$= -\frac{1}{t^2}.$$

■

In Section 1.3 we proved that the inclination ϕ of a line l is related to its slope m by the equation $\tan \phi = m$ (equation 1.13). When l is the tangent line to a curve $y = f(x)$ at point (x_0, y_0), as shown in Figure 3.10, its slope is $f'(x_0)$.

FIGURE 3.10

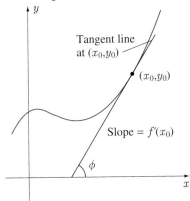

Hence, the inclination ϕ of the tangent line to a curve at a point (x_0, y_0) is given by the equation

$$\tan \phi = f'(x_0). \qquad (3.3)$$

EXAMPLE 3.5

Find the inclination of the tangent line to the curve $y = (x-1)/(x+2)$ at the point $(4, 1/2)$.

SOLUTION In Example 3.3, we calculated the slope of the tangent line to this curve at any point as

$$\frac{dy}{dx} = \frac{3}{(x+2)^2}.$$

According to equation 3.3, the inclination ϕ of the tangent line at $(4, 1/2)$ is given by

$$\tan\phi = \frac{3}{(4+2)^2} = \frac{1}{12}.$$

The angle with tangent equal $1/12$ is $\phi = 0.083$ radians.

EXERCISES 3.1

In Exercises 1–10 use equation 3.2 to find $f'(x)$.

1. $f(x) = x + 2$
2. $f(x) = 3x^2 + 5$
3. $f(x) = 1 + 2x - x^2$
4. $f(x) = x^3 + 2x^2$
5. $f(x) = x^4 + 4x - 12$
6. $f(x) = \dfrac{x+4}{x-5}$
7. $f(x) = \dfrac{x^2+2}{x+3}$
8. $f(x) = x^2(x+2)$
9. $f(x) = \dfrac{3x-2}{4-x}$
10. $f(x) = \dfrac{x^2-x+1}{x^2+x+1}$

In Exercises 11–14 find the specified rate of change.

11. The rate of change of the circumference C of a circle with respect to its radius r.

12. The rate of change of the area A of a circle with respect to its radius r.

13. The rate of change of the area A of a sphere with respect to its radius r.

14. The rate of change of the volume V of a sphere with respect to its radius r.

Answer Exercises 15–19 by drawing a graph of the function. Do not calculate the required derivative.

15. What is $f'(x)$ if $f(x) = 2x - 4$?

16. What is $f'(x)$ if $f(x) = mx + b$, where m and b are constants?

17. What is $f'(0)$ if $f(x) = x^2$?

18. What is $f'(1)$ if $f(x) = (x-1)^2$?

19. What is $f'(0)$ if $f(x) = x^{1/3}$?

In Exercises 20–23 find the equation of the tangent line to the curve at the given point.

20. $y = x^2 + 3$ at $(1, 4)$
21. $y = 3 - 2x - x^2$ at $(4, -21)$
22. $y = 1/x^2$ at $(2, 1/4)$
23. $y = (x+1)/(x+2)$ at $(0, 1/2)$

In Exercises 24–27 find the inclination of the tangent line to the curve at the given point.

24. $y = x^2$ at $(1, 1)$
25. $y = x^3 - 6x$ at $(2, -4)$
26. $y = 1/x^2$ at $(2, 1/4)$
27. $y = 1/(x+1)$ at $(0, 1)$

In Exercises 28–32 find $f'(x)$.

28. $f(x) = x^8$
29. $f(x) = \sqrt{x+1}$
30. $f(x) = \dfrac{1}{(x-2)^4}$
31. $f(x) = \dfrac{1}{\sqrt{x-3}}$
32. $f(x) = x\sqrt{x+1}$

In Exercises 33 and 34 find the specified rate of change.

33. The rate of change of the radius r of a circle with respect to its area A.

34. The rate of change of the volume V of a sphere with respect to its area A.

35. Find the angle between the tangent lines to the curves $y = x^2$ and $x = y^2$ at their point of intersection.

36. Find $f'(x)$ if $f(x) = |x|$.

37. A sphere of radius R has a uniform charge distribution of ρ coulombs per cubic metre. The electrostatic potential V at a distance r from the centre of the sphere is defined by

$$V = f(r) = \begin{cases} \dfrac{\rho}{6\epsilon_0}(3R^2 - r^2) & 0 \leq r \leq R \\ \dfrac{R^3\rho}{3\epsilon_0 r} & r > R, \end{cases}$$

where ϵ_0 is a constant. Sketch a graph of this function. Does $f(r)$ appear to have a derivative at the surface of the sphere? That is, does $f'(R)$ exist? Prove your conjecture using Definition 3.1.

38. Repeat Exercise 37 for the magnitude E of the electrostatic field:

$$E = f(r) = \begin{cases} \dfrac{\rho r}{3\epsilon_0} & 0 \leq r \leq R \\ \dfrac{\rho R^3}{3\epsilon_0 r^2} & r > R. \end{cases}$$

39. Find $f'(x)$ if $f(x) = x^{1/3}$.

40. Let $f(x)$ be a function with the property that $f(x+z) = f(x)f(z)$ for all x and z, and be such that $f(0) = f'(0) = 1$. Prove that $f'(x) = f(x)$ for all x.

SECTION 3.2

Rules for Differentiation

Since calculus plays a key role in many branches of applied science, we need to differentiate many types of functions: polynomials, rational functions, trigonometric functions, exponential and logarithmic functions, etc. To use equation 3.2 each time would be extremely laborious, and we might soon doubt the usefulness of differentiation. Fortunately, however, we can develop a number of rules for taking derivatives that eliminate the necessity of using the definition each time. We state each of these formulas in the form of a theorem.

Theorem 3.1

If $f(x) = c$, where c is a constant, then $f'(x) = 0$.

Proof By equation 3.2,

$$f'(x) = \lim_{h \to 0} \frac{f(x+h) - f(x)}{h} = \lim_{h \to 0} \frac{c - c}{h} = 0.$$

In short, the derivative of a constant function is zero;

$$\frac{d}{dx}(c) = 0. \tag{3.4}$$

Theorem 3.2

If $f(x) = x$, then $f'(x) = 1$.

Proof With equation 3.2,

$$f'(x) = \lim_{h \to 0} \frac{f(x+h) - f(x)}{h} = \lim_{h \to 0} \frac{(x+h) - x}{h} = \lim_{h \to 0} \frac{h}{h} = 1.$$

In short,

$$\frac{d}{dx}(x) = 1. \tag{3.5}$$

Note that graphs of the functions $f(x) = c$ and $f(x) = x$ in Figure 3.11 confirm the results of equations 3.4 and 3.5. The tangent line to $y = c$ always has slope zero, whereas the tangent line to $y = x$ always has slope equal to one.

FIGURE 3.11

Theorem 3.3 *If $f(x) = x^n$, where n is a positive integer, then $f'(x) = nx^{n-1}$.*

Proof Equation 3.2 gives

$$\frac{dy}{dx} = f'(x) = \lim_{h \to 0} \frac{f(x+h) - f(x)}{h} = \lim_{h \to 0} \frac{(x+h)^n - x^n}{h}.$$

If we expand $(x+h)^n$ by means of the binomial theorem, we have

$$\frac{dy}{dx} = \lim_{h \to 0} \frac{1}{h}\left[x^n + nx^{n-1}h + \frac{n(n-1)}{2}x^{n-2}h^2 + \cdots + nxh^{n-1} + h^n - x^n\right].$$

The first and last terms in brackets cancel, and dividing h into the remaining terms gives

$$\frac{dy}{dx} = \lim_{h \to 0}\left[nx^{n-1} + \frac{n(n-1)}{2}x^{n-2}h + \cdots + h^{n-1}\right] = nx^{n-1}.$$

In short,

$$\frac{d}{dx}(x^n) = nx^{n-1}. \tag{3.6}$$

This is called the **power rule** for differentiation. Although we have proved the power rule only for n a positive integer, it is in fact true for every real number n. We will assume that equation (3.6) can be used for any real number n, and will prove this more general result in Section 3.9. For example,

$$\frac{d}{dx}(x^3) = 3x^2 \quad \text{and} \quad \frac{d}{dx}\left(x^{1/3}\right) = \frac{1}{3}x^{-2/3}.$$

Theorem 3.4 *If $g(x) = cf(x)$, where c is a constant, then*

$$g'(x) = cf'(x). \tag{3.7a}$$

Proof By equation 3.2 and Theorem 2.1,

$$g'(x) = \lim_{h \to 0} \frac{g(x+h) - g(x)}{h} = \lim_{h \to 0} \frac{cf(x+h) - cf(x)}{h}$$
$$= c \lim_{h \to 0} \frac{f(x+h) - f(x)}{h} = cf'(x).$$

Thus, for $y = f(x)$, we may write that

$$\frac{d}{dx}(cy) = c\frac{dy}{dx}. \tag{3.7b}$$

Theorem 3.5 *If $p(x) = f(x) + g(x)$, then*

$$p'(x) = f'(x) + g'(x). \tag{3.8a}$$

Proof Equation 3.2 gives

$$p'(x) = \lim_{h \to 0} \frac{p(x+h) - p(x)}{h}$$

$$= \lim_{h \to 0} \frac{[f(x+h) + g(x+h)] - [f(x) + g(x)]}{h}$$

$$= \lim_{h \to 0} \left[\frac{f(x+h) - f(x)}{h} + \frac{g(x+h) - g(x)}{h} \right]$$

$$= f'(x) + g'(x).$$

In short, if we set $u = f(x)$ and $v = g(x)$,

$$\frac{d}{dx}(u+v) = \frac{du}{dx} + \frac{dv}{dx}, \qquad (3.8b)$$

or, in words, the derivative of a sum is the sum of the derivatives.

We now use these formulas to calculate derivatives in the following examples.

EXAMPLE 3.6

Find dy/dx if

(a) $y = x^4$ (b) $y = 3x^6 - x^{-2}$ (c) $y = \dfrac{x^4 - 6x^2}{3x^3}$

SOLUTION (a) By equation 3.6,

$$\frac{dy}{dx} = 4x^3.$$

(b) Formula 3.8 allows us to differentiate each term separately; and by formulas 3.7 and 3.6 it follows that

$$\frac{dy}{dx} = 3(6x^5) - (-2x^{-3}) = 18x^5 + 2x^{-3}.$$

(c) If we write y in the form $y = (x/3) - 2x^{-1}$, then we can proceed as in (b):

$$\frac{dy}{dx} = \frac{1}{3}(1) - 2(-x^{-2}) = \frac{1}{3} + \frac{2}{x^2}.$$

EXAMPLE 3.7

If $f(x) = 3x^4 - 2$, evaluate $f'(1)$.

SOLUTION Since $f'(x) = 12x^3$, it follows that $f'(1) = 12$. Geometrically, 12 is the slope of the tangent line to the curve $y = 3x^4 - 2$ at the point $(1, 1)$ in Figure 3.12. Algebraically, the result implies that at $x = 1$, y changes 12 times as fast as x.

FIGURE 3.12

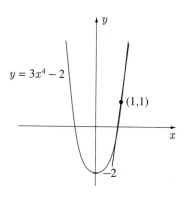

EXAMPLE 3.8 Find the equation of the tangent line to the curve $y = f(x) = x^3 + 5x$ at the point $(1, 6)$.

SOLUTION Since $f'(x) = 3x^2 + 5$, the slope of the tangent line to the curve at $(1, 6)$ is
$$f'(1) = 3(1)^2 + 5 = 8.$$

Using point-slope formula 1.10 for a straight line, we obtain for the equation of the tangent line at $(1, 6)$
$$y - 6 = 8(x - 1) \quad \text{or} \quad 8x - y = 2.$$

The line through $(1, 6)$ perpendicular to the tangent line in Figure 3.13 is called the **normal line** to the curve at $(1, 6)$.

FIGURE 3.13

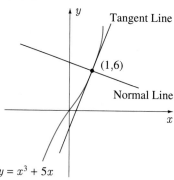

Since two lines are perpendicular only if their slopes are negative reciprocals (see equation 1.12), the normal line at $(1, 6)$ must have slope $-1/8$. The equation of the normal line to $y = x^3 + 5x$ at $(1, 6)$ is therefore
$$y - 6 = -\frac{1}{8}(x - 1) \quad \text{or}$$
$$x + 8y = 49.$$

> **EXAMPLE 3.9**
>
> Find points on the curve $y = x^4/4 - 2x^3/3 - x^2/2 + x$ at which the slope of the tangent line is -1.
>
> **SOLUTION** The slope of the tangent line to the curve is -1 whenever
>
> $$-1 = \frac{dy}{dx} = x^3 - 2x^2 - x + 1.$$
>
> When the -1 is transposed to the right side, and the cubic polynomial is factored, the result is
>
> $$0 = x^3 - 2x^2 - x + 2 = (x-1)(x+1)(x-2).$$
>
> This equation has solutions $x = 1, -1,$ and 2. Thus the points at which the slope of the tangent line is -1 are $(1, 1/12), (-1, -7/12)$, and $(2, -4/3)$. ∎

In applications, independent and dependent variables of functions $y = f(x)$ represent physical quantities and have units associated with them. Units for the derivative $f'(x)$ are units of y divided by units of x. We see this from equation 3.2 where units of the quotient $[f(x+h) - f(x)]/h$ are clearly units of y divided by units of x, and the limit of this quotient as $h \to 0$ does not alter these units. For example, if x measures length in metres and y measures mass in kilograms, the units of dy/dx are kilograms per metre (kilograms divided by metres).

Increment Notation

We have used the letter h to represent a small change in x when calculating derivatives. An alternative notation is sometimes more suggestive. A small change in x, called an **increment** in x, is often denoted by Δx. It is pronounced "delta x" since Δ is the capital letter delta in the Greek alphabet. When x is given an increment Δx in the function $y = f(x)$, the corresponding change or increment in y is denoted by Δy. It is equal to

$$\Delta y = f(x + \Delta x) - f(x). \tag{3.9}$$

With this notation, definition 3.2 takes the form

$$f'(x) = \lim_{\Delta x \to 0} \frac{\Delta y}{\Delta x}. \tag{3.10a}$$

The notation dy/dx for the derivative fits very nicely with increment notation,

$$\frac{dy}{dx} = \lim_{\Delta x \to 0} \frac{\Delta y}{\Delta x}; \tag{3.10b}$$

the derivative of y with respect to x is the limit of the change in y divided by the change in x as the change in x approaches zero. We use this notation in the following example.

EXAMPLE 3.10

Use equation 3.10 to calculate the derivative of $y = f(x) = 3x^2 - 2x$ and check your answer using the differentiation rules discussed at the beginning of this section.

SOLUTION Since

$$\Delta y = f(x + \Delta x) - f(x)$$
$$= [3(x + \Delta x)^2 - 2(x + \Delta x)] - [3x^2 - 2x]$$
$$= 3[x^2 + 2x\Delta x + (\Delta x)^2] - 2x - 2\Delta x - 3x^2 + 2x$$
$$= \Delta x(6x - 2 + 3\Delta x),$$

equation 3.10 gives

$$\frac{dy}{dx} = \lim_{\Delta x \to 0} \frac{\Delta y}{\Delta x} = \lim_{\Delta x \to 0} \frac{\Delta x(6x - 2 + 3\Delta x)}{\Delta x} = \lim_{\Delta x \to 0} (6x - 2 + 3\Delta x) = 6x - 2.$$

Rules 3.8, 3.7, and 3.6 for differentiation of $3x^2 - 2x$ also yield $6x - 2$.

EXERCISES 3.2

In Exercises 1–20 find $f'(x)$.

1. $f(x) = 2x^2 - 3$
2. $f(x) = 3x^3 + 4x + 5$
3. $f(x) = 10x^2 - 3x$
4. $f(x) = 4x^5 - 10x^3 + 3x$
5. $f(x) = 1/x^2$
6. $f(x) = 2/x^3$
7. $f(x) = 5x^4 - 3x^3 + 1/x$
8. $f(x) = -\dfrac{1}{2x^2} + \dfrac{3}{x^4}$
9. $f(x) = x^{10} - \dfrac{1}{x^{10}}$
10. $f(x) = 5x^4 + \dfrac{1}{4x^5}$
11. $f(x) = 5x^{-4} + \dfrac{1}{4x^{-5}}$
12. $f(x) = \sqrt{x}$
13. $f(x) = \dfrac{3}{x^2} + \dfrac{2}{\sqrt{x}}$
14. $f(x) = \dfrac{1}{x^{3/2}} + x^{3/2}$
15. $f(x) = 2x^{1/3} - 3x^{2/3}$
16. $f(x) = \pi x^\pi$
17. $f(x) = (x^2 + 2)^2$
18. $f(x) = (4x^6 - x^2)/x^5$
19. $f(x) = x^{5/3} - x^{2/3} + 3$
20. $f(x) = (2x + 5)^3$

In Exercises 21–24 find equations for the tangent and normal lines to the curve at the point indicated. In each case, sketch the curve and lines.

21. $y = x^2 - 2x + 5$ at $(2, 5)$
22. $y = \sqrt{x} + 5$ at $(4, 7)$
23. $y = 2x^3 - 3x^2 - 12x$ at $(2, -20)$
24. $x = \sqrt{y + 1}$ at $(3, 8)$

25. Find the points on the curve $y = x^4/4 - 2x^3/3 - 19x^2/2 + 22x$ at which the slope of the tangent line is 2.

26. An electronics firm manufactures a device called a "gadget". The money that it receives from the sale of x gadgets is the total revenue and is denoted by $R(x)$. The total cost $C(x)$ of producing these x gadgets also depends on x. The total profit realized on the sale is

$$P(x) = R(x) - C(x).$$

The derivatives of these functions, $R'(x)$, $C'(x)$, and $P'(x)$ are called respectively, marginal revenue, marginal cost, and marginal profit. Calculate these marginal functions if

$$R(x) = 75x - \frac{7x^2}{100} \quad \text{and}$$

$$C(x) = 50x - \frac{x^2}{30} - \frac{x^3}{60\,000},$$

both functions being defined for $1 \leq x \leq 800$.

27. Sketch a graph of the function $x = f(t) = t^3 - 8t^2$. Find the value(s) of t at which the tangent line to this curve is parallel to the line $x = 6t - 3$.

28. At what point(s) on the curve $y = x^3 + x^2 - 22x + 20$ does the tangent line pass through the origin?

29. At what point(s) on the parabola $y = x^2$ does the normal line pass through the point $(2,5)$? Can you suggest an application of this result?

30. Show that the sum of the x- and y-intercepts of the tangent line to the curve $\sqrt{x} + \sqrt{y} = \sqrt{a}$ is always equal to a.

31. Prove that
$$\frac{d}{dx}(ax+b)^n = an(ax+b)^{n-1}$$

when a and b are constants.

32. Find a formula for
$$\frac{d}{dx}|x|^n,$$

when $n > 1$ is an integer.

33. Find the two points on the curve $y = x(1 + 2x - x^3)$ that share a common tangent line.

SECTION 3.3

Differentiability and Continuity

Many functions fail to have a derivative at isolated points. We examine such behaviour in this section. For example, consider the function $f(x) = |x|$ shown in Figure 3.14.

It is clear that for $x > 0$, $f'(x) = 1$, and for $x < 0$, $f'(x) = -1$. At $x = 0$, however, there is a problem. If $f(x)$ is to have a derivative at $x = 0$, it must be given by

$$\lim_{h \to 0} \frac{f(0+h) - f(0)}{h} = \lim_{h \to 0} \frac{|h|}{h}.$$

But this limit does not exist since

$$\lim_{h \to 0^-} \frac{|h|}{h} = -1 \quad \text{and} \quad \lim_{h \to 0^+} \frac{|h|}{h} = 1.$$

Consequently, $f(x) = |x|$ does not have a derivative at $x = 0$.

The same conclusion can be drawn at any point at which the graph of a function takes an abrupt change in direction. Such a point is often called a *corner*. As a result, the function in Figure 3.15 does not have a derivative at $x = a,b,c$, or d. It is also true that a function cannot have a derivative at a point where the function is discontinuous (see, for example, the discontinuities in Figure 2.14). This result is an immediate consequence of the following theorem.

FIGURE 3.14

FIGURE 3.15

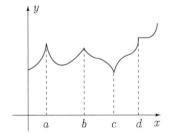

Theorem 3.6

If a function $f(x)$ has a derivative at $x = a$, then $f(x)$ is continuous at $x = a$.

Proof To prove this theorem we show that $\lim_{x \to a} f(x) = f(a)$, the condition that defines continuity of $f(x)$ at $x = a$ (Definition 2.1 in Section 2.4). The derivative $f'(a)$ is defined by

$$f'(a) = \lim_{h \to 0} \frac{f(a+h) - f(a)}{h}.$$

No one will argue with the fact that

$$0 = \lim_{h \to 0} h.$$

But note that if we multiply the left-hand and right-hand sides of these two equations, we obtain

$$0 \cdot f'(a) = \left[\lim_{h \to 0} h\right] \left[\lim_{h \to 0} \frac{f(a+h) - f(a)}{h}\right].$$

Using part (iv) of Theorem 2.1, we may bring the limits on the right together,

$$0 = \lim_{h \to 0} [f(a+h) - f(a)], \quad \text{or}$$

$$\lim_{h \to 0} f(a+h) = f(a).$$

But this is just another way of saying that

$$\lim_{x \to a} f(x) = f(a),$$

(see Exercises 34 in Section 2.4).

As a corollary to this theorem we have the following.

Corollary If $f(x)$ is discontinuous at $x = a$, then $f'(a)$ does not exist.

We introduced this section by showing that the derivative of $|x|$ is 1 when $x > 0$; it is -1 when $x < 0$; and it does not exist at $x = 0$. These can be combined into the simple formula

$$\frac{d}{dx}|x| = \frac{|x|}{x}. \tag{3.11}$$

It is straightforward to generalize 3.11 and obtain the derivative of $|f(x)|$ at any point at which $f(x) \neq 0$ and $f'(x)$ exists. When $f(x) > 0$, we may write

$$\frac{d}{dx}|f(x)| = \frac{d}{dx}f(x) = f'(x).$$

On the other hand, when $f(x) < 0$, we have

$$\frac{d}{dx}|f(x)| = \frac{d}{dx}[-f(x)] = -f'(x).$$

Both of these results are contained in the one equation

$$\frac{d}{dx}|f(x)| = \frac{|f(x)|}{f(x)} f'(x). \tag{3.12}$$

This then is the derivative of $|f(x)|$ at any point at which $f(x) \neq 0$. The exceptional case when $f(x) = 0$ is discussed in Exercise 20.

Right-hand and Left-hand Derivatives

When a function $f(x)$ is defined on a closed interval $b \leq x \leq c$, equation 3.2 can be used to calculate $f'(x)$ only at points in the open interval $b < x < c$. For instance, it is impossible to evaluate

$$f'(b) = \lim_{h \to 0} \frac{f(b+h) - f(b)}{h}$$

since $f(x)$ is not defined for $x < b$ and therefore $f(b+h)$ is not defined for $h < 0$. When a function $f(x)$ is defined only to the right of a point, we define a right-hand derivative at the point; and when $f(x)$ is defined only to the left of a point, we define its left-hand derivative.

> **Definition 3.2**
>
> The **right-hand derivative** of $f(x)$ with respect to x is defined as
>
> $$f'_+(x) = \lim_{h \to 0^+} \frac{f(x+h) - f(x)}{h}, \qquad (3.13\text{a})$$
>
> provided the limit exists. The **left-hand derivative** of $f(x)$ is
>
> $$f'_-(x) = \lim_{h \to 0^-} \frac{f(x+h) - f(x)}{h}, \qquad (3.13\text{b})$$
>
> if the limit exists.

When a function has a derivative at a point x, its right- and left-hand derivatives both exist at x and are equal to $f'(x)$. It is possible, however, for a function to have both a left-hand and a right-hand derivative at a point but not a derivative. An example of this is the absolute value function $f(x) = |x|$ at $x = 0$ (see Figure 3.14). Its right-hand derivative at $x = 0$ is equal to 1 and its left-hand derivative there is -1.

When a function has a derivative at $x = a$, we say that it is **differentiable** at $x = a$. When it has a derivative at every point in some interval, we say that it is differentiable on that interval. In the event that the interval is closed, $b \le x \le c$, we understand that derivatives at $x = b$ and $x = c$ mean right-hand and left-hand derivatives respectively.

> **EXAMPLE 3.11**
>
> Is the function
>
> $$g(x) = \begin{cases} x \sin(1/x) & x \ne 0 \\ 0 & x = 0 \end{cases}$$
>
> differentiable at $x = 0$?
>
> **SOLUTION** We encountered the function $x \sin(1/x)$ in Example 2.8 of Section 2.1 and drew its graph (except for the point at $x = 0$) in Figure 2.6. According to equation 3.2, the derivative of $g(x)$ at $x = 0$ is
>
> $$g'(0) = \lim_{h \to 0} \frac{g(0+h) - g(0)}{h},$$
>
> provided the limit exists. When we substitute from the definition of $g(x)$ for $g(h)$ and $g(0)$, this limit takes the form
>
> $$g'(0) = \lim_{h \to 0} \frac{h \sin(1/h) - 0}{h} = \lim_{h \to 0} \sin\left(\frac{1}{h}\right).$$
>
> Since this limit does not exist (see Example 2.8), $g(x)$ is not differentiable at $x = 0$.

EXERCISES 3.3

In Exercises 1–6 determine whether the function has a right-hand derivative, a left-hand derivative, and a derivative at the given value of x.

1. $f(x) = |x - 5|$ at $x = 5$
2. $f(x) = x^{3/2}$ at $x = 0$
3. $f(x) = |x - 5|^3$ at $x = 5$
4. $f(x) = \operatorname{sgn} x$ at $x = 0$ (see Exercise 36 in Section 2.4)
5. $f(x) = (x^2 - 1)/(x - 1)$ at $x = 1$
6. $f(x) = H(x - a)$ at $x = a$ (see Exercise 38 in Section 2.4)
7. Does it make any difference in Exercise 6 if we define $H(a) = 0$?
8. Does it make any difference in Exercise 6 if we define $H(a) = 1$?
9. Does it make any difference in Exercise 4 if $\operatorname{sgn} x$ does not have a value at $x = 0$?

In Exercises 10–13 determine whether the statement is true or false.

10. If a function $f(x)$ has a derivative at $x = a$, then its graph has a tangent line at the point $(a, f(a))$.
11. If a function $f(x)$ has a tangent line at a point $(a, f(a))$, then it has a derivative at $x = a$.
12. If a function does not have a tangent line at $(a, f(a))$, then it does not have a derivative at $x = a$.
13. If a function $f(x)$ does not have a derivative at $x = a$, then it does not have a tangent line at $(a, f(a))$.

In Exercises 14–16 show algebraically that $f'(0)$ does not exist. Sketch a graph of the function.

14. $f(x) = x^{1/3}$
15. $f(x) = x^{2/3}$
16. $f(x) = x^{1/4}$
17. Sketch a graph of the function $f'(x)$ when $f(x) = [[x]]$ is the greatest integer function (Exercise 36 in Section 1.5).
18. Is the function $f(x) = x|x|$ differentiable at $x = 0$?
19. Find $f'(x)$ if $f(x) = |x| + |x - 1|$. Sketch graphs of $f(x)$ and $f'(x)$.
20. If $f(x)$ is a differentiable function, does $|f(x)|$ have a derivative at points where $f(x) = 0$? Hint: Draw some pictures.
21. If $\lim_{x \to \infty} f(x) = L$ (so that $y = L$ is a horizontal asymptote for the graph of the function), is it necessary that $\lim_{x \to \infty} f'(x) = 0$?
22. Is the function
$$f(x) = \begin{cases} x^2 \sin(1/x) & x \neq 0 \\ 0 & x = 0 \end{cases}$$
differentiable at $x = 0$?
23. For what values of the real number n is the function
$$f(x) = \begin{cases} x^n \sin(1/x) & x \neq 0 \\ 0 & x = 0 \end{cases}$$
differentiable at $x = 0$?
24. Is the function
$$f(x) = \begin{cases} x^2 & x \text{ a rational number} \\ 0 & x \text{ an irrational number} \end{cases}$$
differentiable at $x = 0$?
25. Show that the derivative of the function in Exercise 22 is not continuous at $x = 0$.

SECTION 3.4

Product and Quotient Rules

In this section we add two more formulas to those of Section 2 for calculating derivatives. The first is a rule for differentiating a function that is the product of two other functions.

> **Theorem 3.7**
>
> If $p(x) = f(x)g(x)$, where $f(x)$ and $g(x)$ are differentiable, then
>
> $$p'(x) = f(x)g'(x) + f'(x)g(x). \qquad (3.14\text{a})$$

Proof By equation 3.2,

$$\begin{aligned} p'(x) &= \lim_{h \to 0} \frac{p(x+h) - p(x)}{h} \\ &= \lim_{h \to 0} \frac{f(x+h)g(x+h) - f(x)g(x)}{h}. \end{aligned}$$

To organize this quotient further, we add and subtract the quantity $f(x+h)g(x)$ in the numerator:

$$\begin{aligned} p'(x) &= \lim_{h \to 0} \left\{ \frac{[f(x+h)g(x+h) - f(x+h)g(x)] + [f(x+h)g(x) - f(x)g(x)]}{h} \right\} \\ &= \lim_{h \to 0} \left[f(x+h) \frac{g(x+h) - g(x)}{h} + g(x) \frac{f(x+h) - f(x)}{h} \right] \\ &= f(x)g'(x) + g(x)f'(x). \end{aligned}$$

In taking the limit of the first term, we have used the fact that $\lim_{h \to 0} f(x+h) = f(x)$, which follows from continuity of $f(x)$ (see Theorem 3.6 and Exercise 34 in Section 2.4).

This result is called the **product rule** for differentiation. If we set $u = f(x)$ and $v = g(x)$, then the product rule may also be expressed in the form

$$\frac{d}{dx}(uv) = u\frac{dv}{dx} + v\frac{du}{dx}. \qquad (3.14\text{b})$$

We use increment notation to prove the **quotient rule**. Use of h in place of Δx works equally well.

> **Theorem 3.8**
>
> If $p(x) = f(x)/g(x)$, where $f(x)$ and $g(x)$ are differentiable, then
>
> $$p'(x) = \frac{g(x)f'(x) - f(x)g'(x)}{[g(x)]^2}. \qquad (3.15\text{a})$$

Proof Using equation 3.10,

$$\begin{aligned} p'(x) &= \lim_{\Delta x \to 0} \frac{p(x + \Delta x) - p(x)}{\Delta x} \\ &= \lim_{\Delta x \to 0} \frac{1}{\Delta x} \left[\frac{f(x + \Delta x)}{g(x + \Delta x)} - \frac{f(x)}{g(x)} \right] \\ &= \lim_{\Delta x \to 0} \frac{f(x + \Delta x)g(x) - g(x + \Delta x)f(x)}{\Delta x\, g(x) g(x + \Delta x)}. \end{aligned}$$

To simplify this limit we add and subtract $f(x)g(x)$ in the numerator:

$$p'(x) = \lim_{\Delta x \to 0} \frac{[f(x+\Delta x)g(x) - f(x)g(x)] - [g(x+\Delta x)f(x) - f(x)g(x)]}{\Delta x g(x)g(x+\Delta x)}$$

$$= \lim_{\Delta x \to 0} \frac{1}{g(x)g(x+\Delta x)} \left\{ g(x) \left[\frac{f(x+\Delta x) - f(x)}{\Delta x} \right] - f(x) \left[\frac{g(x+\Delta x) - g(x)}{\Delta x} \right] \right\}$$

$$= \frac{1}{[g(x)]^2} \left\{ g(x) \lim_{\Delta x \to 0} \frac{f(x+\Delta x) - f(x)}{\Delta x} - f(x) \lim_{\Delta x \to 0} \frac{g(x+\Delta x) - g(x)}{\Delta x} \right\}$$

$$= \frac{g(x)f'(x) - f(x)g'(x)}{[g(x)]^2}.$$

If we set $u = f(x)$ and $v = g(x)$, then quotient rule 3.15a can be expressed in the form

$$\frac{d}{dx}\left(\frac{u}{v}\right) = \frac{v\dfrac{du}{dx} - u\dfrac{dv}{dx}}{v^2}. \qquad (3.15\text{b})$$

It makes no difference which term in product rule 3.14 is written first; it does make a difference in the quotient rule. Do not interchange the terms in 3.15b.

EXAMPLE 3.12

For the following three functions, find $f'(x)$ in simplified form:

(a) $f(x) = (x^2 + 2)(x^4 + 5x^2 + 1)$ (b) $f(x) = \dfrac{\sqrt{x}}{3x^2 - 2}$ (c) $f(x) = \dfrac{1}{(5x^3 - 2)^2}$

SOLUTION (a) With product rule 3.14a,

$$f'(x) = (x^2 + 2)\frac{d}{dx}(x^4 + 5x^2 + 1) + (x^4 + 5x^2 + 1)\frac{d}{dx}(x^2 + 2)$$
$$= (x^2 + 2)(4x^3 + 10x) + (x^4 + 5x^2 + 1)(2x)$$
$$= 6x^5 + 28x^3 + 22x.$$

(b) With quotient rule 3.15,

$$f'(x) = \frac{(3x^2 - 2)\left(\dfrac{1}{2\sqrt{x}}\right) - \sqrt{x}(6x)}{(3x^2 - 2)^2}$$

$$= \frac{\dfrac{3x^2 - 2 - 12x^2}{2\sqrt{x}}}{(3x^2 - 2)^2}$$

$$= -\frac{9x^2 + 2}{2\sqrt{x}(3x^2 - 2)^2}.$$

(c) If we express $f(x)$ in the form

$$f(x) = \left(\frac{1}{5x^3 - 2}\right)\left(\frac{1}{5x^3 - 2}\right),$$

and use the product rule,

$$f'(x) = \frac{1}{5x^3 - 2} \frac{d}{dx}\left(\frac{1}{5x^3 - 2}\right) + \frac{1}{5x^3 - 2} \frac{d}{dx}\left(\frac{1}{5x^3 - 2}\right)$$
$$= \frac{2}{5x^3 - 2} \frac{d}{dx}\left(\frac{1}{5x^3 - 2}\right).$$

The quotient rule now gives

$$f'(x) = \frac{2}{5x^3 - 2}\left[\frac{(5x^3 - 2)(0) - 1(15x^2)}{(5x^3 - 2)^2}\right]$$
$$= \frac{-30x^2}{(5x^3 - 2)^3}.$$

Differentiation in part (c) of this example was somewhat cumbersome considering the simplicity of the function. Imagine the difficulty had the power been 12 instead of 2. Fortunately, in Section 3.6 we develop a very simple rule for differentiation of functions like this.

EXAMPLE 3.13

Sketch a graph of the function $y = f(x) = (x - 1)/(x + 2)$. Calculate dy/dx and show qualitatively that the graph agrees with your calculation.

SOLUTION The x- and y-intercepts of the graph are 1 and $-1/2$ respectively. Since $f(x)$ is discontinuous at $x = -2$, we calculate

$$\lim_{x \to -2^-} f(x) = \infty \quad \text{and} \quad \lim_{x \to -2^+} f(x) = -\infty$$

This information, together with the following limits leads to the graph in Figure 3.16,

$$\lim_{x \to -\infty} f(x) = 1^+ \quad \text{and} \quad \lim_{x \to \infty} f(x) = 1^-.$$

Using quotient rule 3.15b, we find

$$\frac{dy}{dx} = \frac{(x+2)(1) - (x-1)(1)}{(x+2)^2} = \frac{3}{(x+2)^2}.$$

The sketch in Figure 3.16 and dy/dx agree that:

(a) The slope of the curve is always positive;

(b) The slope becomes larger and larger as x approaches -2; i.e.,

$$\lim_{x \to -2} f'(x) = \infty;$$

(c) The slope approaches zero as x approaches $\pm\infty$; i.e.,

$$\lim_{x \to \pm\infty} f'(x) = 0.$$

FIGURE 3.16

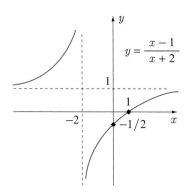

EXAMPLE 3.14

Sketch a graph of the function $f(x) = (x-4)/x^2$.

SOLUTION The x-intercept of the graph is $x = 4$. Since the function is discontinuous at $x = 0$, we calculate

$$\lim_{x \to 0^-} f(x) = -\infty \quad \text{and} \quad \lim_{x \to 0^+} f(x) = -\infty.$$

Limits as $x \to \pm\infty$ are

$$\lim_{x \to -\infty} f(x) = 0^- \quad \text{and} \quad \lim_{x \to \infty} f(x) = 0^+.$$

This information, shown in Figure 3.17(a), would lead us to suspect that the graph should be completed as the dashed lines indicate. To show that this is indeed the case, we calculate

$$f'(x) = \frac{x^2(1) - (x-4)(2x)}{x^4} = \frac{8-x}{x^3}.$$

Since

$$\begin{aligned} f'(x) &< 0 \quad \text{for } x < 0, \\ f'(x) &> 0 \quad \text{for } 0 < x < 8, \\ f'(x) &= 0 \quad \text{for } x = 8, \\ f'(x) &< 0 \quad \text{for } x > 8, \end{aligned}$$

our suspicions were correct, and the final graph is shown in Figure 3.17(b).

FIGURE 3.17

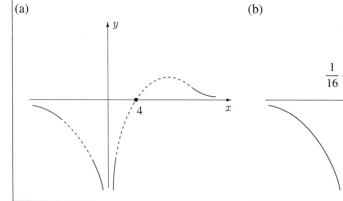

Examples 3.13 and 3.14 illustrate that derivatives can be helpful in sketching graphs. A complete discussion of the application of derivatives to curve sketching will be given in Section 4.5.

Angle Between Intersecting Curves

The angle θ between two curves which intersect at a point (x_0, y_0) (Figure 3.18) is defined as the angle between the tangent lines to the curves at (x_0, y_0). This can be calculated using formula 1.14 once slopes of the tangent lines are known. In the event that $\theta = \pi/2$ (Figure 3.19), the curves are said to be **orthogonal** or perpendicular at (x_0, y_0). Should $\theta = 0$, the curves are said to be tangent at (x_0, y_0) (Figure 3.20).

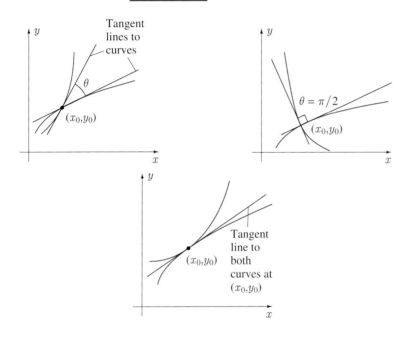

FIGURE 3.18

FIGURE 3.19

FIGURE 3.20

EXAMPLE 3.15

Find the angle between the curves $y = 6 - x^2$ and $y = 3x/(x+1)$ at their point of intersection.

SOLUTION The curves intersect at the point, or points, with x-coordinate satisfying

$$6 - x^2 = \frac{3x}{x+1}.$$

When we multiply both sides by $x + 1$, and simplify, the result is

$$0 = x^3 + x^2 - 3x - 6 = (x-2)(x^2 + 3x + 3).$$

Since the only (real) solution of this equation is $x = 2$, the point of intersection of the curves is $(2, 2)$ (Figure 3.21).

FIGURE 3.21

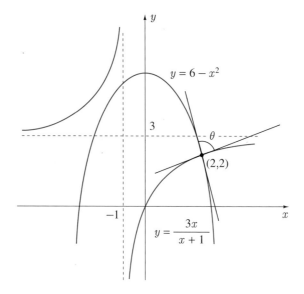

The slope of $y = f(x) = 6 - x^2$ at $(2, 2)$ is

$$f'(2) = -2x\big|_{x=2} = -4.$$

When a quantity is suffixed at the subscript level with an expression like $|x = 2$, it means that the quantity is to be evaluated at $x = 2$. The slope of $y = g(x) = 3x/(x+1)$ at this point is

$$g'(2) = \frac{(x+1)(3) - 3x(1)}{(x+1)^2}\bigg|_{x=2} = \frac{1}{3}.$$

Using formula 1.14 with $m_1 = -4$ and $m_2 = 1/3$, the angle θ between the curves at $(2, 2)$ is given by

$$\tan \theta = \left| \frac{-4 - 1/3}{1 + (-4)(1/3)} \right| = 13.$$

Hence, $\theta = 1.494$ radians. ∎

EXERCISES 3.4

In Exercises 1–16 find $f'(x)$ in simplified form.

1. $f(x) = (x^2 + 2)(x + 3)$

2. $f(x) = (2 - x^2)(x^2 + 4x + 2)$

3. $f(x) = \dfrac{x}{3x + 2}$

4. $f(x) = \dfrac{x^2}{4x^2 - 5}$

5. $f(x) = \dfrac{x^2}{2x - 1}$

6. $f(x) = \dfrac{x^3}{4x^2 + 1}$

7. $f(x) = \sqrt{x}(x + 1)$

8. $f(x) = x\sqrt{x + 1}$

9. $f(x) = \dfrac{2x^2 - 5}{3x + 4}$

10. $f(x) = \dfrac{x + 5}{2x^2 - 1}$

11. $f(x) = \dfrac{x(x + 1)}{1 - 3x}$

12. $f(x) = \dfrac{x^2 + 2x + 3}{x^2 - 5x + 1}$

13. $f(x) = \dfrac{x^{1/3}}{1 - \sqrt{x}}$

14. $f(x) = \dfrac{\sqrt{x} + 2x}{\sqrt{x} - 4}$

15. $f(x) = \dfrac{1}{x^3 - 3x^2 + 2x + 5}$

16. $f(x) = \dfrac{x^3 + 3x^2 + 3x + 10}{(x + 1)^3}$

17. Find equations for the tangent and normal lines to the curve $y = (x + 3)/(x - 4)$ at the point $(1, -4/3)$. Sketch the curve and lines.

18. (a) Sketch a graph of the function $f(x) = x^2/(x^2 + x - 2)$.

    (b) Find $f'(x)$ and show that it agrees with your sketch.

19. If the total cost of producing x items of a commodity is given by the equation

$$C(x) = ax\left(\frac{x+b}{x+c}\right),$$

where a, b, and c are constants, show that the marginal cost $C'(x)$ is

$$a\left[1 + \frac{c(b-c)}{(x+c)^2}\right].$$

20. Find a rule for the derivative of the product of three functions $f(x)$, $g(x)$, and $h(x)$.

In Exercises 21–24 find the angle (or angles) between the curves at their point (or points) of intersection.

21. $y = x^2$, $x + y = 2$ 22. $y = x^2$, $y = 1 - x^2$
23. $y = x(x-5)$, $y = x\sqrt{4-x}$
24. $y = 2x + 2$, $y = x^2/(x-1)$

In Exercises 25 and 26 show that the curves are orthogonal.

25. $x - 2y + 1 = 0$, $y = 2 - x^2$
26. $y = 3 - x^2$, $4y - 7 = x^2$

In Exercises 27 and 28 show that the curves are tangent at the indicated point.

27. $y = x - 2x^2$, $y = x^3 + 2x$ at $(-1, -3)$
28. $y = x^3$, $y = x^2 + x - 1$ at $(1, 1)$

29. (a) The equation $x + 2y = C_1$, where C_1 is a constant, represents a *one-parameter family of curves*. For each value of C_1, the parameter, a different curve in the family is obtained. Draw curves corresponding to $C_1 = -2, -1, 0, 1, 2$.

    (b) Draw a few curves from the one-parameter family $2x - y = C_2$ on the graph in (a).

    (c) Two families of curves are said to be *orthogonal trajectories* if every curve from one family intersects every curve from the other family orthogonally. Are the families in (a) and (b) orthogonal trajectories?

30. For what value of k are the one-parameter families $2x - 3y = C_1$ and $x + ky = C_2$ orthogonal trajectories?

31. (a) A manufacturer's profit from the sale of x kilograms of a commodity per week is given by

$$P(x) = \frac{3x - 200}{x + 400}.$$

Sketch a graph of this function.

    (b) The average profit per kilogram when x kilograms are sold is given by

$$p(x) = \frac{P(x)}{x}.$$

If a point (x, P) on the total profit curve is joined to the origin, the slope of this line is the average profit $p(x)$ for that x. Use this idea to find the sales level for highest average profit.

32. Find all points on the curve $y = (5 - x)/(6 + x)$ at which the tangent line passes through the origin.

33. Show that the ellipse $b^2x^2 + a^2y^2 = a^2b^2$ and the hyperbola $d^2x^2 - c^2y^2 = c^2d^2$ intersect orthogonally if $a^2 - b^2 = c^2 + d^2$.

SECTION 3.5
Higher-Order Derivatives

When $y = f(x)$ is a function of x, its derivative $f'(x)$ is also a function of x. We can therefore take the derivative of the derivative to get what is called the **second derivative** of the function. This can be repeated over and over again. For instance, if $y = f(x) = x^3 + 1/x$, then

$$\frac{dy}{dx} = 3x^2 - \frac{1}{x^2}.$$

We denote the second derivative of y with respect to x by d^2y/dx^2 or $f''(x)$:

$$f''(x) = \frac{d^2y}{dx^2} = \frac{d}{dx}\left(\frac{dy}{dx}\right) = 6x + \frac{2}{x^3}.$$

Similarly,
$$f'''(x) = \frac{d^3y}{dx^3} = 6 - \frac{6}{x^4}.$$

This is called the **third derivative** or the derivative of order three. Clearly we can continue the differentiation process indefinitely to produce derivatives of any positive integer order whatsoever.

EXAMPLE 3.16

Find a formula for the nth derivative of $y = x/(x+1)$.

SOLUTION When $n = 1$,
$$\frac{dy}{dx} = \frac{(x+1)(1) - x(1)}{(x+1)^2} = \frac{1}{(x+1)^2}.$$

When $n = 2$,
$$\frac{d^2y}{dx^2} = \frac{d}{dx}\left[\frac{1}{(x+1)^2}\right] = \frac{(x+1)^2(0) - \frac{d}{dx}(x+1)^2}{(x+1)^4}.$$

According to Exercise 31 in Section 3.2,
$$\frac{d}{dx}(x+1)^n = n(x+1)^{n-1}.$$

Thus,
$$\frac{d^2y}{dx^2} = \frac{-2(x+1)}{(x+1)^4} = \frac{-2}{(x+1)^3}.$$

When $n = 3$,
$$\frac{d^3y}{dx^3} = \frac{(x+1)^3(0) + 2(3)(x+1)^2}{(x+1)^6} = \frac{2 \cdot 3}{(x+1)^4}.$$

When $n = 4$,
$$\frac{d^4y}{dx^4} = \frac{(x+1)^4(0) - 2(3)(4)(x+1)^3}{(x+1)^8} = \frac{-2 \cdot 3 \cdot 4}{(x+1)^5}.$$

It is fairly clear that the pattern emerging is
$$\frac{d^ny}{dx^n} = \frac{(-1)^{n+1} n!}{(x+1)^{n+1}},$$

where the notation $n!$, called "n factorial," represents the product of the first n positive integers: $n! = 1(2)(3) \cdots (n-1)(n)$.

18. (a) Sketch a graph of the function $f(x) = x^2/(x^2 + x - 2)$.

    (b) Find $f'(x)$ and show that it agrees with your sketch.

19. If the total cost of producing x items of a commodity is given by the equation

    $$C(x) = ax\left(\frac{x+b}{x+c}\right),$$

    where a, b, and c are constants, show that the marginal cost $C'(x)$ is

    $$a\left[1 + \frac{c(b-c)}{(x+c)^2}\right].$$

20. Find a rule for the derivative of the product of three functions $f(x)$, $g(x)$, and $h(x)$.

In Exercises 21–24 find the angle (or angles) between the curves at their point (or points) of intersection.

21. $y = x^2$, $x + y = 2$ 22. $y = x^2$, $y = 1 - x^2$
23. $y = x(x-5)$, $y = x\sqrt{4-x}$
24. $y = 2x + 2$, $y = x^2/(x-1)$

In Exercises 25 and 26 show that the curves are orthogonal.

25. $x - 2y + 1 = 0$, $y = 2 - x^2$
26. $y = 3 - x^2$, $4y - 7 = x^2$

In Exercises 27 and 28 show that the curves are tangent at the indicated point.

27. $y = x - 2x^2$, $y = x^3 + 2x$ at $(-1, -3)$
28. $y = x^3$, $y = x^2 + x - 1$ at $(1, 1)$

29. (a) The equation $x + 2y = C_1$, where C_1 is a constant, represents a *one-parameter family of curves*. For each value of C_1, the parameter, a different curve in the family is obtained. Draw curves corresponding to $C_1 = -2, -1, 0, 1, 2$.

    (b) Draw a few curves from the one-parameter family $2x - y = C_2$ on the graph in (a).

    (c) Two families of curves are said to be *orthogonal trajectories* if every curve from one family intersects every curve from the other family orthogonally. Are the families in (a) and (b) orthogonal trajectories?

30. For what value of k are the one-parameter families $2x - 3y = C_1$ and $x + ky = C_2$ orthogonal trajectories?

31. (a) A manufacturer's profit from the sale of x kilograms of a commodity per week is given by

    $$P(x) = \frac{3x - 200}{x + 400}.$$

    Sketch a graph of this function.

    (b) The average profit per kilogram when x kilograms are sold is given by

    $$p(x) = \frac{P(x)}{x}.$$

    If a point (x, P) on the total profit curve is joined to the origin, the slope of this line is the average profit $p(x)$ for that x. Use this idea to find the sales level for highest average profit.

32. Find all points on the curve $y = (5 - x)/(6 + x)$ at which the tangent line passes through the origin.

33. Show that the ellipse $b^2 x^2 + a^2 y^2 = a^2 b^2$ and the hyperbola $d^2 x^2 - c^2 y^2 = c^2 d^2$ intersect orthogonally if $a^2 - b^2 = c^2 + d^2$.

SECTION 3.5

Higher-Order Derivatives

When $y = f(x)$ is a function of x, its derivative $f'(x)$ is also a function of x. We can therefore take the derivative of the derivative to get what is called the **second derivative** of the function. This can be repeated over and over again. For instance, if $y = f(x) = x^3 + 1/x$, then

$$\frac{dy}{dx} = 3x^2 - \frac{1}{x^2}.$$

We denote the second derivative of y with respect to x by d^2y/dx^2 or $f''(x)$:

$$f''(x) = \frac{d^2y}{dx^2} = \frac{d}{dx}\left(\frac{dy}{dx}\right) = 6x + \frac{2}{x^3}.$$

Similarly,
$$f'''(x) = \frac{d^3y}{dx^3} = 6 - \frac{6}{x^4}.$$

This is called the **third derivative** or the derivative of order three. Clearly we can continue the differentiation process indefinitely to produce derivatives of any positive integer order whatsoever.

EXAMPLE 3.16 Find a formula for the nth derivative of $y = x/(x+1)$.

SOLUTION When $n = 1$,
$$\frac{dy}{dx} = \frac{(x+1)(1) - x(1)}{(x+1)^2} = \frac{1}{(x+1)^2}.$$

When $n = 2$,
$$\frac{d^2y}{dx^2} = \frac{d}{dx}\left[\frac{1}{(x+1)^2}\right] = \frac{(x+1)^2(0) - \frac{d}{dx}(x+1)^2}{(x+1)^4}.$$

According to Exercise 31 in Section 3.2,
$$\frac{d}{dx}(x+1)^n = n(x+1)^{n-1}.$$

Thus,
$$\frac{d^2y}{dx^2} = \frac{-2(x+1)}{(x+1)^4} = \frac{-2}{(x+1)^3}.$$

When $n = 3$,
$$\frac{d^3y}{dx^3} = \frac{(x+1)^3(0) + 2(3)(x+1)^2}{(x+1)^6} = \frac{2 \cdot 3}{(x+1)^4}.$$

When $n = 4$,
$$\frac{d^4y}{dx^4} = \frac{(x+1)^4(0) - 2(3)(4)(x+1)^3}{(x+1)^8} = \frac{-2 \cdot 3 \cdot 4}{(x+1)^5}.$$

It is fairly clear that the pattern emerging is
$$\frac{d^ny}{dx^n} = \frac{(-1)^{n+1} n!}{(x+1)^{n+1}},$$

where the notation $n!$, called "n factorial," represents the product of the first n positive integers: $n! = 1(2)(3) \cdots (n-1)(n)$. ∎

EXAMPLE 3.17

Find a formula for $d^2(uv)/dx^2$ if $u = f(x)$ and $v = g(x)$.

SOLUTION Since
$$\frac{d}{dx}(uv) = v\frac{du}{dx} + u\frac{dv}{dx},$$
then
$$\frac{d^2}{dx^2}(uv) = \frac{d}{dx}\left(v\frac{du}{dx} + u\frac{dv}{dx}\right)$$
$$= v\frac{d^2u}{dx^2} + \frac{dv}{dx}\frac{du}{dx} + u\frac{d^2v}{dx^2} + \frac{du}{dx}\frac{dv}{dx}$$
$$= v\frac{d^2u}{dx^2} + 2\frac{du}{dx}\frac{dv}{dx} + u\frac{d^2v}{dx^2}.$$

EXAMPLE 3.18

How many derivatives does $f(x) = x^{8/3}$ have at $x = 0$?

SOLUTION Since
$$f'(x) = \frac{8}{3}x^{5/3}, \quad f''(x) = \left(\frac{8}{3}\right)\left(\frac{5}{3}\right)x^{2/3}, \quad f'''(x) = \left(\frac{8}{3}\right)\left(\frac{5}{3}\right)\left(\frac{2}{3}\right)x^{-1/3},$$
and $f'''(0)$ is not defined, $f(x)$ has only a first and second derivative at $x = 0$.

EXERCISES 3.5

In Exercises 1–10 find the indicated derivative.

1. $f''(x)$ if $f(x) = x^3 + 5x^4$
2. $f'''(x)$ if $f(x) = x^3 - 3x^2 + 2x + 1$
3. $f''(2)$ if $f(x) = (x+1)(x^3 + 3x + 2)$
4. $f'''(1)$ if $f(x) = x^4 - 3x^2 + 1/x$
5. $f''(x)$ if $f(x) = (x+1)/\sqrt{x}$
6. $f'''(t)$ if $f(t) = t^3 - 1/t^3$
7. d^9y/dx^9 if $y = x^{10}$
8. $f''(u)$ if $f(u) = \sqrt{u}/(u+1)$
9. d^6t/dx^6 if $t = x/(2x-6)$
10. $f''(x)$ if $f(x) = x/(\sqrt{x}+1)$
11. When a beam of length L and uniform mass m kilograms per metre is supported as shown in Figure 3.22, it bends under its own weight.

FIGURE 3.22

Beam

It can be shown that the deflection $y = f(x)$ from the horizontal must satisfy the equation
$$(EI)\frac{d^2y}{dx^2} = -\frac{mg}{2}(L-x)^2 + \frac{3}{8}mgL(L-x),$$
where E, I, and g are constants. Show that
$$y = f(x) = -\frac{mg}{48EI}(2x^4 - 5Lx^3 + 3L^2x^2)$$
satisfies this equation.

12. Steady-state temperature T in a region bounded by two concentric spheres of radii a and b (where $a < b$) must satisfy the equation

$$\frac{d^2 T}{dr^2} + \frac{2}{r}\frac{dT}{dr} = 0,$$

where r is the radial distance from the common centre of the spheres.

(a) Show that for any constants c and d, the function

$$T = f(r) = c + \frac{d}{r}$$

satisfies the equation.

(b) If temperatures on the two spheres are maintained at constant values T_a and T_b, calculate c and d in terms of a, b, T_a, and T_b.

13. Find a and b in order that the function $y = f(x) = ax^3 + bx^2$ has its second derivative equal to zero when $x = 1$ and $y = 2$.

14. Find constants a, b, c, and d in order that the function $y = f(x) = ax^3 + bx^2 + cx + d$ has its first derivative equal to 4 when $x = 1$ and $y = 0$, and its second derivative equal to 5 when $x = 2$ and $y = 4$.

15. (a) Find a formula for $d^3(uv)/dx^3$ if $u = f(x)$ and $v = g(x)$.

(b) On the basis of part (a) and Example 3.17, could you hazard a guess at a formula for $d^4(uv)/dx^4$; i.e., do you see the pattern emerging? Exercise 18 asks you to verify the correct formula.

16. If n is a positive integer, find a formula for

$$\frac{d^n}{dx^n}\left(\frac{2x}{4-x^2}\right).$$

Hint: Decompose $2x/(4 - x^2)$ into two simpler fractions and proceed as in Example 3.16.

17. Evaluate

$$\frac{d^{2n}}{dx^{2n}}(x^2 - 1)^n$$

for n a positive integer.

18. If n is a positive integer, and u and v are functions of x, show by mathematical induction that

$$\frac{d^n}{dx^n}(uv) = \sum_{r=0}^{n}\binom{n}{r}\left(\frac{d^r u}{dx^r}\right)\left(\frac{d^{n-r} v}{dx^{n-r}}\right),$$

where $d^0 u/dx^0 = u$, and $\binom{n}{r}$ are the binomial coefficients

$$\binom{n}{r} = \frac{n!}{r!(n-r)!}.$$

(If you are not familiar with sigma notation, see Section 6.1. Mathematical induction is discussed in Appendix C.)

SECTION 3.6

The Chain Rule and the Extended Power Rule

With the differentiation rules in Sections 3.2 and 3.4, we can differentiate any polynomial whatsoever. For instance, if $f(x) = x^3 - 3x^2 + 2x + 1$, then $f'(x) = 3x^2 - 6x + 2$. Consider the polynomial $f(x) = (2x^2 - 3)^8$, which has been conveniently factored for us. To find its derivative, we could expand $(2x^2 - 3)^8$, by the binomial theorem say, then differentiate, and finally simplify. We could also consider using the product rule over and over and over again. Thus, in spite of the fact that the rules of Sections 3.2 and 3.4 permit differentiation of $(2x^2 - 3)^8$, they are not convenient to use. Even more disturbing would be differentiation of, say, $g(x) = 1/(3x^2 + 8)^{12}$. In this section, we obtain results which enable us to differentiate much wider classes of functions, which include $(2x^2 - 3)^8$ and $1/(3x^2 + 8)^{12}$.

Suppose that y is defined as a function of u by

$$y = f(u) = \frac{u}{u+1},$$

and u, in turn, is defined as a function of x by

$$u = g(x) = \frac{\sqrt{x}}{x+2}.$$

These equations imply that y is a function of x; indeed, y is the composition of f and g,

$$y = (f \circ g)(x) = f(g(x)) = \frac{\frac{\sqrt{x}}{x+2}}{\frac{\sqrt{x}}{x+2} + 1}.$$

After some algebraic simplification we find that

$$y = \frac{\sqrt{x}}{\sqrt{x} + x + 2},$$

and we can therefore calculate

$$\frac{dy}{dx} = \frac{(\sqrt{x} + x + 2)\left(\frac{1}{2\sqrt{x}}\right) - \sqrt{x}\left(\frac{1}{2\sqrt{x}} + 1\right)}{(\sqrt{x} + x + 2)^2}.$$

This can be reduced to

$$\frac{dy}{dx} = \frac{2 - x}{2\sqrt{x}(\sqrt{x} + x + 2)^2}.$$

But notice that if we differentiate the original functions,

$$\frac{dy}{du} = \frac{(u+1)(1) - u(1)}{(u+1)^2} = \frac{1}{(u+1)^2} \quad \text{and}$$

$$\frac{du}{dx} = \frac{(x+2)\left(\frac{1}{2\sqrt{x}}\right) - \sqrt{x}(1)}{(x+2)^2} = \frac{2-x}{2\sqrt{x}(x+2)^2}.$$

The product of these derivatives is

$$\frac{dy}{du}\frac{du}{dx} = \frac{2-x}{2\sqrt{x}(x+2)^2(u+1)^2} = \frac{2-x}{2\sqrt{x}(x+2)^2\left(\frac{\sqrt{x}}{x+2}+1\right)^2} = \frac{2-x}{2\sqrt{x}(\sqrt{x}+x+2)^2}.$$

Consequently, for this example, we can write

$$\frac{dy}{dx} = \frac{dy}{du}\frac{du}{dx}.$$

The derivative of a composite function $y = f(g(x))$ can always be calculated by the above formula, according to the following theorem.

Theorem 3.9

(The chain rule) If $y = f(u)$ and $u = g(x)$ are differentiable functions, then the derivative of the composite function $y = f(g(x))$ is

$$\frac{dy}{dx} = \frac{dy}{du}\frac{du}{dx}. \qquad (3.16)$$

Increment notation is particularly useful in proving the chain rule.

Proof By equation 3.10, the derivative of the composite function is

$$\frac{dy}{dx} = \lim_{\Delta x \to 0} \frac{\Delta y}{\Delta x} = \lim_{\Delta x \to 0} \frac{f(g(x + \Delta x)) - f(g(x))}{\Delta x}.$$

Now a change Δx in x produces a change $g(x + \Delta x) - g(x)$ in u. If we denote this change by Δu, then it in turn produces the change $f(u + \Delta u) - f(u)$ in y. We may write

$$\frac{dy}{dx} = \lim_{\Delta x \to 0} \frac{f(u + \Delta u) - f(u)}{\Delta x}.$$

Since $f(u)$ is differentiable, its derivative exists and is defined by

$$f'(u) = \lim_{\Delta u \to 0} \frac{f(u + \Delta u) - f(u)}{\Delta u}.$$

An equivalent way to express the fact that this limit is $f'(u)$ is to say that

$$\frac{f(u + \Delta u) - f(u)}{\Delta u} = f'(u) + \epsilon,$$

where ϵ must satisfy the condition $\lim_{\Delta u \to 0} \epsilon = 0$. We may write

$$f(u + \Delta u) - f(u) = [f'(u) + \epsilon]\Delta u,$$

and if we substitute this into the second expression for dy/dx above, we obtain

$$\frac{dy}{dx} = \lim_{\Delta x \to 0} \frac{[f'(u) + \epsilon]\Delta u}{\Delta x} = \lim_{\Delta x \to 0} \left\{ [f'(u) + \epsilon]\frac{\Delta u}{\Delta x} \right\}.$$

But

$$\lim_{\Delta x \to 0} \frac{\Delta u}{\Delta x} = \lim_{\Delta x \to 0} \frac{g(x + \Delta x) - g(x)}{\Delta x} = \frac{du}{dx}.$$

Furthermore, since $g(x)$ is differentiable, it is continuous (Theorem 3.6), and this implies that $\Delta u \to 0$ as $\Delta x \to 0$. Consequently, $\lim_{\Delta x \to 0} \epsilon = \lim_{\Delta u \to 0} \epsilon = 0$, and these results give

$$\frac{dy}{dx} = f'(u)\frac{du}{dx} = \frac{dy}{du}\frac{du}{dx}.$$

This result is called the **chain rule** for the derivative of a composite function. It expresses the derivative of a composite function as the product of the derivatives of the functions in the composition. From the point of view of rates of change, the chain rule seems quite reasonable. It states that if a variable is defined in terms of a second variable, which is in turn defined in terms of a third variable, then the rate of change of the first variable with respect to the third is the rate of change of the first with respect to the second multiplied by the rate of change of the second with respect to the third. For example, if car A travels twice as fast as car B, and car B travels three times as fast as car C, then car A travels six times as fast as car C.

It is essential to understand the difference between the derivatives dy/dx and dy/du in 3.16. The second, dy/du, is the derivative of y regarded as a function of u, the given function $y = f(u)$. On the other hand, dy/dx is the derivative of y with respect to x, the derivative of the composite function $f(g(x))$.

EXAMPLE 3.19

Find dy/dt at $t = 4$ when $y = x^2 - x$ and $x = \sqrt{t}/(t+1)$.

SOLUTION By the chain rule

$$\frac{dy}{dt} = \frac{dy}{dx}\frac{dx}{dt} = (2x - 1)\left[\frac{(t+1)(1/2)t^{-1/2} - \sqrt{t}(1)}{(t+1)^2}\right].$$

When $t = 4$, we find $x = \sqrt{4}/(4+1) = 2/5$. Once again we use the notation $\frac{dy}{dt}|_{t=4}$ to represent dy/dt evaluated at $t = 4$,

$$\frac{dy}{dt}\bigg|_{t=4} = [2(2/5) - 1]\left[\frac{(4+1)(1/2)(4)^{-1/2} - \sqrt{4}}{(4+1)^2}\right] = \frac{3}{500}.$$

■

The Extended Power Rule

When the function $f(u)$ in Theorem 3.9 is a power function, the chain rule gives what we call the **extended power rule**, often considered the most important differentiation formula of calculus.

Corollary When $u = g(x)$ is *differentiable*,

$$\frac{d}{dx}u^n = nu^{n-1}\frac{du}{dx}. \tag{3.17}$$

It is essentially power rule 3.6 in Section 3.2 with an extra factor du/dx to account for the fact that what is under the power (u) is not just x, it is a function of x. In the special case that u is equal to x, 3.17 reduces to 3.6. It is important not to confuse these rules. Rule 3.6 can be used only for x^n; if anything other than x is raised to a power, formula 3.17 should be used. The most common error in using 3.17 is to forget the du/dx. With 3.17 it is a simple matter to differentiate the functions in the first paragraph of this section:

$$\frac{d}{dx}(2x^2 - 3)^8 = 8(2x^2 - 3)^7\frac{d}{dx}(2x^2 - 3) = 8(2x^2 - 3)^7(4x) = 32x(2x^2 - 3)^7 \quad \text{and}$$

$$\frac{d}{dx}\left[\frac{1}{(3x^2 + 8)^{12}}\right] = \frac{d}{dx}(3x^2 + 8)^{-12} = -12(3x^2 + 8)^{-13}\frac{d}{dx}(3x^2 + 8)$$
$$= \frac{-12}{(3x^2 + 8)^{13}}(6x) = \frac{-72x}{(3x^2 + 8)^{13}}.$$

Notice that in neither of these examples do you see the letter u, although rule 3.17 is stated in terms of u's. We could have introduced u, in the first example, say, by setting $u = 2x^2 - 3$, and proceeded as follows: With $u = 2x^2 - 3$,

$$\frac{d}{dx}(2x^2 - 3)^8 = \frac{d}{dx}u^8 = \frac{d}{du}(u^8)\frac{du}{dx} = 8u^7(4x) = 32xu^7 = 32x(2x^2 - 3)^7.$$

But this is unnecessary; with an understanding of 3.17, we should proceed directly to the derivatives without defining an intermediate variable. With a little practice, the writing should be shortened even more. For example, calculation of the derivative of $(2x^2-3)^8$ should appear as

$$\frac{d}{dx}(2x^2-3)^8 = 8(2x^2-3)^7(4x) = 32x(2x^2-3)^7.$$

Here is an example to illustrate power rules within power rules within power rules.

EXAMPLE 3.20

Find the derivative of the function $f(x) = \sqrt{1 + \left(2 + \sqrt{1-3x}\right)^2}$.

SOLUTION If we write the function in the form

$$f(x) = \left[1 + \left(2 + \sqrt{1-3x}\right)^2\right]^{1/2},$$

then

$$f'(x) = \frac{1}{2}\left[1 + \left(2 + \sqrt{1-3x}\right)^2\right]^{-1/2} \frac{d}{dx}\left[1 + \left(2 + \sqrt{1-3x}\right)^2\right]$$

$$= \frac{1}{2\sqrt{1 + \left(2 + \sqrt{1-3x}\right)^2}}\left[2\left(2 + \sqrt{1-3x}\right)\frac{d}{dx}\left(2 + \sqrt{1-3x}\right)\right]$$

$$= \frac{2 + \sqrt{1-3x}}{\sqrt{1 + \left(2 + \sqrt{1-3x}\right)^2}}\left[\frac{1}{2}(1-3x)^{-1/2}\frac{d}{dx}(1-3x)\right]$$

$$= \frac{2 + \sqrt{1-3x}}{2\sqrt{1-3x}\sqrt{1 + \left(2 + \sqrt{1-3x}\right)^2}}(-3)$$

$$= \frac{-3\left(2 + \sqrt{1-3x}\right)}{2\sqrt{1-3x}\sqrt{1 + \left(2 + \sqrt{1-3x}\right)^2}}.$$

EXERCISES 3.6

In Exercises 1–8 use the chain rule to find dy/dx.

1. $y = t^2 + \frac{1}{t}$, $\quad t = x^2 + 1$

2. $y = \frac{u}{u+1}$, $\quad u = \sqrt{x} + 1$

3. $y = (u^2 + 1)(u + 1)$, $\quad u = \frac{1}{\sqrt{x-4}}$

4. $y = \frac{s}{s^2 - 2}$, $\quad s = x^2 - 2x + 3$

5. $y = (v^2 + v)(\sqrt{v} + 1)$, $\quad v = \frac{x}{x^2 - 1}$

6. $y = \frac{t+3}{t-4}$, $\quad t = \frac{x-2}{x+1}$

7. $y = \frac{t^2 + 3}{t-4}$, $\quad t = (3x+2)(x^2+4x)$

8. $y = u^2\left(1 + \sqrt{u}\right)$, $\quad u = \dfrac{x+1}{x - x^2}$

In Exercises 9–36 find dy/dx, where $y = f(x)$.

9. $f(x) = x(x^3 + 3)^4$

10. $f(x) = x\sqrt{x+1}$

11. $f(x) = x^2(2x+1)^2$

12. $f(x) = \dfrac{x}{\sqrt{2x+1}}$

13. $f(x) = (x+2)^2(x^2+3)$

14. $f(x) = \dfrac{(2x-1)^2}{3x+5}$

15. $f(x) = \dfrac{3x+5}{(2x-1)^2}$

16. $f(x) = x^3(2-5x^2)^{1/3}$

17. $f(x) = \dfrac{x^3}{(2-5x^2)^{1/3}}$

18. $f(x) = (x+1)^2(3x+1)^3$

19. $f(x) = \dfrac{x^{1/3}}{1 - \sqrt{x}}$

20. $f(x) = \dfrac{\sqrt{2-3x}}{x^2}$

21. $f(x) = \left(\dfrac{x^3-1}{2x^3+1}\right)^4$

22. $f(x) = \sqrt[4]{\dfrac{2-x}{2+x}}$

23. $f(x) = (x^3 - 2x^2)^3(x^4 - 2x)^5$

24. $f(x) = (x+5)^4\sqrt{1+x^3}$

25. $f(x) = \dfrac{x\sqrt{1-x^2}}{(3+x)^{1/3}}$

26. $f(x) = x(x+5)^4\sqrt{1+x^3}$

27. $f(x) = \dfrac{x^2(x^3+3)^2}{(x-2)(x+5)^2}$

28. $f(x) = x\sqrt{1 + x\sqrt{1+x}}$

29. $y = t^2 + \dfrac{1}{t^3}$, $\quad t = \sqrt{4-x^2}$

30. $y = (2s - s^2)^{1/3}$, $\quad s = \dfrac{1}{x^2+5}$

31. $y = \dfrac{v^2}{v^3-1}$, $\quad v = x\sqrt{x^2-1}$

32. $y = \dfrac{u}{u+5}$, $\quad u = \dfrac{\sqrt{x-1}}{x}$

33. $y = u^4(u^3 - 2u)^2$, $\quad u = \sqrt{x - 2x^2}$

34. $y = t + \sqrt{t + \sqrt{t}}$, $\quad t = \dfrac{x^2+1}{x^2-1}$

35. $y = \left(\dfrac{v^2+1}{1-v^3}\right)^3$, $\quad v = \dfrac{1}{x^3+3x^2+2}$

36. $y = \dfrac{\sqrt{k}}{1+k+k^2}$, $\quad k = x(x^2+5)^5$

37. If $y = f(u)$, $u = g(s)$, and $s = h(x)$, show that

$$\dfrac{dy}{dx} = \dfrac{dy}{du}\dfrac{du}{ds}\dfrac{ds}{dx}.$$

38. When an electrostatic charge $q = 5 \times 10^{-6}$ coulombs is at a distance r metres from a stationary charge $Q = 3 \times 10^{-6}$ coulombs, the magnitude of the force of repulsion of Q on q is

$$F = \dfrac{Qq}{4\pi\epsilon_0 r^2} \text{ newtons,}$$

where $\epsilon_0 = 8.85 \times 10^{-12}$. If q is moved directly away from Q at 2 m/s, how fast is F changing when $r = 2$ m?

39. When a mass m of 5 kg is r metres from the centre of the earth, the magnitude of the force of attraction of the earth on m is

$$F = \dfrac{GmM}{r^2} \text{ newtons,}$$

where M is the mass of the earth and $G = 6.67 \times 10^{-11}$. If m is falling at 100 km/h when it is 5 km above the surface of the earth, how fast is F changing? The mean density of the earth is 5.52×10^3 kg/m³, and its mean radius is 6370 km.

40. Show that if an equation can be solved for both y in terms of x, $y = f(x)$, and x in terms of y, $x = g(y)$, then

$$\dfrac{dy}{dx} = \dfrac{1}{\dfrac{dx}{dy}}.$$

41. Prove that the derivative of an odd function is an even function, and that the derivative of an even function is an odd function. Recall Definition 1.9 in Section 1.6.

42. Determine whether the following very simple proof of the chain rule has a flaw? If Δu denotes the change in

SECTION 3.6 THE CHAIN RULE AND THE EXTENDED POWER RULE

$u = g(x)$ resulting from a change Δx in x, and Δy is the change in $y = f(u)$ resulting from Δu, then

$$\frac{dy}{dx} = \lim_{\Delta x \to 0} \frac{\Delta y}{\Delta x} = \lim_{\Delta x \to 0} \frac{\Delta y}{\Delta u} \frac{\Delta u}{\Delta x}$$
$$= \left(\lim_{\Delta x \to 0} \frac{\Delta y}{\Delta u} \right) \left(\lim_{\Delta x \to 0} \frac{\Delta u}{\Delta x} \right)$$
$$= \left(\lim_{\Delta u \to 0} \frac{\Delta y}{\Delta u} \right) \left(\lim_{\Delta x \to 0} \frac{\Delta u}{\Delta x} \right)$$
$$= \frac{dy}{du} \frac{du}{dx}.$$

In Exercises 43–46 find d^2y/dx^2.

43. $y = v^2 + v$, $v = \dfrac{x}{x+1}$

44. $y = (u+1)^3 - \dfrac{1}{u}$, $u = x + \sqrt{x+1}$

45. $y = \sqrt{t-1}$, $t = (x + x^2)^2$

46. $y = \dfrac{s}{s+6}$, $s = \dfrac{\sqrt{x}}{1+\sqrt{x}}$

In Exercises 47–54 assume that $f(u)$ is a differentiable function of u. Find the derivative of the given function with respect to x in as simplified a form as possible. Then set $f(u) = u^3 - 2u$ and simplify further.

47. $f(2x + 3)$ **48.** $[f(3 - 4x)]^2$

49. $f(1 - x^2)$ **50.** $f(x + 1/x)$

51. $f(f(x))$ **52.** $\sqrt{3 - 4[f(1-3x)]^2}$

53. $\dfrac{f(-x)}{3 + 2f(x^2)}$ **54.** $f(x - f(x))$

In Exercises 55–58 show that the families of curves are orthogonal trajectories (see question 29 in Exercises 3.4). Sketch both families of curves.

55. $y = mx$, $x^2 + y^2 = r^2$

56. $y = ax^2$, $x^2 + 2y^2 = c^2$

57. $x^2 - y^2 = C_1$, $xy = C_2$

58. $2x^2 + 3y^2 = C^2$, $y^2 = ax^3$

59. If $y = f(u)$ and $u = g(x)$, use the chain rule to show that
$$\frac{d^2y}{dx^2} = \frac{d^2u}{dx^2} \frac{dy}{du} + \frac{d^2y}{du^2} \left(\frac{du}{dx} \right)^2.$$

60. Use the result of Exercise 59 to find d^2y/dx^2 at $x = 1$ when
$$y = (u+1)^3, \quad u = 3x - \frac{2}{x^2}.$$

61. Find the rate of change of $y = f(x) = x^9 + x^6$ with respect to x^3.

62. Find the rate of change of $y = f(x) = \sqrt{1-x^2}$ with respect to $x/(x+1)$.

63. If $y = f(u)$ and $u = g(x)$, show that
$$\frac{d^3y}{dx^3} = \frac{d^3u}{dx^3} \frac{dy}{du} + 3 \frac{d^2y}{du^2} \frac{d^2u}{dx^2} \frac{du}{dx} + \frac{d^3y}{du^3} \left(\frac{du}{dx} \right)^3.$$

64. At what point(s) on the hyperbola $x^2 - 16y^2 = 16$ do tangent lines pass through the point $(2, 3)$?

65. Generalize the result of equation 3.12 to find a formula for
$$\frac{d}{dx}|f(x)|^n, \quad \text{for } n > 1 \text{ an integer}.$$

66. The curve $x^2y^2 = (x + 1)^2(4 - x^2)$ is called a *conchoid of Nicomedes*. Sketch the curve and find points where its tangent line is horizontal.

SECTION 3.7

Implicit Differentiation

We say that y is defined **explicitly** as a function of x if the dependence of y on x is given in the form

$$y = f(x). \tag{3.18}$$

Examples are $y = x^2$, $y = 3x + \sin x$, and $y = 1/(x+1)$. In each case, the dependent variable stands alone on the left side of the equation. The differentiation rules in Theorems 3.1 to 3.5, 3.7, and 3.8 are applicable to explicitly defined functions.

An equation in x and y may define y as a function of x even when it is not in form (3.18). Equations for which y has not been separated out are often written in the generic form

$$F(x, y) = 0. \tag{3.19}$$

The notation $F(x, y)$ is used to denote an expression that depends on two variables x and y. For example, the volume of a right circular cylinder depends on its radius r and its height h; in particular, $V = \pi r^2 h$. In such a case we write that $V = F(r, h) = \pi r^2 h$. Examples of equations of form 3.19 are $y - x^3 = 0$ and $y^3 - 3y - 2x = 0$. Equation 3.19 is said to define y **implicitly** as a function of x in some domain D, if for each x in D, there is one, and only one, value of y for which (x, y) satisfies 3.19. The equation $y - x^3 = 0$ defines y implicitly as a function of x for all (real) x. The explicit definition of the function is $y = x^3$. The equation $y^3 - 3y - 2x = 0$ does not define y as a function of x. For each x in the interval $-1 < x < 1$, there are three solutions of the equation for y. For $x < -1$ and $x > 1$, the equation has only one value of y corresponding to each x. This is most easily seen from the graph of the curve in Figure 3.23. We shall have more to say about the equation $y^3 - 3y - 2x = 0$ later in this section.

FIGURE 3.23

Each of the following equations defines y as a function of x, and does so implicitly:

$$y + x^2 - 2x + 5 = 0;$$
$$2y + x^2 + y^2 - 25 = 0, \quad y \geq -1;$$
$$x + y^5 + x^2 + y = 0.$$

It is easy to solve the first equation for the explicit definition of the function,

$$y = -x^2 + 2x - 5,$$

and find its derivative,

$$\frac{dy}{dx} = -2x + 2.$$

We can also obtain an explicit definition of the function defined by the second equation, but not so simply. If we write

$$y^2 + 2y + (x^2 - 25) = 0,$$

and use quadratic formula 1.17, we obtain

$$y = \frac{-2 \pm \sqrt{4 - 4(x^2 - 25)}}{2} = -1 \pm \sqrt{26 - x^2}.$$

Since y is required to be greater than or equal to -1, we must choose

$$y = -1 + \sqrt{26 - x^2}.$$

This is the explicit definition of the function, and it is now easy to find the derivative of y with respect to x:

$$\frac{dy}{dx} = \frac{1}{2\sqrt{26-x^2}}(-2x) = \frac{-x}{\sqrt{26-x^2}}.$$

The third equation is quite different, for it is impossible to solve this equation for an explicit definition of the function. Does this mean that it is also impossible to obtain dy/dx? The answer is no! To see this, we differentiate both sides of the equation with respect to x, keeping in mind that y is a function of x. Only the term in y^5 presents any difficulty, but its derivative can be calculated using extended power rule 3.17,

$$1 + 5y^4 \frac{dy}{dx} + 2x + \frac{dy}{dx} = 0.$$

We can now solve this equation for dy/dx by grouping the two terms in dy/dx on one side of the equation, and transposing the remaining two terms:

$$(5y^4 + 1)\frac{dy}{dx} = -1 - 2x.$$

Division by $5y^4 + 1$ gives the required derivative,

$$\frac{dy}{dx} = -\frac{2x+1}{5y^4+1}.$$

This process of differentiating an equation that implicitly defines a function is called **implicit differentiation**. It could also have been used in each of the first two examples. If $y + x^2 - 2x + 5 = 0$ is differentiated with respect to x, we have

$$\frac{dy}{dx} + 2x - 2 = 0,$$

from which, as before

$$\frac{dy}{dx} = -2x + 2.$$

When $2y + x^2 + y^2 - 25 = 0$ is differentiated with respect to x, we find

$$2\frac{dy}{dx} + 2x + 2y\frac{dy}{dx} = 0,$$

from which,

$$(2 + 2y)\frac{dy}{dx} = -2x,$$

and therefore

$$\frac{dy}{dx} = \frac{-x}{y+1}.$$

Although this result appears different from the previous expression for dy/dx, when we recall that $y = -1 + \sqrt{26-x^2}$, we find

$$\frac{dy}{dx} = \frac{-x}{-1+\sqrt{26-x^2}+1} = \frac{-x}{\sqrt{26-x^2}}.$$

These examples illustrate that if implicit differentiation is used to obtain a derivative, then the result may depend on y as well as x. Naturally if we require the derivative at a certain value of x, then the y-value used is determined by the original equation defining y implicitly as a function of x.

EXAMPLE 3.21

Assuming that y is defined implicitly as a function of x by the equation

$$x^3 y^3 + x^2 y + 2x = 12,$$

find dy/dx when $x = 1$.

SOLUTION When we differentiate both sides of the equation with respect to x, using the product rule on the first two terms, we find

$$3x^2 y^3 + 3x^3 y^2 \frac{dy}{dx} + 2xy + x^2 \frac{dy}{dx} + 2 = 0.$$

Thus,

$$(3x^3 y^2 + x^2)\frac{dy}{dx} = -(2 + 2xy + 3x^2 y^3) \quad \text{and}$$

$$\frac{dy}{dx} = -\frac{2 + 2xy + 3x^2 y^3}{3x^3 y^2 + x^2}.$$

When $x = 1$ is substituted into the given equation defining y as a function of x, the result is

$$0 = y^3 + y - 10 = (y - 2)(y^2 + 2y + 5),$$

and the only solution of this equation is $y = 2$. We now substitute $x = 1$ and $y = 2$ into the formula for dy/dx to calculate the derivative at $x = 1$,

$$\frac{dy}{dx}\bigg|_{x=1} = -\frac{2 + 2(1)(2) + 3(1)^2(2)^3}{3(1)^3(2)^2 + (1)^2} = -\frac{30}{13}.$$

∎

Implicit differentiation can also be used to find second and higher order derivatives of functions that are defined implicitly. Calculations can be messy, but principles are the same. For example, when the equation $x^5 + y^3 + y^2 = 1$ defines y implicitly as a function of x, it is straightforward to calculate that

$$\frac{dy}{dx} = \frac{-5x^4}{3y^2 + 2y}.$$

To find the second derivative $d^2 y/dx^2$, we differentiate both sides of this equation with respect to x. We use the quotient rule on the right, and when differentiating the denominator we once again keep in mind that y is a function of x. The result is

$$\frac{d^2 y}{dx^2} = \frac{(3y^2 + 2y)(-20x^3) + 5x^4 \left(6y \frac{dy}{dx} + 2\frac{dy}{dx}\right)}{(3y^2 + 2y)^2}.$$

We now replace dy/dx by its expression in terms of x and y,

$$\frac{d^2 y}{dx^2} = \frac{-20x^3(3y^2 + 2y) + 5x^4(6y + 2)\left(\frac{-5x^4}{3y^2 + 2y}\right)}{(3y^2 + 2y)^2},$$

and bring the two terms in the numerator to a common denominator,

$$\frac{d^2 y}{dx^2} = \frac{\dfrac{-20x^3(3y^2 + 2y)^2 - 25x^8(6y + 2)}{(3y^2 + 2y)}}{(3y^2 + 2y)^2}$$

$$= -\frac{20x^3(3y^2+2y)^2 + 25x^8(6y+2)}{(3y^2+2y)^3}.$$

It would not be a pleasant task to proceed to the third derivative of y with respect to x.

In the above examples we seem to have adopted the principle that both sides of an equation can be differentiated with respect to the same variable. But can all equations be differentiated? The answer is an emphatic no! For example, if we differentiate both sides of the equation $4x = 2x$, we obtain the ludicrous result that $4 = 2$. Obviously, then, this equation cannot be differentiated with respect to x.

Possibly the reason differentiation fails in the above example is that the equation contains only one variable, namely, x. Perhaps a more reasonable question might be: "Can every equation containing two variables be differentiated?" To answer this question we consider the situation of a ship heading northeast from a pier. Let us choose east as the positive x-direction and north as the positive y-direction, both originating from the pier (Figure 3.24). We will find the rate of change of the y-coordinate of the ship with respect to its x-coordinate when the ship is 2 km from the pier. Obviously the answer is 1, since the x- and y-coordinates change at the same rate. But consider the following argument.

FIGURE 3.24

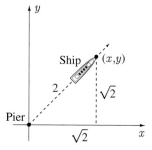

When the ship is 2 km from the pier, $x = y = \sqrt{2}$, and therefore

$$x^2 + y^2 = 4.$$

If we differentiate this equation with respect to x, we find that

$$2x + 2y\frac{dy}{dx} = 0, \quad \text{or} \quad \frac{dy}{dx} = -\frac{x}{y},$$

and with $x = y = \sqrt{2}$,

$$\frac{dy}{dx} = -1.$$

Differentiation of the equation $x^2 + y^2 = 4$, which contains two variables, has again led to an erroneous result.

These two examples have certainly illustrated that not all equations can be differentiated. What then distinguishes an equation that can be differentiated from one that cannot? The answer is that an equation can be differentiated with respect to a variable only if the equation is valid for a continuous range of values of that variable. The equation $4x = 2x$ cannot be differentiated because it is true only for $x = 0$. The equation $x^2 + y^2 = 4$ in the above example cannot be differentiated because *in that example* it is valid only when the ship is 2 km from the pier; i.e., only when $x = y = \sqrt{2}$. In other situations it might be acceptable to differentiate $x^2 + y^2 = 4$. For example, if at position $(\sqrt{2}, \sqrt{2})$, the ship is following the circular path defined by the equation $x^2 + y^2 = 4$, then it is indeed correct to differentiate $x^2 + y^2 = 4$ to find $dy/dx = -1$. This is an extremely important principle, and we will return to it many times. To emphasize it once again, *we may differentiate an equation with respect to a variable only if the equation is valid for a continuous range of values of that variable.*

Each of the three equations at the beginning of this section was given as defining y as a function of x, and as such defines the function for some domain of values for x. Differentiation of these equations was therefore acceptable according to the above principle.

When we are told that equation 3.19 defines y as a function of x, implicit differentiation leads to the derivative dy/dx. But how can we tell whether an equation $F(x,y) = 0$ defines y implicitly as a function of x? Additionally, given that $F(x,y) = 0$ does define y as a function of x, how do we know that the expression for dy/dx obtained by implicit differentiation is a valid representation of the derivative of the function? After all, functions do not always have derivatives at all points in their domains.

Answers to these questions are intimately related. To see how, suppose an equation $F(x,y) = 0$ in x and y is satisfied by a point (x_0, y_0), and when the equation is differentiated implicitly it leads to a quotient for dy/dx,

$$\frac{dy}{dx} = \frac{P(x,y)}{Q(x,y)}.$$

Mathematicians have shown that if $Q(x_0, y_0) \neq 0$, then the equation $F(x,y) = 0$ defines y implicitly as a function of x for some open interval containing x_0, and the derivative of this function at x_0 is $P(x_0, y_0)/Q(x_0, y_0)$. Proofs are usually given in advanced books on mathematical analysis. When $Q(x_0, y_0) = 0$, two possibilities exist. First, the equation $F(x,y) = 0$ might not define y as a function of x in an interval around x_0. Second, the equation might define a function, but the function does not have a derivative at x_0. To illustrate consider first the equation $y^3 - 3y - 2x = 0$, which we introduced earlier in this section. The curve defined by this equation is shown in Figure 3.23, and it clearly illustrates that the equation does not define y as a function of x. It does define the three functions of x in Figure 3.25. Suppose we differentiate the equation with respect to x,

$$3y^2 \frac{dy}{dx} - 3\frac{dy}{dx} - 2 = 0,$$

and solve for dy/dx,

$$\frac{dy}{dx} = \frac{2}{3(y^2 - 1)}.$$

This derivative is obviously undefined at the points $(-1, 1)$ and $(1, -1)$, and these are precisely the points which separate the original curve into three parts. It is impossible to find a portion of the curve around either of these points which defines y as a function of x. At any other point on the curve, the formula $dy/dx = (2/3)(y^2 - 1)^{-1}$ is a valid representation for the derivative for whichever function of Figure 3.25 contains the point.

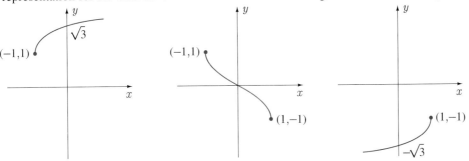

FIGURE 3.25

Consider now the equation $x = y^3$, shown graphically in Figure 3.26. Implicit differentiation leads to

$$\frac{dy}{dx} = \frac{1}{3y^2}.$$

Clearly, y is a function of x for all x, but dy/dx is undefined at $(0,0)$ since the tangent line is vertical at this point.

FIGURE 3.26

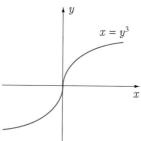

EXAMPLE 3.22

Sketch the curve defined by the equation $y^2 = x^2 - x^4$. Use implicit differentiation to find dy/dx and discuss what happens when the point $(0,0)$ is substituted into the result.

SOLUTION The curve is symmetric about both the x- and y-axes. To draw that portion in the first quadrant, we write $y = x\sqrt{1-x^2}$ and multiply ordinates of $y = x$ and $y = \sqrt{1-x^2}$ (Figure 3.27(a)). The entire curve is shown in Figure 3.27(b).

Implicit differentiation of $y^2 = x^2 - x^4$ gives

$$2y\frac{dy}{dx} = 2x - 4x^3,$$

from which

$$\frac{dy}{dx} = \frac{x(1-2x^2)}{y}.$$

This result is undefined at $(0,0)$, and Figure 3.27(b) indicates why. The equation does not define y as a function of x around $x = 0$.

FIGURE 3.27

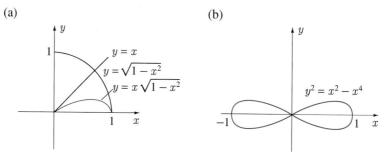

(a)   (b)

EXERCISES 3.7

In Exercises 1–10 find dy/dx wherever y is defined as a function of x.

1. $y^4 + y = 4x^3$
2. $x^4 + y^2 + y^3 = 1$
3. $xy + 2x = 4y^2 + 2$
4. $2x^3 - 3xy^4 + 5xy - 10 = 0$
5. $x + xy^5 + x^2y^3 = 3$
6. $(x+y)^2 = 2x$
7. $x(x-y) - 4y^3 = 2x + 5$
8. $\sqrt{x+y} + y^2 = 12x^2 + y$
9. $\sqrt{1+xy} - xy = 15$
10. $\dfrac{x}{x+y} - \dfrac{y}{x} = 4$

11. Find equations for the tangent and normal lines to the

curve $xy^2 + y^3 = 2$ at the point $(1, 1)$.

In Exercises 12–15 find d^2y/dx^2 wherever y is defined as a function of x.

12. $x^2 + y^3 + y = 1$
13. $2x^2 - y^3 = 4 - xy$
14. $y^2 + 2y = 5x$
15. $(x + y)^2 = x$

16. Find dy/dx at $x = 1$ when $x^3y + xy^3 = 2$.

In Exercises 17–20 find dy/dx and d^2y/dx^2.

17. $(x + y)^2 = x^2 + y^2$
18. $x^2y^3 + 2x + 4y = 5$
19. $xy^2 - 3x^2y = x + 1$
20. $x = y\sqrt{1 - y^2}$

21. Find dy/dx when $x = 0$ if $\sqrt{1 - xy} + 3y = 4$.

22. Find dy/dx when $x = 1$ if $x^2y^3 + xy = 2$.

23. Find dy/dx and d^2y/dx^2 when $x = 2$ if $y^5 + (x-2)y = 1$.

24. Find dy/dx and d^2y/dx^2 when $y = 1$ if $x^2 + 2xy + 3y^2 = 2$.

25. Find point(s) on the curve $xy^2 + x^2y = 16$ at which the slope of the tangent line is equal to zero.

26. Find point(s) on the curve $x^2 + y^{2/3} = 2$ at which the second derivative is equal to zero.

27. If a thermal nuclear reactor is built in the shape of a right circular cylinder of radius r and height h, then according to neutron diffusion theory, r and h must satisfy the equation

$$\left(\frac{2.4048}{r}\right)^2 + \left(\frac{\pi}{h}\right)^2 = k = \text{constant}.$$

Find dr/dh.

28. The elasticity of a function $y = f(x)$ is defined as

$$\frac{Ey}{Ex} = \frac{x}{y}\frac{dy}{dx}.$$

Calculate elasticity for the function defined by each of the following equations:

(a) $f(x) = x\left(\dfrac{x+1}{x+2}\right)$
(b) $x = \dfrac{400y + 200}{3 - y}$

29. Show that the elasticity of a function is equal to one if and only if the tangent line to its graph passes through the origin.

30. Find that point $P(a, b)$ on the first-quadrant part of the ellipse $2x^2 + 3y^2 = 14$ at which the tangent line at P is perpendicular to the line joining P and $(2, 5)$.

31. Show that the equation of the tangent line to the hyperbola $b^2x^2 - a^2y^2 = a^2b^2$ at the point (x_0, y_0) is $b^2xx_0 - a^2yy_0 = a^2b^2$.

32. Prove that for any circle $(x - h)^2 + (y - k)^2 = r^2$,

$$\left|\frac{d^2y/dx^2}{[1 + (dy/dx)^2]^{3/2}}\right| = \frac{1}{r}.$$

33. A solution passes through a conical filter 24 cm deep and 16 cm across the top into a cylindrical vessel of diameter 12 cm. Find an equation relating the depth h of solution in the filter and depth H of solution in the cylinder. What is the rate of change of h with respect to H?

34. If x objects are sold at a price of $r(x)$ per object, the total revenue is $R(x) = xr(x)$. Find the marginal revenue $R'(x)$ if price is defined implicitly by the equation

$$x = 4a^3 - 3ar^2 + r^3,$$

where $a > 0$ is a constant, and $0 < x < 4a^3$.

35. The general polynomial of degree n is

$$a_0 + a_1x + a_2x^2 + \cdots + a_nx^n,$$

where $a_0, a_1, \ldots, a_n$ are constants. Show that two polynomials of degree n can be equal for all x,

$$a_0 + a_1x + a_2x^2 + \cdots + a_nx^n = b_0 + b_1x + b_2x^2 + \cdots + b_nx^n,$$

if and only if $a_0 = b_0, a_1 = b_1, \ldots, a_n = b_n$.

36. (a) Find $f'(0)$ if $y = f(x)$ is defined implicitly as a function of x by

$$x\sqrt{1 + 2y} = x^2 - y.$$

(b) Show that by squaring this equation

$$x^2 + 4x^2y = x^4 + y^2.$$

Differentiate this equation with respect to x to find $f'(0)$. Do you have any difficulties? Explain.

37.–40. Use implicit differentiation to redo questions 55–58 in Exercises 3.6.

41. (a) Use implicit differentiation to find dy/dx if $y^2 = x^2 - 4x^4$.

(b) Can you calculate dy/dx at $x = 0$ using the result of part (a)? Sketch the curve $y^2 = x^2 - 4x^4$ in order to explain this difficulty.

42. (a) Find dy/dx if $\sqrt{x} + \sqrt{y} = 1$, where $0 \leq x \leq 1$, defines y implicitly as a function of x.

    (b) Sketch the curve $\sqrt{|x|} + \sqrt{|y|} = 1$.

43. What is dy/dx if $x^2 + 4x + y^2 + 6y + 15 = 0$?

44. Find dy/dx if $y = u/\sqrt{u^2 - 1}$ and u is defined implicitly as a function of x by $x^2 u^2 + \sqrt{u^2 - 1} = 4$.

45. Given that the equations

    $$y^4 + yv^3 = 3,$$
    $$x^2 v + 3xv^2 = 2x^3 y + 1$$

    define y as a function of v and v as a function of x, find dy/dx in terms of $x, y,$ and v.

46. Show that any function defined implicitly by the equation

    $$\frac{x^2}{y^3} - x = C,$$

    where C is a constant, satisfies the equation

    $$3x^2 \frac{dy}{dx} + y^4 = 2xy.$$

47. Show that any function defined implicitly by the equation

    $$2x^2 - 3y = Cx^2 y^3,$$

    where C is a constant, satisfies the equation

    $$(xy - x^3)\frac{dy}{dx} + y^2 = 0.$$

48. Verify power rule 3.6 in the case that n is a rational number.

49. Find points on the curve $xy^2 + x^2 y = 2$ at which the slope of the tangent line is equal to one.

50. Show that the families of curves $y^2 = x^3/(a - x)$ and $(x^2 + y^2)^2 = b(2x^2 + y^2)$ are orthogonal trajectories.

51. The curve described by the equation $x^3 + y^3 = 3axy$, where $a > 0$ is a constant, is called the *folium of Descartes* (Figure 3.28). Find points where the slope of the tangent line is equal to -1.

FIGURE 3.28

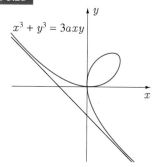

$x^3 + y^3 = 3axy$

52. (a) When a point (x, y) moves in the xy-plane so that the product of its distances from the points $(a, 0)$ and $(-a, 0)$ is always equal to a constant which we denote by c^2 ($c > a$), the curve that it follows is called the *ovals of Cassini*. Verify that the equation for these ovals is $(x^2 + y^2 + a^2)^2 = c^4 + 4a^2 x^2$.

    (b) Show that when $c < \sqrt{2}a$, there are six points on the ovals at which the tangent line is horizontal, but when $c \geq \sqrt{2}a$, there are only two such points.

SECTION 3.8

Derivatives of the Trigonometric Functions

The trigonometric functions, their properties, and the identities that they satisfy are discussed in Appendix A. We assume in this section and the remainder of the book that the reader is familiar with this material. For those readers who feel comfortable with their grasp of trigonometry and do not intend reading the appendix, we emphasize one extremely important convention in calculus. Arguments of trigonometric functions are always measured in radians, never degrees. With this in mind we prove the following theorem.

Theorem 3.10

$$\lim_{\theta \to 0} \frac{\sin \theta}{\theta} = 1. \qquad (3.20)$$

Proof If θ is a positive acute angle as shown in Figure 3.29, then

area of triangle BOP < area of sector BOA < area of triangle OBT.

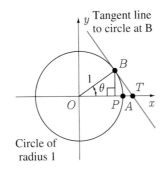

FIGURE 3.29

With θ expressed in radians, the area of a sector of a circle is $(r^2 \theta)/2$. Consequently,

$$\frac{1}{2} \|BP\| \|OP\| < \frac{1}{2} (1)^2 \theta < \frac{1}{2} \|OB\| \|BT\|.$$

When we multiply each term in these inequalitites by 2 and express lengths of the line segments in terms of θ,

$$(\sin \theta)(\cos \theta) < \theta < (1)(\tan \theta).$$

Division by $\sin \theta$ gives

$$\cos \theta < \frac{\theta}{\sin \theta} < \frac{1}{\cos \theta},$$

and when each term is inverted, the inequality signs are reversed, so

$$\frac{1}{\cos \theta} > \frac{\sin \theta}{\theta} > \cos \theta.$$

We now take limits as $\theta \to 0^+$. Since $\cos \theta \to 1$ and $1/\cos \theta \to 1$, it follows that $(\sin \theta)/\theta$ which is between $1/\cos \theta$ and $\cos \theta$ must also approach 1,

$$\lim_{\theta \to 0^+} \frac{\sin \theta}{\theta} = 1.$$

When $\theta < 0$, we set $\phi = -\theta$, in which case

$$\lim_{\theta \to 0^-} \frac{\sin \theta}{\theta} = \lim_{\phi \to 0^+} \frac{\sin(-\phi)}{-\phi} = \lim_{\phi \to 0^+} \frac{\sin \phi}{\phi} = 1.$$

Since left-hand and right-hand limits of $(\sin \theta)/\theta$ are both equal to 1, it follows that

$$\lim_{\theta \to 0} \frac{\sin \theta}{\theta} = 1.$$

With this result we can now find the derivative of the sine function. Derivatives of the other five trigonometric functions follow easily.

Theorem 3.11

$$\frac{d}{dx}\sin x = \cos x \qquad (3.21)$$

Proof If we set $f(x) = \sin x$ in equation 3.2 and use the trigonometric identity

$$\sin A - \sin B = 2\cos\left(\frac{A+B}{2}\right)\sin\left(\frac{A-B}{2}\right),$$

we obtain

$$\begin{aligned}
f'(x) &= \lim_{h \to 0} \frac{f(x+h) - f(x)}{h} \\
&= \lim_{h \to 0} \frac{\sin(x+h) - \sin x}{h} \\
&= \lim_{h \to 0} \frac{1}{h}\left[2\cos\left(\frac{2x+h}{2}\right)\sin\left(\frac{h}{2}\right)\right] \\
&= \lim_{h \to 0} \left[\cos\left(x + \frac{h}{2}\right)\frac{\sin(h/2)}{h/2}\right] \\
&= \lim_{h \to 0} \cos\left(x + \frac{h}{2}\right) \lim_{h/2 \to 0} \frac{\sin(h/2)}{h/2} \\
&= \cos x.
\end{aligned}$$

It is a simple application of the chain rule to prove the following corollary.

Corollary If $u = f(x)$ is a differentiable function,

$$\frac{d}{dx}\sin u = \cos u \frac{du}{dx}. \qquad (3.22)$$

EXAMPLE 3.23

Find derivatives for the following functions:

(a) $f(x) = \sin 3x$ (b) $f(x) = \sin^3 4x$

SOLUTION (a) According to formula 3.22,

$$f'(x) = \cos 3x \frac{d}{dx}(3x) = 3\cos 3x.$$

(b) For this function we must use extended power rule 3.17 before 3.22,

$$f'(x) = 3\sin^2 4x \frac{d}{dx}\sin 4x = 3\sin^2 4x \cos 4x \frac{d}{dx}(4x) = 12\sin^2 4x \cos 4x.$$

Since the cosine function can be expressed in terms of the sine function, it is straightforward to find its derivative.

Theorem 3.12

$$\frac{d}{dx}\cos x = -\sin x \qquad (3.23)$$

Proof Since $\cos x$ can always be expressed in the form $\cos x = \sin(\pi/2 - x)$, it follows that

$$\frac{d}{dx}\cos x = \frac{d}{dx}\sin\left(\frac{\pi}{2} - x\right) = -\cos\left(\frac{\pi}{2} - x\right) = -\sin x.$$

Corollary If $u = f(x)$ is a differentiable function,

$$\frac{d}{dx}\cos u = -\sin u \frac{du}{dx}. \qquad (3.24)$$

Derivatives of the other trigonometric functions are obtained by expressing them in terms of the sine and cosine functions. For the tangent function, we use the chain rule and the quotient rule to calculate

$$\frac{d}{dx}\tan u = \frac{d}{du}\left(\frac{\sin u}{\cos u}\right)\frac{du}{dx} = \frac{\cos u(\cos u) - \sin u(-\sin u)}{\cos^2 u}\frac{du}{dx} = \frac{1}{\cos^2 u}\frac{du}{dx}.$$

Consequently,

$$\frac{d}{dx}\tan u = \sec^2 u \frac{du}{dx}. \qquad (3.25)$$

A similar calculation gives

$$\frac{d}{dx}\cot u = -\csc^2 u \frac{du}{dx}. \qquad (3.26)$$

For the secant function, we obtain

$$\frac{d}{dx}\sec u = \frac{d}{du}\left(\frac{1}{\cos u}\right)\frac{du}{dx} = (-1)(\cos u)^{-2}(-\sin u)\frac{du}{dx} = \frac{1}{\cos u}\frac{\sin u}{\cos u}\frac{du}{dx} \qquad \text{or}$$

$$\frac{d}{dx}\sec u = \sec u \tan u \frac{du}{dx}. \qquad (3.27)$$

Similarly,

$$\frac{d}{dx}\csc u = -\csc u \cot u \frac{du}{dx}. \qquad (3.28)$$

Notice the relationship between derivatives of the *cofunctions* — cosine is the cofunction of sine, cotangent of tangent, and cosecant of secant — and corresponding derivatives for the functions. Each function is replaced by its cofunction, and a negative sign is added.

EXAMPLE 3.24

Find dy/dx if y is defined as a function of x in each of the following:

(a) $y = \sin 2x$ (b) $y = 4 \sec(2x^3 + 5)$ (c) $y = \tan^2 4x$
(d) $y = 2 \csc 3x^2 + 5x \sin x$ (e) $x^2 \tan y + y \sin x = 5$

SOLUTION

(a) $\dfrac{dy}{dx} = (\cos 2x)\dfrac{d}{dx}(2x) = 2 \cos 2x$

(b) $\dfrac{dy}{dx} = 4 \sec(2x^3 + 5) \tan(2x^3 + 5) \dfrac{d}{dx}(2x^3 + 5)$
$= 24x^2 \sec(2x^3 + 5) \tan(2x^3 + 5)$

(c) $\dfrac{dy}{dx} = 2 \tan 4x \dfrac{d}{dx} \tan 4x = 2 \tan 4x (4 \sec^2 4x) = 8 \tan 4x \sec^2 4x$

(d) $\dfrac{dy}{dx} = -2 \csc 3x^2 \cot 3x^2 \dfrac{d}{dx}(3x^2) + 5 \sin x + 5x \cos x$
$= -12x \csc 3x^2 \cot 3x^2 + 5 \sin x + 5x \cos x$

(e) If we differentiate the equation (implicitly) with respect to x, we obtain

$$2x \tan y + x^2 \sec^2 y \dfrac{dy}{dx} + \dfrac{dy}{dx} \sin x + y \cos x = 0.$$

When we solve this equation for dy/dx, the result is

$$\dfrac{dy}{dx} = -\dfrac{2x \tan y + y \cos x}{x^2 \sec^2 y + \sin x}.$$

∎

EXAMPLE 3.25

Find values of x for which the derivative of the function $f(x) = \cos(x + 1/x)$ is equal to zero.

SOLUTION For

$$0 = f'(x) = -\left(1 - \dfrac{1}{x^2}\right) \sin\left(x + \dfrac{1}{x}\right),$$

there are two possibilities:

$$1 - \dfrac{1}{x^2} = 0 \quad \text{or} \quad \sin\left(x + \dfrac{1}{x}\right) = 0.$$

The first gives $x = \pm 1$. The second implies that

$$x + \dfrac{1}{x} = n\pi,$$

where n is an integer. Multiplication by x leads to the quadratic equation

$$x^2 - n\pi x + 1 = 0,$$

with solutions

$$x = \frac{n\pi \pm \sqrt{n^2\pi^2 - 4}}{2}.$$

Clearly, we must choose $n \neq 0$ (else $n^2\pi^2 - 4 < 0$).

EXERCISES 3.8

In Exercises 1–30 find dy/dx.

1. $y = 2\sin 3x$
2. $y = \cos x - 4\sin 5x$
3. $y = \sin^2 x$
4. $y = \tan^{-3} 3x$
5. $y = \sec^4 10x$
6. $y = \csc(4 - 2x)$
7. $y = \sin^2(3 - 2x^2)$
8. $y = x\cot x^2$
9. $y = \dfrac{\sin 2x}{\cos 5x}$
10. $y = \dfrac{x\sin x}{x+1}$
11. $y = \sqrt{\sin 3x}$
12. $y = (1 + \tan^3 x)^{1/4}$
13. $y = \sin^3 x + \cos x$
14. $y = \sin 2x \cos 2x$
15. $2\sin y + 3\cos x = 1$
16. $x\cos y - y\cos x = 3$
17. $4\sin^2 x - 3\cos^3 y = 1$
18. $\tan(x + y) = y$
19. $x + \sec xy = 5$
20. $x^3 y + \tan^2 y = 3x$
21. $x = y^3 \csc^3 y$
22. $y = \cos(\tan x)$
23. $y = x^3 - x^2\cos x + 2x\sin x + 2\cos x$
24. $y = \sin^4 x^2 - \cos^4 x^2$
25. $y = u^3 \sec u, \quad u = x\tan(x+1)$
26. $y = \sqrt{3 - \sec v}, \quad v = \tan\sqrt{x}$
27. $y = \sqrt{t^2 + 1}, \quad t = \sin(\sin x)$
28. $y = (1 + \sec^3 u)^{1/3}, \quad u = \sqrt{1 + \cos x^2}$
29. $y = \dfrac{\sin^2 x}{1 + \cos^3 x}$
30. $y = \dfrac{1 + \tan^3(3x^2 - 4)}{x^2 \sin x}$

In Exercises 31 and 32 find d^2y/dx^2.

31. $\sin y = x^2 + y$
32. $\tan y = x + xy$

In Exercises 33–38 evaluate the limit if it exists.

33. $\lim\limits_{x \to 0} \dfrac{\tan x}{x}$
34. $\lim\limits_{x \to 0} \dfrac{1 - \cos x}{x}$
35. $\lim\limits_{x \to 0} \dfrac{\sin 2x}{x}$
36. $\lim\limits_{x \to \infty} \dfrac{\sin(2/x)}{\sin(1/x)}$
37. $\lim\limits_{x \to 0} \dfrac{(x+1)^2 \sin x}{3x^3}$
38. $\lim\limits_{x \to \pi/2} \dfrac{\cos x}{(x - \pi/2)^2}$

39. Sketch graphs of the functions $f(x) = |\sin x|$ and $g(x) = \sin|x|$. Where do these functions fail to be differentiable?

40. Does $\lim_{x \to 0} g'(x)$ exist for the function $g(x)$ in Example 3.11 of Section 3.3?

41. The angular displacement of the pendulum in Figure 3.30 at time t is given by

$$\theta = f(t) = A\cos(\omega t + \phi), \quad t \geq 0,$$

where A, ω, and ϕ are constants. Show that $\theta = f(t)$ satisfies the equation

$$\frac{d^2\theta}{dt^2} + \omega^2 \theta = 0.$$

FIGURE 3.30

42. When the mass m in Figure 3.31 moves vertically on the end of the spring, its displacement y must satisfy the equation

$$m\frac{d^2 y}{dt^2} + ky = 0,$$

where $k > 0$ is the constant of elasticity for the spring. Show that the function

$$y = f(t) = A\sin\left(\sqrt{\frac{k}{m}}\,t\right) + B\cos\left(\sqrt{\frac{k}{m}}\,t\right)$$

satisfies this equation for any constants A and B.

FIGURE 3.31

43. If the voltage source labeled E in the circuit of Figure 3.32 is suddenly short-circuited by switching the switch labeled S, the current I in the circuit thereafter must satisfy the equation

$$L\frac{d^2 I}{dt^2} + \frac{1}{C}I = 0,$$

where L and C are the sizes of the inductor and capacitor, respectively, and t is time. Show that

$$I = f(t) = A\cos\left(\frac{t}{\sqrt{LC}}\right) + B\sin\left(\frac{t}{\sqrt{LC}}\right)$$

satisfies this equation for any constants A and B whatsoever.

FIGURE 3.32

44. Find

$$\frac{d}{dx}|\sin x|.$$

45. If n is a positive integer, find formulas for

$$\frac{d^n}{dx^n}(\sin x) \quad \text{and} \quad \frac{d^n}{dx^n}(\cos x).$$

46. Show that the function

$$f(x) = \begin{cases} x^2 \sin(1/x) & x \neq 0 \\ 0 & x = 0 \end{cases}$$

has a derivative at $x = 0$, but $f'(x)$ is not continuous at $x = 0$.

47. Any cross-section of the reflector in a car headlight is in the form of a parabola

$$y^2 = 4x, \quad 0 \leq x \leq 2,$$

with the bulb at the point $(1, 0)$ called the *focus* of the parabola. It is a principle of optics that all light rays from the bulb are reflected by the mirror so that the angle between the incident ray and the normal to the mirror is equal to the angle between the reflected ray and the normal. Show that all rays are reflected parallel to the x-axis.

SECTION 3.9

Derivatives of the Exponential and Logarithm Functions

Exponential functions abound in applied mathematics, and much of this is due to the fact that the derivative of an exponential function is itself, varied by a constant in some cases. We obtain derivatives of exponential and logarithm functions in this section, but many of the applications which rely on these derivatives are delayed until Section 9.2. Exponential and logarithm functions and their properties are discussed in Appendix B. We assume that the reader is familiar with this material. To differentiate logarithm and exponential functions, we must return to the definition of the derivative (see equation 3.2 in Section 3.1). We can apply the definition to differentiate either $\log_a x$ or a^x, and once one of these derivatives is obtained, implicit differentiation can be used to find the other. We choose to differentiate $\log_a x$.

> **Theorem 3.13**
>
> The derivative of the logarithm function $\log_a x$ is
>
> $$\frac{d}{dx} \log_a x = \frac{1}{x} \log_a e, \qquad (3.29\text{a})$$
>
> where
>
> $$e = \lim_{z \to \infty} \left(1 + \frac{1}{z}\right)^z = \lim_{v \to 0} (1+v)^{1/v}. \qquad (3.29\text{b})$$

Proof By equation 3.2,

$$\begin{aligned}
\frac{d}{dx} \log_a x &= \lim_{h \to 0} \frac{\log_a(x+h) - \log_a x}{h} \\
&= \lim_{h \to 0} \frac{1}{h} \left[\log_a\left(\frac{x+h}{x}\right)\right] \qquad \left[\text{since } \log_a b - \log_a c = \log_a(b/c)\right] \\
&= \lim_{h \to 0} \frac{1}{h} \left[\log_a\left(1 + \frac{h}{x}\right)\right] \\
&= \frac{1}{x} \lim_{h \to 0} \frac{x}{h} \left[\log_a\left(1 + \frac{h}{x}\right)\right] \\
&= \frac{1}{x} \lim_{h \to 0} \log_a \left(1 + \frac{h}{x}\right)^{x/h} \qquad \left[\text{since } c \log_a b = \log_a b^c\right].
\end{aligned}$$

Since the logarithm function is continuous, we may interchange the limit and logarithm operations (see Theorem 2.3 in Section 2.4),

$$\frac{d}{dx} \log_a x = \frac{1}{x} \log_a \left[\lim_{h \to 0} \left(1 + \frac{h}{x}\right)^{x/h}\right].$$

But

$$\lim_{h \to 0} \left(1 + \frac{h}{x}\right)^{x/h} = \lim_{h/x \to 0} \left(1 + \frac{h}{x}\right)^{x/h} = e,$$

and therefore

$$\frac{d}{dx} \log_a x = \frac{1}{x} \log_a e.$$

If we choose e as the base of logarithms, that is, set $a = e$, derivative 3.29 simplifes to

$$\frac{d}{dx} \log_e x = \frac{1}{x}.$$

Because of the simplicity of this result, logarithms to base e are used more extensively than those to any other base. With the notation $\ln x = \log_e x$ for logarithms to base e, we may state the following corollary.

Corollary 1

$$\frac{d}{dx} \ln x = \frac{1}{x} \qquad (3.30)$$

The chain rule gives the next two results.

Corollary 2 *If $u(x)$ is a differentiable function of x, then*

$$\frac{d}{dx}\log_a u = \frac{1}{u}\frac{du}{dx}\log_a e. \tag{3.31}$$

Corollary 3 *If $u(x)$ is differentiable, then*

$$\frac{d}{dx}\ln u = \frac{1}{u}\frac{du}{dx}. \tag{3.32}$$

We now use implicit differentiation to find the derivative of the exponential function.

Theorem 3.14

$$\frac{d}{dx}a^x = a^x \ln a \tag{3.33}$$

Proof If we set $y = a^x$, then $x = \log_a y$, and we can differentiate (implicitly) with respect to x using 3.31,

$$1 = \frac{1}{y}\frac{dy}{dx}\log_a e.$$

Thus,

$$\frac{dy}{dx} = \frac{y}{\log_a e} = \frac{a^x}{\log_a e}.$$

When it is necessary to change bases of logarithms, the rule is

$$\log_b x = (\log_a x)(\log_b a),$$

(see equation B.8 in Appendix B). If we set $x = b = e$ in this equation, the result is

$$\log_e e = (\log_a e)(\log_e a) \qquad \text{or} \qquad 1 = (\log_a e)(\ln a).$$

Consequently, $1/\log_a e = \ln a$, and our differentiation formula becomes

$$\frac{dy}{dx} = a^x \ln a.$$

Certainly it is again clear that should we use exponentials to base e, the differentiation formula in Theorem 3.14 simplifies.

Corollary 1

$$\frac{d}{dx}e^x = e^x \tag{3.34}$$

The exponential function e^x is therefore its own derivative. In fact, for any constant C whatsoever, the function Ce^x differentiates to give itself. In Chapter 5 we shall see that Ce^x is the only function that is its own derivative.

Corollary 2 *For a differentiable function $u(x)$,*

$$\frac{d}{dx}a^u = a^u \frac{du}{dx}\ln a. \tag{3.35}$$

Corollary 3 *For a differentiable function $u(x)$,*

$$\frac{d}{dx}e^u = e^u \frac{du}{dx}. \tag{3.36}$$

EXAMPLE 3.26

Find dy/dx if y is defined as a function of x in each of the following:

(a) $y = 2^{3x}$ (b) $y = \log_{10}(3x^2 + 4)$ (c) $y = x^2 e^{-2x}$
(d) $y = (\ln x)/x$ (e) $y = \sqrt{1 + e^{2x}}$

SOLUTION (a) Using 3.35,

$$\frac{dy}{dx} = 2^{3x} \frac{d}{dx}(3x) \ln 2 = (3 \ln 2) 2^{3x}.$$

(b) With 3.31,

$$\frac{dy}{dx} = \frac{1}{3x^2 + 4} \frac{d}{dx}(3x^2 + 4) \log_{10} e = \frac{6x}{3x^2 + 4} \log_{10} e.$$

(c) The product rule and 3.36 give

$$\frac{dy}{dx} = 2xe^{-2x} + x^2 e^{-2x}(-2) = 2x(1 - x)e^{-2x}.$$

(d) The quotient rule and 3.30 yield

$$\frac{dy}{dx} = \frac{x(1/x) - \ln x}{x^2} = \frac{1 - \ln x}{x^2}.$$

(e) With extended power rule 3.17 and 3.36,

$$\frac{dy}{dx} = \frac{1}{2\sqrt{1 + e^{2x}}} \frac{d}{dx}(1 + e^{2x}) = \frac{2e^{2x}}{2\sqrt{1 + e^{2x}}} = \frac{e^{2x}}{\sqrt{1 + e^{2x}}}.$$

EXAMPLE 3.27

Find values of x for which the first derivative of the function $f(x) = x^2 \ln x$ is equal to zero, and values of x for which the second derivative is equal to zero.

SOLUTION The first derivative of $f(x)$ vanishes when

$$0 = f'(x) = 2x \ln x + x^2 \left(\frac{1}{x}\right) = x(2 \ln x + 1).$$

Since the function is undefined at $x = 0$, we set

$$2 \ln x + 1 = 0, \quad \text{or} \quad \ln x = -\frac{1}{2}.$$

Thus, $x = e^{-1/2} = 1/\sqrt{e}$.

The second derivative is equal to zero when

$$0 = f''(x) = (2 \ln x + 1) + x\left(\frac{2}{x}\right) = 2 \ln x + 3.$$

The only solution of this equation is $x = e^{-3/2}$.

In Section 3.2 we proved power rule 3.6 only in the case that n was a nonnegative integer. It is now easy to prove it for all real n, at least when $x > 0$. To do this we write x^n in the form $x^n = e^{n \ln x}$ (see equation B.13a). Formulas 3.36 and 3.30 give

$$\frac{d}{dx} x^n = \frac{d}{dx} e^{n \ln x} = e^{n \ln x} \frac{d}{dx} (n \ln x) = x^n \left(\frac{n}{x}\right) = nx^{n-1}.$$

A discussion of the power rule in the case that $x \leq 0$ depends on the value of n (see Exercise 39).

EXERCISES 3.9

In Exercises 1–30, y is defined as a function of x. Find dy/dx in as simplified a form as possible.

1. $y = 3^{2x}$
2. $y = \ln(3x^2 + 1)$
3. $y = \log_{10}(2x + 1)$
4. $y = e^{1-2x}$
5. $y = xe^{2x}$
6. $y = x \ln x$
7. $y = e^{2 \ln x}$
8. $y = \log_{10}(3 - 4x)$
9. $y = \ln(\sin x)$
10. $y = \ln(3 \cos x)$
11. $y = x \ln(x + 1)$
12. $y = x^2 + x^3 e^{4x}$
13. $y = \dfrac{e^{1-x}}{1-x}$
14. $y = \sin(e^{2x})$
15. $y = \ln(\ln x)$
16. $y = e^{-2x} \sin 3x$
17. $y = \ln(x^2 e^{4x})$
18. $y = \dfrac{e^x - e^{-x}}{e^x + e^{-x}}$
19. $y = e^{-x} \ln x$
20. $y = x[\sin(\ln x) - \cos(\ln x)]$
21. $y = e^{\sin u}$, $u = e^{1/x}$
22. $y = \ln(\cos v)$, $v = \sin^2 x$
23. $\ln(x + y) = x^2 y$
24. $xe^y + x^2 \ln y + y \sin x = 0$
25. $y = \ln(\sec x + \tan x)$
26. $y = x\sqrt{x^2 + 1} - \ln\left(x + \sqrt{x^2 + 1}\right)$
27. $y = \ln\left(x + 4 + \sqrt{8x + x^2}\right)$
28. $y = x - \dfrac{1}{4} \ln(1 + 5e^{4x})$
29. $e^{xy} = (x + y)^2$
30. $e^{1/x} + e^{1/y} = \dfrac{1}{x} + \dfrac{1}{y}$

31. A spring with constant 5 N/m is attached to a fixed wall on one end and a 1 kg mass m on the other (Figure 3.33).

FIGURE 3.33

We choose a coordinate system with x positive to the right and with $x = 0$ at the centre of mass of m when the spring is in the unstretched position. At time $t = 0$, m is pulled 1 m to the left of $x = 0$ and given a speed of 3 m/s to the right. During the subsequent motion, a frictional force equal in newtons to twice the velocity of the mass acts on m. It can be shown that if $x(t)$ is the position of m, then $x(t)$ must satisfy the equation

$$\frac{d^2 x}{dt^2} + 2 \frac{dx}{dt} + 5x = 0$$

and the initial conditions $x(0) = -1$ and $x'(0) = 3$. Verify that $x(t) = e^{-t}(\sin 2t - \cos 2t)$ satisfies the equation and initial conditions.

32. Two long parallel rectangular loops lying in the same plane have lengths l and L and widths w and W, respectively (Figure 3.34).

FIGURE 3.34

If the loops do not overlap, and the distance between the near sides is s, the mutual inductance between the loops is

$$M = \frac{\mu_0 l}{2\pi} \ln\left[\frac{s+W}{s\left(1 + \frac{W}{s+w}\right)}\right]$$

where $\mu_0 > 0$ is a constant. Show that the derivative of M as a function of s is negative for $s > 0$.

33. Two parallel wires carrying currents I (Figure 3.35) create a magnetic field.

FIGURE 3.35

The flux Φ of this field through the loop of dimensions h and w is

$$\Phi = \frac{\mu_0 h I}{2\pi} \ln\left[\frac{R(r+w)}{r(R+w)}\right],$$

where $\mu_0 > 0$ is a constant. Sketch a graph of Φ first as a function of h and then as a function of w.

34. Show that the function

$$y = f(x) = Ax + B\left[\frac{x}{2}\ln\left(\frac{x-1}{x+1}\right) + 1\right]$$

satisfies the equation

$$(1-x^2)\frac{d^2y}{dx^2} - 2x\frac{dy}{dx} + 2y = 0$$

for any constants A and B.

35. The equation $x = e^y - e^{-y}$ defines y implicitly as a function of x. Find dy/dx in terms of x in two ways:
    (a) Solve first for an explicit definition for $y = f(x)$.
    (b) Differentiate implicitly with respect to x.

36. The current I in the RC-circuit of Figure 3.36 must satisfy the equation

$$R\frac{dI}{dt} + \frac{I}{C} = \frac{dE}{dt}.$$

FIGURE 3.36

If $E = E_0 \sin\omega t$, where E_0 and ω are constants, show that a solution is

$$I = f(t) = Ae^{-t/(RC)} + \frac{E_0}{Z}\sin(\omega t - \phi),$$

where A is any constant whatsoever, and

$$Z = \sqrt{R^2 + \frac{1}{\omega^2 C^2}}, \qquad \tan\phi = -\frac{1}{\omega CR}.$$

37. The current I in the LR-circuit of Figure 3.37 must satisfy the equation

$$L\frac{dI}{dt} + RI = E.$$

FIGURE 3.37

If E is as in Exercise 36, show that a solution is

$$I = f(t) = Ae^{-Rt/L} + \frac{E_0}{Z}\sin(\omega t - \phi),$$

where A is any constant, and

$$Z = \sqrt{R^2 + \omega^2 L^2}, \qquad \tan\phi = \frac{\omega L}{R}.$$

38. The current in the RCL-circuit of Figure 3.38 must satisfy the equation

$$L\frac{d^2 I}{dt^2} + R\frac{dI}{dt} + \frac{I}{C} = \frac{dE}{dt}.$$

FIGURE 3.38

If E is as in Exercise 36, show that a solution is

$$I = f(t) = e^{-Rt/(2L)}(A\cos\nu t + B\sin\nu t) + \frac{E_0}{Z}\sin(\omega t - \phi),$$

where A and B are any constants whatsoever, and

$$\nu = \sqrt{\frac{1}{LC} - \frac{R^2}{4L^2}}, \quad Z = \sqrt{R^2 + \left(\omega L - \frac{1}{\omega C}\right)^2},$$

$$\tan\phi = \frac{\omega L - 1/(\omega C)}{R}.$$

39. (a) The function $f(x) = x^n$ is not defined for $x < 0$ if n is irrational, that is, is not rational. Verify power rule 3.6 for $x < 0$ if n is rational and x^n is defined.

(b) Discuss the derivative of $f(x) = x^n$ at $x = 0$.

SECTION 3.10

Logarithmic Differentiation

Compare the three functions

$$x^n, \qquad a^x, \qquad x^x.$$

The first, in which the exponent is constant and the base is variable, is called a **power function**. The second, in which the base is constant and the exponent is variable, is called an **exponential function**. We have differentiation rules for power functions and exponential functions, namely

$$\frac{d}{dx}x^n = nx^{n-1} \quad\text{and}\quad \frac{d}{dx}a^x = a^x \ln a,$$

respectively. The third function x^x, which we consider only for $x > 0$, is neither a power nor an exponential function; therefore, we cannot use either of the above formulas to find its derivative. Instead, we set $y = f(x) = x^x$, and take natural logarithms,

$$\ln y = x\ln x.$$

Implicit differentiation with respect to x now gives

$$\frac{1}{y}\frac{dy}{dx} = \ln x + \frac{x}{x} = \ln x + 1.$$

Consequently,

$$\frac{dy}{dx} = y(\ln x + 1) \quad\text{or}\quad \frac{d}{dx}x^x = x^x(\ln x + 1).$$

This process of taking logarithms and then differentiating is called **logarithmic differentiation**.

EXAMPLE 3.28

Find the derivative of the function $f(x) = x^{\sin x}$ when $x > 0$.

SOLUTION If we set $y = x^{\sin x}$, then

$$\ln y = \sin x \ln x.$$

Differentiation with respect to x gives

$$\frac{1}{y}\frac{dy}{dx} = \cos x \ln x + \frac{1}{x}\sin x,$$

and this can be solved for dy/dx,

$$\frac{dy}{dx} = x^{\sin x}\left(\cos x \ln x + \frac{1}{x}\sin x\right).$$

∎

Logarithmic differentiation can also be used to differentiate complicated products or quotients, as in the following example.

EXAMPLE 3.29

Find the derivative of the function

$$y = f(x) = \frac{x^3(x^2 + 1)^{2/3}}{\sin^3 x}$$

on the interval $0 < x < \pi$. Extend the result to other values of x.

SOLUTION When $0 < x < \pi$, we take natural logarithms of both sides of the definition for y:

$$\ln y = 3\ln x + \frac{2}{3}\ln(x^2 + 1) - 3\ln(\sin x).$$

Differentiation with respect to x gives

$$\frac{1}{y}\frac{dy}{dx} = \frac{3}{x} + \frac{2}{3}\frac{2x}{x^2 + 1} - \frac{3}{\sin x}\cos x,$$

and therefore

$$\frac{dy}{dx} = y\left[\frac{3}{x} + \frac{4x}{3(x^2 + 1)} - 3\cot x\right] = \frac{x^3(x^2 + 1)^{2/3}}{\sin^3 x}\left[\frac{3}{x} + \frac{4x}{3(x^2 + 1)} - 3\cot x\right].$$

When x is not in the interval $0 < x < \pi$, this derivation may not be valid. For instance, when $x < 0$, it is not acceptable to write $\ln x$, and when x is in the interval $\pi < x < 2\pi$, the term $\ln(\sin x)$ is not defined. These difficulties are easily overcome by first taking absolute values,

$$|y| = \frac{|x|^3(x^2 + 1)^{2/3}}{|\sin x|^3}.$$

Logarithms now give

$$\ln|y| = 3\ln|x| + \frac{2}{3}\ln(x^2+1) - 3\ln|\sin x|.$$

To differentiate this equation, we use equation 3.12 which states that when $f(x)$ is differentiable, and $f(x) \neq 0$,

$$\frac{d}{dx}|f(x)| = \frac{|f(x)|}{f(x)}f'(x).$$

When this is combined with formula 3.32, we obtain

$$\frac{d}{dx}\ln|f(x)| = \frac{1}{|f(x)|}\frac{d}{dx}|f(x)| = \frac{1}{|f(x)|}\frac{|f(x)|}{f(x)}f'(x) = \frac{f'(x)}{f(x)}. \quad (3.37)$$

Application of this result to the equation in $\ln|y|$ gives

$$\frac{1}{y}\frac{dy}{dx} = \frac{3}{x} + \frac{2}{3}\frac{2x}{x^2+1} - 3\frac{\cos x}{\sin x}.$$

This equation is identical to that obtained for dy/dx when $0 < x < \pi$, but its derivation here shows that it is valid even when x is not in the interval $0 < x < \pi$. ∎

EXERCISES 3.10

In Exercises 1–24 use logarithmic differentiation to find $f'(x)$.

1. $f(x) = x^{-x}$, $x > 0$
2. $f(x) = x^{4\cos x}$, $x > 0$
3. $f(x) = x^{4x}$, $x > 0$
4. $f(x) = (\sin x)^x$, $0 < x < \pi$
5. $f(x) = \left(1 + \dfrac{1}{x}\right)^x$, $x > 0$
6. $f(x) = \left(1 + \dfrac{1}{x}\right)^{x^2}$, $x > 0$
7. $f(x) = \left(\dfrac{1}{x}\right)^{1/x}$, $x > 0$
8. $f(x) = \left(\dfrac{2}{x}\right)^{3/x}$, $x > 0$
9. $f(x) = (\sin x)^{\sin x}$, $0 < x < \pi$
10. $f(x) = (\ln x)^{\ln x}$, $x > 1$
11. $f(x) = (x^2 + 3x^4)^3(x^2+5)^4$
12. $f(x) = \dfrac{\sqrt{x}(1+2x^2)}{\sqrt{1+x^2}}$
13. $f(x) = x\sqrt[3]{1-\sin x}$
14. $f(x) = (x^2+3x)^3(x^2+5)^4$
15. $f(x) = x^2 e^{4x}$
16. $f(x) = x^{3/2}e^{-2x}$
17. $f(x) = x^2 \ln x$
18. $f(x) = \dfrac{e^x}{\ln(x-1)}$
19. $f(x) = (x^3+3)^3(x^2-2x)$
20. $f(x) = \dfrac{\sqrt{x}(1-x^2)}{\sqrt{1+x^2}}$
21. $f(x) = \dfrac{x^2-1}{x\sqrt{1-4\tan^2 x}}$
22. $f(x) = x^3(x^2-4x)\sqrt{1+x^3}$
23. $f(x) = \dfrac{\sin^3 3x}{\tan^5 2x}$

24. $f(x) = \dfrac{\sin 2x \sec 5x}{(1 - 2\cot x)^3}$

25. If $u(x)$ is positive for all x, find a formula for the derivative of u^u with respect to x.

26. If a company sells a certain commodity at price r, the market demands

$$x = r^a e^{-b(r+c)},$$

items per week, where $r > a/b$, and a, b, and c are positive constants.

(a) Show that the demand increases as the price decreases.

(b) Calculate the elasticity of demand defined by

$$\dfrac{Er}{Ex} = \dfrac{x}{r}\dfrac{dr}{dx}.$$

SECTION 3.11

Rolle's Theorem and the Mean Value Theorems

Certain results in calculus are immediately seen to be important. For example, the power, product, and quotient rules, which eliminate the necessity of using equation 3.2 to calculate derivatives are clearly indispensable. Even the algebraic and geometric interpretations of the derivative itself are recognized as useful. Through various examples and exercises of this chapter, we have hinted at the variety and quantity of applications of the derivative. These will be dealt with at length in Chapter 4. Other results in calculus, especially those of a theoretical nature, are regarded as less important, or even unimportant, often because it is not obvious how they will be used. In this section we consider three very important theorems, without which we would encounter serious difficulty in treating many of the topics in the remainder of this book. The first theorem is needed to prove the second, and the second leads immediately to the third.

Theorem 3.15

(Rolle's Theorem) Suppose $f(x)$ is continuous on the closed interval $a \leq x \leq b$ and has a first derivative at each point in the open interval $a < x < b$. If $f(a) = f(b)$, then there exists at least one point c in the open interval at which $f'(c) = 0$.

For the function in Figure 3.39 there are two possible choices for c.

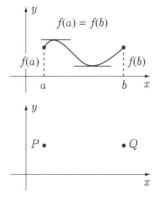

FIGURE 3.39

FIGURE 3.40

Geometrically, Rolle's theorem seems quite evident. Begin with two points, say P and Q in Figure 3.40, which have the same y-coordinate. Now try to join these points by a curve which never has a horizontal tangent line while satisfying the following two conditions:

(a) do not lift pencil from page — continuity of $f(x)$;

(b) the curve must have a tangent line at all points, and this tangent line must not be vertical — $f'(x)$ exists at all points.

It is impossible; therefore a point c where $f'(c) = 0$ must exist.

To verify this theorem directly requires Theorem 4.3 in Section 4.6. Since the latter result is quoted without proof, it seems as reasonable to accept Rolle's theorem on obvious geometric grounds as to base a proof on a theorem which itself is stated without proof. But for those who would like to see a proof based on Theorem 4.3 in Section 4.6, see Exercise 10 in that section.

Rolle's theorem can be used to prove the following result.

Theorem 3.16 (Cauchy's generalized mean value theorem) *Suppose functions $f(x)$ and $g(x)$ are continous on the closed interval $a \leq x \leq b$ and differentiable on the open interval $a < x < b$. If $g'(x) \neq 0$ on $a < x < b$, then there exists at least one point c in the open interval for which*

$$\frac{f(b) - f(a)}{g(b) - g(a)} = \frac{f'(c)}{g'(c)}. \tag{3.38}$$

Proof First note that $g(b) - g(a)$ cannot equal zero. If it did, then $g(a)$ would equal $g(b)$, and Rolle's theorem applied to $g(x)$ on the interval $a \leq x \leq b$ would imply the existence of a point c at which $g'(c) = 0$, contrary to the given assumption. To prove the theorem, we construct a function $h(x)$ to satisfy the conditions of Rolle's theorem. Specifically, we consider

$$h(x) = f(x) - f(a) - \frac{f(b) - f(a)}{g(b) - g(a)}[g(x) - g(a)].$$

Since $f(x)$ and $g(x)$ are continuous for $a \leq x \leq b$, so too is $h(x)$. In addition,

$$h'(x) = f'(x) - \frac{f(b) - f(a)}{g(b) - g(a)} g'(x);$$

therefore, $h'(x)$ exists for $a < x < b$. Finally, since $h(a) = h(b) = 0$, we may conclude from Rolle's theorem that there exists a number c such that $a < c < b$, and

$$0 = h'(c) = f'(c) - \frac{f(b) - f(a)}{g(b) - g(a)} g'(c);$$

that is,

$$\frac{f'(c)}{g'(c)} = \frac{f(b) - f(a)}{g(b) - g(a)}.$$

The next theorem states an important special case of this result that occurs when $g(x) = x$.

Theorem 3.17 (The mean value theorem) *Suppose $f(x)$ is continuous on the closed interval $a \leq x \leq b$ and has a first derivative at each point of the open interval $a < x < b$. Then there exists at least one point c in the open interval for which*

$$f'(c) = \frac{f(b) - f(a)}{b - a}. \tag{3.39}$$

From a geometric point of view 3.39 seems as obvious as Rolle's theorem. Figure 3.41 illustrates that the quotient $[f(b) - f(a)]/(b - a)$ is the slope of the line l joining the points $(a, f(a))$ and $(b, f(b))$ on the graph $y = f(x)$. The mean value theorem states that there is at least one point c between a and b at which the tangent line is parallel to l. In Figure 3.41 there are clearly two such points. Algebraically, 3.39 states that at some point between a and b, the instantaneous rate of change of the function $f(x)$ is equal to its average rate of change over the interval $a \leq x \leq b$. Similar interpretations of Theorem 3.16 are given in Section 10.6.

FIGURE 3.41

EXAMPLE 3.30

Find all values of c satisfying the mean value theorem for the function $f(x) = x^3 - 4x$ on the interval $-1 \leq x \leq 3$.

SOLUTION Since $f(x)$ is differentiable, and therefore continuous, at each point in $-1 \leq x \leq 3$, we can indeed apply the mean value theorem and claim the existence of at least one number c in $-1 < x < 3$ such that

$$f'(c) = \frac{f(3) - f(-1)}{3 - (-1)};$$

that is,

$$3c^2 - 4 = \frac{15 - 3}{4} = 3.$$

Consequently, $c = \pm\sqrt{7/3}$. Since $-\sqrt{7/3} < -1$, the only value of c in the interval $-1 < x < 3$ is $c = \sqrt{7/3}$ (Figure 3.42).

FIGURE 3.42

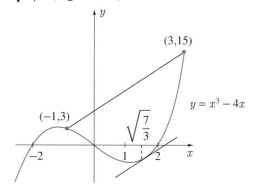

EXAMPLE 3.31

A traffic plane (Figure 3.43) measures the time that it takes a car to travel between points A and B as 15 seconds, and radios this information to a patrol car. What is the maximum speed at which the police officer can claim the car was traveling between A and B?

FIGURE 3.43

SOLUTION The average speed of the car between A and B is $500/15 = 100/3$ m/s. According to the mean value theorem, the instantaneous speed of the car must also have been $100/3$ m/s at least once. This is the maximum speed attributable to the car between A and B. It may have been traveling much faster at some points, but from the information given, no speed greater than $100/3$ m/s can be claimed by the officer.

EXERCISES 3.11

In Exercises 1–14 decide whether the mean value theorem can be applied to the function on the interval. If it cannot, explain why not. If it can, find all values of c in the interval that satisfy equation 3.39.

1. $f(x) = x^2 + 2x$, $-3 \le x \le 2$
2. $f(x) = 4 + 3x - 2x^2$, $1 \le x \le 3$
3. $f(x) = x + 5$, $2 \le x \le 3$
4. $f(x) = |x|$, $-1 \le x \le 1$
5. $f(x) = |x|$, $0 \le x \le 1$
6. $f(x) = x^3 + 2x^2 - x - 2$, $-3 \le x \le 2$
7. $f(x) = x^3 + 2x^2 - x - 2$, $-1 \le x \le 2$
8. $f(x) = (x+2)/(x-1)$, $2 \le x \le 4$
9. $f(x) = (x+1)/(x+2)$, $-3 \le x \le 2$
10. $f(x) = x^2/(x+3)$, $-2 \le x \le 3$
11. $f(x) = \sin x$, $0 \le x \le 2\pi$
12. $f(x) = \ln(2x+1)$, $0 \le x \le 2$
13. $f(x) = e^{-x}$, $-1 \le x \le 1$
14. $f(x) = \sec x$, $0 \le x \le \pi$

In Exercises 15–18 decide whether Cauchy's generalized mean value theorem can be applied to the function on the interval. If it cannot, explain why not. If it can, find all values of c in the interval that satisfy equation 3.38.

15. $f(x) = x^2$, $g(x) = x$, $1 \le x \le 2$
16. $f(x) = x + 1$, $g(x) = |x|^{3/2}$, $-1 \le x \le 1$
17. $f(x) = x^2 + 3x - 1$, $g(x) = x^3 + 5x + 4$, $0 \le x \le 2$
18. $f(x) = x/(x+1)$, $g(x) = x/(x-1)$, $-3 \le x \le -2$
19. Show that if $|f'(x)| \le M$ on $a \le x \le b$, then

$$|f(b) - f(a)| \le M(b - a).$$

20. Show that the value c that satisfies the mean value theorem for any quadratic function $f(x) = dx^2 + ex + g$ on any interval $a \le x \le b$ whatsoever is $c = (a+b)/2$.

21. Use the mean value theorem to show that

$$|\sin a - \sin b| \le |a - b|$$

for all real a and b. Is the same inequality valid for the cosine function?

22. Let $f(x)$ and $g(x)$ be two functions that are differentiable at each point of the interval $a \leq x \leq b$. Prove that if $f(a) = g(a)$ and $f(b) = g(b)$, then there exists c in the open interval $a < x < b$ for which $f'(c) = g'(c)$.

23. Verify that, for a cubic polynomial $f(x) = dx^3 + ex^2 + gx + h$ defined on any interval $a \leq x \leq b$, the values of c that satisfy equation 3.39 are equidistant from $x = -e/(3d)$.

SUMMARY

In this chapter we defined the derivative of a function $y = f(x)$ as

$$\frac{dy}{dx} = f'(x) = \lim_{h \to 0} \frac{f(x+h) - f(x)}{h}.$$

Algebraically it is the instantaneous rate of change of y with respect to x; geometrically it is the slope of the tangent line to the graph of $f(x)$. To eliminate the necessity of using this definition over and over again, we derived the sum, product, quotient, and power rules:

$$\frac{d}{dx}(u+v) = \frac{du}{dx} + \frac{dv}{dx},$$

$$\frac{d}{dx}(uv) = u\frac{dv}{dx} + v\frac{du}{dx},$$

$$\frac{d}{dx}\left(\frac{u}{v}\right) = \frac{v\frac{du}{dx} - u\frac{dv}{dx}}{v^2},$$

$$\frac{d}{dx}(u^n) = nu^{n-1}\frac{du}{dx}.$$

These four simple rules are fundamental to all calculus.

When a function $y = f(x)$ is defined implicitly by some equation $F(x, y) = 0$, we use implicit differentiation to find its derivative. We differentiate each term in the equation with respect to x, and then solve the resulting equation for dy/dx. We pointed out that care must be taken in differentiating equations. An equation can be differentiated with respect to a variable only if it is valid for a continuous range of values of that variable.

The chain rule defines the derivative of a composite function $y = f(g(x))$ as the product of the derivatives of $y = f(u)$ and $u = g(x)$:

$$\frac{dy}{dx} = \frac{dy}{du}\frac{du}{dx}.$$

These rules and techniques form the basis for the rest of differential calculus. When they are combined with the derivatives of the trigonometric functions, the exponential and logarithm functions, and the functions of Chapter 8, we will be well prepared to handle those applications of calculus that involve differentiation. Derivative formulas for the trigonometric, exponential and logarithm functions are listed below:

$$\frac{d}{dx}\sin u = \cos u \frac{du}{dx}, \qquad \frac{d}{dx}\cos u = -\sin u \frac{du}{dx},$$

$$\frac{d}{dx}\tan u = \sec^2 u \frac{du}{dx}, \qquad \frac{d}{dx}\cot u = -\csc^2 u \frac{du}{dx},$$

$$\frac{d}{dx}\sec u = \sec u \tan u \frac{du}{dx}, \qquad \frac{d}{dx}\csc u = -\csc u \cot u \frac{du}{dx},$$

$$\frac{d}{dx}\log_a u = \frac{1}{u}\frac{du}{dx}\log_a e, \qquad \frac{d}{dx}\ln u = \frac{1}{u}\frac{du}{dx},$$

$$\frac{d}{dx}a^u = a^u \frac{du}{dx}\ln a, \qquad \frac{d}{dx}e^u = e^u \frac{du}{dx}.$$

We completed the chapter by using Rolle's theorem to prove two mean value theorems. When $f(x)$ and $g(x)$ are continuous for $a \leq x \leq b$ and differentiable for $a < x < b$, Cauchy's generalized mean value theorem guarantees the existence of at least one point c between a and b such that

$$\frac{f(b) - f(a)}{g(b) - g(a)} = \frac{f'(c)}{g'(c)},$$

provided also that $g'(x) \neq 0$ for $a < x < b$. When $g(x) = x$, we obtain as a special case, the mean value theorem

$$f'(c) = \frac{f(b) - f(a)}{b - a}.$$

Key Terms and Formulas

In reviewing this chapter, you should be able to define or discuss the following key terms:

Average rate of change
Instantaneous rate of change
Derivative
Tangent line
Derivative function
Power rule
Normal line
Increment
Differentiable
Right-hand derivative
Left-hand derivative
Product rule

Quotient rule
Angle between curves
Orthogonal
Chain rule
Extended power rule
Explicit definition of a function
Implicit definition of a function
Implicit differentiation
Logarithmic differentiation
Rolle's theorem
Generalized mean value theorem
Mean value theorem

REVIEW EXERCISES

In Exercises 1–48 assume that y is defined as a function of x and find dy/dx in as simplified a form as possible.

1. $y = x^3 + \dfrac{1}{x^2}$
2. $y = 3x^2 + 2x + \dfrac{1}{x}$
3. $y = 2x - \dfrac{1}{3x^2} + \dfrac{1}{\sqrt{x}}$
4. $y = x^{1/3} - \dfrac{2}{3}x^{5/3}$
5. $y = x(x^2 + 5)^4$
6. $y = (x^2 + 2)^2(x^3 - 3)^3$
7. $y = \dfrac{3x^2}{x^3 - 5}$
8. $y = \dfrac{3x - 2}{x + 5}$
9. $y = \dfrac{x^2 + 2x + 2}{x^2 + 2x - 1}$
10. $y = \dfrac{4x}{x^2 + 5x - 2}$
11. $xy + 3y^3 = x + 1$
12. $\dfrac{x}{y} + \dfrac{y}{x} = x$
13. $x^2 y^2 - 3y \sin x = 14$
14. $x^2 y + y\sqrt{1 + x} = 3$
15. $y = \tan^3(3x + 2)$
16. $y = \sec^2(1 - 4x)$
17. $y = \dfrac{\sin 2x}{\cos 3x}$
18. $y = \sec(\tan 2x)$
19. $y = x^2 \cos x^2$
20. $y = \sin^2 x \cos^2 x$
21. $y = u^2 - 2u, \quad u = (1 + 2x)^{5/3}$
22. $y = t + \cos 2t, \quad t = x - \cos 2x$
23. $y = \sqrt{1 - t^3}, \quad t = \sqrt{1 + x^2}$

24. $y = v\cos^2 v$, $v = \sqrt{1-x^2}$

25. $y = \sqrt{1 + \sqrt{1+x}}$

26. $x = e^{2y}$

27. $\dfrac{2xy}{3x+4} = x^2 + 2$

28. $y = (x^2+1)\ln(x^2-1)$

29. $x\sin y + 2xy = 4$

30. $5\cos(x-y) = 1$

31. $x = \sqrt{1 + x\cot y^2}$

32. $x = \sqrt{\dfrac{4+y}{4-y}}$

33. $y = \sqrt{\dfrac{4+x^2}{4-x^2}}$

34. $y = \dfrac{x^2\sqrt{1-x}}{x+5}$

35. $y = \sqrt{7 - \sqrt{7 - \sqrt{x}}}$

36. $\dfrac{x}{x+y} = \dfrac{y}{x-y}$

37. $y = \sqrt{\dfrac{4+t}{4-t}}$, $t = \tan x$

38. $x = \dfrac{y^2 - 2y}{y^3 + 4y + 6}$

39. $y = \dfrac{x^3 - 6x^2 + 12x - 8}{x^2 - 4x + 4}$

40. $y = \dfrac{x-2}{\sqrt{x} - \sqrt{2}}$

41. $y = x^{2x}$

42. $y = (\cos x)^x$, $0 < x < \pi/2$

43. $y = \dfrac{e^x}{e^x + 1}$

44. $y = \log_{10}(\log_{10} x)$

45. $y = e^x \ln x$

46. $x = e^y + e^{-y}$

47. $xye^{xy} = 1$

48. $x^2 y + \ln(x+y) = x + 2$

In Exercises 49–52 find equations for the tangent and normal lines to the curve at the point indicated.

49. $y = x^3 + 3x - 2$ at $(1, 2)$

50. $y = \dfrac{1}{x+5}$ at $(0, 1/5)$

51. $y = \cos 2x$ at $(\pi/2, -1)$

52. $y = \dfrac{x^2 + 3x}{2x - 5}$ at $(1, -4/3)$

In Exercises 53–55 find d^2y/dx^2 assuming that y is defined implicitly as a function of x.

53. $x^2 - y^2 + 2(x-y) = y^3$

54. $(x-y)^2 = 3xy$

55. $\sin(x+y) = x$

56. Sketch a graph of $f(x) = \sin x^2$. Is is periodic?

57. Find all points on the curve $y = x^3 + x^2$ at which the tangent line passes through the origin.

58. Find that point on the curve $x = y^2 - 4$ at which the normal line passes through the point $(-6, 7)$. What application could be made of this result?

59. (a) How many functions with domain $-1 \le x \le 1$ are defined implicitly by the equation $x^2 + y^2 = 1$?

    (b) How many continuous functions with domain $-1 \le x \le 1$ are defined implicitly by the equation?

60. What is the rate of change of the area A of an equilateral triangle with respect to its side length l?

61. In a heated house, the temperature varies as the thermostat continually engages and disengages the furnace. Suppose that at the thermostat, the temperature T in degress Celsius over a four-hour time interval $0 \le t \le 4$ is given by

$$T = f(t) = 20 + 3\sin(4\pi t - \pi/2).$$

(a) Sketch a graph of $f(t)$.

(b) How many times is the furnace on during the four-hour period?

(c) What is the maximum time rate of change of temperature?

62. The curve defined by the equation $(x^2 + y^2)^2 = x^2 - y^2$ and shown in Figure 3.44 is called a *lemniscate*. Find the four points at which the tangent line is horizontal.

FIGURE 3.44

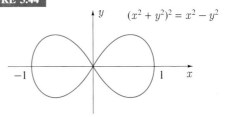

63. Find all points c in the interval $3 \le x \le 6$ that satisfy equation (3.39) when $f(x) = x^3 + 3x - 2$.

64. Find all points c in the interval $-1 \le x \le 1$ that satisfy equation (3.38) when $f(x) = 3x^2 - 2x + 4$ and $g(x) = x^3 + 2x$.

CHAPTER 4

Applications of Differentiation

In chapter 3 we hinted at some of the applications of the derivative; in this chapter we discuss them in detail.

Equations must be solved in almost every area of applied mathematics, and most equations do not have analytic solutions; solutions must be approximated to some specified accuracy. In Section 4.1 we show that derivatives provide one of the most potent methods for doing this. Much of this chapter (Sections 4.2–4.6) is devoted to the topic of optimization and its theory and applications. The first derivative of a function is used to determine where the values of a function are increasing or decreasing, and where the graph of a function attains a maximum or minimum value relative to nearby points. The second derivative determines intervals on which the first derivative is increasing or decreasing, and the transition points between these intervals. Such information can be invaluable in drawing the graph of a function. In Section 4.5 we combine these techniques with the tools from Chapters 1 and 2 for a final discussion on curve sketching. Sections 4.6–4.8, with their wealth of applied problems, show the real power of calculus. In Section 4.6 we illustrate the simplicity that calculus brings to solving applied maxima-minima problems; in Section 4.7 we use the results from Sections 4.2– 4.6 to develop a deeper understanding for the already familiar notions of velocity and acceleration; and in Section 4.8 we use the interpretation of the derivative as a rate of change to investigate interdependences of related quantities in a wide variety of applications. The calculation of many otherwise intractable limits becomes relatively straightforward with L'Hôpital's rule in Section 4.9. In Section 4.10 we discuss differentials, quantities essential to the topic of integration which begins in Chapter 5. To close the chapter in Section 4.11, we offer suggestions on how to organize and solve complicated multistage problems.

SECTION 4.1

Newton's Iterative Procedure for Solving Equations

In almost every area of application of mathematics, it is necessary to solve equations. When the problem is to solve one equation in one unknown, the equation can be expressed in the form

$$f(x) = 0, \qquad (4.1)$$

where $f(x)$ is usually a differentiable function of x. It might be a polynomial, a trigonometric function, an exponential or a logarithm function, or a complicated combination of these. The equation may have one solution, or many solutions, and these solutions are also called roots of the equation, or zeros of the function. Few equations can be solved by formula. Even when $f(x)$ is a polynomial, the only simple

FIGURE 4.1

solution formula is the quadratic formula which solves the equation for second degree polynomials. Functions other than polynomials rarely have formulas for their zeros.

Since we can seldom find a formula for solving 4.1, a different approach must be taken. In this section we discuss one of many techniques for approximating solutions to equations. To illustrate, consider finding the depth of water that half fills (by volume) the hemispherical tank in Figure 4.1. It can be shown (see Exercise 15 in Section 7.4) that when the water is x metres deep, the volume of water in the tank is given by

$$\frac{\pi}{3}(15x^2 - x^3).$$

Since the volume of one-half the tank is one-quarter that of a sphere of radius 5, it follows that the tank is half full when x satisfies the equation

$$\frac{1}{4}\left[\frac{4}{3}\pi(5)^3\right] = \frac{\pi}{3}(15x^2 - x^3),$$

and this equation simplifies to

$$f(x) = x^3 - 15x^2 + 125 = 0.$$

How do we solve this equation when we know that the solution we want is somewhere around $x = 3$? One of many possible methods is called **Newton's iterative procedure**.

A sketch of $f(x)$ between $x = 3$ and $x = 3.5$ is shown in Figure 4.2. The curvature has been exaggerated in order to more clearly depict the following geometric construction. If we set $x_1 = 3$, then x_1 is an approximation to the solution of the equation — not a good approximation, but an approximation nonetheless. Suppose we draw the tangent line to $y = f(x)$ at $(x_1, f(x_1))$. If x_2 is the point of intersection of this tangent line with the x-axis, it is clear that x_2 is a better approximation to the solution of the equation than x_1. If we draw the tangent line to $y = f(x)$ at $(x_2, f(x_2))$, its intersection point x_3 with the x-axis is an even better approximation. Continuation of this process leads to a succession of numbers $x_1, x_2, x_3, \ldots$, each of which is closer to the solution of the equation $f(x) = 0$ than the preceding numbers. This procedure for finding a better approximation to the solution of an equation is called Newton's iterative procedure (or the Newton-Raphson iterative procedure). We say that the numbers $x_1, x_2, x_3, \ldots$ converge to the root of the equation.

FIGURE 4.2

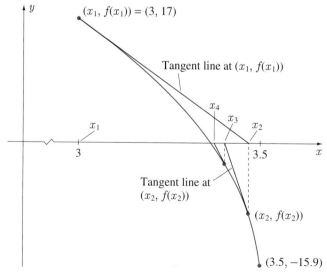

What we need now is an algebraic formula by which to calculate the numbers $x_1, x_2, x_3, \ldots$. The equation of the tangent line to $y = f(x)$ at $(x_1, f(x_1))$ is

$$y - f(x_1) = f'(x_1)(x - x_1).$$

To find the point of intersection of this tangent line with the x-axis, we set $y = 0$,

$$-f(x_1) = f'(x_1)(x - x_1),$$

and solve for x,

$$x = x_1 - \frac{f(x_1)}{f'(x_1)}.$$

But the point of intersection of the tangent line at $(x_1, f(x_1))$ with the x-axis is the second approximation x_2; that is,

$$x_2 = x_1 - \frac{f(x_1)}{f'(x_1)}.$$

To find x_3 we repeat this procedure with x_2 replacing x_1; the result is

$$x_3 = x_2 - \frac{f(x_2)}{f'(x_2)}.$$

As we repeat this process over and over again, the following formula for the $(n+1)^{\text{th}}$ approximation x_{n+1} in terms of the nth approximation x_n emerges

$$x_{n+1} = x_n - \frac{f(x_n)}{f'(x_n)}. \qquad (4.2)$$

This formula defines each approximation in Newton's iterative procedure in terms of its predecessor.

The derivative of $f(x) = x^3 - 15x^2 + 125$ in the above example is $f'(x) = 3x^2 - 30x$, and formula 4.2 becomes

$$x_{n+1} = x_n - \frac{x_n^3 - 15x_n^2 + 125}{3x_n^2 - 30x_n}.$$

Calculation of the first five approximations gives

$$x_2 = x_1 - \frac{x_1^3 - 15x_1^2 + 125}{3x_1^2 - 30x_1} = 3 - \frac{3^3 - 15(3)^2 + 125}{3(3)^2 - 30(3)} = 3.26984;$$

$$x_3 = x_2 - \frac{x_2^3 - 15x_2^2 + 125}{3x_2^2 - 30x_2} = 3.26352;$$

$$x_4 = x_3 - \frac{x_3^3 - 15x_3^2 + 125}{3x_3^2 - 30x_3} = 3.2635182;$$

$$x_5 = x_4 - \frac{x_4^3 - 15x_4^2 + 125}{3x_4^2 - 30x_4} = 3.2635182.$$

Newton's iterative procedure has therefore produced 3.263518 as an approximate solution to the equation $x^3 - 15x^2 + 125 = 0$. In spite of the fact that we have written six decimals in this final answer, our analysis in no way guarantees this degree of accuracy; we have simply judged on the basis that $x_4 = x_5$ that 3.263518 might be accurate to six decimals. The following theorem enables us to confirm this accuracy.

Theorem 4.1

(**Intermediate Value Theorem**) If $f(a)f(b) < 0$ for a function $f(x)$ continuous on $a \leq x \leq b$, then there exists at least one number c between a and b for which $f(c) = 0$.

Figure 4.3 illustrates the situation. The condition $f(a)f(b) < 0$ requires that one of $f(a)$ and $f(b)$ be positive and the other be negative. We have shown $f(a) > 0$ and $f(b) < 0$. As $f(x)$ is continuous for $a \leq x \leq b$, its graph can be traced between a and b without lifting pencil from paper. But then it must cross the x-axis at least once, and such a point is c in the theorem.

FIGURE 4.3

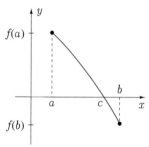

To show that 3.263518 is a root of the equation $f(x) = x^3 - 15x^2 + 125 = 0$ accurate to six decimals, we calculate $f(3.2635175) = 4.77 \times 10^{-5}$ and $f(3.2635185) = -1.83 \times 10^{-5}$. Since these function values are opposite in sign, the intermediate value theorem guarantees a root between 3.2635175 and 3.2635185. Consequently, when rounded to six decimals, the root must be 3.263518.

EXAMPLE 4.1

Use Newton's iterative procedure to find the only positive root of the equation

$$3x^4 + 15x^3 - 125x - 1500 = 0$$

accurate to 5 decimals. (This equation will be encountered in Exercise 44 of Section 4.6.)

SOLUTION
If we set $f(x) = 3x^4 + 15x^3 - 125x - 1500$, then a quick calculation shows that $f(4) = -272$ and $f(5) = 1625$. Thus the required solution is between $x = 4$ and $x = 5$. The nth approximation predicted by Newton's method is defined by

$$x_{n+1} = x_n - \frac{f(x_n)}{f'(x_n)} = x_n - \frac{3x_n^4 + 15x_n^3 - 125x_n - 1500}{12x_n^3 + 45x_n^2 - 125}.$$

When we choose $x_1 = 4$ ($f(4)$ is closer to 0 than $f(5)$), we find

$$x_2 = 4.19956, \qquad x_3 = 4.187268, \qquad x_4 = 4.1872187, \qquad x_5 = 4.1872187.$$

Since

$$f(4.187215) = -5.73 \times 10^{-3} \qquad \text{and} \qquad f(4.187225) = 9.72 \times 10^{-3},$$

it follows that the root is $x = 4.18722$, accurate to five decimals.

In the following example we again use Newton's iterative procedure to approximate the solution of an equation but specify the required accuracy in a different way.

EXAMPLE 4.2

Use Newton's method to find the smallest root of the cubic equation $x^3 - 3x + 1 = 0$ with an error no greater than 10^{-5}.

SOLUTION

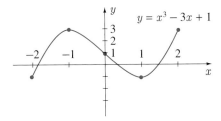

FIGURE 4.4

To find an estimate of the smallest root of the equation, we sketch a quick graph of the function $f(x) = x^3 - 3x + 1$ using the five points in Figure 4.4. The required root is clearly between -2 and -1. We take $x_1 = -2$ and use Newton's iterative procedure to define the nth approximation

$$x_{n+1} = x_n - \frac{f(x_n)}{f'(x_n)} = x_n - \frac{x_n^3 - 3x_n + 1}{3x_n^2 - 3}.$$

The next four approximations are

$x_2 = -1.88889, \qquad x_3 = -1.87945, \qquad x_4 = -1.879385, \qquad x_5 = -1.8793852.$

To determine the root to the required accuracy, we must find an interval of length 2×10^{-5} inside of which the root is known to lie. Since

$$f(-1.87939) = -3.6 \times 10^{-5} \quad \text{and} \quad f(-1.87937) = 1.16 \times 10^{-4},$$

such an interval is $-1.87939 < x < -1.87937$. Because the required solution must lie in this interval, it follows that -1.87938 is a solution with error no greater than 10^{-5}. ∎

Of the many techniques for approximating solutions to equations, Newton's method is the most popular. It converges to a solution a of $f(x) = 0$ provided $f'(a) \neq 0$ and x_1 is chosen sufficiently close to a (see Exercise 33 in Section 11.2). The condition $f'(a) \neq 0$ is included so that the denominator in 4.2 does not approach 0 as $x_n \to a$. However, $f(x_n)$ ususally approaches 0 faster than $f'(x_n)$ so that even when $f'(a) = 0$, Newton's method is successful. It is impossible to indicate how close x_1 must be to a in order to guarantee convergence to a. In some examples x_1 can be any number whatsoever, but we are also aware of examples where $|x_1 - a|$ must be less than 0.01 (see Exercise 31).

When x_1 is not sufficiently close to a, Newton's iterative procedure may not converge, or may converge to a solution other than expected. For instance, were we to attempt to approximate the largest root in Example 4.2, and inadvertently choose $x_1 = 1$, we could not find the second approximation: algebraically because $3x_1^2 - 3 = 0$, and geometrically because at $x_1 = 1$, the tangent line is horizontal, and does not intersect the x-axis. The function $f(x)$ in Figure 4.5 has zeros near $x = 1/2$ and $x = 1$. Were we to attempt to find the smaller zero using an initial approximation $x_1 = 0$, we would find the second approximation x_2 to be larger than 1. Further iterations would then converge to the zero near $x = 1$, not the zero near $x = 1/2$. We conclude, therefore, that the initial approximation in Newton's iterative procedure is most important. A poor choice for x_1 may lead to numbers that either converge to the wrong root or do not converge at all.

FIGURE 4.5

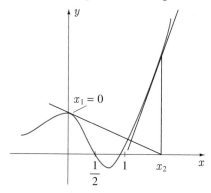

It is sometimes difficult to determine the number of solutions to an equation $f(x) = 0$ and initial approximations to these solutions. Even the roughest sketch of the function $f(x)$ can be instrumental in making such decisions. This was done in our introductory example concerning the hemispherical tank and again in Example 4.2. For some equations, it proves advantageous to arbitrarily separate the function $f(x)$ into two parts and write equation 4.1 in the form $g(x) = h(x)$. Solutions of this equation can be interpreted as x-coordinates of points of intersection of the curves $y = g(x)$ and $y = h(x)$. Sketches of these curves may quickly yield the number of solutions of the equation and their approximate values. For example, given the equation $2 \sin x - (x-1)^2 = 0$, we write $2 \sin x = (x-1)^2$ and sketch the curves $y = 2 \sin x$ and $y = (x-1)^2$ (Figure 4.6). Clearly there are two points of intersection, one between $x = 0$ and $x = 1$, and the other between $x = 2$ and $x = 3$.

FIGURE 4.6

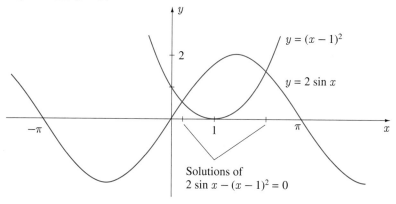

EXERCISES 4.1

In Exercises 1–16 use Newton's iterative procedure to find approximations to all roots of the equation accurate to 6 decimal places. In each case, we suggest making a sketch in order to obtain an initial approximation to each root.

1. $x^2 + 3x + 1 = 0$
2. $x^2 - x - 4 = 0$
3. $x^3 + x - 3 = 0$
4. $x^3 - x^2 + x - 22 = 0$
5. $x^3 - 5x^2 - x + 4 = 0$
6. $x^5 + x - 1 = 0$
7. $x^4 + 3x^2 - 7 = 0$
8. $\dfrac{x+1}{x-2} = x^2 + 1$
9. $x - 10 \sin x = 0$
10. $\sec x = \dfrac{2}{1+x^4}$
11. $(x+1)^2 = \sin 4x$
12. $(x+1)^2 = 5 \sin 4x$
13. $x + 4 \ln x = 0$
14. $x \ln x = 6$
15. $e^x + e^{-x} = 10x$
16. $x^2 - 4e^{-2x} = 0$

In Exercises 17–24 use Newton's iterative procedure to find approximations to all roots of the equation with an error no greater than that specified.

17. $x^3 - 5x - 1 = 0$, 10^{-3}
18. $x^4 - x^3 + 2x^2 + 6x = 0$, 10^{-4}
19. $\dfrac{x}{x+1} = x^2 + 2$, 10^{-5}
20. $(x+1)^2 = x^3 - 4x$, 10^{-3}
21. $(x+1)^2 = 5\sin 4x$, 10^{-3}
22. $\cos^2 x = x^2 - 1$, 10^{-4}
23. $x + (\ln x)^2 = 0$, 10^{-3}
24. $e^{3x} + e^x = 4$, 10^{-4}

In Exercises 25–28 find all points of intersection for the curves accurate to 4 decimals.

25. $y = x^3$, $y = x + 5$
26. $y = (x+1)^2$, $y = x^3 - 4x$
27. $y = x^4 - 20$, $y = x^3 - 2x^2$
28. $y = \dfrac{x}{x+1}$, $y = x^2 + 2$

29. When the beam in Figure 4.7 vibrates vertically, there are certain frequencies of vibration, called natural frequencies.

FIGURE 4.7

They are solutions of the equation
$$\tan x = \dfrac{e^x - e^{-x}}{e^x + e^{-x}}$$
divided by 20π. Find the two smallest frequencies correct to 4 decimals.

30. A stone of mass 100 gm is thrown vertically upward with speed 20 m/s. Air exerts a resistive force on the stone proportional to its speed, and has magnitude 0.1 N when the speed of the stone is 10 m/s. It can be shown that the height y above the projection point attained by the stone is given by
$$y = -98.1t + 1181\left(1 - e^{-t/10}\right) \text{ m},$$
where t is time (measured in seconds with $t = 0$ at the instant of projection).

(a) The time taken for the stone to return to its projection point can be obtained by setting $y = 0$ and solving the equation for t. Do so (correct to 2 decimals).

(b) When air resistance is neglected, the formula for y is
$$y = 20t - 4.905t^2 \text{ m}.$$
What is elapsed time in this case from the instant the stone is projected until it returns to the projection point?

31. A uniform hydro cable $P = 80$ m long with mass per unit length $\rho = 0.5$ kg/m is hung from two supports at the same level $L = 70$ m apart (Figure 4.8).

FIGURE 4.8

The tension T in the cable at its lowest point must satisfy the equation
$$\dfrac{\rho g P}{T} = e^{\rho g L/(2T)} - e^{-\rho g L/(2T)},$$
where $g = 9.81$. If we set $z = \rho g/(2T)$, then z must satisfy
$$2Pz = e^{Lz} - e^{-Lz}.$$
Solve this equation for z and hence find T correct to 1 decimal.

32. Planck's law for the energy density E of blackbody radiation at $1000\,°K$ states that
$$E = E(\lambda) = \dfrac{k\lambda^{-5}}{e^{c/\lambda} - 1}$$
where $k > 0$ is a constant and $c = 0.0014386$. This function is shown in Figure 4.9.

FIGURE 4.9

The value of λ at which E is a maximum must satisfy the equation

$$(5\lambda - c)e^{c/\lambda} - 5\lambda = 0.$$

Find this value of λ correct to 7 decimals.

33. Let $f(x)$ be a continuous function with domain and range both equal to the interval $[a, b]$. Show that there is at least one value of x in (a, b) for which $f(x) = x$.

34. (a) Use the intermediate value theorem to prove that when the domain of a continuous function is an interval, so also is its range. Hint: Use the idea that a set S of points on the y-axis constitutes an interval if for any two points c and d in S, the points $c < y < d$ are all in S.

    (b) If the domain is an open interval, is the range an open interval?

35. Use the Intermediate Value Theorem to prove that at any given time there is a pair of points directly opposite each other on the equator of the earth that have exactly the same temperature. Hint: Take the equator to be the circle $x^2 + y^2 = r^2$. Let $f(x)$ be the temperature on the upper semicircle and $g(x)$ be the temperature on the lower semicircle. Consider the function $F(x) = f(x) - g(-x)$.

36. A marathoner runs the 26 odd miles from point A to point B starting at 7:00 a.m. Saturday morning. Starting at 7:00 a.m. Sunday morning he runs the course again, but this time from point B to point A. Prove that there is a point on the course that he passed at exactly the same time on both days.

SECTION 4.2

Increasing and Decreasing Functions

Many mathematical concepts have their origin in very intuitive ideas. In this section we analyze the intuitive idea of one quantity "getting larger" and another "getting smaller". To describe what it means for a quantity to be increasing (getting larger) or decreasing (getting smaller), we first suppose that the quantity is represented by some function $f(x)$ of a variable x. The mathematical definition for $f(x)$ to be increasing or decreasing is as follows.

Definition 4.1

A function $f(x)$ is said to be **increasing** on an interval I if for all $x_1 > x_2$ in I

$$f(x_1) > f(x_2). \tag{4.3}$$

A function $f(x)$ is said to be **decreasing** on I if for all $x_1 > x_2$ in I

$$f(x_1) < f(x_2). \tag{4.4}$$

The continuous function in Figure 4.10 is increasing on the intervals

$$a \leq x \leq b, \qquad c \leq x \leq d, \qquad e \leq x \leq f,$$

and decreasing on the intervals

$$b \leq x \leq c, \qquad d \leq x \leq e.$$

When the function has points of discontinuity, such as in Figure 4.11, the situation is somewhat more complicated. This function is increasing on the intervals

$$a \leq x < b, \qquad d \leq x < e, \qquad e < x \leq f,$$

and decreasing for

$$b < x < d.$$

Pay special attention to whether end points of each interval are included.

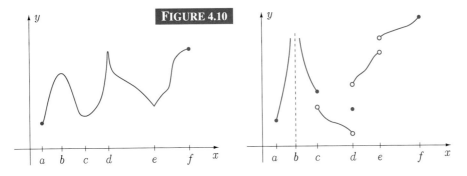

FIGURE 4.10

FIGURE 4.11

For continuous functions the graph in Figure 4.10 indicates that the sign of $f'(x)$ determines whether a function is increasing or decreasing on an interval. A precise statement of the situation for continuous functions is as follows.

Increasing and Decreasing Test for Continuous Functions

(i) A continuous function $f(x)$ is increasing on an interval I if on I

$$f'(x) \geq 0 \tag{4.5}$$

and is equal to zero at only a finite number of points.

(ii) A continuous function $f(x)$ is decreasing on an interval I if on I

$$f'(x) \leq 0 \tag{4.6}$$

and is equal to zero at only a finite number of points.

In conditions 4.5 and 4.6 we permit $f'(x)$ to vanish at only a finite number of points and do not, therefore, allow it to vanish on an interval. The reason for this is illustrated in Figures 4.12 and 4.13. In Figure 4.12, $f'(x)$ is equal to zero on the interval $a \leq x \leq b$, and certainly $f(x)$ is not increasing on any interval that contains these points. The function $f(x)$ in Figure 4.13 has $f'(0) = 0 = f'(1)$, and yet $f(x)$ is increasing on the interval $a \leq x \leq b$.

FIGURE 4.12

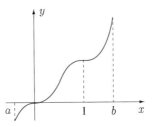

FIGURE 4.13

With the mean value theorem from Section 3.11, it is easy to verify this test. We prove (i); the proof of (ii) is similar. Suppose $f'(x) \geq 0$ on an interval $a \leq x \leq b$ and $d_1, d_2, \ldots, d_n$ are the points at which $f'(x) = 0$. Take any two points x_1 and x_2 in the interval $d_1 \leq x \leq d_2$, where $x_2 > x_1$. By the mean value theorem there exists a point c between x_1 and x_2 such that

$$f'(c) = \frac{f(x_2) - f(x_1)}{x_2 - x_1} \quad \text{or} \quad f(x_2) - f(x_1) = f'(c)(x_2 - x_1).$$

Since $x_2 > x_1$ and $f'(c) > 0$, it follows that $f(x_2)$ must be greater than $f(x_1)$. In other words, $f(x)$ is increasing on the interval $d_1 \leq x \leq d_2$. But by this argument $f(x)$

is increasing on each of the intervals $a \leq x \leq d_1$, $d_1 \leq x \leq d_2$, ..., $d_{n-1} \leq x \leq d_n$, $d_n \leq x \leq b$, and therefore must be increasing for $a \leq x \leq b$.

According to the above test, the sign of $f'(x)$ dictates where a function $f(x)$ is increasing and where it is decreasing. Sign diagrams introduced in Section 1.1 are useful in analyzing positivity and negativity of $f'(x)$.

EXAMPLE 4.3

Find intervals on which the following functions are increasing and decreasing:

(a) $f(x) = 2x^3 + 3x^2 - 12x + 4$ (b) $f(x) = \dfrac{x}{x+1}$

SOLUTION (a) Since $f'(x) = 6x^2 + 6x - 12 = 6(x-1)(x+2)$, the sign diagram in Figure 4.14 indicates that $f'(x) \geq 0$ when $x \leq -2$ and when $x \geq 1$, and that $f'(x) \leq 0$ when $-2 \leq x \leq 1$. Consequently, $f(x)$ is increasing on the intervals $x \leq -2$ and $x \geq 1$, and decreasing on $-2 \leq x \leq 1$.

FIGURE 4.14

(b) Because

$$f'(x) = \frac{(x+1)(1) - x(1)}{(x+1)^2} = \frac{1}{(x+1)^2},$$

$f'(x)$ is always positive, except at $x = -1$ where it does not exist. We conclude that $f(x)$ is increasing on the intervals $x < -1$ and $x > -1$. The function is not, however, increasing for all x [since, for example, $f(0) < f(-2)$]. The graph of the function in Figure 4.15 clarifies the situation.

FIGURE 4.15

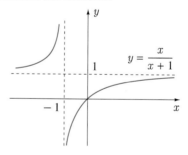

EXERCISES 4.2

In Exercises 1–24 determine intervals on which the function is increasing and decreasing.

1. $f(x) = 2x - 3$

2. $f(x) = 4 - 5x$

3. $f(x) = x^2 - 3x + 4$

4. $f(x) = -2x^2 + 5x$

5. $f(x) = 3x^2 + 6x - 2$

6. $f(x) = 5 + 2x - 4x^2$

7. $f(x) = 2x^3 - 18x^2 + 48x + 1$

8. $f(x) = x^3 + 6x^2 + 12x + 5$

9. $f(x) = 4x^3 - 18x^2 + 1$

10. $f(x) = 4 - 18x - 9x^2 - 2x^3$

11. $f(x) = 3x^4 + 4x^3 - 24x + 2$

12. $f(x) = 3x^4 - 4x^3 + 24x^2 - 48x$

13. $f(x) = x^4 - 4x^3 - 8x^2 + 48x + 24$

14. $f(x) = x^5 - 5x + 2$

15. $f(x) = x + \dfrac{1}{x}$

16. $f(x) = x^2 + \dfrac{1}{x^2}$

17. $f(x) = \dfrac{x}{2 - x}$

18. $f(x) = \dfrac{x^2 + 4}{x^2 - 1}$

19. $f(x) = \dfrac{x^3}{x + 1}$

20. $f(x) = |x^2 - 1| + 1$

21. $f(x) = xe^{-x}$

22. $f(x) = x^2 e^{-x}$

23. $f(x) = \ln(x^2 + 5)$

24. $f(x) = x \ln x$

25. If the price of a certain gadget is set at r, then the market demands x gadgets per week, where

$$x = 4a^3 - 3ar^2 + r^3,$$

where $a > 0$ is a constant, and $0 < r < 2a$. Show that the price function $r = f(x)$ defined implicitly by this equation is a decreasing function.

In Figure 4.16(a) we have sketched the graph of a function $f(x)$ and in Figure 4.16(b), the graph of $f'(x)$. Note that $f'(x) \geq 0$ when $f(x)$ is increasing and $f'(x) \leq 0$ when $f(x)$ is decreasing. In addition, the corner in $f(x)$ at $x = 2$ is reflected in the discontinuity in $f'(x)$. In Exercises 26–33 draw similar graphs for the function and its derivative.

FIGURE 4.16

(a)

(b)

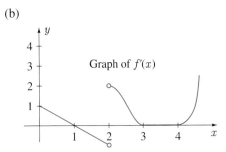

26. $f(x) = x^2 + 2x$

27. $f(x) = x^4 - x^2$

28. $f(x) = \dfrac{x - 1}{x + 2}$

29. $f(x) = \dfrac{x - 4}{x^2}$

30. $f(x) = \dfrac{|x|}{x}$

31. $f(x) = |x^2 - 4|$

32. $f(x) = \dfrac{|x^2 - 4|}{x - 2}$

33. $f(x) = \sin 3x$

 In Exercises 34–37 find (accurate to 4 decimals) intervals on which the function is increasing and decreasing.

34. $f(x) = x^4 + 2x^2 - 6x + 5$

35. $f(x) = 3x^4 - 20x^3 - 24x^2 + 48x$

36. $f(x) = x^2 \sin x$, $-\pi \leq x \leq \pi$

37. $f(x) = \tan x - x(x + 2)$

38. Show that $\sin x < x$ for all $x > 0$. Hint: Calculate $f(0)$ and $f'(x)$ for $f(x) = x - \sin x$.

39. Show that $\cos x > 1 - x^2/2$ for all $x > 0$. Hint: See the technique of Exercise 38.

40. Use the result of Exercise 39 to prove that for $x > 0$

$$\sin x > x - \dfrac{x^3}{6}.$$

41. Use the result of Exercise 40 to prove that for $x > 0$

$$\cos x < 1 - \dfrac{x^2}{2} + \dfrac{x^4}{24}.$$

42. Use the technique of Exercise 38 to verify that for $x > 0$

$$\dfrac{1}{\sqrt{1 + 3x}} > 1 - \dfrac{3x}{2}.$$

43. If $f(x)$ and $g(x)$ are differentiable and increasing on an interval I, is $f(x)g(x)$ increasing on I?

44. If positive functions $f(x)$ and $g(x)$ are differentiable and increasing on an interval I, is $f(x)g(x)$ increasing on I?

SECTION 4.3

Relative Maxima and Minima

One of the most important applications of calculus is in the field of optimization, the study of maxima and minima. In this section we begin discussions of this topic, which continue through to Section 4.6. Fundamental to discussions on optimization are critical points.

Definition 4.2

A **critical point** of a function is a point in the domain of the function at which the first derivative either is equal to zero or does not exist.

Specifically, $x = c$ is a critical point for $f(x)$ if $f'(c) = 0$ or $f'(c)$ does not exist, but in the latter $f(c)$ must exist. Interpreting derivatives as slopes of tangent lines, we can state that a critical point of a function is a point at which the graph of the function has a horizontal tangent line, a vertical tangent line, or no tangent line at all. For example, the eight points a through h on the x-axis in Figure 4.17 are all critical. At a, b, c, and d, the tangent line is horizontal, at e and f, the tangent line is vertical, and at g and h, there is no tangent line.

FIGURE 4.17

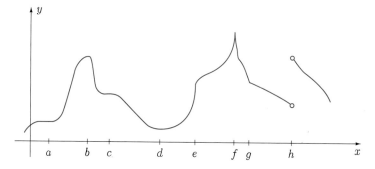

When the domain of a function $f(x)$ is a closed interval $a \leq x \leq b$, the end point $x = a$ is critical if the right-hand derivative of $f(x)$ at $x = a$ is equal to zero or does not exist. Likewise, $x = b$ is critical if the left-hand derivative vanishes or does not exist there. This is consistent with the definition of differentiability in Section 3.3. The function in Figure 4.18 has domain $1 \leq x \leq 3$. Both end points are critical — $x = 1$ because $f'_+(1)$ does not exist, and $x = 3$ because $f'_-(3) = 0$.

FIGURE 4.18

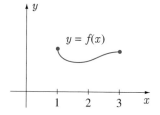

The function in Figure 4.17 is discontinuous at $x = h$, and we know that a function cannot have a derivative at a point of discontinuity (Theorem 3.6). Does this mean that every point of discontinuity of a function is critical? Not necessarily. The function must be defined at a point for that point to be critical. Thus, points of discontinuity are critical only if the function is defined at the point. In the remainder of this section we consider only critical points at which a function is continuous and which are not end points of its domain of definition.

EXAMPLE 4.4

Find critical points for the following functions:

(a) $f(x) = x^3 - 7x^2 + 11x + 6$ (b) $f(x) = \dfrac{x^2}{x^3 - 1}$ (c) $f(x) = x \ln x$

SOLUTION (a) For critical points we first solve

$$0 = f'(x) = 3x^2 - 14x + 11 = (3x - 11)(x - 1),$$

and obtain $x = 11/3$ and $x = 1$. These are the only critical points since there are no points where $f'(x)$ does not exist.

(b) For critical points we calculate

$$f'(x) = \frac{(x^3 - 1)(2x) - x^2(3x^2)}{(x^3 - 1)^2} = \frac{-x(x^3 + 2)}{(x^3 - 1)^2}.$$

Clearly $f'(x) = 0$ when $x = 0$, and when $x^3 + 2 = 0$, which implies that $x = -2^{1/3}$. The derivative does not exist when $x^3 - 1 = 0$; i.e., when $x = 1$. But $f(x)$ is not defined at $x = 1$ either, and therefore $x = 1$ is not a critical point. There are only two critical points: $x = 0$ and $x = -2^{1/3}$.

(c) For critical points we first solve

$$0 = f'(x) = \ln x + \frac{x}{x} = \ln x + 1.$$

The only solution of this equation is $x = 1/e$. Since $f'(x)$ exists for all x in the domain of $f(x)$, namely, $x > 0$, the function has no other critical points. ∎

At the critical points b and f in Figure 4.19, the graph of the function has "high" points.

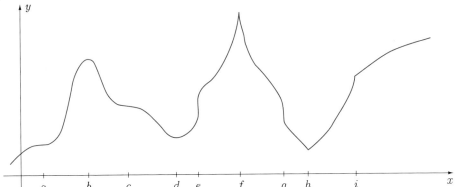

FIGURE 4.19

Definition 4.3

A function $f(x)$ is said to have a **relative maximum** $f(x_0)$ at $x = x_0$ if there exists an open interval I containing x_0 such that for all x in I

$$f(x) \leq f(x_0). \tag{4.7}$$

Since such intervals can be drawn around $x = b$ and $x = f$, relative or local maxima occur at these points. At a relative or local maximum, the graph of the function is highest relative to nearby points.

Critical points d and h in Figure 4.19, where the graph has "low" points, are described in a similar definition.

> **Definition 4.4**
>
> A function $f(x)$ is said to have a **relative minimum** $f(x_0)$ at $x = x_0$ if there exists an open interval I containing x_0 such that for all x in I
>
> $$f(x) \geq f(x_0). \tag{4.8}$$

Relative (or local) minima therefore occur at $x = d$ and $x = h$.

The critical points $x = a$ and $x = c$, where $f'(x) = 0$ and $x = e$ and $x = g$, where $f'(x)$ does not exist will be discussed in Section 4.4.

At $x = i$, the graph takes an abrupt change in direction. The function has a left-hand derivative and a right-hand derivative, but $f'(i)$ does not exist. In Section 3.3 we called the corresponding point on the curve a corner. Corners can sometimes be relative extrema ($x = h$ yields a corner and a relative minimum).

Relative maxima and minima represent high and low points on the graph of a function relative to points near them. It is not coincidence in Figure 4.19 that the two relative maxima and the two relative minima occur at critical points. According to the following theorem, this is always the case.

> **Theorem 4.2**
>
> Relative maxima and relative minima of a function must occur at critical points of the function.

Proof To verify this we prove that at any point at which the derivative $f'(x)$ of a function $f(x)$ exists and is not zero, it is impossible for $f(x)$ to have a relative extremum, that is, a relative maximum or a relative minimum. Suppose that at some point $x = a$, the derivative $f'(a)$ exists and is positive (Figure 4.20),

$$0 < f'(a) = \lim_{h \to 0} \frac{f(a+h) - f(a)}{h}.$$

According to Exercise 30 in Section 2.5, there exists an open interval $I : b < x < c$ around $x = a$ in which

$$\frac{f(a+h) - f(a)}{h} > 0.$$

This implies that when $h > 0$ (and $a + h$ is in I), $f(a + h) - f(a)$ must also be positive, and therefore $f(a + h) > f(a)$. But when $h < 0$ (and $a + h$ is in I), $f(a + h) - f(a)$ must be negative, and therefore $f(a + h) < f(a)$. There is an interval $b < x < a$ in which $f(x) < f(a)$, and an interval $a < x < c$ in which $f(x) > f(a)$. Thus, $x = a$ cannot yield a relative extremum.

A similar proof holds when $f'(a) < 0$.

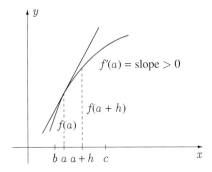

FIGURE 4.20

Although every relative extremum of a function must occur at a critical point, not all critical points give relative extrema. For continuous functions, there is a simple test to determine whether a critical point gives a relative maximum or a relative minimum. To understand this test, consider the critical points in Figure 4.19 as you read the statements.

First-derivative Test for Relative Extrema of Continuous Functions

(i) If $f'(x)$ (slope of a graph) changes from a positive quantity to a negative quantity as x increases through a critical point, then that critical point yields a relative maximum of $f(x)$.

(ii) If $f'(x)$ changes from a negative quantity to a positive quantity as x increases through a critical point, then that critical point yields a relative minimum of $f(x)$.

We prove part (i) of this test and leave verification of (ii) to the reader. Let $x = c$ be a critical point of $f(x)$ and let $f'(x)$ change from a positive quantity to a negative quantity as x increases through c. The fact that $f'(x)$ is positive to the left of c indicates that $f(x)$ must be increasing on some interval $a < x \leq c$ (Figure 4.21). Likewise, for $f'(x)$ to be negative to the right of c, $f(x)$ must be decreasing on some interval $c \leq x < b$. It follows that the value of $f(x)$ at $x = c$ must be larger than its value at any other point in the interval $a < x < b$.

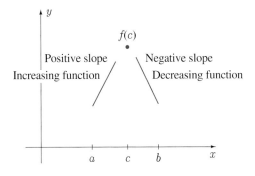

FIGURE 4.21

According to the first derivative test, a critical point at which $f(x)$ is continuous yields a relative maximum or a relative minimum when $f'(x)$ changes sign as x passes through the critical point. Sign diagrams are once again useful in making these decisions.

EXAMPLE 4.5

Find critical points for the following functions and determine whether they yield relative maxima or relative minima for the function.

(a) $f(x) = 2x^3 - 9x^2 - 24x + 6$ (b) $f(x) = \dfrac{x^2}{x^3 - 1}$ (c) $f(x) = x^{5/3} - x^{2/3}$

SOLUTION (a) For critical points we first solve

$$0 = f'(x) = 6x^2 - 18x - 24 = 6(x - 4)(x + 1),$$

and obtain $x = -1$ and $x = 4$. These are the only critical points as there are no points at which $f'(x)$ is undefined. The sign diagram in Figure 4.22 illustrates that as x increases through -1, $f'(x)$ changes from positive to negative; therefore $x = -1$ yields a relative maximum. Since $f'(x)$ changes from negative to positive as x increases through 4, $x = 4$ yields a relative minimum.

FIGURE 4.22

(b) In Example 4.4 we found that

$$f'(x) = \frac{-x(x^3 + 2)}{(x^3 - 1)^2},$$

and that the only critical points are $x = 0$ and $x = -2^{1/3}$. Because the denominator is always positive, the sign of $f'(x)$ is completely determined by the sign of $-x(x^3 + 2)$. This is analyzed in Figure 4.23. Since $f'(x)$ changes from negative to positive as x increases through $-2^{1/3}$, $x = -2^{1/3}$ gives a relative minimum. A relative maximum exists at $x = 0$ since $f'(x)$ changes from positive to negative as x increases through 0.

FIGURE 4.23

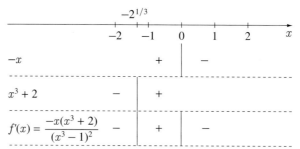

(c) For critical points we first solve

$$0 = f'(x) = \frac{5}{3}x^{2/3} - \frac{2}{3}x^{-1/3} = \frac{5x - 2}{3x^{1/3}}.$$

Clearly, $x = 2/5$ is a critical point since $f'(2/5) = 0$. In addition, because $f'(0)$ does not exist, but $f(0)$ does $(= 0)$, it follows that $x = 0$ is also a critical point. From the sign diagram in Figure 4.24, we see that $f'(x)$ changes from negative to positive as x increases through $x = 2/5$, which implies a relative minumum at this value of x. Since $f(x)$ is continuous at $x = 0$, and $f'(x)$ changes from positive to negative as x increases through 0, a relative maximum occurs at $x = 0$.

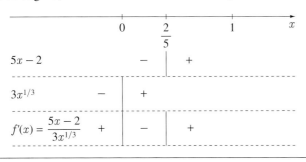

FIGURE 4.24

It is perhaps evident that were we required to sketch the graph of a function, it would be very helpful to know the function's critical points and whether these points give relative maxima or relative minima. This is indeed true, but before we consider the entire topic of curve sketching in Section 4.5, we introduce one further topic in Section 4.4.

EXERCISES 4.3

In Exercises 1–40 find all critical points of the function and determine whether they yield relative maxima or relative minima.

1. $f(x) = x^2 - 2x + 6$

2. $f(x) = 2x^3 + 15x^2 + 24x + 1$

3. $f(x) = x^4 - 4x^3/3 + 2x^2 - 24x$

4. $f(x) = (x - 1)^5$

5. $f(x) = \dfrac{x + 1}{x^2 + 8}$

6. $f(x) = \dfrac{x^2 + 1}{x - 1}$

7. $f(x) = \dfrac{x}{\sqrt{1 - x}}$

8. $f(x) = \dfrac{x^3}{x^4 + 1}$

9. $f(x) = \sin^2 x$

10. $f(x) = x^{1/3}$

11. $f(x) = x^3/3 - x^2/2 - 2x$

12. $f(x) = x^3 - 6x^2 + 12x + 9$

13. $f(x) = 3x^4 - 16x^3 + 18x^2 + 2$

14. $f(x) = x + \dfrac{1}{x}$

15. $f(x) = 2x^3 - 15x^2 + 6x + 4$

16. $f(x) = |x| + x$

17. $f(x) = (x - 1)^{2/3}$

18. $f(x) = \dfrac{(x - 1)^3}{(x + 1)^4}$

19. $f(x) = (x + 2)^3(x - 4)^3$

20. $f(x) = x + 2\sin x$

21. $f(x) = x^{1/5} + x$

22. $f(x) = x^2 + \dfrac{25x^2}{(x - 2)^2}$

23. $f(x) = \left(\dfrac{x + 8}{x}\right)\sqrt{x^2 + 100}$

24. $f(x) = \dfrac{1 + x + x^2 + x^3}{1 + x^3}$

25. $f(x) = x^{5/4} - x^{1/4}$

26. $f(x) = \dfrac{x^2}{x^2 - 4}$

27. $f(x) = \dfrac{x^2}{x^3 - 1}$

28. $f(x) = \dfrac{(2x-1)(x-8)}{(x-1)(x-4)}$

29. $f(x) = \sin^2 x \cos x$, $\quad 0 \le x \le 2\pi$

30. $f(x) = \sin x + \cos x$

31. $f(x) = 2\csc x - \cot x$, $\quad 0 < x < \pi/2$

32. $f(x) = \csc x + 8 \sec x$, $\quad 0 < x < \pi/2$

33. $f(x) = \dfrac{\tan x}{x}$

34. $f(x) = x + \sin^2 x$, $\quad 0 < x < 2\pi$

35. $f(x) = e^{1/x}$

36. $f(x) = x \ln x$

37. $f(x) = x^2 \ln x$

38. $f(x) = xe^{-2x}$

39. $f(x) = xe^{-x^2}$

40. $f(x) = x^2 e^{3x}$

41. Does the function $f(x) = x^2$ defined only for $0 \le x \le 1$ have a relative minimum of $f(0) = 0$ and a relative maximum of $f(1) = 1$?

42. If $f(x) = x^2$ is defined only for $0 \le x \le 1$, are $x = 0$ and $x = 1$ critical points?

In Exercises 43–50 determine whether the statement is true or false.

43. Points of discontinuity of a function are critical if, and only if, the function is defined at the point of discontinuity.

44. When a function is defined only on the interval $a \le x \le b$, the ends $x = a$ and $x = b$ must yield relative maxima or minima for the function.

45. A function is discontinuous at a point if, and only if, it has no derivative at the point.

46. A function can have a relative maximum and a relative minimum at the same point.

47. If the derivative of a function changes sign when passing through a point, the point must yield a relative extrema for the function.

48. If a function has two relative maxima, it must have a relative minimum between them.

49. It is possible for every point in the domain of a function to be critical.

50. On an interval of finite length, a nonconstant function can have only a finite number of critical points at which its derivative is equal to zero.

51. Verify that the function $f(x) = x^3 + \cos x$ has two critical points, one of which is $x = 0$. Find the other critical point correct to three decimal places.

52. (a) Use multiplication of ordinates to sketch a graph of the function $f(x) = e^{-x} \sin x$, $x \ge 0$.

   (b) For what values of x does $f(x)$ have relative extrema? Does your graph in (a) agree?

In Exercises 53–56 find all critical points of the function correct to four decimals. Determine whether each critical point yields a relative maximum or a relative minimum.

53. $f(x) = x^4 + 6x^2 + 4x + 1$

54. $f(x) = x^4 - 10x^2 - 4x + 5$

55. $f(x) = x^3 - 2\cos x$

56. $f(x) = \dfrac{x^2 - 4}{(x^2 - 5x + 4)^2}$

57. The distance from the origin to any point (x, y) on the curve $C: y = f(x)$ is given by $D = \sqrt{x^2 + y^2}$. Show that if $P(x_0, y_0)$ is a point on C for which D has a relative extremum, then the line OP is perpendicular to the tangent line to C at P. Assume that $f(x)$ is differentiable, and that the curve does not pass through the origin.

58. An experiment is performed n times, resulting in n measurements $x_1, x_2, \ldots, x_n$ of the variable x. To obtain the best estimate of the value of x, we choose that value $\overline{x}$ which makes the sum of the squares of the differences between the x_i and $\overline{x}$ as small as possible; i.e., we choose $\overline{x}$ to minimize

$$S(x) = (x - x_1)^2 + (x - x_2)^2 + \cdots + (x - x_n)^2.$$

Show that $\overline{x}$ is the ordinary arithmetic mean of the x_i:

$$\overline{x} = \frac{1}{n}(x_1 + x_2 + \cdots + x_n).$$

59. A 200-gram mass is attached to a spring with elasticity constant 20 N/m (newtons per metre) in Figure 4.25.

FIGURE 4.25

If the mass is pulled 10 cm below the position where it would hang motionless ($y = 0$), and given an upward velocity of 2 m/s, then its position as a function of time t (in seconds) is given by

$$y(t) = -\frac{1}{10}\cos(10t) + \frac{1}{5}\sin(10t) \text{ m}.$$

Find the maximum distance from $y=0$ attained by the mass.

60. An inductor L, a resistor R, and a capacitor C are connected in series with a generator, producing an oscillatory voltage $E = E_0 \cos\omega t$, for $t \geq 0$ (Figure 4.26).

FIGURE 4.26

If L, C, R, E_0, and ω are all constants, the steady-state current I in the circuit is given by

$$I = \frac{E_0}{Z}\cos(\omega t - \phi),$$

where

$$Z = \sqrt{R^2 + \left(\omega L - \frac{1}{\omega C}\right)^2} \quad \text{and} \quad \tan\phi = \frac{\omega L - 1/(\omega C)}{R}.$$

Find the value of ω that makes the amplitude E_0/Z of the current a maximum. What is the maximum amplitude?

61. A landing approach is to be shaped generally as shown in Figure 4.27.

FIGURE 4.27

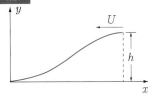

The following conditions are imposed on the approach pattern:

(a) The altitude must be h when descent commences;
(b) Touchdown must occur at $x = 0$;
(c) A constant horizontal speed U must be maintained throughout;
(d) At no time must vertical acceleration in absolute value exceed some fixed positive constant k.

Find a cubic polynomial satisfying these requirements.

62. The equation $(x^2 + y^2 + x)^2 = x^2 + y^2$ describes a *cardioid* (Figure 4.28). Find the maximum y-coordinate for points on the curve.

FIGURE 4.28

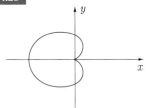

63. The equation $(x^2 + y^2)^2 = x^2 y$ describes a *bifolium* (Figure 4.29). Find the points on the curve farthest from the origin.

FIGURE 4.29

SECTION 4.4

Concavity and Points of Inflection

In the first three sections of this chapter we concentrated on the first derivative of a function. In this section we turn our attention to the second derivative.

Consider the function in Figure 4.30. If we draw tangent lines to the graph at the five points c_1, c_2, c_3, c_4, and c_5, it is clear that the slope is greater at c_5 than it is at c_4, greater at c_4 than at c_3, and so on. In fact, given any two points x_1 and x_2 in the interval $a < x < b$, where $x_2 > x_1$, the slope at x_2 is greater than at x_1,

$$f'(x_2) > f'(x_1).$$

What we are saying is that the function $f'(x)$ is increasing on the interval $a < x < b$.

FIGURE 4.30

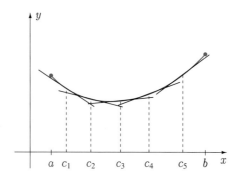

The first derivative of the function $f(x)$ in Figure 4.31 is decreasing on the interval $b < x < c$. In Figure 4.32, we have pieced together the functions in Figures 4.30 and 4.31 to form a function which has $f'(x)$ increasing on $a < x \leq b$ and decreasing on $b \leq x < c$. Since the left-hand derivative $f'_-(b)$ in Figure 4.30 and the right-hand derivative $f'_+(b)$ in Figure 4.31 are the same, it follows that $f'(b)$ exists in Figure 4.32, and that is why b is included in the intervals in which $f'(x)$ is increasing and decreasing. We give names to intervals on which the first derivative of a function is increasing and decreasing, and points which separate such intervals in the following definition.

FIGURE 4.31

FIGURE 4.32

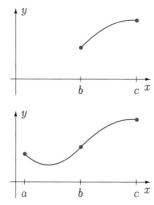

Definition 4.5

The graph of a function $f(x)$ is said to be **concave upward** on an interval I if $f'(x)$ is increasing on I, and **concave downward** on I if $f'(x)$ is decreasing on I. Points on a graph that separate intervals of opposite concavity are called **points of inflection**.

The graph in Figure 4.33 is concave upward on the intervals

$$b \leq x \leq c, \qquad d \leq x < e, \qquad e < x \leq f,$$

and concave downward for

$$a < x \leq b, \qquad c \leq x \leq d, \qquad f \leq x < g, \qquad g < x < h.$$

The points $x = e$ and $x = g$ are not included in these intervals because $f'(x)$ is not defined at these points. Points on the curve corresponding to $x = b$, c, d, and f are points of inflection; those at $x = e$ and g are not. They do not separate intervals of opposite concavity.

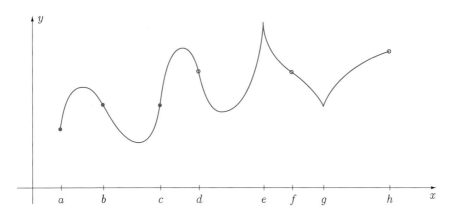

FIGURE 4.33

The tangent line has a peculiar property at a point of inflection. To see it, draw the tangent line at the four points of inflection in Figure 4.33. The tangent line actually crosses from one side of the curve to the other (something many readers think a tangent line should not do).

The critical points $x = a$ and $x = c$ in Figure 4.19, where $f'(x) = 0$, and the critical points $x = e$ and $x = g$ where $f'(x)$ does not exist, do not yield relative extrema for $f(x)$. Critical points of these types are illustrated again in Figures 4.34 and 4.35. The tangent line at $x = a$ is horizontal in Figure 4.34 ($f'(a) = 0$). Since the graph is concave downward to the left of $x = a$ and concave upward to the right, the point $(a, f(a))$ is a point of inflection. We call it a **horizontal point of inflection**. The point $(a, f(a))$ in Figure 4.35 is also a point of inflection. As the tangent line at this point is vertical, we call it a **vertical point of inflection**.

FIGURE 4.34

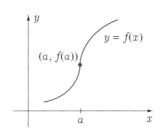

FIGURE 4.35

In Section 4.2 we stated that a function is increasing on an interval if its first derivative is greater than or equal to zero on that interval, and equal to zero at only a finite number of points; it is decreasing if its first derivative is less than or equal to zero. Furthermore, points that separate intervals on which the function is increasing and decreasing are points at which the first derivative is either equal to zero (and changes sign) or does not exist. We can use these ideas to locate analogously points of inflection and intervals on which the graph of a function is concave upward and concave downward.

Test for Concavity

(i) The graph of a function $f(x)$ is concave upward on an interval I if on I

$$f''(x) \geq 0 \qquad (4.9)$$

and is equal to zero at only a finite number of points.

(ii) The graph of a function $f(x)$ is concave downward on an interval I if on I

$$f''(x) \leq 0 \qquad (4.10)$$

and is equal to zero at only a finite number of points.

(iii) Points of inflection occur where

$$f''(x) = 0 \qquad (4.11)$$

and $f''(x)$ changes sign, or perhaps also where $f''(x)$ does not exist.

Consequently, just as the first derivative of a function is used to determine its relative extrema and intervals on which it is increasing and decreasing, the second derivative is used to find points of inflection and intervals on which the function is concave upward and concave downward. This parallelism between the first and second derivative is reiterated in the following table.

First Derivative

1. A relative extremum occurs at a point where $f'(x) = 0$ and $f'(x)$ changes sign as x increases through the point.

2. A relative extremum may also occur at a point where $f'(x)$ does not exist.

3. $f(x)$ is increasing on an interval if $f'(x) \geq 0$; $f(x)$ is decreasing on an interval if $f'(x) \leq 0$. $f'(x)$ may be equal to zero at only a finite number of points in either case.

Second Derivative

1. A point of inflection occurs at a point where $f''(x) = 0$ and $f''(x)$ changes sign as x increases through the point.

2. A point of inflection may also occur at a point where $f''(x)$ does not exist.

3. $f(x)$ is concave upward on an interval if $f''(x) \geq 0$; $f(x)$ is concave downward on an interval if $f''(x) \leq 0$. $f''(x)$ may be equal to zero at only a finite number of points in either case.

These discussions lead to what is often called the *second-derivative test* for determining whether a critical point at which $f'(x) = 0$ yields a relative maximum or a relative minimum.

Second-derivative Test for Relative Extrema

(i) If $f'(x_0) = 0$ and $f''(x_0) > 0$, then $x = x_0$ yields a relative minimum for $f(x)$.

(ii) If $f'(x_0) = 0$ and $f''(x_0) < 0$, then $x = x_0$ yields a relative maximum for $f(x)$.

(iii) If $f'(x_0) = 0$ and $f''(x_0) = 0$, then no conclusion can be made.

In (i), $f'(x_0) = 0$ implies that $x = x_0$ is a critical point with a horizontal tangent line; $f''(x_0) > 0$ implies that around $x = x_0$ the curve is concave upward, and $x = x_0$ therefore yields a relative minimum. Similarly, $f''(x_0) < 0$ at a critical point in (ii) implies that the curve is concave downward and therefore has a relative maximum.

To show that the test fails in case (iii), consider the three functions $f(x) = x^4$, $f(x) = -x^4$, and $f(x) = x^3$ shown in Figures 4.36, 4.37, and 4.38 respectively. For each function $f'(0) = f''(0) = 0$; yet the first has a relative minimum, the second a relative maximum, and the third a horizontal point of inflection at $x = 0$.

FIGURE 4.36

FIGURE 4.37

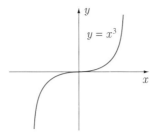

FIGURE 4.38

The second-derivative test very quickly determines the nature of a critical point x_0 when $f'(x_0) = 0$ and $f''(x_0) \neq 0$. For instance, the second derivative of the function $f(x) = 2x^3 - 9x^2 - 24x + 6$ in Example 4.5(a) of Section 4.3 is $f''(x) = 12x - 18$. Since $f''(-1) = -30$, the critical point $x = -1$ yields a relative maximum, and because $f''(4) = 30$, $x = 4$ gives a relative minimum.

A more refined test to determine the nature of a critical point x_0 when $f'(x_0) = f''(x_0) = 0$ is discussed in Exercise 34.

EXAMPLE 4.6

For each of the following functions, find all points of inflection and intervals on which the graph of the function is concave upward and concave downward.

(a) $f(x) = \dfrac{2}{3}x^3 - 6x^2 + 16x + 1$ (b) $f(x) = x^{6/5} + x^{1/5}$

SOLUTION (a) For points of inflection we first solve

$$0 = f''(x) = \frac{d}{dx}(2x^2 - 12x + 16) = 4x - 12,$$

and obtain $x = 3$. Since $f''(x)$ changes sign as x passes through 3, $x = 3$ gives the point of inflection $(3, 13)$. The graph $y = f(x)$ is concave downward for $x \leq 3$ (since $f''(x) \leq 0$) and concave upward for $x \geq 3$ ($f''(x) \geq 0$). Since $f''(x)$ is defined for all x, there can be no other points of inflection.

(b) For points of inflection we first solve

$$0 = f''(x) = \frac{d}{dx}\left(\frac{6}{5}x^{1/5} + \frac{1}{5}x^{-4/5}\right) = \frac{6}{25}x^{-4/5} - \frac{4}{25}x^{-9/5} = \frac{6x - 4}{25x^{9/5}},$$

and find $x = 2/3$. Since $f''(x)$ changes sign as x passes through $2/3$ (Figure 4.39), $x = 2/3$ gives a point of inflection with $f(2/3) \approx 1.5$.

We also note that $f'(x)$ and $f''(x)$ do not exist at $x = 0$, although $f(0)$ does ($= 0$). Since $f''(x)$ changes sign as x passes through 0, $(0, 0)$ must also be a point of inflection. Finally, the sign diagram in Figure 4.39 indicates that the graph $y = f(x)$ is concave upward for $x < 0$ and $x \geq 2/3$ and concave downward for $0 < x \leq 2/3$.

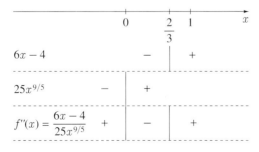

FIGURE 4.39

In Section 4.6 we will see the importance of critical points in the solution of applied maxima-minima problems. In some problems the significance of inflection points is equally important. As an example, consider how an infectious disease spreads through a population of susceptible individuals. To simplify the problem, we assume that all individuals are equally susceptible and that the disease is introduced to the population by one infective. Contact between individuals spreads the disease slowly at first, then more rapidly as more individuals become infected, and finally, when most of the population has become infected, the spread tapers off. To describe this process in more mathematical terms, we denote by N the total number of individuals in the population, and by $x(t)$ the number of infectives at any given time t. Since the disease is spread through contacts between infectives and noninfectives, the rate at which individuals become infected depends on both the number of infectives x and the number of noninfectives $N - x$ at any given time. In practice, the rate of infection, represented by dx/dt, is often assumed proportional to the product of x and $N - x$:

$$\frac{dx}{dt} = kx(N - x),$$

where k is a constant. Given that this equation describes the rate at which the disease spreads, it is possible to determine the number of infectives $x(t)$ present at any given time. We cannot solve this problem completely at this point, but we can obtain some information on its graph.

First, we note that dx/dt is always nonnegative so that the graph of $x(t)$ is increasing. Second, since points of inflection on the graph are defined by a vanishing second derivative, we set

$$0 = x''(t) = \frac{d}{dt}[kNx - kx^2] = kN\frac{dx}{dt} - 2kx\frac{dx}{dt} = k(N - 2x)\frac{dx}{dt}.$$

But $k \neq 0$, and $dx/dt = 0$ only when the population is either totally infected or not at all infected; therefore, a point of inflection occurs only when $N - 2x = 0$ or $x = N/2$, that is, when the population is half infected. Since $x''(t) > 0$ when $x < N/2$ and $x''(t) < 0$ when $x > N/2$, the graph of $x(t)$ is concave upward for $x < N/2$ and concave downward for $x > N/2$. These are shown in Figure 4.40; we have chosen a time convention of $t = 0$ when the first infective is introduced. In other words, we have completely determined the shape of the graph of $x(t)$. What remains is the insertion of a time scale and this we shall do in Section 9.7.

FIGURE 4.40

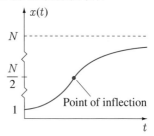

At this point the significance of the point of inflection at $x = N/2$ is clear. Since the slope of the curve (dx/dt) is increasing for $x < N/2$ and decreasing for $x > N/2$, it follows that when $x = N/2$, the rate of spread of the disease is at its peak.

EXERCISES 4.4

In Exercises 1–14 determine where the graph of the function is concave upward, is concave downward, and has points of inflection.

1. $f(x) = x^3 - 3x^2 - 3x + 5$
2. $f(x) = 3x^4 + 4x^3 - 24x + 2$
3. $f(x) = x^2 + \dfrac{1}{x^2}$
4. $f(x) = \dfrac{x^2 + 4}{x^2 - 1}$
5. $f(x) = x + \cos x$, $|x| < 2\pi$
6. $f(x) = x^2 - \sin x$
7. $f(x) = x^2 - 2\sin x$
8. $f(x) = x^2 - 4\sin x$, $|x| < 2\pi$
9. $f(x) = x \ln x$
10. $f(x) = x^2 \ln x$
11. $f(x) = e^{1/x}$
12. $f(x) = xe^{-2x}$
13. $f(x) = x^2 e^{3x}$
14. $f(x) = x^2 - e^{-x}$

In Exercises 15–22 use the second-derivative test to determine whether critical points where $f'(x) = 0$ yield relative maxima or relative minima.

15. $f(x) = x^3 - 3x^2 - 3x + 5$
16. $f(x) = x + \dfrac{1}{x}$
17. $f(x) = 3x^4 - 16x^3 + 18x^2 + 2$
18. $f(x) = x^{5/4} - x^{1/4}$
19. $f(x) = x \ln x$
20. $f(x) = x^2 \ln x$
21. $f(x) = xe^{2x}$
22. $f(x) = x^2 e^{-2x}$

23. Is the graph of a function concave upward, concave downward, both, or neither on an interval I if on I its second derivative is always equal to zero?

24. Prove that the curve $y = 2\cos x$ passes through all points of inflection of the curve $y = x \sin x$.

25. Show that every cubic polynomial has exactly one point of inflection on its graph.

Sketch the graph of a function $f(x)$ that is defined everywhere on the interval $I : a \leq x \leq b$, and that possesses the properties in Exercises 26–31. In each case assume that $a < x_0 < b$.

26. $f(x)$ is increasing on I and discontinuous at x_0.

27. $f''(x) \geq 0$ on I and $f(x)$ is not concave upward on I.

28. $f(x)$ is decreasing on I, $f(x)$ is continuous at x_0, but $f'(x)$ is discontinuous at x_0.

29. $f(x)$ is increasing on $a \leq x \leq x_0$, increasing on $x_0 < x \leq b$, but not increasing on I.

30. $f(x)$ is concave downward on $a \leq x < x_0$, concave downward on $x_0 < x \leq b$, but not concave downward on I.

31. $f(x)$ is increasing on $a \leq x \leq x_0$, concave downward on $a \leq x < x_0$, decreasing on $x_0 \leq x \leq b$, and concave upward on $x_0 < x \leq b$.

32. If $f(x)$ and $g(x)$ are twice differentiable and concave upward on an interval I, is $f(x)g(x)$ concave upward on I?

33. If the functions in Exercise 32 are also increasing on I, is $f(x)g(x)$ concave upward on I?

34. The second-derivative test fails to classify a critical point x_0 of a function $f(x)$ as yielding a relative maximum, a relative minimum, or a horizontal point of inflection if $f'(x_0) = f''(x_0) = 0$. The following test can be used in such cases. Suppose $f(x)$ has derivatives of all orders in an open interval around x_0, and the first n derivatives all vanish at x_0, but the $(n+1)^{\text{th}}$ derivative at x_0 is not zero. If we denote the nth derivative of $f(x)$ at x_0 by $f^{(n)}(x_0)$, these conditions are

$$0 = f'(x_0) = f''(x_0) = \cdots = f^{(n)}(x_0), \quad f^{(n+1)}(x_0) \neq 0.$$

Then:

(i) if n is even, $f(x)$ has a horizontal point of inflection at x_0;

(ii) if n is odd and $f^{(n+1)}(x_0) > 0$, $f(x)$ has a relative minimum at x_0;

(ii) if n is odd and $f^{(n+1)}(x_0) < 0$, $f(x)$ has a relative maximum at x_0.

A proof of this result requires the use of material from Chapter 11 and is therefore delayed until that time (see Exercise 10 in Section 11.9). Note that the second-derivative test is the special case when $n = 1$. Use this test to determine whether critical points of the following functions yield relative maxima, relative minima, or horizontal points of inflection.

(a) $f(x) = (x^2 - 1)^3$ (b) $f(x) = x^2\sqrt{1-x}$

35. Show that the points of inflection of $f(x) = (k - x)/(x^2 + k^2)$, where k is a constant, all lie on a straight line.

36. Prove that if a cubic polynomial has both a relative maximum and a relative minimum, then the point of inflection between these extrema is the midpoint of the line segment joining them.

37. (a) In Exercise 22 of Section 3.5, the derivative of the function

$$f(x) = \begin{cases} x^2 \sin\left(\dfrac{1}{x}\right), & x \neq 0 \\ 0 & x = 0 \end{cases}$$

is zero at $x = 0$; that is, $f'(0) = 0$. It follows that the graph of the function has a horizontal tangent line at $(0,0)$. Sketch the graph and the tangent line.

(b) Show that $f''(0)$ does not exist. Hint: See Exercise 25 in Section 3.3.

(c) Is the point $(0,0)$ a relative maximum, a relative minimum, a horizontal point of inflection, or none of these?

SECTION 4.5

Curve Sketching

Some years ago, one of the major applications of differentiation was to draw graphs of functions. With computers and advanced electronic calculators that provide visual displays of these graphs, many users of mathematics feel that this application is now obsolete. Such is not the case, and for a number of reasons. First, plotting programs in calculators and computers have built in smoothing routines which eliminate anomalies on curves. This may not be appropriate. For example, a corner on the graph of a function represents a point where the function does not have a derivative. A plotting program which "rounds" the corner is misleading; it removes the evidence that the function has a point where its derivative does not exist. Second, calculators may suggest, but do not offer conclusive evidence, that a graph has vertical, horizontal, or "slanted" asymptotes, and discontinuities are not handled well by calculators. Finally, the electronic display may suggest that a point on a curve is a horizontal or vertical point of inflection, but cannot be a conclusive argument.

What we suggest is that although the use of calculus in curve sketching may not require the stress that it once enjoyed, there are definitely situations in which it is essential.

Curve sketching, as a topic, began in Sections 1.3 and 1.4 when we used the basic idea of a table of values. Following this, we illustrated in Section 1.5 that addition and multiplication of ordinates often yield rough but very quick sketches. We also saw that symmetries of a curve about the x- and y-axes can be helpful in many cases. We used limits in Section 2.4 to discover the behaviour of the graph of a function near a point of discontinuity or for very large positive and negative values of the independent variable. Each of these techniques has its value and none should be neglected.

In this chapter we have defined critical points, relative extrema, increasing and decreasing functions, concavity, and points of inflection, and each of these concepts can add information to a sketch. The natural question to ask at this point is: "Given a curve to sketch, which of the many tools that are now available should we use?" There is no simple answer to this question, except to say that we should use those tools that give us a suitable sketch with the least amount of effort. But a sketch suitable for one problem may not be suitable for another. For some problems, a very rough sketch might be sufficient, in which case we might need to use only addition or multiplication of ordinates; in other problems, a fairly detailed graph might be appropriate, and we would need further information such as relative extrema and points of inflection. In this section we illustrate how further detail can be added to a sketch by using more and more tools on a specific example. At the same time, we see how the various techniques complement each other.

EXAMPLE 4.7

Sketch a graph of the function $f(x) = (x-4)/x^2$.

SOLUTION With an x-intercept equal to 4, and the limits

$$\lim_{x \to 0^-} f(x) = -\infty, \qquad \lim_{x \to 0^+} f(x) = -\infty, \qquad \lim_{x \to -\infty} f(x) = 0^-, \qquad \lim_{x \to \infty} f(x) = 0^+,$$

we begin our sketch as shown in Figure 4.41(a). This information would lead us to suspect that the graph should be completed as shown in Figure 4.41(b). To verify this we find critical points for $f(x)$:

$$0 = f'(x) = \frac{x^2(1) - (x-4)(2x)}{x^4} = \frac{x(8-x)}{x^4} = \frac{8-x}{x^3}.$$

Clearly, $x = 8$ is the only critical point of the function, and the first derivative test indicates that it gives a relative maximum. Our sketch is as shown in Figure 4.41(c).

Figure 4.41(c) makes it clear that there is a point of inflection to the right of $x = 8$, which we could pinpoint with $f''(x)$:

$$f''(x) = \frac{x^3(-1) - (8-x)(3x^2)}{x^6} = \frac{2(x-12)}{x^4}.$$

Since $f''(12) = 0$ and $f''(x)$ changes sign as x passes through 12, $(12, 1/18)$ is the point of inflection. Figure 4.41(d) contains the final graph.

FIGURE 4.41

(a)

(b)

(c)

(d)
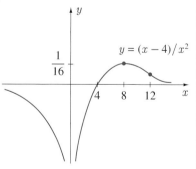

EXAMPLE 4.8

Sketch a graph of the function $f(x) = x + \sin x$.

SOLUTION We might try addition of ordinates for the curves $y = x$ and $y = \sin x$ in Figure 4.42(a). This could result in the sketches in Figures 4.42(b),(c), or (d). To decide which of these curves is correct, we consider the critical points of $f(x)$, defined by

$$0 = f'(x) = 1 + \cos x.$$

The solutions of this equation are $x = (2n+1)\pi$, where n is an integer. Since $f'(x)$ does not change sign as x passes through these points, they cannot yield relative maxima or minima. They must give horizontal points of inflection, and the graph in Figure 4.42(c) is correct.

FIGURE 4.42

(a)

(b)

(c)

(d)
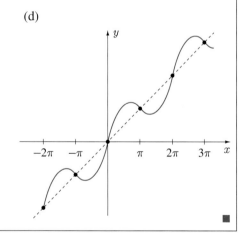

EXAMPLE 4.9

Sketch a graph of the function $f(x) = x^{5/3} - x^{2/3}$.

SOLUTION We first note that the three points $(0,0)$, $(1,0)$, and $(-1,-2)$ are on the graph. Next we add the facts that

$$\lim_{x \to -\infty} f(x) = -\infty \quad \text{and} \quad \lim_{x \to \infty} f(x) = \infty,$$

as shown in Figure 4.43(a). We cannot be sure that the concavity is as indicated, but this will be verified shortly. To find critical points of $f(x)$, we first solve

$$0 = f'(x) = \frac{5}{3}x^{2/3} - \frac{2}{3}x^{-1/3} = \frac{5x-2}{3x^{1/3}}.$$

The only solution is $x = 2/5$, and because $f'(x)$ changes from negative to positive as x increases through $2/5$, we have a relative minimum of $f(2/5) = (2/5)^{5/3} - (2/5)^{2/3} = -0.33$. Since $f'(x)$ does not exist at $x = 0$, this is also a critical point, and we calculate that

$$\lim_{x \to 0^-} f'(x) = \infty \quad \text{and} \quad \lim_{x \to 0^+} f'(x) = -\infty.$$

This information, along with the relative minimum, is shown in Figure 4.43(b). We now join these parts smoothly to produce the sketch in Figure 4.43(c). To verify that the concavity is as indicated we calculate

$$f''(x) = \left(\frac{5}{3}\right)\left(\frac{2}{3}\right)x^{-1/3} - \left(\frac{2}{3}\right)\left(-\frac{1}{3}\right)x^{-4/3} = \frac{2(5x+1)}{9x^{4/3}}.$$

Clearly there is only one point of inflection, namely when $x = -1/5$. The sign of $f''(x)$ on each side of this point indicates that the concavity is indeed as shown in Figure 4.43(c), and the final sketch is in Figure 4.43(d).

FIGURE 4.43

(a)

(b)

(c)

(d)

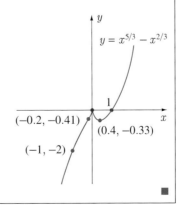

EXAMPLE 4.10

Sketch a graph of the function $f(x) = (x+2)^2/x^2$.

SOLUTION We write

$$f(x) = \frac{(x+2)^2}{x^2} = \left(\frac{x+2}{x}\right)^2 = \left(1 + \frac{2}{x}\right)^2.$$

A sketch of $y = 2/x$ is shown in Figure 4.44(a), and $y = 1 + 2/x$ requires only a shift of 1 unit in the y-direction. If we now square all ordinates of the curve $y = 1 + 2/x$, we obtain the sketch of $y = (1 + 2/x)^2$ in Figure 4.44(b). The only information that could be added is the exact position of the point of inflection to the left of $x = -2$.

FIGURE 4.44

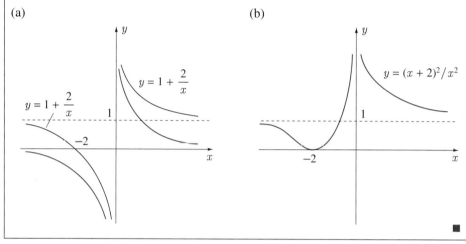

EXAMPLE 4.11

Sketch a graph of the function $f(x) = (x^2 + 1)/(x-1)$.

SOLUTION If we divide $x^2 + 1$ by $x - 1$, we find

$$f(x) = (x+1) + \frac{2}{x-1}.$$

Sketches of $y = x+1$ and $y = 2/(x-1)$ are shown in Figure 4.45(a), and by addition of ordinates, we obtain the graph in Figure 4.45(b).

To add the positions of the relative extrema we find critical points. Since

$$f'(x) = \frac{(x-1)(2x) - (x^2+1)(1)}{(x-1)^2} = \frac{x^2 - 2x - 1}{(x-1)^2},$$

we set

$$x^2 - 2x - 1 = 0.$$

Critical points are therefore

$$x = \frac{2 \pm \sqrt{4+4}}{2} = 1 \pm \sqrt{2}.$$

Our sketch in Figure 4.45(b) clearly indicates that $x = 1 + \sqrt{2}$ gives a relative minimum and $x = 1 - \sqrt{2}$ a relative maximum. This could also be verified by the first-derivative test. Would you consider the second-derivative test? The final graph is shown in Figure 4.45(c). As $x \to \pm\infty$, the graph gets arbitrarily close to the line $y = x + 1$. We call this line a **slanted asymptote**.

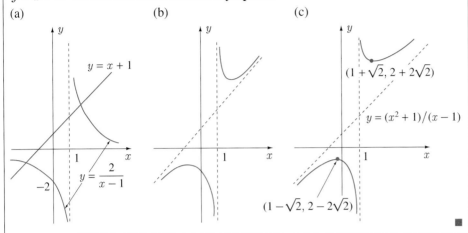

FIGURE 4.45

For rational functions, such as that in Example 4.11, slanted asymptotes occur when the numerator has degree exactly one higher than the denominator.

EXERCISES 4.5

In Exercises 1–10 find all relative maxima and minima for the function and points of inflection for its graph. Use this information to sketch a graph of the function.

1. $f(x) = \dfrac{x^3}{3} - \dfrac{x^2}{2} - 2x$

2. $f(x) = x^3 - 6x^2 + 12x + 9$

3. $f(x) = 3x^4 - 16x^3 + 18x^2 + 2$

4. $f(x) = x - 3x^{1/3}$

5. $f(x) = 2x^3 - 15x^2 + 6x + 4$

6. $f(x) = \dfrac{x}{x^2 + 4}$

7. $f(x) = (x - 2)^3 (x + 2)$

8. $f(x) = x^{2/3}(8 - x)$

9. $f(x) = \dfrac{(x + 2)^2}{x^3}$

10. $f(x) = 2x^{3/2} - 9x + 12x^{1/2}$

In Exercises 11–20 sketch a graph of the function.

11. $f(x) = \dfrac{x^3 + 16}{x}$

12. $f(x) = \dfrac{x^2 + x + 1}{x}$

13. $f(x) = x + \dfrac{1}{x}$

14. $f(x) = \dfrac{x^3}{x^2 - 4}$

15. $f(x) = \dfrac{2x^2}{x^2 - 8x + 12}$

16. $f(x) = \dfrac{x^2 + 1}{x^2 - 1}$

17. $f(x) = \dfrac{(x - 1)^3}{(x + 1)^4}$

18. $f(x) = (x - 1)^{2/3}$

19. $f(x) = (x + 2)^3 (x - 4)^3$

20. $f(x) = x + 2 \sin x$

21. A voltage source with constant electromotive force E and constant internal resistance r maintains a current I through a circuit with resistance R (Figure 4.46), where

$$E = I(r + R).$$

The power (work per unit time) necessary to maintain this current in r and R is given by

$$P = I^2(r + R) = I^2 r + I^2 R.$$

(a) If we define $P_R = I^2 R$ and $P_r = I^2 r$ as the power dissipated in R and r respectively, sketch P_R and P_r as functions of R.

(b) Sketch a graph of $P(R) = P_R(R) + P_r(R)$.

FIGURE 4.46

22. The rate of photosynthesis P in a leaf depends on the intensity of light I on the leaf according to

$$P = \frac{MI}{I+K} - R,$$

where $M > R$ and K are all positive constants. Sketch a graph of this function.

In Exercises 23–34 sketch a graph of the function.

23. $f(x) = x/(x^2 + 3)$
24. $f(x) = x^4 + 10x^3 + 6x^2 - 64x + 5$
25. $f(x) = x^4 - 2x^3 + 2x$
26. $f(x) = \dfrac{x^2 - 8}{x - 5}$
27. $f(x) = \dfrac{x^2}{x^2 - 4}$
28. $f(x) = x^{5/4} - x^{1/4}$
29. $f(x) = |x^2 - 9| + 2$
30. $f(x) = \dfrac{(2x - 1)(x - 8)}{(x - 1)(x - 4)}$
31. $f(x) = \sin^2 x \cos x$, $0 \le x \le 2\pi$
32. $f(x) = \sin x + \cos x$
33. $f(x) = \dfrac{x}{\sqrt{1-x}}$
34. $f(x) = \left(\dfrac{x+8}{x}\right)\sqrt{x^2 + 100}$

35. (a) A company produces x kilograms of a commodity per day at a total cost of

$$C(x) = \frac{x^2}{300}\left(\frac{x + 100}{x + 300}\right) + 60, \quad 1 \le x \le 200.$$

Show that the cost graph is always concave upward. Sketch the graph.

(b) The average production cost per kilogram when x kilograms are produced is given by $c(x) = C(x)/x$. Sketch a graph of this function.

(c) Show that the output at which the average cost is least satisfies the equation

$$(x + 300)^2(x^2 - 18\,000) - 60\,000\,x^2 = 0.$$

Do this in two ways:
(i) by finding critical points for $c(x)$;
(ii) by noting that the average cost is the slope of the line joining a point $(x, C(x))$ to the origin; hence minimum average cost occurs when the tangent line to the $C(x)$ graph passes through the origin.

36. The function

$$f(x) = \frac{1}{\sqrt{2\pi}\sigma} e^{-(x-\mu)^2/(2\sigma^2)}$$

where $\sigma > 0$ and μ are constants is called the *normal probability density function*. Sketch its graph.

37. When a drug is injected into the blood at time $t = 0$, it is sometimes assumed that the concentration of the drug thereafter is given by the function

$$f(t) = k\left(e^{-at} - e^{-bt}\right)$$

where k, a, and b are positive constants ($b > a$). Sketch a graph of this function indicating any relative extrema and points of inflection.

38. Sketch a graph of the function

$$f(x) = \frac{1 + x + x^2 + x^3}{1 + x^3}.$$

SECTION 4.6

Absolute Maxima and Minima

Seldom in applications of maxima and minima theory do we hear questions such as: What are the relative maxima of this quantity, or what are the relative minima of that quantity? More likely it is: What is the biggest of these, the smallest of those, the best way to do this, the cheapest way to do that? In other words, when a function which represents a measureable quantity is to be maximized (or minimized), it is not relative maxima (or minima) that are required; a different kind of extrema is involved—considering all points in the domain of the function, what is the largest (or smallest) value of the function?

Definition 4.6

The **absolute maximum** of a function $f(x)$ on an interval I is $f(x_0)$ if x_0 is in I and if for all x in I

$$f(x) \leq f(x_0); \qquad (4.12)$$

$f(x_0)$ is said to be the **absolute minimum** of $f(x)$ on I if for all x in I

$$f(x) \geq f(x_0). \qquad (4.13)$$

For the function in Figure 4.47, the absolute maximum of $f(x)$ on the interval $a \leq x \leq b$ is $f(c)$, and the absolute minimum is $f(d)$. For the function in Figure 4.48, the absolute maximum on $a \leq x \leq b$ is $f(b)$, and the absolute minimum is $f(a)$. For the function in Figure 4.49, the absolute maximum on $a \leq x \leq b$ is $f(b)$, and the absolute minimum is $f(c)$. Note that we speak of absolute maxima and minima (absolute extrema) of a function only on some specified interval; that is, we do not ask for the absolute maximum or minimum of a function $f(x)$ without specifying the interval I.

FIGURE 4.47

FIGURE 4.48

FIGURE 4.49

Each of the functions in Figures 4.47–4.49 is continuous on a closed interval $a \leq x \leq b$. The following theorem asserts that every such function has absolute extrema. For a proof of this result, the interested reader should consult books on advanced analysis.

Theorem 4.3

A function that is continuous on a closed interval must attain an absolute maximum and an absolute minimum on that interval.

The conditions of this theorem are sufficient to guarantee existence of absolute extrema; that is, if a function is continuous on a closed interval, then it *must* have absolute extrema on that interval. However, they are not necessary. If these conditions are not met, the function may or may not have absolute extrema. For instance, if the function in Figure 4.47 is confined to the open interval $a < x < b$, it still attains its absolute extrema at $x = c$ and $x = d$. On the other hand, the function in Figure 4.48 does not have absolute extrema on the open interval $a < x < b$. The function in Figure 4.50 is not continuous on the closed interval $a \leq x \leq d$; it has no absolute maximum on this interval, but it does have absolute minimum $f(a)$. This function is not continuous on $d \leq x \leq b$, but it has absolute maximum $f(d)$ and absolute minimum $f(b)$.

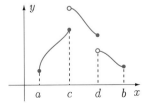
FIGURE 4.50

Absolute extrema of the functions in Figures 4.47–4.49 always occur either at a critical point of the function or at an end of the interval $a \leq x \leq b$. This result is always true for continuous functions.

Theorem 4.4

A function that is continuous on a closed interval always attains its absolute extrema either at critical points or at the ends of the interval.

Proof Let $f(x)$ be a continuous function on a closed interval $a \leq x \leq b$. To verify the theorem, we show that $f(x)$ cannot have an absolute extremum at any point x_0 in $a < x < b$ that is not critical. If $a < x_0 < b$ and x_0 is not critical, then x_0 cannot yield a relative extremum for $f(x)$ (see Theorem 4.2). Definitions 4.3 and 4.4 now imply that in every open interval around x_0, $f(x)$ takes on values that are larger than $f(x_0)$ and values that are smaller than $f(x_0)$. It follows that $f(x_0)$ cannot possibly be an absolute extremum for $f(x)$ on $a \leq x \leq b$.

Because of this theorem, there is a simple procedure to determine the absolute extrema of a *continuous* function $f(x)$ on a *closed* interval $a \leq x \leq b$.

Finding Absolute Extrema for a Continuous Function $f(x)$ on a Closed Interval $a \leq x \leq b$

(i) Find all critical points $x_1, x_2, \ldots, x_n$ of $f(x)$ in $a < x < b$.

(ii) Evaluate
$$f(a), \quad f(x_1), \quad f(x_2), \quad \ldots, \quad f(x_n), \quad f(b).$$

(iii) The absolute maximum of $f(x)$ on $a \leq x \leq b$ is the largest of the numbers in (ii); the absolute minimum is the smallest of these numbers.

Note that it is not necessary to classify the critical points of $f(x)$ as yielding relative maxima, relative minima, horizontal points of inflection, vertical points of inflection, corners, or none of these. We need only evaluate $f(x)$ at $x = a$, $x = b$, and at its critical points. The largest and smallest of these numbers must be the absolute extrema.

EXAMPLE 4.12

Find the absolute extrema for the function $f(x) = (x^2 + 2x + 21)/(x + 5)$ on the interval $0 \leq x \leq 6$.

SOLUTION For critical points of $f(x)$, we first solve

$$0 = f'(x) = \frac{(x+5)(2x+2) - (x^2 + 2x + 21)(1)}{(x+5)^2}$$
$$= \frac{x^2 + 10x - 11}{(x+5)^2} = \frac{(x+11)(x-1)}{(x+5)^2}.$$

The only critical point between $x = 0$ and $x = 6$ is $x = 1$. Since

$$f(0) = \frac{21}{5}, \quad f(1) = 4, \quad f(6) = \frac{69}{11},$$

the absolute maximum is $69/11$ and the absolute minimum is 4. ∎

In many applied problems the function is continuous, but its domain is not a finite interval. In this case we replace evaluation of the function at its end points with limits as the independent variable becomes very large negatively, positively, or both.

EXAMPLE 4.13

What are the absolute extrema of the function

$$f(x) = \frac{(2-x)^2}{x^2}, \quad x \geq \frac{1}{2}.$$

SOLUTION For critical points of $f(x)$, we first solve

$$0 = f'(x) = \frac{x^2(-2)(2-x) - (2-x)^2(2x)}{x^4} = \frac{4(x-2)}{x^3}.$$

The only solution is $x = 2$. Furthermore, $f'(x)$ exists for all $x > 1/2$. We now evaluate

$$f(1/2) = 9, \quad f(2) = 0, \quad \lim_{x \to \infty} f(x) = 1.$$

Maximum and minimum values of $f(x)$ are therefore 9 and 0 respectively. ∎

We now show how absolute extrema can be used to solve applied problems.

EXAMPLE 4.14

A rectangular field is to be fenced on three sides with 1000 m of fencing (the fourth side being a straight river's edge). Find the dimensions of the field in order that the area be as large as possible.

FIGURE 4.51

SOLUTION Since the area of the field is to be maximized, we first define a function representing this area. The area of a field of width w and length l (Figure 4.51) is

$$A = lw.$$

This function represents the area of a field with arbitrary length l and arbitrary width w. Our problem states that only 1000 m of fencing are available for the three sides; therefore, l and w must satisfy the equation

$$2w + l = 1000.$$

With this equation we can express A completely in terms of w (or l):

$$A(w) = w(1000 - 2w) = 1000w - 2w^2.$$

To maximize the area of the field, we must therefore maximize the function $A(w)$. But what are the values of w under consideration? Clearly, w cannot be negative, and in order to satisfy the restriction $2w + l = 1000$, w cannot exceed 500. The physical problem has now been modeled mathematically. Find the absolute maximum of the (continuous) function $A(w)$ on the (closed) interval $0 \leq w \leq 500$.

We first find the critical points of $A(w)$ by solving

$$0 = A'(w) = 1000 - 4w;$$

the only solution is $w = 250$. Since $A'(w)$ exists for all $0 \leq w \leq 500$, we evaluate $A(w)$ at $w = 250$ and the end points:

$$A(0) = 0, \qquad A(500) = 0, \qquad A(250) = 1000(250) - 2(250)^2 > 0.$$

We conclude that the largest possible area is obtained when the width of the field is 250 m and its length is 500 m. ∎

EXAMPLE 4.15

Pop cans to hold 300 mL are made in the shape of right-circular cylinders. Find dimensions of the can that minimize its surface area.

SOLUTION The surface area of the can (Figure 4.52(a)) consists of a circular top, an identical bottom, and a rectangular piece formed into the side of the can. The total area of these three pieces is

$$A = 2\pi r^2 + 2\pi r h,$$

where we measure r and h in centimetres (see Figure 4.52(b)). Since the can must hold 300 mL of pop, it follows that

$$\pi r^2 h = 300 \qquad \text{or} \qquad h = \frac{300}{\pi r^2}.$$

This equation can be used to express A completely in terms of r,

$$A(r) = 2\pi r^2 + 2\pi r \left(\frac{300}{\pi r^2}\right).$$

For what values of r is $A(r)$ defined? The radius of the can must be positive and therefore $r > 0$. How large can r be? In effect, it can be as large as desired. The height of the cylinder can always be chosen sufficiently small when r is large to satisfy the volume condition $\pi r^2 h = 300$.

FIGURE 4.52

The can may not be esthetically pleasing for large r and small h, but no mathematical difficulties occur in this situation. To solve the problem we must therefore maximize

$$A(r) = 2\pi r^2 + \frac{600}{r}, \qquad r > 0.$$

For critical points of $A(r)$, we solve

$$0 = A'(r) = 4\pi r - \frac{600}{r^2} \qquad \text{or} \qquad 4\pi r^3 - 600 = 0.$$

The only solution of this equation is $r = (150/\pi)^{1/3}$. Were $A(r)$ defined on a closed interval we would evaluate it at the critical point and the ends of the interval. With a continuous function on the interval $r > 0$, we replace evaluation at the ends of the interval with limits as r approaches the ends:

$$\lim_{r \to 0^+} A(r) = \infty, \qquad A\left((150/\pi)^{1/3}\right) > 0, \qquad \lim_{r \to \infty} A(r) = \infty.$$

Thus, minimum surface area occurs when the radius of the can is $(150/\pi)^{1/3}$ cm and its height, obtained from the fact that $h = 300/(\pi r^2)$, is $2(150/\pi)^{1/3}$ cm. ∎

On the basis of these two examples, we suggest the following steps in solving applied maxima-minima problems:

1. Sketch a diagram illustrating the situation.
2. Identify the quantity that is to be maximized or minimized, choose a letter to represent it, and find an expression for this quantity.
3. If necessary, use information in the problem to rewrite the expression in 2 as a function of only one variable.
4. Determine the domain of the function in 3.
5. Maximize or minimize the function in 3 on the domain in 4.
6. Interpret the maximum or minimum values in terms of the original problem.

Step 2 is crucial. Do not consider subsidiary information in the problem until the quantity to be maximized or minimized is clearly identified, labelled, and an expression found for it. Only then should other information be considered. For instance, in Example 4.14, the restriction $2w + l = 1000$ for the length of fencing available was not introduced until the expression $A = lw$ had been identified. Likewise in Example 4.15, volume condition $\pi r^2 h = 300$ was introduced after surface area $A = 2\pi r^2 + 2\pi rh$.

We now take a look at three additional examples.

EXAMPLE 4.16

A lighthouse is 6 km off shore and a cabin on the straight shoreline is 9 km from the point on the shore nearest the lighthouse. If a man rows at a rate of 3 km/h, and walks at a rate of 5 km/h, where should he beach his boat in order to get from the lighthouse to the cabin as quickly as possible? Repeat the problem in the case that the cabin is 4 km along the shoreline.

SOLUTION Figure 4.53 illustrates the path followed by the man when he beaches the boat x kilometres from the point on land closest to the lighthouse. His travel time t for this path is the sum of his time t_1 in the water and his time t_2 on land. Since velocities are constant, each of these times may be calculated by dividing distance by velocity; that is, $t_1 = \sqrt{x^2 + 36}/3$ and $t_2 = (9-x)/5$. To minimize t for some value of x between 0 and 9, we minimize the function

$$t(x) = t_1 + t_2 = \frac{\sqrt{x^2 + 36}}{3} + \frac{9 - x}{5}, \quad 0 \leq x \leq 9.$$

For critical points of $t(x)$ we solve

$$0 = t'(x) = \frac{x}{3\sqrt{x^2+36}} - \frac{1}{5} \quad \text{or} \quad 5x = 3\sqrt{x^2+36}.$$

Thus,
$$25x^2 = 9(x^2 + 36),$$

from which we accept only the positive solution $x = 9/2$. Since $t'(x)$ exists for all x, and

$$t(0) = \frac{\sqrt{36}}{3} + \frac{9}{5} = \frac{19}{5},$$
$$t(9) = \frac{\sqrt{81+36}}{3} = \sqrt{13},$$
$$t\left(\frac{9}{2}\right) = \frac{\sqrt{81/4+36}}{3} + \frac{9-9/2}{5} = \frac{17}{5} < \sqrt{13},$$

the boat should therefore be beached 4.5 km from the cabin.

When the cabin is 4 km along the shoreline, then

$$t(x) = \frac{\sqrt{x^2+36}}{3} + \frac{4-x}{5}, \quad 0 \leq x \leq 4.$$

Critical points are again given by the equation

$$\frac{x}{3\sqrt{x^2+36}} - \frac{1}{5} = 0,$$

but the solution $x = 9/2$ must be rejected since it does not fall in the interval $0 \leq x \leq 4$. Travel time must therefore be minimized by heading directly to shore or directly to the cabin. Since

$$t(0) = \frac{\sqrt{36}}{3} + \frac{4}{5} = \frac{14}{5},$$
$$t(4) = \frac{\sqrt{16+36}}{3} = \frac{2\sqrt{13}}{3} < \frac{14}{5},$$

the boat should head directly toward the cabin.

FIGURE 4.53

EXAMPLE 4.17

A wall of a building is to be braced by a beam that must pass over a parallel wall 5 metres high and 2 metres from the building (Figure 4.54). Find the length of the shortest beam that can be used.

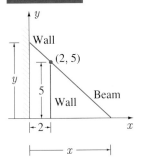

FIGURE 4.54

SOLUTION If we introduce coordinates as shown in Figure 4.54, we see that the shortest beam is the shortest of all line segments that stretch from points on the positive x-axis to points on the positive y-axis, and pass through the point $(2,5)$. The length of the line segment that has intercepts x and y is given by

$$D = \sqrt{x^2 + y^2}.$$

To express D in terms of only one variable, we must find an equation relating x and y. Since slopes of the line segments joining $(0, y)$ to $(x, 0)$ and $(2, 5)$ to $(x, 0)$ must be identical, we have

$$\frac{y - 0}{0 - x} = \frac{0 - 5}{x - 2} \quad \text{or} \quad y = \frac{5x}{x - 2}.$$

Thus,

$$D = \sqrt{x^2 + \frac{25 x^2}{(x - 2)^2}}.$$

Since D is a minimum when D^2 is a minimum, we consider minimizing

$$D^2 = f(x) = x^2 + \frac{25 x^2}{(x - 2)^2}, \quad x > 2.$$

For critical points of $f(x)$, we solve

$$0 = f'(x) = 2x + \frac{50x}{(x-2)^2} - \frac{50x^2}{(x-2)^3}$$

$$= \frac{2x[(x-2)^3 + 25(x-2) - 25x]}{(x-2)^3}$$

$$= \frac{2x}{(x-2)^3}[(x-2)^3 - 50].$$

Since x cannot be equal to zero, we set

$$(x - 2)^3 = 50,$$

and this implies that $x = 2 + 50^{1/3}$. Once again since $f(x)$ is defined on the infinite interval $x > 2$, we evaluate it at the critical point and take limits as $x \to 2^+$ and $x \to \infty$:

$$\lim_{x \to 2^+} f(x) = \infty, \quad f(2 + 50^{1/3}) > 0, \quad \lim_{x \to \infty} f(x) = \infty.$$

We conclude that D^2, and therefore D, must be minimized for $x = 2 + 50^{1/3}$, and the shortest beam has length

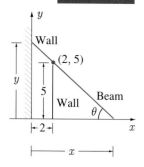

FIGURE 4.55

$$\sqrt{f(2+50^{1/3})} = (2+50^{1/3})\sqrt{1+\frac{25}{50^{2/3}}} = 9.6 \text{ m}.$$

We could also have solved this problem by introducing the angle θ in Figure 4.55. The length D of the beam can be expressed in terms of θ by

$$D = x \sec \theta.$$

Since $x - 2 = 5 \cot \theta$, it follows that

$$D = g(\theta) = \sec \theta (2 + 5 \cot \theta), \quad 0 < \theta < \pi/2.$$

For critical points of this function, we set

$$0 = g'(\theta) = \sec \theta \tan \theta (2 + 5 \cot \theta) + \sec \theta (-5 \csc^2 \theta).$$

When this equation is expressed in the form

$$\frac{\cos \theta}{\sin^2 \theta}(2 \tan^3 \theta - 5) = 0,$$

we see that the only critical point is given by

$$\tan \theta = \left(\frac{5}{2}\right)^{1/3}.$$

The angle between 0 and $\pi/2$ satisfying this equation is 0.9358 radians. Since

$$\lim_{\theta \to 0^+} g(\theta) = \infty, \quad g(0.9358) = 9.6, \quad \lim_{\theta \to \pi/2^-} g(\theta) = \infty,$$

it follows once again that the shortest beam has length 9.6 m. ∎

EXAMPLE 4.18

A printer contracts to print 200 000 copies of a membership card. It costs $10 per hour to run the press, and the press produces 1000 impressions per hour. The printer is at liberty to choose the number of set types per impression to a limit of 40, each set type costing $2. If x set types are chosen, then each impression yields x cards. How many set types should be used?

SOLUTION The printer will choose that number of set types which will minimize production costs. If x set types are used, then the cost for the set types themselves is $2x$ dollars. With this number of set types, the press prints $1000x$ cards per hour. In order to produce 200 000 cards, the press must therefore run for $200\,000/(1000x)$ hours at a cost of $[10][200\,000/(1000x)]$ dollars. The total cost of producing the cards when x set types are used is therefore given by

$$C(x) = 2x + \frac{2000}{x}.$$

Now, $C(x)$ is different than most functions that we have encountered; it is defined only for integer values between 1 and 40 inclusively since the printer must use at least one set type, cannot use more than 40, and cannot use a fraction of a set type. To minimize $C(x)$ we could therefore evaluate it at the first 40 positive integers, and pick the smallest such number. Then there would be no need for calculus at all.

On the other hand, if this were a different problem in which $C(x)$ were again only defined for integer values, but, say, a billion of them, we might not be so complacent about avoiding calculus. But it is impossible to differentiate a function that is defined only for integers. To remedy this, we extend the domain of definition of $C(x)$ from integer values to all real numbers between 1 and 40, and minimize $C(x)$ on this interval, $1 \leq x \leq 40$.

Our customary procedure is to first find critical points of $C(x)$ by setting

$$0 = C'(x) = 2 - \frac{2000}{x^2}.$$

The only solution of this equation in the interval $1 \leq x \leq 40$ is $x = 10\sqrt{10} = 31.6$. As a function of a continuous variable x, $C(x)$ has one critical point, namely, 31.6, at which its derivative is zero. But this is an unacceptable number to the printing problem since x must be an integer. We therefore do not evaluate $C(x)$ at 31.6, but rather at the integer values 31 and 32 on either side; that is, we calculate

$$C(1) = 2002, \qquad C(31) = 126.52, \qquad C(32) = 126.50, \qquad C(40) = 130.$$

The printer should therefore use 32 set types. ∎

EXERCISES 4.6

In Exercises 1–9 find absolute extrema for the function on the interval, if they exist.

1. $f(x) = x^3 - x^2 - 5x + 4, \quad -2 \leq x \leq 3$
2. $f(x) = \dfrac{x-4}{x+1}, \quad 0 \leq x \leq 10$
3. $f(x) = x + \dfrac{1}{x}, \quad \dfrac{1}{2} \leq x \leq 5$
4. $f(x) = x - 2\sin x, \quad 0 \leq x \leq 4\pi$
5. $f(x) = x\sqrt{x+1}, -1 \leq x \leq 1$
6. $f(x) = \dfrac{12}{x^2 + 2x + 2}, \quad x < 0$
7. $f(x) = \dfrac{x+1}{x-1}, \quad x > 1$
8. $f(x) = \dfrac{x}{x^2+3}, \quad x > 0$
9. $f(x) = \dfrac{(2x-1)(x-8)}{(x-1)(x-4)}, \quad x \leq -2$

10. Use Theorem 4.3 to prove Theorem 3.15 in Section 3.11.

11. One end of a uniform beam of length L is built into a wall, and the other end is simply supported (Figure 4.56). If the beam has constant mass per unit length m, its deflection y from the horizontal at a distance x from the built-in end is given by

$$(48\,EI)y = -mg(2x^4 - 5Lx^3 + 3L^2x^2),$$

where E and I are constants depending on the material and cross-section of the beam, and $g > 0$ is a constant. How far from the built-in end does maximum deflection occur?

FIGURE 4.56

12. An agronomist wishes to fence eight rectangular plots for experimentation as shown in Figure 4.57. If each plot must contain 9000 m², find the minimum amount of fencing that can be used.

FIGURE 4.57

13. A closed box is to have length equal to three times its width and total surface area of 30 m². Find the dimensions that produce maximum volume.

14. A square box with open top is to have a volume of 6000 L. Find the dimensions of the box that minimize the amount of material used.

15. A cherry orchard has 255 trees, each of which produces on the average 25 baskets of cherries. For each additional tree planted, the yield per tree decreases by one-twelfth of a basket. How many more trees will produce a maximum crop?

16. A high school football field is being designed to accommodate a 400 m track. The track runs along the length of the field and has semicircles beyond the end zones with diameters equal to the width of the field. If the end zones must be 10 m deep, find the dimensions of the field that maximize playing area (not including the end zones).

17. When a manufacturing company sells x objects per month, it sets the price r of each object at

$$r(x) = 100 - \frac{x^2}{10\,000}.$$

The total cost C of producing the x objects per month is

$$C(x) = \frac{x^2}{10} + 2x + 20.$$

Find the number of objects the company should sell per month in order to realize maximum profits.

18. The base of an isosceles triangle, which is not one of the equal sides, has length b, and its altitude has length a. Find the area of the largest rectangle that can be placed inside the triangle if one of the sides of the rectangle must lie on the base of the triangle.

19. A manufacturer builds cylindrical metal cans that hold 1000 cm³. There is no waste involved in cutting material for the curved surface of the can. However, each circular end piece is cut from a square piece of metal, leaving four waste pieces. Find the height and radius of the can that uses the least amount of metal, including all waste materials.

20. Sides AB and AC of an isosceles triangle have equal length. The base BC has length $2a$, as does the altitude AD from A to BC. Find the height of a point P on AD at which the sum of the distances AP, BP, and CP is a minimum.

21. Two poles are driven into the ground 3 m apart. One pole protrudes 2 m above the ground and the other pole 1 m above the ground. A single piece of rope is attached to the top of one pole, passed through a loop on the ground, pulled taut, and attached to the top of the other pole. Where should the loop be placed in order that the rope be as short as possible?

22. In designing pages for a book, a publisher decides that the rectangular printed region on each page must have area 150 cm². If the page must have 2.5 cm margins on each side and 3.75 cm margins at top and bottom, find the dimensions of the page of smallest possible area.

23. Find the points on the hyperbola $y^2 - x^2 = 9$ closest to $(4, 0)$.

24. Find the point on the parabola $y = x^2$ closest to $(-2, 5)$.

25. A light source is to be placed directly above the centre of a circular area of radius r (Figure 4.58). The illumination at any point on the edge of the circle is directly proportional to the cosine of the angle θ and inversely proportional to the square of the distance d from the source. Find the height h above the circle at which illumination on the edge of the table is maximized.

FIGURE 4.58

26. Among all line segments that stretch from points on the positive x-axis to points on the positive y-axis and pass through the point $(2, 5)$, find that one which makes with the positive x- and y-axes the triangle with least possible area.

27. A long piece of metal one metre wide is to be bent in two places to form a spillway so that its cross-section is an isosceles trapezoid (Figure 4.59). Find the angle θ at which the bends should be formed in order to obtain maximum possible flow along the spillway if the lengths of AB, BC, and CD are all one-third metre.

FIGURE 4.59

28. The strength of a rectangular beam varies as the product of its width and the square of its depth. Find the dimensions of the strongest rectangular beam that can be cut from a circular log of radius 25 cm.

29. At noon a ship S_1 is 20 km north of ship S_2. If S_1 sails south at 6 km/h, and S_2 east at 8 km/h, find when the two ships are closest together.

30. A military courier is located on a desert 6 km from a point P, which is the point on a long straight road nearest him. He is ordered to report to a point Q on the road. If we assume that he can travel at a rate of 14 km/h on the desert and a rate of 50 km/h on the road, find the point where he should reach the road in order to get to Q in the least possible time when:
    (a) Q is 3 km from P;
    (b) Q is 1 km from P.

31. Find the area of the largest rectangle that can be inscribed inside a circle of radius r.

32. Among all rectangles that can be inscribed inside the ellipse $b^2 x^2 + a^2 y^2 = a^2 b^2$ and have sides parallel to the axes, find the one with largest area.

33. Among all rectangles that can be inscribed inside the ellipse $b^2 x^2 + a^2 y^2 = a^2 b^2$ and have sides parallel to the axes, find the one with largest perimeter.

34. Find the area of the largest rectangle that has one side on the x-axis and two vertices on the curve $y = e^{-x^2}$.

35. Find the volume of the largest right circular cylinder that can be inscribed in a sphere of radius r.

36. Find the volume of the largest right circular cylinder that can be inscribed inside a right circular cone of radius r and height h.

37. Two corridors, one 3 m wide and the other 6 m wide, meet at right angles. Find the length of the longest beam that can be transported horizontally around the corner. Ignore the dimensions of the beam.

38. Repeat Exercise 37 taking into account that the beam has square cross-section $1/3$ m on each side.

39. A bee's cell is always constructed in the shape of a regular hexagonal cylinder open on one end and a trihedral apex at the other (Figure 4.60).

    FIGURE 4.60

    It can be shown that the total area of the nine faces is given by

    $$A = 6xy + \frac{3}{2}x^2(\sqrt{3}\csc\theta - \cot\theta).$$

    Find the angle that minimizes A.

40. A rental company buys a new machine for p dollars, which it then rents to customers. If the company keeps the machine for t years (before replacing it), the average replacement cost per year for the t years is p/t. During these t years, the company must make repairs on the machine, the number n depending on t as given by $n(t) = t^\alpha/\beta$, where $\alpha > 1$ and $\beta > 0$ are constants. If r is the average cost per repair, then the average maintenance cost per year over the life of the machine is nr/t. The total yearly expense associated with the machine if it is kept for t years is therefore

    $$C(t) = \frac{p}{t} + \frac{nr}{t}.$$

    Find the optimum time at which to replace the machine.

41. Discuss Exercise 40 in the cases that $\alpha = 1$ and $0 < \alpha < 1$.

42. The illumination at a point is inversely proportional to the square of the distance from the light source and directly proportional to the intensity of the source. If two light sources of intensities I_1 and I_2 are a distance d apart, at what point on the line segment joining the sources is the sum of their illuminations a minimum — relative to all other points on the line segment?

43. A window is in the form of a rectangle surmounted by a semicircle with diameter equal to the width of the window. If the rectangle is of clear glass while the semicircle is of coloured glass that transmits only half as much light per unit area as the clear glass, and if the total perimeter is fixed, find the proportions of the rectangular and semicircular part of the window that admit the most light.

44. A company wishes to construct a storage tank in the form of a rectangular parallelepiped with a square horizontal cross-section. The volume of the tank must be 100 m³.
    (a) If material for the sides and top costs $1.25 per square metre, and material for the bottom costs $4.75 per square metre, find the dimensions that minimize material costs.
    (b) Repeat (a) if the 12 edges must be welded at a cost of $7.50 per metre of weld.

45. When an unloaded die is thrown, there is a probability of $1/6$ that it will come up "two". If the die is loaded, on the other hand, the probability that a "two" will appear is not $1/6$, but is some number p between zero and one. To find p we could roll the die a large number of times, say n, and count the number of times "two" appears, say, m. It seems reasonable that an estimate for p is m/n. Mathematicians define a likelihood function, which for the present situation turns out to be

$$f(x) = \frac{n!}{m!(n-m)!} x^m (1-x)^{n-m}.$$

The value of x that maximizes $f(x)$ on the interval $0 \leq x \leq 1$ is called the maximum likelihood estimate of p. Show that this estimate is m/n.

46. An underground pipeline is to be constructed between two cities A and B (Figure 4.61).

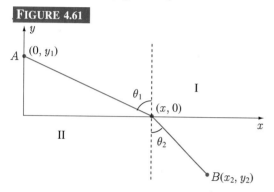

FIGURE 4.61

An analysis of the substructure indicates that construction costs per kilometre in regions I($y > 0$) and II($y < 0$) are c_1 and c_2 respectively. Show that the total construction cost is minimized when x is chosen so that

$$c_1 \sin \theta_1 = c_2 \sin \theta_2.$$

47. A submarine is sailing on the surface due east at a rate of s km/h. It is to pass 1 km north of a point of land on an island at midnight. Soldiers on the island wishing to escape the enemy plan to intercept the submarine by rowing a rubber raft in a straight-line course at a rate of v km/h ($v < s$). What is the last instant that they can leave the island and expect to make contact with the submarine?

48. A packing company wishes to form the 1 m by 2 m piece of cardboard in Figure 4.62(a) into a box as shown in Figure 4.62(b). Cuts are to be made along solid lines and folds along dotted lines, and two sides are to be taped together as shown. If the outer flaps on top and bottom must meet in the centre but the inner flaps need not, find the dimensions of the box holding the most volume. How far apart will the inner flaps be?

FIGURE 4.62

(a)

(b) Bottom is same as top

Tape

49. The cost of fuel per hour for running a ship varies directly as the cube of the speed, and is $\$B$ per hour when the speed is b kilometres per hour. There are also fixed costs of $\$A$ per hour. Find the most economical speed at which to make a trip.

50. A paper drinking cup in the form of a right-circular cone can be made from a circular piece of paper by removing a sector and joining edges OA and OB, as shown in Figure 4.63. If the radius of the circle is R, what choice of θ yields a cup of maximum volume?

FIGURE 4.63

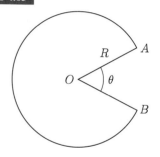

51. A truck driver estimates that the cost per kilometre of operating his rig is $40 + (50 - v)^2/10$ cents, where v is his average speed in kilometres per hour. In order to cover this cost he charges $20 per hour for rental of his truck. For his driving he charges $10 per hour. For a 1000 km nonstop haul, what average speed would you recommend to the driver?

52. Prove that the minimum distance from a point (x_1, y_1) to a line $Ax + By + C = 0$ is

$$\frac{|Ax_1 + By_1 + C|}{\sqrt{A^2 + B^2}}.$$

53. A window is in the form of a rectangle surmounted by a semicircle with diameter equal to the width of the rectangle. The rectangle is of clear glass costing $\$a$ per unit area, while the semicircle is of coloured glass costing $\$b$ per unit area. The coloured glass transmits only a fraction p ($0 < p < 1$) as much light per unit area as the clear glass. In addition, the curved portion of the window is surmounted by a special frame at a cost of $c per unit length. If the total cost of the window must not exceed $A, find the dimensions of the window that admit the most light.

54. Find the point on the ellipse $4x^2 + 9y^2 = 36$ closest to $(4, 13\sqrt{5}/6)$.

55. The frame for a kite is to be made from six pieces of wood as shown in Figure 4.64. The four outside pieces have predetermined lengths a and b ($b > a$). Yet to be cut are the two diagonal pieces. How long should they be in order to make the area of the kite as large as possible?

FIGURE 4.64

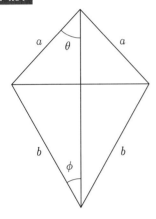

56. Suppose that the line $Ax + By + C = 0$ intersects the parabola $y = ax^2 + bx + c$ at points P and Q (Figure 4.65). Find the point R on the parabola between P and Q that maximizes the area of triangle PQR.

FIGURE 4.65

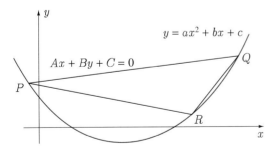

57. There are n red stakes securely driven into the ground in a straight line. A blue stake is to be added to the line, and each red stake is to be joined to the blue one by a string. Where should the blue stake be placed in order that the total length of all strings be as small as possible?

58. When blood flows through a vein or artery, it encounters resistance due to friction with the walls of the blood vessel and the viscosity of the blood itself. *Poiseuille's law* for laminar blood flow states that for a circular vessel, resistance R to blood flow is proportional to the length L of the blood vessel and inversely proportional to the fourth power of its radius r:

$$R = k\frac{L}{r^4},$$

where k is a constant. Figure 4.66 shows a blood vessel of radius r_1 from A to B and a branching vessel from D to C of radius $r_2 < r_1$.

FIGURE 4.66

The resistance encountered by the blood in flowing from A to C is given by

$$R = k\frac{L_1}{r_1^4} + k\frac{L_2}{r_2^4},$$

where L_1 and L_2 are the lengths of AD and DC respectively.

(a) If B is assumed to the right of E, show that R can be expressed in terms of θ as

$$R = f(\theta) = \frac{k}{r_1^4}(X - Y\cot\theta) + \frac{k}{r_2^4}Y\csc\theta,$$

where X and Y are the lengths of AE and CE respectively.

(b) Show that $f(\theta)$ has only one critical point $\bar{\theta}$ in the range $0 < \theta < \pi/2$, and $\bar{\theta}$ is defined by

$$\cos\bar{\theta} = \frac{r_2^4}{r_1^4}.$$

(c) Verify that $\bar{\theta}$ yields a relative minimum for $f(\theta)$.

(d) Show that

$$f(\bar{\theta}) = \frac{kX}{r_1^4} + \frac{kY}{r_2^4}\sqrt{1 - \left(\frac{r_2}{r_1}\right)^8} < f(\pi/2).$$

(e) Does $\bar{\theta}$ provide an absolute minimum for $f(\theta)$?

59. If the fencing in Example 4.14 is to form the arc of a circle, what is the maximum possible area?

60. The upper left corner of a piece of paper a units wide and b units long ($b > a$) is folded to the right edge as shown in Figure 4.67. Calculate length x in order that the length y of the fold be a minimum.

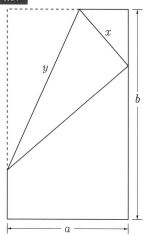

FIGURE 4.67

SECTION 4.7

Velocity and Acceleration

One of the most important applications of calculus and one that can be traced back to its founder, Sir Isaac Newton, is the field of kinematics — the study of the relationships among displacement, velocity, and acceleration. In this section we treat kinematics of one-dimensional motion, but do so only from a differentiation point of view; given the position of a particle moving along a straight line, find its velocity and acceleration and use these quantities to describe the motion of the particle. In Section 5.2 we reverse these operations; beginning with the acceleration, we find velocity and position. This process is essential to physics and engineering, where acceleration of a particle is determined by the forces acting on it.

In Section 3.1 we indicated that when the position of a particle moving along the x-axis is known as a function of time t, say, $x(t)$, its **instantaneous velocity** is the derivative of $x(t)$ with respect to t,

$$v(t) = \frac{dx}{dt}. \qquad (4.14)$$

The **instantaneous acceleration** of the particle is defined as the rate of change of velocity with respect to time:

$$a(t) = \frac{dv}{dt} = \frac{d^2 x}{dt^2}. \qquad (4.15)$$

In actual fact, $x(t)$, $v(t)$, and $a(t)$ are the x-components of the displacement, velocity, and acceleration vectors, respectively. In the absence of a complete discussion of vectors, we omit the terms "vector" and "component", and simply call $x(t)$, $v(t)$, and $a(t)$ displacement, velocity, and acceleration.

When distance is measured in metres and time in seconds, velocity is measured in metres per second (m/s). Since acceleration is the time derivative of velocity, its units must be units of velocity divided by units of time, i.e., metres per second per second (m/s/s). Usually we shorten this by saying "metres per second squared," and write m/s^2.

Suppose that the position of a particle moving along the x-axis is

$$x(t) = t^3 - 27t^2 + 168t + 46, \quad t \geq 0,$$

where x is measured in metres and t in seconds, and we wish to describe the motion of the particle. The best way to do this is to draw a graph of the displacement function. With calculus we can then show how geometric properties of the graph reflect important features about the velocity and acceleration of the particle. The velocity and acceleration of the particle are quickly calculated to be

$$v(t) = \frac{dx}{dt} = 3t^2 - 54t + 168 = 3(t-4)(t-14) \quad \text{m/s},$$

$$a(t) = \frac{d^2x}{dt^2} = 6t - 54 = 6(t-9) \quad \text{m/s}^2.$$

Ignoring the physical interpretations of dx/dt and d^2x/dt^2 as velocity and acceleration for the moment, and concentrating only on the fact that they are the first and second derivatives of the function $x(t)$, we immediately find that $x(t)$ has a relative maximum of $x(4) = 350$ and a relative minimum of $x(14) = -150$ at the critical points $t = 4$ and $t = 14$. Add to this a point of inflection at $(9, 100)$ and we obtain the graph of $x(t)$ in Figure 4.68.

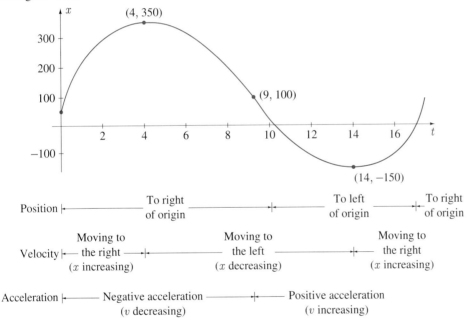

FIGURE 4.68

Let us now discuss what this graph tells us about the motion of the particle. Ordinates of this graph represent horizontal distances of the particle from the origin $x = 0$. When an ordinate is positive, the particle is that distance to the right of the origin; when an ordinate is negative, the particle is that distance to the left of the origin. For instance, at time $t = 0$, we calculate $x = 46$ and therefore the particle begins 46 m to the right of $x = 0$. At time $t = 4$, it is 350 m to the right of $x = 0$, and at $t = 14$, it is 150 m to the left of the origin.

The slope of the graph represents velocity of the particle. When slope is positive, namely, in the intervals $0 < t < 4$ and $t > 14$, the particle is moving to the right along the x-axis; for $4 < t < 14$, velocity is negative, indicating that the particle is moving to the left. At times $t = 4$ and $t = 14$, the particle is instantaneously at rest.

The concavity of the graph reflects the sign of the acceleration. For $0 < t < 9$, the graph is concave downward, its slope is decreasing. Physically this means that it has a negative acceleration; that is, its velocity is decreasing. For $t > 9$, the graph is concave upward, its slope is increasing. Physically, acceleration is positive; that is, velocity is increasing. At the point of inflection $(9, 100)$, the acceleration changes sign. Notice that the acceleration is not zero at $t = 4$ and $t = 14$. In spite of the fact that the velocity is zero at these times, the acceleration does not vanish. You might ask yourself what feature of the graph would reflect coincident zeros for velocity and acceleration (see Exercise 21).

Speed is defined to be the magnitude of velocity. It represents how fast the particle is moving without regard for direction. For instance, at time $t = 0$, the velocity and speed are both 168 m/s, whereas at time $t = 10$, velocity is -72 m/s and speed is 72 m/s. Geometrically, speed is represented by the slope of the graph without regard for sign. Do not confuse velocity and speed; they are closely related, but they are not interchangeable.

With the above ideas in mind, let us detail the history of the particle's motion. At time $t = 0$, it begins 46 m to the right of the origin, moving to the right with velocity (or speed) 168 m/s. Since the acceleration is negative (to the left), the particle is slowing down (both velocity and speed are decreasing), until at time $t = 4$ s, it comes to an instantaneous stop 350 m to the right of the origin. Because the acceleration continues to be negative, the particle moves to the left, its velocity decreasing, but its speed increasing. At time $t = 9$ s, when the particle is 100 m to the right of the origin, the acceleration changes sign. At this instant, the velocity has attained a (relative) minimum value, but speed is a (relative) maximum. With acceleration to the right for $t > 9$ s, the particle continues to move left, but slows down, its velocity increasing until at time $t = 14$ s, when it once again comes to a stop 150 m to the left of the origin. For time $t > 14$ s, it moves to the right, picking up speed, and passes through the origin just before $t = 17$ s.

Further analysis of the interdependences of displacement, velocity, and acceleration is contained in the following example.

EXAMPLE 4.19

The position of a particle moving along the x-axis is given by the function

$$x(t) = 3t^4 - 32t^3 + 114t^2 - 144t + 40, \quad 0 \le t \le 5,$$

where x is measured in metres and t in seconds. Answer the following questions concerning its motion:

(a) What is its velocity and speed at $t = 1/2$ s?

(b) When is its acceleration increasing?

(c) Is the velocity increasing or decreasing at $t = 2$ s?

(d) Is the particle speeding up or slowing down at $t = 2$ s?

(e) What is the maximum velocity of the particle in the time interval $0 \le t \le 2$?

(f) What is the maximum distance the particle ever attains from the origin?

SOLUTION Taking into consideration that there are six parts to this example, which examine much of the particle's motion, a graph of the displacement function could be very helpful. However, had only one or two parts of the example been required, a graphical approach might not have been particularly efficient. Let us take the approach here that each question had been asked in isolation, and that we shall find answers using the displacement, velocity and acceleration functions. The velocity and acceleration are

$$v(t) = 12t^3 - 96t^2 + 228t - 144 = 12(t-1)(t-3)(t-4) \text{ m/s},$$
$$a(t) = 12(3t^2 - 16t + 19) \text{ m/s}^2.$$

(a) The velocity at $t = 1/2$ is $v(1/2) = -105/2$ m/s, and the speed is $105/2$ m/s.

(b) The acceleration is increasing when its derivative is nonnegative. Since $da/dt = 12(6t - 16) = 24(3t - 8)$, acceleration is increasing for $8/3 \leq t \leq 5$.

(c) Since $a(2) = -12$ m/s², the velocity is decreasing at $t = 2$ s.

(d) Since $v(2) = 24$ m/s, the particle is moving to the right. But according to part (c), its acceleration is to the left at this time. As a result, it must be slowing down.

(e) To maximize $v(t)$ on $0 \leq t \leq 2$, we first find critical points of $v(t)$; that is, solve for zeros of the acceleration

$$0 = 12(3t^2 - 16t + 19).$$

Of the two solutions $(8 \pm \sqrt{7})/3$, only $(8 - \sqrt{7})/3$ is in the given interval $0 \leq t \leq 2$. Since

$$v(0) = -144, \quad v\left((8 - \sqrt{7})/3\right) = 25.35, \quad v(2) = 24,$$

maximum velocity is 25.35 m/s occuring at $t = (8 - \sqrt{7})/3$ s.

(f) Critical points for $x(t)$ occur at $t = 1, 3,$ and 4, where velocity is zero. Since

$$x(0) = 40, \quad x(1) = -19, \quad x(3) = 13, \quad x(4) = 8, \quad x(5) = 45,$$

maximum distance from the origin is 45 m. ∎

EXERCISES 4.7

1. Figure 4.69 shows the graph of the displacement function $x(t)$ of a particle moving along the x-axis during the time interval $0 \leq t \leq 6$. Answer the following questions, estimating values from the graph when necessary.

FIGURE 4.69

(a) Is the particle to the left or right of the origin at times $t = 1$ and $t = 4$?

(b) By drawing a horizontal line at $x = 2$ determine whether the particle is more often to the right or to the left of the point $x = 2$.

(c) What is the maximum distance from the origin achieved by the particle?

(d) What is the maximum distance from the point $x = 3$ achieved by the particle?

(e) Is the particle moving to the right or to the left at times $t = 1/2$ and $t = 3$?

(f) How many times does the particle change direction?

(g) Is the velocity greater at $t = 7/2$ or at $t = 9/2$?

(h) Is the speed greater at $t = 7/2$ or at $t = 9/2$?

(i) Is the acceleration positive or negative at $t = 1$ and $t = 4$?

(j) How many times does the acceleration change sign?

2. Repeat Exercise 1 for the graph in Figure 4.70.

FIGURE 4.70

In Exercises 3–12 find the velocity and acceleration of an object that moves along the x-axis with the given position function. In each exercise, discuss the motion, including in your discussion a graph of the function $x(t)$. Assume that x is measured in metres and t in seconds.

3. $x(t) = 2t + 5, \quad t \geq 5$

4. $x(t) = t^2 - 7t + 6, \quad t \geq 0$

5. $x(t) = t^2 + 5t + 10, \quad t \geq 1$

6. $x(t) = -2t^3 + 2t^2 + 16t - 1, \quad t \geq 0$

7. $x(t) = t^3 - 9t^2 + 15t + 3$, $t \geq 2$

8. $x(t) = 3\cos 4t$, $t \geq 0$

9. $x(t) = 1/t$, $t \geq 1$

10. $x(t) = t + 4/t$, $t \geq 1$

11. $x(t) = (t-4)/t^2$, $t \geq 2$

12. $x(t) = t^2\sqrt{1-t}$, $0 \leq t \leq 1$

13. An object moving along the x-axis has position given by

$$x(t) = t^3 - 9t^2 + 24t + 1,$$

where x is measured in metres and $t \geq 0$ is time in seconds. Determine at $t = 1$ s:
   (a) the position of the object;
   (b) whether the object is moving to the right or to the left;
   (c) the acceleration of the object;
   (d) whether the object's speed is increasing or decreasing.

14. The position of a particle moving along the x-axis is measured in metres to be

$$x(t) = t^3 - 9t^2 + 15t - 2,$$

where $t \geq 0$ is time in seconds.
   (a) What is the velocity and acceleration of the particle?
   (b) Is the particle ever instantaneously at rest?
   (c) Is the particle speeding up or slowing down at $t = 4$ s?
   (d) When is the velocity equal to -10 m/s?
   (e) Is the particle's acceleration increasing or decreasing at $t = 3$ s?

15. A particle moves along the x-axis in such a way that its position as a function of time t is given by

$$x(t) = t \sin t, \quad 0 \leq t \leq 9.$$

Determine graphically each of the following.
   (a) The number of times the particle reverses direction.
   (b) The number of times the particle has zero acceleration.
   (c) The direction of motion at time $t = 6$.

16. Can the position curve $x = x(t)$ of a realistic particle moving along the x-axis be represented by a function $x(t)$ that has a discontinuity? Explain.

In Exercises 17–20 we have defined the position function of an object moving along the x-axis. Sketch graphs of the position, velocity, and acceleration functions. Also draw a graph of speed as a function of time.

17. $x(t) = t^2 + 2t - 15$, $t \geq 0$

18. $x(t) = t^3 - 3t$, $t \geq 0$

19. $x(t) = t + 4/t$, $t \geq 1$

20. $x(t) = 5\sin t$, $t \geq 0$

21. What feature on the displacement graph would indicate a time when the velocity and acceleration are simultaneously zero?

22. Are critical points for the velocity function the same as those for the speed function? Explain using the graphs from Exercise 20.

23. What is the maximum speed of the particle in Example 4.19 over the time interval $0 \leq t \leq 2$?

24. In many velocity and acceleration problems it is more convenient to express acceleration in terms of a derivative with respect to position, as opposed to a derivative with respect to time. Show that acceleration can be written in the form

$$a = v\frac{dv}{dx}.$$

25. When an object moves with constant acceleration a along the x-axis, its position as a function of time t must be of the form

$$x = x(t) = \frac{1}{2}at^2 + bt + c,$$

where b and c are constants.
   (a) If the object is at positions x_1 and x_2 at times t_1 and t_2, what is its average velocity over the time interval $t_1 \leq t \leq t_2$?
   (b) At what time in this time interval is the instantaneous velocity equal to the average velocity? Where is the object at this time? Is it at the midpoint of the interval between x_1 and x_2, is it closer to x_1, or is it closer to x_2? Assume in this part of the problem that $a > 0$ and that the velocity of the object is positive at time t_1.

In Exercises 26–29 assume that a particle moves along the x-axis in such a way that its position, velocity, and acceleration are continuous functions on the interval $a \leq t \leq b$. Discuss the validity of each statement.

26. When position has a relative maximum so also does the absolute value of the distance from the origin to the particle.

27. When position has a relative minimum so also does the absolute value of the distance from the origin to the particle.

28. When velocity has a relative minimum so does speed.

29. When velocity has a relative maximum so does speed.

SECTION 4.8
Related Rates

Many interesting and practical problems involving rates of change are commonly referred to as *related rate problems*. In these problems, two or more quantities are related to each other and rates at which some of them change are known. It is required to find rates at which the others change. Related rate problems deal almost exclusively with rates of change of quantities with respect to time. To solve these problems we first consider three examples. These will suggest the general procedures by which all related rate problems can be analyzed. We shall then discuss two somewhat more complicated problems.

EXAMPLE 4.20

A man 2 m tall walks directly away from a streetlight that is 8 m high at the rate of $3/2$ m/s. How fast is the length of his shadow changing?

FIGURE 4.71

SOLUTION When x denotes the distance between the man and the lightpost (Figure 4.71), the fact that he walks directly away from the light at $3/2$ m/s means that x is changing at a rate of $3/2$ m/s; that is, $dx/dt = 3/2$ m/s. If s represents the length of the man's shadow, then we are searching for ds/dt. Similar triangles in Figure 4.71 enable us to relate s and x,

$$\frac{x+s}{s} = \frac{8}{2},$$

and this equation can be solved for s in terms of x:

$$s = \frac{x}{3}.$$

Now, s and x are each functions of time t,

$$s = f(t) \quad \text{and} \quad x = g(t),$$

although we have not calculated the exact form of these functions. Indeed, the essence of the related rate problem is to find ds/dt without ever knowing $f(t)$ explicitly. To do this we note that since the equation $s = x/3$ is valid at any time t when the man is walking away from the light, we may differentiate with respect to t to obtain

$$\frac{ds}{dt} = \frac{1}{3}\frac{dx}{dt}.$$

This equation relates the known rate $dx/dt = 3/2$ with the unknown rate ds/dt. It follows that

$$\frac{ds}{dt} = \left(\frac{1}{3}\right)\left(\frac{3}{2}\right) = \frac{1}{2},$$

and the man's shadow is therefore getting longer at the rate of $1/2$ m/s. ∎

Knowing dx/dt in this example, we have calculated ds/dt, and have done so without finding s explicitly as a function of time t. This is the essence of a related rate problem. Since dx/dt is a constant value, it is quite easy to find s as a function of t, and hence ds/dt. Indeed, if we choose time $t = 0$ when the man starts to walk away from the streetlight, then his distance from the light at any given time is $x = 3t/2$ metres. Combine this with the fact that $s = x/3$ and we may write

$$s = \frac{1}{3}\left(\frac{3t}{2}\right) = \frac{t}{2} \text{ m}.$$

With this explicit formula for s, it is clear that $ds/dt = 1/2$. What is important to realize is that the solution in this paragraph is possible only because the man walks at a constant rate. Were his speed not constant, it might be impossible to find s explicitly in terms of t. The next example illustrates this point in that the given rate is known only at one instant in time.

EXAMPLE 4.21

A ladder leaning against a house (Figure 4.72) is prevented from moving by a young child. Suddenly, something distracts the child and he releases the ladder. The ladder begins slipping down the wall of the house, picking up speed as it falls. If the top end of the ladder is moving at 1 m/s when the lower end is 15 m from the house, how fast is the foot of the ladder moving away from the house at this instant?

FIGURE 4.72

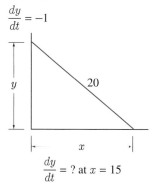

SOLUTION Figure 4.72 indicates that when y denotes the height of the top of the ladder above the ground, then $dy/dt = -1$ m/s when $x = 15$ m (the negative sign because y is decreasing). We emphasize here that $dy/dt = -1$ *only* when $x = 15$. What is required is dx/dt when $x = 15$ m. Because the triangle in the figure is right-angled, we may write

$$x^2 + y^2 = 20^2,$$

and this equation is valid at any time during which the ladder is slipping. If we differentiate with respect to time t, using extended power rule 3.17,

$$2x\frac{dx}{dt} + 2y\frac{dy}{dt} = 0.$$

When $x = 15$, we calculate that $y = \sqrt{400 - 225} = 5\sqrt{7}$, and therefore at this instant

$$15\frac{dx}{dt} + 5\sqrt{7}(-1) = 0.$$

This yields $dx/dt = \sqrt{7}/3$, and we can say that when the foot of the ladder is 15 m from the wall, it is moving away from the wall at $\sqrt{7}/3$ m/s.

EXAMPLE 4.22

A tank in the form of a right circular cone with altitude 6 m and base radius 3 m (Figure 4.73) is being filled with water at a rate of 4000 L/min. How fast is the surface of the water rising when the depth is 3 m?

FIGURE 4.73

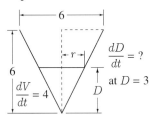

SOLUTION Figure 4.73, which illustrates a cross-section of the tank, indicates that when the depth of water in the tank is D, the volume V of water is

$$V = \frac{1}{3}\pi r^2 D.$$

Of the three variables V, r, and D in this equation, we are concerned only with V and D, since dV/dt is given and dD/dt is what we want. This suggests that we eliminate r using similar triangles. Since $r/D = 3/6$, we have $r = D/2$; therefore,

$$V = \frac{1}{3}\pi \left(\frac{D}{2}\right)^2 D = \frac{1}{12}\pi D^3.$$

Because this result is valid for all time t during the filling process, we can differentiate with respect to t, once again using extended power rule 3.17:

$$\frac{dV}{dt} = \frac{1}{4}\pi D^2 \frac{dD}{dt}.$$

Since $dV/dt = 4$ m³/min which is converted from the rate of 4000 L/min since the litre is not an acceptable unit of measure for volume, we find that when $D = 3$,

$$4 = \frac{1}{4}\pi(3)^2 \frac{dD}{dt},$$

from which $dD/dt = 16/(9\pi)$. The surface is therefore rising at a rate of $16/(9\pi)$ m/min. ∎

Examples 4.20–4.22 illustrate the following general procedure for solving related-rate problems:

1. Sketch a diagram illustrating all given information, especially given rates of change and desired rates of change.

2. Find an equation valid for all time (in some interval about the instant in question) that involves only variables whose rates of change are given or required.

3. Differentiate the equation in 2 and solve for the required rate.

Steps 1 and 3 are usually quite straightforward; step 2, on the other hand, may tax your ingenuity. To find the equation in the appropriate variables, it may be necessary to introduce and substitute for additional variables. Finding these substitutions requires you to analyze the problem very closely.

Be careful not to substitute numerical data that represent the instant at which the derivative is required before differentiation has taken place. Numerical data must be substituted after differentiation. For instance, in Example 4.22, the radius r of the surface of the water when $D = 3$ is $3/2$. If we substitute this into $V = \pi r^2 D/3$, we obtain a function $V = \pi(3/2)^2 D/3 = 3\pi D/4$ which is valid only when $D = 3$. It cannot therefore be differentiated; only equations that are valid for a range of values of t can be differentiated with respect to t.

We now apply this procedure to two further examples. The first is an extension of Example 4.20.

EXAMPLE 4.23

A man 2 m tall walks along the edge of a straight road 10 m wide. On the other edge of the road stands a streetlight 8 m high. If the man walks at $3/2$ m/s, how fast is his shadow lengthening when he is 10 m from the point directly opposite the light?

FIGURE 4.74

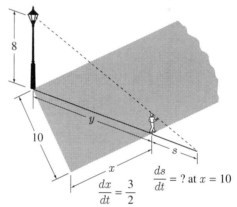

SOLUTION First we draw Figure 4.74 wherein the man's speed is represented as the time rate of change of his distance x from the point on his side of the road directly opposite the light. What is required is the rate of change ds/dt of the length of his shadow when $x = 10$. To find an equation relating x and s, we first use similar (vertical) triangles to write

$$\frac{y+s}{s} = \frac{8}{2} = 4,$$

from which

$$y = 3s,$$

an equation that relates s to y, rather than s to x. However, since

$$y^2 = x^2 + 100,$$

we substitute to obtain

$$9s^2 = x^2 + 100.$$

The derivative of this equation with respect to time t gives

$$18s\frac{ds}{dt} = 2x\frac{dx}{dt}.$$

When $x = 10$, we obtain $s = \sqrt{100 + 100}/3 = 10\sqrt{2}/3$, and, at this instant,

$$18\left(\frac{10\sqrt{2}}{3}\right)\frac{ds}{dt} = 2(10)\left(\frac{3}{2}\right).$$

Thus,
$$\frac{ds}{dt} = \frac{30}{60\sqrt{2}} = \frac{\sqrt{2}}{4},$$
and the man's shadow is therefore lengthening at the rate of $\sqrt{2}/4$ m/s. ∎

It is worthwhile noting in this example that were we to solve for ds/dt before substituting $x = 10$ and $s = 10\sqrt{2}/3$, then we would have
$$\frac{ds}{dt} = \frac{x}{9s}\frac{dx}{dt}.$$
Since dx/dt is always equal to $3/2$, and $s = \sqrt{x^2 + 100}/3$, we can write
$$\frac{ds}{dt} = \frac{x}{3\sqrt{x^2 + 100}}\left(\frac{3}{2}\right) = \frac{x}{2\sqrt{x^2 + 100}},$$
a general formula for ds/dt. The limit of this rate as x becomes very large is
$$\lim_{x \to \infty} \frac{ds}{dt} = \lim_{x \to \infty} \frac{x}{2\sqrt{x^2 + 100}} = \frac{1}{2}.$$
But for very large x, the man essentially walks directly away from the light, and this answer, as we might expect, is identical to that in Example 4.20.

EXAMPLE 4.24

One end of a rope is tied to a box. The other end is passed over a pulley 5 m above the floor and tied at a level 1 m above the floor to the back of a truck. If the rope is taut and the truck moves at $1/2$ m/s, how fast is the box rising when the truck is 3 m from the plumb line through the pulley?

FIGURE 4.75

SOLUTION In Figure 4.75, we have represented the speed of the truck as the rate of change of length x, $dx/dt = 1/2$ m/s. What is required is dy/dt when $x = 3$. To find an equation relating x and y, we first use the fact that the length z of rope between pulley and truck is the hypotenuse of a right-angled triangle with sides of lengths x and 4,
$$z^2 = x^2 + 16.$$
For an equation relating y and z, we note that the length of the rope, call it L, remains constant, and is equal to the sum of z and $5 - y$,
$$L = z + (5 - y).$$

These two equations can be combined into

$$(L - 5 + y)^2 = x^2 + 16,$$

and differentiation with respect to time t now gives

$$2(L - 5 + y)\frac{dy}{dt} = 2x\frac{dx}{dt}.$$

When $x = 3$, we may write that

$$(L - 5 + y)^2 = 9 + 16 = 25.$$

We could solve this equation for y (in terms of L), but it is really not y that is needed to obtain dy/dt from the previous equation. It is $L - 5 + y$, and this is clearly equal to 5. Thus, when $x = 3$, we have

$$2(5)\frac{dy}{dt} = 2(3)\left(\frac{1}{2}\right);$$

that is, $dy/dt = 3/10$, and the box is rising at a rate of $3/10$ m/s.

The reader should compare this example with the problem in Exercise 2. They may appear similar, but are really quite different.

EXERCISES 4.8

1. A convertible is travelling along a straight highway at 100 km/h. A child in the car accidentally releases a helium-filled balloon, which then rises vertically at 10 m/s. How fast are the child and balloon separating 4 seconds after the balloon is released?

2. A rope passes over a pulley and one end is attached to a cart as shown in Figure 4.76. If the rope is pulled vertically downward at 2 m/s, how fast is the cart moving when $s = 6$ m?

FIGURE 4.76

3. A light is on the ground 20 m from a building. A man 2 m tall walks from the light directly toward the building at 3 m/s. How fast is the length of his shadow on the building changing when he is 8 m from the building?

4. A funnel in the shape of a right circular cone is 15 cm across the top and 30 cm deep. A liquid is flowing in at the rate of 80 mL/s and flowing out at 15 mL/s. At what rate is the surface of the liquid rising when the liquid fills the funnel to a depth of 20 cm?

5. A water tank is in the form of a right circular cylinder of diameter 3 m and height 3 m on top of a right circular cone of diameter 3 m and height 1 m. If water is being drawn from the bottom at the rate of 1 L/min, how fast is the water level falling when
   (a) it is 1 m from the top of the tank?
   (b) it is 3.5 m from the top of the tank?

6. A point P moves along the curve $y = x^2 + x + 4$ where x and y are measured in metres. Its x-coordinate decreases at 2 m/s. If the perpendicular from P to the x-axis intersects this axis at point Q, how fast is the area of the triangle with vertices P, Q, and the origin changing when the x-coordinate of P is 2 m?

7. Water is being pumped into a swimming pool which is 10 m wide, 20 m long, 1 m deep at the shallow end, and 3 m deep at the deep end. If the water level is rising at 1 cm/min when the depth is 1 m at the deep end, at what rate is water being pumped into the pool?

8. Boyle's law for a perfect gas states that the pressure exerted by the gas on its containing vessel is inversely

proportional to the volume occupied by the gas. If when the volume is 10 L and the pressure is 50 N/m², the volume is increasing at 1/2 L/s, find the rate of change of the pressure of the gas.

9. A woman driving 100 km/h along a straight highway notes that the shadow of a cloud is keeping pace with her. What can she conclude about the speed of the cloud?

10. A fisherman is trolling at a rate of 2 m/s with his lure 100 m behind the boat and on the surface. Suddenly a fish strikes and dives vertically at a rate of 3 m/s. If the fisherman permits the line to run freely and it always remains straight, how fast is the line being played out when the reel is 50 m from its position at the time of the strike?

11. Air expands adiabatically in accordance with the law $PV^{7/5}$ = constant. If, at a given time, the volume V is 100 L and the pressure P is 40 N/cm², at what rate is the pressure changing when the volume is decreasing at 1 L/s?

12. Sand is poured into a right-circular cylinder of radius 1/2 m along its axis (Figure 4.77). Once sand completely covers the bottom, a right circular cone is formed on the top.

FIGURE 4.77

(a) If 0.02 m³ of sand enters the container every minute, how fast is the top of the sand pile rising?
(b) How fast is the sand rising along the side of the cylinder?

13. A balloon has the shape of a right-circular cylinder of radius r and length l with a hemisphere at each end of radius r. The balloon is being filled at a rate of 10 mL/s in such a way that l increases twice as fast as r. Find the rate of change of r when $r = 8$ cm and $l = 20$ cm.

14. An oval racetrack has a straight stretch 100 m long and two semicircles, each of radius 50 m (Figure 4.78). Car 1, on the infield, moves along the x-axis from O to B. It accelerates from rest at O, attains a speed of 10 m/s at C, and maintains this speed along CB. Car 2 travels along the quarter oval $ADEB$. It is at D when Car 1 is at C, and between D and B, Car 2 maintains the same rate of change of its x-coordinate as does Car 1.

FIGURE 4.78

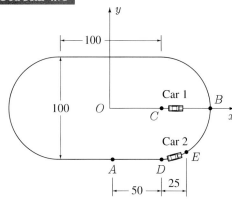

(a) Find a formula for the rate of change of the y-coordinate of Car 2 between D and B.
(b) How fast is the y-coordinate of Car 2 changing when it is at point E?
(c) If the cars collide at B, which car suffers the most damage?

15. A ship is 1 km north of a pier and is travelling N30°E at 3 km/h. A second ship is 3/4 km east of the pier and is travelling east at 7 km/h. How fast are the ships separating?

16. The circle in Figure 4.79 represents a long-playing record which is rotating clockwise at 100/3 revolutions per minute. A bug is walking away from the centre of the record directly toward point P on the rim of the record at 1 cm/s. When the bug is at position R, 10 cm from O, angle θ is $\pi/4$ radians. Find the rate at which the distance from the bug to the fixed point Q is changing when the bug is at R.

FIGURE 4.79

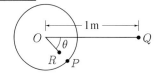

17. Eight skaters form a "whip". Show that the seventh person on the whip travels twice as fast as the fourth person.

18. A particle moves counterclockwise around a circle of radius 5 cm centred at the origin making 4 revolutions each second. How fast is the particle moving away from the point with coordinates $(5, 6)$ when it is at position $(-3, 4)$?

19. Two people (A and B in Figure 4.80) walk along opposite sides of a road 10 m wide. A walks to the right at 1 m/s, and B walks to the left at 2 m/s. A third person C walks along a sidewalk 5 m from the road in

such a way that B is always on the line joining A and C. Find the speed of C.

FIGURE 4.80

20. Let $P(x, y)$ be a point on the first quadrant portion of the hyperbola $x^2 - y^2 = 1$. Let R be the foot of the perpendicular from P to the x-axis, and $Q(x^*, 0)$ be the x-intercept of the normal line to the hyperbola at P.
    (a) Show that $x^* = 2x$.
    (b) If P moves along the hyperbola so that its x-coordinate is decreasing at 3 units per unit time, how fast is the area of triangle QPR changing when $x = 4$?

21. A solution passes from a conical filter 24 cm deep and 16 cm across the top into a cylindrical container of diameter 12 cm. When the depth of solution in the filter is 12 cm, its level is falling at the rate of 1 cm/min. How fast is the level of solution rising in the cylinder at this instant?

22. A light is at the top of a pole 25 m high and a ball is dropped at the same height from a point 10 m from the light. How fast is the shadow of the ball moving along the ground 1 second later? The distance fallen by the ball t seconds after it has been dropped is $d = 4.905t^2$ metres.

23. A point moves along the parabola $y = x^2 - 3x$ (x and y measured in metres) in such a way that its x-coordinate changes at the rate of 2 m/s. How fast is its distance from the point $(1, 2)$ changing when it is at $(4, 4)$?

24. Repeat Exercise 23 given that the parabola is replaced by the curve $(x + y)^2 = 16x$.

25. The volume of wood in the trunk of a tree is sometimes calculated by considering it as a frustrum of a right-circular cone (Figure 4.81).
    (a) Verify that the volume of the trunk is
    $$V = \frac{1}{3}\pi h(R^2 + rR + r^2).$$
    (b) Suppose that at the present time the radii of the top and bottom are $r = 10$ and $R = 50$ cm, and the height is $h = 30$ m. If the tree continues to grow so that ratios r/R and r/h always remain the same as they are now, and R increases at a rate of $1/2$ cm/year, how fast will the volume be changing in 2 years?

FIGURE 4.81

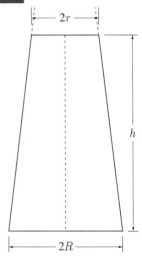

26. Sand is poured into a right-circular cone of radius 2 m and height 3 m along its axis (Figure 4.82).

FIGURE 4.82

The sand forms two cones of equal height h, one inverted on top of the other.
    (a) If 0.02 m³ of sand enters the container every minute, how fast is the top of the pile rising when it is just level with the top of the container?
    (b) How fast is the sand rising along the side of the container at this instant?

27. Figure 4.83 shows a rod OB rotating counterclockwise in the xy-plane about O at a rate of 2 revolutions per second. Attached to OB is a rod AB, and A is confined to sliding horizontally along the x-axis. We are given that OB has length l metres, and AB has length L metres ($L > 2l$).

FIGURE 4.83

(a) Find the velocity of A as a function of x and θ.
(b) Find the acceleration of A.

28. If to the mechanism in Figure 4.83 we add a rod AC (Figure 4.84), where C is confined to sliding vertically, find the velocity of C in terms of x, θ, and y.

FIGURE 4.84

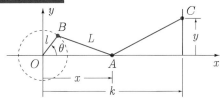

29. In Figure 4.85 the boy's feet make 1 revolution per second around a sprocket of radius R metres. The chain travels around a sprocket of radius r metres on the back wheel, which itself has radius $\overline{R}$ metres. If a stone imbedded in the tire becomes dislodged, how fast is it travelling when it leaves the tire?

FIGURE 4.85

30. (a) If θ is the angle formed by the minute and hour hands of a clock, what is the time rate of change of θ (in radians per minute)?

(b) If the lengths of the hands on the clock are 10 cm and 7.5 cm, find the rate at which their tips approach each other at 3:00 (i.e., find dz/dt in Figure 4.86).

(c) Repeat (b) but replace time 3:00 with 8:05.

FIGURE 4.86

31. If the sides of a triangle have lengths a, b, and c, its area is given by

$$A = \sqrt{s(s-a)(s-b)(s-c)},$$

where $s = (a+b+c)/2$ is one-half its perimeter. If the length of each side increases at a rate of 1 cm/min, how fast is A changing when $a = 3$ cm, $b = 4$ cm, and $c = 5$ cm?

SECTION 4.9

Indeterminate Forms and L'Hôpital's Rule

Derivatives are instantaneous rates of change, defined as limits of average rates of change. As a result, it was necessary to discuss limits in Chapter 2 prior to the introduction of derivatives in Section 3.1. Now that we have derivatives, it may seem quite surprising that they can be used to evaluate certain limits.

The Indeterminate Form 0/0

If we let x approach zero in numerator and denominator of the limit

$$\lim_{x \to 0^+} \frac{\sqrt{1+x} - 1}{\sqrt{x}}, \qquad (4.16)$$

we find that
$$\lim_{x \to 0^+} \left(\sqrt{1+x} - 1\right) = 0 \quad \text{and} \quad \lim_{x \to 0^+} \sqrt{x} = 0.$$

We say that limit 4.16 is of the **indeterminate form** $0/0$. Likewise, the limit

$$\lim_{x \to 2} \frac{x^3 - x^2 - 8x + 12}{x^2 - 4x + 4} \tag{4.17}$$

is of the indeterminate form $0/0$ since both numerator and denominator approach zero as x approaches 2.

In Section 2.1 we evaluated limit 4.16 by rationalizing the numerator (see Example 2.6). Factoring numerator and denominator in limit 4.17 gives

$$\lim_{x \to 2} \frac{x^3 - x^2 - 8x + 12}{x^2 - 4x + 4} = \lim_{x \to 2} \frac{(x-2)^2(x+3)}{(x-2)^2} = \lim_{x \to 2} (x+3) = 5.$$

These two examples illustrate the "trickery" to which we resorted in Chapter 2 in order to evaluate limits. With Cauchy's generalized mean value theorem from Section 3.11, however, we can prove a result called L'Hôpital's rule, which makes evaluation of many limits of the indeterminate form $0/0$ quite simple.

Theorem 4.5

(L'Hôpital's rule) Let $f(x)$ and $g(x)$ be functions that are differentiable on an open interval I except possibly at the point $x = a$ in I. Suppose further that $g'(x) \neq 0$ in I, except possibly at a. If $\lim_{x \to a} f(x) = 0$ and $\lim_{x \to a} g(x) = 0$, and if

$$\lim_{x \to a} \frac{f'(x)}{g'(x)} = L,$$

then

$$\lim_{x \to a} \frac{f(x)}{g(x)} = L.$$

Proof We define two functions $F(x)$ and $G(x)$ that are identical to $f(x)$ and $g(x)$ on I but have value zero at $x = a$:

$$F(x) = \begin{cases} f(x) & x \neq a \\ 0 & x = a, \end{cases} \qquad G(x) = \begin{cases} g(x) & x \neq a \\ 0 & x = a. \end{cases}$$

Since $F'(x) = f'(x)$ and $G'(x) = g'(x)$ for x in I except possibly at $x = a$, $F(x)$ and $G(x)$ are therefore differentiable on I except possibly at $x = a$. In addition, differentiability of a function implies continuity of the function (Theorem 3.6), so that $F(x)$ and $G(x)$ must certainly be continuous on I except possibly at $x = a$. But

$$\lim_{x \to a} F(x) = \lim_{x \to a} f(x) = 0 = F(a),$$

and the same is true for $G(x)$; hence, $F(x)$ and $G(x)$ are continuous for all x in I. Consequently, $F(x)$ and $G(x)$ are identical to $f(x)$ and $g(x)$ in every respect, except that they have been assigned a value at $x = a$ to guarantee their continuity there. This extra condition permits us to apply Cauchy's generalized mean value theorem to $F(x)$ and $G(x)$ on the interval between a and x, so long as x is in I. We can conclude that there exists a number c between a and x such that

$$\frac{F(x) - F(a)}{G(x) - G(a)} = \frac{F'(c)}{G'(c)},$$

or, since $F(a) = G(a) = 0$,
$$\frac{F(x)}{G(x)} = \frac{F'(c)}{G'(c)}.$$

Since x and c are points in I, we can also write that $F(x) = f(x)$, $F'(c) = f'(c)$, $G(x) = g(x)$, $G'(c) = g'(c)$, and therefore

$$\frac{f(x)}{g(x)} = \frac{F(x)}{G(x)} = \frac{F'(c)}{G'(c)} = \frac{f'(c)}{g'(c)}.$$

If we now let x approach a, then c must also approach a since it is always between a and x. Consequently,

$$\lim_{x \to a} \frac{f(x)}{g(x)} = \lim_{c \to a} \frac{f'(c)}{g'(c)} = \lim_{x \to a} \frac{f'(x)}{g'(x)},$$

and if

$$L = \lim_{x \to a} \frac{f'(x)}{g'(x)},$$

it follows that

$$\lim_{x \to a} \frac{f(x)}{g(x)} = L.$$

This theorem is also valid if L is replaced by ∞ or $-\infty$. The only difference in the proof is to make the same change in the last sentence.

Theorem 4.5 is also valid if $x \to a$ is replaced by either a right-hand limit, $x \to a^+$, or a left-hand limit, $x \to a^-$. The only difference in these cases is that interval I is replaced by open intervals $a < x < b$ and $b < x < a$ respectively, and the proofs are almost identical. In addition, the following theorem indicates that $x \to a$ can be replaced by $x \to \infty$ (or $x \to -\infty$).

Theorem 4.6

(L'Hôpital's rule) Let $f(x)$ and $g(x)$ be functions that are differentiable for some interval $x > b > 0$, and suppose that $g'(x) \neq 0$ on this interval. If $\lim_{x \to \infty} f(x) = 0$ and $\lim_{x \to \infty} g(x) = 0$, and if

$$\lim_{x \to \infty} \frac{f'(x)}{g'(x)} = L \quad (\text{or } \pm\infty),$$

then

$$\lim_{x \to \infty} \frac{f(x)}{g(x)} = L \quad (\text{or } \pm\infty).$$

In other words, L'Hôpital's rule applies to any type of limit that yields the indeterminate form $0/0$ (be it $x \to a$, $x \to a^+$, $x \to a^-$, $x \to \infty$, or $x \to -\infty$). A common error when using L'Hôpital's rule is to differentiate $f(x)/g(x)$ with the quotient rule and then take the limit of the resulting derivative. L'Hôpital's rule calls for the limit of $f'(x)/g'(x)$; $f(x)$ and $g(x)$ are differentiated separately.

If we use L'Hôpital's rule on limit 4.16, we find

$$\lim_{x \to 0^+} \frac{\sqrt{1+x} - 1}{\sqrt{x}} = \lim_{x \to 0^+} \frac{\frac{1}{2\sqrt{1+x}}}{\frac{1}{2\sqrt{x}}} = \lim_{x \to 0^+} \frac{\sqrt{x}}{\sqrt{1+x}} = 0.$$

For limit 4.17, we have

$$\lim_{x \to 2} \frac{x^3 - x^2 - 8x + 12}{x^2 - 4x + 4} = \lim_{x \to 2} \frac{3x^2 - 2x - 8}{2x - 4},$$

which is still a limit of the indeterminate form $0/0$. Note that this is a conditional equation; that is, it says that the limit on the left is equal to the limit on the right, provided the limit on the right exists. If we apply L'Hôpital's rule a second time, to the limit on the right, we obtain

$$\lim_{x \to 2} \frac{x^3 - x^2 - 8x + 12}{x^2 - 4x + 4} = \lim_{x \to 2} \frac{6x - 2}{2} = 5.$$

EXAMPLE 4.25

Evaluate the following limits:

(a) $\lim_{x \to -3} \dfrac{3 + x}{\sqrt{3} - \sqrt{-x}}$
(b) $\lim_{x \to 0} \dfrac{\tan x}{x}$

(c) $\lim_{x \to 4} \dfrac{x - 4}{x^2 - 8x + 16}$
(d) $\lim_{x \to 4} \dfrac{x^3 - 4x^2 + 9x - 36}{x^2 + 5}$

SOLUTION (a) Since we have the indeterminate form $0/0$, we use L'Hôpital's rule to write

$$\lim_{x \to -3} \frac{3 + x}{\sqrt{3} - \sqrt{-x}} = \lim_{x \to -3} \frac{1}{\dfrac{1}{2\sqrt{-x}}} = \lim_{x \to -3} 2\sqrt{-x} = 2\sqrt{3}.$$

(b) If we use L'Hôpital's rule, we have

$$\lim_{x \to 0} \frac{\tan x}{x} = \lim_{x \to 0} \frac{\sec^2 x}{1} = 1.$$

(c) By L'Hôpital's rule,

$$\lim_{x \to 4} \frac{x - 4}{x^2 - 8x + 16} = \lim_{x \to 4} \frac{1}{2x - 8}.$$

Since

$$\lim_{x \to 4^+} \frac{1}{2x - 8} = \infty \quad \text{and} \quad \lim_{x \to 4^-} \frac{1}{2x - 8} = -\infty,$$

we conclude that

$$\lim_{x \to 4^+} \frac{x - 4}{x^2 - 8x + 16} = \infty \quad \text{and} \quad \lim_{x \to 4^-} \frac{x - 4}{x^2 - 8x + 16} = -\infty.$$

(d) This limit is not of the indeterminate form $0/0$ since $\lim_{x \to 4} (x^2 + 5) = 21$; thus we cannot use L'Hôpital's rule. Since $\lim_{x \to 4} (x^3 - 4x^2 + 9x - 36) = 0$,

$$\lim_{x \to 4} \frac{x^3 - 4x^2 + 9x - 36}{x^2 + 5} = 0.$$

∎

Had we used L'Hôpital's rule in part (d) of Example 4.25, we would have obtained an incorrect answer:

$$\lim_{x \to 4} \frac{x^3 - 4x^2 + 9x - 36}{x^2 + 5} = \lim_{x \to 4} \frac{3x^2 - 8x + 9}{2x} = \frac{25}{8}.$$

In other words, L'Hôpital's rule is not to be used indiscriminately; *it must be used only on the indeterminate forms for which it is designed.*

The Indeterminate Form ∞/∞

The limit

$$\lim_{x \to \infty} \frac{1 + \sqrt{x-1}}{2x + 5}$$

is said to be of the **indeterminate form** ∞/∞ since numerator and denominator become increasingly large as $x \to \infty$. Theorems 4.5 and 4.6 for L'Hôpital's rule can be adapted to this indeterminate form also; hence we calculate that

$$\lim_{x \to \infty} \frac{1 + \sqrt{x-1}}{2x + 5} = \lim_{x \to \infty} \frac{\frac{1}{2\sqrt{x-1}}}{2} = \lim_{x \to \infty} \frac{1}{4\sqrt{x-1}} = 0.$$

EXAMPLE 4.26

Evaluate the following limits, if they exist:

(a) $\displaystyle\lim_{x \to \infty} \frac{x^2}{e^x}$ (b) $\displaystyle\lim_{x \to -\infty} \frac{\sqrt{2x^2 + 3x + 2}}{1 - x}$

SOLUTION (a) Since this limit exhibits the indeterminate form ∞/∞, we use L'Hôpital's rule to write

$$\lim_{x \to \infty} \frac{x^2}{e^x} = \lim_{x \to \infty} \frac{2x}{e^x}.$$

Since this limit is still of the form ∞/∞, we use L'Hôpital's rule again,

$$\lim_{x \to \infty} \frac{x^2}{e^x} = \lim_{x \to \infty} \frac{2}{e^x} = 0.$$

(b) By L'Hôpital's rule,

$$\lim_{x \to -\infty} \frac{\sqrt{2x^2 + 3x + 2}}{1 - x} = \lim_{x \to -\infty} \frac{\frac{4x + 3}{2\sqrt{2x^2 + 3x + 2}}}{-1} = \lim_{x \to -\infty} \frac{4x + 3}{2\sqrt{2x^2 + 3x + 2}}.$$

This limit is also of the indeterminate form ∞/∞. Further applications of L'Hôpital's rule do not lead to a simpler form for the limit. Thus, L'Hôpital's rule does not prove advantageous on this limit. It is better to divide numerator and denominator by x,

$$\lim_{x \to -\infty} \frac{\sqrt{2x^2 + 3x + 2}}{1 - x} = \lim_{x \to -\infty} \frac{\frac{\sqrt{2x^2 + 3x + 2}}{x}}{\frac{1}{x} - 1} = \lim_{x \to -\infty} \frac{-\sqrt{2 + \frac{3}{x} + \frac{2}{x^2}}}{\frac{1}{x} - 1} = \sqrt{2}.$$

∎

The Indeterminate Form $0 \cdot \infty$

The limits
$$\lim_{x \to \infty} xe^{-2x} \quad \text{and} \quad \lim_{x \to 0^+} x^2 \ln x$$

are said to be of the **indeterminate form** $0 \cdot \infty$. L'Hôpital's rule can again be used if we first rearrange the limits into one of the forms $0/0$ or ∞/∞:

$$\lim_{x \to \infty} xe^{-2x} = \lim_{x \to \infty} \frac{x}{e^{2x}} = \lim_{x \to \infty} \frac{1}{2e^{2x}} = 0;$$

$$\lim_{x \to 0^+} x^2 \ln x = \lim_{x \to 0^+} \frac{\ln x}{\frac{1}{x^2}} = \lim_{x \to 0^+} \frac{\frac{1}{x}}{\frac{-2}{x^3}} = \lim_{x \to 0^+} \left(-\frac{x^2}{2}\right) = 0.$$

Note that had we converted the second limit into the $0/0$ form, we would have had

$$\lim_{x \to 0^+} x^2 \ln x = \lim_{x \to 0^+} \frac{x^2}{\frac{1}{\ln x}} = \lim_{x \to 0^+} \frac{2x}{\frac{-1}{x(\ln x)^2}} = \lim_{x \to 0^+} -2x^2 (\ln x)^2.$$

Although this is correct, the limit on the right is more difficult to evaluate than the original. In other words, we must be judicious in converting a limit from the $0 \cdot \infty$ indeterminate form to either $0/0$ or ∞/∞.

EXAMPLE 4.27

Evaluate the following limits if they exist:

(a) $\displaystyle\lim_{x \to \pi/2} (x - \pi/2) \sec x$ (b) $\displaystyle\lim_{x \to 0^+} xe^{1/x}$

SOLUTION (a) $\displaystyle\lim_{x \to \pi/2} (x - \pi/2) \sec x = \lim_{x \to \pi/2} \frac{x - \pi/2}{\cos x} = \lim_{x \to \pi/2} \frac{1}{-\sin x} = -1$

(b) $\displaystyle\lim_{x \to 0^+} xe^{1/x} = \lim_{x \to 0^+} \frac{e^{1/x}}{1/x} = \lim_{x \to 0^+} \frac{e^{1/x}(-1/x^2)}{-1/x^2} = \lim_{x \to 0^+} e^{1/x} = \infty.$ ∎

The Indeterminate Forms 0^0, 1^∞, ∞^0, and $\infty - \infty$

Various other indeterminate forms arise in the evaluation of limits, and many of these can be reduced to the $0/0$ and ∞/∞ forms by introducing logarithms. In particular, the limits

$$\lim_{x \to 0^+} x^x, \quad \lim_{x \to \infty} \left(1 + \frac{1}{x}\right)^{x^2}, \quad \lim_{x \to \pi/2^-} (\sec x)^{\cos x} \quad \text{and} \quad \lim_{x \to \pi/2} (\sec x - \tan x) \quad (4.18)$$

are said to display the indeterminate forms 0^0, 1^∞, ∞^0, and $\infty - \infty$, respectively. To evaluate $\lim_{x \to 0^+} x^x$, we set
$$L = \lim_{x \to 0^+} x^x$$
and take natural logarithms of both sides,

$$\ln L = \ln \left(\lim_{x \to 0^+} x^x\right).$$

As the logarithm function is continuous, we may interchange the limit and logarithm operations (see Theorem 2.3),

$$\ln L = \lim_{x \to 0^+} (\ln x^x) = \lim_{x \to 0^+} x \ln x = \lim_{x \to 0^+} \frac{\ln x}{\frac{1}{x}}.$$

We are now in a position to use L'Hôpital's rule:

$$\ln L = \lim_{x \to 0^+} \frac{\frac{1}{x}}{\frac{-1}{x^2}} = \lim_{x \to 0^+} (-x) = 0.$$

Exponentiation of both sides of $\ln L = 0$ now gives $L = e^0 = 1$; that is,

$$\lim_{x \to 0^+} x^x = 1.$$

For the second limit in 4.18 we again set

$$L = \lim_{x \to \infty} \left(1 + \frac{1}{x}\right)^{x^2}$$

and take natural logarithms:

$$\ln L = \ln \left[\lim_{x \to \infty} \left(1 + \frac{1}{x}\right)^{x^2}\right] = \lim_{x \to \infty} \left[\ln \left(1 + \frac{1}{x}\right)^{x^2}\right]$$

$$= \lim_{x \to \infty} \left[x^2 \ln \left(1 + \frac{1}{x}\right)\right] = \lim_{x \to \infty} \left[\frac{\ln\left(\frac{x+1}{x}\right)}{\frac{1}{x^2}}\right].$$

By L'Hôpital's rule, we have

$$\ln L = \lim_{x \to \infty} \left[\frac{\frac{x}{x+1}\left(\frac{-1}{x^2}\right)}{\frac{-2}{x^3}}\right] = \lim_{x \to \infty} \frac{x^2}{2(x+1)} = \infty.$$

Consequently,

$$L = \lim_{x \to \infty} \left(1 + \frac{1}{x}\right)^{x^2} = \infty.$$

In the third limit of 4.18 we set

$$L = \lim_{x \to \pi/2^-} (\sec x)^{\cos x},$$

in which case

$$\ln L = \ln \left[\lim_{x \to \pi/2^-} (\sec x)^{\cos x}\right] = \lim_{x \to \pi/2^-} [\cos x \ln (\sec x)]$$

$$= \lim_{x \to \pi/2^-} \left[\frac{\ln (\sec x)}{\sec x}\right] = \lim_{x \to \pi/2^-} \left[\frac{\frac{1}{\sec x} \sec x \tan x}{\sec x \tan x}\right]$$

$$= \lim_{x \to \pi/2^-} \cos x = 0.$$

Thus,
$$L = \lim_{x \to \pi/2^-} (\sec x)^{\cos x} = e^0 = 1.$$

Finally, the last limit in 4.18 is evaluated by rewriting it in the $0/0$ form,

$$\lim_{x \to \pi/2} (\sec x - \tan x) = \lim_{x \to \pi/2} \left(\frac{1 - \sin x}{\cos x} \right) = \lim_{x \to \pi/2} \frac{-\cos x}{-\sin x} = 0.$$

EXAMPLE 4.28

Sketch a graph of the function $f(x) = x \ln x$.

SOLUTION For critical points of $f(x)$, we solve

$$0 = f'(x) = \frac{x}{x} + \ln x.$$

The only critical point is $x = 1/e$. Since $f''(x) = 1/x$, it follows that $f''(1/e) = e > 0$, and $x = 1/e$ yields a relative minimum of $f(1/e) = -1/e$. Furthermore, since $f''(x)$ is never equal to zero, there are no points of inflection on the graph, which is always concave upward. This information, along with the point $(1, 0)$, is shown in Figure 4.87(a). To complete the sketch, we use l'Hôpital's rule to calculate

$$\lim_{x \to 0^+} x \ln x = \lim_{x \to 0^+} \frac{\ln x}{\frac{1}{x}} = \lim_{x \to 0^+} \frac{\frac{1}{x}}{-\frac{1}{x^2}} = \lim_{x \to 0^+} (-x) = 0^-,$$

and add the extra information that

$$\lim_{x \to 0^+} f'(x) = -\infty.$$

The final graph is shown in Figure 4.87(b).

FIGURE 4.87

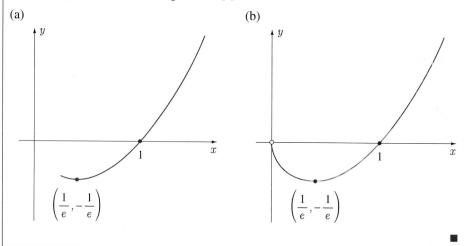

EXERCISES 4.9

In Exercises 1–42 evaluate the limit if it exists.

1. $\lim_{x \to 0} \dfrac{x^2 + 3x}{x^3 + 5x^2}$
2. $\lim_{x \to 3} \dfrac{x^2 - 9}{x - 3}$
3. $\lim_{x \to -\infty} \dfrac{x^3 + 3x - 2}{x^2 + 5x + 1}$
4. $\lim_{x \to \infty} \dfrac{2x^2 + 3x}{5x^3 + 4}$
5. $\lim_{x \to 5} \dfrac{x^2 - 10x + 25}{x^3 - 125}$
6. $\lim_{x \to 1} \dfrac{1}{(x - 1)^2}$
7. $\lim_{x \to \infty} \dfrac{\sqrt{x^2 + 1}}{2x + 5}$
8. $\lim_{x \to -\infty} \dfrac{\sin x}{x}$
9. $\lim_{x \to \infty} \dfrac{\sin(2/x)}{\sin(1/x)}$
10. $\lim_{x \to \pi/2} \dfrac{\cos x}{(x - \pi/2)^2}$
11. $\lim_{x \to 1^+} \dfrac{(1 - 1/x)^3}{\sqrt{x - 1}}$
12. $\lim_{x \to \infty} \dfrac{\sin(1/x)}{1/x^2}$
13. $\lim_{x \to 9^-} \dfrac{\sqrt{x} - 3}{\sqrt{9 - x}}$
14. $\lim_{x \to 0} \dfrac{\sqrt{5 + x} - \sqrt{5 - x}}{x}$
15. $\lim_{x \to 0} \dfrac{x - \sin x}{x^3}$
16. $\lim_{x \to a} \dfrac{x^n - a^n}{x - a}$
17. $\lim_{x \to 0} \dfrac{(1 - \cos x)^2}{3x^2}$
18. $\lim_{x \to 0} \dfrac{\tan x}{x}$
19. $\lim_{x \to 0} \dfrac{\sin 3x}{\tan 2x}$
20. $\lim_{x \to 1} \dfrac{\left(1 - \sqrt{2 - x}\right)^{3/2}}{x - 1}$
21. $\lim_{x \to 0} \dfrac{\sqrt{x + 1} - \sqrt{2x + 1}}{\sqrt{3x + 4} - \sqrt{2x + 4}}$
22. $\lim_{x \to 0} \dfrac{(1 - \cos x)^2}{3x^4}$
23. $\lim_{x \to \infty} x \sin\left(\dfrac{1}{x}\right)$
24. $\lim_{x \to 2} \dfrac{(x - 2)^{10}}{\left(\sqrt{x} - \sqrt{2}\right)^{10}}$
25. $\lim_{x \to 0} \left(\dfrac{4}{x^2} - \dfrac{2}{1 - \cos x}\right)$
26. $\lim_{x \to -\infty} xe^x$
27. $\lim_{x \to \infty} x^2 e^{-4x}$
28. $\lim_{x \to -\infty} x \sin\left(\dfrac{4}{x}\right)$
29. $\lim_{x \to 0} x \cot x$
30. $\lim_{x \to 0} \csc x (1 - \cos x)$
31. $\lim_{x \to 0^+} (\sin x)^x$
32. $\lim_{x \to 0^+} x^{\sin x}$
33. $\lim_{x \to \infty} \left(\dfrac{x + 5}{x + 3}\right)^x$
34. $\lim_{x \to 0} (1 + x)^{\cot x}$
35. $\lim_{x \to \infty} x^{1/x}$
36. $\lim_{x \to 0^+} |\ln x|^{\sin x}$
37. $\lim_{x \to 0^+} xe^{1/x}$
38. $\lim_{x \to 0} (\tan x - \csc x)$
39. $\lim_{x \to 0} (\csc x - \cot x)$
40. $\lim_{x \to 1} \left(\dfrac{x}{\ln x} - \dfrac{1}{x \ln x}\right)$
41. $\lim_{x \to 1} \left(\dfrac{x}{x - 1} - \dfrac{1}{\ln x}\right)$
42. $\lim_{x \to 0} \left(\dfrac{1}{x^2} - \dfrac{1}{\sin^2 x}\right)$

In Exercises 43–54 sketch a graph of the function.

43. $f(x) = xe^{-2x}$
44. $f(x) = x^2 e^{3x}$
45. $f(x) = xe^{-x^2}$
46. $f(x) = e^{1/x}$
47. $f(x) = \dfrac{\ln x}{x}$
48. $f(x) = x^2 \ln x$
49. $f(x) = xe^{1/x}$
50. $f(x) = \dfrac{x^2}{\ln x}$
51. $f(x) = x^x, \quad x > 0$
52. $f(x) = x^{10} e^{-x}$
53. $f(x) = e^{-x} \ln x, \quad x > 0$
54. $f(x) = 2 \csc x - \cot x, \quad 0 < x < \pi/2$

55. Evaluate
$$\lim_{x \to \infty} \left(\dfrac{x + a}{x + b}\right)^{cx}$$
for any constants a, b, and c.

56. The indeterminate forms 0^0, 1^∞, and ∞^0 are often evaluated by introducing logarithms. Show that the limit
$$\lim_{x \to \infty} (x - \ln x)$$
can be evaluated by introducing exponentials.

57. When an electrostatic field E is applied to a gaseous or liquid polar dielectric, a net dipole moment P per unit volume is set up, where
$$P(E) = \dfrac{e^E + e^{-E}}{e^E - e^{-E}} - \dfrac{1}{E}.$$
Show that $\lim_{E \to 0^+} P(E) = 0$.

58. Planck's law for the energy density ψ of blackbody radiation states that
$$\psi = \psi(\lambda) = \dfrac{k\lambda^{-5}}{e^{c/\lambda} - 1},$$
where k and c are positive constants and λ is the wavelength of the radiation.
(a) Show that
$$\lim_{\lambda \to 0^+} \psi(\lambda) = 0 \quad \text{and} \quad \lim_{\lambda \to \infty} \psi(\lambda) = 0.$$

(b) Show that $\psi(\lambda)$ has one critical point which must satisfy the equation

$$(5\lambda - c)e^{c/\lambda} = 5\lambda.$$

Find the critical point accurate to seven decimals when $c = 0.001\,438\,6$.

(c) Sketch a graph of the function $\psi(\lambda)$.

59. The following limit arises in the calculation of the electric field intensity for a half-wave antenna:

$$\lim_{\theta \to 0} f(\theta),$$

where

$$f(\theta) = \sin\theta \left\{ \frac{\sin[\pi/2(\cos\theta - 1)]}{\cos\theta - 1} + \frac{\sin[\pi/2(\cos\theta + 1)]}{\cos\theta + 1} \right\}.$$

Evaluate this limit by first showing that $f(\theta)$ can be written in the form

$$f(\theta) = \frac{2\cos\left(\dfrac{\pi}{2}\cos\theta\right)}{\sin\theta},$$

and then using L'Hôpital's rule.

60. Find all values of a, b, and c for which

$$\lim_{x \to 0} \frac{e^{ax} - bx - \cos(x + cx^2)}{2x^3 + 5x^2} = 5.$$

61. (a) Sketch a graph of the function

$$f(x) = \begin{cases} e^{-1/x^2} & x \neq 0 \\ 0 & x = 0. \end{cases}$$

(b) Show that for every positive integer n,

$$\lim_{x \to 0} \frac{e^{-1/x^2}}{x^n} = 0.$$

(c) Prove by mathematical induction that $f^{(n)}(0) = 0$, where $f^{(n)}(0)$ is the nth derivative of $f(x)$ evaluated at $x = 0$.

SECTION 4.10

Differentials

In Section 3.1 we pointed out that the notation dy/dx for the derivative of a function $y = f(x)$ should not be considered a quotient. In this section, we change this. It is not that we lied to you in Chapter 3; it's just that we were not prepared to tell you the whole story. Besides, nothing in Chapters 3 and 4 would have been any easier had we regarded dy/dx as "dy divided by dx". Beginning in Chapter 5, however, it is essential that we be able to do this, and therefore in this section we define "differentials" dx and dy so that dy/dx can be regarded as a quotient.

When we use the notation

$$\frac{dy}{dx} = \lim_{\Delta x \to 0} \frac{f(x + \Delta x) - f(x)}{\Delta x}$$

for the derivative of a function $y = f(x)$, we call Δx an increment in x. It represents a change in the value of the independent variable from some value x to another value $x + \Delta x$. This change can be positive or negative depending on whether we want $x + \Delta x$ to be larger or smaller than x. When the independent variable changes from x to $x + \Delta x$, the dependent variable changes by an amount Δy where

$$\Delta y = f(x + \Delta x) - f(x). \qquad (4.19)$$

In other words, Δy is the change in y resulting from the change Δx in x. For the function in Figure 4.88, Δy is positive when Δx is positive, and Δy is negative when Δx is negative.

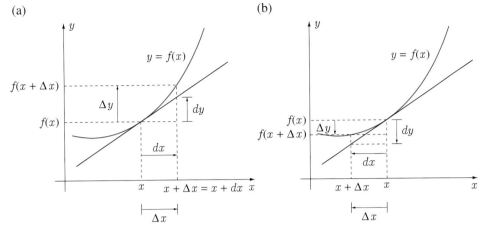

FIGURE 4.88

For example, when $y = x^2 - 2x$, the change Δy in y when x is changed from 3 to 3.2 is
$$\Delta y = [(3.2)^2 - 2(3.2)] - [3^2 - 2(3)] = 0.84.$$
The function increases by 0.84 when x increases from 3 to 3.2.

For purposes of integration, a topic which begins in Chapter 5 and continues in every chapter thereafter, an alternative notation for an increment in the independent variable x is more suggestive.

> **Definition 4.7**
>
> An increment Δx in the independent variable x is denoted by
> $$dx = \Delta x, \quad (4.20)$$
> and when written as dx, it is called the **differential** of x.

The differential dx is synonymous with the increment Δx; it represents a change in x (in most applications a very small change). The differential of the dependent variable is *not* synonymous with Δy.

> **Definition 4.8**
>
> The **differential** of $y = f(x)$, corresponding to the differential dx in x, is denoted by dy and is defined by
> $$dy = f'(x)\,dx. \quad (4.21)$$

The difference between Δy and dy is most easily seen in Figure 4.88. We know that Δy is the exact change in the function $y = f(x)$ when x is changed by an amount Δx or dx. It is the difference in the height of the curve at x and $x + \Delta x = x + dx$. Now the slope of the tangent line to the graph at the point (x, y) is $f'(x)$. Definition 4.8 indicates that dy can be interpreted as the difference in the height of this tangent line at x and at $x + dx$. In other words, dy is the change in y corresponding to the change dx in x if we follow the tangent line to $y = f(x)$ at (x, y) rather than the curve itself.

Figure 4.88 also suggests that when dx is very small (close to zero), dy is approximately equal to Δy; that is,
$$dy \approx \Delta y \quad \text{when } dx \approx 0.$$

This is also illustrated in the following numerical example.

EXAMPLE 4.29

Find Δy and dy for the function $y = f(x) = \sqrt{x^2 + 1}$ when $x = 2$ and $dx = 0.1$.

SOLUTION According to equation 4.19, the change in y as x increases from 2 to 2.1 is

$$\Delta y = f(2.1) - f(2) = \sqrt{(2.1)^2 + 1} - \sqrt{2^2 + 1} = 0.08987.$$

Since $f'(x) = x/\sqrt{x^2 + 1}$, the differential of y for $x = 2$ and $dx = 0.1$ is

$$dy = f'(2)(0.1) = \frac{2}{\sqrt{2^2 + 1}}(0.1) = 0.08944.$$

The difference between dy and Δy is therefore 0.00043, a difference of 43 parts in 8987. ∎

Before the invention of electronic calculators, differentials were used to approximate a function near points at which it was easily evaluated. For example, imagine trying to evaluate the function $f(x) = x^{1/3}$ at $x = 126$ without a calculator. We could use differentials to approximate $126^{1/3}$ as follows:

$$126^{1/3} = f(126) = f(125) + \Delta y \approx f(125) + dy$$
$$= 5 + f'(5)(1) = 5 + \frac{1}{3(125)^{2/3}} = 5 + \frac{1}{75} = \frac{376}{75}.$$

With the advent of the electronic calculator, problems of this type exist, but it is ridiculous to attack them with differentials. Differentials can be useful when we examine changes in a function without specifying values for the independent variable. Very prominent in this context are the "relative" and "percentage" changes.

When a quantity y undergoes a change Δy, then its *relative change* is defined as

$$\frac{\Delta y}{y}, \qquad (4.22)$$

and its *percentage change* is given by

$$100\frac{\Delta y}{y}. \qquad (4.23)$$

Sometimes relative and percentage changes are more important than actual changes. To illustrate this, consider the function $V = 4\pi r^3/3$ which represents the volume of a sphere. If the radius of the sphere is increased from 0.10 m to 0.11 m, then the change in the volume of the sphere is

$$\Delta V = \frac{4}{3}\pi(0.11)^3 - \frac{4}{3}\pi(0.10)^3 = 4.4\pi \times 10^{-4} \text{ m}^3.$$

This is not a very large quantity, but in relation to the original size of the sphere, we have a relative change of

$$\frac{\Delta V}{V} = \frac{4.4\pi \times 10^{-4}}{4\pi(0.10)^3/3} = 0.33,$$

and a percentage change of
$$100 \frac{\Delta V}{V} = 33\%.$$

Suppose the same increase of 0.01 m is applied to a sphere with radius 100 m. The change in the volume is
$$\Delta V = \frac{4}{3}\pi(100.01)^3 - \frac{4}{3}\pi(100)^3 = 4.0\pi \times 10^2 \text{ m}^3.$$

This is quite a large change in volume, but the relative change is
$$\frac{\Delta V}{V} = \frac{4.0\pi \times 10^2}{4\pi(100)^3/3} = 3.0 \times 10^{-4},$$

and the percentage change is
$$100 \frac{\Delta V}{V} = 0.03\%.$$

Although the change 400π in V when $r = 100$ is much larger than the change 0.00044π when $r = 0.1$, the relative and percentage changes are much smaller when $r = 100$. We see, then, that in certain cases, it may be relative and percentage changes that are significant rather than actual changes.

In the above example, when $r = 100$ m, the change $dr = 0.01$ is certainly small compared to r. We should therefore be able to use the differential $dV = V'(r)\,dr = 4\pi r^2\,dr$ to approximate ΔV. With $r = 100$ and $dr = 0.01$, we have
$$dV = 4\pi(100)^2(0.01) = 4.0\pi \times 10^2 \text{ m}^3.$$

(To two significant figures, dV is equal to ΔV.) Thus relative change in V is approximately equal to dV/V and percentage change $100\,dV/V$. So for small changes in an independent variable, the differential of the dependent variable may be used in place of its increment in the calculation of relative and percentage changes; that is, equations (4.22) and (4.23) can be replaced by
$$\frac{dy}{y} \quad \text{and} \quad 100\frac{dy}{y}.$$

We do this in the following example.

EXAMPLE 4.30

When a pendulum swings, the frequency (number of cycles per second) of its oscillations is given by
$$f = h(l) = 2\pi\sqrt{\frac{g}{l}},$$
where l is the length of the pendulum and $g > 0$ is a constant. If the length of the pendulum is increased by $\frac{1}{4}\%$, calculate the approximate percentage change in f.

SOLUTION   The approximate change in f is given by
$$df = h'(l)\,dl = 2\pi\sqrt{g}\left(-\frac{1}{2}\right)l^{-3/2}\,dl = -\pi\sqrt{g}\,\frac{dl}{l^{3/2}};$$

hence the approximate percentage change in f is
$$100\frac{df}{f} = 100\left(\frac{-\pi\sqrt{g}\,dl}{l^{3/2}}\right)\left(\frac{l^{1/2}}{2\pi\sqrt{g}}\right) = -\frac{1}{2}\left(100\frac{dl}{l}\right).$$

But because l increases by $\frac{1}{4}\%$, it follows that $100(dl/l) = 1/4$, and

$$100\frac{df}{f} = -\frac{1}{8}.$$

The frequency therefore changes by $-\frac{1}{8}\%$, the negative sign indicating that because l increases, f decreases. ∎

The differential dy cannot always be used as an approximation for the actual change Δy in a function $y = f(x)$. Sometimes it cannot be used even when dx is very close to zero. For example, if $f(x) = 2x^3 + 9x^2 - 24x + 6$, then

$$dy = f'(x)\, dx = (6x^2 + 18x - 24)\, dx.$$

If x is changed from 1 to 1.01, then the approximate change in y as predicted by the differential is

$$dy = (6 + 18 - 24)(0.01) = 0.$$

In fact, for any dx whatsoever, we find that $dy = 0$. Geometrically speaking, we can see why. Since $f'(1) = 0$, $x = 1$ is a critical point of $f(x)$ (Figure 4.89); therefore dy, which is the tangent line approximation to Δy, will always be zero. We cannot use differentials to approximate function changes at critical points.

FIGURE 4.89

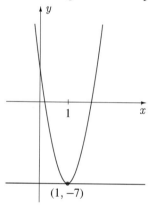

$(1, -7)$

The differential for the function $y = f(x) = x^{100}$ is

$$dy = 100x^{99}\, dx.$$

If x is changed by 1% from $x = 1$ to $x = 1.01$, then $dx = 0.01$ and

$$dy = 100(1)^{99}(0.01) = 1.0.$$

The actual change in y is

$$\Delta y = (1.01)^{100} - 1^{100} = 1.7.$$

We would hardly regard this dy as a very good approximation for Δy even though the change in x is only one percent.

This latter example begs the question. "How small in general must dx be in order that dy be a reasonable approximation for Δy?" This is not a simple question to answer. We will discuss approximations in more detail in Chapter 11, and then be able to answer the more important question, "How good an approximation to Δy is dy?" After all, it is not much use to say that $dy \approx \Delta y$ if we cannot say to how many decimal places the approximation is accurate. Suffice it to say now that use of the differential dy to approximate Δy is to be regarded with some reservation. This is not to say that differentials are useless. We will see in Chapters 5–7 that differentials are indispensable to the topic of integration.

We make one last comment before leaving this section. If equation 4.21 is divided by differential dx, then

$$\frac{dy}{dx} = f'(x).$$

Now, the left side of this equation is the differential of y divided by the differential of x. The quotient of differentials dy and dx is equal to the derivative $f'(x)$. The entity dy/dx can henceforth be regarded either as "the derivative of y with respect to x", or "dy divided by dx", whichever is appropriate for the discussion at hand.

EXERCISES 4.10

In Exercises 1–10 find dy in terms of x and dx.

1. $y = x^2 + 3x - 2$
2. $y = \dfrac{x+1}{x-1}$
3. $y = \sqrt{x^2 - 2x}$
4. $y = \sin(x^2 + 2) - \cos x$
5. $y = x^{1/3} - x^{5/3}$
6. $y = x^3 \sqrt{3 - 4x^2}$
7. $y = x^2 \sin x$
8. $y = \dfrac{x^3 - 3x^2 + 3x + 5}{x^2 - 2x + 1}$
9. $y = \sqrt{1 + \sqrt{1-x}}$
10. $y = \dfrac{x^2(x-2)}{x^3 + 5x}$

11. The momentum M and kinetic energy K of a mass m moving with speed v are given by

$$M = mv \quad \text{and} \quad K = \frac{1}{2}mv^2.$$

If v is changed by 1%, what are approximate percentage changes in M and K?

12. The magnitude of the gravitational force of attraction between two point masses m and M is given by

$$F = \frac{GmM}{r^2},$$

where $G > 0$ is a constant, and r is the distance between the masses. If r changes by 2%, by how much does F change approximately?

13. Under adiabatic expansion, a gas obeys the law

$$PV^{7/5} = \text{a constant},$$

where P is pressure and V is volume. If the pressure is increased by 2%, find the approximate percentage change in the volume.

14. The magnitude of the gravitational force of attraction between two point masses m and M is defined in Exercise 12. If the earth is considered a perfect sphere (radius 6.37×10^6 m), then this law predicts a gravitational attraction of $9.81 m$ N on a mass m on its surface. Use differentials to determine the height above the surface of the earth at which the gravitational attraction decreases to $9.80 m$ N.

SECTION 4.11

Problem Solving

Differentiation is a powerful tool for solving problems in many fields — physics, chemistry, engineering, psychology, medecine, and economics, to name a few. Many of these applications involve related rate and applied maxima-minima problems, and are examples of *word* problems. In this section we give some suggestions on how to handle word problems, which may be helpful in other mathematical problems as well. As we follow these suggestions through the examples, we illustrate the logical thought processes that lead from problem to solution.

Three suggestions important to many mathematical problems, and in particular to those that are calculus-based, are:

Suggestion 1: Make a diagram whenever possible to illustrate all given information.

Suggestion 2: Identify the nature of the problem. Is it a related rate problem, a maxima-minima problem, or some other kind of problem. There are always key words in the statement of the problem to help make this decision. Find them.

This suggestion is extremely important because further steps in the solution are intimately connected to the nature of the problem. However, most problems will also follow the next suggestion.

Suggestion 3: Choose letters to represent key variables and find appropriate formulas or equations involving these variables.

Problem 1 The slope and y-intercept of the line $y = mx + b$ are such that the line intersects the parabola $y = x^2$ in two points P and Q. A point R on the parabola between P and Q is joined to P and Q to create a triangle PQR. Where should R be chosen in order that the area of the triangle be as large as possible?

Certainly we can follow Suggestion 1 and begin with Figure 4.90(a). To identify the nature of the problem (Suggestion 2), we key on the last four words "as large as possible"; they identify that we have a maximum problem. Find the position of R in order to maximize the area of triangle PQR. According to Suggestion 3 we should choose a letter, say A, to represent the area of triangle PQR, and now find a formula for A. What should A be expressed in terms of? Since we want to find the position of R that maximizes A, A should be expressed in terms of R. How do we do this? By giving coordinates to R. If its x-coordinate is denoted by x, then its y-coordinate is x^2 (since it is on the parabola). Notice that we have not assigned coordinates to P and Q; we will do this only if it becomes necessary. We now express A in terms of x since x identifies the position of R. Because the area of a triangle is one-half the product of its base and height, we drop a perpendicular from R to PQ to intersect PQ at S (Figure 4.90(b)). Then

$$A = \frac{1}{2}||PQ||\,||RS||.$$

A key point to realize here is that P and Q are fixed, and therefore so also is length $||PQ||$ of the base of the triangle. It is not necessary to find a formula for $||PQ||$; simply regard it as a constant.

FIGURE 4.90 (a) (b)

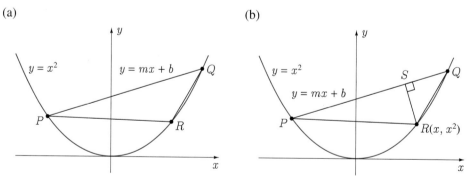

We must find $||RS||$ in terms of x. The simplest way is to use the distance formula developed in Exercise 45 of Section 1.3 and Exercise 52 of Section 4.6. With the equation of the line expressed in the form $mx - y + b = 0$, the distance from $R(x, x^2)$ to S is

$$||RS|| = \frac{|mx - y + b|}{\sqrt{m^2 + 1}} = \frac{|mx - x^2 + b|}{\sqrt{m^2 + 1}}.$$

Thus,

$$A = \frac{1}{2}||PQ||\frac{|mx - x^2 + b|}{\sqrt{m^2 + 1}}.$$

Are we ready to differentiate? No. This is an absolute maximum problem — among all points on the parabola between P and Q, find the one that maximizes A. It follows that x must be restricted to the interval $x_P \leq x \leq x_Q$ where x_P and x_Q are x-coordinates of P and Q. Our mathematical problem therefore is to find the absolute maximum of the function

$$A(x) = \frac{\|PQ\|}{2\sqrt{m^2+1}} |mx - x^2 + b|, \quad x_P \leq x \leq x_Q.$$

To find critical points of $A(x)$, we first set $A'(x) = 0$, using differentiation formula 3.12,

$$0 = A'(x) = \frac{\|PQ\|}{2\sqrt{m^2+1}} \frac{|mx - x^2 + b|}{mx - x^2 + b}(m - 2x).$$

The only solution of this equation is $x = m/2$. We should now evaluate A at x_P, at the critical point $m/2$, and at x_Q. Since the triangle degenerates to a straight line when R is at either P or Q (that is, when $x = x_P$ or $x = x_Q$), it follows that $x = m/2$ must yield a maximum for A. Thus, R should be chosen as the point $(m/2, m^2/4)$.

Problem 2 A hemispherical tank of radius 3 m has a light on its upper edge as in Figure 4.91(a). A stone falls vertically along the axis of symmetry of the tank, and when it is 1 m from the bottom of the tank it is falling at 2 m/s. How fast is its shadow moving along the surface of the tank at this instant?

Although Figure 4.91(a) would seem to fullfil Suggestion 1, there is really no need to draw a 3-dimensional diagram. The shadow follows a circular path down the side of the tank which we have shown in Figure 4.91(b).

(a)

FIGURE 4.91

(b)

(c)

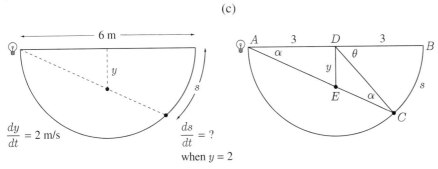

The nature of the problem is identified in the last sentence "How fast is the shadow moving". It is a related rate problem; we want the rate at which the shadow is moving, given the rate at which the stone is falling.

To follow Suggestion 3, we introduce the distance s along the circle as the distance that the shadow has travelled since the stone passed the diameter of the circle. Its rate of change with respect to time t, namely ds/dt, is what is required. To express the rate at which the stone is falling as a derivative, we let y be the distance from the diameter of the circle to the stone. Then $dy/dt = 2$ m/s. These facts have been added to Figure 4.91(b).

For a related rate problem, we must find an equation relating y and s. First, arc length along a circle is the product of the radius of the circle and the angle θ that it subtends at the centre of the circle. We therefore name some key points in Figure 4.91(c), join C and D, and write

$$s = 3\theta.$$

We wanted an equation relating s and y, not s and θ. We must therefore relate y and θ. Since triangle ADC is isosceles, angles DAC and DCA are equal — called α in Figure 4.91(c), and $\theta = 2\alpha$. Hence

$$s = 3(2\alpha) = 6\alpha.$$

From triangle ADE, we may write $y = 3\tan\alpha$. Combine this with $s = 6\alpha$, and we finally relate s and y,

$$y = 3\tan\left(\frac{s}{6}\right).$$

Differentiation with respect to t gives

$$\frac{dy}{dt} = 3\sec^2\left(\frac{s}{6}\right)\left(\frac{1}{6}\frac{ds}{dt}\right).$$

When the stone is 1 m from the bottom of the tank, $y = 2$ m, and therefore

$$2 = 3\tan\left(\frac{s}{6}\right).$$

It follows that at this instant

$$\sec^2\left(\frac{s}{6}\right) = 1 + \tan^2\left(\frac{s}{6}\right) = 1 + \left(\frac{2}{3}\right)^2 = \frac{13}{9}.$$

Substitution of this into the equation that relates dy/dt and ds/dt, along with $dy/dt = 2$, yields

$$2 = 3\left(\frac{13}{9}\right)\left(\frac{1}{6}\frac{ds}{dt}\right).$$

Hence, $ds/dt = 36/13$, and the shadow is moving at $36/13$ m/s.

Problem 3 Prove that when $0 < a < b < \pi/2$,

$$\frac{\tan b}{\tan a} > \frac{b}{a}. \tag{4.18}$$

This is not a word problem; it is included to once again illustrate the logic by which we proceed from problem to solution. We do not begin with $0 < a < b < \pi/2$; the important requirement is to verify the tangent inequality. As there is no obvious connection between the two sides of the inequality, perhaps it would be beneficial to rewrite it in the form

$$\frac{1}{b}\tan b > \frac{1}{a}\tan a. \tag{4.19}$$

This looks promising; both sides now have exactly the same form. Effectively, we have multiplied 4.18 by $b^{-1}\tan a$ to obtain 4.19. Inequality 4.19 remains in the same direction only if $b^{-1}\tan a$ is positive, and this is indeed the case because we are considering only values of a and b between 0 and $\pi/2$.

We have noted that both sides of 4.19 have the same form. In fact, they can be obtained by setting $x = b$ and $x = a$ in the function

$$f(x) = \frac{1}{x} \tan x.$$

We can now paraphrase the problem: Show that for any two points a and b in the interval $0 < x < \pi/2$, with $b > a$,

$$f(b) > f(a).$$

We must show that the function $f(x)$ is increasing on the interval $I : 0 < x < \pi/2$. But according to 4.5, this is established if $f'(x) \geq 0$ on I. Using the product rule we calculate

$$f'(x) = -\frac{1}{x^2} \tan x + \frac{1}{x} \sec^2 x = \frac{1}{x^2}(x \sec^2 x - \tan x)$$

$$= \frac{1}{x^2}\left(\frac{x}{\cos^2 x} - \frac{\sin x}{\cos x}\right) = \frac{1}{x^2 \cos^2 x}(x - \sin x \cos x).$$

Since $x^2 \cos^2 x$ is positive on I, the sign of $f'(x)$ is determined by that of $x - \sin x \cos x$. We must therefore show that on I,

$$0 \leq x - \sin x \cos x.$$

This inequality can be simplified if it is multiplied by 2,

$$0 \leq 2x - 2 \sin x \cos x = 2x - \sin 2x \quad \text{or} \quad 2x \geq \sin 2x. \quad (4.20)$$

Inequality 4.18 is verified once 4.20 is established. Graphs of the functions $2x$ and $\sin 2x$ on I are shown in Figure 4.92. Important to this picture are the facts that the line $y = 2x$ and the curve $y = \sin 2x$ both have slope 2 at $x = 0$, but the slope of $y = \sin 2x$ is less than 2 for $x > 0$. Clearly, then, $2x \geq \sin 2x$, and our proof is complete.

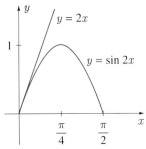

FIGURE 4.92

We have solved Problem 3 backwards. We have replaced inequality 4.18 with 4.19, which is tantamount to showing that $f(x) = x^{-1} \tan x$ is increasing on $I : 0 < x < \pi/2$. But this is established once 4.20 is verified, and Figure 4.92 verifies 4.20.

We could now rewrite the solution to this problem in a very elegant way. We could begin by saying:

"Consider the function $f(x) = x^{-1} \tan x$ on the interval $0 < x < \pi/2$." We could show that $f(x)$ is increasing and 4.19 and 4.18 would immediately follow. This is exactly how the solution would appear in most calculus books and/or solutions manuals. The author would appear brilliant; out of nowhere he would produce the function $f(x) = x^{-1} \tan x$. He actually solved the problem backwards, as above, and then rewrote it.

Problems 1, 2, and 3 have been chosen as involved, multistage problems to emphasize the importance of Suggestions 1, 2, and 3, and to illustrate how to organize our thinking into a logical sequence.

SUMMARY

In this chapter we discussed a number of applications of differentiation, the first of which was Newton's iterative procedure for approximating the roots of equations. It is perhaps the most popular of all approximation methods, because of its speed, simplicity, and accuracy.

In Section 4.3 we defined a critical point of a function as a point in its domain where its first derivative either vanishes or does not exist. Geometrically, this corresponds to a point where the graph of the function has a horizontal tangent line, a vertical tangent line, or no tangent line at all. The first derivative test indicates if critical points yield relative maxima or relative minima. A function is increasing (or decreasing) on an interval if its graph slopes upward to the right (respectively left), and this is characterized by a nonnegative (respectively nonpositive) derivative. It is concave upward (or downward) if its slope is increasing (respectively decreasing), and consequently if its second derivative is nonnegative (respectively nonpositive). Points that separate intervals of opposite concavity are called points of inflection. In Section 4.5 we showed how our curve sketching techniques, which now include addition and multiplication of ordinates, symmetries, limits, relative extrema, and points of inflection, complement one another in drawing graphs.

In Section 4.6 we illustrated that many applied extrema problems require absolute extrema rather than relative extrema. Absolute extrema of a continuous function on a closed interval must occur at either critical points or the ends of the interval. This fact implies that to find the absolute extrema of a continuous function $f(x)$ on a closed interval $a \leq x \leq b$, we evaluate $f(x)$ at its critical points between a and b and at a and b. The largest and smallest of these values are the absolute extrema of $f(x)$ on $a \leq x \leq b$.

When an object moves along a straight line, its velocity and acceleration are the first and second derivatives respectively of its displacement with respect to time. In other words, if we observe straightline motion of an object, and record its position as a function of time, then we can calculate the velocity and acceleration of that object at any instant.

Changes in a number of interrelated quantities usually produce changes in the others—sometimes small, sometimes large. How the rates of change of these variables relate to each other was the subject of Section 4.8. Related-rate problems made us acutely aware of the importance of differentiating an equation with respect to a variable only if the equation is valid for a continuous range of values of that variable.

Cauchy's generalized mean value theorem enabled us to develop L'Hôpital's rule in Section 4.9 for evaluation of various indeterminate forms such as $0/0$, ∞/∞, $0 \cdot \infty$, 0^0, 1^∞, ∞^0, and $\infty - \infty$.

Key Terms and Formulas

In reviewing this chapter, you should be able to define or discuss the following key terms:

Newton's iterative procedure
Increasing function
Decreasing function
Critical point
Relative maximum
Relative minimum
First derivative test
Concave upward
Concave downward
Point of inflection
Vertical point of inflection
Absolute maximum
Absolute minimum
Instantaneous velocity
Instantaneous acceleration
Instantaneous speed
Related rates
Indeterminate form
L'Hôpital's rule
Differential

REVIEW EXERCISES

1. (a) Prove that the area of the isosceles triangle in Figure 4.93 is
$$A = \frac{l^2 \sin \theta}{2}.$$

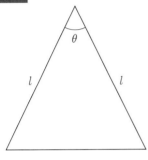

FIGURE 4.93

(b) If the angle θ is increasing at $1/2$ radian per minute, but l remains constant, how fast is the area of the triangle changing? What does your answer predict when $\theta = 0$, $\pi/2$, and π?

(c) When does A change most rapidly and most slowly; that is, when is $|dA/dt|$ largest and smallest?

2. Sketch a graph of each of the following functions, indicating all relative maxima and minima and points of inflection.

(a) $f(x) = 4x^3 + x^2 - 2x + 1$

(b) $f(x) = \dfrac{x^2 - 2x + 4}{x^2 - 2x + 1}$

3. Of all pairs of positive numbers that add to some given constant $c > 0$, find that pair which has the largest product.

4. Solve Exercise 3 for the smallest product.

5. Of all pairs of positive numbers that multiply to some given constant $c > 0$, find that pair which has the smallest sum.

6. Solve Exercise 5 for the largest sum.

7. Two sides of the triangle in Figure 4.94 maintain constant lengths of 3 cm and 4 cm but the length l of the third side decreases at the rate of 1 cm/min. How fast is angle θ changing when l is 4 cm?

FIGURE 4.94

In Exercises 8–19, evaluate the limit if it exists.

8. $\lim\limits_{x \to 0} \dfrac{3x^2 + 2x^3}{3x^3 - 2x^2}$

9. $\lim\limits_{x \to \infty} \dfrac{\sin 3x}{2x}$

10. $\lim\limits_{x \to 4} \dfrac{x^2 - 16}{x - 4}$

11. $\lim\limits_{x \to 0} \dfrac{\sin 3x}{2x}$

12. $\lim\limits_{x \to -\infty} \dfrac{\sin x^2}{2x}$

13. $\lim\limits_{x \to 2^+} \dfrac{\sqrt{x} - 2}{\sqrt{x} - \sqrt{2}}$

14. $\lim\limits_{x \to \infty} x^2 e^{-3x}$

15. $\lim\limits_{x \to 0^+} x^{2x}$

16. $\lim\limits_{x \to 0^+} x^4 \ln x$

17. $\lim\limits_{x \to 0} \dfrac{\sin 2x}{\tan 3x}$

18. $\lim\limits_{x \to \infty} \left(\dfrac{x+1}{x-1}\right)^x$

19. $\lim\limits_{x \to -\infty} xe^x$

20. Use Newton's method to find all critical points for the following functions accurate to six decimals.

(a) $f(x) = x^4 + 3x^2 - 2x + 5$

(b) $f(x) = \dfrac{x^3 + 1}{3x^3 + 5x + 1}$

21. An object moves along the x-axis with its position defined as a function of time t by
$$x = x(t) = t^4 - \frac{44}{3}t^3 + 62t^2 - 84t, \quad t \geq 0.$$

Sketch a graph of this function, indicating times when the velocity and acceleration of the object are equal to zero.

22. If the graph in Figure 4.95 represents the position $x(t)$ of an object moving along the x-axis, what could physically cause the corner at time t_0?

FIGURE 4.95

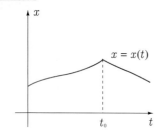

23. If an object moves along the x-axis with constant acceleration, is it possible for a graph of its position function $x(t)$ to have a point of inflection?

24. An open box is formed from a square piece of cardboard (l units long on each side) by cutting out a square at each corner and folding up the sides. What is the maximum possible volume for the box?

25. (a) If the average speed of a car for a trip is 80 km/h, must it at some time have had an instantaneous speed of 80 km/h? Explain.

(b) Must the car at some instant have had a speed of 83 km/h?

26. Sketch graphs of the following functions, indicating all relative maxima and minima and points of inflection:

(a) $f(x) = \dfrac{x^3}{x^2 - 1}$ (b) $f(x) = x^2 + \sin^2 x$

27. If at some instant of time sides a and b of the triangle in Figure 4.96 form a right angle, and if these sides are increasing at equal rates, does it remain a right-angled triangle?

FIGURE 4.96

28. A football team presently sells tickets at prices of $8, $9, and $10 per seat, depending on the position of the seat. At these prices it averages sales of 10 000 at $10, 20 000 at $9, and 30 000 at $8. The team wishes to raise the price of each ticket by the same amount, but feels that for every dollar the price is raised, 10% fewer tickets of each type will be sold. What price increase per ticket will maximize revenue?

29. Repeat Exercise 28 given that the team takes into account the fact that profit from concession sales for each person at the game is fifty cents.

30. If a particle moves away from the origin along the positive x-axis with a constant speed of 10 m/s, how fast is its distance from the curve $y = x^2$ changing when it is at $x = 3$ m?

31. Each evening a cow in a pasture returns to its barn at point B (Figure 4.97). But it always does so by first walking to the river for a drink. If the cow walks at 2 km/h and stops to drink for 2 minutes, what is the minimum time it takes for the cow to get from the pasture to the barn? What is the minimum time if the cow walks twice as fast?

FIGURE 4.97

32. A farmer has 100 hectares to plant in corn and potatoes. Undamaged corn yields p per hectare and potatoes q per hectare. For each crop, the loss due to disease and pests per unit hectare is directly proportional to the area planted. If the farmer plants x hectares of corn, the loss due to disease and pests is equal to ax per hectare. The total loss of corn is therefore ax^2 hectares. Similarly, the total loss of potatoes is by^2, if the area planted in potatoes is y hectares. Find the areas that should be planted in corn and potatoes in order to minimize monetary loss. Substitute sample values for a, b, p, and q to see whether your results look reasonable.

CHAPTER 5

The Indefinite Integral or Antiderivative

In Chapter 3 we introduced the derivative and ways to differentiate functions defined explicitly and implicitly. We then discussed various applications of calculus, including velocity and acceleration, related rates, maxima and minima, Newton's iterative procedure, and L'Hôpital's rule. In this chapter, we reverse the differentiation process. Instead of giving you a function and asking for its derivative, we give you the derivative and ask you to find the function. This process of "backwards differentiation" or "antidifferentiation" has such diverse applications that antidifferentiation is as important to calculus as differentiation. In many problems we find ourselves differentiating at one stage and antidifferentiating at another.

Antidifferentiation is a much more difficult process than differentiation. All but one of the rules for differentiation were developed in Chapter 3; we have a few more functions to differentiate in Chapter 8, and one more rule in Section 10.1. In contrast to this, the list of formulas and techniques for finding antiderivatives is endless. In this chapter we introduce the three simplest, but most important, techniques; in Chapter 9 we discuss many others.

SECTION 5.1

The Reverse Operation of Differentiation

In our discussions on velocity and acceleration in Section 4.7, we showed that when an object moves along the x-axis with its position described by the function $x(t)$, its velocity and acceleration are the first and second derivatives of $x(t)$ with respect to time t:

$$v(t) = \frac{dx}{dt} \quad \text{and} \quad a(t) = \frac{d^2 x}{dt^2}.$$

For example, if $x(t) = t^3 + 3t^2$, then

$$v(t) = 3t^2 + 6t \quad \text{and} \quad a(t) = 6t + 6.$$

When engineers and physicists study the motions of objects, a more common type of problem is to determine the position of an object; that is, the position is not given. What they might know, however, is the acceleration of the object (perhaps through Newton's second law, which states that acceleration is proportional to the resultant force on the object). So the question we must now ask is: If we know the acceleration $a(t)$ of an object as a function of time, can we obtain its velocity and position by reversing the differentiations? In the above example, if we know that

$$a(t) = \frac{dv}{dt} = 6t + 6,$$

can we find the function $v(t)$ that differentiates to give $6t+6$? We know that to arrive at the terms $6t$ and 6 after differentiation, $v(t)$ might have contained the terms $3t^2$ and $6t$. In other words, one possible velocity function that differentiates to give $a(t) = 6t + 6$ is $v(t) = 3t^2 + 6t$. It is not, however, the only one; $v(t) = 3t^2 + 6t + 10$ and $v(t) = 3t^2 + 6t - 22$ also have derivative $6t + 6$. In fact, for any constant C whatsoever, the derivative of

$$v(t) = 3t^2 + 6t + C$$

is $a(t) = 6t + 6$. In Theorem 5.1 we shall show that this velocity function represents all functions that have $6t+6$ as their derivative. We shall also demonstrate how to evaluate the constant C.

For the present, let us set $C = 0$ so that

$$v(t) = \frac{dx}{dt} = 3t^2 + 6t.$$

We now ask what position function $x(t)$ differentiates to give $3t^2 + 6t$. One possibility is $x(t) = t^3 + 3t^2$, and for the same reason as above, so is

$$x(t) = t^3 + 3t^2 + D$$

for any constant D whatsoever.

Thus, by reversing the differentiation operation in this example we have proceeded from the acceleration $a(t) = 6t + 6$ to possible velocity functions $v(t)$, and then to possible position functions $x(t)$. This process of antidifferentiation has applications far beyond velocity and acceleration problems, and we shall see many of them as we progress through this book. We begin our formal study of antidifferentiation with the following definition.

Definition 5.1

A function $F(x)$ is called an **antiderivative** or **indefinite integral** of $f(x)$ on an interval I if on I

$$F'(x) = f(x). \tag{5.1}$$

For example, since

$$\frac{d}{dx}x^4 = 4x^3,$$

we say that x^4 is an antiderivative or an indefinite integral of $4x^3$ for all x. But for any constant C, the function $x^4 + C$ is also an antiderivative of $4x^3$. The following theorem indicates that these are the only antiderivatives of $4x^3$.

Theorem 5.1

If $F(x)$ is an indefinite integral of $f(x)$ on an interval I, then every indefinite integral of $f(x)$ on I is of the form

$$F(x) + C, \quad \text{where } C \text{ is a constant.}$$

Proof Suppose that $F(x)$ and $G(x)$ are two antiderivatives of $f(x)$ on I. If we define a function $D(x) = G(x) - F(x)$, then on I

$$D'(x) = G'(x) - F'(x) = f(x) - f(x) = 0.$$

If x_1 and x_2 are any two points in the interval I, then certainly $D'(x) = 0$ on the interval $x_1 \leq x \leq x_2$. But differentiability of $D(x)$ on $x_1 \leq x \leq x_2$ implies continuity of $D(x)$ thereon also (Theorem 3.6). We may therefore apply the mean value theorem (Theorem 3.15) to $D(x)$ on the interval $x_1 \leq x \leq x_2$, and conclude that there exists a number c between x_1 and x_2 such that

$$D'(c) = \frac{D(x_2) - D(x_1)}{x_2 - x_1}.$$

But because $D'(c) = 0$, it follows that $D(x_1) = D(x_2)$. Since x_1 and x_2 are any two points in I, we conclude that $D(x)$ must have the same value at every point in I; that is, on I,

$$D(x) = G(x) - F(x) = C,$$

where C is a constant. Consequently,

$$G(x) = F(x) + C,$$

and our proof is complete.

Because of this theorem, if we find one antiderivative $F(x)$ of $f(x)$ by any means whatsoever, then we have found every antiderivative of $f(x)$, since every antiderivative can be written as $F(x)$ plus a constant C. As every indefinite integral of $f(x)$ is of the form $F(x) + C$, we call $F(x) + C$ *the* indefinite integral of $f(x)$. The operation of taking the indefinite integral is denoted by

$$\int f(x)\, dx = F(x) + C, \tag{5.2}$$

where the differential dx indicates that antidifferentiation is with respect to x. We call $f(x)$ the **integrand** of the indefinite integral. For example, we write

$$\int x^2\, dx = \frac{x^3}{3} + C \quad \text{and} \quad \int \frac{1}{x^3}\, dx = -\frac{1}{2x^2} + C.$$

The following theorem is fundamental to the calculation of antiderivatives. Its proof is a straightforward exercise in differentiation and the use of Definition 5.1.

Theorem 5.2

If $f(x)$ and $g(x)$ have antiderivatives on an interval I, then on I:

(i) $\quad \displaystyle\int [f(x) + g(x)]\, dx = \int f(x)\, dx + \int g(x)\, dx;$ (5.3)

(ii) $\quad \displaystyle\int k f(x)\, dx = k \int f(x)\, dx, \quad k$ a constant. (5.4)

For example, to evaluate the antiderivative of $2x^3 - 4x$, we write

$$\int (2x^3 - 4x)\,dx = \int 2x^3\,dx + \int -4x\,dx \qquad \text{(by part (i) of Theorem 5.2)}$$

$$= 2\int x^3\,dx - 4\int x\,dx \qquad \text{(by part (ii) of Theorem 5.2)}$$

$$= 2\left(\frac{x^4}{4}\right) - 4\left(\frac{x^2}{2}\right) + C$$

$$= \frac{x^4}{2} - 2x^2 + C.$$

Every differentiation formula developed in Chapter 3 can be expressed as an integration formula. In fact, equations 5.3 and 5.4 are integral counterparts of equations 3.8 and 3.7. Some of the differentiation formulas for trigonometric, exponential, and logarithm functions are restated below as integration formulas.

$$\int \sin x\,dx = -\cos x + C, \qquad (5.5\text{a})$$

$$\int \cos x\,dx = \sin x + C, \qquad (5.5\text{b})$$

$$\int \sec^2 x\,dx = \tan x + C, \qquad (5.5\text{c})$$

$$\int \sec x \tan x\,dx = \sec x + C, \qquad (5.5\text{d})$$

$$\int \csc^2 x\,dx = -\cot x + C, \qquad (5.5\text{e})$$

$$\int \csc x \cot x\,dx = -\csc x + C, \qquad (5.5\text{f})$$

$$\int e^x\,dx = e^x + C, \qquad (5.5\text{g})$$

$$\int a^x\,dx = a^x \log_a e + C, \qquad (5.5\text{h})$$

$$\int \frac{1}{x}\,dx = \ln|x| + C. \qquad (5.5\text{i})$$

The last equation follows immediately from 3.37 with $f(x) = x$.

Perhaps the most important integration formula is the counterpart of power rule 3.6. Since

$$\frac{d}{dx}x^{n+1} = (n+1)x^n,$$

it follows that

$$\int (n+1)x^n\,dx = x^{n+1} + \overline{C}.$$

With property 5.4, the $n+1$ can be removed from the integral and taken to the other side of the equation,

$$\int x^n\,dx = \frac{1}{n+1}x^{n+1} + C, \qquad n \neq -1, \qquad (5.6)$$

where $C = \overline{C}/(n+1)$. As indicated, this result is valid provided $n \neq -1$, but formula 5.5i takes care of this exceptional case.

EXAMPLE 5.1

Evaluate
$$\int \left(2x + \frac{1}{x^2}\right) dx.$$

SOLUTION Using Theorem 5.2 and formula (5.6), we find

$$\int \left(2x + \frac{1}{x^2}\right) dx = 2\int x\, dx + \int \frac{1}{x^2}\, dx = 2\left(\frac{x^2}{2}\right) - \frac{1}{x} + C = x^2 - \frac{1}{x} + C.$$

Where is $x^2 - 1/x + C$, the indefinite integral of $2x + 1/x^2$? It is valid on the intervals $x < 0$ and $x > 0$, but not for all x because the function is not defined at $x = 0$. To make this clear, we should write

$$\int \left(2x + \frac{1}{x^2}\right) dx = \begin{cases} x^2 - \dfrac{1}{x} + C_1 & x < 0 \\ x^2 - \dfrac{1}{x} + C_2 & x > 0 \end{cases}$$

where the constants C_1 and C_2 need not be the same. For brevity we often write

$$\int \left(2x + \frac{1}{x^2}\right) dx = x^2 - \frac{1}{x} + C,$$

thereby suppressing the complete description of the indefinite integral. When the context demands that we distinguish various intervals on which the indefinite integral is defined, we shall be careful to give the extended version. ∎

EXAMPLE 5.2

Find a curve that passes through the point $(1, 5)$ and whose tangent line at each point (x, y) has slope $5x^4 - 3x^2 + 2$.

SOLUTION If $y = f(x)$ is the equation of the curve, then

$$\frac{dy}{dx} = 5x^4 - 3x^2 + 2.$$

If we take antiderivatives of both sides of this equation with respect to x, we obtain

$$\int \frac{dy}{dx}\, dx = \int (5x^4 - 3x^2 + 2)\, dx \quad \text{or}$$

$$y = x^5 - x^3 + 2x + C.$$

Since $(1, 5)$ is a point on the curve, its coordinates must satisfy the equation of the curve:

$$5 = 1^5 - 1^3 + 2(1) + C.$$

Thus $C = 3$, and the required curve is

$$y = x^5 - x^3 + 2x + 3.$$

∎

SECTION 5.1 THE REVERSE OPERATION OF DIFFERENTIATION

In taking the antiderivative of each side of the equation

$$\frac{dy}{dx} = 5x^4 - 3x^2 + 2$$

in the above example, we added an arbitrary constant C to the right-hand side. You might question why we did not add a constant to the left-hand side. Had we done so, the result would have been

$$y + D = x^5 - x^3 + 2x + E.$$

If we then wrote

$$y = x^5 - x^3 + 2x + (E - D)$$

and defined $C = E - D$, we would have obtained exactly the same result. Hence, nothing is gained by adding an arbitrary constant to both sides; a constant on one side is sufficient.

The problem in Example 5.2 was geometric: find the equation of a curve satisfying certain properties. We quickly recast it as the problem of finding the function $y = f(x)$ that satisfies the equation

$$\frac{dy}{dx} = 5x^4 - 3x^2 + 2,$$

subject to the additional condition that $f(1) = 5$. Many applications of calculus lead to problems of this form. Find a function that satisfies an equation which involves the derivative of the function. Such an equation is called a **differential equation**. Differential equations are discussed again in Section 5.4, and before studyng them in detail in Chapter 16, we shall encounter them repeatedly in our discussions. When we solve a differential equation with no subsidiary conditions, say,

$$\frac{dy}{dx} = 4x^2 + 7x,$$

we do not get a function, but rather a one-parameter family of functions. For this differential equation, we obtain

$$y = \frac{4}{3}x^3 + \frac{7}{2}x^2 + C,$$

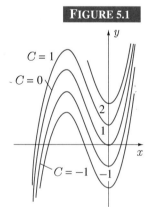

FIGURE 5.1

a one-parameter family of cubic polynomials, C being the parameter. Geometrically, we have the one-parameter family of curves in Figure 5.1. Parameter C represents a vertical shift of one curve relative to another. Note that if a vertical line is drawn at any position x to intersect these curves, then at the points of intersection, every curve has exactly the same slope, namely, $4x^2 + 7x$. The slope of each cubic at $x = 0$ is zero. If an extra condition is added to the differential equation, such as to demand that y be equal to 4 when $x = 2$, then

$$4 = \frac{4}{3}(2)^3 + \frac{7}{2}(2)^2 + C,$$

or, $C = -62/3$. This condition singles out one particular function from the family, namely, $y = 4x^3/3 + 7x^2/2 - 62/3$. Geometrically, it determines that curve in the family which passes through the point $(2, 4)$.

In this chapter, we discuss three basic ways to find antiderivatives. First, some antiderivatives are obvious, and the better you are at differentiation, the more likely this will be. For example, you should have no trouble recognizing that

$$\int 7x^5 \, dx = \frac{7x^6}{6} + C \quad \text{and} \quad \int \frac{1}{x^3} \, dx = \frac{-1}{2x^2} + C.$$

Our second method results from the answer to the following question: How do we check that a function $F(x)$ is an antiderivative of $f(x)$? We differentiate $F(x)$, of course. This simple fact suggests an approach to slightly more complex problems, say,

$$\int (2x+3)^5 \, dx.$$

We might reason that in order to have $2x+3$ raised to power 5 after differentiation, we had $2x+3$ to power 6 before differentiation; that is, a logical "guess" for the antiderivative is $(2x+3)^6$. Differentiation of this function gives

$$\frac{d}{dx}(2x+3)^6 = 6(2x+3)^5(2) = 12(2x+3)^5,$$

and we see that $(2x+3)^6$ is not the correct antiderivative. It has produced $(2x+3)^5$, as required, but it has also given an undesireable factor of 12. We therefore "fix up" our original guess by multiplying it by $1/12$; that is, the correct antiderivative is

$$\int (2x+3)^5 \, dx = \frac{1}{12}(2x+3)^6 + C.$$

This is what we call "guess and fix up": we guess the antiderivative closely enough to have it fixed up. Our guess must be correct within a *multiplicative constant*. We cannot be out by x's. Let us illustrate this with two very similar problems,

$$\int \frac{1}{(5x+2)^5} \, dx \quad \text{and} \quad \int \frac{1}{(5x^2+2)^5} \, dx.$$

For the first problem we might guess at $(5x+2)^{-4}$. Differentiation gives

$$\frac{d}{dx}\left[\frac{1}{(5x+2)^4}\right] = \frac{-4}{(5x+2)^5}(5) = \frac{-20}{(5x+2)^5}.$$

Since we are out by a factor of -20, we fix up the original guess,

$$\int \frac{1}{(5x+2)^5} \, dx = \frac{-1}{20(5x+2)^4} + C.$$

It might seem just as logical to guess at $(5x^2+2)^{-4}$ for the second problem, but differentiation yields

$$\frac{d}{dx}\left[\frac{1}{(5x^2+2)^4}\right] = \frac{-4}{(5x^2+2)^5}(10x) = \frac{-40x}{(5x^2+2)^5}.$$

This time the discrepancy is $-40x$, which involves x. We must not try to fix up the original guess; it must be abandoned totally. This indefinite integral is quite difficult, and we will have to wait for the more powerful techniques of Chapter 9. To emphasize once again, do not try to fix up x's, only constants.

EXAMPLE 5.3

Evaluate the following indefinite integrals:

(a) $\displaystyle\int \cos 3x \, dx$ (b) $\displaystyle\int e^{-2x} \, dx$ (c) $\displaystyle\int \frac{x}{3x^2 - 4} \, dx$

Solutions (a) To obtain $\cos 3x$ after differentiation, we guess $\sin 3x$ as the antiderivative. Since

$$\frac{d}{dx} \sin 3x = 3 \cos 3x,$$

it is necessary to fix up a 3,

$$\int \cos 3x \, dx = \frac{1}{3} \sin 3x + C.$$

(b) With an initial guess of e^{-2x} for the antiderivative based on the fact that the derivative of an exponential function always returns the same exponential, we calculate

$$\frac{d}{dx} e^{-2x} = -2 e^{-2x}.$$

Consequently, we fix up a -2,

$$\int e^{-2x} \, dx = -\frac{1}{2} e^{-2x} + C.$$

(c) Since

$$\frac{d}{dx} \ln(3x^2 - 4) = \frac{1}{3x^2 - 4}(6x),$$

the required antiderivative is

$$\int \frac{x}{3x^2 - 4} \, dx = \frac{1}{6} \ln|3x^2 - 4| + C.$$

We have inserted absolute values as suggested by formula 5.5i. ∎

The third technique for finding antiderivatives is discussed in Section 5.3.

EXERCISES 5.1

In Exercises 1–20 evaluate the indefinite integral.

1. $\displaystyle\int (x^3 - 2x) \, dx$

2. $\displaystyle\int (x^4 + 3x^2 + 5x) \, dx$

3. $\displaystyle\int (2x^3 - 3x^2 + 6x + 6) \, dx$

4. $\displaystyle\int \sin x \, dx$

5. $\displaystyle\int 3 \cos x \, dx$

6. $\displaystyle\int \sqrt{x} \, dx$

7. $\displaystyle\int \left(x^{10} - \frac{1}{x^3} \right) dx$

8. $\displaystyle\int \left(\frac{1}{x^2} - \frac{2}{x^4} \right) dx$

9. $\displaystyle\int \left(x^{3/2} - x^{2/7} \right) dx$

10. $\displaystyle\int \left(\frac{1}{x^2} + \frac{1}{2\sqrt{x}} \right) dx$

11. $\displaystyle\int \left(\frac{4}{x^{3/2}} + 2 x^{1/3} \right) dx$

12. $\displaystyle\int \left(-\frac{1}{2x^2} + 3x^3 \right) dx$

13. $\displaystyle\int \frac{1}{x^\pi} \, dx$

14. $\displaystyle\int \left(2\sqrt{x} + 3x^{3/2} - 5x^{5/2} \right) dx$

15. $\displaystyle\int x^2 (x^2 - 3) \, dx$

16. $\displaystyle\int \sqrt{x}(x + 1) \, dx$

17. $\displaystyle\int \left(\frac{x-2}{x^3}\right) dx$ 18. $\displaystyle\int x^2(1+x^2)^2\, dx$

19. $\displaystyle\int (x^2+1)^3\, dx$ 20. $\displaystyle\int \frac{(x-1)^2}{\sqrt{x}}\, dx$

In Exercises 21–24 find the curve $y = f(x)$ that passes through the given point and whose slope at each point (x, y) is defined by the indicated derivative.

21. $dy/dx = x^2 - 3x + 2$, $(2, 1)$

22. $dy/dx = 2x^3 + 4x$, $(0, 5)$

23. $dy/dx = -2x^4 + 3x^2 + 6$, $(1, 0)$

24. $dy/dx = 2 - 4x + 8x^7$, $(1, 1)$

25. Find the equation of the curve that has a second derivative equal to $6x^2$ and passes through the points $(0, 2)$ and $(-1, 3)$.

26. Find a function $f(x)$ that has a relative maximum $f(2) = 3$ and has a second derivative equal to $-5x$.

27. Is it possible to find a function $f(x)$ that has a relative minimum $f(2) = 3$ and has a second derivative equal to $-5x$?

In Exercises 28–59 use "guess and fix up" to evaluate the antiderivative.

28. $\displaystyle\int \sqrt{x+2}\, dx$ 29. $\displaystyle\int (x+5)^{3/2}\, dx$

30. $\displaystyle\int \sqrt{2-x}\, dx$ 31. $\displaystyle\int \frac{1}{\sqrt{4x+3}}\, dx$

32. $\displaystyle\int (2x-3)^{3/2}\, dx$ 33. $\displaystyle\int (3x+1)^5\, dx$

34. $\displaystyle\int (1-2x)^7\, dx$ 35. $\displaystyle\int \frac{1}{(x+4)^2}\, dx$

36. $\displaystyle\int \frac{1}{(1+3x)^6}\, dx$ 37. $\displaystyle\int x(x^2+1)^3\, dx$

38. $\displaystyle\int x^2(2+3x^3)^7\, dx$ 39. $\displaystyle\int \frac{x}{(2+x^2)^2}\, dx$

40. $\displaystyle\int \cos 2x\, dx$ 41. $\displaystyle\int \cos^2 x \sin x\, dx$

42. $\displaystyle\int 3\sin 2x \cos 2x\, dx$ 43. $\displaystyle\int \sec 12x \tan 12x\, dx$

44. $\displaystyle\int \csc^2 4x\, dx$ 45. $\displaystyle\int e^{4x}\, dx$

46. $\displaystyle\int xe^{-x^2}\, dx$ 47. $\displaystyle\int \frac{e^{3/x}}{x^2}\, dx$

48. $\displaystyle\int e^{4x-3}\, dx$ 49. $\displaystyle\int \frac{1}{3x+2}\, dx$

50. $\displaystyle\int \frac{2}{7-5x}\, dx$ 51. $\displaystyle\int \frac{x}{1-x^2}\, dx$

52. $\displaystyle\int \frac{3x^2}{1-4x^3}\, dx$ 53. $\displaystyle\int 2^x\, dx$

54. $\displaystyle\int 3^{2x}\, dx$ 55. $\displaystyle\int \frac{e^x}{e^x+1}\, dx$

56. $\displaystyle\int \sin x(1+\cos x)^4\, dx$

57. $\displaystyle\int \frac{\cos x}{\sin^3 x}\, dx$

58. $\displaystyle\int e^{2x}(1+e^{2x})^3\, dx$ 59. $\displaystyle\int \frac{\sec^2 x}{\tan^2 x}\, dx$

In Exercises 60–65 find a one-parameter family of functions satisfying the differential equation.

60. $\dfrac{dy}{dx} = x^3 - \dfrac{1}{x^2}$ 61. $\dfrac{dy}{dx} = \sqrt{3-4x}$

62. $\dfrac{dy}{dx} = \dfrac{1}{(3x+5)^{3/2}}$ 63. $\dfrac{dy}{dx} = x^2(2x^3+4)^4$

64. $\dfrac{dy}{dx} = \dfrac{x^3}{(2+3x^4)^2}$ 65. $\dfrac{dy}{dx} = \sin x(1+\cos^2 x)$

66. Find a function $y = f(x)$ that satisfies the differential equation
$$\frac{dy}{dx} = \frac{1}{x^2}$$
and passes through the two points $(1, 1)$ and $(-1, -2)$.

67. (a) Find the indefinite integral of the signum function of Exercise 36 in Section 2.4.

   (b) Prove that $\operatorname{sgn} x$ cannot have an antiderivative on any interval containing $x = 0$.

68. Find the indefinite integral of the Heaviside unit step function of Exercise 38(a) in Section 2.4. Sketch a graph of the indefinite integral function.

SECTION 5.2

Velocity and Acceleration Revisited

In Section 4.7 we discussed relationships among position, velocity, and acceleration from the viewpoint of derivatives. We now consider these relationships through antiderivatives, an approach providing a far more practical viewpoint when it comes to applications.

For motion along the x-axis, velocity is the derivative of position with respect to time:

$v(t) = dx/dt$. We can say, therefore, that position is the antiderivative of velocity:

$$x(t) = \int v(t)\, dt. \tag{5.7}$$

Similarly, as acceleration is the derivative of velocity,

$$v(t) = \int a(t)\, dt. \tag{5.8}$$

Thus, given the acceleration of an object moving along a straight line, we can antidifferentiate to find its velocity and antidifferentiate again for its position. Since each antidifferentiation introduces an arbitrary constant, additional information must be specified in order to evaluate these constants.

EXAMPLE 5.4 The car in Figure 5.2 accelerates from rest when the light turns green. Initially the acceleration is 10 m/s^2, but it decreases linearly to zero after 10 s. Find the velocity and position of the car during this time interval.

SOLUTION Let us set up the coordinate system in Figure 5.2 and choose time $t = 0$ when the car pulls away from the light. Given this time convention, the acceleration of the car as a function of time t (see Figure 5.3) is

$$a(t) = 10 - t, \quad 0 \leq t \leq 10.$$

If $v(t)$ denotes the velocity of the car during the time interval, then

$$\frac{dv}{dt} = 10 - t,$$

FIGURE 5.2 **FIGURE 5.3**

and antidifferentiation gives

$$v(t) = 10t - \frac{t^2}{2} + C.$$

By our time convention, the velocity of the car is zero at time $t = 0$; that is, $v(0) = 0$, and from this we obtain

$$0 = 10(0) - \frac{(0)^2}{2} + C.$$

Consequently, $C = 0$, and
$$v(t) = 10t - \frac{t^2}{2}.$$

This is the velocity of the car during the time interval $0 \leq t \leq 10$, measured in metres per second.

Since the position of the car with respect to its original position at time $t = 0$ is denoted by x,
$$v = \frac{dx}{dt} = 10t - \frac{t^2}{2}.$$

Thus,
$$x(t) = 5t^2 - \frac{t^3}{6} + D.$$

Since $x(0) = 0$, the constant D must also be zero, and the position of the car indicated by its distance in metres from the stoplight, for $0 \leq t \leq 10$, is
$$x(t) = 5t^2 - \frac{t^3}{6}.$$

■

EXAMPLE 5.5

A stone is thrown vertically upward over the edge of a cliff at 25 m/s. When does it hit the base of the cliff if the cliff is 100 m high?

FIGURE 5.4

SOLUTION Let us measure y as positive upward, taking $y = 100$ and $t = 0$ at the point and instant of projection (Figure 5.4). A law of physics states that when an object near the earth's surface is acted on by gravity alone, it experiences an acceleration whose magnitude is 9.81 m/s². If a denotes the acceleration of the stone and v its velocity, then
$$a = \frac{dv}{dt} = -9.81$$

(a is negative since it is in the negative y-direction). To obtain v we take antiderivatives with respect to t:
$$v(t) = -9.81t + C.$$

By our time convention, $v(0) = 25$ so that
$$25 = -9.81(0) + C.$$

Thus, $C = 25$, and
$$v(t) = -9.81t + 25.$$

We now have the velocity of the stone at any given instant. To find the position of the stone we set $v = dy/dt$,
$$\frac{dy}{dt} = -9.81t + 25,$$
and take antiderivatives once again:
$$y(t) = -4.905t^2 + 25t + D.$$

Since $y = 100$ when $t = 0$,
$$100 = -4.905(0)^2 + 25(0) + D.$$

Hence, $D = 100$, and
$$y(t) = -4.905t^2 + 25t + 100.$$

We have found the equation that tells us exactly where the stone is at any given time. To determine when the stone strikes the base of the cliff, we set $y(t) = 0$; that is,
$$0 = -4.905t^2 + 25t + 100,$$
a quadratic equation with solutions
$$t = \frac{-25 \pm \sqrt{25^2 - 4(-4.905)(100)}}{-9.81} = \frac{25 \pm \sqrt{2587}}{9.81} = 7.7 \quad \text{or} \quad -2.6.$$

Since the negative root must be rejected, we find that the stone strikes the base of the cliff after 7.7 seconds. ∎

EXAMPLE 5.6

Two stones are thrown vertically upward one second apart over the edge of the cliff in Example 5.5. The first is thrown at 25 m/s, the second at 20 m/s. Determine if and when they ever pass each other.

FIGURE 5.5

y_1 | y_2
$t = 0$ | $t = 1$
$y_1 = 0$ | $y_2 = 0$
$v_1 = 25$ | $v_2 = 20$

Stone 1 | Stone 2

SOLUTION Let us choose a different coordinate system to that of Example 5.5 by taking y as positive upward with $y = 0$ and $t = 0$ at the point and instant of projection of the first stone (Figure 5.5). The acceleration of stone 1 is
$$a_1 = -9.81 = \frac{dv_1}{dt},$$

from which
$$v_1(t) = -9.81t + C.$$
Since $v_1(0) = 25$, it follows that $25 = C$, and
$$v_1(t) = -9.81t + 25, \quad t \geq 0.$$
Thus,
$$\frac{dy_1}{dt} = -9.81t + 25,$$
from which we have
$$y_1(t) = -4.905t^2 + 25t + D.$$
Since $y_1(0) = 0$, we find that $0 = D$, and
$$y_1(t) = -4.905t^2 + 25t, \quad t \geq 0.$$

The acceleration of stone 2 is also -9.81; hence we have
$$a_2 = -9.81 = \frac{dv_2}{dt},$$
from which
$$v_2(t) = -9.81t + E.$$
Because $v_2(1) = 20$, we must have $20 = -9.81(1) + E$, or $E = 29.81$. Thus,
$$v_2(t) = -9.81t + 29.81, \quad t \geq 1.$$
Consequently,
$$\frac{dy_2}{dt} = -9.81t + 29.81,$$
from which
$$y_2(t) = -4.905t^2 + 29.81t + F.$$
Since $y_2(1) = 0$, it follows that $0 = -4.905(1)^2 + 29.81(1) + F$, or $F = -24.905$. Thus,
$$y_2(t) = -4.905t^2 + 29.81t - 24.905, \quad t \geq 1.$$
The stones will pass each other if y_1 and y_2 are ever equal for the same time t; that is, if
$$-4.905t^2 + 25t = -4.905t^2 + 29.81t - 24.905.$$
The solution of this equation is
$$t = \frac{24.905}{4.81} = 5.2.$$

Because the first stone does not strike the base of the cliff for 7.7 seconds (Example 5.5), it follows that the stones do indeed pass each other 5.2 seconds after the first stone is projected. ∎

An important point to note in Examples 5.4, 5.5, and 5.6 is that the coordinate system and time convention were specified immediately; that is, we decided, and did so at

EXERCISES 5.2

In Exercises 1–8 we have defined acceleration $a(t)$ of an object moving along the x-axis during some time interval and specified the initial conditions $x(0)$ and $v(0)$. Find the velocity $v(t)$ and position $x(t)$ of the object as functions of time.

1. $a(t) = t + 2$, $\quad 0 \leq t \leq 3$; $\quad v(0) = 0, x(0) = 0$
2. $a(t) = 6 - 2t$, $\quad 0 \leq t \leq 3$; $\quad v(0) = 5, x(0) = 0$
3. $a(t) = 6 - 2t$ $\quad 0 \leq t \leq 4$; $\quad v(0) = 5, x(0) = 0$
4. $a(t) = 120t - 12t^2$, $\quad 0 \leq t \leq 10$; $\quad v(0) = 0, x(0) = 4$
5. $a(t) = t^2 + 1$, $\quad 0 \leq t \leq 5$; $\quad v(0) = -1, x(0) = 1$
6. $a(t) = t^2 + 5t + 4$, $\quad 0 \leq t \leq 15$; $\quad v(0) = -2, x(0) = -3$
7. $a(t) = \cos t$, $\quad t \geq 0$; $\quad v(0) = 0, x(0) = 0$
8. $a(t) = 3\sin t$, $\quad t \geq 0$; $\quad v(0) = 1, x(0) = 4$

9. The velocity of an object moving along the x-axis is given in metres per second by
$$v(t) = 3t^2 - 9t + 6$$
where t is time in seconds. If the object starts from position $x = 1$ m at time $t = 0$, answer each of the following questions:
   (a) What is the acceleration of the object at $t = 5$ s?
   (b) What is the position of the object at $t = 2$ s?
   (c) Is the object speeding up or slowing down at $t = 5/4$ s?
   (d) What is the closest the object ever comes to the origin?

10. A particle moving along the x-axis has acceleration
$$a(t) = 6t - 2$$
in metres per second per second for time $t \geq 0$.
   (a) If the particle starts at the point $x = 1$ moving to the left with speed 3 m/s, find its position as a function of time t.
   (b) At what time does the particle have zero velocity (if any)?

11. The acceleration of a particle moving along the x-axis is given in metres per second per second by
$$a(t) = 6t - 15,$$
where $t \geq 0$ is time in seconds.
   (a) If the velocity of the particle at $t = 2$ s is 6 m/s, what is its velocity at $t = 1$ s?
   (b) If the particle is 10 m to the right of the origin at time $t = 0$, what is its position as a function of time t?
   (c) What is the closest the particle ever comes to the origin?

12. A car is sitting at rest at a stoplight. When the light turns green at time $t = 0$, the driver immediately presses the accelerator, imparting an acceleration of $a(t) = 3 - t/5$ m/s^2 to the car for 10 s.
   (a) Where is the car after the 10 s?
   (b) If the driver applies the brakes at $t = 10$ s, and the car experiences a constant deceleration of 2 m/s^2, where and when does the car come to a stop?

13. Find how far a plane will move when landing if, in t seconds after touching the ground, its speed in metres per second is given by $180 - 18t$.

14. A stone is thrown directly upward with an initial speed of 10 m/s. How high will it rise?

15. You are standing on a bridge 25 m above a river. If you wish to drop a stone onto a piece of floating wood, how soon before the wood reaches the appropriate spot should you drop the stone?

16. You are standing at the base of a building and wish to throw a ball to a friend on the roof 20 m above you. With what minimum speed must you throw the ball?

17. A car is travelling at 20 m/s when the brakes are applied. What constant deceleration must the car experience if it is to stop before striking a tree that is 50 m from the car at the instant the brakes are applied? Assume that the car travels in a straight line.

18. The position of an object moving along the x-axis is given by
$$x(t) = t^3 - 6t^2 + 9t - 20, \quad t \geq 0.$$
Sketch graphs of the position, velocity, and acceleration functions for this motion. Pay special attention to the fact that $v(t)$ represents the slope of $x = x(t)$ and $a(t)$ is the slope of $v = v(t)$.

19. Repeat Exercise 18 given that the acceleration of the object is
$$a(t) = 6t - 30, \quad t \geq 0,$$
and $v(0) = -33$ and $x(0) = 400$.

20. You are called on as an expert to testify at a traffic hearing. The question concerns the speed of a car that made an emergency stop with brakes locked and wheels sliding. The skid mark on the road measured 9 m. Assuming that the deceleration of the car was constant and could not exceed the acceleration due to gravity of a freely falling body (and this is indeed a reasonable assumption), what can you say about the speed of the car before the brakes were applied? Are you testifying for the prosecution or the defence?

21. When the brakes of an automobile are applied, they produce a constant deceleration of 5 m/s^2. What is the distance, from the point of application of brakes, required to stop a car travelling at:
    (a) 100 km/hr?
    (b) 50 km/hr?
    (c) What is the ratio of these distances?
    (d) Repeat (a), (b), and (c) given that the reaction time of the driver to get his foot from accelerator to brake is 3/4 s, and distances are calculated taking this reaction time into account.

22. Two trains, one travelling at 100 km/hr and the other at 60 km/hr, are headed toward each other along a straight level track. When they are 2 km apart, each engineer sees the other's train and locks his wheels.
    (a) If the deceleration of each train has magnitude 1/4 m/s^2, determine whether a collision occurs.
    (b) Repeat (a) given that the deceleration is caused by the wheels being reversed rather than locked.
    (c) Illustrate graphically the difference between the situations in (a) and (b).

23. A steel bearing is dropped from the roof of a building. An observer standing in front of a window 1 m high notes that the bearing takes 1/8 s to fall from the top to the bottom of the window. The bearing continues to fall, makes a completely elastic collision with a horizontal sidewalk, and reappears at the bottom of the window 2 s after passing it on the way down. After a completely elastic collision, the bearing will have the same speed at a point going up as it had going down. How tall is the building?

24. In the theory of special relativity, Newton's second law ($F = ma$) is replaced by
$$F = m_0 \frac{d}{dt}\left[\frac{v}{\sqrt{1 - (v^2/c^2)}}\right],$$
where F is the applied force, m_0 the mass of the particle measured at rest, v its speed, and c the speed of light — a constant. Show that if we set $a = dv/dt$, then
$$F = \frac{m_0 a}{[1 - (v^2/c^2)]^{3/2}}.$$
Explain the difference between this law and Newton's second law.

25. Two stones are thrown vertically upward over the edge of a bottomless abyss, the second stone t_0 units of time after the first. The first stone has an initial speed of v_0', and the second an initial speed of v_0''.
    (a) Show that if the stones are ever to pass each other during their motions, two conditions must be satisfied:
    $$gt_0 > v_0' - v_0'' \quad \text{and} \quad v_0' > gt_0/2,$$
    where $g > 0$ is the acceleration due to gravity.
    (b) Show that the first condition is equivalent to the requirement that stone 1 must begin its downward trajectory before stone 2.
    (c) Show that the second condition is equivalent to the requirement that stone 1 must not pass its original projection point before the projection of stone 2.

Change of Variable in the Indefinite Integral

In Section 5.1 we suggested two methods for evaluating indefinite integrals—"recognition" and "guess and fix up". In this section we show how a "change of variable" can often replace a complex integration problem with a simpler one.

Consider the indefinite integral

$$\int x\sqrt{2x+1}\, dx.$$

What is annoying about this integrand is the sum of two terms $2x+1$ under a square root. This can be changed by setting $u = 2x + 1$. As a result, $\sqrt{2x+1} = \sqrt{u}$, and the x in front of the square root is equal to $(u-1)/2$. Now the differential dx is used to indicate antidifferentiation in the problem with respect to x. But surely there must be another reason why we have chosen the differential to denote this. If we regard $x\sqrt{2x+1}\, dx$ as a product of $x\sqrt{2x+1}$ and dx, then perhaps we should obtain an expression for dx in terms of du. To do this we note that since $u = 2x + 1$, then

$$\frac{du}{dx} = 2.$$

As derivatives can be regarded as quotients of differentials, we can rewrite this equation in the form

$$dx = \frac{du}{2}.$$

If we make all these substitutions into the indefinite integral

$$\int x\sqrt{2x+1}\, dx,$$

the result is an integration problem in the variable u, which is easy to evaluate,

$$\int \left(\frac{u-1}{2}\right)\sqrt{u}\, \frac{du}{2} = \frac{1}{4}\int (u^{3/2} - u^{1/2})\, du = \frac{1}{4}\left(\frac{2}{5}u^{5/2} - \frac{2}{3}u^{3/2}\right) + C.$$

If u is now replaced by $2x + 1$, the result is

$$\frac{1}{10}(2x+1)^{5/2} - \frac{1}{6}(2x+1)^{3/2} + C.$$

Differentiation of this function quickly indicates that its derivative is indeed $x\sqrt{2x+1}$, and therefore,

$$\int x\sqrt{2x+1}\, dx = \frac{1}{10}(2x+1)^{5/2} - \frac{1}{6}(2x+1)^{3/2} + C.$$

In this example, the substitution $u = 2x + 1$ replaces a complex integration in x with a simple one in u. Once the problem in u is solved, replacement of u's with x's gives the solution to the original indefinite integral. This method is generally applicable and is justified in the following theorem.

Theorem 5.3

If by the substitution $u = g(x)$, the product $f(x)\,dx$ becomes $h(u)\,du$, and if

$$\int h(u)\,du = H(u) + C,$$

then

$$\int f(x)\,dx = H(g(x)) + C.$$

Proof To prove this theorem we must show that

$$\frac{d}{dx}[H(g(x))] = f(x).$$

If we set $y = H(u)$, then the chain rule gives

$$\frac{dy}{dx} = \frac{dy}{du}\frac{du}{dx} = H'(u)\frac{du}{dx} = h(u)\frac{du}{dx}.$$

But $h(u)\,du = f(x)\,dx$, so that

$$\frac{dy}{dx} = \frac{f(x)\,dx}{dx} = f(x),$$

and the proof is complete.

EXAMPLE 5.7

Evaluate the following indefinite integrals:

(a) $\int \sqrt[5]{2x+4}\,dx$ (b) $\int \frac{x}{\sqrt{x+1}}\,dx$ (c) $\int \sin^3 x \cos^2 x\,dx$

SOLUTION (a) We could use "guess and fix up" in this case. With an initial guess of $(2x+4)^{6/5}$, we calculate

$$\frac{d}{dx}(2x+4)^{6/5} = \frac{6}{5}(2x+4)^{1/5}(2).$$

Thus,

$$\int (2x+4)^{1/5}\,dx = \frac{5}{12}(2x+4)^{6/5} + C.$$

Alternatively, if we set $u = 2x + 4$, then $du = 2\,dx$, and

$$\int (2x+4)^{1/5}\,dx = \int u^{1/5}\frac{du}{2} = \frac{1}{2}\left\{\frac{5}{6}u^{6/5}\right\} + C = \frac{5}{12}(2x+4)^{6/5} + C.$$

(b) If we set $u = x + 1$, then $du = dx$, and

$$\int \frac{x}{\sqrt{x+1}}\,dx = \int \frac{(u-1)}{\sqrt{u}}\,du = \int (u^{1/2} - u^{-1/2})\,du = \frac{2}{3}u^{3/2} - 2u^{1/2} + C$$

$$= \frac{2}{3}(x+1)^{3/2} - 2(x+1)^{1/2} + C.$$

A different substitution is also possible. If we set $u = \sqrt{x+1}$, then

$$du = \frac{1}{2\sqrt{x+1}}\,dx \quad \text{and}$$

$$\int \frac{x}{\sqrt{x+1}}\,dx = \int (u^2 - 1)(2\,du) = 2\left(\frac{u^3}{3} - u\right) + C$$
$$= \frac{2}{3}(x+1)^{3/2} - 2(x+1)^{1/2} + C.$$

(c) If we set $u = \cos x$, then $du = -\sin x\,dx$ and

$$\int \sin^3 x \cos^2 x\,dx = \int \sin^2 x \cos^2 x \sin x\,dx$$
$$= \int (1 - \cos^2 x) \cos^2 x \sin x\,dx$$
$$= \int (1 - u^2) u^2 (-du) = \int (u^4 - u^2)\,du$$
$$= \frac{u^5}{5} - \frac{u^3}{3} + C = \frac{1}{5}\cos^5 x - \frac{1}{3}\cos^3 x + C.$$

∎

EXERCISES 5.3

In Exercises 1–30 evaluate the indefinite integral.

1. $\int (5x + 14)^9\,dx$

2. $\int \sqrt{1 - 2x}\,dx$

3. $\int \frac{1}{(3y - 12)^{1/4}}\,dy$

4. $\int \frac{5}{(5 - 42x)^{1/4}}\,dx$

5. $\int x^2 (3x^3 + 10)^4\,dx$

6. $\int \frac{x}{(x^2 + 4)^2}\,dx$

7. $\int \sin^4 x \cos x\,dx$

8. $\int \frac{x^2}{(x - 2)^4}\,dx$

9. $\int z\sqrt{1 - 3z}\,dz$

10. $\int \frac{x}{\sqrt{2x + 3}}\,dx$

11. $\int \frac{1 + \sqrt{x}}{\sqrt{x}}\,dx$

12. $\int s^3 \sqrt{s^2 + 5}\,ds$

13. $\int \sin^2 x \cos^3 x\,dx$

14. $\int \sqrt{1 - \cos x} \sin x\,dx$

15. $\int \frac{x^3}{(3 - x^2)^3}\,dx$

16. $\int y^2 \sqrt{y - 4}\,dy$

17. $\int \frac{(1 + \sqrt{u})^{1/2}}{\sqrt{u}}\,du$

18. $\int x^8 (3x^3 - 5)^6\,dx$

19. $\int \frac{1 + z^{1/4}}{\sqrt{z}}\,dz$

20. $\int \frac{x + 1}{(x^2 + 2x + 2)^{1/3}}\,dx$

21. $\int \frac{(x - 1)(x + 2)}{\sqrt{x}}\,dx$

22. $\int \frac{\cos^3 x}{(3 - 4\sin x)^4}\,dx$

23. $\int \sqrt{1 + \sin 4t}\, \cos^3 4t\,dt$

24. $\int \sqrt{1 + \sqrt{x}}\,dx$

25. $\int \tan^2 x \sec^2 x\,dx$

26. $\int \tan x \sec^2 x\,dx$

27. $\int \frac{e^{2x}}{e^{2x} + 1}\,dx$

28. $\int \frac{\ln x}{x}\,dx$

29. $\int \frac{1}{x \ln x}\,dx$

30. $\int \frac{x}{(x^2 + 1)[\ln(x^2 + 1)]^2}\,dx$

31. Evaluate

$$\int \sin^3 x \cos^3 x\,dx$$

by:

(a) making the substitution $u = \sin x$;

(b) making the substitution $u = \cos x$.

(c) Verify that these answers are the same.

32. Evaluate
$$\int \sqrt{\frac{x^2}{1+x}}\, dx.$$

In Exercises 33–36 use the suggested change of variable to evaluate the indefinite integral.

33. $\displaystyle\int \frac{\sqrt{4x - x^2}}{x^3}\, dx;$ set $u = 2/x$

34. $\displaystyle\int \frac{\sqrt{x - x^2}}{x^4}\, dx;$ set $u = 1/x$

35. $\displaystyle\int \frac{x}{(5 - 4x - x^2)^{3/2}}\, dx;$ set $u^2 = (5 + x)/(1 - x)$

36. $\displaystyle\int \frac{1}{3(1 - x^2) - (5 + 4x)\sqrt{1 - x^2}}\, dx;$
set $u^2 = (1 - x)/(1 + x)$

37. Show that the substitution $u - x = \sqrt{x^2 + x + 4}$ replaces the integral
$$\int \sqrt{x^2 + x + 4}\, dx$$
with the integral of a rational function of u.

38. Show that the substitution $(x + 1)u = \sqrt{4 + 3x - x^2}$ replaces the integral
$$\int \frac{1}{\sqrt{4 + 3x - x^2}}\, dx$$
with the integral of a rational function of u.

SECTION 5.4

An Introduction to Differential Equations

A differential equation contains derivatives of some unknown function; the equation must be solved for this function. In this section we use calculus to derive mathematical models to describe four applied problems. These models turn out to be differential equations that are reasonably easy to solve.

Before doing so, however, let us express extended power rule 3.17 in an integral form that is particularly useful for differential equations. Equation 3.17 states that when $u(x)$ is a differentiable function of x,
$$\frac{d}{dx}u^n = nu^{n-1}\frac{du}{dx}.$$

It follows, therefore, that the antiderivative of $nu^{n-1}\, du/dx$ is u^n,
$$\int nu^{n-1}\frac{du}{dx}\, dx = u^n + C.$$

Property 5.4 permits us to take the constant outside the integral, and the equation can then be divided by this quantity:
$$\int u^{n-1}\frac{du}{dx}\, dx = \frac{1}{n}u^n + C, \quad n \neq 0.$$

Now n is an arbitrary number in this formula, and therefore there is no loss in generality if we replace each n with $n + 1$. The result is
$$\int u^n \frac{du}{dx}\, dx = \frac{1}{n+1}u^{n+1} + C, \quad n \neq -1. \tag{5.9a}$$

The $n = -1$ case is the integral counterpart of formula 3.37,
$$\int \frac{1}{u}\frac{du}{dx}\, dx = \ln|u| + C. \tag{5.9b}$$

EXAMPLE 5.8

FIGURE 5.6

Find all curves that satisfy the condition that at every point on the curve the normal line passes through the origin.

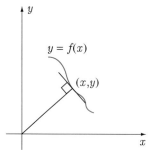

SOLUTION If $y = f(x)$ is the equation of any curve, then the slope of the tangent line at a point (x, y) on this curve is dy/dx, and the slope of the normal line is the negative reciprocal thereof. As derivatives can now be regarded as quotients of differentials, the negative reciprocal of dy/dx is $-dx/dy$. But when the normal line passes through the origin, its slope must also be given by the quotient $(y-0)/(x-0)$ (Figure 5.6). It follows that the required curve $y = f(x)$ must satisfy

$$\frac{y}{x} = -\frac{dx}{dy} \quad \text{or} \quad \frac{dy}{dx} = -\frac{x}{y}.$$

To solve this differential equation, we write

$$y\frac{dy}{dx} = -x$$

and antidifferentiate both sides with respect to x:

$$\int \left(y\frac{dy}{dx}\right) dx = \int -x\, dx.$$

The right integral is obviously $-x^2/2$, and by formula (5.9a), the antiderivative on the left is $y^2/2$. When a constant of integration is added, we have

$$\frac{y^2}{2} = -\frac{x^2}{2} + C.$$

If we set $D^2 = 2C$, then

$$x^2 + y^2 = D^2;$$

thus, the only curves satisfying the given condition are circles centred at the origin.

EXAMPLE 5.9

An upright cylindrical container (Figure 5.7) has height 2 m and radius 1 m. Water flows out through a hole in the bottom with speed in metres per second given by

$$v = \frac{3\sqrt{2gD}}{5},$$

where D (in metres) is the depth of water in the container and $g = 9.81$ m/s^2 is the acceleration due to gravity. This formula indicates that water comes out much more quickly when the container is full than when it is almost empty. Find a formula for D as a function of time if the tank is originally full and if the area of the hole is 3 cm^2.

FIGURE 5.7

SOLUTION If V is the volume of water in the container at any time t, then

$$V = \pi(1)^2 D = \pi D.$$

The derivative

$$\frac{dV}{dt} = \pi \frac{dD}{dt}$$

represents the rate of change of the volume of water in the container, where it is clear that $dV/dt < 0$ since the volume is decreasing. But this rate of change must equal the rate at which the water exits through the hole. Since the area of the hole is 0.0003 m^2, and water exits with speed v m/s, water leaves the container at a rate of $0.0003\, v$ m^3/s. Consequently,

$$-\pi \frac{dD}{dt} = 0.0003\, v = 0.00018\sqrt{2gD},$$

and the unknown function $D(t)$ must satisfy the differential equation

$$\frac{1}{\sqrt{D}} \frac{dD}{dt} = -\frac{0.00018\sqrt{2g}}{\pi}.$$

Antiderivatives with respect to t yield

$$\int \frac{1}{\sqrt{D}} \frac{dD}{dt}\, dt = \int \frac{-0.00018\sqrt{2g}}{\pi}\, dt.$$

Since formula (5.9a) can be used on the left integral, and the integrand on the right is a constant, we obtain

$$2\sqrt{D} = -\frac{0.00018\sqrt{2g}}{\pi}t + C.$$

If we choose time $t = 0$ when the container is full ($D = 2$), substitution of these values into the above equation gives

$$2\sqrt{2} = C.$$

The solution of the differential equation for $D(t)$ is defined implicitly by the equation

$$2\sqrt{D} = -\frac{0.00018\sqrt{2g}}{\pi}t + 2\sqrt{2}.$$

By solving this equation for D in terms of t, we obtain the explicit function

$$D(t) = \left(\sqrt{2} - \frac{0.00009\sqrt{2g}}{\pi}t\right)^2.$$

This solution tells us that we can determine how long it takes for the container to empty. If we set $D = 0$, we obtain

$$0 = \left(\sqrt{2} - \frac{0.00009\sqrt{2g}}{\pi}t\right)^2.$$

Thus,

$$t = \frac{\pi}{0.00009\sqrt{g}} = \frac{\pi}{0.00009\sqrt{9.81}} = 11\,145,$$

and the container therefore empties in 186 minutes.

EXAMPLE 5.10

When a mothball is exposed to the air, it slowly evaporates. If the mothball remains spherical at all times, and its radius after one year is only half of its original radius, when will it completely disappear?

SOLUTION The size of the mothball can be measured in various ways—by its volume, its surface area, or its radius. Since we were given information about its radius, let us denote by $r(t)$ the radius of the mothball as a function of time t, and choose time $t = 0$ when the mothball is originally exposed to the air. If we let R represent the original radius of the mothball, then we can say that $r(0) = R$ and $r(1) = R/2$. This is the only information about $r(t)$ that we were given, and it certainly is not enough to solve the problem.

We need a statement on the instantaneous rate of evaporation of the mothball. Without such a statement, we must make our own reasonable assumption about how fast the naphthalene evaporates from the surface of the mothball. Since evaporation takes place only at the surface, let us assume that the amount of naphthalene evaporating per unit time is proportional to the surface area of the mothball at that time. If V and A represent volume and area of the mothball at any time, then this assumption becomes

$$\frac{dV}{dt} = kA,$$

where k is a constant. We have four variables in the problem: t, r, A, and V. By substituting $V = 4\pi r^3/3$ and $A = 4\pi r^2$, we can eliminate V and A:

$$\frac{d}{dt}\left(\frac{4}{3}\pi r^3\right) = 4\pi r^2 k,$$

and this equation reduces to

$$\frac{dr}{dt} = k.$$

Now we have a differential equation for $r(t)$, with solution

$$r(t) = kt + C.$$

Using the conditions $r(0) = R$ and $r(1) = R/2$, we find $C = R$ and $k = -R/2$. Consequently,

$$r(t) = R - \frac{Rt}{2} = R\left(1 - \frac{t}{2}\right).$$

The mothball completely disappears when $r = 0$, and this occurs in $t = 2$ years. ∎

EXAMPLE 5.11

A tank originally contains 1000 L of water, in which 5 kg of salt has been dissolved.
  (a) If a brine mixture containing 2 kg of salt for each 100 L of solution is poured into the tank at 10 mL/s, find the amount of salt in the tank as a function of time.
  (b) If at the same time brine is being added, the mixture in the tank is being drawn off at 5 mL/s, find the amount of salt in the tank as a function of time. Assume that the mixture is stirred constantly.

SOLUTION (a) Suppose we let $S(t)$ represent the number of grams of salt in the tank at time t. If we choose time $t = 0$ at the instant the brine mixture begins entering the original solution, then $S(0) = 5000$. Since 10 mL of mixture enter the tank each second, and each millilitre contains 0.02 gm of salt, it follows that 0.2 gm of salt enter the tank each second. Consequently, after t seconds, $0.2t$ grams of salt have been added to the tank, and the total amount of salt in the solution is, in grams,

$$S(t) = 5000 + 0.2t.$$

(b) Once again we let $S(t)$ represent the number of grams of salt in the tank at time t. Its derivative dS/dt, the rate of change of $S(t)$, is the difference between how fast salt is being added to the tank in the brine, and how fast salt is being removed as the mixture is removed. As in part (a), salt is being added at the constant rate of 0.2 gm/s. The rate at which salt leaves the tank, on the other hand, is not constant; it depends on the concentration of salt in the tank. Since the tank originally contained 10^6 mL of solution, and a net 5 mL enters each second, the amount of solution in the tank at time t is $10^6 + 5t$ mL. It follows that the concentration of salt in the solution at time t is, in grams per millilitre,

$$\frac{S}{10^6 + 5t},$$

where $S = S(t)$ is the amount of salt in the tank at that time. As solution is being drawn off at the rate of 5 mL/s, the rate at which salt leaves the tank is, in grams per second,

$$\frac{S}{10^6 + 5t}(5).$$

The net rate at which salt enters the tank is therefore

$$0.2 - \frac{5S}{10^6 + 5t}.$$

Since dS/dt also represents the rate of change of the amount of salt in the tank, we set

$$\frac{dS}{dt} = \frac{1}{5} - \frac{5S}{10^6 + 5t}.$$

To find $S(t)$ we must solve this differential equation subject to the condition that $S(0) = 5000$. We can clear the term $10^6 + 5t$ from the denominator if we multiply the equation by this factor:

$$(10^6 + 5t)\frac{dS}{dt} + 5S = \frac{1}{5}(10^6 + 5t).$$

But note that the left-hand side of this differential equation is the derivative of the product of S and $10^6 + 5t$; that is,

$$\frac{d}{dt}[S(10^6 + 5t)] = (10^6 + 5t)\frac{dS}{dt} + 5S.$$

Consequently, we may express the differential equation in the form

$$\frac{d}{dt}[S(10^6 + 5t)] = \frac{1}{5}(10^6 + 5t),$$

and take antiderivatives with respect to t on both sides:

$$S(10^6 + 5t) = \frac{1}{5}\int (10^6 + 5t)\,dt = \frac{1}{50}(10^6 + 5t)^2 + C.$$

Since $S(0) = 5000$, we find

$$5000(10^6) = \frac{1}{50}(10^6)^2 + C,$$

and hence, $C = -15 \times 10^9$. Thus, $S(t)$ is defined implicitly by

$$S(10^6 + 5t) = \frac{1}{50}(10^6 + 5t)^2 - 15 \times 10^9,$$

or explicitly by

$$S(t) = \frac{1}{50}(10^6 + 5t) - \frac{15 \times 10^9}{10^6 + 5t}.$$

A sketch of this function is shown in Figure 5.8. It is asymptotic to the line $S = (10^6 + 5t)/50$. ∎

FIGURE 5.8

EXERCISES 5.4

1. A boy lives 6 km from school. He decides to walk to school at a speed that is always proportional to the square of his distance from the school. If he is half-way to school after one hour, find his distance from school at any time. How long does it take him to reach school?

2. A diving board of length L and uniform mass per unit length m is fixed horizontally (Figure 5.9).

   FIGURE 5.9

   If a diver of mass M stands on the end of the board, the deflection $y = f(x)$ of the board must satisfy the differential equation

   $$(EI)\frac{d^2y}{dx^2} = Mg(L-x) + \frac{mg}{2}(L-x)^2,$$

   where E and I are constants depending on the shape and material of the board, and $g < 0$ is the acceleration due to gravity.
   (a) Find $y = f(x)$.
   (b) What is the maximum deflection of the board from the horizontal?

3. The water trough in Figure 5.10 is 4 m long. Its cross-section is an isosceles triangle with a half-metre base and a half-metre altitude.

   FIGURE 5.10

   Water leaks out through a hole of area 1 cm² in the bottom with speed in metres per second given by $v = \sqrt{gD/2}$, where D (in metres) is the depth of water in the trough, and $g > 0$ is the acceleration due to gravity. Find how long a full trough takes to empty.

4. The y-intercept of the normal line to a curve at any point is always equal to 2. If the curve passes through the point $(3, 4)$, find its equation.

5. A tank has 100 L of solution containing 4 kg of sugar. A mixture with 10 gm of sugar per litre of solution is added at 200 mL/min. At the same time, 100 mL of well-stirred mixture is removed each minute. Find the amount of sugar in the tank as a function of time.

6. A tank in the form of an inverted right-circular cone of height H and radius R is filled with water. Water escapes through a hole of cross-sectional area A at the vertex with speed $v = c\sqrt{2gD}$, where $0 < c < 1$ is a constant called the discharge coefficient, $g > 0$ is the acceleration due to gravity, and D is the depth of water in the tank. Find a formula for the time the tank takes to empty.

7. A spring of negligible mass and elasticity constant $k > 0$ is attached to a wall at one end and a mass M at the other (Figure 5.11).

   FIGURE 5.11

   The mass is free to slide horizontally along a frictionless surface. If $x = 0$ is taken as the position of M when the spring is unstretched and M is set into motion, the differential equation describing the position $x(t)$ of M is

   $$\frac{d^2x}{dt^2} = -\frac{k}{M}x.$$

   If motion is initiated by imparting a speed v_0 in the positive x-direction to M at position $x = 0$, find the velocity of M as a function of position. Note that

   $$\frac{d^2x}{dt^2} = \frac{dv}{dt} = \frac{dv}{dx}\frac{dx}{dt} = v\frac{dv}{dx}.$$

8. When a mass m falls under the influence of gravity alone, it experiences an acceleration d^2r/dt^2 described by

   $$m\frac{d^2r}{dt^2} = -\frac{GmM}{r^2},$$

   where M is the mass of the earth, G is a positive constant, and r is the distance from m to the centre of the earth (Figure 5.12). If m is dropped from a height h above the surface of the earth, find its velocity when it strikes the earth. What is the maximum attainable speed of m?

FIGURE 5.12

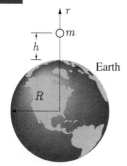

9. A container in the form of an inverted right-circular cone of radius 4 cm and height 10 cm is full of water. Water evaporates from the surface at a rate proportional to the area of the surface. If the water level drops 1 cm in the first 5 days, how long will it take for the water to evaporate completely?

10. Allometry is the quantitative study of the size of one part of a biological object in relation to the size of another part. For example, a researcher might be interested in the rate at which a child's eyes grow in relation to the rate of growth of his or her head. It is clear that the head grows more rapidly than the eyes, but just how much more rapidly? If $x(t)$ and $y(t)$ represent volumes of the eye and head respectively, at any given time t, then $(1/x)\,dx/dt$ and $(1/y)\,dy/dt$ are called *relative growth rates*. Numerous experimental results have suggested that these relative growth rates are proportional

$$\frac{1}{y}\frac{dy}{dt} = k\frac{1}{x}\frac{dx}{dt}.$$

This is called the *allometric law*. If we eliminate t from this equation, we obtain a differential equation for y as a function of x:

$$\frac{dy}{dx} = k\frac{y}{x}.$$

Solve this differential equation. (Hint: Multiply by x^k).

11. In psychology, it is sometimes assumed that if R represents the reaction of a subject to an amount S of stimulus, then relative increases of R and S are proportional. In other words, if S is changed at rate dS/dt, and the resulting rate of change of R is dR/dt, then

$$\frac{1}{R}\frac{dR}{dt} = k\frac{1}{S}\frac{dS}{dt}.$$

Although this is often called the Brentano-Stevens law, it is essentially the allometric law of Exercise 10 in a different setting. Find R as a function of S.

12. Three boys of the same height carry a board of length L and uniform mass per unit length m horizontally as shown in Figure 5.13.

FIGURE 5.13

The weight of the board causes it to bend, and its shape is the same on either side of the middle boy. Between the first two boys, for $0 \leq x \leq L/2$, the displacement of the board from the horizontal $y = f(x)$ must satisfy the differential equation

$$(EI)\frac{d^2y}{dx^2} = Ax + \frac{mg}{2}x^2,$$

where E and I are constants depending on the cross section and material of the board, A is the weight supported by the boy at $x = 0$, and $g < 0$ is the acceleration due to gravity.

(a) Solve this equation along with the conditions $f(0) = f(L/2) = 0$ to find $f(x)$ as a function of x, E, I, m, g, and A.

(b) Find a condition that permits determination of A.

13. The steady-state temperature T in a region bounded by two concentric spheres of radii 1 m and 2 m must satisfy the differential equation

$$\frac{d^2T}{dr^2} + \frac{2}{r}\frac{dT}{dr} = 0,$$

where r is the radial distance from the common centre of the spheres. If temperatures on the inner and outer spheres are maintained at $10\,°C$ and $20\,°C$ respectively, find the temperature distribution $T(r)$ between the spheres.

14. A room with volume 100 m³ initially contains 0.1% carbon dioxide. Beginning at time $t = 0$, fresher air containing 0.05% carbon dioxide flows into the room at 5 m³/min. The well-mixed air in the room flows out at the same rate. Find the amount of carbon dioxide in the room as a function of time. What is the limit of the function as $t \to \infty$?

15. Find all curves such that if the normal is drawn at any point (x, y) on the curve, then that part of the normal between (x, y) and the x-axis is bisected by the y-axis.

16. It is sometimes assumed that the density ρ of the atmosphere is related to height h above sea level according to the differential equation

$$\frac{d\rho}{dh} = -\frac{\rho^{2-\delta}}{k\delta},$$

where $\delta > 1$ and $k > 0$ are constants.

(a) Show that if ρ_0 is density at sea level, then

$$\rho^{\delta-1} = -\frac{h}{k}\left(\frac{\delta-1}{\delta}\right) + \rho_0^{\delta-1}.$$

(b) If air pressure P and density are related by the equation $P = k\rho^\delta$, prove that

$$P^{1-1/\delta} = P_0^{1-1/\delta} - [(1 - 1/\delta)\rho_0 P_0^{-1/\delta}]h,$$

where P_0 is air pressure at sea level.

(c) Show that the effective height of the atmosphere is

$$\frac{\delta P_0}{(\delta-1)\rho_0}.$$

SUMMARY

The indefinite integral or antiderivative of a function $f(x)$ is a family of functions, each of which has $f(x)$ as its first derivative. If we can find one antiderivative of $f(x)$, then every other antiderivative in the family differs from this one by at most an additive constant. According to Theorem 5.2, the antiderivative of a sum of two functions is the sum of their antiderivatives, and multiplicative constants may be bipassed when finding antiderivatives

$$\int [f(x) + g(x)]\,dx = \int f(x)\,dx + \int g(x)\,dx; \qquad \int cf(x)\,dx = c\int f(x)\,dx.$$

The most important integration formula is for powers of x,

$$\int x^n\,dx = \begin{cases} \dfrac{1}{n+1}x^{n+1} + C & n \neq -1 \\ \ln|x| + C & n = -1. \end{cases}$$

Other integration formulas that arise from differentiations of trigonometric and exponential functions are listed in equations 5.5.

In this chapter we have studied three ways to evaluate indefinite integrals. (Others will follow in Chapter 9.) First, due to our expertise in differentiation, some antiderivatives are immediately recognizable. Second, sometimes an antiderivative can be guessed to within a multiplicative constant, and then fixed up. Third, a change of variable can often replace a complex integration problem with a simpler one.

Integration plays a fundamental role in kinematics. Since velocity is the derivative of position, and acceleration is the derivative of velocity, it follows that position is the antiderivative of velocity, and velocity is the antiderivative of acceleration,

$$v(t) = \int a(t)\,dt \quad \text{and} \quad x(t) = \int v(t)\,dt.$$

Many physical systems are modelled by differential equations, and the solution of a differential equation usually involves one or more integrations.

Key Terms and Formulas

In reviewing this chapter, you should be able to define or discuss the following key terms:

Indefinite integral
Antiderivative
Integrand

Differential equation
Change of variable

REVIEW EXERCISES

In Exercises 1–24 evaluate the indefinite integral.

1. $\int (3x^3 - 4x^2 + 5)\, dx$
2. $\int \left(\dfrac{1}{x^5} + 2x - \dfrac{1}{x^3}\right) dx$
3. $\int (2x^2 - 3x + 7x^6)\, dx$
4. $\int \left(\dfrac{1}{x^2} - 2\sqrt{x}\right) dx$
5. $\int \sqrt{x-2}\, dx$
6. $\int x(1 + 3x^2)^4\, dx$
7. $\int \left(\sqrt{x} - \dfrac{1}{\sqrt{x}}\right) dx$
8. $\int \left(\dfrac{x^2 + 5}{\sqrt{x}}\right) dx$
9. $\int \dfrac{1}{(x+5)^4}\, dx$
10. $\int \left(\dfrac{\sqrt{x}}{x^2} - \dfrac{15}{\sqrt{x}}\right) dx$
11. $\int \sin 3x\, dx$
12. $\int x\sqrt{1 - x^2}\, dx$
13. $\int x \cos x^2\, dx$
14. $\int x^2 (1 - 2x^2)^2\, dx$
15. $\int x\sqrt{1 + x}\, dx$
16. $\int \dfrac{x}{\sqrt{2 - x}}\, dx$
17. $\int \dfrac{1}{(1 + x)^2}\, dx$
18. $\int (2 + \sqrt{x})^2\, dx$
19. $\int \dfrac{1}{\sqrt{x}(2 + \sqrt{x})^2}\, dx$
20. $\int \sin^4 x \cos x\, dx$
21. $\int e^{3-5x}\, dx$
22. $\int x e^{-4x^2}\, dx$
23. $\int \dfrac{e^x - 1}{e^{2x}}\, dx$
24. $\int \dfrac{1}{5x \ln x}\, dx$

25. Find the curve $y = f(x)$ for which $f''(x) = x^2 + 1$, and that passes through the point $(1, 1)$ with slope 4.

26. Find the curve $y = f(x)$ for which $f''(x) = 12x^2$, and that passes through the two points $(1, 4)$ and $(-1, -3)$.

27. Find the curve $y = f(x)$ for which $f''(x) = 24x^2 + 6x$, and that is tangent to the line $y = 4x + 4$ at $(1, 8)$.

28. A boy lives 6 km from school. He decides to walk to school at a speed that is always proportional to the square root of his distance from the school. If he is halfway to school after one hour, find his distance from school at any time. How long does it take him to reach school?

29. If a ball is thrown vertically upward with a speed of 30 m/s, how high will it rise?

30. A stone is thrown vertically downward over the edge of a bridge 50 m above a river. If the stone strikes the water in 2.2 s, what was its initial speed?

In Exercises 31–38 evaluate the indefinite integral.

31. $\int \dfrac{1}{\sqrt{1 + \sqrt{x}}}\, dx$
32. $\int \dfrac{x}{\sqrt{1 + x} + 1}\, dx$
33. $\int \dfrac{\sin x}{\sqrt{4 + 3\cos x}}\, dx$
34. $\int x^8 (3 - 2x^3)^6\, dx$
35. $\int \dfrac{(2 + x)^4}{x^6}\, dx$
36. $\int \sin^3 x \cos^3 x\, dx$
37. $\int \dfrac{1}{x\sqrt{1 + 3\ln x}}\, dx$
38. $\int \tan x\, dx$

39. A graph of the acceleration $a(t)$ of an object is shown in Figure 5.14. Find its velocity $v(t)$ and position $x(t)$ in the time interval $0 \le t \le 15$ if $v(0) = 0 = x(0)$, and sketch graphs of each.

FIGURE 5.14

40. Find the equation of the curve that passes through the point $(1, 1)$ such that the slope of the tangent line at any point (x, y) is half the square of the slope of the line from the origin to (x, y).

41. The equation $y = x^3 + C$ describes a family of cubics where C represents the distance of its horizontal point of inflection above the x-axis. Find the equation of the curve that passes through the point $(1, 1)$ and intersects each of these curves at right angles.

CHAPTER 6

The Definite Integral

There are two aspects to calculus: differentiation and integration. We dealt with differentiation and its applications in Chapters 3 and 4. Integration in Chapter 5 was synonymous with antidifferentiation. In this chapter we investigate a new type of integral called the "definite integral". Before doing that, we introduce sigma notation, a compact notation for sums of terms, particularly useful for definite integrals. In Section 6.2 we discuss four different types of problems that motivate the concept of the definite integral. In subsequent sections we develop the definition for the integral and discuss various ways to evaluate it. Section 6.6 presents an application of the definite integral — finding the average value of a function.

The definite integral is, by definition, very different from the indefinite integral, yet the two are intimately related through the Fundamental Theorem of Integral Calculus (Sections 6.4 and 6.5).

SECTION 6.1 Sigma Notation

One of the most important notations in calculus, *sigma notation*, is used to represent a sum of terms, all of which are similar in form. For example, the six terms in the sum

$$\frac{1}{1+2^2} + \frac{2}{1+3^2} + \frac{3}{1+4^2} + \frac{4}{1+5^2} + \frac{5}{1+6^2} + \frac{6}{1+7^2}$$

are all formed in the same way: each is an integer divided by 1 plus the square of the next integer. If k represents an integer, we can say that every term has the form $k/[1+(k+1)^2]$; the first term is obtained by setting $k=1$, the second by setting $k=2$, and so on. As $k/[1+(k+1)^2]$ represents each and every term in the sum, we can describe the sum in words by saying, "Assign k in $k/[1+(k+1)^2]$ the integer values between 1 and 6, inclusively, and add the resulting numbers together." The notation used to represent this statement is

$$\sum_{k=1}^{6} \frac{k}{1+(k+1)^2}.$$

The Σ symbol is the Greek letter sigma which, in this case, means "sum". Summed are expressions of the form $k/[1+(k+1)^2]$, and the "$k=1$" and "6" indicate that every integer from 1 to 6 is substituted into $k/[1+(k+1)^2]$. We call $k/[1+(k+1)^2]$ the **general term** of the sum, since it represents each and every term therein. The letter k is called the **index of summation** or **variable of summation**, and 1 and 6 are called the **limits of summation**. Any letter may be used to represent the index of summation; most commonly used are i, j, k, l, m, and n.

Here is another example:

$$\sum_{n=5}^{15} n^3 = 5^3 + 6^3 + 7^3 + 8^3 + 9^3 + 10^3 + 11^3 + 12^3 + 13^3 + 14^3 + 15^3.$$

In summing a large number of terms, it is quite cumbersome to write them all down. One way around this difficulty is to write the first few terms to indicate the pattern by which the terms are formed, three dots, and the last term. For example, to indicate the sum of the cubes of the positive integers less than or equal to 100, we write

$$1^3 + 2^3 + 3^3 + 4^3 + 5^3 + \cdots + 100^3,$$

where the dots indicate that all numbers between 5^3 and 100^3 are to be filled in according to the same pattern suggested. Obviously, it would be preferable to express the sum in sigma notation

$$\sum_{k=1}^{100} k^3,$$

which is both precise and compact.

EXAMPLE 6.1 Write each of the following sums in sigma notation:

(a) $\dfrac{1}{2 \cdot 3} + \dfrac{4}{3 \cdot 4} + \dfrac{9}{4 \cdot 5} + \dfrac{16}{5 \cdot 6} + \cdots + \dfrac{169}{14 \cdot 15}$;

(b) $\dfrac{16}{\sqrt{2}} + \dfrac{32}{\sqrt{3}} + \dfrac{64}{\sqrt{4}} + \dfrac{128}{\sqrt{5}} + \cdots + \dfrac{4096}{\sqrt{10}}$.

SOLUTION (a) We examine each term carefully to determine the pattern. If we use i as the index of summation, the general term is $i^2/[(i+1)(i+2)]$. Consequently,

$$\frac{1}{2 \cdot 3} + \frac{4}{3 \cdot 4} + \frac{9}{4 \cdot 5} + \frac{16}{5 \cdot 6} + \cdots + \frac{169}{14 \cdot 15} = \sum_{i=1}^{13} \frac{i^2}{(i+1)(i+2)}.$$

(b) If n is chosen as the index of summation, then

$$\frac{16}{\sqrt{2}} + \frac{32}{\sqrt{3}} + \frac{64}{\sqrt{4}} + \frac{128}{\sqrt{5}} + \cdots + \frac{4096}{\sqrt{10}} = \sum_{n=4}^{12} \frac{2^n}{\sqrt{n-2}}.$$

The representations for the sums in Example 6.1 in terms of sigma notation are not unique. In fact, there is an infinite number of representations for each sum. Consider, for example, the sum represented by

$$\sum_{i=1}^{9} \frac{2^{i+3}}{\sqrt{i+1}}.$$

If we write out some of the terms in the summation, we find

$$\sum_{i=1}^{9} \frac{2^{i+3}}{\sqrt{i+1}} = \frac{16}{\sqrt{2}} + \frac{32}{\sqrt{3}} + \frac{64}{\sqrt{4}} + \frac{128}{\sqrt{5}} + \cdots + \frac{4096}{\sqrt{10}},$$

the same sum as that in Example 6.1(b). This sum can also be represented by

$$\sum_{j=-2}^{6} \frac{2^{j+6}}{\sqrt{j+4}} \quad \text{and} \quad \sum_{m=15}^{23} \frac{2^{m-11}}{\sqrt{m-13}}.$$

We can transform any one of these representations into any other by making a change of variable of summation. For example, if in the summation of Example 6.1(b) we set $i = n - 3$, then $n = i + 3$, and

$$\frac{2^n}{\sqrt{n-2}} = \frac{2^{i+3}}{\sqrt{i+1}}.$$

For the limits, we find that $i = 1$ when $n = 4$, and $i = 9$ when $n = 12$. It follows that

$$\sum_{n=4}^{12} \frac{2^n}{\sqrt{n-2}} = \sum_{i=1}^{9} \frac{2^{i+3}}{\sqrt{i+1}}.$$

Similarly, the changes $j = n - 6$ and $m = n + 11$ transform

$$\sum_{n=4}^{12} \frac{2^n}{\sqrt{n-2}}$$

into

$$\sum_{j=-2}^{6} \frac{2^{j+6}}{\sqrt{j+4}} \quad \text{and} \quad \sum_{m=15}^{23} \frac{2^{m-11}}{\sqrt{m-13}}.$$

EXAMPLE 6.2 Change each of the following summations into representations that are initiated with the integer 1:

(a) $\sum_{i=4}^{26} \frac{i^{2/3}}{i^2 + i + 1}$ (b) $\sum_{j=-3}^{102} \frac{j^2 + 2j + 5}{\sin(j + 5)}$

SOLUTION (a) To initiate the summation at 1, we want $n = 1$ when $i = 4$, so we set $n = i - 3$. Then $i = n + 3$, and by substitution, we have

$$\sum_{i=4}^{26} \frac{i^{2/3}}{i^2 + i + 1} = \sum_{n=1}^{23} \frac{(n+3)^{2/3}}{(n+3)^2 + (n+3) + 1} = \sum_{n=1}^{23} \frac{(n+3)^{2/3}}{n^2 + 7n + 13}.$$

(b) In this case, we set $n = j + 4$. Then, $j = n - 4$, and

$$\sum_{j=-3}^{102} \frac{j^2 + 2j + 5}{\sin(j + 5)} = \sum_{n=1}^{106} \frac{(n-4)^2 + 2(n-4) + 5}{\sin(n + 1)} = \sum_{n=1}^{106} \frac{n^2 - 6n + 13}{\sin(n + 1)}.$$ ∎

If we examine the results of Example 6.2 and the summations immediately preceding this example, we soon come to realize that there is a very simple way to change variables. To illustrate, consider once again

$$\sum_{n=4}^{12} \frac{2^n}{\sqrt{n-2}}.$$

Should we wish to initiate the summation with 1, rather than 4, we lower both limits by 3. To compensate, we replace each n in the general term by $n+3$; the result is

$$\sum_{n=1}^{9} \frac{2^{n+3}}{\sqrt{(n+3)-2}} = \sum_{n=1}^{9} \frac{2^{n+3}}{\sqrt{n+1}}.$$

Likewise, for simplicity in the summation

$$\sum_{n=1}^{10} \frac{(n+4)^2}{e^{n+4}},$$

it would be advisable to lower each n in the general term by 4. This can be done provided we raise each limit by 4,

$$\sum_{n=1}^{10} \frac{(n+4)^2}{e^{n+4}} = \sum_{n=5}^{14} \frac{n^2}{e^n}.$$

Every summation represented in sigma notation is of the form

$$\sum_{i=m}^{n} f(i), \qquad (6.1)$$

where m and n are integers ($n > m$), and $f(i)$ is some function of the index of summation i. In Example 6.2(a), $f(i) = i^{2/3}/(i^2 + i + 1)$, $m = 4$, and $n = 26$; in Example 6.1(a), $f(i) = i^2/[(i+1)(i+2)]$, $m = 1$, and $n = 13$. The following properties of sigma notation are easily proved by writing out each summation.

Theorem 6.1

If $f(i)$ and $g(i)$ are functions of i, and m and n are positive integers such that $n > m$, then

$$\sum_{i=m}^{n} [f(i) + g(i)] = \sum_{i=m}^{n} f(i) + \sum_{i=m}^{n} g(i); \qquad (6.2)$$

$$\sum_{i=m}^{n} cf(i) = c \sum_{i=m}^{n} f(i), \qquad (6.3)$$

if c is a constant independent of i.

Compare this theorem with Theorem 5.2 in Section 5.1; notice the similarities between properties of summations in sigma notation and those of indefinite integrals.

We emphasize that sigma notation is simply a concise symbolism used to represent a sum of terms; it does not evaluate the sum. In the following, we develop formulas for certain sums of terms.

Summation $\sum_{i=1}^{n} i$ represents the sum of the first n integers:

$$\sum_{i=1}^{n} i = 1 + 2 + 3 + 4 + \cdots + (n-1) + n.$$

If we write the terms on the right in reverse order, we have

$$\sum_{i=1}^{n} i = n + (n-1) + (n-2) + \cdots + 4 + 3 + 2 + 1.$$

Addition of these two equations gives us

$$2\sum_{i=1}^{n} i = (n+1) + (n+1) + (n+1) + \cdots + (n+1) + (n+1) = n(n+1).$$

Consequently,

$$\sum_{i=1}^{n} i = \frac{n(n+1)}{2}. \tag{6.4}$$

This result can be used to develop formulas for the sums of the squares, cubes, etc., of the positive integers. To find the sum of the squares of the first n integers, we note that by expansion and simplification

$$i^3 - (i-1)^3 = 3i^2 - 3i + 1$$

for any integer i whatsoever. It follows that

$$\sum_{i=1}^{n} [i^3 - (i-1)^3] = \sum_{i=1}^{n} (3i^2 - 3i + 1).$$

But if we write the left-hand side in full, we find

$$\sum_{i=1}^{n} [i^3 - (i-1)^3] = [1^3 - 0^3] + [2^3 - 1^3] + [3^3 - 2^3] + \cdots [n^3 - (n-1)^3].$$

Most of these terms cancel one another, leaving only n^3; that is,

$$\sum_{i=1}^{n} [i^3 - (i-1)^3] = n^3.$$

Thus,

$$n^3 = \sum_{i=1}^{n} (3i^2 - 3i + 1)$$

$$= 3\sum_{i=1}^{n} i^2 - 3\sum_{i=1}^{n} i + \sum_{i=1}^{n} 1 \quad \text{(using Theorem 6.1)}$$

$$= 3\sum_{i=1}^{n} i^2 - 3\frac{n(n+1)}{2} + n \quad \text{(using formula 6.4)}.$$

We can solve this equation for $\sum_{i=1}^{n} i^2$:

$$\sum_{i=1}^{n} i^2 = \frac{1}{3}\left[n^3 + \frac{3n(n+1)}{2} - n\right] = \frac{n(n+1)(2n+1)}{6}. \tag{6.5}$$

A similar procedure beginning with the identity $i^4 - (i-1)^4 = 4i^3 - 6i^2 + 4i - 1$ yields

$$\sum_{i=1}^{n} i^3 = \frac{n^2(n+1)^2}{4}. \qquad (6.6)$$

These results can also be established independently of one another by mathematical induction.

EXERCISES 6.1

In Exercises 1–10 express the sum in sigma notation. Initiate the summation with the integer 1.

1. $2 \cdot 3 + 3 \cdot 4 + 4 \cdot 5 + 5 \cdot 6 + \cdots + 99 \cdot 100$

2. $\dfrac{1}{2} + \dfrac{2}{4} + \dfrac{3}{8} + \dfrac{4}{16} + \dfrac{5}{32} + \cdots + \dfrac{10}{1024}$

3. $\dfrac{16}{14+15} + \dfrac{17}{15+16} + \dfrac{18}{16+17} + \cdots + \dfrac{199}{197+198}$

4. $1 + \sqrt{2} + \sqrt{3} + 2 + \sqrt{5} + \sqrt{6} + \sqrt{7} + \sqrt{8} + 3 + \cdots + 121$

5. $1 + \dfrac{1}{2} + \dfrac{1}{2 \cdot 3} + \dfrac{1}{2 \cdot 3 \cdot 4} + \cdots + \dfrac{1}{2 \cdot 3 \cdot 4 \cdot 5 \cdots 16}$

6. $-2 + 3 - 4 + 5 - 6 + 7 - 8 + \cdots - 1020$

7. $\dfrac{2 \cdot 3}{1 \cdot 4} + \dfrac{6 \cdot 7}{5 \cdot 8} + \dfrac{10 \cdot 11}{9 \cdot 12} + \dfrac{14 \cdot 15}{13 \cdot 16} + \cdots + \dfrac{414 \cdot 415}{413 \cdot 416}$

8. $\dfrac{\tan 1}{2} + \dfrac{\tan 2}{1 + 2^2} + \dfrac{\tan 3}{1 + 3^2} + \dfrac{\tan 4}{1 + 4^2} + \cdots + \dfrac{\tan 225}{1 + 225^2}$

9. $4^3 + 5^2 + 6 + 1 + \dfrac{1}{8} + \dfrac{1}{9^2} + \cdots + \dfrac{1}{25^{18}}$

10. $0.9 + 0.99 + 0.999 + \cdots + 0.999999999$

In Exercises 11–15 verify by a change of variable of summation that the two summations are identical.

11. $\displaystyle\sum_{n=1}^{24} \dfrac{n^2}{2n+1}$ $\displaystyle\sum_{i=4}^{27} \dfrac{i^2 - 6i + 9}{2i - 5}$

12. $\displaystyle\sum_{k=2}^{101} \dfrac{3k - k^2}{\sqrt{k+5}}$ $\displaystyle\sum_{m=0}^{99} \dfrac{2 - m - m^2}{\sqrt{7+m}}$

13. $\displaystyle\sum_{n=5}^{20} (-1)^n \dfrac{2^n}{n^2 + 1}$ $\displaystyle\sum_{j=1}^{16} 16(-1)^j \dfrac{2^j}{j^2 + 8j + 17}$

14. $\displaystyle\sum_{i=0}^{37} \dfrac{3^{3i}}{i!}$ $\displaystyle\sum_{m=2}^{39} \dfrac{3^{3m}}{729(m-2)!}$

15. $\displaystyle\sum_{r=15}^{225} \dfrac{1}{r^2 - 10r}$ $\displaystyle\sum_{n=10}^{220} \dfrac{1}{n^2 - 25}$

In Exercises 16–21 use Theorem 6.1 and the formulas 6.4–6.6 to evaluate the sum.

16. $\displaystyle\sum_{n=1}^{12} (3n + 2)$

17. $\displaystyle\sum_{j=1}^{21} (2j^2 + 3j)$

18. $\displaystyle\sum_{m=1}^{n} (4m - 2)^2$

19. $\displaystyle\sum_{k=2}^{29} (k^3 - 3k^2)$

20. $\displaystyle\sum_{n=1}^{25} (n + 5)(n - 4)$

21. $\displaystyle\sum_{i=1}^{n} i(i - 3)^2$

22. Verify Theorem 6.1.

23. Find a formula for

$$\sum_{k=1}^{n} \dfrac{1}{k(k+1)}.$$

Hint:

$$\dfrac{1}{k(k+1)} = \dfrac{1}{k} - \dfrac{1}{k+1}.$$

24. If $f(x)$ is a function of x, defined for all x, simplify the sum

$$\sum_{i=1}^{n} [f(i) - f(i-1)].$$

25. Prove formula 6.6.

26. Is

$$\sum_{i=1}^{n} [f(i)g(i)] \text{ equal to } \left[\sum_{i=1}^{n} f(i)\right]\left[\sum_{i=1}^{n} g(i)\right]?$$

27. A geometric series is a sum of terms of the form

$$a + ar + ar^2 + ar^3 + \cdots + ar^{n-1}.$$

There is a first term a, and every term thereafter is obtained by multiplying the preceding term by r (called the *common ratio*).

(a) If S_n represents this sum, express S_n in sigma notation.

(b) Prove that
$$S_n = \frac{a(1-r^n)}{1-r}.$$

Use the formula in Exercise 27 to sum the geometric series in Exercises 28–31.

28. $\dfrac{1}{8} + \dfrac{1}{16} + \dfrac{1}{32} + \dfrac{1}{64} + \cdots + \dfrac{1}{1\,048\,576}$

29. $1 - \dfrac{1}{3} + \dfrac{1}{9} - \dfrac{1}{27} + \dfrac{1}{81} - \cdots - \dfrac{1}{19\,683}$

30. $40(0.99) + 40(0.99)^2 + 40(0.99)^3 + \cdots 40(0.99)^{15}$

31. $\sqrt{0.99} + 0.99 + (0.99)^{3/2} + (0.99)^2 + \cdots + (0.99)^{10}$

32. Prove that
$$\left|\sum_{i=1}^{n} f(i)\right| \leq \sum_{i=1}^{n} |f(i)|.$$

33. Express the following summation in sigma notation:
$$1 + \frac{1}{2} - \frac{1}{4} - \frac{1}{8} + \frac{1}{16} + \frac{1}{32} - \frac{1}{64} - \frac{1}{128} + \cdots + \frac{1}{4096}.$$

SECTION 6.2
The Need for the Definite Integral

In this section we consider four inherently different problems: on area, on volume, on blood flow, and on work; but we shall see that a common method of solution exists for all four of them. This common theme leads to the definition of the definite integral in Section 6.3.

Problem 1

We first consider the problem of finding the area A in Figure 6.1(a). At present, we have formulas for areas of very few geometric shapes — rectangles, triangles, polygons, and circles, and the shape in Figure 6.1(a) is not one of them. We can, however, find an approximation to the area by constructing rectangles in the following way. Between a and b we pick $n-1$ points $x_1, x_2, x_3, \ldots, x_{n-1}$ on the x-axis such that

$$a = x_0 < x_1 < x_2 < \cdots < x_{n-1} < x_n = b.$$

We draw vertical lines through each of these $n-1$ points to intersect the curve $y = f(x)$ and form rectangles, as shown in Figure 6.1(b).

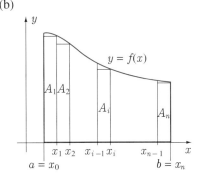

FIGURE 6.1

If we denote by A_i the area of the ith such rectangle ($i = 1, \ldots, n$), then

$$A_i = (\text{height of rectangle})(\text{width of rectangle})$$
$$= f(x_i)(x_i - x_{i-1}).$$

The sum of these n rectangular areas is an approximation to the required area; that is, A is approximately equal to

$$\sum_{i=1}^{n} A_i = \sum_{i=1}^{n} f(x_i)(x_i - x_{i-1}).$$

If we let the number of these rectangles get larger and larger (and at the same time require each to have a smaller and smaller width that eventually approaches zero), the approximation appears to get better and better. In fact, we expect that

$$A = \lim_{n \to \infty} \sum_{i=1}^{n} A_i = \lim_{n \to \infty} \sum_{i=1}^{n} f(x_i)(x_i - x_{i-1}). \tag{6.7}$$

Problem 2

If the area in Figure 6.1(a) is rotated about the x-axis, it traces out a volume V. This is certainly not a standard shape for which we have a volume formula; therefore, to calculate V we first find an approximation. We take the rectangles in Figure 6.1(b) (which approximate the area) and rotate them about the x-axis (Figure 6.2).

FIGURE 6.2

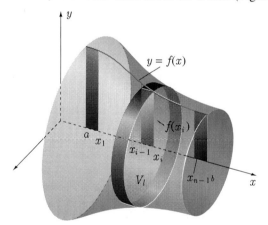

Each A_i traces out a disc of volume V_i ($i = 1, \ldots, n$), where

$$\begin{aligned} V_i &= \text{(surface area of disc)(thickness of disc)} \\ &= \pi(\text{radius of disc})^2(\text{thickness of disc}) \\ &= \pi[f(x_i)]^2(x_i - x_{i-1}). \end{aligned}$$

An approximation to the required volume is then

$$\sum_{i=1}^{n} V_i = \sum_{i=1}^{n} \pi[f(x_i)]^2(x_i - x_{i-1}).$$

Again we feel intuitively that as the number of discs becomes larger and larger (and the width of each approaches zero), the approximation becomes better and better, and

$$V = \lim_{n \to \infty} \sum_{i=1}^{n} V_i = \lim_{n \to \infty} \sum_{i=1}^{n} \pi[f(x_i)]^2(x_i - x_{i-1}). \tag{6.8}$$

Problem 3

When blood flows through a vein or artery, it encounters resistance due to friction with the walls of the blood vessel and due to the viscosity of the blood itself. As a result, the velocity of the blood is not constant across a cross-section of the vessel; blood flows more quickly near the centre of the vessel than near its walls. It has been shown that for laminar blood flow in a vessel of circular cross-section (Figure 6.3), the velocity of blood is given by

$$v = v(r) = c(R^2 - r^2), \quad 0 \leq r \leq R,$$

where $c > 0$ is a constant, R is the radius of the blood vessel, and r is radial distance measured from the centre of the vessel. We wish to find the rate of blood flow through the vessel, that is, the volume of blood flowing through the cross-section per unit time.

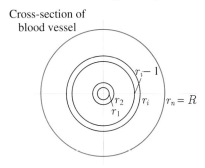

FIGURE 6.3

If v were constant over the cross-section, then flow per unit time would be the product of v and the cross-sectional area. Unfortunately this is not the case, but we can still use the idea that flow is velocity multiplied by area. We divide the cross-section into rings with radii

$$0 = r_0 < r_1 < r_2 < \cdots < r_{n-1} < r_n = R.$$

Over the ith ring, the variation in v is small and v can be approximated by $v(r_i)$. The flow through the ith ring can therefore be approximated by

$$F_i = (\text{area of ring})(\text{velocity at outer radius of ring})$$
$$= (\pi r_i^2 - \pi r_{i-1}^2) v(r_i).$$

An approximation to the required flow F is the sum of these F_i:

$$\sum_{i=1}^{n} F_i = \sum_{i=1}^{n} (\pi r_i^2 - \pi r_{i-1}^2) v(r_i),$$

and it seems reasonable that as the number of rings increases, so too does the accuracy of the approximation; that is, we anticipate that

$$F = \lim_{n \to \infty} \sum_{i=1}^{n} F_i = \lim_{n \to \infty} \sum_{i=1}^{n} (\pi r_i^2 - \pi r_{i-1}^2) v(r_i). \tag{6.9}$$

Problem 4

A spring is fixed horizontally into a wall at one end, and the other end is free. Consider finding the work done in stretching the spring three centimetres by pulling on its free end (Figure 6.4(a)).

FIGURE 6.4

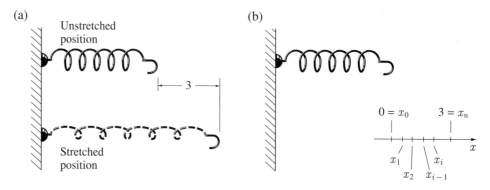

Let us choose an x-axis positive to the right with $x = 0$ at the position of the free end of the spring when it is in the unstretched position (Figure 6.4(b)). In order to calculate the work to stretch the spring, we must know something about the forces involved. It has been shown experimentally that the force F that must be exerted on the free end of the spring in order to maintain a stretch x in the spring is proportional to x:

$$F = F(x) = kx,$$

where $k > 0$ is a constant. This then is the force that will perform the work. The basic definition of work W done by a constant force F acting along a straight line segment of length d is $W = Fd$. Unfortunately, our force $F(x)$ is not constant; it depends on x, and we cannot therefore simply multiply force by distance. What we can find, however, is an approximation to the required work by dividing the length between $x = 0$ and $x = 3$ into n subintervals by $n - 1$ points $x_1, x_2, \ldots, x_{n-1}$ such that

$$0 = x_0 < x_1 < x_2 < \ldots < x_{n-1} < x_n = 3.$$

When the spring is stretched between x_{i-1} and x_i, the force necessary to maintain this stretch does not vary greatly and can be approximated by $F(x_i)$. It follows then that the work necessary to stretch the spring from x_{i-1} to x_i is approximately equal to

$$W_i = F(x_i)(x_i - x_{i-1}).$$

As a result, an approximation to the total work required to pull the free end of the spring from $x = 0$ to $x = 3$ is

$$\sum_{i=1}^{n} W_i = \sum_{i=1}^{n} F(x_i)(x_i - x_{i-1}).$$

Once again we expect that as n becomes indefinitely large, this approximation approaches the required work W, and

$$W = \lim_{n \to \infty} \sum_{i=1}^{n} W_i = \lim_{n \to \infty} \sum_{i=1}^{n} F(x_i)(x_i - x_{i-1}). \qquad (6.10)$$

Each of these four problems on area, volume, blood flow, and work has been tackled in the same way, and the method can be described qualitatively.

The quantity to be calculated, say W, cannot be obtained for the object G given because no formula exists. As a result, n smaller objects, say G_i, are constructed. The G_i are chosen in such a way that the quantity W can be calculated, exactly or approximately, for each G_i, say W_i. Then an approximation for W is

$$\sum_{i=1}^{n} W_i.$$

If the number of G_i is increased indefinitely, this approximation becomes more and more accurate and

$$W = \lim_{n \to \infty} \sum_{i=1}^{n} W_i.$$

It is this "limit-summation" process that we shall discuss throughout the remainder of the chapter. We begin in Section 6.3 with a mathematical description of the process, and by doing so, we obtain a unified approach to the whole idea. We then discover that there is a very simple way to calculate these limits. At that point we will be ready to use the technique in a multitude of applications, including the four problems in this section.

SECTION 6.3

The Definite Integral

The four problems of Section 6.2 have a common theme: the limit of a summation. By means of a summation we approximated some quantity (area, volume, blood flow, work), and the limit led, at least intuitively, to an exact value for the quantity. In this section, we investigate the mathematics of the limit-summation — but only its mathematics. We concentrate here on what a definite integral is, and how to evaluate it, while interpretation of the definite integral as area, volume, work, etc. is made in Chapter 7.

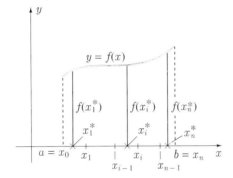

FIGURE 6.5

To define the definite integral of a function $f(x)$ on an interval $a \leq x \leq b$ (Figure 6.5), we divide the interval into n subintervals by any $n-1$ points:

$$a = x_0 < x_1 < x_2 < \cdots < x_i < \cdots < x_{n-1} < x_n = b.$$

Next we choose in each subinterval $x_{i-1} \leq x \leq x_i$ any point x_i^* whatsoever, and evaluate $f(x_i^*)$. We now form the sum

$$f(x_1^*)(x_1 - x_0) + f(x_2^*)(x_2 - x_1) + \cdots + f(x_n^*)(x_n - x_{n-1}) = \sum_{i=1}^{n} f(x_i^*)(x_i - x_{i-1})$$

$$= \sum_{i=1}^{n} f(x_i^*) \Delta x_i,$$

where we have set $\Delta x_i = x_i - x_{i-1}$. We denote by $\|\Delta x_i\|$ the length of the longest of the n subintervals,

$$\|\Delta x_i\| = \max_{i=1,\ldots,n} |\Delta x_i|.$$

It is often called the **norm** of the particular partition of $a \leq x \leq b$ into the subintervals Δx_i. With this notation, we are ready to define the definite integral of $f(x)$. It is the limit of the above summation as the number of subintervals becomes increasingly large and every subinterval shrinks to a point. An easier way to say this is to take the limit as the norm of the partition approaches zero. In other words, we define the **definite integral** of $f(x)$ with respect to x from $x = a$ to $x = b$ as

$$\int_a^b f(x)\,dx = \lim_{||\Delta x_i|| \to 0} \sum_{i=1}^n f(x_i^*)\Delta x_i, \qquad (6.11)$$

provided the limit exists. If the limit exists, but is dependent on the choice of subdivision Δx_i or star-points x_i^*, then the definite integral is of little use. We stipulate, therefore, that in order for the definite integral to exist, the limit of the sum in equation 6.11 must be independent of the manner of subdivision of the interval $a \le x \le b$ and choice of star-points in the subintervals. At first sight this requirement might seem rather severe, since we must now check that all subdivisions and all choices of star-points lead to the same limit before concluding that the definite integral exists. Fortunately, however, the following theorem indicates that, for continuous functions, this is unnecessary. A proof of this theorem can be found in advanced books on mathematical analysis.

Theorem 6.2 *If a function $f(x)$ is continuous on a finite interval $a \le x \le b$, then the definite integral of $f(x)$ with respect to x from $x = a$ to $x = b$ exists.*

For a continuous function, the definite integral exists, and any choice of subdivision and star-points leads to its correct value through the limiting process. We call $f(x)$ on the left-hand side of equation 6.11 the integrand, and a and b the **lower** and **upper limits of integration**, respectively. The sum $\sum_{i=1}^n f(x_i^*)\Delta x_i$ is called a **Riemann sum**, and because of this, definite integral (6.11) is also called the **Riemann integral**. The integral was named after German mathematician G.F.B. Riemann (1826-1866), who introduced the notion of the definite integral as a sum.

EXAMPLE 6.3 Evaluate the definite integral

$$\int_0^1 x^2\,dx.$$

SOLUTION Since $f(x) = x^2$ is continuous on the interval $0 \le x \le 1$, the definite integral exists, and we may choose any subdivision and star-points in its evaluation. The simplest partition is into n equal subintervals of length $1/n$ by the points (Figure 6.6)

$$x_i = \frac{i}{n}, \qquad i = 0, \ldots, n.$$

We choose for star-points the right end of each subinterval; that is, in $x_{i-1} \le x \le x_i$, we choose $x_i^* = x_i = i/n$. Then, by equation 6.11, we have

$$\int_0^1 x^2\,dx = \lim_{||\Delta x_i|| \to 0} \sum_{i=1}^n f(x_i^*)\Delta x_i = \lim_{||\Delta x_i|| \to 0} \sum_{i=1}^n (x_i^*)^2 \Delta x_i.$$

Since all subintervals have equal length $\Delta x_i = 1/n$, the norm of the partition is $||\Delta x_i|| = 1/n$, and taking the limit as $||\Delta x_i|| \to 0$ is tantamount to letting $n \to \infty$. Thus,

$$\int_0^1 x^2 \, dx = \lim_{n \to \infty} \sum_{i=1}^{n} \left(\frac{i}{n}\right)^2 \left(\frac{1}{n}\right)$$

$$= \lim_{n \to \infty} \frac{1}{n^3} \sum_{i=1}^{n} i^2.$$

If we now use equation 6.5 for the sum of the squares of the first n positive integers, we obtain

$$\int_0^1 x^2 \, dx = \lim_{n \to \infty} \frac{1}{n^3} \frac{n(n+1)(2n+1)}{6} = \frac{1}{3}.$$

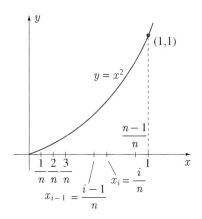

FIGURE 6.6

This example illustrates that even for a simple function such as $f(x) = x^2$, evaluation of the definite integral by equation 6.11 is not really simple. In fact, had we not known formula 6.5 for the sum of the squares of the integers, we would not have been able to complete the calculation. Imagine the magnitude of the problem were the integrand equal to $f(x) = x(x+1)^{-2/3}$. In other words, if definite integrals are to be at all useful, we must find a simpler way to evaluate them. This we do in Section 6.4, but in order to stress the definite integral as a limit-summation, we consider one more example.

EXAMPLE 6.4

Evaluate the definite integral

$$\int_{-1}^{1} (5x - 2) \, dx.$$

SOLUTION Since $f(x) = 5x - 2$ is continuous on the interval $-1 \leq x \leq 1$, the definite integral exists, and we may choose any partition and star-points in its evaluation. For n equal subdivisions of length $2/n$, we use points (Figure 6.7)

$$x_i = -1 + \frac{2i}{n}, \qquad i = 0, \ldots, n.$$

If we choose the right end of each subinterval as a star-point, that is, $x_i^* = x_i = -1 + 2i/n$, then equation 6.11 gives

$$\int_{-1}^{1}(5x-2)\,dx = \lim_{\|\Delta x_i\|\to 0}\sum_{i=1}^{n}f(x_i^*)\Delta x_i$$

$$= \lim_{\|\Delta x_i\|\to 0}\sum_{i=1}^{n}(5x_i^*-2)\Delta x_i.$$

Once again all subintervals have equal length $\Delta x_i = 2/n$, and therefore we may replace $\|\Delta x_i\| \to 0$ with $n \to \infty$,

$$\int_{-1}^{1}(5x-2)\,dx = \lim_{n\to\infty}\sum_{i=1}^{n}\left[5\left(-1+\frac{2i}{n}\right)-2\right]\left(\frac{2}{n}\right) = \lim_{n\to\infty}\sum_{i=1}^{n}\left[\frac{20i-14n}{n^2}\right].$$

Applying Theorem 6.1, we can break the summation into two parts, and take constants outside each summation to obtain

$$\int_{-1}^{1}(5x-2)\,dx = \lim_{n\to\infty}\left[\sum_{i=1}^{n}\frac{20i}{n^2}-\sum_{i=1}^{n}\frac{14}{n}\right]$$

$$= \lim_{n\to\infty}\left[\frac{20}{n^2}\sum_{i=1}^{n}i - \frac{14}{n}\sum_{i=1}^{n}1\right]$$

$$= \lim_{n\to\infty}\left[\frac{20}{n^2}\frac{n(n+1)}{2} - \frac{14}{n}n\right] \quad \text{(using formula 6.4)}$$

$$= \lim_{n\to\infty}\left[\frac{10-4n}{n}\right]$$

$$= -4.$$

FIGURE 6.7

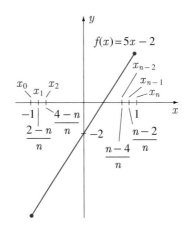

Note that in Example 6.3 the value of the definite integral is positive, and in Example 6.4 it is negative. The value of the definite integral in Exercise 8 is zero. In other words, the value of a definite integral can be positive, negative, or zero, depending on the limits and the integrand.

EXERCISES 6.3

In Exercises 1–8 use equation 6.11 to evaluate the definite integral.

1. $\int_0^1 x \, dx$
2. $\int_0^2 3x \, dx$
3. $\int_0^1 (3x + 2) \, dx$
4. $\int_0^2 x^3 \, dx$
5. $\int_1^2 (x^2 + 2x) \, dx$
6. $\int_{-1}^0 (-x + 1) \, dx$
7. $\int_{-1}^1 x^2 \, dx$
8. $\int_{-1}^1 x^3 \, dx$

9. Evaluate the definite integral

$$\int_{-1}^1 x^{15} \, dx.$$

10. (a) Consider the definite integral

$$\int_0^1 2^x \, dx.$$

Show that when the interval $0 \le x \le 1$ is subdivided into n equal subintervals, and star-points are chosen as right-hand end points in each subinterval, equation 6.11 leads to

$$\int_0^1 2^x \, dx = \lim_{n \to \infty} \frac{1}{n} \sum_{i=1}^n 2^{i/n}.$$

(b) Use the formula in question 27(b) from Exercises 6.1 to express the summation in closed form,

$$\int_0^1 2^x \, dx = \lim_{n \to \infty} \frac{2^{1/n}}{n(2^{1/n} - 1)}.$$

(c) Use L'Hôpital's rule to evaluate this limit, and hence find the value of the definite integral.

11. Use the technique of Exercise 10 to evaluate

$$\int_1^3 e^x \, dx.$$

12. Use the formula

$$\sum_{i=1}^n \sin i\theta = \frac{\sin \frac{(n+1)\theta}{2} \sin \frac{n\theta}{2}}{\sin \frac{\theta}{2}}$$

to evaluate

$$\int_0^\pi \sin x \, dx.$$

13. Use the formula

$$\sum_{i=1}^n \cos i\theta = \frac{\cos \frac{(n+1)\theta}{2} \sin \frac{n\theta}{2}}{\sin \frac{\theta}{2}}$$

to evaluate

$$\int_0^{\pi/2} \cos x \, dx.$$

14. In this exercise we evaluate

$$\int_a^b x^k \, dx$$

for $b > a > 0$ and any $k \ne -1$.

(a) Let $h = (b/a)^{1/n}$ and subdivide the interval $a \le x \le b$ into n subintervals by the points

$$x_0 = a, \quad x_1 = ah, \quad x_2 = ah^2, \ldots, x_i = ah^i, \ldots,$$
$$x_n = ah^n = b.$$

Show that with the choice of $x_i^* = x_i$ in the ith subinterval $x_{i-1} \le x \le x_i$, equation 6.11 gives

$$\int_a^b x^k \, dx = a^{k+1} \lim_{n \to \infty} \left[\left(\frac{h-1}{h}\right) \sum_{i=1}^n (h^{k+1})^i \right].$$

(b) Use the result of Exercise 27 in Section 6.1 to write the summation in closed form, and hence show that

$$\int_a^b x^k \, dx = (b^{k+1} - a^{k+1}) \lim_{n \to \infty} \frac{\left(\frac{b}{a}\right)^{k/n} \left[\left(\frac{b}{a}\right)^{1/n} - 1\right]}{\left(\frac{b}{a}\right)^{(k+1)/n} - 1}.$$

(c) Use L'Hôpital's rule to evaluate this limit, and hence obtain

$$\int_a^b x^k \, dx = \frac{b^{k+1} - a^{k+1}}{k + 1}.$$

SECTION 6.4

The Fundamental Theorem of Integral Calculus

In Section 6.3 we demonstrated how to evaluate definite integrals by using their definition in equation 6.11. Integrands x^2 and $5x - 2$ in Examples 6.3 and 6.4 are very simple polynomials, as are the integrands in questions 1–9 of Exercise set 6.3, but in spite of this, calculations were frequently laborious, and invariably required formulas 6.4–6.6. We promised a very simple technique that would replace these calculations, and this is the substance of the *fundamental theorem of integral calculus*.

Theorem 6.3

(The fundamental theorem of integral calculus) If $f(x)$ is continuous on the interval $a \leq x \leq b$, and $F(x)$ is an antiderivative of $f(x)$ thereon, then

$$\int_a^b f(x)\, dx = F(b) - F(a). \tag{6.12}$$

Proof Since $f(x)$ is continuous on $a \leq x \leq b$, the definite integral of $f(x)$ from $x = a$ to $x = b$ exists and is defined by equation 6.11, where we are at liberty to choose the Δx_i and x_i^* in any way whatsoever. For any choice of Δx_i, a convenient choice for the x_i^* can be found by applying the mean value theorem n times to $F(x)$, once on each subinterval $x_{i-1} \leq x \leq x_i$. This is possible since $F'(x) = f(x)$ is continuous for $a \leq x \leq b$. The mean value theorem states that for each subinterval, there exists at least one point c_i between x_{i-1} and x_i such that

$$\frac{F(x_i) - F(x_{i-1})}{x_i - x_{i-1}} = F'(c_i), \qquad i = 1, \ldots, n.$$

But $F'(c_i) = f(c_i)$, so that

$$F(x_i) - F(x_{i-1}) = f(c_i)\Delta x_i, \qquad i = 1, \ldots, n.$$

If we now choose $x_i^* = c_i$, then

$$f(x_i^*)\Delta x_i = F(x_i) - F(x_{i-1}),$$

and equation 6.11 gives

$$\int_a^b f(x)\, dx = \lim_{||\Delta x_i|| \to 0} \sum_{i=1}^n f(x_i^*)\Delta x_i = \lim_{||\Delta x_i|| \to 0} \sum_{i=1}^n [F(x_i) - F(x_{i-1})].$$

When we write out all terms in the summation, we find that many cancellations take place,

$$\int_a^b f(x)\, dx = \lim_{||\Delta x_i|| \to 0} \{[F(x_1) - F(x_0)] + [F(x_2) - F(x_1)] + \cdots + [F(x_n) - F(x_{n-1})]\}$$

$$= \lim_{||\Delta x_i|| \to 0} \{F(x_n) - F(x_0)\}$$

$$= \lim_{||\Delta x_i|| \to 0} \{F(b) - F(a)\}$$

$$= F(b) - F(a).$$

If we introduce the notation

$$\{F(x)\}_a^b$$

to represent the difference $F(b) - F(a)$, then Theorem 6.3 can be expressed in the form

$$\int_a^b f(x)\,dx = \left\{\int f(x)\,dx\right\}_a^b. \tag{6.13}$$

This is a fantastic result. No longer is it necessary to consider limits of summations in order to evaluate definite integrals. We simply find an antiderivative of the integrand, substitute $x = b$ and $x = a$, and subtract. For instance, to evaluate the definite integral in Example 6.3, we easily write

$$\int_0^1 x^2\,dx = \left\{\frac{x^3}{3}\right\}_0^1 = \frac{1}{3} - 0 = \frac{1}{3}.$$

Note that had we used *the* indefinite integral $x^3/3 + C$ for x^2, we would have had

$$\int_0^1 x^2\,dx = \left\{\frac{x^3}{3} + C\right\}_0^1 = \left\{\frac{1}{3} + C\right\} - C = \frac{1}{3}.$$

Because the arbitrary constant always vanishes in the evaluation of definite integrals, we omit the constant in this context.

Similarly, for Example 6.4, we obtain

$$\int_{-1}^1 (5x - 2)\,dx = \left\{\frac{5x^2}{2} - 2x\right\}_{-1}^1 = \left\{\frac{5}{2} - 2\right\} - \left\{\frac{5}{2} + 2\right\} = -4.$$

EXAMPLE 6.5

Evaluate the following definite integrals:

(a) $\int_1^2 (3x^2 - x + 4)\,dx$ (b) $\int_{-2}^4 \sqrt{x + 4}\,dx$

SOLUTION

(a) $\int_1^2 (3x^2 - x + 4)\,dx = \left\{x^3 - \frac{x^2}{2} + 4x\right\}_1^2 = \{8 - 2 + 8\} - \left\{1 - \frac{1}{2} + 4\right\} = \frac{19}{2}$

(b) $\int_{-2}^4 \sqrt{x + 4}\,dx = \left\{\frac{2}{3}(x + 4)^{3/2}\right\}_{-2}^4 = \frac{2}{3}(8)^{3/2} - \frac{2}{3}(2)^{3/2} = \frac{28\sqrt{2}}{3}$

EXAMPLE 6.6

Evaluate

$$\int_1^3 \left(\frac{1}{x^2} + 3x^3\right)dx.$$

SOLUTION Since $-1/x + 3x^4/4$ is an antiderivative for $1/x^2 + 3x^3$,

$$\int_1^3 \left(\frac{1}{x^2} + 3x^3\right)dx = \left\{-\frac{1}{x} + \frac{3x^4}{4}\right\}_1^3 = \left\{-\frac{1}{3} + \frac{243}{4}\right\} - \left\{-1 + \frac{3}{4}\right\} = \frac{182}{3}.$$

EXAMPLE 6.7

Can the following definite integral be evaluated with Theorem 6.3,

$$\int_{-1}^{1} \frac{1}{x^2} \, dx \,?$$

SOLUTION No! Theorem 6.3 requires the integrand $1/x^2$ to be continuous on the interval $-1 \leq x \leq 1$, and this is not the case. The function is discontinuous at $x = 0$. ∎

There is a difficulty with Theorem 6.3. It is subtle, but important. The theorem states that, to evaluate definite integrals of continuous functions, we use antiderivatives. But how do we know that continuous functions have antiderivatives? We don't yet. This fact will be established in Section 6.5 when we verify an alternative version of the fundamental theorem. You might ask why this alternative version is not proved first. Would it not be more logical to first establish existence of antiderivatives, and then use this fact to prove Theorem 6.3? From a logic point of view, the answer is yes. However, from a practical point of view, Theorem 6.3 is infinitely more useful than its alternative version, and we therefore want to give it every possible emphasis. To prove the alternative version first would detract from the importance and simplicity of Theorem 6.3.

Before we move on to the alternative version of the fundamental theorem, we present the following theorems which describe properties of the definite integral.

Theorem 6.4

If $f(x)$ is continuous on $a \leq x \leq b$, then:

(i) $\quad \int_{b}^{a} f(x)\, dx = -\int_{a}^{b} f(x)\, dx;$ \hfill (6.14)

(ii) $\quad \int_{a}^{b} f(x)\, dx = \int_{a}^{c} f(x)\, dx + \int_{c}^{b} f(x)\, dx.$ \hfill (6.15)

Theorem 6.5

If $f(x)$ and $g(x)$ are continuous on $a \leq x \leq b$, then:

(i) $\quad \int_{a}^{b} [f(x) + g(x)]\, dx = \int_{a}^{b} f(x)\, dx + \int_{a}^{b} g(x)\, dx;$ \hfill (6.16)

(ii) $\quad \int_{a}^{b} kf(x)\, dx = k \int_{a}^{b} f(x)\, dx,$ \hfill (6.17)

when k is a constant.

Properties 6.16 and 6.17 are analogous to 5.3 and 5.4 for indefinite integrals. Theorems 6.4 and 6.5 can be proved using either Theorem 6.3 or equation 6.11.

We can also establish the following property.

Theorem 6.6

When $f(x)$ is continuous on $a \leq x \leq b$ and $m \leq f(x) \leq M$ on this interval,

$$m(b-a) \leq \int_a^b f(x)\, dx \leq M(b-a). \tag{6.18}$$

Proof By equation 6.11, we can write

$$\int_a^b f(x)\, dx = \lim_{\|\Delta x_i\| \to 0} \sum_{i=1}^n f(x_i^*)\Delta x_i$$

$$\leq \lim_{\|\Delta x_i\| \to 0} \sum_{i=1}^n M \Delta x_i$$

$$= M \lim_{\|\Delta x_i\| \to 0} \sum_{i=1}^n \Delta x_i$$

$$= M \lim_{\|\Delta x_i\| \to 0} [(x_1 - a) + (x_2 - x_1) + (x_3 - x_2) + \cdots + (b - x_{n-1})]$$

$$= M \lim_{\|\Delta x_i\| \to 0} (b-a)$$

$$= M(b-a).$$

A similar proof establishes the inequality involving m.

EXAMPLE 6.8

Use Theorem 6.6 to find a maximum possible value for

$$\int_1^4 \frac{\sin x^2}{1+x^2}\, dx.$$

SOLUTION Clearly $\sin x^2 \leq 1$ for all x, and on the interval $1 \leq x \leq 4$, the largest value of $1/(1+x^2)$ is $1/2$. Consequently, $(\sin x^2)/(1+x^2) \leq 1/2$ for $1 \leq x \leq 4$, and by Theorem 6.6,

$$\int_1^4 \frac{\sin x^2}{1+x^2}\, dx \leq \frac{1}{2}(4-1) = \frac{3}{2}.$$

EXERCISES 6.4

In Exercises 1–34 evaluate the definite integral.

1. $\int_3^4 (x^3 + 3)\, dx$

2. $\int_1^3 (x^2 - 2x + 3)\, dx$

3. $\int_{-1}^1 (4x^3 + 2x)\, dx$

4. $\int_{-3}^{-1} \frac{1}{x^2}\, dx$

5. $\int_4^2 \left(x^2 + \frac{3}{x^3}\right) dx$

6. $\int_0^{\pi/2} \sin x\, dx$

7. $\int_{-1}^1 (x^2 - 1 - x^4)\, dx$

8. $\int_{-1}^{-2} \left(\frac{1}{x^2} - 2x\right) dx$

9. $\int_1^2 (x^4 + 3x^2 + 2)\, dx$

10. $\int_0^1 x(x^2 + 1)\, dx$

11. $\int_0^1 x^2(x^2 + 1)^2\, dx$

12. $\int_0^{2\pi} \cos 2x \, dx$

13. $\int_1^3 \frac{x^2+3}{x^2} \, dx$

14. $\int_0^1 (x^{2.2} - x^\pi) \, dx$

15. $\int_{-1}^1 x^2(x^3 - x) \, dx$

16. $\int_3^4 \frac{(x^2-1)^2}{x^2} \, dx$

17. $\int_1^2 \left(\sqrt{x} - \frac{1}{\sqrt{x}}\right) dx$

18. $\int_{-2}^3 (x-1)^3 \, dx$

19. $\int_2^4 \frac{(x^2-1)(x^2+1)}{x^2} \, dx$

20. $\int_0^{\pi/4} 3 \cos x \, dx$

21. $\int_0^{\pi/4} \sec^2 x \, dx$

22. $\int_{\pi/2}^{\pi} \sin x \cos x \, dx$

23. $\int_{\pi/6}^{\pi/3} \csc^2 3x \, dx$

24. $\int_{-\pi/4}^{\pi/4} \sec x \tan x \, dx$

25. $\int_0^2 2^x \, dx$

26. $\int_{-1}^2 e^x \, dx$

27. $\int_0^1 e^{3x} \, dx$

28. $\int_{-3}^{-2} \frac{1}{x} \, dx$

29. $\int_1^3 \frac{(x+1)^2}{x} \, dx$

30. $\int_0^1 3^{4x} \, dx$

31. $\int_0^5 |x| \, dx$

32. $\int_0^4 x|x+1| \, dx$

33. $\int_{-5}^5 |x| \, dx$

34. $\int_{-2}^1 x|x+1| \, dx$

35. If $v(t) = 3t^2 - 6t - 105$ represents the velocity (in metres per second) of a particle moving along the x-axis, evaluate and interpret physically:

   (a) $\int_0^{12} v \, dt$ (b) $\int_0^{12} |v| \, dt$

In Exercises 36–41 use Theorem 6.6 to find maximum and minimum values for the integral.

36. $\int_0^{\pi/4} \frac{\sin x}{1 + x^2} \, dx$

37. $\int_0^{\pi/2} \frac{\sin x}{1 + x} \, dx$

38. $\int_0^{\pi} \frac{\sin x}{2 + x^2} \, dx$

39. $\int_{\pi/4}^{\pi/2} \frac{\sin 2x}{10 + x^2} \, dx$

40. $\int_0^1 (1 + 4x^4) \cos(x^2) \, dx$

41. $\int_1^3 \sqrt{4 + x^3} \, dx$

SECTION 6.5

An Alternative Version of the Fundamental Theorem

The fundamental theorem of integral calculus in Section 6.4 allows us to use antiderivatives to evaluate definite integrals. To complete the picture, we now show that in some sense, definite integrals can be regarded as antiderivatives.

When $f(t)$ is continuous for $a \leq t \leq b$, its definite integral

$$\int_a^b f(t) \, dt$$

is a number. If b is changed but a is kept fixed, the value of the definite integral changes; for each value of b, a new value for the definite integral. In other words, the value of the definite integral is a function of its upper limit. Suppose we replace b by x, and denote the resulting function by $F(x)$,

$$F(x) = \int_a^x f(t) \, dt.$$

We now show that the derivative of $F(x)$ is $f(x)$; that is, $F(x)$ is an antiderivative of $f(x)$.

Theorem 6.7

(Alternative Version of the Fundamental Theorem) When $f(x)$ is continuous for $a \leq x \leq b$, the function

$$F(x) = \int_a^x f(t)\, dt \qquad (6.19)$$

is differentiable for $a \leq x \leq b$, and $F'(x) = f(x)$.

Proof If x is any point in the open interval $a < x < b$, then h can always be chosen sufficiently small that $x + h$ is also in the interval $a < x < b$. By definition 3.1, the derivative of $F(x)$ at this x is defined as

$$\begin{aligned}
F'(x) &= \lim_{h \to 0} \frac{F(x+h) - F(x)}{h} \\
&= \lim_{h \to 0} \frac{1}{h} \left[\int_a^{x+h} f(t)\, dt - \int_a^x f(t)\, dt \right] \\
&= \lim_{h \to 0} \frac{1}{h} \left[\int_a^{x+h} f(t)\, dt + \int_x^a f(t)\, dt \right] \quad \text{(by property 6.14)} \\
&= \lim_{h \to 0} \frac{1}{h} \int_x^{x+h} f(t)\, dt \quad \text{(by property 6.15)}.
\end{aligned}$$

According to property 6.18,

$$mh \leq \int_x^{x+h} f(t)\, dt \leq Mh$$

where m and M are the absolute minimum and absolute maximum of $f(x)$ on the interval between x and $x + h$. Division by h gives

$$m \leq \frac{1}{h} \int_x^{x+h} f(t)\, dt \leq M.$$

Consider what happens as we let $h \to 0$. The limit of the middle term is $F'(x)$. Furthermore, the numbers m and M must approach one another, and in the limit must both be equal to $f(x)$. They are minimum and maximum values of $f(x)$ on the interval between x and $x + h$, and h is approaching zero. We conclude therefore that

$$F'(x) = \lim_{h \to 0} \int_x^{x+h} f(t)\, dt = f(x).$$

This argument can also be used to establish that $F(x)$ has a right-hand derivative $f(a)$ at $x = a$, and a left-hand derivative $f(b)$ at $x = b$.

Recall that Theorem 6.3 assumed the existence of antiderivatives of continuous functions. Theorem 6.7 establishes this fact, and states that definite integral 6.19 with a variable upper limit is one such antiderivative. In practice, we do not use 6.19 to find antiderivatives, but the theorem does establish existence of antiderivatives of continuous functions.

Symbolically, we may write the result of Theorem 6.7 in the form

$$\frac{d}{dx}\int_a^x f(t)\,dt = f(x). \qquad (6.20)$$

The lower limit in this result need not be a; it can also be b,

$$\frac{d}{dx}\int_b^x f(t)\,dt = f(x).$$

The lower limit can also be any number c between a and b,

$$\frac{d}{dx}\int_c^x f(t)\,dt = f(x).$$

EXAMPLE 6.9

Evaluate

$$\frac{d}{dx}\int_0^x \sqrt{1-t^2}\,dt.$$

SOLUTION According to 6.20,

$$\frac{d}{dx}\int_0^x \sqrt{1-t^2}\,dt = \sqrt{1-x^2}.$$

This is valid for $-1 \leq x \leq 1$.

EXAMPLE 6.10

Evaluate

$$\frac{d}{dx}\int_1^{2x^2} \frac{\sin t}{1+t^2}\,dt.$$

SOLUTION For this problem we set $u = 2x^2$, and invoke the chain rule,

$$\frac{d}{dx}\int_1^{2x^2} \frac{\sin t}{1+t^2}\,dt = \frac{d}{dx}\int_1^u \frac{\sin t}{1+t^2}\,dt = \left[\frac{d}{du}\int_1^u \frac{\sin t}{1+t^2}\,dt\right]\frac{du}{dx}$$

$$= \frac{\sin u}{1+u^2}(4x) = \frac{4x\sin(2x^2)}{1+4x^4}.$$

EXAMPLE 6.11

Evaluate

$$\frac{d}{dx}\int_x^5 (1+t^3)^{2/3}\,dt.$$

SOLUTION We can solve this problem by reversing the limits on the integral,

$$\frac{d}{dx}\int_x^5 (1+t^3)^{2/3}\,dt = -\frac{d}{dx}\int_5^x (1+t^3)^{2/3}\,dt = -(1+x^3)^{2/3}.$$

In the following example, the variable x appears in both limits.

EXAMPLE 6.12

Evaluate
$$\frac{d}{dx} \int_{x^2}^{2x} \cos(2t^3 + 1)\, dt.$$

SOLUTION Since the integrand is continuous for all real numbers, property 6.15 permits us to write

$$\int_{x^2}^{2x} \cos(2t^3 + 1)\, dt = \int_{x^2}^{a} \cos(2t^3 + 1)\, dt + \int_{a}^{2x} \cos(2t^3 + 1)\, dt,$$

for any real number a whatsoever. To find the derivative of the second integral on the right, we set $v = 2x$ and use the chain rule (as in Example 6.10), and for the derivative of the first integral on the right, we reverse the limits (as in Example 6.11) and then use the chain rule with $u = x^2$:

$$\frac{d}{dx} \int_{x^2}^{2x} \cos(2t^3 + 1)\, dt = -\frac{d}{dx} \int_{a}^{x^2} \cos(2t^3 + 1)\, dt + \frac{d}{dx} \int_{a}^{2x} \cos(2t^3 + 1)\, dt$$

$$= \left[-\frac{d}{du} \int_{a}^{u} \cos(2t^3 + 1)\, dt \right] \frac{du}{dx} + \left[\frac{d}{dv} \int_{a}^{v} \cos(2t^3 + 1)\, dt \right] \frac{dv}{dx}$$

$$= -2x \cos(2u^3 + 1) + 2\cos(2v^3 + 1)$$

$$= -2x \cos(2x^6 + 1) + 2\cos(16x^3 + 1).$$

EXERCISES 6.5

In Exercises 1–20 differentiate the definite integral with respect to x.

1. $\displaystyle\int_{0}^{x} (3t^2 + t)\, dt$

2. $\displaystyle\int_{1}^{x} \frac{1}{\sqrt{t^2 + 1}}\, dt$

3. $\displaystyle\int_{x}^{2} \sin(t^2)\, dt$

4. $\displaystyle\int_{x}^{-1} t^3 \cos t\, dt$

5. $\displaystyle\int_{0}^{3x} (2t - t^4)^2\, dt$

6. $\displaystyle\int_{1}^{2x} \sqrt{t + 1}\, dt$

7. $\displaystyle\int_{4}^{3x^2} \sin(3t + 4)\, dt$

8. $\displaystyle\int_{-2}^{5x+4} \sqrt{t^3 + 1}\, dt$

9. $\displaystyle\int_{x}^{2x} (3\sqrt{t} - 2t)\, dt$

10. $\displaystyle\int_{4x}^{4x+4} \left(t^3 - \frac{1}{\sqrt{t}} \right) dt$

11. $\displaystyle\int_{-2x}^{x} \tan(3t + 1)\, dt$

12. $\displaystyle\int_{-x^2}^{-2x^2} \sec(1 - t)\, dt$

13. $\displaystyle\int_{0}^{\sin x} \cos(t^2)\, dt$

14. $\displaystyle\int_{\cos x}^{\sin x} \frac{1}{\sqrt{t + 1}}\, dt$

15. $\displaystyle\int_{0}^{2\sqrt{x}} \sqrt{t}\, dt$

16. $\displaystyle\int_{\sqrt{x}}^{2\sqrt{x}} \sqrt{t}\, dt$

17. $\displaystyle\int_{1}^{x^2} t^2 e^{4t}\, dt$

18. $\displaystyle\int_{x}^{2} \ln(t^2 + 1)\, dt$

19. $\displaystyle\int_{x}^{2x} t \ln t\, dt$

20. $\displaystyle\int_{-2x}^{3x} e^{-4t^2}\, dt$

21. Verify that when $a(x)$ and $b(x)$ are differentiable functions of x,

$$\frac{d}{dx} \int_{a(x)}^{b(x)} f(t)\, dt = f[b(x)] \frac{db}{dx} - f[a(x)] \frac{da}{dx}.$$

Is equation 6.19 a special case of this result?

22.–28. Use the result of Exercise 21 to redo Exercises 8, 10, 12, 14, 16, 18, and 20.

SECTION 6.6

Average Values

The average value of two numbers c and d is defined as $(c+d)/2$. The average value of a set of n numbers $y_1, y_2, \ldots, y_n$ is $(y_1 + y_2 + \cdots + y_n)/n$. We would like to extend this idea to define the average value of a function $f(x)$ over an interval $a \leq x \leq b$. By beginning with some simple functions we can see eventually how average values should be defined.

The function $f(x)$ in Figure 6.8(a) is equal to 1 for the first third of the interval shown, 2 for the second third, and 3 for the last third. We suggest that its average value over the interval $0 \leq x \leq 6$ is 2. The function in Figure 6.8(b) takes on the same function values, namely, 1, 2, and 3, but not on the same subintervals. The fact that it has value 2 for $2 < x \leq 5$ and value 3 for $5 < x \leq 6$ would indicate that its average value should be somewhat less than the average value of 2 for the function in Figure 6.8(a). The average value of the function in Figure 6.8(c) should be even less than that in Figure 6.8(b). These three functions suggest that two factors are important when considering average values of functions: values that the function takes on, and lengths of the intervals on which they take these values. Perhaps what should be done to calculate average values for these functions is to add together the products obtained by multiplying each of the function values by the length of the interval in which it has this value, and then divide this sum by the length of the overall interval.

FIGURE 6.8

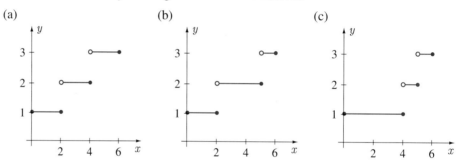

For the function in Figure 6.8(a) this would yield an average value of

$$\frac{1}{6-0}[1(2-0) + 2(4-2) + 3(6-4)] = 2;$$

for the function in Figure 6.8(b),

$$\frac{1}{6-0}[1(2-0) + 2(5-2) + 3(6-5)] = \frac{11}{6};$$

and for the function in Figure 6.8(c),

$$\frac{1}{6-0}[1(4-0) + 2(5-4) + 3(6-5)] = \frac{3}{2}.$$

This procedure is applicable only to a function whose domain can be subdivided into a finite number of subintervals inside each of which the function has a constant value. Such a function is said to be *piecewise constant*. What shall we do for functions which are not piecewise constant? By rephrasing the above procedure, it will become obvious. The same three average values are obtained if we adopt the following approach [illustrated for the function $f(x)$ in Figure 6.8(c)]. Divide the interval $0 \leq x \leq 6$ into three subintervals $0 < x \leq 4$, $4 < x \leq 5$, and $5 < x \leq 6$. Pick a point in each subinterval; call the points $x_1^*, x_2^*,$ and x_3^*. Evaluate $f(x)$ at each point and multiply by the length of the subinterval in which the point is found. Add these results, and divide by the length of the interval to obtain the average value,

$$\frac{1}{6}[f(x_1^*)(4-0) + f(x_2^*)(5-4) + f(x_3^*)(6-5)] = \frac{1}{6}[1(4) + 2(1) + 3(1)] = \frac{3}{2}.$$

But this is the procedure used to define the definite integral of a function; it lacks the limit because the function is piecewise constant. In other words, for a function $f(x)$ which has a value at every point in the interval $a \leq x \leq b$ (but is not necessarily piecewise constant), we define its average value as

$$\text{Average Value} = \frac{1}{b-a}\int_a^b f(x)\,dx = \frac{1}{b-a}\lim_{\|\Delta x_i\|\to 0}\sum_{i=1}^n f(x_i^*)\Delta x_i. \quad (6.21)$$

EXAMPLE 6.13

What is the average value of the function $f(x) = x^2$ on the interval $0 \leq x \leq 2$?

SOLUTION By equation 6.21,

$$\text{Average Value} = \frac{1}{2}\int_0^2 x^2\,dx = \frac{1}{2}\left\{\frac{x^3}{3}\right\}_0^2 = \frac{1}{2}\left(\frac{8}{3}\right) = \frac{4}{3}.$$

EXAMPLE 6.14

Find the average value of $f(x) = \sin x$ on the intervals: (a) $0 \leq x \leq \pi/2$ (b) $0 \leq x \leq \pi$ (c) $0 \leq x \leq 2\pi$.

SOLUTION We calculate average values on these intervals as:

(a) $\dfrac{1}{\pi/2}\displaystyle\int_0^{\pi/2} \sin x\,dx = \dfrac{2}{\pi}\{-\cos x\}_0^{\pi/2} = \dfrac{2}{\pi}[0+1] = \dfrac{2}{\pi};$

(b) $\dfrac{1}{\pi}\displaystyle\int_0^{\pi} \sin x\,dx = \dfrac{1}{\pi}\{-\cos x\}_0^{\pi} = \dfrac{1}{\pi}[1+1] = \dfrac{2}{\pi};$

(c) $\dfrac{1}{2\pi}\displaystyle\int_0^{2\pi} \sin x\,dx = \dfrac{1}{2\pi}\{-\cos x\}_0^{2\pi} = \dfrac{1}{2\pi}[-1+1] = 0.$

The graph of $\sin x$ in Figure 6.9 also suggests that the average values in (a) and (b) should be the same, and that the average value in (c) should be zero.

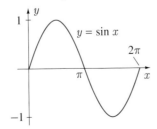

FIGURE 6.9

EXAMPLE 6.15 What is the average value of a linear function?

SOLUTION The average value of the linear function $f(x) = mx + c$ over an interval $a \leq x \leq b$ is

$$\frac{1}{b-a}\int_a^b (mx+c)\,dx = \frac{1}{b-a}\left\{\frac{mx^2}{2} + cx\right\}_a^b$$
$$= \frac{1}{b-a}\left[\left(\frac{mb^2}{2} + cb\right) - \left(\frac{ma^2}{2} + ca\right)\right]$$
$$= \frac{1}{b-a}\left[\frac{m}{2}(b^2 - a^2) + c(b-a)\right]$$
$$= \frac{m}{2}(b+a) + c$$
$$= m\left(\frac{b+a}{2}\right) + c.$$

This is the value of the function midway between $x = a$ and $x = b$. ∎

Theorem 6.7 can be used to establish the next theorem.

Theorem 6.8 (Mean Value Theorem for Definite Integrals) If $f(x)$ is continuous for $a \leq x \leq b$, then there exists at least one number c between a and b such that

$$\int_a^b f(x)\,dx = (b-a)f(c). \qquad (6.22)$$

Proof By Theorem 6.7, the function

$$F(x) = \int_a^x f(t)\,dt$$

is continuous for $a \leq x \leq b$, and has derivative $f(x)$ for $a < x < b$. Mean Value Theorem 3.15 applied to $F(x)$ guarantees at least one number c between a and b such that

$$F(b) - F(a) = (b-a)F'(c).$$

By substitution,

$$\int_a^b f(t)\,dt - \int_a^a f(t)\,dt = (b-a)f(c).$$

Since the second integral vanishes, we have

$$\int_a^b f(x)\,dx = (b-a)f(c).$$

By writing equation 6.22 in the form

$$f(c) = \frac{1}{b-a}\int_a^b f(x)\,dx,$$

Theorem 6.8 states that the function must take on its average value at least once in the interval.

EXAMPLE 6.16

Find all values of c satisfying Theorem 6.8 for the function $f(x) = x^2$ on the interval $0 \leq x \leq 1$.

SOLUTION Substituting $a = 0$ and $b = 1$ in equation 6.22 gives

$$(1-0)c^2 = \int_0^1 x^2\, dx = \left\{\frac{x^3}{3}\right\}_0^1 = \frac{1}{3} - 0 = \frac{1}{3}.$$

Of the two solutions $c = \pm 1/\sqrt{3}$ for this equation, only the positive one is between 0 and 1. Thus, only $c = 1/\sqrt{3}$ satisfies Theorem 6.8 for the function $f(x) = x^2$ on the interval $0 \leq x \leq 1$. ■

EXAMPLE 6.17

Find all values of c satisfying Theorem 6.8 for the function $f(x) = \sin x$ on the interval $\pi/4 \leq x \leq 3\pi/4$.

SOLUTION Substituting $a = \pi/4$ and $b = 3\pi/4$ in equation 6.22 gives

$$\left(\frac{3\pi}{4} - \frac{\pi}{4}\right)\sin c = \int_{\pi/4}^{3\pi/4} \sin x\, dx = \{-\cos x\}_{\pi/4}^{3\pi/4} = \frac{1}{\sqrt{2}} + \frac{1}{\sqrt{2}} = \sqrt{2}.$$

Thus,

$$\sin c = \frac{2\sqrt{2}}{\pi}.$$

There are two angles between $\pi/4$ and $3\pi/4$ with a sine equal to $2\sqrt{2}/\pi$, namely $c = 1.12$ and $c = 2.02$. ■

EXERCISES 6.6

In Exercises 1–20 find the average value of the function over the interval.

1. $f(x) = x^2 - 2x, \quad 0 \leq x \leq 2$
2. $f(x) = x^3 - x, \quad -1 \leq x \leq 1$
3. $f(x) = x^3 - x, \quad 0 \leq x \leq 1$
4. $f(x) = x^4, \quad 1 \leq x \leq 2$
5. $f(x) = \sqrt{x+1}, \quad 0 \leq x \leq 1$
6. $f(x) = \sqrt{x+1}, \quad -1 \leq x \leq 1$
7. $f(x) = x^4 - 1, \quad 0 \leq x \leq 1$
8. $f(x) = x^4 - 1, \quad 0 \leq x \leq 2$
9. $f(x) = \cos x, \quad -\pi/2 \leq x \leq \pi/2$
10. $f(x) = \cos x, \quad 0 \leq x \leq \pi/2$
11. $f(x) = |x|, \quad -2 \leq x \leq 2$
12. $f(x) = |x|, \quad 0 \leq x \leq 2$
13. $f(x) = |x^2 - 4|, \quad 0 \leq x \leq 3$
14. $f(x) = |x^2 - 4|, \quad -3 \leq x \leq 3$
15. $f(x) = \operatorname{sgn} x, \quad -1 \leq x \leq 1$ (see question 36 in Exercises 2.4)
16. $f(x) = \operatorname{sgn} x, \quad -1 \leq x \leq 3$
17. $f(x) = H(x-1), \quad 0 \leq x \leq 2$ (see question 38 in Exercises 2.4)
18. $f(x) = H(x-4), \quad 0 \leq x \leq 2$
19. $f(x) = [[x]], \quad 0 \leq x \leq 3$ (see question 36 in Exercises 1.5)
20. $f(x) = [[x]], \quad 0 \leq x \leq 3.5$

21. If a particle moving along the x-axis is at position x_1 at time t_1 and at position x_2 at time t_2, its average velocity over this time interval is

$$\frac{x_2 - x_1}{t_2 - t_1}.$$

Verify that this is the same as the average of the velocity function over the time interval $t_1 \leq t \leq t_2$.

22. The velocity v of blood flowing through a circular vein or artery of radius R at a distance r from the centre of the blood vessel is

$$v(r) = c(R^2 - r^2).$$

(See Problem 3 in Section 6.2.) What is the average value of $v(r)$ with respect to r?

In Exercises 23–30 find all values of c satisfying equation 6.22 for the function $f(x)$ on the specified interval.

23. $f(x) = 2x - x^2$, $\quad 0 \leq x \leq 2$

24. $f(x) = x^3 - 8x$, $\quad -2 \leq x \leq 2$

25. $f(x) = \cos x$, $\quad 0 \leq x \leq \pi/2$

26. $f(x) = \cos x$, $\quad 0 \leq x \leq \pi$

27. $f(x) = \sqrt{x+1}$, $\quad 1 \leq x \leq 3$

28. $f(x) = x^2(x+1)$, $\quad 0 \leq x \leq 1$

29. $f(x) = x\sqrt{x^2+1}$, $\quad 0 \leq x \leq 2$

30. $f(x) = 1/x^2 + 1/x^3$, $\quad 1 \leq x \leq 2$

SECTION 6.7

Change of Variable in the Definite Integral

The fundamental theorem of integral calculus indicates that to evaluate the definite integral

$$\int_{-4}^{-1} \frac{x}{\sqrt{x+5}}\, dx$$

we should first find an indefinite integral for $x/\sqrt{x+5}$. To do this we set $u = x+5$, in which case $du = dx$, and

$$\int \frac{x}{\sqrt{x+5}}\, dx = \int \frac{u-5}{\sqrt{u}}\, du = \int (u^{1/2} - 5u^{-1/2})\, du$$

$$= \frac{2}{3} u^{3/2} - 10 u^{1/2} + C = \frac{2}{3}(x+5)^{3/2} - 10(x+5)^{1/2} + C.$$

Consequently,

$$\int_{-4}^{-1} \frac{x}{\sqrt{x+5}}\, dx = \left\{ \frac{2}{3}(x+5)^{3/2} - 10(x+5)^{1/2} \right\}_{-4}^{-1}$$

$$= \left[\frac{2}{3}(4)^{3/2} - 10(4)^{1/2} \right] - \left[\frac{2}{3}(1)^{3/2} - 10(1)^{1/2} \right]$$

$$= -\frac{16}{3}.$$

An alternative approach, that often turns out to be less work, is to make the change of variable $u = x + 5$ directly in the definite integral. In this case we again replace

$$\frac{x}{\sqrt{x+5}}\, dx \quad \text{by} \quad \frac{u-5}{\sqrt{u}}\, du.$$

In addition, we replace the limits $x = -4$ and $x = -1$ by those values of u that correspond to these values of x, namely, $u = 1$ and $u = 4$ respectively. We then obtain

$$\int_{-4}^{-1} \frac{x}{\sqrt{x+5}}\, dx = \int_{1}^{4} \frac{u-5}{\sqrt{u}}\, du = \int_{1}^{4} (u^{1/2} - 5u^{-1/2})\, du$$

288   CHAPTER 6   THE DEFINITE INTEGRAL

$$= \left\{\frac{2}{3}u^{3/2} - 10\,u^{1/2}\right\}_1^4$$

$$= \left[\frac{2}{3}(4)^{3/2} - 10(4)^{1/2}\right] - \left[\frac{2}{3}(1)^{3/2} - 10(1)^{1/2}\right]$$

$$= -\frac{16}{3}.$$

That this method is generally acceptable is proved in the following theorem.

Theorem 6.9

Suppose $f(x)$ is continuous on $a \leq x \leq b$, and we set $x = g(u)$, where $a = g(\alpha)$ and $b = g(\beta)$. Then

$$\int_a^b f(x)\,dx = \int_\alpha^\beta f(g(u))\,g'(u)\,du, \qquad (6.23)$$

if $g'(u)$ is continuous on $\alpha \leq u \leq \beta$, and if when u is between α and β, $g(u)$ is between a and b.

Proof If we define

$$F(x) = \int f(x)\,dx \quad \text{and} \quad G(u) = \int f(g(u))\,g'(u)\,du,$$

then by Theorem 6.3,

$$\int_a^b f(x)\,dx = F(b) - F(a) \quad \text{and} \quad \int_\alpha^\beta f(g(u))\,g'(u)\,du = G(\beta) - G(\alpha).$$

So we must show that

$$F(b) - F(a) = G(\beta) - G(\alpha).$$

Now the chain rule applied to $y = F(x)$, $x = g(u)$ yields

$$\frac{dy}{du} = \frac{dy}{dx}\frac{dx}{du} = f(x)g'(u);$$

that is,

$$\frac{d}{du}F(g(u)) = f(g(u))\,g'(u).$$

Since this is an identity in u, we may integrate with respect to u from $u = \alpha$ to $u = \beta$:

$$\int_\alpha^\beta \frac{d}{du}F(g(u))\,du = \int_\alpha^\beta f(g(u))\,g'(u)\,du.$$

Thus,

$$\{F(g(u))\}_\alpha^\beta = \{G(u)\}_\alpha^\beta \quad \text{or}$$

$$G(\beta) - G(\alpha) = F(g(\beta)) - F(g(\alpha)) = F(b) - F(a).$$

EXAMPLE 6.18

Evaluate

$$\int_2^4 \frac{\sqrt{x}}{(\sqrt{x}-1)^4}\,dx.$$

SOLUTION If we set $u = \sqrt{x} - 1$, then $du = [1/(2\sqrt{x})]\,dx$, and

$$\int_2^4 \frac{\sqrt{x}}{(\sqrt{x}-1)^4}\,dx = \int_{\sqrt{2}-1}^1 \frac{u+1}{u^4} 2(u+1)\,du$$

$$= 2\int_{\sqrt{2}-1}^1 \left(\frac{1}{u^2} + \frac{2}{u^3} + \frac{1}{u^4}\right)du$$

$$= 2\left\{-\frac{1}{u} - \frac{1}{u^2} - \frac{1}{3u^3}\right\}_{\sqrt{2}-1}^1$$

$$= -2\left[1 + 1 + \frac{1}{3}\right] + 2\left[\frac{1}{\sqrt{2}-1} + \frac{1}{(\sqrt{2}-1)^2} + \frac{1}{3(\sqrt{2}-1)^3}\right]$$

$$= 21.2.$$

EXERCISES 6.7

In Exercises 1–22 evaluate the definite integral.

1. $\int_1^2 x(3x^2 - 2)^4\,dx$
2. $\int_0^1 z\sqrt{1-z}\,dz$
3. $\int_{-1}^0 \frac{x}{\sqrt{x+3}}\,dx$
4. $\int_{\pi/4}^{\pi/3} \cos^5 x \sin x\,dx$
5. $\int_1^3 x^3\sqrt{9-x^2}\,dx$
6. $\int_{-5}^6 \frac{x}{\sqrt{x^2-12}}\,dx$
7. $\int_4^5 y^2\sqrt{y-4}\,dy$
8. $\int_{1/2}^1 \sqrt{\frac{x^2}{1+x}}\,dx$
9. $\int_1^4 \frac{\sqrt{1+\sqrt{u}}}{\sqrt{u}}\,du$
10. $\int_{-2}^1 \frac{x+1}{(x^2+2x+2)^{1/3}}\,dx$
11. $\int_3^4 \frac{x^2}{(x-2)^4}\,dx$
12. $\int_0^{\pi/6} \sqrt{2+3\sin x}\cos x\,dx$
13. $\int_{\pi/4}^{\pi/2} \frac{\sin^3 x}{(1+\cos x)^4}\,dx$
14. $\int_1^4 \frac{(x+1)(x-1)}{\sqrt{x}}\,dx$
15. $\int_4^9 \sqrt{1+\sqrt{x}}\,dx$
16. $\int_{-1/2}^1 \sqrt{\frac{x^2}{1+x}}\,dx$
17. $\int_{-1}^1 \frac{|x|}{(x+2)^3}\,dx$
18. $\int_{-1}^1 \left|\frac{x}{(x+2)^3}\right|\,dx$
19. $\int_0^1 x^2 e^{x^3}\,dx$
20. $\int_1^2 \frac{(\ln x)^2}{x}\,dx$
21. $\int_2^4 \frac{1}{x \ln x}\,dx$
22. $\int_{-\pi/4}^{\pi/4} \frac{\sec^2 x}{\sqrt{4+3\tan x}}\,dx$

23. Show that if $f(x)$ is an odd function, then

$$\int_{-a}^a f(x)\,dx = 0;$$

and that if $f(x)$ is an even function,

$$\int_{-a}^a f(x)\,dx = 2\int_0^a f(x)\,dx.$$

In Exercises 24–26 use the suggested substitution to evaluate the definite integral.

24. $\int_1^3 \frac{1}{x^{3/2}\sqrt{4-x}}\,dx \qquad u = \frac{1}{\sqrt{x}}$

25. $\int_{-6}^{-1} \frac{\sqrt{x^2 - 6x}}{x^4}\,dx \qquad u = \frac{1}{x}$

26. $\int_{-4}^0 \frac{x}{(5-4x-x^2)^{3/2}}\,dx \qquad u^2 = \frac{1-x}{5+x}$

27. Show that

$$\int_0^\pi \frac{\cos^2[(\pi/2)\cos\theta]}{\sin\theta}\,d\theta = \frac{\pi}{2}\int_0^{2\pi} \frac{1-\cos\phi}{\phi(2\pi-\phi)}\,d\phi.$$

This integral is used in calculating radiated power from a half-wave antenna.

SUMMARY

The definite integral of a continuous function $f(x)$ from $x = a$ to $x = b$ is a number, one that depends on the function $f(x)$ and the limits a and b. We have defined the definite integral by subdividing the interval $a \leq x \leq b$ into n parts by $n + 1$ points $a = x_0 < x_1 < \cdots < x_{n-1} < x_n = b$, and choosing a point x_i^* in each subinterval $x_{i-1} \leq x \leq x_i$. The definite integral is then the limit of the summation

$$\int_a^b f(x)\, dx = \lim_{||\Delta x_i|| \to 0} \sum_{i=1}^n f(x_i^*) \Delta x_i,$$

where $\Delta x_i = x_i - x_{i-1}$. Since all calculations in this limit take place on the x-axis, we regard the definite integral as an integration along the x-axis from $x = a$ to $x = b$.

The fundamental theorem of integral calculus allows us to calculate definite integrals using antiderivatives,

$$\int_a^b f(x)\, dx = \left\{ \int f(x)\, dx \right\}_a^b.$$

This presupposes that a continuous function has an antiderivative, as verified in Section 6.5. It was shown that when the definite integral is given a variable upper limit x, the resulting function

$$F(x) = \int_a^x f(t)\, dt$$

is an antiderivative of $f(x)$, that is, $F'(x) = f(x)$.

The average value of a function $f(x)$ over an interval $a \leq x \leq b$ is defined as

$$\frac{1}{b-a} \int_a^b f(x)\, dx.$$

The mean value theorem for definite integrals guarantees the existence of at least one number c between a and b, at which $f(x)$ takes on its average value,

$$f(c) = \frac{1}{b-a} \int_a^b f(x)\, dx.$$

Evaluation of a definite integral with a complex integrand can sometimes be simplified with an appropriate change of variable.

Key Terms and Formulas

In reviewing this chapter, you should be able to define or discuss the following key terms:

Sigma notation
Definite integral
Norm of a partition
Fundamental theorem

Average value of a function
Mean value theorem for integrals
Change of variable

REVIEW EXERCISES

In Exercises 1–20 evaluate the definite integral.

1. $\int_0^3 (x^2 + 3x - 2)\, dx$ 2. $\int_{-1}^1 (x^2 - x^4)\, dx$

3. $\int_{-1}^1 (x^3 - 3x)\, dx$ 4. $\int_0^2 (x^2 - 2x)\, dx$

5. $\int_1^2 (x+1)^2\, dx$ 6. $\int_{-3}^{-2} \frac{1}{x^2}\, dx$

7. $\int_4^9 \left(\frac{1}{\sqrt{x}} - \sqrt{x}\right) dx$ 8. $\int_0^\pi \cos x\, dx$

9. $\int_{-1}^1 x(x+1)^2\, dx$ 10. $\int_1^2 x^2(x^2+3)\, dx$

11. $\int_0^3 \sqrt{x+1}\, dx$ 12. $\int_1^5 x\sqrt{x^2-1}\, dx$

13. $\int_1^4 \left(\frac{\sqrt{x}+1}{\sqrt{x}}\right) dx$ 14. $\int_{-1}^0 x\sqrt{x+1}\, dx$

15. $\int_1^2 \frac{x^2+1}{(x+1)^4}\, dx$ 16. $\int_{-4}^{-2} x^2\sqrt{2-x}\, dx$

17. $\int_0^{\pi/4} \frac{\cos x}{(1+\sin x)^2}\, dx$

18. $\int_2^3 x(1+2x^2)^4\, dx$

19. $\int_1^8 \frac{(1+x^{1/3})^2}{x^{2/3}}\, dx$ 20. $\int_{-4}^4 |x+2|\, dx$

21. Use equation 6.11 to evaluate the following definite integrals:

(a) $\int_0^2 (x-5)\, dx$ (b) $\int_0^3 (x^2+3)\, dx$

22. Prove that each of the following answers is incorrect, but do so without evaluating the definite integral. Think about what the definite integral represents.

(a) $\int_0^4 (x^2 + 3x)\, dx = -3$ (b) $\int_{-3}^{-2} \frac{1}{x}\, dx = 5$

In Exercises 23–26 find the average value of the function on the interval.

23. $f(x) = \sqrt{x+4}$, $0 \le x \le 1$

24. $f(x) = 1/x^2 - x$, $-2 \le x \le -1$

25. $f(x) = x\sqrt{x+1}$, $0 \le x \le 1$

26. $f(x) = \cos^3 x \sin^2 x$, $0 \le x \le \pi/2$

In Exercises 27–32 differentiate the integral with respect to x.

27. $\int_1^x t\sqrt{t^3+1}\, dt$ 28. $\int_x^{-3} t^2(t+1)^3\, dt$

29. $\int_1^{x^2} \sqrt{t^2+1}\, dt$ 30. $\int_{2x}^4 t\cos t\, dt$

31. $\int_{2x+3}^{1-x} \frac{1}{t^2+1}\, dt$ 32. $\int_{-x^2}^{x^2} \sin^2 t\, dt$

In Exercises 33–40 evaluate the definite integral.

33. $\int_0^1 \frac{x^3}{(x^2+1)^{3/2}}\, dx$ 34. $\int_0^{\pi/6} \frac{\cos^3 x}{\sqrt{1+\sin x}}\, dx$

35. $\int_{-1}^1 x^3\sqrt{1-x^2}\, dx$ 36. $\int_{-1}^2 \left|\frac{x}{\sqrt{3+x}}\right| dx$

37. $\int_{-1}^2 x^2(4-x^3)^5\, dx$ 38. $\int_1^5 \frac{6x^2+8x+2}{\sqrt{x^3+2x^2+x}}\, dx$

39. $\int_1^2 \frac{x-25}{\sqrt{x}-5}\, dx$ 40. $\int_0^1 \frac{1}{\sqrt{2+x}+\sqrt{x}}\, dx$

CHAPTER 7

Applications of the Definite Integral

In Chapter 6 we defined the definite integral of a function $f(x)$ as the limit of a summation

$$\int_a^b f(x)\,dx = \lim_{\|\Delta x_i\| \to 0} \sum_{i=1}^{n} f(x_i^*) \Delta x_i.$$

In this chapter we use this equation to evaluate certain geometric and physical quantities. Each of these quantities is expressed as a limit-summation of the form in this equation. The limit-summation can immediately be interpreted as a definite integral. But then the definite integral can be evaluated by means of the corresponding indefinite integral.

SECTION 7.1

Area

We have formulas for areas of very few shapes, such as squares, rectangles, triangles, and polygons of any shape, since they can be divided into rectangles and triangles. We also know that the area of a circle is π times the square of its radius (but do you remember where this formula came from?). Now consider finding the area in Figure 7.1(a). Each of us has intuitive ideas about this area. In this section we take these intuitive ideas of what the area *ought to be*, and make a precise mathematical definition of what the area *is*.

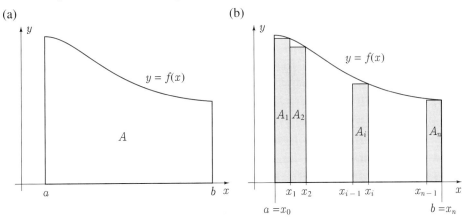

FIGURE 7.1

Recall from Problem 1 of Section 6.2 that the area in Figure 7.1(a) can be approximated by rectangles. Specifically, we partition the interval $a \le x \le b$ into

n parts by points
$$a = x_0 < x_1 < x_2 < \cdots < x_{n-1} < x_n = b,$$
and construct rectangles as shown in Figure 7.1(b). The area A_i of the ith rectangle is
$$A_i = f(x_i)(x_i - x_{i-1}) = f(x_i)\Delta x_i,$$
and an approximation to the required area A is therefore
$$\sum_{i=1}^{n} A_i = \sum_{i=1}^{n} f(x_i)\Delta x_i.$$

If we were to let the number of these rectangles get larger and larger, and at the same time require each to have smaller and smaller width that eventually approaches zero, we feel that a better and better approximation would be obtained. In fact, were we to take the limit as the norm $||\Delta x_i||$ of the partition approaches zero, we would get what we think is A. Since we have as yet no formal definition for the area of odd-shaped figures, we take this opportunity to make our own. We define the area A in Figure 7.1(a) as

$$A = \lim_{||\Delta x_i|| \to 0} \sum_{i=1}^{n} f(x_i)\Delta x_i. \tag{7.1}$$

Area A then has been defined as the limit of a sum of rectangular areas.

The right-hand side of equation 7.1 is strikingly similar to the definition of the definite integral of $f(x)$ from $x = a$ to $x = b$ (equation 6.11),

$$\int_a^b f(x)\,dx = \lim_{||\Delta x_i|| \to 0} \sum_{i=1}^{n} f(x_i^*)\Delta x_i. \tag{6.11}$$

The only difference is the absence of *'s in 7.1. But when we recall that in 6.11, x_i^* may be chosen as any point in the subinterval $x_{i-1} \le x \le x_i$, we see that by choosing $x_i^* = x_i$,

$$\int_a^b f(x)\,dx = \lim_{||\Delta x_i|| \to 0} \sum_{i=1}^{n} f(x_i)\Delta x_i.$$

It follows that area A of Figure 7.1(a) may be calculated by means of the definite integral

$$A = \int_a^b f(x)\,dx. \tag{7.2}$$

It is important to realize that equation 7.2 does not imply that a definite integral should always be thought of as an area; on the contrary, we will find that definite integrals can represent many other quantities. What we have said is that the area in Figure 7.1(a) is defined by limit 7.1. But this limit may also be interpreted as the definite integral of $f(x)$ with respect to x from $x = a$ to $x = b$; hence the area may be calculated by definite integral 7.2. Since definite integrals can be evaluated using antiderivatives, it seems that we have a very simple way to find areas.

It is simple to extend this result to the problem of finding the area in Figure 7.2. With our interpretation of equation 7.2, we can state that the area under the curve $y = f(x)$, above the x-axis, and between the vertical lines $x = a$ and $x = b$ is given by

$$A_2 = \int_a^b f(x)\,dx.$$

Since the area under the curve $y = g(x)$ is given similarly by

$$A_1 = \int_a^b g(x)\,dx,$$

it follows that the required area is

$$A = A_2 - A_1 = \int_a^b f(x)\,dx - \int_a^b g(x)\,dx = \int_a^b [f(x) - g(x)]\,dx. \quad (7.3)$$

We now have two formulas for finding areas: equation 7.2 for the area under a curve, above the x-axis, and between two vertical lines; and equation 7.3 for the area between two curves and two vertical lines. Note that 7.2 is a special case of 7.3. At this point we could solve a number of area problems using these two results, but to do so would strongly suggest that integration should be approached from a "formula" point of view; and if there is any point of view that we wish to adopt, it is completely the opposite. By the end of this chapter we hope to have developed a sufficiently clear understanding of the limit-summation process that use of integration in situations other than those discussed here will be straightforward.

FIGURE 7.2

FIGURE 7.3

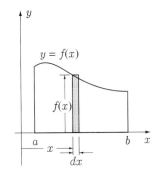

To illustrate how we arrive at the correct definite integral for area problems without memorizing either 7.2 or 7.3, consider again finding the area in Figure 7.1(a). We draw a rectangle of width dx at position x as shown in Figure 7.3. The area of this rectangle is

$$f(x)\,dx.$$

We visualize this rectangle as a representative for a large number of such rectangles between a and b. To find the required area we add together all such rectangular areas and take the limit as their widths approach zero. But this is precisely the concept of the definite integral, so that we write for the limit-summation process

$$A = \int_a^b f(x)\,dx.$$

The limits a and b identify the x-positions of the first and last rectangles, respectively.

FIGURE 7.4

Similarly, for the area in Figure 7.4 we draw a rectangle of width dx and length $f(x) - g(x)$ and, therefore, of area

$$[f(x) - g(x)]\,dx.$$

To add the areas of all such rectangles between a and b, and at the same time to take the limit as their widths approach zero, we once again use the definite integral,

$$A = \int_a^b [f(x) - g(x)]\,dx.$$

For area problems then, we start with the area of a representative rectangle and proceed to the required area by summation with the definite integral. We might write symbolically that

EXAMPLE 7.1

Find the area enclosed by the curves $y = x^2$ and $y = x^3$.

SOLUTION The area of the representative rectangle in Figure 7.5 is

$$(x^2 - x^3)\,dx;$$

hence,

$$A = \int_0^1 (x^2 - x^3)\,dx = \left\{ \frac{x^3}{3} - \frac{x^4}{4} \right\}_0^1 = \frac{1}{12}.$$

FIGURE 7.5

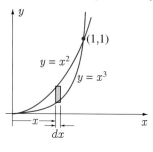

In the next two examples, the expression for the area of the representative rectangle varies within the required area. In such cases, we must set up different integrals corresponding to the different representative rectangles.

EXAMPLE 7.2

Find the area bounded by the x-axis and the curve $y = x^3 - x$.

SOLUTION

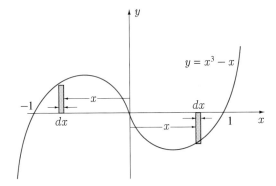

FIGURE 7.6

The area of any rectangle between $x = -1$ and $x = 0$ (Figure 7.6) is

$$[(x^3 - x) - 0]\, dx,$$

whereas the area of a rectangle between $x = 0$ and $x = 1$ is

$$[0 - (x^3 - x)]\, dx.$$

Consequently,

$$A = \int_{-1}^{0} (x^3 - x)\, dx + \int_{0}^{1} (-x^3 + x)\, dx$$

$$= \left\{ \frac{x^4}{4} - \frac{x^2}{2} \right\}_{-1}^{0} + \left\{ -\frac{x^4}{4} + \frac{x^2}{2} \right\}_{0}^{1}$$

$$= -\left(\frac{1}{4} - \frac{1}{2} \right) + \left(-\frac{1}{4} + \frac{1}{2} \right)$$

$$= \frac{1}{2}.$$

We could have saved ourselves some calculations in this example by noting that because of the symmetry of the diagram ($x^3 - x$ is an odd function), the two areas are identical. Hence, we could find the left area and double it,

$$A = 2\int_{-1}^{0} (x^3 - x)\, dx = 2\left\{ \frac{x^4}{4} - \frac{x^2}{2} \right\}_{-1}^{0} = -2\left(\frac{1}{4} - \frac{1}{2} \right) = \frac{1}{2}.$$

EXAMPLE 7.3

Find the area of the triangle with edges $y = x$, $y = -x/2$, and $y = 5x - 44$.

SOLUTION

FIGURE 7.7

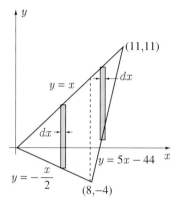

Areas of representative rectangles to the left and right of $x = 8$ (Figure 7.7) are, respectively,

$$[x - (-x/2)]\, dx \quad \text{and} \quad [x - (5x - 44)]\, dx;$$

therefore

$$A = \int_0^8 [x - (-x/2)]\, dx + \int_8^{11} [x - (5x - 44)]\, dx$$

$$= \frac{3}{2} \int_0^8 x\, dx + \int_8^{11} (44 - 4x)\, dx$$

$$= \frac{3}{2} \left\{ \frac{x^2}{2} \right\}_0^8 + \left\{ 44x - 2x^2 \right\}_8^{11}$$

$$= 48 + (484 - 242) - (352 - 128)$$

$$= 66.$$

We see from the above examples that the length of a representative rectangle in equation 7.4 as "upper y minus lower y" is valid whether the rectangle is in the first quadrant (Figure 7.5), the second and fourth quadrants (Figure 7.6), or partially in the first and partially in the fourth (Figure 7.7). In fact it is valid for rectangles in all quadrants.

FIGURE 7.8

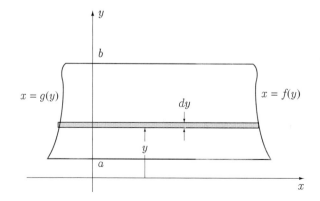

There is nothing special about vertical rectangles. Sometimes it is more convenient to subdivide an area into horizontal rectangles. For example, to find the area in Figure 7.8, we draw a representative rectangle at position y of width dy. Its length is $[f(y) - g(y)]$, and therefore its area is

$$[f(y) - g(y)]\,dy.$$

Adding over all rectangles gives

$$A = \int_a^b [f(y) - g(y)]\,dy. \qquad (7.5)$$

Corresponding to 7.4, we could write that for a subdivision into horizontal rectangles

$$A = \int \begin{matrix} \text{y-position of} \\ \text{last rectangle} \\ \\ \text{y-position of} \\ \text{first rectangle} \end{matrix} \left[\begin{matrix} x\text{-coordinate} \\ \text{of right end} \\ \text{of rectangle} \end{matrix} - \begin{matrix} x\text{-coordinate} \\ \text{of left end} \\ \text{of rectangle} \end{matrix} \right] dy. \qquad (7.6)$$

EXAMPLE 7.4

Find the area bounded by the curves $y = \sqrt{x + 14}$, $x = \sqrt{y}$, and $y = 0$.

SOLUTION Subdivision of the area (Figure 7.9) into vertical rectangles results in two integrations: one to the left and the other to the right of the y-axis. On the other hand, throughout the required area, the area of a horizontal rectangle is

$$[\sqrt{y} - (y^2 - 14)]\,dy,$$

and therefore

$$A = \int_0^4 \left(\sqrt{y} - y^2 + 14\right) dy = \left\{ \frac{2}{3} y^{3/2} - \frac{1}{3} y^3 + 14 y \right\}_0^4 = \frac{16}{3} - \frac{64}{3} + 56 = 40.$$

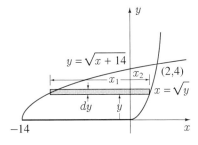

FIGURE 7.9

In choosing between horizontal and vertical rectangles, two objectives must be considered:

1. to minimize the number of integrations; and
2. to obtain simple definite integrals.

For instance, by choosing one type of rectangle we may obtain only one definite integral, but that may be very difficult to evaluate. If the other type of rectangle leads to two simple definite integrals, then it would be wise to choose the two simple integrals.

EXAMPLE 7.5

Find the area enclosed by the curves

$$y = \frac{x}{\sqrt{x^2 - 16}}, \qquad y = \frac{x^2}{15}, \qquad y = -x^2, \qquad x = 6, \qquad x \geq 0.$$

SOLUTION Examination of Figure 7.10 indicates that a choice of horizontal rectangles would necessitate three definite integrals. In addition, they would require us to solve the equations $y = x^2/15$ and $y = x/\sqrt{x^2 - 16}$ for x in terms of y. We therefore opt for vertical rectangles.

$$\begin{aligned}
A &= \int_0^5 \left[\frac{x^2}{15} - (-x^2)\right] dx + \int_5^6 \left[\frac{x}{\sqrt{x^2-16}} - (-x^2)\right] dx \\
&= \frac{16}{15} \int_0^5 x^2\, dx + \int_5^6 \left[\frac{x}{\sqrt{x^2-16}} + x^2\right] dx \\
&= \frac{16}{15} \left\{\frac{x^3}{3}\right\}_0^5 + \left\{\sqrt{x^2-16} + \frac{x^3}{3}\right\}_5^6 \\
&= \frac{16}{15}\left(\frac{125}{3}\right) + \left(\sqrt{20} + 72\right) - \left(3 + \frac{125}{3}\right) \\
&= 76.2.
\end{aligned}$$

FIGURE 7.10

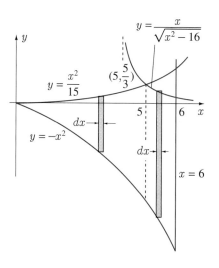

Recall the mean value theorem for definite integrals (Theorem 6.8),

$$\int_a^b f(x)\, dx = (b-a) f(c).$$

It has a very simple interpretation in terms of area — at least when $f(x) \geq 0$ for $a \leq x \leq b$. Since $f(x) \geq 0$ for $a \leq x \leq b$, the definite integral may be interpreted as the area under the curve $y = f(x)$, above the x-axis, and between vertical lines at $x = a$ and $x = b$ (Figure 7.11). The right side is the area of the rectangle of width $b - a$ and height $f(c)$ shaded in Figure 7.11. The mean value theorem guarantees at least one point c between a and b for which the rectangular area is equal to the area under the curve. For the curve in Figure 7.11, there is exactly one such point c; for the curve in Figure 7.12, there are three choices for c.

FIGURE 7.11

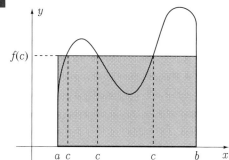

FIGURE 7.12

EXERCISES 7.1

In Exercises 1–16 find the area bounded by the curves.

1. $y^2 = 4x$, $x^2 = 4y$
2. $y = x^3 + 8$, $y = 4x + 8$
3. $yx^2 = 4$, $y = 5 - x^2$
4. $x = y(y - 2)$, $x + y = 12$
5. $x = 4y - 4y^2$, $y = x - 3$, $y = 1$, $y = 0$
6. $y = e^{3x}$, $x = 1$, $x = 2$, $y = -x$
7. $12y = 7 - x^2$, $y = 1/(2x)$
8. $y = \sqrt{x + 4}$, $y = (x + 4)^2/8$
9. $y = (x - 1)^5$, $x = 0$, $y = 0$
10. $y = \sin x$ $(0 \le x \le \pi)$, $y = 0$
11. $y = x^5 - x$, $y = 0$
12. $x + y = 1$, $x + y = 5$, $y = 2x + 1$, $y = 2x + 6$
13. $x = \sec^2 y$, $y = 0$, $y = \pi/4$, $x = 0$
14. $xy = e$, $y = x^2$, $y = 2$
15. $y = e^2(2 - x^2)$, $y = e^{2x}$, $y = e^{-2x}$
16. $x = |y| + 1$, $x + (y - 1)^2 = 4$, $y = 0$ (above the x-axis)

17. (a) Find the area A bounded by the curves
$$y = 16 - x^2, \quad y = 0, \quad x = 1, \quad x = 3.$$
In the remainder of this problem we approximate A by rectangles and show that the accuracy of the approximation increases as the number of rectangles increases. To do this you will need a large graph of the function $f(x) = 16 - x^2$ on the interval $1 \le x \le 3$.

(b) On the graph draw two rectangles of equal width (one unit) and with heights determined in the same way as in Figure 7.1(b). If A_2 denotes the sum of the areas of these two rectangles, what is A_2? What is the error in the approximation of A by A_2? Illustrate this error on the graph.

(c) Repeat part (b) with four rectangles all of width one-half, denoting the sum of the areas of the four rectangles by A_4. Show on the graph the extra precision of A_4 over A_2.

(d) Repeat part (b) with eight rectangles of equal width, denoting the sum of the areas of the eight rectangles by A_8. Show on the graph the extra precision of A_8 over A_4.

This discussion is continued in Exercise 46.

18. Repeat Exercise 17 with the area bounded by the curves
$$y = x^3 + 1, \quad x = 1, \quad x = 3, \quad y = 0.$$

In Exercises 19–22 set up (but do not evaluate) definite integral(s) for the area bounded by the curves.

19. $x = 1/\sqrt{4 - y^2}$, $4x = -y^2$, $y = -1$, $y = 1$
20. $x^2 + y^2 = 4$, $x^2 + y^2 = 4x$ (interior to both)
21. $x^2 + y^2 = 4$, $x^2 + y^2 = 6x$ (interior to both)
22. $x^2 + y^2 = 16$, $x = y^2$ (smaller area)

23. Find the area bounded by $y^2 = 4ax$ and $x^2 = 4ay$, where $a > 0$ is a constant.

In Exercises 24–35 find the area bounded by the curves.

24. $y = x/\sqrt{x + 3}$, $x = 1$, $x = 6$, $y = -x^2$
25. $x = y^2 + 2$, $x = -(y - 4)^2$, $y = -x + 4$, $y = 0$
26. $y = x^3 - x$, $x + y + 1 = 0$, $x = \sqrt{y + 1}$

27. $y = \left| \dfrac{x}{(x-2)^3} \right|$, $y = 0$, $x = -1$, $x = 1$

28. $x = 2ye^{-y^2}$, $y = x$

29. $y = \sin^3 x$, $y = 1/8$, $0 \le x \le 2\pi$

30. $y = \ln x^2$, $y = 1 - x^2$, $y = 1$

31. $|x|^{1/2} + |y|^{1/2} = 1$

32. $y^2 = x^2(4 - x^2)$

33. $y^2 = x^4(9 + x^2)$

34. $y^2 = x^2(x^2 - 4)$, $x = 5$

35. $(2x - y)^2 = x^3$, $x = 4$

 In Exercises 36–43, it is necessary to use a calculator or computer to find points of intersection of the curves. Find the area bounded by the curves (to 3 decimal places).

36. $y = x^3 + 3x^2 + 2x + 1$, $x = 0$, $y = 0$

37. $y = x^3 - 4x$, $y = 2 - x - x^2$

38. $y = x^4 - 5x^2 + 5$, $y = 0$

39. $y = e^x$, $y = 2 - x^2$

40. $y = \cos x$, $4y = x + 2$

41. $y = \dfrac{2}{x+2}$, $y = x^3 + 3x - 1$, $x = 0$

42. $y = x^3 - 3x^2 + 4x - 2$, $x = 4 - y^2$

43. $x = y^3 - y^2 - 2y$, $2y = x^2 - 1$

44. For what values of m do the curves

$$y = \dfrac{x}{3x^2 + 1} \quad \text{and} \quad y = mx$$

bound an area? Find the area.

45. Find a point (a, b) on the curve $y = x/\sqrt{x^2 + 1}$ such that the area bounded by this curve, the x-axis, and the line $x = a$ is equal to twice the area bounded by the curve, the y-axis, and the line $y = b$.

46. If 2^n ($n = 1, 2, \ldots$) rectangles (all of equal width) are drawn in Exercise 17 to approximate A, and A_{2^n} denotes the sum of the areas of these rectangles, show that

$$A_{2^n} = A - \left[\dfrac{1}{2^{n-2}} + \dfrac{1}{6}\left(\dfrac{3}{2^{n-3}} + \dfrac{1}{2^{2n-3}} \right) \right],$$

and hence that

$$\lim_{n \to \infty} A_{2^n} = A.$$

47. Show that curves

$$y = \dfrac{x^3}{x^4 + 16} \quad \text{and} \quad 204y = 13x^2 - 1$$

bound two areas. Find the larger area.

48. Let P be a point on the cubic curve $y = f(x) = ax^3$. Let the tangent line at P intersect $y = f(x)$ again at Q, and let A be the area of the region bounded by $y = f(x)$ and the line PQ. Let B be the area of the region defined in the same way by starting with Q instead of P. Show that B is 16 times as large as A.

49. Prove that the result in Exercise 48 is valid for any cubic $y = f(x) = ax^3 + bx^2 + cx + d$.

SECTION 7.2

Volumes of Solids of Revolution

In Section 6.2 we discussed the idea of rotating flat areas about coplanar lines to produce volumes of solids of revolution. To find the volume generated when the area in Figure 7.1(a) is revolved about the x-axis, we again approximate the area by n rectangles as in Figure 7.1(b). If each of these rectangles is rotated about the x-axis, then n discs are formed. Since the radius of the ith disc is $f(x_i)$ (Figure 7.13), its volume is given by

$$V_i = \pi [f(x_i)]^2 (x_i - x_{i-1}) = \pi [f(x_i)]^2 \Delta x_i.$$

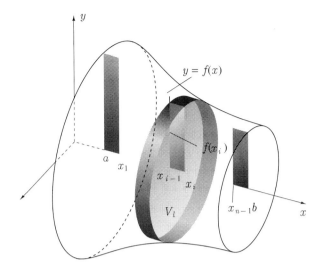

FIGURE 7.13

An approximation to the required volume V is therefore

$$\sum_{i=1}^{n} V_i = \sum_{i=1}^{n} \pi [f(x_i)]^2 \Delta x_i.$$

If we let the number of rectangles get larger and larger, and at the same time require the width of each to approach zero, it seems reasonable that we will obtain a better approximation. Furthermore, were we to take the limit as the norm of the partition approaches zero, we should get what we think is V. Since we have as yet no formal definition for such volumes, we make our own. We define

$$V = \lim_{\|\Delta x_i\| \to 0} \sum_{i=1}^{n} \pi [f(x_i)]^2 \Delta x_i. \tag{7.7}$$

But we can interpret the right-hand side of this definition as the definite integral of $\pi [f(x)]^2$ with respect to x from $x = a$ to $x = b$; that is,

$$\int_a^b \pi [f(x)]^2 \, dx = \lim_{\|\Delta x_i\| \to 0} \sum_{i=1}^{n} \pi [f(x_i)]^2 \Delta x_i.$$

Consequently, we can calculate the volume of the solid of revolution generated by rotating the area in Figure 7.1(a) about the x-axis, by means of the definite integral

$$V = \int_a^b \pi [f(x)]^2 \, dx. \tag{7.8}$$

The volume of the solid of revolution has been defined by limit 7.7, but for evaluation of this limit we use definite integral 7.8.

To avoid memorizing 7.8 as a formula, we use the technique introduced in Section 7.1. For the volume obtained by rotating the area in Figure 7.1(a) about the x-axis, we construct a rectangle of width dx at position x, as shown in either Figure 7.3 or Figure 7.14. When this rectangle is rotated around the x-axis, the volume of the disc generated is

$$\pi [f(x)]^2 \, dx,$$

where dx is the thickness of the disc and $\pi[f(x)]^2$ is the area of its flat surface. This disc is pictured as representing a large number of such discs between a and b. We find the required volume by adding the volumes of all such discs and taking the limit as their widths approach zero. But this is precisely the concept of the definite integral, so that we write for the limit-summation process

$$V = \int_a^b \pi[f(x)]^2\, dx,$$

where the limits $x = a$ and $x = b$ identify the x-positions of first and last discs respectively. This is called the **disc method** for finding the volume of a solid of revolution.

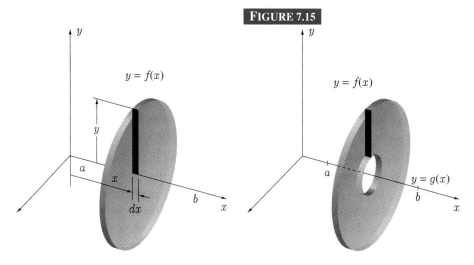

FIGURE 7.14

FIGURE 7.15

A slightly more general problem is that of finding the volume of the solid of revolution generated by rotating an area bounded by two curves and two vertical lines as shown in Figure 7.15 about the x-axis. If a representative rectangle of width dx at position x is rotated around the x-axis, the volume formed is a **washer**. Since outer and inner radii of this washer are $f(x)$ and $g(x)$ respectively, its volume is

$$\{\pi[f(x)]^2 - \pi[g(x)]^2\}\, dx.$$

To add the volumes of all such washers between a and b, and at the same time take limits as their widths approach zero, we use the definite integral

$$V = \int_a^b \{\pi[f(x)]^2 - \pi[g(x)]^2\}\, dx. \tag{7.9}$$

EXAMPLE 7.6 Prove that the volume of a sphere of radius r is $4\pi r^3/3$.

SOLUTION A sphere of radius r is formed when the semicircle $x^2 + y^2 \leq r^2$ ($y \geq 0$) is rotated around the x-axis. The volume of the representative disc generated by rotating the rectangle in Figure 7.16 about the x-axis is

$$\pi y^2\, dx = \pi(r^2 - x^2)\, dx.$$

Since the volume formed by the left quarter circle is the same as for the right quarter circle, we calculate the volume generated by the right quarter and double the result:

$$V = 2\int_0^r \pi(r^2 - x^2)\,dx = 2\pi\left\{r^2 x - \frac{x^3}{3}\right\}_0^r = \frac{4}{3}\pi r^3.$$

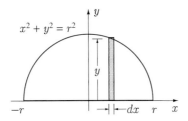

FIGURE 7.16

EXAMPLE 7.7

Prove that the volume of a right-circular cone of base radius r and height h is $\pi r^2 h/3$.

SOLUTION The cone can be generated by rotating the triangle in Figure 7.17 around the x-axis. The volume of the representative disc formed by rotating the rectangle shown is given by

$$\pi y^2\,dx = \pi\left(\frac{rx}{h}\right)^2 dx;$$

hence,

$$V = \int_0^h \frac{\pi r^2 x^2}{h^2}\,dx = \frac{\pi r^2}{h^2}\left\{\frac{x^3}{3}\right\}_0^h = \frac{1}{3}\pi r^2 h.$$

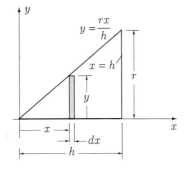

FIGURE 7.17

EXAMPLE 7.8

Find the volume of the solid of revolution obtained by rotating the area bounded by the curves $y = 1 - x^2$ and $y = 4 - 4x^2$ about (a) the x-axis and (b) the line $y = -1$.

SOLUTION (a) When the rectangle in Figure 7.18 is rotated around the x-axis, the volume of the washer formed is

$$(\pi y_2^2 - \pi y_1^2)\, dx = \pi[(4-4x^2)^2 - (1-x^2)^2]\, dx = 15\pi(1 - 2x^2 + x^4)\, dx.$$

FIGURE 7.18

Because of the symmetry of the area, we rotate only the right half and double the result:

$$V = 2\int_0^1 15\pi(1 - 2x^2 + x^4)\, dx = 30\pi\left\{x - \frac{2x^3}{3} + \frac{x^5}{5}\right\}_0^1 = 16\pi.$$

(b) When the rectangle in Figure 7.19 is rotated about the line $y = -1$, inner and outer radii of the washer are r_1 and r_2 respectively. Now r_1 is a length in the y-direction, and in Section 7.1 we learned that to calculate lengths in the y-direction, we take upper y minus lower y. Hence, $r_1 = (1-x^2) - (-1) = 2 - x^2$. Similarly, $r_2 = (4-4x^2) - (-1) = 5 - 4x^2$. The volume of the washer is therefore

$$(\pi r_2^2 - \pi r_1^2)\, dx = [\pi(5-4x^2)^2 - \pi(2-x^2)^2]\, dx = 3\pi(7 - 12x^2 + 5x^4)\, dx.$$

Consequently,

$$V = 2\int_0^1 3\pi(7 - 12x^2 + 5x^4)\, dx = 6\pi\left\{7x - 4x^3 + x^5\right\}_0^1 = 24\pi.$$

FIGURE 7.19

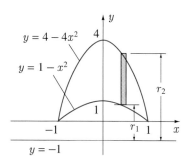

We can also rotate horizontal rectangles about vertical lines to produce discs and washers as illustrated in the next two examples.

EXAMPLE 7.9

Find the volume of the solid of revolution when the area enclosed by the curves $y = \ln x$, $y = 0$, $y = 1$, and $x = 0$ is revolved around the y-axis.

SOLUTION When the rectangle in Figure 7.20 is rotated around the y-axis, the volume of the disc formed is

$$\pi x^2\, dy = \pi(e^y)^2\, dy = \pi e^{2y}\, dy.$$

The required volume is therefore

$$V = \int_0^1 \pi e^{2y}\, dy = \pi \left\{ \frac{1}{2} e^{2y} \right\}_0^1 = \frac{\pi}{2}(e^2 - 1).$$

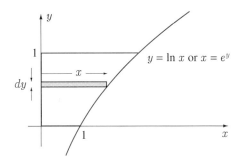

FIGURE 7.20

EXAMPLE 7.10

Find the volume of the solid of revolution obtained by rotating the area enclosed by the curves $y = x^2 - 1$ and $y = 0$ about the line $x = 5$.

SOLUTION When the horizontal rectangle in Figure 7.21 is rotated around the line $x = 5$, the volume of the washer formed is

$$(\pi r_2^2 - \pi r_1^2)\, dy = \pi[(5 - x_2)^2 - (5 - x_1)^2]\, dy.$$

Since x_1 and x_2 are x-coordinates of points on the curve $y = x^2 - 1$, we solve this equation for $x = \pm\sqrt{y+1}$. Since $x_1 > 0$ and $x_2 < 0$, we set $x_1 = \sqrt{y+1}$ and $x_2 = -\sqrt{y+1}$. Thus, the volume of the washer can be expressed as

$$\pi\left[\left(5 + \sqrt{y+1}\right)^2 - \left(5 - \sqrt{y+1}\right)^2 \right] dy = 20\pi\sqrt{y+1}\, dy.$$

The required volume is therefore

$$V = \int_{-1}^0 20\pi\sqrt{y+1}\, dy = 20\pi\left\{ \frac{2}{3}(y+1)^{3/2} \right\}_{-1}^0 = \frac{40\pi}{3}.$$

FIGURE 7.21

The solution of Example 7.10 by washers is not as straightforward as in previous examples. For some problems, washers are totally inappropriate. Consider, for example, rotating the area in Figure 7.22 about the x-axis. Each of the rectangles shown yields a washer with a volume formula different from the others. As a result, use of washers requires six definite integrals and only one of these is easy to set up. Determination of this volume seems to lend itself to the use of horizontal rather than vertical rectangles.

To see that this is indeed true, we divide the interval $a \leq y \leq b$ into n parts by the points

$$a = y_0 < y_1 < y_2 < \cdots < y_{n-1} < y_n = b.$$

In each subinterval $y_{i-1} \leq y \leq y_i$, we find the midpoint

$$y_i^* = \frac{y_{i-1} + y_i}{2}$$

and construct a rectangle of length $f(y_i^*) - g(y_i^*)$ and width $y_i - y_{i-1}$ as shown in Figure 7.23. When this i^{th} rectangle is rotated around the x-axis, a cylindrical shell is formed. Since the length of the shell is $f(y_i^*) - g(y_i^*)$, and its inner and outer radii are y_{i-1} and y_i respectively, its volume is

$$(\pi y_i^2 - \pi y_{i-1}^2)[f(y_i^*) - g(y_i^*)] = \pi(y_i + y_{i-1})(y_i - y_{i-1})[f(y_i^*) - g(y_i^*)]$$
$$= 2\pi y_i^*[f(y_i^*) - g(y_i^*)]\Delta y_i,$$

where $\Delta y_i = y_i - y_{i-1}$. When we add the volumes of all such shells, we obtain an approximation to the required volume V:

$$\sum_{i=1}^{n} 2\pi y_i^*[f(y_i^*) - g(y_i^*)]\Delta y_i.$$

If we let the number of rectangles get larger and larger, and at the same time require the width of each to approach zero, it seems reasonable to expect that we will get a better approximation. As the norm $\|\Delta y_i\|$ approaches zero, the limit should yield what we think is V. We therefore define the required volume as

$$V = \lim_{\|\Delta y_i\| \to 0} \sum_{i=1}^{n} 2\pi y_i^*[f(y_i^*) - g(y_i^*)]\Delta y_i. \tag{7.10}$$

FIGURE 7.22

FIGURE 7.23

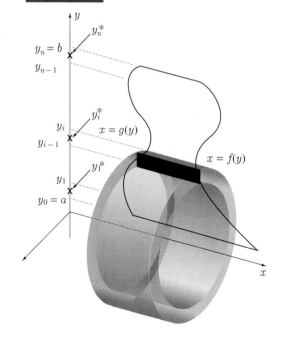

But the right-hand side of this definition may be interpreted as the definite integral of $2\pi y[f(y) - g(y)]$ with respect to y from $y = a$ to $y = b$; that is,

$$\int_a^b 2\pi y[f(y) - g(y)]\, dy = \lim_{||\Delta y_i|| \to 0} \sum_{i=1}^n 2\pi y_i^*[f(y_i^*) - g(y_i^*)]\Delta y_i.$$

Consequently, the volume of the solid of revolution generated by rotating the area in Figure 7.22 around the x-axis is given by the definite integral

$$V = \int_a^b 2\pi y[f(y) - g(y)]\, dy. \tag{7.11}$$

In practice we develop 7.11 by drawing a rectangle of width dy at position y as shown in Figure 7.24(a). When this rectangle is rotated about the x-axis, the volume of the cylindrical shell generated is approximately

$$2\pi y[f(y) - g(y)]\, dy.$$

We obtain this by picturing the shell as being cut along the rectangle and opened up into a slab with dimensions dy, $f(y) - g(y)$, and $2\pi y$ as in Figure 7.24(b). The thickness dy of the shell corresponds to the thickness of the slab; the length $f(y) - g(y)$ of the shell corresponds to that of the slab; and the inner circumference $2\pi y$ of the shell corresponds to the width of the slab. If we now add the volumes of all such cylindrical shells, and at the same time take the limit as their widths approach zero, we obtain the required volume. But this limit-summation defines the definite integral

$$\int_a^b 2\pi y[f(y) - g(y)]\, dy.$$

It follows that

$$V = \int_a^b 2\pi y[f(y) - g(y)]\, dy.$$

The method described is called the **cylindrical-shell method** for finding the volume of a solid of revolution.

(a)

(b)

FIGURE 7.24

EXAMPLE 7.11

Find the volume of the solid of revolution when the area enclosed by $x = 2y - y^2$ and the y-axis is revolved around the x-axis.

SOLUTION The volume of the cylindrical shell formed by rotating the rectangle in Figure 7.25 around the x-axis is approximately

$$(2\pi y)(x)\,dy = 2\pi y(2y - y^2)\,dy = 2\pi(2y^2 - y^3)\,dy.$$

Hence,

$$V = \int_0^2 2\pi(2y^2 - y^3)\,dy = 2\pi \left\{ \frac{2y^3}{3} - \frac{y^4}{4} \right\}_0^2 = 2\pi \left(\frac{16}{3} - 4 \right) = \frac{8\pi}{3}.$$

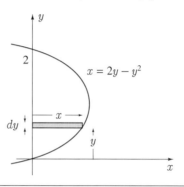

FIGURE 7.25

EXAMPLE 7.12

Use cylindrical shells to calculate the volume in Example 7.10.

SOLUTION When we rotate the rectangle in Figure 7.26 around the line $x = 5$, the volume of the cylindrical shell is approximately

$$2\pi(5 - x)(0 - y)\,dx = 2\pi(5 - x)(1 - x^2)\,dx = 2\pi(x^3 - 5x^2 - x + 5)\,dx.$$

The total volume is therefore

$$V = \int_{-1}^1 2\pi(x^3 - 5x^2 - x + 5)\,dx = 2\pi \left\{ \frac{x^4}{4} - \frac{5x^3}{3} - \frac{x^2}{2} + 5x \right\}_{-1}^1$$

$$= 2\pi \left(\frac{1}{4} - \frac{5}{3} - \frac{1}{2} + 5 \right) - 2\pi \left(\frac{1}{4} + \frac{5}{3} - \frac{1}{2} - 5 \right) = \frac{40\pi}{3}.$$

This is the result obtained by the washer method in Example 7.10.

FIGURE 7.26

310   CHAPTER 7   APPLICATIONS OF THE DEFINITE INTEGRAL

Even though the area in this example is symmetric about the y-axis, the volume generated by the right half is less than that generated by the left. For this reason we cannot integrate from $x = 0$ to $x = 1$ and double the result. If rotation were performed about the x-axis, we could indeed integrate over either half of the interval and double. Finally, for rotation about the y-axis, we would integrate over only one half of the interval and neglect the other half in order to eliminate duplications.

EXAMPLE 7.13

Find the volume of the solid of revolution obtained by rotating the area enclosed by the curves $y = x^3$, $y = -x$, and $y = 1$ about the line $y = -2$.

SOLUTION When we rotate the rectangle in Figure 7.27 around the line $y = -2$, the volume of the representative cylindrical shell is approximately

$$2\pi(y+2)(x_2 - x_1)\,dy = 2\pi(y+2)(y^{1/3} + y)\,dy.$$

The total volume is therefore

$$V = \int_0^1 2\pi(y+2)(y^{1/3} + y)\,dy = 2\pi \int_0^1 (y^{4/3} + 2y^{1/3} + y^2 + 2y)\,dy$$

$$= 2\pi \left\{ \frac{3}{7}y^{7/3} + \frac{3}{2}y^{4/3} + \frac{y^3}{3} + y^2 \right\}_0^1 = 2\pi \left(\frac{3}{7} + \frac{3}{2} + \frac{1}{3} + 1 \right) = \frac{137\pi}{21}.$$

FIGURE 7.27

FIGURE 7.28

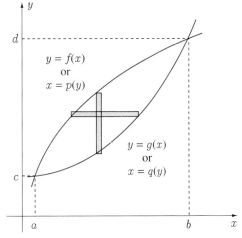

Equations 7.9 and 7.11 provide two methods for calculating volumes of solids of revolution: washers and shells. Where applicable both methods give the same results. We illustrate this for the area bounded by the curves in Figure 7.28. As indicated in the figure, each curve defines y as a function of x, and x as a function of y. When this area is rotated around the x-axis, cylindrical shells can be used to calculate the volume of the solid of revolution

$$V = \int_c^d 2\pi y[q(y) - p(y)]\,dy = 2\pi \int_c^d yq(y)\,dy - 2\pi \int_c^d yp(y)\,dy. \quad (7.12)$$

Suppose we make the change of variable $x = q(y)$ in the first integral. When this equation is solved for y in terms of x, the result is $y = g(x)$, and therefore $dy = g'(x)\,dx$. Because $x = a$ when $y = c$ and $x = b$ when $y = d$, we obtain

$$\int_c^d yq(y)\,dy = \int_a^b g(x)xg'(x)\,dx.$$

Now, the product rule for differentiation gives

$$\frac{d}{dx}\left\{x[g(x)]^2\right\} = 2xg(x)g'(x) + [g(x)]^2,$$

and therefore

$$g(x)xg'(x) = \frac{1}{2}\frac{d}{dx}\left\{x[g(x)]^2\right\} - \frac{1}{2}[g(x)]^2.$$

So,

$$\int_c^d yq(y)\,dy = \int_a^b \left\{\frac{1}{2}\frac{d}{dx}\left\{x[g(x)]^2\right\} - \frac{1}{2}[g(x)]^2\right\}dx$$

$$= \left\{\frac{1}{2}x[g(x)]^2\right\}_a^b - \frac{1}{2}\int_a^b [g(x)]^2\,dx$$

$$= \frac{1}{2}b[g(b)]^2 - \frac{1}{2}a[g(a)]^2 - \frac{1}{2}\int_a^b [g(x)]^2\,dx$$

$$= \frac{1}{2}bd^2 - \frac{1}{2}ac^2 - \frac{1}{2}\int_a^b [g(x)]^2\,dx.$$

Similarly, the change of variable $x = p(y)$ on the second integral in 7.12 leads to

$$\int_c^d yp(y)\,dy = \frac{1}{2}bd^2 - \frac{1}{2}ac^2 - \frac{1}{2}\int_a^b [f(x)]^2\,dx.$$

Substitution of these into 7.12 gives

$$V = 2\pi\left\{\frac{1}{2}bd^2 - \frac{1}{2}ac^2 - \frac{1}{2}\int_a^b [g(x)]^2\,dx\right\} - 2\pi\left\{\frac{1}{2}bd^2 - \frac{1}{2}ac^2 - \frac{1}{2}\int_a^b [f(x)]^2\,dx\right\}$$

$$= -\pi\int_a^b [g(x)]^2\,dx + \pi\int_a^b [f(x)]^2\,dx$$

$$= \int_a^b \left\{\pi[f(x)]^2 - \pi[g(x)]^2\right\}dx.$$

But this is precisely the integral obtained when the washer method is used to find the volume.

EXERCISES 7.2

In Exercises 1–12 use the disc or washer method to find the volume of the solid of revolution obtained by rotating the area bounded by the curves about the line.

1. $x^2 + y^2 = 36$, about $y = 0$
2. $y^2 = 5 - x$, $x = 0$, about $x = 0$
3. $y = x^2 + 4$, $y = 2x^2$, about $y = 0$
4. $x - y^2 = 16$, $x = 20$, about $x = 0$
5. $x - 1 = y^2$, $x = 5$, about $x = 1$
6. $x + y + 1 = 0$, $2y = x - 2$, $y = 0$, about $y = 0$
7. $y = 4x^2 - 4x$, $y = x^3$, about $y = -2$
8. $x = 2y - y^2 - 2$, $x = -5$, about $x = 0$
9. $y^2 = 5 - x$, $x = 0$, about $x = 6$
10. $y = x^2 - 2x$, $y = 2x - x^2$, about $y = 2$
11. $y = \csc x$, $y = 0$, $x = \pi/4$, $x = 3\pi/4$, about $y = 0$
12. $y = \ln(x + 1)$, $y = 1$, $x = 0$, about $x = 0$

In Exercises 13–24 use the cylindrical shell method to find the volume of the solid of revolution obtained by rotating the area bounded by the curves about the line.

13. $y = 1 - x^3$, $x = 0$, $y = 0$, about $x = 0$
14. $y = -\sqrt{4 - x}$, $x = 0$, $y = 0$, about $y = 0$
15. $y = (x - 1)^2$, $y = 1$, about $x = 0$
16. $x + y = 4$, $y = 2\sqrt{x - 1}$, $y = 0$, about $y = 0$
17. $y = 3x - x^2$, $y = x^2 - 3x$, about $x = 4$
18. $y = 2 - |x|$, $y = 0$, about $y = -1$
19. $x = y^3$, $x = 2 - y^2$, $y = 0$, about $y = 1$
20. $y = x^2$, $y = -x^2$, $x = -1$, about $x = -1$
21. $y = -\sqrt{9 - x}$, $x = 0$, $y = 0$, about $y = 0$
22. $y = x$, $xy = 9$, $x + y = 10$, $(x \geq y)$, about $x = 0$
23. $y = 0$, $(x + 1)y = \sin x$, $0 \leq x \leq 2\pi$, about $x = -1$
24. $y = 10 - x^2$, $x^2 y = 9$, about $y = 0$

In Exercises 25–32 use the most appropriate method to find the volume of the solid of revolution obtained by rotating the area bounded by the curves about the line.

25. $(x^2 + 1)^2 y = 4$, $y = 1$, about $x = 0$
26. $y = (x - 1)^2 - 4$, $5y = 12x$, $x = 0$, $(x \geq 0)$, about $x = 0$
27. $x^2 - y^2 = 5$, $9y = x^2 + 9$, $9y + x^2 + 9 = 0$, $(-3 \leq x \leq 3)$, about $x = 0$
28. $y = |x^2 - 1|$, $x = -2$, $x = 2$, $y = -1$, about $y = -1$
29. $y = x^2 - 2$, $y = 0$, about $y = -1$
30. $x = \sqrt{4 + 12y^2}$, $x - 20y = 24$, $y = 0$, about $y = 0$
31. $y = (x + 1)^{1/4}$, $y = -(x + 1)^2$, $x = 0$, about $x = 0$
32. $y = x^4 - 3$, $y = 0$, about $y = -1$

In Exercises 33–36 a calculator or computer is needed to find the points of intersection of the curves. Find the volume of the solid of revolution when the area bounded by the curves is rotated about the line correct to three decimal places.

33. $y = x^3 - x$, $y = \sqrt{x}$, about $x = 0$
34. $y = e^{-2x}$, $y = 4 - x^2$, about $y = -1$
35. $y = \dfrac{1}{\sqrt{x - 1}}$, $y = 16 - x^2$, about $y = -1$
36. $y = \sqrt{4 - x}$, $y = x^3 + 1$, about $y = 0$

37. A tapered rod of length L has circular cross sections. If the radii of its ends are a and b, what is the volume of the rod?

38. During one revolution an airplane propeller displaces an amount of air that can be calculated as a volume of a solid of revolution. If the area yielding the volume is that bounded by the curves $64x = y(y - 4)$ and $64x = y(4 - y)$, and is rotated about the x-axis, calculate the volume of air displaced.

39. If a sphere of radius r is sliced a distance h from its centre, show that the volume of the smaller piece is

$$V = \frac{\pi}{3}(r - h)^2(2r + h).$$

40. An embankment is to be built around the circular wading pool in Figure 7.29(a). Figure 7.29(b) shows a cross-section of the embankment.
   (a) Find a cubic polynomial $y = ax^3 + bx^2 + cx + d$ to fit the three points $(0, 0)$, $(4, 2)$, and $(6, 0)$.
   (b) Determine the amount of fill required to build the embankment.

FIGURE 7.29
(a)

(b)

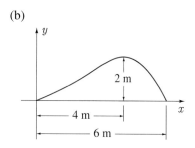

41. Find the volume of the donut obtained by rotating the circle in Figure 7.30 about the y-axis.

FIGURE 7.30

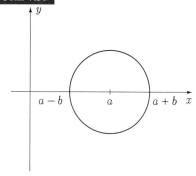

42. A cylindrical hole is bored through the centre of a sphere, the length of the hole being L. Show that no matter what the radius of the sphere, the volume of the sphere that remains is always the same and equal to the volume of a sphere of diameter L.

43. Water half fills a cylindrical pail of radius a and height L. When the pail is rotated about its axis of symmetry with angular speed ω (Figure 7.31), the surface of the water assumes a parabolic shape, the cross-section of which is given by

$$y = H + \frac{\omega^2 x^2}{2g},$$

where $g > 0$ is the acceleration due to gravity and H is a constant. Find the speed ω in terms of L, a, and g, at which water spills over the top.

FIGURE 7.31

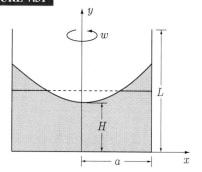

44. A right circular cone of height H and base radius r has its vertex at the centre of a sphere of radius R ($R < H$). Find that part of the volume of the sphere inside the cone.

45. Devise a way to calculate the volume of the solid of revolution when the area in the first quadrant bounded by

$$y = 1 - x^2, \quad x = 0, \quad y = 0$$

is rotated about the line $y = x + 1$. Hint: You may want to use the distance formula in question 45 of Exercises 1.3.

SECTION 7.3

Lengths of Curves

Formula 1.7 defines the length of the straight-line segment joining two points $P(x_1, y_1)$ and $Q(x_2, y_2)$. The formula $s = r\theta$ gives the length of the arc of a circle of radius r subtended by an angle θ at the centre of the circle. In this section we derive a result that will theoretically enable us to find the length of any curve.

Consider finding the length of the curve $C : y = f(x)$ in Figure 7.32(a) joining points $A(a, f(a))$ and $B(b, f(b))$. To find its length L we begin by approximating C with a series of straight-line segments. Specifically, we choose $n - 1$ consecutive points on C

between A and B,
$$A = P_0, P_1, P_2, \ldots, P_{n-1}, P_n = B,$$
and join each P_{i-1} to P_i ($i = 1, \ldots, n$) by means of a straight-line segment, as in Figure 7.32(b). If the coordinates of P_i are denoted by (x_i, y_i), then the length of the line segment joining P_{i-1} and P_i is
$$\|P_{i-1} P_i\| = \sqrt{(x_i - x_{i-1})^2 + (y_i - y_{i-1})^2}.$$
For a large number of these line segments, it is reasonable to approximate L by the sum of the lengths of the segments:
$$\sum_{i=1}^{n} \|P_{i-1} P_i\| = \sum_{i=1}^{n} \sqrt{(x_i - x_{i-1})^2 + (y_i - y_{i-1})^2}.$$
In fact, as we increase n and at the same time decrease the length of each segment, we expect the approximation to become more and more accurate. We therefore define
$$L = \lim_{\|\Delta x_i\| \to 0} \sum_{i=1}^{n} \|P_{i-1} P_i\| = \lim_{\|\Delta x_i\| \to 0} \sum_{i=1}^{n} \sqrt{(x_i - x_{i-1})^2 + (y_i - y_{i-1})^2}. \quad (7.13)$$

(a) (b)

FIGURE 7.32

If we set $\Delta x_i = x_i - x_{i-1}$, and note that $y_{i-1} = f(x_{i-1})$ and $y_i = f(x_i)$, we have
$$L = \lim_{\|\Delta x_i\| \to 0} \sum_{i=1}^{n} \sqrt{(\Delta x_i)^2 + [f(x_i) - f(x_{i-1})]^2}$$
$$= \lim_{\|\Delta x_i\| \to 0} \sum_{i=1}^{n} \sqrt{1 + \left[\frac{f(x_i) - f(x_{i-1})}{\Delta x_i}\right]^2} \, \Delta x_i.$$

In order that 7.13 be a useful definition from a calculational point of view, we must find a convenient way to evaluate this limit-summation. To do this we assume that $f'(x)$ exists at every point in the interval $a \leq x \leq b$, and apply the mean value theorem to $f(x)$ on each subinterval $x_{i-1} \leq x \leq x_i$ (Figure 7.33). The theorem guarantees the existence of at least one point x_i^* between x_{i-1} and x_i such that
$$f'(x_i^*) = \frac{f(x_i) - f(x_{i-1})}{x_i - x_{i-1}} = \frac{f(x_i) - f(x_{i-1})}{\Delta x_i}.$$
Consequently, using these points x_i^*, we can express the length of C in the form
$$L = \lim_{\|\Delta x_i\| \to 0} \sum_{i=1}^{n} \sqrt{1 + [f'(x_i^*)]^2} \, \Delta x_i. \quad (7.14)$$

But the right-hand side of this equation is the definition of the definite integral of the function $\sqrt{1 + [f'(x)]^2}$ with respect to x from $x = a$ to $x = b$:

$$\int_a^b \sqrt{1 + [f'(x)]^2} \, dx = \lim_{\|\Delta x_i\| \to 0} \sum_{i=1}^n \sqrt{1 + [f'(x_i^*)]^2} \, \Delta x_i.$$

Thus we can evaluate the length of C by the definite integral

$$L = \int_a^b \sqrt{1 + \left(\frac{dy}{dx}\right)^2} \, dx. \qquad (7.15)$$

FIGURE 7.33

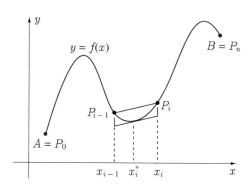

In this derivation we assumed that $f'(x)$ was defined at each point in the interval $a \leq x \leq b$, necessary for the mean value theorem to apply. But in order to guarantee existence of the definite integral in 7.15, Theorem 6.2 requires continuity of the integrand $\sqrt{1 + (dy/dx)^2}$. Consequently, to ensure that the length of a curve can be calculated by means of 7.15, we assume that $f(x)$ has a continuous first derivative on the interval $a \leq x \leq b$.

EXAMPLE 7.14

Find the length of the curve $y = x^{3/2}$ from $(1, 1)$ to $(2, 2\sqrt{2})$ (Figure 7.34).

FIGURE 7.34

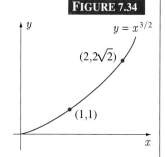

SOLUTION Since

$$\frac{dy}{dx} = \frac{3}{2} x^{1/2},$$

equation 7.15 gives

$$L = \int_1^2 \sqrt{1 + \left(\frac{3}{2} x^{1/2}\right)^2} \, dx = \int_1^2 \sqrt{1 + \frac{9}{4} x} \, dx$$

$$= \left\{ \frac{8}{27} \left(1 + \frac{9}{4} x\right)^{3/2} \right\}_1^2 = \frac{8}{27} \left[\left(\frac{11}{2}\right)^{3/2} - \left(\frac{13}{4}\right)^{3/2} \right] = \frac{1}{27}(22^{3/2} - 13^{3/2}).$$

FIGURE 7.35

The result in 7.15 is clearly useful when the equation of the curve is expressed in the form $y = f(x)$. If the curve is defined in the form $x = g(y)$ (Figure 7.35), a similar analysis gives

$$L = \int_c^d \sqrt{1 + \left(\frac{dx}{dy}\right)^2} \, dy. \qquad (7.16)$$

For example, we may write the equation of the curve in Example 7.14 in the form $x = y^{2/3}$, in which case $dx/dy = (2/3)y^{-1/3}$, and

$$L = \int_1^{2\sqrt{2}} \sqrt{1 + \left(\frac{2}{3}y^{-1/3}\right)^2}\, dy = \int_1^{2\sqrt{2}} \sqrt{1 + \frac{4}{9}y^{-2/3}}\, dy$$

$$= \int_1^{2\sqrt{2}} \sqrt{\frac{4 + 9y^{2/3}}{9y^{2/3}}}\, dy = \int_1^{2\sqrt{2}} \frac{\sqrt{4 + 9y^{2/3}}}{3y^{1/3}}\, dy.$$

If we set $u = 4 + 9y^{2/3}$, then $du = 6y^{-1/3}\, dy$, and

$$L = \frac{1}{3}\int_{13}^{22} \sqrt{u}\, \frac{du}{6} = \frac{1}{18}\left\{\frac{2}{3}u^{3/2}\right\}_{13}^{22} = \frac{1}{27}(22^{3/2} - 13^{3/2}).$$

Integrals 7.15 and 7.16 can both be interpreted geometrically. This interpretation will be very useful when we study parametric equations of curves in Section 10.1, and line integrals in Chapter 15. Figure 7.36 shows the curve C of Figure 7.32(a). At a point (x,y) on C, we draw the tangent line to the curve. If dx is a short length along the x-axis at position x, the length of that part of the tangent line between vertical lines at x and $x + dx$ is given by

$$\sqrt{(dx)^2 + (dy)^2}.$$

This length along the tangent line closely approximates the length of the curve between the same two vertical lines, the approximation being more accurate the shorter the length dx. If we picture a large number of these tangential line segments between a and b, we can find the total length of C by adding together all such lengths and taking the limit as each approaches zero. But this process is represented by the definite integral, and we therefore write

$$L = \int_{x=a}^{x=b} \sqrt{(dx)^2 + (dy)^2} \quad \text{or} \quad L = \int_{y=c}^{y=d} \sqrt{(dx)^2 + (dy)^2}. \quad (7.17)$$

The first integral is chosen when it is more convenient to integrate with respect to x, in which case dx is taken outside the square root. We set

$$L = \int_a^b \sqrt{1 + \left(\frac{dy}{dx}\right)^2}\, dx,$$

which is equation 7.15. The second corresponds to a more convenient integral with respect to y, in which case dy is taken outside the square root. We obtain

$$L = \int_c^d \sqrt{1 + \left(\frac{dx}{dy}\right)^2}\, dy,$$

which is equation 7.16.

FIGURE 7.36

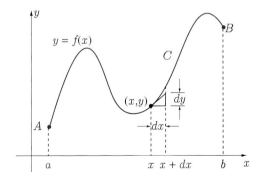

EXAMPLE 7.15

Find the length of the curve $24xy = x^4 + 48$ between $(2, 4/3)$ and $(3, 43/24)$.

SOLUTION It is easier to express y in terms of x (rather than x in terms of y),

$$y = \frac{x^4 + 48}{24x} = \frac{x^3}{24} + \frac{2}{x}.$$

We now use the fact that small lengths along the curve are approximated by

$$\sqrt{(dx)^2 + (dy)^2} = \sqrt{1 + \left(\frac{dy}{dx}\right)^2}\,dx = \sqrt{1 + \left(\frac{x^2}{8} - \frac{2}{x^2}\right)^2}\,dx$$

$$= \sqrt{1 + \frac{x^4}{64} - \frac{1}{2} + \frac{4}{x^4}}\,dx = \sqrt{\frac{1}{64x^4}(x^8 + 32x^4 + 256)}\,dx$$

$$= \frac{1}{8x^2}\sqrt{(x^4 + 16)^2}\,dx = \frac{x^4 + 16}{8x^2}\,dx.$$

FIGURE 7.37

The total length along the curve is therefore (Figure 7.37)

$$L = \int_2^3 \frac{x^4 + 16}{8x^2}\,dx = \frac{1}{8}\int_2^3 \left(x^2 + \frac{16}{x^2}\right)dx$$

$$= \frac{1}{8}\left\{\frac{x^3}{3} - \frac{16}{x}\right\}_2^3 = \frac{9}{8}.$$

EXERCISES 7.3

In Exercises 1–10 find the length of the curve.

1. $8x^2 = y^3$ from $(1, 2)$ to $(2\sqrt{2}, 4)$
2. $3y = 2(x^2 + 1)^{3/2}$ from $(-2, 10\sqrt{5}/3)$ to $(-1, 4\sqrt{2}/3)$
3. $x = 2(y - 2)^{3/2}$ from $(0, 2)$ to $(2, 3)$
4. $y = (x - 1)^{3/2}$ from $(2, 1)$ to $(10, 27)$
5. $y = \frac{x^3}{4} + \frac{1}{3x}$ from $(1, 7/12)$ to $(2, 13/16)$
6. $y = (e^x + e^{-x})/2$ between the lines $x = 0$ and $x = 1$
7. $y = \frac{x^4}{4} + \frac{1}{8x^3}$ from $(2, 257/64)$ to $(1/2, 65/64)$
8. $36xy = x^4 + 108$ from $(3, 7/4)$ to $(4, 91/36)$
9. $x = \frac{y^7}{20} + \frac{1}{7y^5}$ between the lines $y = -1$ and $y = -2$
10. $y = x^5/5 + 1/(12x^3)$ from $(1, 17/60)$ to $(2, 3077/480)$

In Exercises 11–20 set up (but do not evaluate) a definite integral for the length of the curve.

11. $y = x^2$ from $(0, 0)$ to $(1, 1)$
12. $y = 3x^2 - 4x$ from $(1, -1)$ to $(2, 4)$
13. $x^2 - y^2 = 1$ from $(1, 0)$ to $(2, \sqrt{3})$
14. $x^2 - y^2 = 1$ from $(2, -\sqrt{3})$ to $(3, 2\sqrt{2})$
15. $y = \sin x$ from $(0, 0)$ to $(0, \pi)$
16. $y = \ln(\cos x)$ between the lines $x = 0$ and $x = \pi/4$
17. $y = \ln x$ from $(1, 0)$ to any other point on the curve
18. $8y^2 = x^2(1 - x^2)$ (complete length)
19. $x = y^2 - 2y$ from $(0, 0)$ to $(0, 2)$
20. $x^2/4 + y^2/9 = 1$ (complete length)
21. Find the length of the curve $3x = \sqrt{y}(y - 3)$ from $(-2/3, 1)$ to $(2/3, 4)$.
22. Find the length of the curve $x^{2/3} + y^{2/3} = 1$.

23. If n is any number other than 1 or -1, and $0 < a < b$, find the length of the curve

$$y = \frac{x^{n+1}}{n+1} + \frac{1}{4(n-1)x^{n-1}}$$

between $x = a$ and $x = b$.

24. If n is any number greater than $1/2$, and $0 < a < b$, find the length of the curve

$$y = \frac{x^{2n+1}}{4(2n-1)} + \frac{1}{(2n+1)x^{2n-1}}$$

between $x = a$ and $x = b$.

SECTION 7.4

Work

When a body moves a distance d along a straight line l under the action of a constant force F, which acts in the same direction as the motion (Figure 7.38), the work done on the body by F is defined as

$$W = Fd. \tag{7.18}$$

FIGURE 7.38

Forces often vary in either magnitude or direction, or both, and when such forces act on the body, calculation of the work done is not as simple as using equation (7.18). In this section we show how we can use definite integrals to calculate work done by forces that always act along the direction of motion, but do not have constant magnitude. Specifically, let us consider a particle that moves along the x-axis from $x = a$ to $x = b$, where $b > a$, under the action of some force. Suppose the force always acts in the positive x-direction, but its size, which we denote by $F(x)$, is not constant along the x-axis. To find the work done by the force as the particle moves from $x = a$ to $x = b$, we cannot simply multiply force by distance. (Where would we evaluate $F(x)$?) Instead, we divide the interval $a \le x \le b$ into n subintervals by points (Figure 7.39)

$$a = x_0 < x_1 < x_2 < \cdots < x_{n-1} < x_n = b.$$

FIGURE 7.39

In each subinterval $x_{i-1} \le x \le x_i$ we choose a point x_i^*. If lengths of the subintervals are small and $F(x)$ is continuous, then $F(x)$ does not vary greatly over any given subinterval. In such a situation, we may approximate $F(x)$ by a constant force $F(x_i^*)$ on each subinterval $x_{i-1} \le x \le x_i$. It follows that the work done by the force over the ith subinterval is approximately

$$F(x_i^*)(x_i - x_{i-1}) = F(x_i^*)\Delta x_i.$$

Further, an approximation to the total work done by the force as the particle moves from $x = a$ to $x = b$ is

$$\sum_{i=1}^{n} F(x_i^*) \Delta x_i.$$

As n becomes larger and each Δx_i approaches zero, this approximation becomes better, and we therefore define

$$W = \lim_{||\Delta x_i|| \to 0} \sum_{i=1}^{n} F(x_i^*) \Delta x_i. \tag{7.19}$$

But this limit may also be interpreted as the definite integral of $F(x)$ with respect to x from $x = a$ to $x = b$. Consequently, W may be calculated with the definite integral

$$W = \int_a^b F(x)\,dx. \tag{7.20}$$

It is simple enough to interpret the various parts of 7.20, and in so doing we begin to feel the definite integral at work. The integrand $F(x)$ is the force at position x, and dx is a small distance along the x-axis. The product $F(x)\,dx$ is therefore interpreted as the (approximate) work done by the force along dx. The definite integral then adds over all dx's, beginning at $x = a$ and ending at $x = b$, to give the total work.

What is important about equation 7.20 is not the particular form of the definite integral, but the fact that work done by a force can be evaluated by means of a definite integral. As we solve work problems in this section, we find that the form of the definite integral varies considerably from one problem to another, but the underlying fact remains that each problem is solved with a definite integral.

With the exception of work problems that involve emptying tanks (Example 7.18), we recommend that you always make a diagram illustrating the physical setup at some intermediate stage between start and finish. Determine forces at this position in order to set up the work integral.

In the derivation of 7.20 we assumed continuity of $F(x)$, and this guarantees existence of the definite integral.

EXAMPLE 7.16

Find the work necessary to expand a spring from a stretch of 5 cm to a stretch of 15 cm if a force of 200 N stretches it 10 cm.

SOLUTION Let the spring be stretched in the positive x-direction, and let $x = 0$ correspond to the free end of the spring in the unstretched position in the top half of Figure 7.40. In the bottom half of this figure we have shown the spring stretched to an intermediate position.

FIGURE 7.40

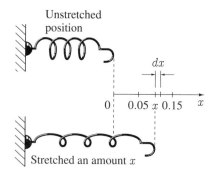

According to Hooke's law, when the spring is stretched an amount x, the restoring force in the spring is proportional to x:

$$F_s = -kx,$$

where $k > 0$ is a constant. The negative sign indicates that the force is in the negative x-direction. Since $F_s = -200$ N when $x = 0.10$ m,

$$-200 = -k(0.10),$$

it follows that $k = 2000$ N/m. The force required to counteract the restoring force of the spring when it is stretched an amount x is therefore

$$F(x) = 2000\,x.$$

The work done by this force in stretching the spring from position x a further amount dx is approximately

$$2000\,x\,dx,$$

and hence the total work to increase the stretch from 5 cm to 15 cm is

$$W = \int_{0.05}^{0.15} 2000\,x\,dx = 2000 \left\{ \frac{x^2}{2} \right\}_{0.05}^{0.15} = 20.$$

The total work done is therefore 20 N-m (or joules). ∎

EXAMPLE 7.17

A cable of length 100 m and mass 300 kg hangs vertically from the top of a building. What work is required to lift the entire cable to the top of the building?

SOLUTION

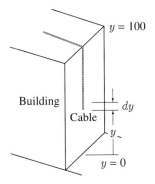

FIGURE 7.41

When the cable has been lifted to an intermediate point where its lower end is y metres above its original position (Figure 7.41), the length of cable still hanging is $100 - y$ m. Since each metre of cable has mass 3 kg, the mass of $100 - y$ m of cable is $3(100 - y)$ kg. It follows that the force that must be exerted to overcome gravity on this much cable is $9.81(3)(100 - y)$ N. The work that this force does in raising the end of the cable an additional amount dy is approximately

$$9.81(3)(100 - y)\,dy,$$

and the total work to raise the entire cable is therefore

$$W = \int_0^{100} 9.81(3)(100-y)\,dy = 29.43\left\{100y - \frac{y^2}{2}\right\}_0^{100} = 147\,150 \text{ J}.$$

■

EXAMPLE 7.18

A tank in the form of an inverted right-circular cone of depth 10 m and radius 4 m is full of water. Find the work required to pump the water to a level 1 m above the top of the tank.

SOLUTION

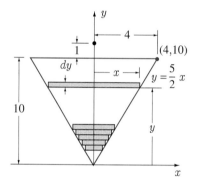

FIGURE 7.42

A cross-section of the tank is shown in Figure 7.42. Suppose we approximate the water in the tank with circular discs formed by rotating the rectangle shown around the y-axis. The force of gravity on the representative disc at position y is its volume $\pi x^2\,dy$, multiplied by the density of water (1000 kg/m^3), multiplied by the acceleration due to gravity (-9.81),

$$-9.81(1000)\pi x^2\,dy.$$

It is negative because gravity acts in the negative y-direction. The work done by an equal and opposite force in lifting the disc to a level 1 m above the top of the tank (a distance $11 - y$) is approximately

$$(11-y)9.81(1000)\pi x^2\,dy.$$

The total work to empty the tank is therefore

$$W = \int_0^{10} (11-y)9810\pi x^2\,dy.$$

To express x in terms of y, we note that x and y are the coordinates of points on the straight line through the origin and the point $(4, 10)$. Since the equation of this line is $y = 5x/2$, we obtain

$$W = 9810\pi \int_0^{10} (11-y)\left(\frac{2y}{5}\right)^2 dy = 1569.6\pi \int_0^{10}(11y^2 - y^3)\,dy$$

$$= 1569.6\pi \left\{\frac{11y^3}{3} - \frac{y^4}{4}\right\}_0^{10} = 5.75 \times 10^6 \text{ J.}$$

Two observations are noteworthy:

1. In Examples 7.16 and 7.17 the differential represents distance moved and the integrand represents force, and we applied equation 7.20 directly. In Example 7.18 the differential is part of the force, and the distance moved is part of the integrand. This is why we stated earlier that it is not advisable to use equation 7.20 as a formula. The integrand is not always force and the differential is not always distance moved.

2. In each of these examples we set up the coordinate system; it was not given. We can use any coordinate system whatsoever; but once we have chosen our coordinates, we must refer everything to that system. For instance, were we to use the coordinate system in Figure 7.43 for Example 7.18, the definite integral would be

$$W = \int_{-10}^{0} (1-y)\, 9810\pi \left[\frac{2}{5}(10+y)\right]^2 dy.$$

Evaluation of this definite integral would again lead to the result $W = 5.75 \times 10^6$ J.

FIGURE 7.43

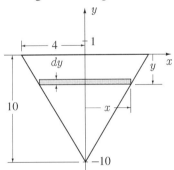

EXERCISES 7.4

1. A spring requires a 10 N force to stretch it 3 cm. Find the work to increase the stretch of the spring from 5 cm to 7 cm.

2. Find the work to increase the stretch of the spring in Exercise 1 from 7 cm to 9 cm.

3. A cage of mass M kilograms is to be lifted from the bottom of a mine shaft h metres deep. If the mass of the cable used to hoist the cage is m kilograms per metre, find the work done.

4. A uniform cable of length 50 m and mass 100 kg hangs vertically from the top of a building 100 m high. How much work is required to get 10 m of the cable on top of the building?

5. A 2 m chain of mass 20 kg lies on the floor. If friction between floor and chain is ignored, how much work is required to lift one end of the chain 2 m straight up?

6. How much work is required to lift the end of the chain in Exercise 5 only 1 m?

7. How much work is required to lift the end of the chain in Exercise 5 a distance of 4 m?

8. A 5 m chain of mass 15 kg hangs vertically. It is required to lift the lower end of the chain 5 m so that it is level with the upper end. Calculate the work done using each of the coordinate systems in Figure 7.44.

FIGURE 7.44

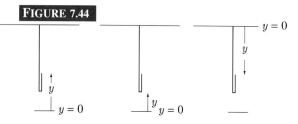

9. A tank filled with water has the form of a paraboloid of revolution with vertical axis (Figure 7.45). If the depth of the tank is 12 m and the diameter of the top is 8 m, find the work in pumping the water to the top of the tank.

FIGURE 7.45

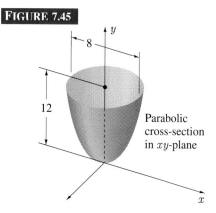

Parabolic cross-section in xy-plane

10. In Exercise 9, find the work to empty the tank through an outlet 2 m above the top of the tank.

11. The ends of a trough are isosceles triangles with width 2 m and depth 1 m. The trough is 5 m long and it is full of water. How much work is required to lift all of the water to the top of the trough?

12. How much work is required to lift the water in the trough of Exercise 11 to a height of 2 m above the top of the trough?

13. A rectangular swimming pool full of water is 25 m long and 10 m wide. The depth is 3 m for the first 10 m of length, decreasing linearly to 1 m at the shallow end. How much work is required to lower the level of the water in the pool by 1/2 m?

14. How much work is required to empty the pool of Exercise 13 over its edge?

15. The force of repulsion between two point charges of like sign, one of size q and the other of size Q, has magnitude
$$F = \frac{qQ}{4\pi\epsilon_0 r^2},$$

where ϵ_0 is a constant and r is the distance between the charges. If Q is placed at the origin, and q is moved along the x-axis from $x = 2$ to $x = 5$, find the work done by the electrostatic force.

16. Two positive charges q_1 and q_2 are placed at positions $x = 5$ and $x = -2$ on the x-axis. A third positive charge q_3 is moved along the x-axis from $x = 1$ to $x = -1$. Find the work done by the electrostatic forces of q_1 and q_2 on q_3 (see Exercise 15).

17. If the chain in Exercise 5 is stretched out straight on the floor, and if the coefficient of friction between floor and chain is 0.01, what work is required to raise one end of the chain 2 m? The force of friction on that part of the chain on the floor is 0.01 times the weight of the chain on the floor.

18. Two similar springs, each 1 m long in an unstretched position, have spring constant k N/m. The springs are joined together at P (Figure 7.46), and their free ends are fastened to two posts 4 m apart. What work is done in moving the midpoint P a distance b m to the right?

FIGURE 7.46

19. If the force of a crossbow (in newtons) is proportional to the draw (in metres), and it is 200 N at a full draw of 50 cm, what is the speed of a 20-gm arrow from a fully drawn crossbow? Hint: The work necessary to draw the bow is converted into kinetic energy K, where $K = mv^2/2$ (m is the mass of the arrow and v is the speed of the arrow).

20. A bucket of water with mass 100 kg is on the ground attached to one end of a cable with mass per unit length 5 kg/m. The other end of the cable is attached to a windlass 100 m above the bucket. If the bucket is raised at a constant speed, water runs out through a hole in the bottom at a constant rate to the extent that the bucket has mass 80 kg when it reaches the top. To further complicate matters, a pigeon of mass 2 kg lands on the bucket when it is 50 m above the ground. He immediately begins taking a bath, splashing water over the side of the bucket at the rate of 1 kg/m. Find the work done by the windlass in raising the bucket 100 m.

21. A hemispherical tank with diametral plane at the top has radius 5 m.

   (a) Show that if the depth of oil in the tank is h metres when the tank is one-half full by volume, then h must satisfy the equation
   $$h^3 - 15h^2 + 125 = 0.$$
   Find h to four decimals.

(b) Find the work to empty the half-full tank if the oil has density 750 kg/m³, and the outlet is at the top of the tank.

22. Newton's universal law of gravitation states that the force of attraction between two point masses m and M has magnitude
$$F = \frac{GmM}{r^2},$$
where r is the distance between the masses and $G = 6.67 \times 10^{-11}$ Nm²/kg² is a constant.

(a) If M represents the mass of the earth and we regard it as a point mass concentrated at its centre, show that Newton's universal law of gravitation at the earth's surface reduces to $F = mg$, where $g = 9.82$ m/s². Assume for the calculation of M that the earth is a sphere with radius 6370 km and mean density 5.52×10^3 kg/m³.

(b) Use the original law $F = GmM/r^2$ with the earth regarded as a point mass to calculate the work required to lift a mass of 10 kg from the earth's surface to a height of 10 km.

(c) Calculate the work in (b) using the constant gravitational force $F = mg$ in (a). Is there a significant difference?

23. A right circular cylinder with horizontal axis has radius r metres and length h metres. If it is full of oil with density ρ kilograms per cubic metre, how much work is required to empty the tank through an outlet at its top?

24. A gas is confined in a cylinder, closed on one end with a piston on the other. If the temperature is held constant, then $PV = C$, where P is the pressure of the gas in the cylinder, V is its volume, and C is a constant. The force needed to move the piston is $F = PA$, where A is the area of the face of the piston. If the radius of the cylinder is 3 units, find the work done in moving the piston from a point 10 units to a point 5 units from the closed end.

25. A gas is confined in a cylinder, closed on one end with a piston on the other. If the gas expands adiabatically, then the pressure P and volume V of the gas obey the law
$$PV^{7/5} = C,$$
where C is a constant. The force on the piston face due to the pressure of the gas is given by $F = PA$, where A is the area of the face of the piston. Show that if the piston is moved so as to reduce the volume occupied by the gas from V_0 to $V_0/2$, then the work done by the piston is
$$\frac{5}{2}\left(2^{2/5} - 1\right)CV_0^{-2/5}.$$

26. A particle is moved along the x-axis by a force $F(x)$ in the x-direction. Use Newton's second law to show that if $F(x)$ is the total resultant force on the particle, then the work done by $F(x)$ is equal to the change in kinetic energy of the particle.

27. A diving bell of mass 10 000 kilograms is attached to a chain of mass 5 kg/m, and the bell sits on the bottom of the ocean 100 m below the surface (Figure 7.47). How much work is required to lift the bell to deck level 5 m above the surface? Take into account the fact that when the bell and chain are below the surface, they weigh less than when above. The apparent loss in weight is equal to the weight of water displaced by the bell and chain. Assume that the bell is a perfect cube, 2 m on each side; therefore, when completely submerged, it displaces 8 m³ of water. Assume also that the chain displaces 1 litre of water per metre of length.

FIGURE 7.47

28. Two springs with constants k_1 and k_2 are joined together, and then one end is fastened to a wall (Figure 7.48). It is shown in physics that when a horizontal force is applied to the free end, the ratio of the stretches s_1 and s_2 in the two springs is inversely proportional to the ratio of their spring constants:
$$\frac{s_1}{s_2} = \frac{k_2}{k_1}.$$

Show that if the free end is moved so as to produce a total stretch in the springs of L, the amount of work done is
$$\frac{k_1 k_2 L^2}{k_1 + k_2}.$$

FIGURE 7.48

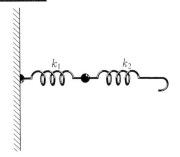

SECTION 7.5

Fluid Pressure

When an object is immersed in a fluid, it is acted on by fluid forces. These forces are independent of the object, and are therefore a property of the fluid itself. They always act perpendicular to the surface of the submerged object. We use the concept of pressure to describe these fluid forces. We define **pressure** at a point in a fluid as the *magnitude of the force per unit area* that would act on a surface at that point in the fluid. Because pressure is the magnitude of the fluid force at a point, it is therefore a positive quantity.

FIGURE 7.49

Experience suggests that pressure depends on two factors: depth below the surface of the fluid and the type of fluid itself. In order to discover the precise dependence, we consider a small horizontal disc of the fluid (Figure 7.49). Suppose we denote by $P(y)$ the functional dependence of pressure P on depth y. Then the pressure at the bottom of the disc is $P(y)$ and the pressure at the top of the disc is $P(y + \Delta y)$. If the fluid is stationary, then the sum of all vertical forces on the disc must be zero. There are three vertical forces acting on the disc: fluid forces on its top and bottom faces, and gravity. Since pressure $P(y)$, which is force per unit area, is the same at all points on the bottom of the disc, it follows that the force on the bottom of the disc must be $AP(y)$; that is, the product of the area A of the bottom of the disc and pressure at points on the bottom of the disc. Similarly, the fluid force on the top of the disc is $-AP(y + \Delta y)$; it is negative because it is in the negative y-direction. Finally, if ρ is the density of the fluid (mass per unit volume), the force of gravity on the disc is $-9.81\rho(A\Delta y)$. Since the sum of these three forces must be zero, we set

$$AP(y) - AP(y + \Delta y) - 9.81\rho(A\Delta y) = 0.$$

Rearrangement of this equation yields

$$\frac{P(y + \Delta y) - P(y)}{\Delta y} = -9.81\rho,$$

and if we take limits of both sides as $\Delta y \to 0$, we obtain

$$\frac{dP}{dy} = -9.81\rho.$$

This differential equation for $P(y)$ is immediately integrable:

$$P(y) = -9.81\rho y + C.$$

Since fluid pressure at the surface of the fluid is equal to zero ($P(0) = 0$), C must be equal to zero; hence,

$$P = -9.81\rho y. \qquad (7.21)$$

Since $-y$ is a measure of depth d below the surface of the fluid, we have shown that

$$P = 9.81 \rho d, \qquad (7.22)$$

where d is always taken as positive.

An illuminating interpretation of formula 7.22 is suggested by Figure 7.50. Above a point at depth d below the surface of a fluid, we consider a column of fluid of unit cross-sectional area. The weight of this column of fluid is its volume multiplied by 9.81ρ:

$$W = 9.81 \rho V.$$

But V is the product of the length of the column, d, and the (unit) cross-sectional area; that is, $V = (1)d = d$, and hence

$$W = 9.81 \rho d. \qquad (7.23)$$

A comparison of equations 7.22 and 7.23 suggests that pressure at any point is precisely the weight of a column of fluid of unit cross-sectional area above that point.

FIGURE 7.50

FIGURE 7.51

We now consider the problem of determining total force on one side of a flat vertical plate (Figure 7.51), when it is immersed in a fluid of density ρ. If the area of the plate is subdivided into horizontal rectangles of width dy, then the pressure at each point of this rectangle is approximately $P = -9.81\rho y$. This is an approximation because slight variations in pressure do occur over vertical displacements within the rectangle. It follows that the force on the representative rectangle is approximately equal to its area multiplied by $-9.81\rho y$,

$$-9.81 \rho y (x_2 - x_1) \, dy = -9.81 \rho y [f(y) - g(y)] \, dy.$$

Total force on the plate is found by adding the forces on all such rectangles and taking the limit as their widths approach zero. We obtain the required force, therefore, as

$$F = \int_a^b -9.81 \rho y [f(y) - g(y)] \, dy. \qquad (7.24)$$

Once again we do not suggest that equation 7.24 be memorized because it is closely connected with the choice of coordinates in Figure 7.51. For a different coordinate system, the definite integral would be correspondingly different (see Example 7.19). What is important is the procedure: subdivide the area of the plate into horizontal rectangles, find the force on a representative rectangle, and finally add over all rectangles with the definite integral.

Note that vertical rectangles cannot be used without further discussion since it is not evident how to calculate the force on such a rectangle. Consideration of vertical rectangles is given in Exercise 8.

EXAMPLE 7.19

The vertical face of a dam is parabolic with breadth 100 m and height 50 m. Find the total force due to fluid pressure on the face.

SOLUTION If we set up the coordinate system in Figure 7.52, we see that the edge of the dam has an equation of the form $y = kx^2$. Since $(50, 50)$ is a point on this curve, it follows that $k = 1/50$, and $y = x^2/50$. The area of the representative rectangle shown is $2x\,dy$, and it is at a depth $50 - y$ below the surface of the water. Since the density of water is 1000 kg/m^3, the force on the representative rectangle is approximately equal to

$$(9.81)(1000)(50 - y)2x\,dy = 19\,620(50 - y)\sqrt{50y}\,dy.$$

The total force on the dam must therefore be

$$F = \int_0^{50} 19\,620(50 - y)5\sqrt{2}\,y^{1/2}\,dy = 98\,100\sqrt{2}\int_0^{50}(50y^{1/2} - y^{3/2})\,dy$$

$$= 98\,100\sqrt{2}\left\{\frac{100y^{3/2}}{3} - \frac{2y^{5/2}}{5}\right\}_0^{50} = 6.54 \times 10^8 \text{ N}.$$

FIGURE 7.52

EXAMPLE 7.20

A tank in the form of a right-circular cylinder of radius 2 m and length 10 m lies on its side. If it is half filled with oil of density ρ kilograms per cubic metre, find the force on each end of the tank.

SOLUTION The force on the representative rectangle in Figure 7.53 is approximately equal to the pressure $9.81\rho(-y)$ multiplied by the area $2x\,dy$ of the rectangle,

$$9.81\rho(-y)2x\,dy = -19.62\,\rho y\sqrt{4 - y^2}\,dy.$$

The total force on each end of the tank is therefore

$$F = \int_{-2}^0 -19.62\,\rho y\sqrt{4 - y^2}\,dy = 19.62\,\rho\left\{\frac{1}{3}(4 - y^2)^{3/2}\right\}_{-2}^0 = 52.32\,\rho \text{ N}.$$

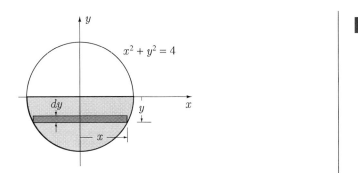

FIGURE 7.53

EXERCISES 7.5

1. A tropical fish tank has length 1 m, width 0.5 m, and depth 0.5 m. Find the force due to water pressure on each of the sides and bottom when the tank is full.

2. The vertical surface of a dam exposed to the water of a lake has the shape shown in Figure 7.54. Find the force of the water on the face of the dam.

   FIGURE 7.54

3. The vertical end of a water trough is an isosceles triangle with width 2 m and depth 1 m. Find the force of the water on each end when the trough is half full by volume.

4. A square plate, 2 m on each side, has one diagonal vertical. If it is one-half submerged in water, what is the force due to water pressure on each side of the plate?

5. A cylindrical oil tank of radius r and height h has its axis vertical. If the density of the oil is ρ, find the force on the bottom of the tank when it is full.

6. A rectangular swimming pool full of water is 25 m long and 10 m wide. The depth is 3 m for the first 10 m at the deep end, decreasing linearly to 1 m at the shallow end. Find the force due to the weight of the water on each of the sides and ends of the pool.

7. The vertical face of a dam across a river has the shape of a parabola 36 m across the top and 9 m deep at the centre. What is the force that the river exerts on the dam if the water is 0.5 m from the top?

8. Show that the force due to fluid pressure on the vertical rectangle in Figure 7.55 is
   $$F = \frac{9.81\rho}{2} h(y_1^2 - y_2^2),$$
   where ρ is the density of the fluid.

   FIGURE 7.55

9. A flat plate in the shape of a trapezoid is submerged vertically in a fluid with density ρ. The plate has two parallel vertical sides of lengths 6 and 8 and a third side of length 5 that is perpendicular to the parallel sides and at a depth of 1 below the surface of the fluid (Figure 7.56). Find the force due to fluid pressure on each side of the plate, using both horizontal and vertical rectangles (see Exercise 8 for vertical rectangles).

   FIGURE 7.56

   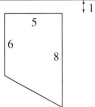

10. The base of a triangular plate, of length a, lies in the surface of a fluid of density ρ. The third vertex of the triangle is at depth b below the surface (Figure 7.57). Show that the force due to fluid pressure on each side of the plate is $9.81\rho ab^2/6$, no matter what the shape of the triangle.

FIGURE 7.57

11. Set up (but do not evaluate) a definite integral(s) to find the force due to water pressure on each side of the flat vertical plate in Figure 7.58.

FIGURE 7.58

12. In Exercise 6, find the force due to water pressure on each part of the bottom of the pool.

13. The bow of a landing barge (Figure 7.59) consists of a rectangular flat plate A metres wide and B metres long. When the barge is stationary, this plate makes an angle of $\pi/6$ radians with the surface of the water. Find the maximum force of the water on the bow.

FIGURE 7.59

14. A water tank is to be built in the form of a rectangular box with all six sides welded along their joins (Figure 7.60). Sides $ABCD$ and $EFGH$ are to be square, and all sides except $ABCD$ are supported from the outside. If the maximum force that $ABCD$ can withstand is 20 000 newtons, what is the largest cross-section that can be built, assuming that at some stage the tank will be full?

FIGURE 7.60

15. A cylindrical oil tank of radius r and length h has its axis horizontal. If the density of the oil is ρ, find the force on each end of the tank when it is half full.

16. A right-circular cylinder of radius r and height h is immersed in a fluid of density ρ with its axis vertical. Show that the buoyant force on the cylinder due to the pressure of the fluid is equal to the weight of the fluid displaced. This is known as *Archimedes' principle*, and is valid for an object of any shape.

17. The lower half of a cubical tank 2 m on each side is occupied by water, and the upper half by oil (density 0.90 gm/cm^3).
  (a) What is the force on each side of the tank due to the pressure of the water and the oil?
  (b) If the oil and water are stirred to create a uniform mixture, does the force on each side change? If not explain why not. If so, by how much does it increase or decrease?

18. Archimedes' principle states that the buoyant force on an object when immersed or partially immersed in a fluid is equal to the weight of the fluid displaced by the object.
  (a) Show that if an object floats partially submerged in water, the percentage of the volume of the object above water is
  $$100\frac{\rho_w - \rho_o}{\rho_w},$$
  where ρ_o and ρ_w are the densities of the object and water respectively.
  (b) If the densities of ice and water are 915 and 1000 kg/m^3 respectively, show that only 8.5% of the volume of an iceberg is above water.

19. If a full tube of mercury is inverted in a large container of mercury, the level of mercury in the tube will fall, but it will stabilize at a point higher than that in the container (Figure 7.61). This is due to the fact that air pressure acts on the surface of the mercury in the container but not on the surface of the mercury in the tube. The extra column of mercury, of height h, creates a force at A that counteracts the atmospheric pressure transmitted through the mercury in the container to the tube so that the total pressure at A is equal to the total pressure at B.

FIGURE 7.61

(a) Show that if the density of mercury is 13.6 gm/mL, then the atmospheric pressure at the surface of the mercury in the container is $1.33 \times 10^5 h$ N/m², provided h is measured in metres.

(b) If h is measured as 761 millimetres, find the atmospheric pressure.

20. (a) A block of wood (density 0.40 gm/cm³) is cubical (0.25 m on each side). If it floats in water, how deep is its lowest point below the surface? Refer to Archimedes' principle in Exercise 18.

(b) Repeat part (a) given that the block is a sphere of radius 0.25 m.

21. A tank is to be built in the form of a right-circular cylinder with horizontal axis. The ends are to be joined to the cylindrical side by a continuous weld. One end of the tank and the cylindrical side are supported from the outside. The remaining unsupported end can withstand a total force of 40 000 N less 1000 N for each metre of weld on that end. What is the maximum radius for the tank if it is to hold a fluid weighing 1.019×10^3 N/m³?

22. Find the ratio L/R such that the forces due to fluid pressure on the rectangular and semicircular parts of the plate in Figure 7.62 are equal.

FIGURE 7.62

SECTION 7.6

Centres of Mass and Centroids

Everyone is acquainted with the action of a teeter-totter or seesaw. Two children of unequal masses can pass many hours rocking, provided the child with greater mass sits closer to the fulcrum. In this section we discuss the mathematics of the seesaw. This requires a definition of "moments of masses", and moments lead to the idea of the "centre of mass" of distributions of masses, lumped or continuous.

To discuss the mathematics of a seesaw, we consider in Figure 7.63 a uniform seesaw of length $2L$ balanced at its centre, with a child of mass m at one end. If a second child of equal mass is placed at the other end, the ideal seesaw situation is created. If, however, the mass of the second child is $M > m$, then this child must be placed somewhat closer to the fulcrum. To find the exact position, we must determine what might be called *rocking power* of a mass. A little experimentation shows that when $M = 2m$, M must be placed halfway between the end and the fulcrum; when $M = 3m$, M must be placed a distance $L/3$ from the fulcrum; and in general, when $M = am$ ($a > 1$), M must be placed L/a from the fulcrum. Now the rocking power of the child of mass m is constant, and for each mass $M = am$ we have found an equal and opposite rocking power if M is placed at L/a.

FIGURE 7.63

Clearly, then, rocking power depends on both mass and distance from the fulcrum. A little thought shows that the mathematical quantity that remains constant for the various masses $M = am$ is the product of M and distance to the fulcrum; in each case, this product is $(am)(L/a) = mL$, the same product as for the child of mass m. It would appear then, that rocking power should be defined as the product of mass and distance. We do this in the following definition, and at the same time give rocking power a new name.

Definition 7.1

The **first moment** of a point mass m about a point P is the product md where d is the directed distance from P to m.

If directed distances to the right of point P in Figure 7.64 are chosen as positive and distances to the left are negative, then d_1 is positive and d_2 is negative. Mass m_1 has a positive first moment $m_1 d_1$ about P, and m_2 has a negative first moment $m_2 d_2$.

FIGURE 7.64

In Figure 7.65, we have placed five children of masses m_1, m_2, m_3, m_4, and m_5 on the same seesaw. A sixth child of mass m_6 is to be placed somewhere on the seesaw so that all six children form the ideal seesaw.

FIGURE 7.65

To find the appropriate position for the sixth child, we let x be the directed distance from the fulcrum to the point where this child should be placed. The total first moment of all six children about the fulcrum, choosing distances to the right as positive and to the left as negative, is

$$m_1(-L) + m_2(-3L/4) + m_3(-L/2) + m_4(L/4) + m_5(L) + m_6(x).$$

We regard this as the resultant first moment of all six children attempting to turn the seesaw — clockwise if the moment is positive, counterclockwise if the moment is negative. Balance occurs if this resultant first moment is zero:

$$0 = -m_1 L - \frac{3}{4} m_2 L - \frac{1}{2} m_3 L + \frac{1}{4} m_4 L + m_5 L + m_6 x.$$

We may solve this equation for the position of m_6:

$$x = \frac{L}{4m_6}(4m_1 + 3m_2 + 2m_3 - m_4 - 4m_5).$$

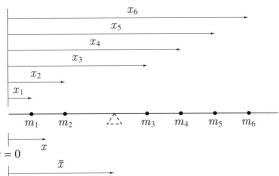

FIGURE 7.66

We now turn this problem around and place the six children at distances from the left end as shown in Figure 7.66. If the mass of the seesaw itself is neglected, where should the fulcrum be placed in order to create the ideal seesaw? To solve this problem, we let the distance that the fulcrum should be placed from the left end be represented by $\bar{x}$. In order for balance to occur, the total first moment of all six children about the fulcrum must vanish; hence,

$$0 = m_1(x_1 - \bar{x}) + m_2(x_2 - \bar{x}) + m_3(x_3 - \bar{x}) + m_4(x_4 - \bar{x}) + m_5(x_5 - \bar{x}) + m_6(x_6 - \bar{x}).$$

The solution of this equation is

$$\bar{x} = \frac{m_1 x_1 + m_2 x_2 + m_3 x_3 + m_4 x_4 + m_5 x_5 + m_6 x_6}{m_1 + m_2 + m_3 + m_4 + m_5 + m_6} = \frac{1}{M}\sum_{i=1}^{6} m_i x_i, \quad (7.25)$$

where $M = \sum_{i=1}^{6} m_i$ is the total mass of all six children. This point $\bar{x}$ at which the fulcrum creates a balancing position is called the *centre of mass* for the six children. It is a point where masses to the right are balanced by masses to the left.

In the remainder of this section we extend the idea of a centre of mass of point masses along a line (the seesaw) to the centre of mass of a distribution of point masses in a plane, and then to the centre of mass of a continuous distribution of mass. Our first step is to define the first moment of a point mass in a plane about a line in the plane (Figure 7.67).

FIGURE 7.67

> **Definition 7.2**
>
> The first moment of a mass m about a line l is md where d is the directed distance from l to m.

Once again directed distances are used in calculating first moments, and therefore distances on one side of the line must be chosen as positive and distances on the other side as negative. For vertical and horizontal lines, there is a natural convention for doing this. Distances to the right of a vertical line are chosen as positive, and distances to the left are negative. Distance upward from a horizontal line are positive, and distances downward are negative. In particular, when a mass m is located at position (x, y) in the xy-plane, its first moments about the x- and y-axes are my and mx, respectively.

The first moments of a system of n point masses $m_1, m_2, \ldots, m_n$ located at points $(x_1, y_1), (x_2, y_2), \ldots, (x_n, y_n)$ respectively (Figure 7.68(a)) about the x- and y-axes are defined as the sums of the first moments of the individual masses about these lines:

$$\text{First moment of system about } x\text{-axis} = \sum_{i=1}^{n} m_i y_i \quad \text{and} \quad (7.26\,\text{a})$$

$$\text{First moment of system about } y\text{-axis} = \sum_{i=1}^{n} m_i x_i. \quad (7.26\,\text{b})$$

FIGURE 7.68

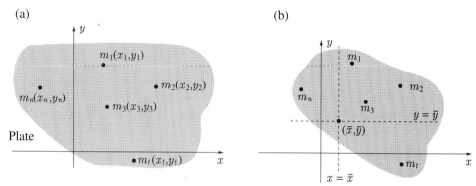

What is the physical meaning of these first moments? Do they, for instance, play the same role that first moments did for the seesaw? To see this we imagine that each point mass is embedded in a thin plastic plate in the xy-plane. The plate itself is massless and extends to include all n masses m_i. Picture now that the plate is horizontal, and a sharp edge is placed along the y-axis. Does the plate rotate about this edge, or does it balance? It is clear that the plate balances if the first moment of the system about the y-axis is equal to zero, and rotates otherwise, the direction depending on whether the first moment is positive or negative. Similarly, the plate balances on a sharp edge placed along the x-axis only if the first moment of the system about the x-axis vanishes. In general, the plate balances along any straight edge only if the first moment of the system about that edge vanishes.

We defined the centre of mass for a given distribution of children on the seesaw as the point at which to place the fulcrum in order to obtain balance. Analogously, we define the *centre of mass* $(\bar{x}, \bar{y})$ of the distribution of point masses in Figure 7.68(a) as the position to place a sharp point in order to obtain balance. Remember that the plastic itself is massless and only the point masses m_i can create moments. Balance occurs at a point $(\bar{x}, \bar{y})$ only if balance occurs about every straight line through $(\bar{x}, \bar{y})$. In particular, balance must occur about the lines $x = \bar{x}$ and $y = \bar{y}$ parallel to the y- and x-axes as in Figure 7.68(b). Since balance occurs about $x = \bar{x}$ only if the total first moment of the system about this line vanishes, we obtain the condition that

$$0 = m_1(x_1 - \bar{x}) + m_2(x_2 - \bar{x}) + \cdots + m_n(x_n - \bar{x}),$$

which can be solved for $\bar{x}$,

$$\bar{x} = \frac{1}{M} \sum_{i=1}^{n} m_i x_i, \quad (7.27)$$

where $M = \sum_{i=1}^{n} m_i$ is the total mass of the system. Note that this equation is identical to 7.25. Similarly, for balance about $y = \overline{y}$, we find that $\overline{y}$ must be

$$\overline{y} = \frac{1}{M} \sum_{i=1}^{n} m_i y_i. \tag{7.28}$$

We have obtained a unique point $(\overline{x}, \overline{y})$ based on conditions of balance about the lines $x = \overline{x}$ and $y = \overline{y}$. Does this necessarily imply that balance occurs about every straight line through $(\overline{x}, \overline{y})$? The answer is yes, and the proof is found in Exercise 34.

We have now proved that every planar point mass distribution has a centre of mass $(\overline{x}, \overline{y})$, where $\overline{x}$ and $\overline{y}$ are defined by equations 7.27 and 7.28. Our derivation has shown that the first moment of the system about any line through $(\overline{x}, \overline{y})$ must be equal to zero. Are there any other significant features of this point? To answer this question we note that if a fictitious particle of mass M (the total mass of the system) is located at the centre of mass $(\overline{x}, \overline{y})$, then its first moment about the y-axis is $M\overline{x}$. But from equation 7.27, we have

$$M\overline{x} = \sum_{i=1}^{n} m_i x_i, \tag{7.29}$$

and we conclude that the first moment of this fictitious particle M about the y-axis is exactly the same as the first moment of the system about the y-axis. Similarly, the first moment of M about the x-axis is $M\overline{y}$, and from 7.28,

$$M\overline{y} = \sum_{i=1}^{n} m_i y_i. \tag{7.30}$$

Thus, the centre of mass of a system of point masses m_i is a point at which a single particle of mass $M = \sum_{i=1}^{n} m_i$ has the same first moments about the x- and y-axes as the system. It can be shown further (Exercise 34) that the first moment of M about any line is the same as the first moment of the system about that line.

In summary, we defined the centre of mass of a system of point masses as a balance point. We found as a result that the centre of mass is a point at which a single particle of mass equal to the total mass of the system has the same first moment about any line as the system itself. This argument is reversible. Were we to define the centre of mass as a point to place the mass of the system for equivalent first moments, it would be a balance point. In other words, we have two equivalent definitions of the centre of mass of a system of point masses—a balance point or an equivalent point for first moments.

We now make the transition from a discrete system of particles to a continuous distribution of mass in the form of a thin plate of constant mass per unit area ρ (Figure 7.69). In order to find the mass of the plate, we proceed in exactly the same way that we did for areas. We divide the plate into vertical rectangles, the mass in a representative rectangles of width dx at position x being

$$\rho[f(x) - g(x)]\,dx.$$

To find the total mass of the plate, we add over all such rectangles of ever-diminishing widths to obtain

$$M = \int_a^b \rho[f(x) - g(x)]\,dx. \tag{7.31}$$

Based on our discussion for systems of point masses, we define the centre of mass of a continuous distribution of mass as that point $(\overline{x}, \overline{y})$ where a particle of mass M

has the same first moments about the x- and y-axes as the distribution. In algebraic terms, we note that $M\bar{x}$ is the first moment about the y-axis of a particle of mass M at $(\bar{x}, \bar{y})$. To this we must equate the first moment of the original distribution about the y-axis. Now each point in the representative rectangle in Figure 7.69 is approximately the same distance x from the y-axis — approximately, because the rectangle does have finite, though very small, width. The first moment, then, of this rectangle about the y-axis is approximately

$$x\rho[f(x) - g(x)]\, dx.$$

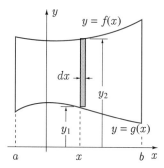

FIGURE 7.69

We find the first moment of the entire plate about the y-axis by adding first moments of all such rectangles and taking the limit as their widths approach zero. But once again this is the process defined by the definite integral, and we obtain, therefore, for the first moment of the plate about the y-axis

$$\int_a^b x\rho[f(x) - g(x)]\, dx.$$

Consequently,

$$M\bar{x} = \int_a^b x\rho[f(x) - g(x)]\, dx, \qquad (7.32)$$

and this equation can be solved for $\bar{x}$ once M and the integral on the right have been evaluated. Equation 7.32 represents for continuous distributions what equation 7.27 does for discrete distributions.

To find $\bar{y}$ we must equate the product $M\bar{y}$ to the first moment of the plate about the x-axis. If we consider the representative rectangle in Figure 7.69 we see that not all points therein are the same distance from the x-axis. To circumvent this problem, we consider all of the mass of the rectangle to be concentrated at its centre of mass. Since the centre of mass is the midpoint of the rectangle — a point distant $[f(x) + g(x)]/2$ from the x-axis, it follows that the first moment of this rectangle about the x-axis is

$$\frac{1}{2}[f(x) + g(x)]\rho[f(x) - g(x)]\, dx.$$

The total first moment of the plate about the x-axis is the definite integral of this expression, and we set

$$M\bar{y} = \int_a^b \frac{1}{2}[f(x) + g(x)]\rho[f(x) - g(x)]\, dx. \qquad (7.33)$$

This equation is used to evaluate $\bar{y}$.

Equations 7.32 and 7.33 can be memorized as formulas for $\bar{x}$ and $\bar{y}$, but it is far easier to perform the above operations mentally and arrive at these equations. Besides, for various shapes of plates, we might use horizontal rectangles or combinations of horizontal and vertical rectangles, and in such cases 7.32 and 7.33 would have to be modified.

On the basis of definitions 7.32 and 7.33 for $(\bar{x}, \bar{y})$, we can show that the first moment of M at $(\bar{x}, \bar{y})$ about any line is the same as the first moment of the plate about that same line. This implies that the plate balances along any line through $(\bar{x}, \bar{y})$ and therefore at $(\bar{x}, \bar{y})$.

EXAMPLE 7.21

Find the centre of mass of a thin plate of constant mass per unit area ρ if its edges are defined by the curves

$$y = 2 - x^2, \quad y = 0, \quad x = 0, \quad x \geq 0.$$

SOLUTION

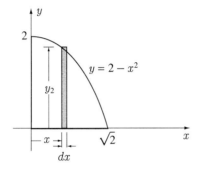

FIGURE 7.70

Using vertical rectangles (Figure 7.70), we find

$$M = \int_0^{\sqrt{2}} \rho y \, dx = \rho \int_0^{\sqrt{2}} (2 - x^2) \, dx = \rho \left\{ 2x - \frac{x^3}{3} \right\}_0^{\sqrt{2}} = \frac{4\sqrt{2}\rho}{3}.$$

If $(\bar{x}, \bar{y})$ is the centre of mass of the plate, then $M\bar{x}$ is the first moment of the single particle of mass M about the y-axis. This must be equated to the first moment of the plate about the y-axis. Since $x\rho(2 - x^2) \, dx$ is approximately the first moment of the rectangle in Figure 7.70 about the y-axis, the following integral gives the first moment of the plate about the y-axis,

$$\int_0^{\sqrt{2}} x\rho(2 - x^2) \, dx = \rho \int_0^{\sqrt{2}} (2x - x^3) \, dx = \rho \left\{ x^2 - \frac{x^4}{4} \right\}_0^{\sqrt{2}} = \rho.$$

Hence we set $M\bar{x} = \rho$, and solve this equation for $\bar{x}$,

$$\bar{x} = \frac{\rho}{M} = \rho \frac{3}{4\sqrt{2}\rho} = \frac{3}{4\sqrt{2}}.$$

To find $\bar{y}$, we calculate the first moment of the plate about the x-axis. Since the centre of mass of the rectangle in Figure 7.70 is $y/2$ units above the x-axis, it follows that the first moment of this rectangle about the x-axis is $(y/2)\rho y \, dx$. When we integrate this to find the first moment of the plate about the x-axis, and equate it to $M\bar{y}$, the result is

$$M\bar{y} = \int_0^{\sqrt{2}} \frac{y}{2}\rho y\, dx = \frac{\rho}{2}\int_0^{\sqrt{2}}(2-x^2)^2\, dx = \frac{\rho}{2}\int_0^{\sqrt{2}}(4-4x^2+x^4)\, dx$$

$$= \frac{\rho}{2}\left\{4x - \frac{4x^3}{3} + \frac{x^5}{5}\right\}_0^{\sqrt{2}} = \frac{16\sqrt{2}\rho}{15}.$$

Thus,
$$\bar{y} = \frac{16\sqrt{2}\rho}{15}\frac{3}{4\sqrt{2}\rho} = \frac{4}{5}.$$

EXAMPLE 7.22

Find the centre of mass of a thin plate of constant mass per unit area ρ if its edges are defined by the curves

$$y = 2x - x^2, \qquad y = x^2 - 4.$$

SOLUTION

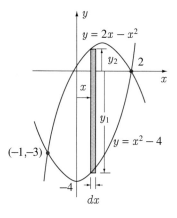

FIGURE 7.71

Using vertical rectangles (Figure 7.71),

$$M = \int_{-1}^{2} \rho(y_2 - y_1)\, dx = \rho\int_{-1}^{2}[(2x - x^2) - (x^2 - 4)]\, dx$$

$$= \rho\int_{-1}^{2}(4 + 2x - 2x^2)\, dx = \rho\left\{4x + x^2 - \frac{2x^3}{3}\right\}_{-1}^{2} = 9\rho.$$

If $(\bar{x}, \bar{y})$ is the centre of mass of the plate, then first moments of the plate and M about the y-axis give

$$M\bar{x} = \int_{-1}^{2} x\rho(y_2 - y_1)\, dx = \rho\int_{-1}^{2}(4x + 2x^2 - 2x^3)\, dx$$

$$= \rho\left\{2x^2 + \frac{2x^3}{3} - \frac{x^4}{2}\right\}_{-1}^{2} = \frac{9\rho}{2}.$$

Thus,
$$\bar{x} = \frac{9\rho}{2}\frac{1}{9\rho} = \frac{1}{2}.$$

To find $\bar{y}$, we use moments about the x-axis to write

$$M\bar{y} = \int_{-1}^{2} \frac{1}{2}(y_1 + y_2)\rho(y_2 - y_1)\, dx = \frac{\rho}{2}\int_{-1}^{2}(y_2^2 - y_1^2)\, dx$$

$$= \frac{\rho}{2}\int_{-1}^{2}[(2x - x^2)^2 - (x^2 - 4)^2]\, dx = 2\rho\int_{-1}^{2}(-x^3 + 3x^2 - 4)\, dx$$

$$= 2\rho\left\{-\frac{x^4}{4} + x^3 - 4x\right\}_{-1}^{2} = -\frac{27\rho}{2}.$$

Consequently,
$$\bar{y} = -\frac{27\rho}{2}\frac{1}{9\rho} = -\frac{3}{2}.$$

■

EXAMPLE 7.23

Find the first moments of a thin plate of constant mass per unit area ρ about the lines (a) $y = 0$, (b) $y = -2$, and (c) $x = -2$, if its edges are defined by the curves

$$x = |y|^3, \qquad x = 2 - y^2.$$

SOLUTION (a) Since the mass is distributed symmetrically about the x-axis (Figure 7.72), the first moment about $y = 0$ is zero.

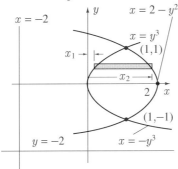

FIGURE 7.72

(b) Since the centre of mass of the plate is on the x-axis, its first moment about the line $y = -2$ is

$$(2)(\text{mass of plate}) = 2(2)(\text{mass of plate above the }x\text{-axis})$$

$$= 4\int_0^1 \rho(x_2 - x_1)\, dy = 4\rho\int_0^1(2 - y^2 - y^3)\, dy$$

$$= 4\rho\left\{2y - \frac{y^3}{3} - \frac{y^4}{4}\right\}_0^1$$

$$= \frac{17\rho}{3}.$$

(c) The first moment of the plate about the line $x = -2$ is twice the first moment of its upper half. Since the x-coordinate of the centre of mass of the representative rectangle in Figure 7.72 is $(x_1 + x_2)/2$, the distance from the line $x = -2$ to the centre of mass of the rectangle is $2 + (x_1 + x_2)/2$. The first moment of the horizontal rectangle about $x = -2$ is therefore

$$\left(2 + \frac{x_1 + x_2}{2}\right) \rho(x_2 - x_1)\, dy,$$

and the first moment of the plate is

$$2 \int_0^1 \left(2 + \frac{x_1 + x_2}{2}\right) \rho(x_2 - x_1)\, dy = \rho \int_0^1 [4(x_2 - x_1) + (x_2^2 - x_1^2)]\, dy$$

$$= 4\rho \int_0^1 (x_2 - x_1)\, dy$$

$$+ \rho \int_0^1 [(2 - y^2)^2 - (y^3)^2]\, dy$$

$$= \frac{17\rho}{3} + \rho \int_0^1 (4 - 4y^2 + y^4 - y^6)\, dy$$

$$= \frac{17\rho}{3} + \rho \left\{4y - \frac{4y^3}{3} + \frac{y^5}{5} - \frac{y^7}{7}\right\}_0^1$$

$$= \frac{881\rho}{105}.$$

∎

It has become apparent through our discussions and examples that in calculating the centre of mass of a thin plate with constant mass per unit area ρ, ρ is really unnecessary. As a constant it is taken out of each integration and cancels in the final division. The location of the centre of mass depends only on the geometric shape of the plate, and for this reason we could replace all references to mass by area. In particular, the mass of the plate M can be replaced by its area A, first moments (of mass) $M\bar{x}$ and $M\bar{y}$ can be replaced by first moments (of area) $A\bar{x}$ and $A\bar{y}$, and equations 7.32 and 7.33 then take the form

$$A\bar{x} = \int_a^b x[f(x) - g(x)]\, dx, \tag{7.34}$$

$$A\bar{y} = \int_a^b \frac{1}{2}[f(x) + g(x)][f(x) - g(x)]\, dx. \tag{7.35}$$

It is customary when using first moments of area to call $(\bar{x}, \bar{y})$ the centroid of the area rather than the centre of mass of the plate, simply because all references to mass have been deleted. We emphasize, however, that the statements in this paragraph apply only when mass per unit area is constant.

EXERCISES 7.6

In Exercise 1–5 find the centre of mass of the thin plate with constant mass per unit area.

1.

Parabolic

2.

Semiparabolic

3.

Parabolic spandrel

4.

Quarter-circle

5.

Semicircle

In Exercise 6–13 find the centroid of the area bounded by the curves.

6. $y = x^2 - 1$, $y = -x^2 - 2x - 1$
7. $y = \sqrt{|x|}$, $y = 2 - x^2$
8. $y = x^3$, $x = y^3$
9. $y = x$, $y = 2x$, $2y = x + 3$
10. $x = y^2 - 2y$, $x + y = 12$
11. $y = \sqrt{2-x}$, $x + y = 2$
12. $x = 4y - 4y^2$, $y = x - 3$, $y = 1$, $y = 0$
13. $x^3 y = 8$, $y = 9 - x^3$

14. The edges of a thin plate with constant mass per unit area ρ are defined by the curves
$$y = |x|^{1/2}, \quad y = x + 2, \quad y = 2 - x.$$
Find its first moment about the line $x = -5$.

15. Show that the centroids of areas A and B in Figure 7.73 have coordinates:
$$\overline{x}_A = \left(\frac{n+1}{n+2}\right)a, \quad \overline{y}_A = \left(\frac{n+1}{4n+2}\right)b,$$
$$\overline{x}_B = \left(\frac{n+1}{2n+4}\right)a, \quad \overline{y}_B = \left(\frac{n+1}{2n+1}\right)b.$$

FIGURE 7.73

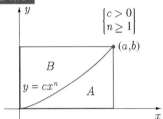

16. Find the centre of mass of the seesaw in Figure 7.66 if the mass of the seesaw is not neglected. Assume that it has uniform mass per unit length ρ and length $2L$.

In Exercises 17–19 find the centroid of the area bounded by the curves.

17. $x = \sqrt{y+2}$, $y = x$, $y = 0$
18. $y + x^2 = 0$, $x = y + 2$, $x + y + 2 = 0$, $y = 2$ (above $y + x^2 = 0$)
19. $y = \sqrt{2-x}$, $15y = x^2 - 4$

20. Find the centre of mass of the thin plate in Figure 7.74 if it has constant mass per unit area.

FIGURE 7.74

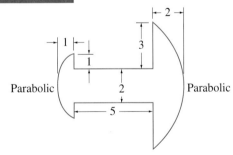

21. Find the first moment about the line $x + y = 1$ for the area defined by the curves
$$x + 2 = y^2, \quad y = x.$$

22. If an area A can be subdivided into n subareas A_i ($i = 1, \ldots, n$) such that the centroid of each A_i is $(\bar{x}_i, \bar{y}_i)$, show that the centroid $(\bar{x}, \bar{y})$ of A is given by

$$\bar{x} = \frac{1}{A}\sum_{i=1}^{n} A_i \bar{x}_i, \qquad \bar{y} = \frac{1}{A}\sum_{i=1}^{n} A_i \bar{y}_i.$$

In Exercises 23–26 use the technique suggested in Exercise 22 to find the centroid of the area.

23.

24.

25.

26.

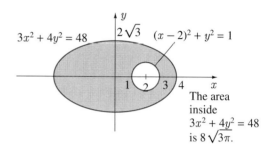

The area inside $3x^2 + 4y^2 = 48$ is $8\sqrt{3}\pi$.

In Exercises 27–30 find coordinates of the centroid of the area accurate to three decimal places.

27. $y = x^3 + 3x^2 + 2x + 1$, $x = 0$, $y = 0$

28. $y = x^4 - 5x^2 + 5$, $y = 0$ (above the x-axis)

29. $x = y^3 - y^2 - 2y$, $2y = x^2 - 1$

30. $y = x^3 - x$, $y = \sqrt{x}$

31. Prove the following *theorem of Pappus*: If a plane area is revolved about a coplanar axis not crossing the area, the volume generated is equal to the product of the area and the circumference of the circle described by the centroid of the area.

32. Use the result of Exercise 31 to find the volume of the donut in Exercise 41 of Section 7.2.

33. Use the result of Exercise 31 to find the volume in Exercise 45 of Section 7.2.

34. (a) Show that the first moment of the system of point masses in Figure 7.68(b) about any line is the same as the first moment of a point mass $M = \sum_{i=1}^{n} m_i$ at $(\bar{x}, \bar{y})$ about that line. Hint: See Exercise 45 in Section 1.3 for a formula for the distance from a point to a line.

(b) Does it follow that the first moment of the system about any line through $(\bar{x}, \bar{y})$ is zero?

35. A thin flat plate of area A is immersed vertically in a fluid of density ρ. Show that the total force due to fluid pressure on each side of the plate is equal to the product of 9.81ρ, A, and the depth of the centroid of the plate below the surface of the fluid.

36. Show that centroid of a triangle with vertices (x_1, y_1), (x_2, y_2), and (x_3, y_3) is

$$\bar{x} = \frac{x_1 + x_2 + x_3}{3}, \qquad \bar{y} = \frac{y_1 + y_2 + y_3}{3}.$$

SECTION 7.7

Moments of Inertia

Newton's second law $F = ma$ is fundamental to the study of translational motion of bodies. For rotational motion of bodies, its counterpart is $\tau = I\alpha$ where τ is torque, α is angular acceleration, and I is the "moment of inertia" of the body. The kinetic energy of a body of mass m moving with velocity v is $mv^2/2$. The kinetic energy of a body rotating with angular velocity ω is $I\omega^2/2$. Thus, for rotational motion, there is a quantity called the "moment of inertia" of a body which is analogous to mass in translational equations. In this section we define and calculate moments of inertia.

To define moments of inertia of bodies, we begin with the moment of inertia of a point mass.

Definition 7.3

The **moment of inertia** or **second moment** of a point mass m about a line l (Figure 7.67) is the product md^2 where d is the directed distance from l to m.

In particular, if m is at position (x, y) in the xy-plane, its moments of inertia about the x- and y-axes are my^2 and mx^2 respectively. For a system of n particles of masses m_1, m_2, ..., m_n located at points (x_1, y_1), (x_2, y_2), ..., (x_n, y_n) as in Figure 7.68(a), the moments of inertia of the system about the x- and y-axes are the sums of the moments of inertia of the particles about the x- and y-axes:

$$\text{Moment of inertia about } x\text{-axis} = \sum_{i=1}^{n} m_i y_i^2, \qquad (7.36\,\text{a})$$

$$\text{Moment of inertia about } y\text{-axis} = \sum_{i=1}^{n} m_i x_i^2. \qquad (7.36\,\text{b})$$

The transition from the discrete case to a continuous distribution in the form of a thin plate with constant mass per unit area ρ is not always so simple as for first moments. First consider the moment of inertia of the plate in Figure 7.69 in Section 7.6 about the y-axis. The mass of the representative rectangle shown is

$$\rho[f(x) - g(x)]\,dx,$$

and each point of the rectangle is approximately the same distance x from the y-axis. The moment of inertia, then, of this rectangle about the y-axis is approximately

$$x^2 \rho[f(x) - g(x)]\,dx.$$

The moment of inertia of the plate about the y-axis is found by adding moments of inertia of all such rectangles and taking the limit as their widths approach zero. But again this process defines a definite integral, and therefore the moment of inertia of the plate in Figure 7.69 about the y-axis is

$$I = \int_a^b x^2 \rho[f(x) - g(x)]\,dx. \qquad (7.37)$$

EXAMPLE 7.24

Find the moment of inertia about the y-axis of a thin plate with constant mass per unit area ρ if its edges are defined by the curves

$$y = x^3, \qquad y = \sqrt{2-x}, \qquad x = 0.$$

SOLUTION

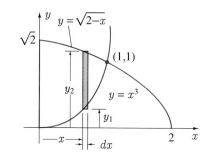

FIGURE 7.75

Since the moment of inertia of the vertical rectangle in Figure 7.75 is

$$x^2 \rho(y_2 - y_1)\, dx = \rho x^2 \left(\sqrt{2-x} - x^3\right) dx,$$

the moment of inertia of the plate is

$$I = \int_0^1 \rho x^2 \left(\sqrt{2-x} - x^3\right) dx = \rho \int_0^1 x^2 \sqrt{2-x}\, dx - \rho \int_0^1 x^5\, dx.$$

In the first integral we set $u = 2 - x$, in which case $du = -dx$, and

$$I = \rho \int_2^1 (2-u)^2 u^{1/2}(-du) - \rho \left\{\frac{x^6}{6}\right\}_0^1 = \rho \int_1^2 (4u^{1/2} - 4u^{3/2} + u^{5/2})\, du - \frac{\rho}{6}$$

$$= \rho \left\{\frac{8u^{3/2}}{3} - \frac{8u^{5/2}}{5} + \frac{2u^{7/2}}{7}\right\}_1^2 - \frac{\rho}{6} = \frac{256\sqrt{2} - 319}{210}\rho.$$

∎

EXAMPLE 7.25

Find the moment of inertia about the line $y = -1$ of a thin plate of constant mass per unit area ρ if its edges are defined by the curves

$$x = y^2, \qquad x = 2y.$$

SOLUTION Since the directed distance from the line $y = -1$ to all points in the horizontal rectangle in Figure 7.76 is approximately $y + 1$, the moment of inertia of the rectangle about the line $y = -1$ is approximately $(y+1)^2 \rho(x_2 - x_1)\, dy$. It follows that the moment of inertia of the plate is

$$I = \int_0^2 (y+1)^2 \rho(x_2 - x_1)\, dy = \rho \int_0^2 (y+1)^2 (2y - y^2)\, dy$$

$$= \rho \int_0^2 (-y^4 + 3y^2 + 2y)\, dy = \rho \left\{-\frac{y^5}{5} + y^3 + y^2\right\}_0^2 = \frac{28\rho}{5}.$$

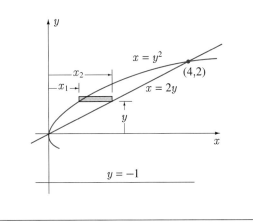

FIGURE 7.76

In Examples 7.24 and 7.25, and in the discussion leading to equation (7.37), we chose rectangles that had lengths parallel to the line about which we required the moment of inertia. This is not just coincidence; the use of perpendicular rectangles is more complicated. Consider, for instance, finding the moment of inertia about the x-axis of the plate in Example 7.24. To use the vertical rectangles in Figure 7.75, we first require the moment of inertia of such a rectangle about the x-axis. The mass of the rectangle, as in Example 7.24, must now be multiplied by the square of the distance from the x-axis to the rectangle. Unfortunately, different points in the rectangle are at different distances. One suggestion might be to concentrate all of the mass of the rectangle at its centre of mass and use the distance from the x-axis to the centre of mass. This is *incorrect*. The centre of mass is a point at which mass can be concentrated if we are discussing first moments. We are discussing second moments. (In Exercise 12 we show that the moment of inertia of the rectangle in Figure 7.77 cannot be obtained by concentrating its mass at its centre of mass.) What are we to do then? To use this type of rectangle, we must first develop a formula for its moment of inertia. To obtain this formula, let us consider the moment of inertia about the x-axis of the rectangle of width h and length $y_2 - y_1$ in Figure 7.77. If we subdivide this rectangle into smaller rectangles of width dy, the moment of inertia of the tiny rectangle about the x-axis is approximately

$$y^2 \rho h \, dy.$$

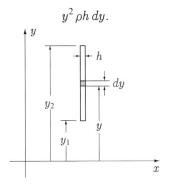

FIGURE 7.77

The moment of inertia of the long, vertical rectangle can be obtained by adding over all the tiny rectangles as their widths dy approach zero,

$$\int_{y_1}^{y_2} y^2 \rho h \, dy = \rho h \int_{y_1}^{y_2} y^2 \, dy = \frac{\rho h}{3}(y_2^3 - y_1^3). \quad (7.38)$$

We can use this formula to state that the moment of inertia about the x-axis of the vertical rectangle in Figure 7.75 is

$$\frac{\rho}{3}(y_2^3 - y_1^3) \, dx = \frac{\rho}{3}[(2-x)^{3/2} - x^9] \, dx.$$

The moment of inertia of the plate about the x-axis is therefore

$$I = \int_0^1 \frac{\rho}{3}[(2-x)^{3/2} - x^9]\,dx = \frac{\rho}{3}\left\{-\frac{2}{5}(2-x)^{5/2} - \frac{x^{10}}{10}\right\}_0^1 = \frac{(16\sqrt{2}-5)\rho}{30}.$$

The alternative procedure for this problem is to use horizontal rectangles which are parallel to the x-axis and obtain two definite integrals:

$$I = \int_0^1 y^2 \rho y^{1/3}\,dy + \int_1^{\sqrt{2}} y^2 \rho(2-y^2)\,dy.$$

In summary, we have two methods for determining moments of inertia of thin plates:

1. Choose rectangles parallel to the line about which the moment of inertia is required, in which case only the basic idea of a moment of inertia (mass times distance squared) is needed.

2. Choose rectangles perpendicular to the line about which the moment of inertia is required, in which case formula 7.38, or a similar formula, is needed.

    As for finding centres of mass, we could, in the special case of uniform mass distributions, drop all references to mass and talk about second moments of area about a line.

EXAMPLE 7.26

Find the moment of inertia about the line $x = 2$ of a plate with mass per unit area ρ if its edges are defined by the curves

$$x = 2y - y^2, \qquad x = y^2 - 2y.$$

SOLUTION

FIGURE 7.78

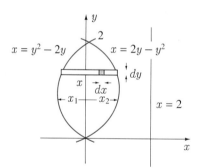

We first divide the plate into horizontal rectangles of width dy, and then subdivide this rectangle into smaller rectangles of width dx (Figure 7.78). Since the directed distance from the line $x = 2$ to the tiny rectangle is $x - 2$, the moment of inertia of the tiny rectangle about $x = 2$ is

$$(x-2)^2 \rho\,dy\,dx.$$

It follows that the moment of inertia of the long horizontal rectangle about $x = 2$ is

$$\int_{x_1}^{x_2} (x-2)^2 \rho\,dy\,dx = \rho\,dy\left\{\frac{1}{3}(x-2)^3\right\}_{x_1}^{x_2} = \frac{\rho}{3}[(x_2-2)^3 - (x_1-2)^3]\,dy.$$

The moment of inertia of the entire plate is now

$$I = \int_0^2 \frac{\rho}{3}[(x_2 - 2)^3 - (x_1 - 2)^3]\,dy$$

$$= \frac{\rho}{3}\int_0^2 [(2y - y^2 - 2)^3 - (y^2 - 2y - 2)^3]\,dy$$

$$= \frac{2\rho}{3}\int_0^2 (24y - 12y^2 + 8y^3 - 12y^4 + 6y^5 - y^6)\,dy$$

$$= \frac{2\rho}{3}\left\{12y^2 - 4y^3 + 2y^4 - \frac{12y^5}{5} + y^6 - \frac{y^7}{7}\right\}_0^2$$

$$= \frac{1184\rho}{105}.$$

∎

EXERCISES 7.7

In Exercises 1–10 the curves define a thin plate with constant mass per unit area ρ. Find its moment of inertia about the line.

1. $y = x^2$, $y = x^3$, about the y-axis
2. $y = x$, $y = 2x + 4$, $y = 0$, about the x-axis
3. $y = x^2$, $2y = x^2 + 4$, about $y = 0$
4. $y = x^2 - 4$, $y = 2x - x^2$, about $x = -2$
5. $xy\sqrt{y^2 + 12} = 1$, $x = 0$, $y = 1$, $y = 1/2$, about $y = 0$
6. $y = |x|^{1/3}$, $y = 2 - |x|^{1/3}$, about $x = 0$
7. $x + y = 2$, $y - x = 2$, $y = 0$, about $x = -2$
8. $x = 1 - y^2$, $x = y^2 - 1$, about $x = -1$
9. $x = 1 - y^2$, $x = y^2 - 1$, about $y = 1$
10. $x = y^2$, $x + y = 2$, about $y = 3$

11. (a) If I_{Ax} and I_{Bx} represent the second moments of area about the x-axis of the areas in Figure 7.73, show that
$$I_{Bx} = 3nI_{Ax}.$$
(b) Show that if I_{Ay} and I_{By} represent the second moments of area about the y-axis, then
$$nI_{Ay} = 3I_{By}.$$

12. What is the product of the mass of the rectangle in Figure 7.77 and the square of the distance from the x-axis to the centre of mass of the rectangle? Is it equal to the expression in equation 7.38?

13. The radius of gyration r of a thin plate with constant mass per unit area about a line is defined by $I = Mr^2$, where M is the mass of the plate and I is its moment of inertia about that line. Find radii of gyration about the x- and y-axes for the plate with edges
$$y = 2x^3, \qquad y + x^3 = 0, \qquad 2y = x + 3.$$
Explain the physical significance of r.

14. Show that the kinetic energy of a long-playing record is equal to one-half the product of the moment of inertia of the record about a line through its centre and perpendicular to its face, and the square of its angular speed.

15. Prove the *parallel axis theorem*: The moment of inertia of a thin plate (with constant mass per unit area) with respect to any coplanar line is equal to the moment of inertia with respect to the parallel line through the centre of mass plus the mass multiplied by the square of the distance between the lines.

16. If a line $x = \tilde{x}$ is drawn through the area in Figure 7.69, what integral represents the second moment of area about this line? What should be the value of $\tilde{x}$ for the smallest possible second moment?

17. The polar moment of inertia of a point mass m at (x, y) is defined as the product of m and the square of its distance from the origin
$$J_0 = m(x^2 + y^2).$$
For the thin plate (with constant mass per unit area) in Figure 7.79, let I_x and I_y be its moments of inertia about the x- and y-axes. Show that

$$J_0 = I_x + I_y.$$

FIGURE 7.79

In Exercises 18–20 the curves define a plate with mass per unit area equal to 2. Find its moment of inertia about the line accurate to three decimals.

18. $y = 1 - x^2$, $\quad x = y^2 \quad$ about $y = 0$

19. $y = x^3 - x$, $\quad y = \sqrt{x} \quad$ about $x = 0$

20. $y = x^3 - x$, $\quad y = \sqrt{x} \quad$ about $y = 0$

21. Find the second moment of area of a rectangle about its diagonal.

SECTION 7.8

Additional Applications

Volumes by Slicing

If we can represent the area of parallel cross-sections of a volume as a function of one variable, we can use a definite integral to calculate the volume. In particular, when we use the disc or washer method to determine the volume of a solid of revolution, parallel cross-sections are circles. In the following example, parallel cross-sections are squares.

EXAMPLE 7.27

A uniformly tapered rod of length 2 m has square cross-sections. If the areas of its ends are 4 and 16 cm² as in Figure 7.80(a), what is the volume of the rod?

FIGURE 7.80

(a)

(b)

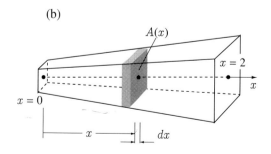

SOLUTION If we define an x-coordinate perpendicular to the square cross-sections as in Figure 7.80(b), then the side lengths of the cross-sections at $x = 0$ and $x = 2$ are 0.02 and 0.04 m respectively. Since the rod is uniformly tapered, the side length of the cross-section at x is

$$0.02 + x\left(\frac{0.04 - 0.02}{2}\right) = 0.02 + 0.01x = \frac{2 + x}{100}.$$

The area of the cross-section at x is therefore

$$A(x) = 10^{-4}(2 + x)^2.$$

If we construct at x a "slab" of cross-sectional area $A(x)$ and width dx, the volume of the slab is

$$A(x)\,dx = 10^{-4}(2 + x)^2\,dx.$$

To obtain the volume of the rod, we add the volumes of all such slabs between $x = 0$ and $x = 2$, and take the limit as their widths approach zero:

$$V = \int_0^2 A(x)\,dx = \int_0^2 10^{-4}(2 + x)^2\,dx$$

$$= 10^{-4}\left\{\frac{(2+x)^3}{3}\right\}_0^2 = \frac{10^{-4}}{3}(64 - 8) = \frac{56}{3} \times 10^{-4}\,\text{m}^3.$$

Area of a Surface of Revolution

If a curve in the xy-plane is rotated about the x- or y-axis (or a line parallel to the x- or y-axis) in order to produce a surface, we can calculate the area of this surface using the ideas of lengths along curves discussed in Section 7.3.

EXAMPLE 7.28

If that part of the parabola $y = x^2$ between $x = 0$ and $x = 1$ is rotated around the y-axis, find the area of the surface of revolution traced out by the curve (Figure 7.81).

FIGURE 7.81

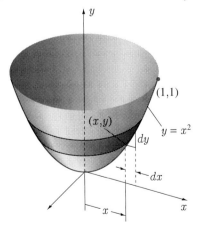

SOLUTION We can approximate length along the parabola corresponding to a change dx in x by the tangential straight-line length:

$$\sqrt{(dx)^2 + (dy)^2} = \sqrt{1 + \left(\frac{dy}{dx}\right)^2}\, dx = \sqrt{1 + (2x)^2}\, dx = \sqrt{1 + 4x^2}\, dx.$$

If this straight-line segment is rotated around the y-axis, each point follows a circular path of radius approximately equal to x; therefore, the area traced out by the line segment is approximately equal to

$$(2\pi x)\left(\sqrt{1 + 4x^2}\, dx\right).$$

We can find the total surface area by adding all such areas, and taking the limit as the widths dx approach zero:

$$A = \int_0^1 2\pi x\sqrt{1 + 4x^2}\, dx = 2\pi \left\{\frac{(1 + 4x^2)^{3/2}}{12}\right\}_0^1 = \frac{(5\sqrt{5} - 1)\pi}{6}.$$

■

EXAMPLE 7.29

Find the area of the surface of revolution traced out by rotating that part of the curve $y = x^3$ between $x = 1$ and $x = 2$ about the x-axis.

SOLUTION We can approximate length along the cubic corresponding to a change dx in x by the tangential straight-line length:

$$\sqrt{(dx)^2 + (dy)^2} = \sqrt{1 + \left(\frac{dy}{dx}\right)^2}\, dx = \sqrt{1 + (3x^2)^2}\, dx = \sqrt{1 + 9x^4}\, dx.$$

FIGURE 7.82

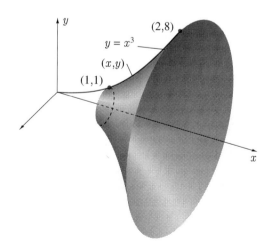

If this straight-line segment is rotated about the x-axis (Figure 7.82), each point follows a circular path of radius approximately equal to y; therefore, the area traced out by the line segment is approximately equal to

$$(2\pi y)\left(\sqrt{1+9x^4}\,dx\right).$$

By adding over all such areas,

$$A = \int_1^2 2\pi y\sqrt{1+9x^4}\,dx = 2\pi\int_1^2 x^3\sqrt{1+9x^4}\,dx$$

$$= 2\pi\left\{\frac{(1+9x^4)^{3/2}}{54}\right\}_1^2 = \frac{(145^{3/2}-10^{3/2})\pi}{27}.$$

■

Rates of Flow

In Problem 3 of Section 6.2 we considered laminar blood flow in a circular vessel. Specifically, the velocity of blood through the cross-section in Figure 7.83(a) is a function of radial distance r from the centre of the vessel:

$$v(r) = c(R^2 - r^2), \quad 0 \le r \le R,$$

where $c > 0$ is a constant and R is the radius of the vessel.

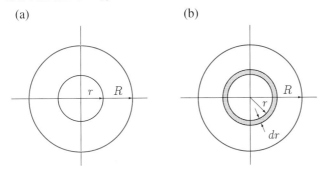

FIGURE 7.83

If we construct, at radius r, a thin ring of width dr as in Figure 7.83(b), then the area of this ring is approximately $(2\pi r)\,dr$. Since v does not vary greatly over this ring, the amount of blood flowing through the ring per unit time is approximately $v(r)$ multiplied by the area of the ring:

$$v(r)(2\pi r\,dr).$$

We can find the total flow through the blood vessel by adding the flows through all rings and taking the limit as the widths dr of the rings approach zero:

$$F = \int_0^R v(r)2\pi r\,dr = 2\pi\int_0^R rc(R^2-r^2)\,dr = 2\pi c\left\{\frac{R^2 r^2}{2}-\frac{r^4}{4}\right\}_0^R = \frac{\pi c R^4}{2}.$$

Repairs, Depreciation, and Replacement

In most businesses, depreciation and eventual replacement of equipment must be considered. Take, for example, a salesman who must decide when to replace his car. When he buys a new car, he knows that its value depreciates quickly at first and less rapidly after a few years. On the other hand, the longer he keeps the car, the more costly repairs become. The problem then is to decide on the optimum time to replace the car. Let us consider a situation in which the rate of depreciation of the car is $f(t)$, and the rate of accumulation of maintenance costs is $g(t)$. We have shown a possible $f(t)$ in Figure 7.84, where the ordinate at time t represents the number of dollars per year the car depreciates at that time. If we draw a rectangle of width dt at time t, then the area of this rectangle, $f(t)\,dt$, represents the amount the car depreciates during the time interval dt. Consequently, the total depreciation of the car from the time of purchase ($t = 0$) to time T is

$$\int_0^T f(t)\,dt.$$

In a similar way, the total maintenance cost from time $t = 0$ to $t = T$ (Figure 7.85) is

$$\int_0^T g(t)\,dt.$$

The cost associated with owning and maintaining the car over a period of T years is

$$\int_0^T f(t)\,dt + \int_0^T g(t)\,dt = \int_0^T [f(t) + g(t)]\,dt,$$

and the average yearly cost is

$$C(T) = \frac{1}{T}\int_0^T [f(t) + g(t)]\,dt.$$

FIGURE 7.84

FIGURE 7.85

The best time to replace the car is that value T which minimizes this average yearly cost. Critical points of $C(T)$ are defined by

$$0 = C'(T) = -\frac{1}{T^2}\int_0^T [f(t) + g(t)]\,dt + \frac{1}{T}\frac{d}{dT}\int_0^T [f(t) + g(t)]\,dt.$$

Equation 6.20 in Section 6.5 can be used to evaluate the derivative of the integral; the result is

$$0 = -\frac{1}{T^2}\int_0^T [f(t) + g(t)]\,dt + \frac{1}{T}[f(T) + g(T)], \qquad \text{or}$$

$$\frac{1}{T}\int_0^T [f(t) + g(t)]\,dt = f(T) + g(T).$$

Since the left side of this equation is $C(T)$, we can say that critical points of $C(T)$ occur when

$$C(T) = f(T) + g(T); \qquad (7.39)$$

that is, when average yearly costs are equal to the rate at which costs are changing. At one of these points — if there is more than one, $C(T)$ will take on its minimum value.

EXAMPLE 7.30

Find the optimum time to replace a car if rates of depreciation and accumulation of maintenance costs are, respectively,

$$f(t) = -\frac{V}{50}(t - 10) \quad \text{and} \quad g(t) = \frac{V}{400}t^2,$$

where V is the value of the new car. Assume that the car must be replaced by the time it has depreciated to no value.

SOLUTION Since the total depreciation of the car in T years is given by

$$\int_0^T f(t)\,dt = \int_0^T -\frac{V}{50}(t-10)\,dt = -\frac{V}{50}\left\{\frac{t^2}{2} - 10t\right\}_0^T = -\frac{V}{100}(T^2 - 20T),$$

it follows that the car completely depreciates when

$$V = -\frac{V}{100}(T^2 - 20T).$$

The only solution of this equation is $T = 10$ years. Do not make the mistake of assuming that complete depreciation occurs when $f(t) = 0$; this is not always true. Complete depreciation occurs when the area under $f(t)$ is equal to the initial value of the car.

The average yearly cost associated with the car if it is kept for T years is

$$\begin{aligned}
C(T) &= \frac{1}{T}\int_0^T [f(t) + g(t)]\,dt \\
&= \frac{1}{T}\int_0^T \left(-\frac{Vt}{50} + \frac{V}{5} + \frac{Vt^2}{400}\right)dt \\
&= \frac{1}{T}\left(-\frac{VT^2}{100} + \frac{VT}{5} + \frac{VT^3}{1200}\right) \\
&= -\frac{VT}{100} + \frac{V}{5} + \frac{VT^2}{1200}, \quad 0 < T \leq 10.
\end{aligned}$$

For critical point of $C(T)$, we solve

$$0 = C'(T) = -\frac{V}{100} + \frac{VT}{600},$$

the solution of which is $T = 6$. We could also have obtained $T = 6$ by solving equation 7.39. Since

$$\lim_{T \to 0^+} C(T) = \frac{V}{5}, \quad C(6) = \frac{17V}{100}, \quad C(10) = \frac{11V}{60},$$

it follows that the car should be replaced in 6 years. ∎

EXERCISES 7.8

1. Verify that the surface area of a sphere of radius r is $4\pi r^2$.

2. Find the area of the curved surface of a right-circular cone of radius r and height h.

3. Calculate the rate of flow of blood through a circular vessel of radius R if the velocity profile is
   (a) $v = f(r) = cR\sqrt{R^2 - r^2}$ and
   (b) $v = f(r) = (c/R^2)(R^2 - r^2)^2$.

4. Find the volume of the pyramid in Figure 7.86.

FIGURE 7.86

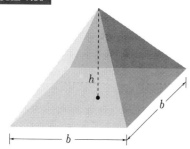

5. The amount of water consumed by a community varies throughout the day, peaking, naturally, around meal hours. During the six-hour period between 12:00 noon ($t = 0$) and 6:00 p.m. ($t = 6$), we find that the number of cubic metres of water consumed per hour at time t is given by the function

$$f(t) = 5000 + 21.65t^2 - 249.7t^3 + 97.52t^4 - 9.680t^5.$$

Find the total consumption during this 6-hour period.

6. The number of bees per unit area at a distance x from a hive is given by

$$\rho(x) = \frac{600\,000}{31\pi R^5}(R^3 + 2R^2 x - Rx^2 - 2x^3), \quad 0 \leq x \leq R,$$

where R is the maximum distance travelled by the bees.
   (a) What is the number of bees in the colony?
   (b) How many bees are within a distance $R/2$ of the hive?

7. If the radius of the blood vessel in Figure 7.83 is reduced to $R/2$ because of arteriosclerosis, the velocity profile is

$$v(r) = c(R^2 - 4r^2), \quad 0 \leq r \leq R/2.$$

What percentage of the normal flow ($\pi cR^4/2$) gets through the hardened vessel?

8. A tree trunk of diameter 50 cm (Figure 7.87) has a wedge cut from it by two planes. The lower plane is perpendicular to the axis of the trunk, and together the planes make an angle $\pi/3$ radians, meeting along a diameter of the circular cross-section of the trunk. Find the volume of the wedge.

FIGURE 7.87

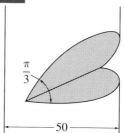

9. Find the area of the surface of revolution formed by rotating that part of the curve $24xy = x^4 + 48$ between $x = 1$ and $x = 2$ about the x-axis.

10. Find the area of the surface of revolution generated by rotating the curve $8y^2 = x^2(1 - x^2)$ about the x-axis.

11. Find the area of the surface of revolution generated by rotating the loop of the curve $9y^2 = x(3 - x)^2$ about (a) the y-axis (b) the x-axis.

12. The base of a solid is the circle $x^2 + y^2 = r^2$, and every plane section perpendicular to the x-axis is an isosceles triangle (Figure 7.88). Find the volume of the solid.

FIGURE 7.88

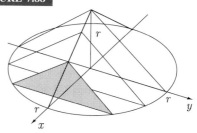

13. Electrons are fired from an electron gun at a target (Figure 7.89). The probability that an electron strikes the target in a ring of unit area at a distance x from the centre of the target is given by

$$p = f(x) = \frac{5}{3\pi R^5}(R^3 - x^3), \quad 0 \leq x \leq R.$$

What percentage of a cascade of electrons hits within a distance r from the centre of the target?

FIGURE 7.89

In Exercises 14-16, $f(t)$ and $g(t)$ represent rate of depreciation and rate of accumulation of maintenance costs for a piece of machinery. Sketch graphs of these functions, and show that the total area under $y = f(t)$ is equal to V (the original value of the machine). Determine the optimum time at which to replace the machine. Sketch a graph of $C(T)$ (the average yearly expense) to illustrate your answer.

14. $f(t) = \begin{cases} \dfrac{V}{10} - \dfrac{Vt}{200} & 0 < t \le 20 \\ 0 & t > 20 \end{cases}$ $g(t) = \dfrac{Vt^2}{4000}$

15. $f(t) = \begin{cases} -\dfrac{V}{50}(t - 10) & 0 < t \le 10 \\ 0 & t > 10 \end{cases}$ $g(t) = \dfrac{Vt}{400}$

16. $f(t) = \begin{cases} \dfrac{V}{10} - \dfrac{6Vt^2}{15\,625} & 0 < t \le 12.5 \\ 0 & t > 12.5 \end{cases}$ $g(t) = \dfrac{Vt^2}{3000}$

17. Two right-circular cylinders, each of radius r, have axes that intersect at right angles. Find the volume common to the two cylinders.

18. What is the volume of air trapped in the attic of a house if the roof has shape shown in Figure 7.90? The peak of the roof is 2 m above the base.

FIGURE 7.90

19. Suppose that at time $t = 0$ the number of people in a certain population is N_0, and the probability that each individual lives to age t is $p(t)$. For time $t > 0$, the birth rate is $r(t)$ individuals per unit time. Show that at time T the number of people in the population is

$$N(T) = N_0 p(T) + \int_0^T p(T - t) r(t)\, dt.$$

20. The depth of water in the container in Figure 7.91 is originally 10 cm. Water evaporates from the surface at a rate proportional to the area of the surface. If the water level drops 1 cm in 5 days, how long does it take for the water to evaporate completely? Note that Exercise 9 in Section 5.4 is this same problem given a container that is a right-circular cone. Now we ask you to repeat the problem with no knowledge of the shape of the container.

FIGURE 7.91

21. An appliance retailer must be concerned with her inventory costs. For example, given that she sells N refrigerators per year, she must decide whether to order the year's supply at one time, in which case a large storage area would be necessary, or to make periodic orders throughout the year, perhaps of $N/12$ at the first of each month. In the latter case she would incur costs due to paper work, delivery charges, etc. Let us suppose that the retailer decides to order in equal-lot sizes, x, at equally spaced intervals, N/x times per year.

(a) If each order has fixed costs of F dollars, plus f dollars for each refrigerator, what are the total yearly ordering costs?

(b) Between successive deliveries, the retailer's stock dwindles from x to 0. If she assumes that the number of refrigerators in stock decreases linearly in time, and the yearly stocking cost per refrigerator is p dollars, what are the yearly stocking costs?

(c) If the retailer's total yearly inventory costs are her ordering costs in part (a) plus her stocking costs in part (b), what value of x minimzes inventory costs?

SECTION 7.9

Improper Integrals

According to the fundamental theorem of integral calculus (Theorem 6.3), definite integrals are evaluated with indefinite integrals,

$$\int_a^b f(x)\,dx = \left\{\int f(x)\,dx\right\}_a^b,$$

provided $f(x)$ is continuous on $a \le x \le b$. In this section we investigate what to do when $f(x)$ is not continuous on $a \le x \le b$, or when either a or b is infinite.

Consider whether a reasonable meaning can be given to the integrals

$$\int_1^\infty \frac{1}{x^2}\,dx \quad \text{and} \quad \int_1^\infty \frac{1}{\sqrt{x}}\,dx.$$

Both integrals are "improper" in the sense that their upper limits are not finite, and we therefore call them **improper integrals**. If $b > 1$, there is no difficulty with the evaluation and possible interpretation of either

$$\int_1^b \frac{1}{x^2}\,dx \quad \text{or} \quad \int_1^b \frac{1}{\sqrt{x}}\,dx.$$

Clearly,

$$\int_1^b \frac{1}{x^2}\,dx = \left\{-\frac{1}{x}\right\}_1^b = 1 - \frac{1}{b} \quad \text{and}$$

$$\int_1^b \frac{1}{\sqrt{x}}\,dx = \left\{2\sqrt{x}\right\}_1^b = 2\sqrt{b} - 2.$$

We can interpret the first integral as the area under the curve $y = 1/x^2$, above the x-axis, and between the vertical lines $x = 1$ and $x = b$ (Figure 7.92). The second integral has exactly the same interpretation but uses the curve $y = 1/\sqrt{x}$ in place of $y = 1/x^2$ (Figure 7.93).

FIGURE 7.92

FIGURE 7.93

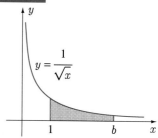

In Table 7.1 we have listed values of these definite integrals corresponding to various values of b. It is clear that the two integrals display completely different characteristics. As b is made very large, the integral of $1/x^2$ is always less than 1, but gets closer to 1 as b increases. In other words,

$$\lim_{b \to \infty} \int_1^b \frac{1}{x^2}\,dx = 1.$$

Geometrically, the area in Figure 7.92 is always less than 1 but approaches 1 as b approaches infinity.

Contrast this with the integral of $1/\sqrt{x}$, which becomes indefinitely large as b increases:

$$\lim_{b \to \infty} \int_1^b \frac{1}{\sqrt{x}}\,dx = \infty;$$

that is, the area in Figure 7.93 can be made as large as desired by choosing b sufficiently large.

TABLE 7.1

b	$\int_1^b \frac{1}{x^2}\,dx$	$\int_1^b \frac{1}{\sqrt{x}}\,dx$
100	0.99	18
10 000	0.9999	198
1 000 000	0.999999	1 998
100 000 000	0.99999999	19 998

On the basis of these calculations, we would like to say that the improper integral of $1/x^2$ has value 1, whereas the improper integral of $1/\sqrt{x}$ has no value. In other words, if these improper integrals are to have values, they should be defined by the limits

$$\int_1^\infty \frac{1}{x^2}\,dx = \lim_{b \to \infty} \int_1^b \frac{1}{x^2}\,dx \quad \text{and} \quad \int_1^\infty \frac{1}{\sqrt{x}}\,dx = \lim_{b \to \infty} \int_1^b \frac{1}{\sqrt{x}}\,dx.$$

In the second case the limit does not exist, and we interpret this to mean that the improper integral does not exist or has no value.

In general, then, improper integrals are defined in terms of limits as follows.

Definition 7.4

If $f(x)$ is continuous for $x \geq a$, we define

$$\int_a^\infty f(x)\,dx = \lim_{b \to \infty} \int_a^b f(x)\,dx, \qquad (7.40)$$

provided the limit exists.

If the limit exists, we say that the improper integral is equal to the limit, or that the improper integral *converges* to the limit. If the limit does not exist, we say that the improper integral has no value, or that it *diverges*.

EXAMPLE 7.31

Determine whether the following improper integrals converge or diverge:

(a) $\displaystyle\int_3^\infty \frac{1}{(x-2)^3}\,dx$ (b) $\displaystyle\int_0^\infty \frac{x^2}{\sqrt{1+x^3}}\,dx$ (c) $\displaystyle\int_1^\infty \frac{2x^2+6}{3x^2+5}\,dx$

SOLUTION (a) Using equation 7.40,

$$\int_3^\infty \frac{1}{(x-2)^3}\,dx = \lim_{b\to\infty} \int_3^b \frac{1}{(x-2)^3}\,dx = \lim_{b\to\infty}\left\{\frac{-1}{2(x-2)^2}\right\}_3^b$$

$$= \lim_{b\to\infty}\left[\frac{1}{2} - \frac{1}{2(b-2)^2}\right] = \frac{1}{2},$$

and the improper integral therefore converges to $\frac{1}{2}$.

(b) Once again 7.40 gives

$$\int_0^\infty \frac{x^2}{\sqrt{1+x^3}}\,dx = \lim_{b\to\infty}\int_0^b \frac{x^2}{\sqrt{1+x^3}}\,dx = \lim_{b\to\infty}\left\{\frac{2}{3}\sqrt{1+x^3}\right\}_0^b$$

$$= \lim_{b\to\infty}\left[\frac{2}{3}\sqrt{1+b^3} - \frac{2}{3}\right] = \infty,$$

and the improper integral therefore diverges.

(c) Although we cannot at this time find an antiderivative for $f(x) = (2x^2 + 6)/(3x^2 + 5)$, we can solve this problem by interpreting the improper integral as an area. The graph of $f(x)$ in Figure 7.94 indicates that the curve is asymptotic to the line $y = \frac{2}{3}$. Clearly, then, the improper integral

$$\int_1^\infty \frac{2x^2 + 6}{3x^2 + 5}\,dx$$

can have no value since the area under the curve is larger than the area of a rectangle of width $\frac{2}{3}$ and infinite length.

FIGURE 7.94

We can define improper integrals with infinite lower limits in exactly the same way as for improper integrals with infinite upper limits. Specifically, if $f(x)$ is continuous for $x \leq b$, we define

$$\int_{-\infty}^b f(x)\,dx = \lim_{a\to-\infty}\int_a^b f(x)\,dx, \tag{7.41}$$

provided the limit exists. For example,

$$\int_{-\infty}^5 \frac{1}{(x-6)^5}\,dx = \lim_{a\to-\infty}\int_a^5 \frac{1}{(x-6)^5}\,dx = \lim_{a\to-\infty}\left\{\frac{-1}{4(x-6)^4}\right\}_a^5$$

$$= \lim_{a\to-\infty}\left[-\frac{1}{4} + \frac{1}{4(a-6)^4}\right] = -\frac{1}{4}.$$

When $f(x)$ is continuous for all x, we define

$$\int_{-\infty}^\infty f(x)\,dx = \lim_{a\to-\infty}\int_a^c f(x)\,dx + \lim_{b\to\infty}\int_c^b f(x)\,dx, \tag{7.42}$$

provided both limits exist. The number c is arbitrary; existence of the limits is independent of c (see Exercise 33). For example, if we choose $c = 0$, then

$$\int_{-\infty}^{\infty} \frac{x}{(1+x^2)^2}\,dx = \lim_{a \to -\infty} \int_a^0 \frac{x}{(1+x^2)^2}\,dx + \lim_{b \to \infty} \int_0^b \frac{x}{(1+x^2)^2}\,dx$$

$$= \lim_{a \to -\infty} \left\{ \frac{-1}{2(1+x^2)} \right\}_a^0 + \lim_{b \to \infty} \left\{ \frac{-1}{2(1+x^2)} \right\}_0^b$$

$$= \lim_{a \to -\infty} \left[\frac{1}{2(1+a^2)} - \frac{1}{2} \right] + \lim_{b \to \infty} \left[\frac{1}{2} - \frac{1}{2(1+b^2)} \right]$$

$$= -\frac{1}{2} + \frac{1}{2} = 0.$$

EXAMPLE 7.32

(a) Is it possible to assign a number to represent the area bounded by the curves $y = 1/x$, $x = 1$, and $y = 0$?

(b) If the region in (a) is rotated around the x-axis, is it possible to assign a number to represent the volume of the solid so generated?

SOLUTION (a) If a number can be assigned to the area of the region bounded by the curves (Figure 7.95), it must be defined by the improper integral

$$A = \int_1^{\infty} \frac{1}{x}\,dx = \lim_{b \to \infty} \int_1^b \frac{1}{x}\,dx = \lim_{b \to \infty} \{\ln|x|\}_1^b = \lim_{b \to \infty} \ln b = \infty.$$

Consequently, we cannot assign an area to the region.

(b) If we use circular discs to calculate the volume in the solid formed by rotating the region in Figure 7.95 around the x-axis, then

$$V = \int_1^{\infty} \pi y^2\,dx = \lim_{b \to \infty} \int_1^b \frac{\pi}{x^2}\,dx = \lim_{b \to \infty} \left\{ \frac{-\pi}{x} \right\}_1^b = \lim_{b \to \infty} \left(\pi - \frac{\pi}{b} \right) = \pi.$$

Thus we should assign to the solid a volume of π cubic units. Does this bother you? The area in Figure 7.95 bounded by the curves $y = 1/x$ the x-axis, and vertical lines at $x = 1$ and $x = b$ can be made larger than any number whatsoever by choosing b sufficiently large. On the other hand, when this area is revolved around the x-axis, no matter how large b is, the volume of the solid of revolution is never greater than π. This may seem to defy your intuition at first, but think about it very carefully.

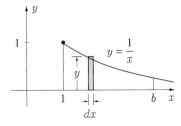

FIGURE 7.95

SECTION 7.9 IMPROPER INTEGRALS

EXAMPLE 7.33

Find the escape velocity of a projectile from the earth's surface.

FIGURE 7.96

Earth

SOLUTION Suppose the projectile is fired from the earth's surface with speed v in the x-direction, as shown in Figure 7.96. If the projectile is to escape the earth's gravitational pull, its initial kinetic energy must be greater than or equal to the work done against gravity as the projectile travels from the earth's surface ($x = R$) to a point where it is free of gravity ($x = \infty$). When the projectile is a distance x from the centre of the earth, the force of attraction on it is

$$F(x) = -\frac{GMm}{x^2},$$

where $G > 0$ is a constant, m is the mass of the projectile, and M is the mass of the earth. This is known as *Newton's universal law of gravitation*. The work done against gravity in travelling from $x = R$ to $x = \infty$ is therefore

$$W = \int_R^\infty \frac{GmM}{x^2}\,dx = \lim_{b \to \infty} \int_R^b \frac{GmM}{x^2}\,dx$$

$$= \lim_{b \to \infty} \left\{-\frac{GmM}{x}\right\}_R^b = \lim_{b \to \infty} \left[\frac{GmM}{R} - \frac{GmM}{b}\right] = \frac{GmM}{R}.$$

Since the initial kinetic energy of the projectile is $mv^2/2$, it escapes the gravitational pull of the earth if

$$\frac{1}{2}mv^2 \geq \frac{GmM}{R};$$

that is, if

$$v \geq \sqrt{\frac{2GM}{R}}.$$

Hence $\sqrt{2GM/R}$ is the escape velocity (and notice that it is independent of the mass of the projectile). If we take the mean radius of the earth as 6370 km, its mean density as $\rho = 5.52 \times 10^3$ kg/m^3, and $G = 6.67 \times 10^{-11}$, we obtain $v = 11.2$ km/s. (Compare this with a 308 Winchester that fires a 150-grain bullet with a muzzle velocity of only 0.81 km/s.) ∎

A second type of improper integral occurs when the integrand is discontinuous at a point or points in the interval of integration. For example, integrands of the improper integrals

$$\int_1^5 \frac{1}{\sqrt{x-1}}\,dx \quad \text{and} \quad \int_1^5 \frac{1}{(x-1)^4}\,dx$$

each have infinite discontinuities at $x = 1$. If c is a number between 1 and 5, it is easy to calculate

$$\int_c^5 \frac{1}{\sqrt{x-1}}\,dx = \left\{2\sqrt{x-1}\right\}_c^5 = 4 - 2\sqrt{c-1} \quad \text{and}$$

$$\int_c^5 \frac{1}{(x-1)^4}\,dx = \left\{\frac{-1}{3(x-1)^3}\right\}_c^5 = \frac{1}{3(c-1)^3} - \frac{1}{192},$$

and one possible interpretation of these definite integrals is the areas in Figures 7.97 and 7.98.

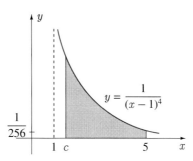

FIGURE 7.97   FIGURE 7.98

If we let c approach 1 from the right in the first integral, we find that

$$\lim_{c \to 1^+} \int_c^5 \frac{1}{\sqrt{x-1}}\,dx = \lim_{c \to 1^+} \left(4 - 2\sqrt{c-1}\right) = 4;$$

and in the second integral,

$$\lim_{c \to 1^+} \int_c^5 \frac{1}{(x-1)^4}\,dx = \lim_{c \to 1^+} \left[\frac{1}{3(c-1)^3} - \frac{1}{192}\right] = \infty.$$

It would seem reasonable then to define the area under the curve $y = 1/\sqrt{x-1}$ above the x-axis, and between the vertical lines $x = 1$ and $x = 5$ as 4 square units. On the other hand, the area bounded by $y = 1/(x-1)^4$, $y = 0$, $x = 5$, and $x = c$ becomes increasingly large as $x = c$ moves closer to $x = 1$, and we cannot therefore assign a value to this area when $c = 1$. In other words, if we define

$$\int_1^5 \frac{1}{\sqrt{x-1}}\,dx = \lim_{c \to 1^+} \int_c^5 \frac{1}{\sqrt{x-1}}\,dx \quad \text{and}$$

$$\int_1^5 \frac{1}{(x-1)^4}\,dx = \lim_{c \to 1^+} \int_c^5 \frac{1}{(x-1)^4}\,dx,$$

then the first improper integral has a value of 4, whereas the second has no value.

These improper integrals are examples of the general situation described in the following definition and illustrated in Figure 7.99.

(a) (b) (c)   FIGURE 7.99

Definition 7.5

If $f(x)$ is continuous at every point in the interval $a \leq x \leq b$ except at $x = a$, then

$$\int_a^b f(x)\, dx = \lim_{c \to a^+} \int_c^b f(x)\, dx, \tag{7.43}$$

provided the limit exists (Figure 7.99(a)). If $f(x)$ is continuous at every point in the interval $a \leq x \leq b$ except $x = b$, then

$$\int_a^b f(x)\, dx = \lim_{c \to b^-} \int_a^c f(x)\, dx, \tag{7.44}$$

provided the limit exists (Figure 7.99(b)). If $f(x)$ is continuous at every point in the interval $a \leq x \leq b$ except $x = d$ where $a < d < b$, then

$$\int_a^b f(x)\, dx = \lim_{c \to d^-} \int_a^c f(x)\, dx + \lim_{c \to d^+} \int_c^b f(x)\, dx, \tag{7.45}$$

provided both limits exist (Figure 7.99(c)).

EXAMPLE 7.34

Determine whether the following improper integrals converge or diverge:

(a) $\displaystyle\int_{-2}^{2} \frac{1}{(x-2)^3}\, dx$ (b) $\displaystyle\int_{-1}^{1} \frac{1}{x^4}\, dx$ (c) $\displaystyle\int_{1}^{5} \frac{x}{\sqrt{x-1}}\, dx$

FIGURE 7.100

FIGURE 7.101

FIGURE 7.102

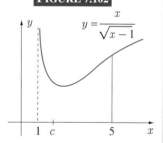

SOLUTION

(a) The graph of the integrand (Figure 7.100) shows a discontinuity at $x = 2$. Hence,

$$\int_{-2}^{2} \frac{1}{(x-2)^3}\, dx = \lim_{c \to 2^-} \int_{-2}^{c} \frac{1}{(x-2)^3}\, dx = \lim_{c \to 2^-} \left\{ \frac{-1}{2(x-2)^2} \right\}_{-2}^{c}$$

$$= \lim_{c \to 2^-} \left[\frac{-1}{2(c-2)^2} + \frac{1}{32} \right] = -\infty.$$

The improper integral therefore diverges.

(b) Since the discontinuity is interior to the interval of integration (Figure 7.101), we use equation 7.45,

$$\int_{-1}^{1} \frac{1}{x^4}\,dx = \lim_{c \to 0^-} \int_{-1}^{c} \frac{1}{x^4}\,dx + \lim_{c \to 0^+} \int_{c}^{1} \frac{1}{x^4}\,dx$$

$$= \lim_{c \to 0^-} \left\{ \frac{-1}{3x^3} \right\}_{-1}^{c} + \lim_{c \to 0^+} \left\{ \frac{-1}{3x^3} \right\}_{c}^{1}$$

$$= \lim_{c \to 0^-} \left[-\frac{1}{3} - \frac{1}{3c^3} \right] + \lim_{c \to 0^+} \left[\frac{1}{3c^3} - \frac{1}{3} \right].$$

Since neither of these limits exists, the improper integral diverges.

(c) Since the integrand is discontinuous only at $x = 1$ (Figure 7.102),

$$\int_{1}^{5} \frac{x}{\sqrt{x-1}}\,dx = \lim_{c \to 1^+} \int_{c}^{5} \frac{x}{\sqrt{x-1}}\,dx.$$

If we set $u = x - 1$, then $du = dx$, and

$$\int \frac{x}{\sqrt{x-1}}\,dx = \int \frac{u+1}{\sqrt{u}}\,du = \int (u^{1/2} + u^{-1/2})\,du$$

$$= \frac{2}{3} u^{3/2} + 2 u^{1/2} + C = \frac{2}{3}(x-1)^{3/2} + 2\sqrt{x-1} + C.$$

Thus,

$$\int_{1}^{5} \frac{x}{\sqrt{x-1}}\,dx = \lim_{c \to 1^+} \left\{ \frac{2}{3}(x-1)^{3/2} + 2\sqrt{x-1} \right\}_{c}^{5}$$

$$= \lim_{c \to 1^+} \left[\frac{28}{3} - \frac{2}{3}(c-1)^{3/2} - 2\sqrt{c-1} \right] = \frac{28}{3},$$

and the improper integral converges. ∎

Various types of improper integrals may occur in the same problem. For example, Figure 7.103 shows that the integral of $f(x) = 1/(x^2 - 1)$ from $x = -3$ to $x = \infty$ involves the use of five limits:

$$\int_{-3}^{\infty} \frac{1}{x^2 - 1}\,dx = \lim_{c \to -1^-} \int_{-3}^{c} \frac{1}{x^2 - 1}\,dx + \lim_{c \to -1^+} \int_{c}^{0} \frac{1}{x^2 - 1}\,dx + \lim_{c \to 1^-} \int_{0}^{c} \frac{1}{x^2 - 1}\,dx$$

$$+ \lim_{c \to 1^+} \int_{c}^{10} \frac{1}{x^2 - 1}\,dx + \lim_{b \to \infty} \int_{10}^{b} \frac{1}{x^2 - 1}\,dx.$$

If just one of these limits fails to exist, then the improper integral diverges.

If the integrand has a finite discontinuity, such as the function $f(x)$ in Figure 7.104, then the improper integral of $f(x)$ from $x = a$ to $x = b$ is defined in terms of two limits,

$$\int_{a}^{b} f(x)\,dx = \lim_{c \to d^-} \int_{a}^{c} f(x)\,dx + \lim_{c \to d^+} \int_{c}^{b} f(x)\,dx,$$

but there is no question in this case that the improper integral converges.

FIGURE 7.103

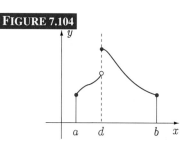

FIGURE 7.104

EXERCISES 7.9

In Exercises 1–18 determine whether the improper integral converges or diverges. Find the value for each convergent integral.

1. $\int_3^\infty \dfrac{1}{(x+4)^2}\,dx$
2. $\int_3^\infty \dfrac{1}{(x+4)^{1/3}}\,dx$
3. $\int_{-\infty}^{-4} \dfrac{x}{\sqrt{x^2-2}}\,dx$
4. $\int_{-\infty}^{-4} \dfrac{x}{(x^2-2)^4}\,dx$
5. $\int_{-\infty}^\infty \dfrac{10^{10}\,x^3}{(x^4+5)^2}\,dx$
6. $\int_{-\infty}^\infty \dfrac{x^3}{(x^4+5)^{1/4}}\,dx$
7. $\int_0^1 \dfrac{1}{(1-x)^{5/3}}\,dx$
8. $\int_0^1 \dfrac{1}{\sqrt{1-x}}\,dx$
9. $\int_1^\infty x\sqrt{x^2-1}\,dx$
10. $\int_2^5 \dfrac{x}{\sqrt{x^2-4}}\,dx$
11. $\int_{-1}^1 \dfrac{x}{(1-x^2)^2}\,dx$
12. $\int_{-\infty}^\infty \dfrac{1}{x^2}\,dx$
13. $\int_0^\infty \dfrac{1}{\sqrt{x}}\,dx$
14. $\int_{-\infty}^{\pi/2} \dfrac{x}{(x^2-4)^2}\,dx$
15. $\int_4^\infty \cos x\,dx$
16. $\int_{-\infty}^\infty \sin x\,dx$
17. $\int_0^\infty \dfrac{x}{\sqrt{x+3}}\,dx$
18. $\int_2^3 \dfrac{x^3}{\sqrt{x^2-4}}\,dx$

19. If $f(x)$ is a continuous odd function, is it necessarily true that
$$\int_{-\infty}^\infty f(x)\,dx = 0?$$

20. (a) Is it possible to assign a number to represent the area bounded by the curves
$$y = 1 - x^{-1/4}, \quad y = 1, \quad x = 1 ?$$
(b) If the region in (a) is rotated about the line $y = 1$, is it possible to assign a number to represent its volume?

21. Repeat Exercise 20 if $y = 1 - x^{-1/4}$ is replaced by $y = 1 - x^{-2/3}$.

22. Repeat Exercise 20 if $y = 1 - x^{-1/4}$ is replaced by $y = 1 - x^{-3}$.

23. A function $f(x)$ qualifies as a *probability density function* (pdf) on the interval $x \geq 0$ if it satisfies two conditions: $f(x) \geq 0$ for $x \geq 0$, and
$$\int_0^\infty f(x)\,dx = 1.$$

(a) Show that each of the following functions qualifies as a pdf:
$$f(x) = \dfrac{6x}{(1+3x^2)^2}; \qquad f(x) = \dfrac{2x}{(1+x)^3}$$

(b) If a variable x has pdf $f(x)$ defined for $x \geq 0$, then the probability that x lies in an interval I is the definite integral of $f(x)$ over I. In particular, the probability that x is greater than or equal to a is
$$P(x \geq a) = \int_a^\infty f(x)\,dx.$$

Calculate $P(x \geq 3)$ for each pdf in part (a).

24. Verify that
$$\int_1^\infty \dfrac{1}{x^p}\,dx$$
converges if $p > 1$ and diverges if $p < 1$.

25. The force of repulsion between two point charges of like sign, one of size q and the other of size unity, has magnitude
$$F = \dfrac{q}{4\pi\epsilon_0 r^2},$$
where ϵ_0 is a constant and r is the distance between the charges. The potential V at any point P due to charge q is defined as the work required to bring the unit charge to P from infinity along the straight line joining q and P. Find a formula for V.

26. Verify that if $f(x)$ is continuous on $a \leq x \leq b$ except for a finite discontinuity at d (Figure 7.104), then the improper integral of $f(x)$ from $x = a$ to $x = b$ must converge.

27. Find the length of the loop of the curve $9y^2 = x(3-x)^2$.

28. Is equation 7.42 equivalent to the following equation?
$$\int_{-\infty}^{\infty} f(x)\,dx = \lim_{a \to \infty} \int_{-a}^{a} f(x)\,dx$$

In Exercises 29–32 determine whether the improper integral converges or diverges. Hint: Compare each integral to a known convergent or divergent integral.

29. $\int_{2}^{\infty} \dfrac{x^2}{\sqrt{x^2-1}}\,dx$ 30. $\int_{1}^{3} \dfrac{x^3}{(27-x^3)^2}\,dx$

31. $\int_{0}^{1} \dfrac{x^2}{\sqrt{1-x^2}}\,dx$ 32. $\int_{-\infty}^{-2} \dfrac{\sqrt{-x}}{(x^2+5)^2}\,dx$

33. Show that existence of the limits in equation 7.42 is independent of the choice of c.

34. One of the most important functions in physics and engineering is the *Dirac delta function* $\delta(x-a)$ introduced in Exercise 24 of Section 2.5 as the limit of the unit pulse function (Figure 7.105):
$$\delta(x-a) = \lim_{\epsilon \to 0} P_\epsilon(x-a).$$

For a continuous function $f(x)$,
$$\int_{-\infty}^{\infty} f(x)\delta(x-a)\,dx = \int_{-\infty}^{\infty} f(x)\left[\lim_{\epsilon \to 0} P_\epsilon(x-a)\right]dx.$$

Assuming that the order of integration and the process of taking the limit as ϵ approaches zero can be interchanged, that is,
$$\int_{-\infty}^{\infty} f(x)\left[\lim_{\epsilon \to 0} P_\epsilon(x-a)\right]dx = \lim_{\epsilon \to 0}\int_{-\infty}^{\infty} f(x)P_\epsilon(x-a)\,dx,$$

show that
$$\int_{-\infty}^{\infty} f(x)\delta(x-a)\,dx = f(a).$$

FIGURE 7.105

SUMMARY

In this chapter we used definite integrals in a wide variety of applications. With differentials and representative elements we were able to avoid memorization of formulas. To summarize the use of representative elements in various applications, consider the region R in Figure 7.106.

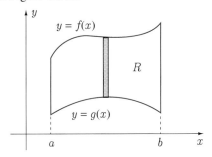

FIGURE 7.106

To find the area of R, we draw at position x a rectangle of length $f(x) - g(x)$ and width dx. The areas $[f(x) - g(x)]\,dx$ of all such rectangles are then added to find the total area:
$$\int_{a}^{b} [f(x) - g(x)]\,dx.$$

If R is rotated around the x-axis to form a solid of revolution, the rectangle generates a washer with volume $\{\pi[f(x)]^2 - \pi[g(x)]^2\}\,dx$, and therefore the total volume is

$$\int_a^b \{\pi[f(x)]^2 - \pi[g(x)]^2\}\,dx.$$

If R is rotated about the y-axis, the rectangle traces out a cylindrical shell of volume $2\pi x[f(x) - g(x)]\,dx$, and the total volume is

$$\int_a^b 2\pi x[f(x) - g(x)]\,dx.$$

First moments of this rectangle about the x- and y-axes are

$$\frac{1}{2}[f(x) + g(x)][f(x) - g(x)]\,dx \quad \text{and} \quad x[f(x) - g(x)]\,dx,$$

and these lead to the definite integrals

$$\int_a^b \frac{1}{2}\{[f(x)]^2 - [g(x)]^2\}\,dx \quad \text{and} \quad \int_a^b x[f(x) - g(x)]\,dx$$

for first moments of area of R about the x- and y-axes. Second moments of area of region R about the x- and y-axes are

$$\int_a^b \frac{1}{3}\{[f(x)]^3 - [g(x)]^3\}\,dx \quad \text{and} \quad \int_a^b x^2[f(x) - g(x)]\,dx,$$

where in the first integral it was necessary to use formula 7.38 for the second moment of the rectangle.

The key, then, to the use of the definite integral is to divide the region into smaller elements, calculate the required quantity for a representative element, and then use a definite integral to add over all elements.

An important point to keep in mind is that for lengths in the y-direction, we always take "upper y minus lower y", and for lengths in the x-direction, we use "larger x minus smaller x". This way many sign errors can be avoided when calculating *lengths*. For instance, note that this rule is used to find the length of a rectangle (in areas, volumes, fluid pressure), the radius of a circle (in volumes, surface area), depth (fluid pressure), and some distances (work).

An improper integral is a definite integral with an infinite limit and/or a point of discontinuity in the interval of integration. The value of an improper integral is defined by first calculating it over a finite interval in which the integrand is continuous, and then letting the limit on the integral approach either infinity or the point of discontinuity.

Key Terms and Formulas

In reviewing this chapter, you should be able to define or discuss the following key terms:

Area
Representative rectangle
Volume of solid of revolution
Disc method
Washer method
Cylindrical shell method
Length of a curve
Work
Fluid pressure
Fluid force

First moment of mass
First moment of area
Centre of mass
Centroid
Moment of inertia
Second moment
Volume by slicing
Area of surface of revolution
Improper integral

REVIEW EXERCISES

In Exercises 1–5, calculate for the region bounded by the curves: (a) its area; (b) the volumes of the solids of revolution obtained by rotating the region about the x- and y-axes; (c) its centroid; and (d) its second moments of area about the x- and y-axes.

1. $y = 9 - x^2$, $y = 0$
2. $y = x^3$, $y = 2 - x$, $x = 0$
3. $x = y^2 - 1$, $x = 4 - 4y^2$
4. $2y = x$, $y + 1 = x$, $y = 0$
5. $2x + y = 2$, $y - 2x = 2$, $x = y + 1$, $y + x + 1 = 0$

In Exercises 6–10 find the volume of the solid of revolution obtained by rotating the area bounded by the curves about the line.

6. $y = |x| + 2$, $y = 3$, about $y = 1$
7. $x = y^4$, $x = 4$, about $x = 4$
8. $y = x^3 - x$, $y = 0$, about $y = 0$
9. $yx^2 = 1$, $x = 1$, $x = 2$, $y = 0$ about $y = -1$
10. $y = x + 2$, $y = \sqrt{4 - 2x}$, $y = 0$, about $y = 1$

In Exercises 11–15 find the first and second moments of area of the region bounded by the curves about the line.

11. $y = 4x^2$, $y = 2 + 2x^2$, about $y = -1$
12. $2x = \sqrt{y}$, $\sqrt{2}x = \sqrt{y - 2}$, $x = 0$, about $x = 1$
13. $3y = 2x + 6$, $2y + x = 4$, $y = 0$, about $y = -2$
14. $3y = 2x + 6$, $2y + x = 4$, $y = 0$, about $y = 2$
15. $y = x^2 - 2x$ ($y \le 0$), $y = 2x$, $y = 4 - 2x$, about $x = 1$

16. A water tank in the form of a right-circular cylinder with radius 1 m and height 3 m has its axis vertical. If it is half full, how much work is required to empty the tank through an outlet at the top of the tank?

17. A car runs off a bridge into a river and submerges. If the tops of its front and rear side windows are 2 m below the surface of the water (Figure 7.107), what is the force due to water pressure on each window?

FIGURE 7.107

18. Find the surface area of the volume of the solid of revolution obtained by rotating the area bounded by the curves

$$2x + y = 2, \quad y - 2x = 2, \quad y = 0$$

about the x-axis.

19. A tank in the form of an inverted right-circular cone of depth 10 m and radius 4 m is full of water. How much work is required to lower the water level in the tank by 2 m by pumping water out through an outlet 1 m above the top of the tank?

20. Show that if it takes W units of work to stretch a spring from equilibrium to a certain stretch, then it requires $3W$ more units of work to double that stretch.

In Exercises 21–28 determine whether the improper integral converges or diverges. Evaluate each convergent integral.

21. $\int_1^\infty \dfrac{1}{x^{3/2}}\, dx$
22. $\int_0^3 \dfrac{1}{\sqrt{3-x}}\, dx$
23. $\int_{-1}^0 \dfrac{1}{(x+1)^2}\, dx$
24. $\int_{-2}^2 \dfrac{x}{\sqrt{4-x^2}}\, dx$
25. $\int_{-\infty}^\infty \dfrac{1}{(x+3)^3}\, dx$
26. $\int_{-\infty}^{-3} \dfrac{1}{\sqrt{-x}}\, dx$
27. $\int_{-6}^\infty x\sqrt{x^2+4}\, dx$
28. $\int_{-\infty}^\infty \dfrac{x}{(x^2-1)^2}\, dx$

29. A uniformly tapered rod of length 1 m has circular cross-sections, the radii of its ends being 1 cm and 2 cm. Find the volume of the rod by:
    (a) using the fact that every cross-section is circular;
    (b) considering the rod as a volume of a solid of revolution.

30. In Exercise 16, how much work is required to empty the tank if its axis is horizontal and the outlet is at the top of the tank?

CHAPTER 8

Further Transcendental Functions and their Derivatives

In Chapter 3 we learned to differentiate functions that involve a finite number of additions, subtractions, multiplications, divisions, and roots. These functions fall into a wider class called *algebraic functions*. A function $y = f(x)$ is said to be **algebraic** if, for all x in its domain, it satisfies an equation of the form

$$P_0(x)y^n + P_1(x)y^{n-1} + \cdots + P_{n-1}(x)y + P_n(x) = 0, \qquad (8.1)$$

where $P_0(x), \ldots, P_n(x)$ are polynomials in x, and n is a positive integer. A polynomial $P(x)$ is itself algebraic since it satisfies $y - P(x) = 0$; that is, it satisfies 8.1 with $n = 1$, $P_0(x) = 1$, and $P_1(x) = -P(x)$. Rational functions $P(x)/Q(x)$ are also algebraic ($n = 1$, $P_0(x) = Q(x)$, and $P_1(x) = -P(x)$). The function $f(x) = x^{1/3}$ is algebraic since it satisfies $y^3 - x = 0$. The equation $y^3 + y = x$ defines y implicitly as a function of x. We cannot find an explicit form for the function, but according to 8.1, the function so defined is algebraic.

A function that is not algebraic is called **transcendental**. The trigonometric, exponential, and logarithm functions which we differentiated in Chapter 3 are transcendental. In this chapter we consider two other classes of transcendental functions: inverse trigonometric and hyperbolic functions.

SECTION 8.1

Inverse Functions

When we speak of a function $y = g(x)$, it is inherent that there is a unique y associated with each x; that is, given a value of x in the domain of $g(x)$, the function associates one — and only one — value y in the range. However, it may happen, and quite often does, that a value of y in the range of the function may be associated with more than one value of x. For example, each $y > 0$ in the range of the function $y = g(x) = x^2$ (Figure 8.1) is associated with two values of x, namely, $x = \pm\sqrt{y}$.

Some functions have the property that each value of y in the range arises from only one x in the domain. For instance, given any value y in the range of the function $y = f(x) = x^3$ in Figure 8.2, there is a unique x such that $y = x^3$, namely, $x = y^{1/3}$. Such a function is said to be "one-to-one". Formally, we say that a function $f(x)$ is **one-to-one** if, for any two values x_1 and x_2 in the domain of $f(x)$, it follows that $f(x_1) \neq f(x_2)$. To repeat, a one-to-one function $f(x)$ has the property that given any y in its range, there is one — and only one — x in its domain for which $y = f(x)$. We can therefore define a function which maps values in the range of $f(x)$ onto values in the domain, a function that maps y onto x if x is mapped by $f(x)$ onto y. We call this function the inverse function of $f(x)$. It reverses the action of $f(x)$. For the function $f(x) = x^3$ in Figure 8.2, the inverse function is the function that takes cube roots of real numbers.

FIGURE 8.1

FIGURE 8.2

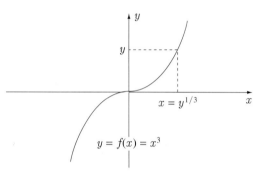

Unless there is a good reason to do otherwise, it is our custom to denote the independent variable of a function by the letter x and the dependent variable by y. For the inverse function of a function $y = f(x)$, there is a natural tendency to denote the independent variable by y and the dependent variable by x since the inverse function maps the range of $f(x)$ onto its domain. Not for obvious reasons, it is better to maintain our usual practice of using x as independent variable and y as dependent variable even for inverse functions. When a function $f(x)$ has an inverse function, we adopt the notation $f^{-1}(x)$ to represent the inverse function. For example, when $f(x) = x^3$, the inverse function is $f^{-1}(x) = x^{1/3}$. Be careful with this notation. Do not interpret the "-1" as a power and write $f^{-1}(x)$ as $1/f(x)$. This is not correct. The notation f^{-1} represents a function, just as tan represents the tangent function and $\sqrt{}$ represents the positive square root function. It represents the inverse function of $f(x)$.

The inverse function $f^{-1}(x)$ of a function $f(x)$ "undoes" what $f(x)$ "does"; it reverses the effect of $f(x)$. For example, the function $f(x) = x^2$, $x \geq 0$ in Figure 8.3 is one-to-one; it squares nonnegative numbers. Its inverse is $f^{-1}(x) = \sqrt{x}$, the positive square root function (Figure 8.4). Squaring an x and then taking the positive square root of the result returns the original x. Likewise, the inverse of $g(x) = x^2$, $x \leq 0$ (Figure 8.5) is the negative square root function $g^{-1}(x) = -\sqrt{x}$ (Figure 8.6).

FIGURE 8.3

FIGURE 8.4

CHAPTER 8

Further Transcendental Functions and their Derivatives

In Chapter 3 we learned to differentiate functions that involve a finite number of additions, subtractions, multiplications, divisions, and roots. These functions fall into a wider class called *algebraic functions*. A function $y = f(x)$ is said to be **algebraic** if, for all x in its domain, it satisfies an equation of the form

$$P_0(x)y^n + P_1(x)y^{n-1} + \cdots + P_{n-1}(x)y + P_n(x) = 0, \qquad (8.1)$$

where $P_0(x), \ldots, P_n(x)$ are polynomials in x, and n is a positive integer. A polynomial $P(x)$ is itself algebraic since it satisfies $y - P(x) = 0$; that is, it satisfies 8.1 with $n = 1$, $P_0(x) = 1$, and $P_1(x) = -P(x)$. Rational functions $P(x)/Q(x)$ are also algebraic ($n = 1$, $P_0(x) = Q(x)$, and $P_1(x) = -P(x)$). The function $f(x) = x^{1/3}$ is algebraic since it satisfies $y^3 - x = 0$. The equation $y^3 + y = x$ defines y implicitly as a function of x. We cannot find an explicit form for the function, but according to 8.1, the function so defined is algebraic.

A function that is not algebraic is called **transcendental**. The trigonometric, exponential, and logarithm functions which we differentiated in Chapter 3 are transcendental. In this chapter we consider two other classes of transcendental functions: inverse trigonometric and hyperbolic functions.

SECTION 8.1

Inverse Functions

When we speak of a function $y = g(x)$, it is inherent that there is a unique y associated with each x; that is, given a value of x in the domain of $g(x)$, the function associates one — and only one — value y in the range. However, it may happen, and quite often does, that a value of y in the range of the function may be associated with more than one value of x. For example, each $y > 0$ in the range of the function $y = g(x) = x^2$ (Figure 8.1) is associated with two values of x, namely, $x = \pm\sqrt{y}$.

Some functions have the property that each value of y in the range arises from only one x in the domain. For instance, given any value y in the range of the function $y = f(x) = x^3$ in Figure 8.2, there is a unique x such that $y = x^3$, namely, $x = y^{1/3}$. Such a function is said to be "one-to-one". Formally, we say that a function $f(x)$ is **one-to-one** if, for any two values x_1 and x_2 in the domain of $f(x)$, it follows that $f(x_1) \neq f(x_2)$. To repeat, a one-to-one function $f(x)$ has the property that given any y in its range, there is one — and only one — x in its domain for which $y = f(x)$. We can therefore define a function which maps values in the range of $f(x)$ onto values in the domain, a function that maps y onto x if x is mapped by $f(x)$ onto y. We call this function the inverse function of $f(x)$. It reverses the action of $f(x)$. For the function $f(x) = x^3$ in Figure 8.2, the inverse function is the function that takes cube roots of real numbers.

FIGURE 8.1

FIGURE 8.2

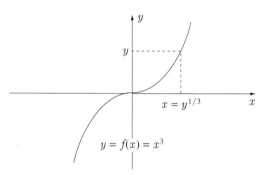

Unless there is a good reason to do otherwise, it is our custom to denote the independent variable of a function by the letter x and the dependent variable by y. For the inverse function of a function $y = f(x)$, there is a natural tendency to denote the independent variable by y and the dependent variable by x since the inverse function maps the range of $f(x)$ onto its domain. Not for obvious reasons, it is better to maintain our usual practice of using x as independent variable and y as dependent variable even for inverse functions. When a function $f(x)$ has an inverse function, we adopt the notation $f^{-1}(x)$ to represent the inverse function. For example, when $f(x) = x^3$, the inverse function is $f^{-1}(x) = x^{1/3}$. Be careful with this notation. Do not interpret the "-1" as a power and write $f^{-1}(x)$ as $1/f(x)$. This is not correct. The notation f^{-1} represents a function, just as tan represents the tangent function and $\sqrt{}$ represents the positive square root function. It represents the inverse function of $f(x)$.

The inverse function $f^{-1}(x)$ of a function $f(x)$ "undoes" what $f(x)$ "does"; it reverses the effect of $f(x)$. For example, the function $f(x) = x^2$, $x \geq 0$ in Figure 8.3 is one-to-one; it squares nonnegative numbers. Its inverse is $f^{-1}(x) = \sqrt{x}$, the positive square root function (Figure 8.4). Squaring an x and then taking the positive square root of the result returns the original x. Likewise, the inverse of $g(x) = x^2$, $x \leq 0$ (Figure 8.5) is the negative square root function $g^{-1}(x) = -\sqrt{x}$ (Figure 8.6).

FIGURE 8.3

FIGURE 8.4

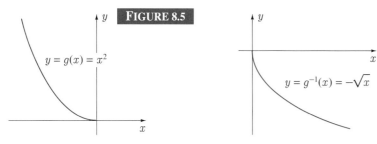

FIGURE 8.5

FIGURE 8.6

The fact that $f^{-1}(x)$ undoes $f(x)$ can be stated algebraically as follows. For each x in the domain of $f(x)$,

$$f^{-1}(f(x)) = x. \tag{8.2}$$

This is the defining relation for an inverse function. We understand that the domain of $f^{-1}(x)$ is the same as the range of $f(x)$ so that $f^{-1}(x)$ operates on all outputs of $f(x)$.

EXAMPLE 8.1

What is the inverse function of $f(x) = (x+1)/(x-2)$.

SOLUTION To find the algebraic definition of $f^{-1}(x)$, we solve the equation $y = (x+1)/(x-2)$ for x in terms of y. First we cross-multiply,

$$x + 1 = y(x - 2);$$

then group terms in x,

$$x(y - 1) = 2y + 1;$$

and finally divide by $y - 1$,

$$x = \frac{2y + 1}{y - 1}.$$

What have we accomplished by solving for x in terms of y? If we take an x in the domain of $f(x)$, then $f(x)$ produces $y = (x+1)/(x-2)$. Now this same x and y satisfy the equation $x = (2y+1)/(y-1)$ because this equation is simply a rearrangement of $y = (x+1)/(x-2)$. Consequently, if we substitute y into the right-hand side of the equation,

$$x = \frac{2y + 1}{y - 1},$$

we obtain the original x. This equation must therefore define the inverse function of $f(x)$. In other words, if we denote the independent variable by x, the inverse function is

$$f^{-1}(x) = \frac{2x + 1}{x - 1}.$$

■

Example 8.1 has illustrated that to find the inverse of a function $y = f(x)$, the equation should be solved for x in terms of y, and then variables should be renamed. If the equation does not have a unique solution for x in terms of y, then $f(x)$ does not have an inverse function. Such is the case for the function $g(x) = x^2$ in Figure 8.1. Solving $y = g(x) = x^2$ for x gives two solutions $x = \pm\sqrt{y}$.

So far our discussion of inverse functions has been algebraic. The geometry of inverse functions is most revealing. You may have noticed a relationship between the graphs of

$f(x)$ and $f^{-1}(x)$ in Figures 8.3 and 8.4, and of $g(x)$ and $g^{-1}(x)$ in Figures 8.5 and 8.6. The inverse function is the mirror image of the function in the line $y = x$; that is, graphs of inverse pairs are symmetric about the line $y = x$. These two examples are not mere coincidence, but as we now show, graphs of inverse functions are always mirror images of each other in the line $y = x$.

Suppose $f^{-1}(x)$ is the inverse of $f(x)$, and that (x, y) is any point on the graph of $y = f(x)$. Since $f^{-1}(x)$ is the inverse of $f(x)$, the result of substituting y into $f^{-1}(x)$ is x; that is, the point (y, x) is on the graph of $f^{-1}(x)$. But the point (y, x) is the mirror image of (x, y) in the line $y = x$. When this is combined with the fact that the domain of $f^{-1}(x)$ is the range of $f(x)$, our conjecture is verified.

This suggests a very simple way to graphically determine the inverse of a function $f(x)$: take its mirror image in the line $y = x$. Note, too, that if the mirror image does not represent a function, then no inverse function for $f(x)$ exists. For example, the mirror image of $y = x^2$, $-\infty < x < \infty$, in the line $y = x$ is shown in Figure 8.7 and does not represent a function. We conclude as before that this function does not have an inverse.

FIGURE 8.7

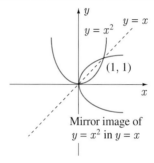

Mirror image of $y = x^2$ in $y = x$

If $y = f^{-1}(x)$ is the mirror image of $y = f(x)$ in the line $y = x$, then $y = f(x)$ is the mirror image of $y = f^{-1}(x)$. This means that $f(x)$ is the inverse of $f^{-1}(x)$, and that $f(x)$ undoes what $f^{-1}(x)$ does:

$$f\left(f^{-1}(x)\right) = x. \qquad (8.3)$$

Geometrically, a function $f(x)$ has an inverse if the reflection of its graph $y = f(x)$ in the line $y = x$ represents a function. But we know that a curve represents a function if every vertical line that intersects it does so at exactly one point. Furthermore, a vertical line intersects the reflected curve at exactly one point only if its horizontal reflection intersects $y = f(x)$ at exactly one point. These two facts enable us to state that a function $f(x)$ has an inverse function if — and only if — every horizontal line that intersects it does so at exactly one point. This is a geometric interpretation of a function being one-to-one. See, for example, the functions in Figures 8.3 and 8.5. Horizontal lines that intersect the curves do so at exactly one point. The function in Figure 8.1, which has no inverse, is intersected in two points by every horizontal line $y = c > 0$. When a function is increasing on an interval I (Figure 8.8), it passes the horizontal line test (and is one-to-one). Likewise, a function which is decreasing on I (Figure 8.9) is one-to-one. A function which is either increasing on an interval I or decreasing on I is said to be **monotonic** on I. What we have shown is the following result.

FIGURE 8.8   **FIGURE 8.9**

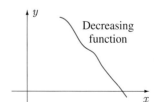

> **Theorem 8.1**
>
> A function which is monotonic on an interval has an inverse function on that interval.

It is important to realize that monotonicity is a sufficient condition for existence of an inverse function; that is, if a function is monotonic, then it has an inverse. It is not, however, a necessary condition. The function in Figure 8.10(a) is not monotonic on the interval $-1 \leq x \leq 1$, but it does have the inverse function in Figure 8.10(b).

FIGURE 8.10

(a)

(b)
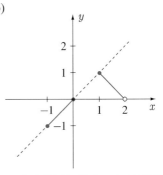

EXAMPLE 8.2

Does the function $f(x) = x^2 - 2x$ have an inverse? Does it have an inverse on the interval $x \geq 1$, and on the interval $x \leq 1$?

SOLUTION The graph of $f(x)$ in Figure 8.11 indicates that $f(x)$ does not have an inverse. When restricted to the interval $x \geq 1$, the function is one-to-one, and does have an inverse. To find it, we solve $y = x^2 - 2x$ for x in terms of y. The quadratic formula applied to

$$x^2 - 2x - y = 0$$

gives

$$x = \frac{2 \pm \sqrt{4 + 4y}}{2} = 1 \pm \sqrt{y + 1}.$$

Since x must be greater than or equal to one, we must choose $x = 1 + \sqrt{y + 1}$. Hence, the inverse of $f(x) = x^2 - 2x$, $x \geq 1$ is

$$f^{-1}(x) = 1 + \sqrt{x + 1}.$$

Similarly, when restricted to the interval $x \leq 1$, $f(x)$ has inverse

$$f^{-1}(x) = 1 - \sqrt{x + 1}.$$

FIGURE 8.11

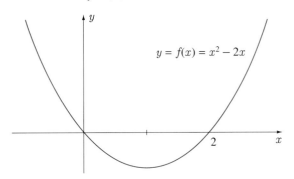

∎

This example illustrates that when a function is not monotonic on an interval I, the interval can usually be subdivided into subintervals on which the function is monotonic, and on each such subinterval the function has an inverse.

EXERCISES 8.1

In Exercises 1–14 determine graphically whether the function has an inverse. Find each inverse function.

1. $f(x) = 2x + 3$
2. $f(x) = \sqrt{x+1}$
3. $f(x) = x^2 + x$
4. $f(x) = \dfrac{x+5}{2x+4}$
5. $f(x) = 1/x$
6. $f(x) = 3x^3 + 2$
7. $f(x) = \sqrt{4 - x^2}$, $0 \le x \le 2$
8. $f(x) = 2x + |x|$
9. $f(x) = x + |x|$
10. $f(x) = \sqrt{1 - x^2}$
11. $f(x) = x^4 + 2x^2 + 2$, $x \le 0$
12. $f(x) = x^2 - 2x + 4$, $x \ge 1$
13. $f(x) = \left(\dfrac{x+2}{x-2}\right)^3$
14. $f(x) = \dfrac{x}{3 + x^2}$

In Exercises 15–20 show that the function does not have an inverse. Subdivide its domain of definition into subintervals on which the function has an inverse, and find the inverse function on each subinterval.

15. $f(x) = x^4$
16. $f(x) = 1/x^4$
17. $f(x) = x^2 + 2x + 3$
18. $f(x) = x^4 + 4x^2 + 2$
19. $f(x) = \dfrac{x^2}{x^2 + 4}$
20. $f(x) = \dfrac{x^4}{x^2 + 4}$

21. If a manufacturing firm sells x objects of a certain commodity per week, it sells them at a price of r per object, and r depends on x, $r = f(x)$. In economic theory this function is usually considered decreasing, as shown in Figure 8.12(a); hence it has an inverse function $x = f^{-1}(r)$. In this function x depends on r, indicating that if the price of the object is set at r, then the market will demand x of them per week. This function is therefore called the *demand function* (Figure 8.12(b)). Find demand functions if

(a) $r = \dfrac{a}{x+b} + c$, a, b, and c positive constants; and

(b) $r = \dfrac{a}{x^2+b} + c$, a, b, and c positive constants.

Sketch the demand function $x = f^{-1}(r)$ and given function $r = f(x)$ in each case.

FIGURE 8.12

(a)

(b)

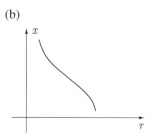

22. Show that the demand function
$$x = f(r) = 4a^3 - 3ar^2 + r^3, \quad 0 < r < 2a$$
has an inverse function $r = f^{-1}(x)$. What is the domain of $f^{-1}(x)$? Sketch its graph.

23. Find the inverse function for
$$f(x) = \dfrac{x^2}{(1+x)^2}, \quad -1 < x \le 0.$$

SECTION 8.2

The Inverse Trigonometric Functions

The function $f(x) = \sin x$, defined for all real x, does not have an inverse. This is evident from the fact that the function is not one-to-one, or that it fails the horizontal line test. By restricting the domain of $\sin x$, however, the function can be made one-to-one, and this can be done in many ways. In particular, when restricted to the domain

$-\pi/2 \leq x \leq \pi/2$, $\sin x$ is one-to-one, and therefore has an inverse function. This function, denoted by

$$y = \text{Sin}^{-1} x, \tag{8.4}$$

and called the **inverse sine function**, is shown in Figure 8.13. The range of the function is

$$-\frac{\pi}{2} \leq \text{Sin}^{-1} x \leq \frac{\pi}{2}. \tag{8.5}$$

The values that fall in this range are called the **principal values** of the inverse sine function. They have resulted from our restriction of the domain of $\sin x$ to this same interval.

An equivalent way to derive the inverse sine function is as follows. The reflection of the graph of $y = \sin x$ in the line $y = x$ does not represent a function, but by restricting the range of values of the reflected curve, we can produce a single-valued function (Figure 8.14). Once again this can be done in many ways, and when we do so by restricting the y-values to the interval $[-\pi/2, \pi/2]$, the resulting function is called the inverse sine function $\text{Sin}^{-1} x$. Note very carefully that $\text{Sin}^{-1} x$ is not the inverse function of $f(x) = \sin x$, because the sine function has no inverse. It is, however, the inverse of the sine function restricted to the domain $-\pi/2 \leq x \leq \pi/2$. It is perhaps then a misnomer to call the function $\text{Sin}^{-1} x$ the inverse sine function, but this has become the accepted terminology.

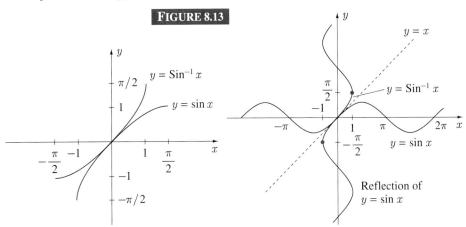

FIGURE 8.13

FIGURE 8.14 Reflection of $y = \sin x$

We now know what the inverse sine function looks like graphically, but what does it mean to say that $y = \text{Sin}^{-1} x$? Certainly given any value of x, we can push a few buttons on an electronic calculator and find $\text{Sin}^{-1} x$ for that x. To use inverse trigonometric functions in practice, we must have a feeling for what they do. To obtain this insight, we note that if (x, y) is a point on the curve $y = \text{Sin}^{-1} x$, then (y, x) is a point on the sine curve; that is,

$$y = \text{Sin}^{-1} x \quad \text{only if} \quad x = \sin y. \tag{8.6}$$

In the latter equation y is an angle and x is the sine of that angle. Thus, when we see $y = \text{Sin}^{-1} x$, we should not read "y equals inverse sine x", but rather "y is an angle whose sine is x." For instance, if $x = \frac{1}{2}$, then $y = \text{Sin}^{-1}\left(\frac{1}{2}\right)$ is read "y is an angle whose sine is $\frac{1}{2}$." Clearly the angle whose sine is $\frac{1}{2}$ is $\pi/6$, and we write $y = \text{Sin}^{-1}\left(\frac{1}{2}\right) = \pi/6$. Remember that we must choose the principal value $\pi/6$; there are many angles that have a sine equal to $\frac{1}{2}$, but the inverse sine function demands that we choose that angle in the range $-\pi/2 \leq y \leq \pi/2$, and state it with a number in this interval. For example, $2\pi + \pi/6$ may represent the same angle with the positive x-axis as $\pi/6$, but the number

$2\pi + \pi/6$ does not lie between $-\pi/2$ and $\pi/2$. In summary, the function $\sin x$ regards x as an angle and assigns to x the sine of the angle; the inverse sine function $\mathrm{Sin}^{-1} x$ regards x as the sine of an angle and assigns to x the angle with that sine.

Another notation that is commonly used for the inverse sine function is $\arcsin x$. One reason the notation $\arcsin x$ is preferable to $\mathrm{Sin}^{-1} x$ is the possible misinterpretation of $\mathrm{Sin}^{-1} x$. Sometimes students regard the "-1" as a power and write $\mathrm{Sin}^{-1} x$ as $1/\sin x$. This is *not* correct, a fact that you were warned about in Section 8.1. Sin^{-1} is the name of a function, just as $\sin$ is the name of the sine function, and $\sqrt{}$ is the notation for the positive square root function. The capital S in Sin^{-1} should also warn you that this is the inverse sine function.

EXAMPLE 8.3

Simplify each of the following expressions:

(a) $\mathrm{Sin}^{-1}(-\sqrt{3}/2)$ (b) $\mathrm{Sin}^{-1}(1)$ (c) $\mathrm{Sin}^{-1}(3)$
(d) $\mathrm{Sin}^{-1}(3/5) + \mathrm{Sin}^{-1}(4/5)$ (e) $\sin\left[\mathrm{Sin}^{-1}(\sqrt{3}/2)\right]$

SOLUTION (a) $\mathrm{Sin}^{-1}(-\sqrt{3}/2)$ asks for the angle whose sine is equal to $-\sqrt{3}/2$. Clearly,
$$\mathrm{Sin}^{-1}(-\sqrt{3}/2) = -\pi/3.$$

(b) $\mathrm{Sin}^{-1}(1) = \pi/2$.
(c) $\mathrm{Sin}^{-1}(3)$ is not defined since the domain of $\mathrm{Sin}^{-1} x$ is $-1 \leq x \leq 1$.
(d) If $\phi = \mathrm{Sin}^{-1}(3/5)$, then ϕ is illustrated in the triangle in Figure 8.15. Since the third side must have length 4, it follows that $\mathrm{Sin}^{-1}(4/5) = \pi/2 - \phi$, and
$$\mathrm{Sin}^{-1}\left(\frac{3}{5}\right) + \mathrm{Sin}^{-1}\left(\frac{4}{5}\right) = \phi + \left(\frac{\pi}{2} - \phi\right) = \frac{\pi}{2}.$$

(e) $\sin\left[\mathrm{Sin}^{-1}(\sqrt{3}/2)\right] = \sin(\pi/3) = \sqrt{3}/2$.

FIGURE 8.15

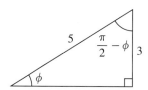

EXAMPLE 8.4

For what values of x are the following valid:

(a) $\sin\left(\mathrm{Sin}^{-1} x\right) = x$ (b) $\mathrm{Sin}^{-1}(\sin x) = x$

SOLUTION These two equations express the fact that the sine function and the inverse sine function are inverses, provided we are careful about domains:
(a) The equation $\sin\left(\mathrm{Sin}^{-1} x\right) = x$ is valid for $-1 \leq x \leq 1$. Given an x in this interval, $\mathrm{Sin}^{-1} x$ finds that angle (in the principal value range) which has x as its sine. Then $\sin\left(\mathrm{Sin}^{-1} x\right)$ takes the sign of this angle. Naturally it returns the original number x.
(b) The composite function $\mathrm{Sin}^{-1}(\sin x)$ is defined for all x, but only on the domain $-\pi/2 \leq x \leq \pi/2$ is it equal to x.

Our analysis of the inverse sine function is now complete. We could repeat a similar discussion five times, once for each of the inverse functions associated with the other five trigonometric functions. However, we shall give an abbreviated version for the inverse cosine function and tabulate results for the remaining four functions.

The reflection of the graph of the function $f(x) = \cos x$ is shown in Figure 8.16, and it does not represent a function. If we restrict the y-values on the reflected curve to

$$0 \leq y \leq \pi, \tag{8.7}$$

then we do obtain a function called the inverse cosine function, denoted by

$$y = \text{Cos}^{-1} x. \tag{8.8}$$

The values in 8.7 are called the principal values of the inverse cosine function. Note again that $\text{Cos}^{-1} x$ is not the inverse function of $\cos x$, but is the inverse function of $f(x) = \cos x$, $0 \leq x \leq \pi$.

When we write $y = \text{Cos}^{-1} x$, we do not read this as "y equals inverse cosine x," but rather as "y is an angle whose cosine is x,", for if $y = \text{Cos}^{-1} x$, then $x = \cos y$.

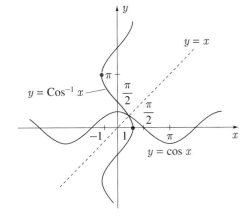

FIGURE 8.16

The remaining four inverse trigonometric functions are shown in Figures 8.17(c)–(f). Lighter curves represent reflections in the line $y = x$ of the trigonometric functions $\tan x$, $\cot x$, $\csc x$, and $\sec x$. Heavier curves represent principal values of the inverse trigonometric functions $\text{Tan}^{-1} x$, $\text{Cot}^{-1} x$, $\text{Csc}^{-1} x$, and $\text{Sec}^{-1} x$. Principal values of the six inverse trigonometric functions are listed in Table 8.1.

TABLE 8.1

Inverse Trigonometric Function	Principal Values
$\text{Sin}^{-1} x$	$-\frac{\pi}{2} \leq y \leq \frac{\pi}{2}$
$\text{Tan}^{-1} x$	$-\frac{\pi}{2} < y < \frac{\pi}{2}$
$\text{Cos}^{-1} x$	$0 \leq y \leq \pi$
$\text{Cot}^{-1} x$	$0 < y < \pi$
$\text{Csc}^{-1} x$	$-\pi < y \leq -\frac{\pi}{2}, \; 0 < y \leq \frac{\pi}{2}$
$\text{Sec}^{-1} x$	$-\pi \leq y < -\frac{\pi}{2}, \; 0 \leq y < \frac{\pi}{2}$

FIGURE 8.17

(a)

(b)

(c)

(d)

(e)

(f)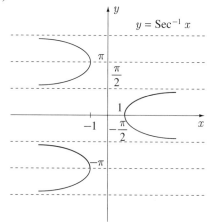

It would be reasonable to ask why principal values of $\operatorname{Sec}^{-1} x$ were not chosen as $0 \leq y < \pi/2$, $\pi/2 < y \leq \pi$. Had they been chosen in this way, they would have been very similar to those of $\operatorname{Cos}^{-1} x$ and $\operatorname{Cot}^{-1} x$. Likewise, why are the principal values of $\operatorname{Csc}^{-1} x$ not $-\pi/2 \leq y < 0$, $0 < y \leq \pi/2$? The answer is that they could have been selected in this way, and some authors do indeed make this choice; it is simply a matter of preference. Each choice does, however, create corresponding changes in later work. Specifically, in the exercises of Section 8.3, if principal values of $\operatorname{Sec}^{-1} x$ and $\operatorname{Csc}^{-1} x$ are selected in this alternative way, then derivatives of these functions are modified correspondingly.

In the remainder of this section we solve a number of problems which illustrate some of the properties of the inverse trigonometric functions and their applications.

EXAMPLE 8.5

Show that
$$\operatorname{Csc}^{-1} x = \begin{cases} \operatorname{Sin}^{-1}\left(\dfrac{1}{x}\right) & x \geq 1 \\ -\pi - \operatorname{Sin}^{-1}\left(\dfrac{1}{x}\right) & x \leq -1. \end{cases}$$

SOLUTION When $x \geq 1$, we set $y = \operatorname{Csc}^{-1} x$, in which case $0 < y \leq \pi/2$. It follows that $x = \csc y$, and
$$\frac{1}{x} = \frac{1}{\csc y} = \sin y.$$

If we apply the inverse sine function to both sides of this equation, the result is
$$\operatorname{Sin}^{-1}\left(\frac{1}{x}\right) = \operatorname{Sin}^{-1}(\sin y) = y,$$

because y is in the principal value range of the inverse sine function. Hence, when $x \geq 1$,
$$\operatorname{Csc}^{-1} x = \operatorname{Sin}^{-1}\left(\frac{1}{x}\right).$$

When $x \leq -1$, we again set $y = \operatorname{Csc}^{-1} x$, and obtain
$$\operatorname{Sin}^{-1}\left(\frac{1}{x}\right) = \operatorname{Sin}^{-1}(\sin y).$$

But in this case the right side is not equal to y, because $-\pi < y \leq -\pi/2$. To remedy this, we note that when $-\pi < y \leq -\pi/2$, we may write $\sin y = \sin(-\pi - y)$. Since $-\pi - y$ is in the principal range for the inverse sine function ($-\pi/2 \leq -\pi - y < 0$), it follows that
$$\operatorname{Sin}^{-1}\left(\frac{1}{x}\right) = \operatorname{Sin}^{-1}(\sin y) = \operatorname{Sin}^{-1}[\sin(-\pi - y)] = -\pi - y = -\pi - \operatorname{Csc}^{-1} x.$$
∎

Similar relations involving the inverse secant and inverse cotangent functions are presented in Exercises 45 and 46.

EXAMPLE 8.6

Show that
$$\operatorname{Sin}^{-1} x = \begin{cases} -\operatorname{Cos}^{-1}\sqrt{1 - x^2} & -1 \leq x < 0 \\ \operatorname{Cos}^{-1}\sqrt{1 - x^2} & 0 \leq x \leq 1. \end{cases}$$

SOLUTION When $0 \leq x \leq 1$, we set $y = \operatorname{Sin}^{-1} x$, in which case $0 \leq y \leq \pi/2$. It follows that $x = \sin y$, and because $\sin^2 y + \cos^2 y = 1$, we have
$$\cos y = \pm\sqrt{1 - \sin^2 y} = \pm\sqrt{1 - x^2}.$$

Since y is an angle in the first quadrant, its cosine must be nonnegative, and therefore
$$\cos y = \sqrt{1 - x^2}.$$

When we apply the inverse cosine function to both sides of this equation, we obtain

$$\text{Cos}^{-1}(\cos y) = y = \text{Cos}^{-1}\sqrt{1-x^2}.$$

When $-1 \leq x < 0$, we continue to set $y = \text{Sin}^{-1} x$, and once again obtain

$$\cos y = \sqrt{1-x^2},$$

because $-\pi/2 \leq y < 0$. Application of the inverse cosine function gives

$$\text{Cos}^{-1}(\cos y) = \text{Cos}^{-1}\sqrt{1-x^2},$$

but the left side is not equal to y because y is not in the principal value range of the inverse cosine function. This is easily adjusted by noting that with $-\pi/2 \leq y < 0$, we have $\cos y = \cos(-y)$. Hence,

$$\text{Cos}^{-1}(\cos y) = \text{Cos}^{-1}[\cos(-y)] = -y = \text{Cos}^{-1}\sqrt{1-x^2};$$

that is,

$$y = -\text{Cos}^{-1}\sqrt{1-x^2}.$$

Similar relations among other inverse trigonometric functions are presented in Exercises 47 and 48. We now consider a few examples wherein inverse trigonometric functions are used to solve equations involving trigonometric functions.

EXAMPLE 8.7 Find all solutions of the equation

$$\sin^2 x + 3\sin x = 2.$$

SOLUTION The quadratic formula applied to the equation

$$(\sin x)^2 + 3(\sin x) - 2 = 0$$

yields

$$\sin x = \frac{-3 \pm \sqrt{9+8}}{2} = \frac{-3 \pm \sqrt{17}}{2}.$$

Since $\sin x$ can only take on values in the interval $-1 \leq \sin x \leq 1$, the possibility $\sin x = (-3 - \sqrt{17})/2$ must be rejected, and

$$\sin x = \frac{-3 + \sqrt{17}}{2}.$$

One solution of this equation is

$$x = \text{Sin}^{-1}\left(\frac{\sqrt{17}-3}{2}\right) = 0.60 \text{ radians.}$$

This is not the only solution, however; there are many other angles with a sine equal to $(\sqrt{17} - 3)/2$. Since $\sin x$ is 2π-periodic, the angles $2n\pi + 0.60$, for n any integer, all have sine equal to $(\sqrt{17} - 3)/2$. Because $\sin(\pi - x) = \sin x$, it follows that $\sin(\pi - 0.60) = (\sqrt{17} - 3)/2$, and therefore $\pi - 0.60$ is another solution. When multiples of 2π are added to this angle, $2n\pi + (\pi - 0.60)$ are also solutions. Thus, the complete set of solutions is

$$2n\pi + 0.60, \quad 2n\pi + (\pi - 0.60),$$

where n is an integer. Figure 8.18 suggests that this set of numbers can be represented more compactly as an initial rotation of $\pi/2$, plus or minus $\pi/2 - 0.60 = 0.97$, and possible multiples of 2π; that is,

$$\frac{\pi}{2} \pm 0.97 + 2n\pi = \left(\frac{4n+1}{2}\right)\pi \pm 0.97.$$

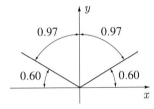

FIGURE 8.18

Often in problems like these, $x = \text{Sin}^{-1}[(\sqrt{17} - 3)/2] = 0.60$ is the only solution given. We can see the reasoning behind this conclusion. There is only one principal value of the inverse sine function, and therefore one solution to the equation. But the equation $\sin x = (\sqrt{17} - 3)/2$ says nothing about the inverse sine function. We have introduced it simply as a convenience. When we obtain the solution 0.60 we must ask whether there are other solutions to the original equation. For the above example, there are other solutions; in some problems, the principal value is the only acceptable solution. Remember, then, if we introduce an inverse trigonometric function into a problem along with its corresponding principal values, we must ask whether there are other possibilities besides the principal values. This point is emphasized again in the following example.

EXAMPLE 8.8

Find all solutions of the equation

$$\tan(\cos x) = \frac{1}{\sqrt{3}}.$$

SOLUTION If we set $y = \cos x$, then $\tan y = 1/\sqrt{3}$, and one solution of this equation is $y = \text{Tan}^{-1}(1/\sqrt{3}) = \pi/6$. But there are many other solutions for y, namely,

$$y = \frac{\pi}{6} + n\pi,$$

where n is an integer. But $y = \cos x$, and $\cos x$ must take on values in the interval $-1 \leq \cos x \leq 1$. There is only one possibility for n, namely, $n = 0$; hence,

$$\cos x = \frac{\pi}{6}.$$

From the solution $x = \text{Cos}^{-1}(\pi/6) = 1.02$ and Figure 8.19, we obtain

$$x = \pm 1.02 + 2n\pi,$$

where n is an integer, as the complete set of solutions.

FIGURE 8.19

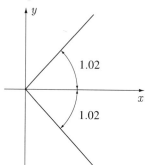

EXAMPLE 8.9

Find constants $R > 0$ and $0 < \phi < \pi$ so that the function $f(x) = 3\cos\omega x - 4\sin\omega x$, where $\omega > 0$ is a constant, can be expressed in the form $R\sin(\omega x + \phi)$ for all real x.

SOLUTION If we expand $R\sin(\omega x + \phi)$ by means of a compound angle formula, and equate it to $f(x)$, we have

$$R[\sin\omega x \cos\phi + \cos\omega x \sin\phi] = 3\cos\omega x - 4\sin\omega x.$$

This equation will be valid for all x if we can find values of R and ϕ so that

$$R\cos\phi = -4 \quad \text{and} \quad R\sin\phi = 3.$$

When we square and add these equations, the result is

$$R^2\cos^2\phi + R^2\sin^2\phi = R^2 = (-4)^2 + (3)^2 = 25.$$

Consequently, $R = 5$, and

$$\cos\phi = -\frac{4}{5} \quad \text{and} \quad \sin\phi = \frac{3}{5}.$$

The only angle in the range $0 < \phi < \pi$ satisfying these equations is $\phi = \text{Cos}^{-1}(-4/5) = 2.50$ radians. Notice that $\text{Sin}^{-1}(3/5)$ does not give this angle. Thus, $f(x)$ can be expressed in the form

$$f(x) = 5\sin(\omega x + 2.50).$$

EXERCISES 8.2

In Exercises 1–16 evaluate the expression (if it has a value).

1. $\text{Tan}^{-1}(-1/3)$
2. $\text{Sin}^{-1}(1/4)$
3. $\text{Sec}^{-1}(\sqrt{3})$
4. $\text{Csc}^{-1}(-2/\sqrt{3})$
5. $\text{Cot}^{-1}(1)$
6. $\text{Cos}^{-1}(3/2)$
7. $\text{Sin}^{-1}(\pi/2)$
8. $\text{Tan}^{-1}(-1)$
9. $\sin\left(\text{Tan}^{-1}\sqrt{3}\right)$
10. $\tan\left(\text{Sin}^{-1}3\right)$
11. $\text{Sin}^{-1}[\tan(1/6)]$
12. $\text{Tan}^{-1}[\sin(1/6)]$
13. $\sec\left[\text{Cos}^{-1}(1/2)\right]$
14. $\text{Sin}^{-1}[\sin(3\pi/4)]$
15. $\sin\left[\text{Sin}^{-1}(1/\sqrt{2})\right]$
16. $\text{Sin}^{-1}\left[\cos\left(\text{Sec}^{-1}(-\sqrt{2})\right)\right]$

In Exercises 17–22 find all solutions of the equation.

17. $\sin x = 1/3$
18. $\tan x = -1.2$
19. $4\sin^2 x - 2\cos^2 x = 1$
20. $4\sin^2 x + 2\cos^2 x = 1$
21. $\cos^2 x - 3\cos x + 1 = 0$
22. $\sin^2 x - 3\sin x - 5 = 0$

In Exercises 23–26 sketch a graph of the function.

23. $f(x) = 2 + \left(\text{Csc}^{-1}x\right)^2$
24. $f(x) = \sqrt{\text{Tan}^{-1}x} + \sqrt{\text{Sec}^{-1}x}$
25. $f(x) = \text{Sin}^{-1}(x-3)$
26. $f(x) = \text{Sin}^{-1}x + \text{Csc}^{-1}x$

27. If ϕ is the angle formed by AB and AO in Figure 4.83, find ϕ as a function of θ.

In Exercises 28–33 find all solutions of the equation.

28. $\sin x \tan^2 x - 3 + \tan^2 x - 3\sin x = 0$
29. $\sin x + \cos x = 1$
30. $\sec(\sin x) = -\sqrt{2}$
31. $\cos\left(\text{Sin}^{-1}x\right) = 1/2$
32. $\sec(\tan x) = -\sqrt{2}$
33. $\text{Cos}^{-1}[\tan(x^2+4)] = 2\pi - 5$

34. Sketch graphs of the following functions:
   (a) $f(x) = \sin\left(\text{Sin}^{-1}x\right)$ (b) $f(x) = \text{Sin}^{-1}(\sin x)$

35. Sketch graphs of the following functions:
   (a) $f(x) = \cos\left(\text{Cos}^{-1}x\right)$ (b) $f(x) = \text{Cos}^{-1}(\cos x)$

36. Express the function $f(x) = 4\sin 2x + \cos 2x$ in the form $R\sin(2x+\phi)$ where $R > 0$ and $0 < \phi < \pi$.

37. Express the function $f(x) = -2\sin 3x + 4\cos 3x$ in the form $R\cos(3x+\phi)$ where $R > 0$ and $0 < \phi < \pi$.

38. Repeat Exercise 36 for $f(x) = -2\sin 2x + 4\cos 2x$.

39. Repeat Exercise 37 for $f(x) = -4\sin 3x + 5\cos 3x$.

40. A pendulum consists of a mass m suspended from a string of length L (Figure 8.20). At time $t = 0$, the mass is pulled through a small angle θ_0 — to the right when $\theta_0 > 0$, and to the left when $\theta_0 < 0$, and given an initial speed v_0 to the right. Its subsequent angular displacements are given by

$$\theta = \theta(t) = \theta_0 \cos\omega t + \frac{v_0}{\omega L}\sin\omega t, \quad t \geq 0,$$

where $\omega = \sqrt{9.81/L}$, provided any resistance due to the air is neglected. Show that $\theta(t)$ can be expressed in the form

$$\theta(t) = \sqrt{\theta_0^2 + \frac{v_0^2}{\omega^2 L^2}}\sin(\omega t + \phi),$$

where $\phi = \text{Tan}^{-1}(\omega L\theta_0/v_0)$.

FIGURE 8.20

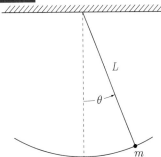

41. What changes, if any, must be made in Exercise 40 if the initial speed v_0 is to the left rather than the right?

42. A mass m is suspended from a spring with constant k (Figure 8.21). If, at time $t = 0$, the mass is given an initial displacement y_0 and an upward speed v_0, its subsequent displacements are given by

$$y = y(t) = y_0 \cos \omega t + \frac{v_0}{\omega} \sin \omega t, \quad t \geq 0,$$

where $\omega = \sqrt{k/m}$, provided any resistance due to the air is neglected. Show that $y(t)$ can be expressed in the form

$$y(t) = \sqrt{y_0^2 + \frac{v_0^2}{\omega^2}} \sin(\omega t + \phi),$$

where $\phi = \text{Tan}^{-1}(\omega y_0 / v_0)$.

FIGURE 8.21

$y = 0$ at equilibrium

43. What changes, if any, must be made in Exercise 42 if the initial speed v_0 is downward rather than upward?

44. An inductance L, resistance R, and capacitance C are connected in series with a generator producing an oscillatory voltage $E = E_0 \cos \omega t$, $t \geq 0$ (Figure 8.22). If L, C, R, $E_0 > 0$, and $\omega > 0$ are all constants, the steady-state current I in the circuit is

$$I = \frac{E_0}{\sqrt{R^2 + \left(\omega L - \frac{1}{\omega C}\right)^2}} \left[R \cos \omega t + \left(\omega L - \frac{1}{\omega C}\right) \sin \omega t \right].$$

Express I in the form

$$I = A \cos(\omega t - \phi),$$

where $A > 0$ and $-\pi/2 \leq \phi \leq \pi/2$.

FIGURE 8.22

45. Prove that

$$\text{Sec}^{-1} x = \begin{cases} \text{Cos}^{-1}\left(\frac{1}{x}\right) & x \geq 1 \\ -\text{Cos}^{-1}\left(\frac{1}{x}\right) & x \leq -1. \end{cases}$$

46. Prove that

$$\text{Cot}^{-1} x = \begin{cases} \text{Tan}^{-1}\left(\frac{1}{x}\right) & x > 0 \\ \pi + \text{Tan}^{-1}\left(\frac{1}{x}\right) & x < 0. \end{cases}$$

47. Verify that

$$\text{Sec}^{-1} x = \begin{cases} \text{Tan}^{-1} \sqrt{x^2 - 1} & x \geq 1 \\ -\pi + \text{Tan}^{-1} \sqrt{x^2 - 1} & x \leq -1. \end{cases}$$

48. Verify that

$$\text{Csc}^{-1} x = \begin{cases} \text{Cot}^{-1} \sqrt{x^2 - 1} & x \geq 1 \\ -\pi + \text{Cot}^{-1} \sqrt{x^2 - 1} & x \leq -1. \end{cases}$$

49. Verify that

$$\text{Cos}^{-1} x = \begin{cases} \text{Sin}^{-1} \sqrt{1 - x^2} & 0 \leq x \leq 1 \\ \pi - \text{Sin}^{-1} \sqrt{1 - x^2} & -1 \leq x < 0. \end{cases}$$

50. In Figure 8.23, a ray of light is incident on a prism at angle i. The light is refracted and leaves the prism at an angle of deviation ψ relative to the incident direction.

FIGURE 8.23

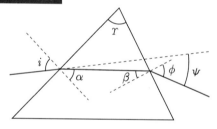

(a) Show that $\gamma = \alpha + \beta$, and use this to derive the expression

$$\psi = i + \phi - \gamma.$$

(b) To express ψ in terms of i and properties of the prism, we use Snell's law, which states that if the prism is in air, then the relationships between the incident and refracted light at the two faces are

$$n \sin \alpha = \sin i \quad \text{and} \quad n \sin \beta = \sin \phi,$$

where n is the refractive index of the prism. Use these equations to show that ψ can be expressed in the form

$$\psi = i - \gamma + \text{Sin}^{-1}\left(\sqrt{n^2 - \sin^2 i} \sin \gamma - \sin i \cos \gamma\right).$$

51. Show that for the prism in Exercise 50 the angle of deviation is a minimum when $i = \phi$. If ψ_m is the minimum angle of deviation, prove that

$$n = \frac{\sin\left[(\psi_m + \gamma)/2\right]}{\sin(\gamma/2)}.$$

In practice, this equation is used to determine the index of refraction of the prism.

52. Verify that if $0 \leq x < 1$, then

$$2\,\text{Tan}^{-1}\sqrt{\frac{1+x}{1-x}} = \pi - \text{Cos}^{-1} x.$$

53. Find intervals on which the function

$$f(x) = x + \sec\left(\frac{x}{2} - 1\right), \quad 0 \leq x \leq 5,$$

is increasing and decreasing.

SECTION 8.3

Derivatives of the Inverse Trigonometric Functions

Derivatives of inverse trigonometric functions are most easily obtained with implicit differentiation. We begin with the inverse sine function.

Theorem 8.2

$$\frac{d}{dx}\text{Sin}^{-1} x = \frac{1}{\sqrt{1-x^2}} \tag{8.9}$$

Proof If we set $y = \text{Sin}^{-1} x$, then $x = \sin y$, and we can differentiate implicitly with respect to x:

$$1 = \cos y \frac{dy}{dx}.$$

Thus,

$$\frac{dy}{dx} = \frac{1}{\cos y}.$$

To express $\cos y$ in terms of x we can proceed in two ways:
  (i) From the trigonometric identity $\sin^2 y + \cos^2 y = 1$, we obtain

$$\cos y = \pm\sqrt{1 - \sin^2 y} = \pm\sqrt{1 - x^2}.$$

We know that $y = \text{Sin}^{-1} x$ and the principal values of $\text{Sin}^{-1} x$ are $-\pi/2 \leq y \leq \pi/2$. Therefore y is an angle in either the first or fourth quadrant. It follows that $\cos y \geq 0$, and we must choose

$$\cos y = \sqrt{1 - x^2}.$$

  (ii) We use the triangle in Figure 8.24 to replace the trigonometric identity. The triangle is designed to fit the equation $x = \sin y$. The third side is then $\sqrt{1 - x^2}$. It follows that

$$\cos y = \sqrt{1 - x^2}.$$

But $\cos y$ is not always positive as indicated here. For example, if y is between $\pi/2$ and π, then $\cos y < 0$. To decide on the correct sign for $\cos y$, we use the discussion of principal values in (i), and obtain again $\cos y = \sqrt{1 - x^2}$.

FIGURE 8.24

Finally, then, we have

$$\frac{dy}{dx} = \frac{1}{\sqrt{1-x^2}}.$$

It is a straightforward application of the chain rule to obtain the following corollary.

Corollary *If $u(x)$ is a differentiable function, then*

$$\frac{d}{dx}\text{Sin}^{-1} u = \frac{1}{\sqrt{1-u^2}}\frac{du}{dx}. \tag{8.10}$$

Next we obtain the derivative of the inverse cosine function.

Theorem 8.3

$$\frac{d}{dx}\text{Cos}^{-1} x = \frac{-1}{\sqrt{1-x^2}} \tag{8.11}$$

Proof If we set $y = \text{Cos}^{-1} x$, then $x = \cos y$, and we can differentiate with respect to x,

$$1 = -\sin y \frac{dy}{dx},$$

and solve for

$$\frac{dy}{dx} = \frac{-1}{\sin y}.$$

The triangle in Figure 8.25, obtained from $x = \cos y$, yields $\sin y = \sqrt{1-x^2}$. Since the principal values of $y = \text{Cos}^{-1} x$ are $0 \leq y \leq \pi$, it follows that $\sin y$ is indeed nonnegative, and therefore

$$\frac{dy}{dx} = \frac{-1}{\sqrt{1-x^2}}.$$

FIGURE 8.25

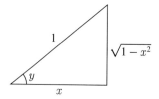

Corollary *If $u(x)$ is differentiable, then*

$$\frac{d}{dx}\text{Cos}^{-1} u = \frac{-1}{\sqrt{1-u^2}}\frac{du}{dx}. \tag{8.12}$$

Derivatives for the other inverse trigonometric functions can be derived in a similar way. We list them below and include derivatives of $\text{Sin}^{-1} u$ and $\text{Cos}^{-1} u$ for completeness:

$$\frac{d}{dx}\text{Sin}^{-1} u = \frac{1}{\sqrt{1-u^2}}\frac{du}{dx}; \qquad (8.10)$$

$$\frac{d}{dx}\text{Cos}^{-1} u = \frac{-1}{\sqrt{1-u^2}}\frac{du}{dx}; \qquad (8.12)$$

$$\frac{d}{dx}\text{Tan}^{-1} u = \frac{1}{1+u^2}\frac{du}{dx}; \qquad (8.13)$$

$$\frac{d}{dx}\text{Cot}^{-1} u = \frac{-1}{1+u^2}\frac{du}{dx}; \qquad (8.14)$$

$$\frac{d}{dx}\text{Sec}^{-1} u = \frac{1}{u\sqrt{u^2-1}}\frac{du}{dx}; \qquad (8.15)$$

$$\frac{d}{dx}\text{Csc}^{-1} u = \frac{-1}{u\sqrt{u^2-1}}\frac{du}{dx}. \qquad (8.16)$$

Note that the derivative of an inverse cofunction is the negative of the derivative of the corresponding inverse function; that is, derivatives of $\text{Cos}^{-1} u$, $\text{Cot}^{-1} u$, and $\text{Csc}^{-1} u$ are the negatives of the derivatives of $\text{Sin}^{-1} u$, $\text{Tan}^{-1} u$, and $\text{Sec}^{-1} u$.

EXAMPLE 8.10

Find dy/dx if y is defined as a function of x in each of the following:

(a) $y = \text{Sec}^{-1}(x^2)$ (b) $y = 2x\text{Cot}^{-1}(3x)$ (c) $y = \left[\text{Sin}^{-1}(x^3)\right]^2$

(d) $y = \dfrac{\text{Sin}^{-1} x}{\text{Cos}^{-1} x}$ (e) $\text{Cos}^{-1}(xy) + x^2 y + 5 = 0$

SOLUTION (a) $\dfrac{dy}{dx} = \dfrac{1}{x^2\sqrt{(x^2)^2 - 1}}\dfrac{d}{dx}(x^2) = \dfrac{2x}{x^2\sqrt{x^4 - 1}} = \dfrac{2}{x\sqrt{x^4 - 1}}$

(b) $\dfrac{dy}{dx} = 2\text{Cot}^{-1}(3x) + 2x\left(\dfrac{-1}{1+9x^2}\right)(3) = 2\text{Cot}^{-1}(3x) - \dfrac{6x}{1+9x^2}$

(c) $\dfrac{dy}{dx} = 2\text{Sin}^{-1}(x^3)\dfrac{d}{dx}\text{Sin}^{-1}(x^3) = 2\text{Sin}^{-1}(x^3)\dfrac{1}{\sqrt{1-x^6}}(3x^2) = \dfrac{6x^2\text{Sin}^{-1}(x^3)}{\sqrt{1-x^6}}$

(d) $\dfrac{dy}{dx} = \dfrac{\text{Cos}^{-1} x\left(\dfrac{1}{\sqrt{1-x^2}}\right) - \text{Sin}^{-1} x\left(\dfrac{-1}{\sqrt{1-x^2}}\right)}{\left(\text{Cos}^{-1} x\right)^2} = \dfrac{\text{Cos}^{-1} x + \text{Sin}^{-1} x}{\sqrt{1-x^2}\left(\text{Cos}^{-1} x\right)^2}$

(e) Differentiation with respect to x gives

$$\frac{-1}{\sqrt{1-x^2 y^2}}\left(y + x\frac{dy}{dx}\right) + 2xy + x^2\frac{dy}{dx} = 0.$$

Thus,

$$\frac{dy}{dx}\left(\frac{-x}{\sqrt{1-x^2 y^2}} + x^2\right) = \frac{y}{\sqrt{1-x^2 y^2}} - 2xy,$$

from which

$$\frac{dy}{dx} = \frac{y - 2xy\sqrt{1-x^2 y^2}}{\sqrt{1-x^2 y^2}}\frac{\sqrt{1-x^2 y^2}}{-x + x^2\sqrt{1-x^2 y^2}} = \frac{y - 2xy\sqrt{1-x^2 y^2}}{-x + x^2\sqrt{1-x^2 y^2}}.$$

∎

EXAMPLE 8.11

Find the derivative of the function $f(x) = \text{Sin}^{-1} x + \text{Cos}^{-1} x$. What is your conclusion?

SOLUTION The derivative of the function is

$$f'(x) = \frac{1}{\sqrt{1-x^2}} + \frac{-1}{\sqrt{1-x^2}} = 0.$$

Since this result is valid for the entire domain $-1 \leq x \leq 1$ of $f(x)$, we can antidifferentiate both sides of the equation,

$$f(x) = C = \text{ a constant}.$$

Since $f(0) = \text{Sin}^{-1}(0) + \text{Cos}^{-1}(0) = 0 + \pi/2 = \pi/2$, it follows that $C = \pi/2$. Consequently,

$$\text{Sin}^{-1} x + \text{Cos}^{-1} x = \frac{\pi}{2}.$$

EXAMPLE 8.12

A road information sign specifying the distance to the next city is 1 m high and sits 5 m above road level (Figure 8.26). If the average motorist's eye level is 1.5 m above ground level, at what distance from the sign do the letters appear tallest?

FIGURE 8.26

FIGURE 8.27

SOLUTION The letters appear tallest when the angle θ that they subtend at the eye (Figure 8.27) is greatest. We assume that the letters are 1 m high. Since

$$\theta = \beta - \alpha = \text{Tan}^{-1}\left(\frac{9/2}{x}\right) - \text{Tan}^{-1}\left(\frac{7/2}{x}\right), \quad 0 < x < \infty,$$

critical points of $\theta(x)$ are given by

$$0 = \frac{d\theta}{dx} = \frac{1}{1 + \left(\frac{9}{2x}\right)^2}\left(-\frac{9}{2x^2}\right) - \frac{1}{1 + \left(\frac{7}{2x}\right)^2}\left(-\frac{7}{2x^2}\right)$$

$$= \frac{-18}{4x^2 + 81} + \frac{14}{4x^2 + 49}.$$

Equivalently,

$$18(4x^2 + 49) = 14(4x^2 + 81),$$

which has only one positive solution, $x = 3\sqrt{7}/2$. Since $\theta(x)$ approaches zero as x approaches zero and as x becomes very large, it follows that θ is maximized when the motorist is $3\sqrt{7}/2$ m from the sign.

EXERCISES 8.3

In Exercises 1–30 y is defined as a function of x. Find dy/dx in as simplified a form as possible.

1. $y = \text{Cos}^{-1}(2x + 3)$
2. $y = \text{Cot}^{-1}(x^2 + 2)$
3. $y = \text{Csc}^{-1}(3 - 4x)$
4. $y = \text{Tan}^{-1}(2 - x^2)$
5. $y = \text{Sec}^{-1}(3 - 2x^2)$
6. $y = x\text{Csc}^{-1}(x^2 + 5)$
7. $y = (x^2 + 2)\text{Sin}^{-1}(2x)$
8. $y = \text{Tan}^{-1}\sqrt{x+2}$
9. $y = \text{Sin}^{-1}\sqrt{1-x^2}$
10. $y = \text{Cot}^{-1}\sqrt{x^2-1}$
11. $y = \left[\text{Tan}^{-1}(x^2)\right]^2$
12. $y = x^2\text{Sec}^{-1}x$
13. $y = \tan(3\text{Sin}^{-1}x)$
14. $y = \text{Cot}^{-1}[(1+x)/(1-x)]$
15. $y = \text{Csc}^{-1}(1/x)$
16. $y = \text{Sin}^{-1}[(1-x)/(1+x)]$
17. $y = \text{Tan}^{-1}(u^2 + 1/u),\ u = \tan(x^2 + 4)$
18. $y = t\text{Cos}^{-1}t,\ t = \sqrt{1-x^2}$
19. $y^2\sin x + y = \text{Tan}^{-1}x$
20. $\text{Sin}^{-1}(xy) = 5x + 2y$
21. $y = \sqrt{x^2 - 1} - \text{Sec}^{-1}x$
22. $y = \dfrac{1}{3}\text{Csc}^{-1}\left(\dfrac{x}{3}\right) - \dfrac{\sqrt{x^2-9}}{x^2}$
23. $y = x\text{Cos}^{-1}\left(\dfrac{x}{2}\right) - \sqrt{4 - x^2}$
24. $y = \dfrac{\text{Csc}^{-1}(3x)}{x} - \dfrac{\sqrt{9x^2 - 1}}{x}$
25. $y = x^2\text{Sec}^{-1}x - \sqrt{x^2 - 1}$
26. $y = x\left(\text{Cos}^{-1}x\right)^2 - 2x - 2\sqrt{1-x^2}$
27. $y = (1 + 9x^2)\text{Cot}^{-1}(3x) + 3x$
28. $y = (x - 2)\sqrt{4x - x^2} + 4\text{Sin}^{-1}\left(\dfrac{x-2}{2}\right)$
29. $y = (2x^2 - 1)\text{Sin}^{-1}x + x\sqrt{1 - x^2}$
30. $y = \text{Tan}^{-1}\left(\dfrac{\sqrt{2}\,x}{\sqrt{1 + x^4}}\right)$

31. Evaluate the derivative of
$$f(x) = \text{Sec}^{-1}x + \text{Cot}^{-1}\sqrt{x^2 - 1}.$$
What is your conclusion?

32. Find all relative extrema and points of inflection for the function
$$f(x) = x\text{Sin}^{-1}x + \sqrt{1 - x^2}.$$
Sketch its graph.

33. (a) Show that if the principal values of $f(x) = \text{Sec}^{-1}x$ are chosen as $0 \le y \le \pi,\ y \ne \pi/2$, then the derivative of the function is
$$\dfrac{d}{dx}\text{Sec}^{-1}x = \dfrac{1}{|x|\sqrt{x^2 - 1}}.$$

(b) What is the derivative of $f(x) = \text{Csc}^{-1}x$ if its principal values are chosen as $-\pi/2 \le y \le \pi/2$, $y \ne 0$?

34. Find the angle between the curves $y = \text{Sin}^{-1}x$ and $y = \text{Cos}^{-1}x$ at their point of intersection.

35. Verify the results in equations 8.13–8.16.

36. When a force F is applied to the object of mass m in Figure 8.28, three other forces act on m: the force of gravity mg ($g > 0$) directly downward, a reactional force of the supporting surface, and a horizontal, frictional force opposing F. The least force F that will overcome friction and produce motion is given by
$$F = \dfrac{\mu mg}{\cos\theta + \mu\sin\theta},$$
where μ is a constant called the *coefficient of static friction*. Find the angle θ for which F is minimal.

FIGURE 8.28

37. Find d^2y/dx^2 if y is defined implicitly as a function of x by
$$\ln(x^2 + y^2) = 2\text{Tan}^{-1}\left(\dfrac{y}{x}\right).$$

38. A runner moves counterclockwise around the track in Figure 8.29 at a rate of 4 m/s. A camera at the centre of the track is placed on a swivel so that it can follow the runner. Find the rate at which the camera turns when (a) the runner is at A and (b) the runner is at B.

FIGURE 8.29

39. A rope with a ring at one end is passed through two fixed rings at the same level (Figure 8.30). The end of the rope without the ring is then passed through the ring at the other end, and a mass m is attached to it. If the rope moves so as to maximize the distance from m to the line through the fixed rings, find angle θ.

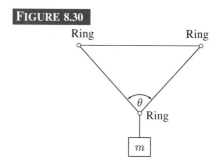

FIGURE 8.30

SECTION 8.4

Hyperbolic Functions and Their Derivatives

Certain combinations of the exponential function occur so often in physical applications that they are given special names. Specifically, half the difference of e^x and e^{-x} is defined as the hyperbolic sine function and half their sum is the hyperbolic cosine function. These functions are denoted as follows:

$$\sinh x = \frac{e^x - e^{-x}}{2} \quad \text{and} \quad \cosh x = \frac{e^x + e^{-x}}{2}. \tag{8.17}$$

The names of these functions and their notations bear a striking resemblance to those for the trigonometric functions, and there are reasons for this. First, the hyperbolic functions $\sinh x$ and $\cosh x$ are related to the curve $x^2 - y^2 = 1$, called the *unit hyperbola*, in much the same way as the trigonometric functions $\sin x$ and $\cos x$ are related to the unit circle $x^2 + y^2 = 1$. We will point out two of these similarities in Examples 8.16 and 8.17. Second, for each identity satisfied by the trigonometric functions, there is a corresponding identity satisfied by the hyperbolic functions — not the same identity, but one very similar. For example, using equations 8.17, we have

$$(\cosh x)^2 - (\sinh x)^2 = \left(\frac{e^x + e^{-x}}{2}\right)^2 - \left(\frac{e^x - e^{-x}}{2}\right)^2$$
$$= \frac{1}{4}\left[(e^{2x} + 2 + e^{-2x}) - (e^{2x} - 2 + e^{-2x})\right]$$
$$= 1.$$

Thus the hyperbolic sine and cosine functions satisfy the identity

$$\cosh^2 x - \sinh^2 x = 1, \tag{8.18}$$

which is reminiscent of the identity $\cos^2 x + \sin^2 x = 1$ for the trigonometric functions.

Just as four other trigonometric functions are defined in terms of $\sin x$ and $\cos x$, four corresponding hyperbolic functions are defined as follows:

$$\tanh x = \frac{\sinh x}{\cosh x} = \frac{e^x - e^{-x}}{e^x + e^{-x}}, \quad \coth x = \frac{\cosh x}{\sinh x} = \frac{e^x + e^{-x}}{e^x - e^{-x}},$$
$$\operatorname{sech} x = \frac{1}{\cosh x} = \frac{2}{e^x + e^{-x}}, \quad \operatorname{csch} x = \frac{1}{\sinh x} = \frac{2}{e^x - e^{-x}}. \tag{8.19}$$

These definitions and 8.18 immediately imply that

$$1 - \tanh^2 x = \operatorname{sech}^2 x, \qquad (8.20\text{a})$$
$$\coth^2 x - 1 = \operatorname{csch}^2 x, \qquad (8.20\text{b})$$

analogous to $1 + \tan^2 x = \sec^2 x$ and $1 + \cot^2 x = \csc^2 x$ respectively.

Graphs of $\cosh x$ and $\sinh x$ in Figures 8.31(a) and (b) are easily obtained by addition and subtraction of ordinates of e^x and e^{-x}. Note that $\cosh x$ is an even function, whereas $\sinh x$ is odd, and that neither is periodic. To sketch $\tanh x$, we use division of ordinates, together with the facts that the function is odd and that $\lim_{x \to \infty} \tanh x = 1$ in Figure 8.31(c). These three graphs then lead to the graphs for $\coth x$, $\operatorname{sech} x$, and $\operatorname{csch} x$ in Figures 8.31(d)–(f).

FIGURE 8.31

(a)

(b)

(c)

(d)

(e)

(f)
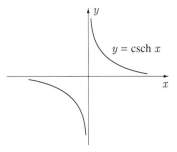

In Appendix A we point out that most trigonometric identities can be derived from the compound angle formulas for $\sin(A \pm B)$ and $\cos(A \pm B)$. It is easy to verify similar formulas for the hyperbolic functions:

$$\sinh(A \pm B) = \sinh A \cosh B \pm \cosh A \sinh B \qquad \text{and} \qquad (8.21\text{a})$$

$$\cosh(A \pm B) = \cosh A \cosh B \pm \sinh A \sinh B. \tag{8.21b}$$

For example, equations 8.17 give

$$\cosh A \cosh B - \sinh A \sinh B = \left(\frac{e^A + e^{-A}}{2}\right)\left(\frac{e^B + e^{-B}}{2}\right)$$
$$- \left(\frac{e^A - e^{-A}}{2}\right)\left(\frac{e^B - e^{-B}}{2}\right)$$
$$= \frac{1}{4}\left[\left(e^{A+B} + e^{A-B} + e^{B-A} + e^{-A-B}\right) - \left(e^{A+B} - e^{A-B} - e^{B-A} + e^{-A-B}\right)\right]$$
$$= \frac{1}{2}\left[e^{A-B} + e^{-(A-B)}\right]$$
$$= \cosh(A - B).$$

With these formulas, we can derive hyperbolic identities analogous to trigonometric identities (A.12)–(A.17):

$$\tanh(A \pm B) = \frac{\tanh A \pm \tanh B}{1 \pm \tanh A \tanh B}, \tag{8.21c}$$

$$\sinh 2A = 2 \sinh A \cosh A, \tag{8.21d}$$

$$\cosh 2A = \cosh^2 A + \sinh^2 A \tag{8.21e}$$

$$= 2 \cosh^2 A - 1 \tag{8.21f}$$

$$= 1 + 2 \sinh^2 A, \tag{8.21g}$$

$$\tanh 2A = \frac{2 \tanh A}{1 + \tanh^2 A}, \tag{8.21h}$$

$$\sinh A \sinh B = \frac{1}{2} \cosh(A + B) - \frac{1}{2} \cosh(A - B), \tag{8.21i}$$

$$\sinh A \cosh B = \frac{1}{2} \sinh(A + B) + \frac{1}{2} \sinh(A - B), \tag{8.21j}$$

$$\cosh A \cosh B = \frac{1}{2} \cosh(A + B) + \frac{1}{2} \cosh(A - B), \tag{8.21k}$$

$$\sinh A + \sinh B = 2 \sinh\left(\frac{A + B}{2}\right) \cosh\left(\frac{A - B}{2}\right), \tag{8.21l}$$

$$\sinh A - \sinh B = 2 \cosh\left(\frac{A + B}{2}\right) \sinh\left(\frac{A - B}{2}\right), \tag{8.21m}$$

$$\cosh A + \cosh B = 2 \cosh\left(\frac{A + B}{2}\right) \cosh\left(\frac{A - B}{2}\right), \tag{8.21n}$$

$$\cosh A - \cosh B = 2 \sinh\left(\frac{A + B}{2}\right) \sinh\left(\frac{A - B}{2}\right). \tag{8.21o}$$

Since the hyperboic sine and cosine are defined in terms of the exponential function, for which we know the derivative, and the remaining hyperbolic functions are defined in terms of the hyperbolic sine and cosine, it follows that calculation of the derivatives of the hyperbolic functions should be straightforward. Indeed, if $u(x)$ is a differentiable function of x, then

$$\frac{d}{dx} \sinh u = \cosh u \, \frac{du}{dx}, \tag{8.22a}$$

$$\frac{d}{dx} \cosh u = \sinh u \, \frac{du}{dx}, \tag{8.22b}$$

$$\frac{d}{dx}\tanh u = \operatorname{sech}^2 u \, \frac{du}{dx}, \tag{8.22c}$$

$$\frac{d}{dx}\coth u = -\operatorname{csch}^2 u \, \frac{du}{dx}, \tag{8.22d}$$

$$\frac{d}{dx}\operatorname{sech} u = -\operatorname{sech} u \tanh u \, \frac{du}{dx}, \tag{8.22e}$$

$$\frac{d}{dx}\operatorname{csch} u = -\operatorname{csch} u \coth u \, \frac{du}{dx}. \tag{8.22f}$$

EXAMPLE 8.13

Find dy/dx if y is defined as a function of x by:

(a) $y = \operatorname{sech}(3x^2)$
(b) $y = \tanh(1 - 4x)$
(c) $y = \cos 2x \sinh 2x$
(d) $y = \cosh(\operatorname{Tan}^{-1} x^2)$

SOLUTION

(a) $\dfrac{dy}{dx} = -\operatorname{sech}(3x^2)\tanh(3x^2)\dfrac{d}{dx}(3x^2)$
$= -6x\operatorname{sech}(3x^2)\tanh(3x^2)$

(b) $\dfrac{dy}{dx} = \operatorname{sech}^2(1-4x)\dfrac{d}{dx}(1-4x) = -4\operatorname{sech}^2(1-4x)$

(c) $\dfrac{dy}{dx} = \cos 2x(2\cosh 2x) + (-2\sin 2x)\sinh 2x$
$= 2(\cos 2x \cosh 2x - \sin 2x \sinh 2x)$

(d) $\dfrac{dy}{dx} = \sinh(\operatorname{Tan}^{-1} x^2)\dfrac{d}{dx}\operatorname{Tan}^{-1} x^2 = \sinh(\operatorname{Tan}^{-1} x^2)\dfrac{2x}{1+x^4}$

■

EXAMPLE 8.14

If a uniform cable is hung between two fixed supports (Figure 8.32), the shape of the curve $y = f(x)$ must satisfy the differential equation

$$\frac{d^2 y}{dx^2} = \frac{\rho g}{H}\sqrt{1 + \left(\frac{dy}{dx}\right)^2},$$

where ρ is the mass per unit length of the cable, $g > 0$ is the acceleration due to gravity, and $H > 0$ is the tension in the cable at its lowest point. Show that a solution of the equation is

$$y = f(x) = \frac{H}{\rho g}\cosh\left(\frac{\rho g x}{H}\right) + C,$$

where C is a constant.

FIGURE 8.32

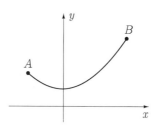

SOLUTION In Exercises 41 of Section 9.5 we derive this solution. For now we simply wish to verify that the hyperbolic cosine is indeed a solution. The first derivative of the function is

$$\frac{dy}{dx} = \sinh\left(\frac{\rho g x}{H}\right),$$

and therefore its second derivative is

$$\frac{d^2 y}{dx^2} = \frac{\rho g}{H} \cosh\left(\frac{\rho g x}{H}\right).$$

On the other hand,

$$\frac{\rho g}{H}\sqrt{1 + \left(\frac{dy}{dx}\right)^2} = \frac{\rho g}{H}\sqrt{1 + \sinh^2\left(\frac{\rho g x}{H}\right)} = \frac{\rho g}{H}\sqrt{\cosh^2\left(\frac{\rho g x}{H}\right)} = \frac{\rho g}{H}\cosh\left(\frac{\rho g x}{H}\right).$$

Thus, the function $y = (H/(\rho g))\cosh(\rho g x/H) + C$ does indeed satisfy the given differential equation. ∎

Example 8.14 shows that the many telephone and hydro wires crisscrossing the country hang in the form of hyperbolic cosines. Engineers often call this curve a *catenary*.

EXAMPLE 8.15

In studying wave guides, the electrical engineer often encounters the differential equation

$$\frac{d^2 y}{dx^2} - ky = 0,$$

where $k > 0$ is a constant. Show that $y = f(x) = A\cosh\sqrt{k}x + B\sinh\sqrt{k}x$ is a solution for any constants A and B.

SOLUTION The first derivative of $y = f(x)$ is

$$\frac{dy}{dx} = \sqrt{k}A\sinh\sqrt{k}x + \sqrt{k}B\cosh\sqrt{k}x,$$

and therefore

$$\frac{d^2 y}{dx^2} = kA\cosh\sqrt{k}x + kB\sinh\sqrt{k}x = k(A\cosh\sqrt{k}x + B\sinh\sqrt{k}x).$$

Thus,

$$\frac{d^2 y}{dx^2} = ky.$$

∎

In Examples 8.16 and 8.17 we illustrate geometric parallels between the trigonometric sine and cosine functions and the hyperbolic sine and cosine functions.

EXAMPLE 8.16

Show that:
(a) every point (x, y) on the unit circle $x^2 + y^2 = 1$ can be expressed in the form $x = \cos t$, $y = \sin t$ for some real number t in the interval $0 \leq t \leq 2\pi$;
(b) every point (x, y) on the right half of the unit hyperbola $x^2 - y^2 = 1$ can be expressed in the form $x = \cosh t$, $y = \sinh t$ for some real number t.

SOLUTION
(a) If t is the angle shown in Figure 8.33, then clearly the coordinates of P are $x = \cos t$ and $y = \sin t$. As angle t ranges for 0 to 2π, P traces the circle exactly once.

FIGURE 8.33

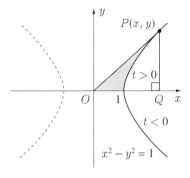

FIGURE 8.34

(b) A sketch of the unit hyperbola $x^2 - y^2 = 1$ is shown in Figure 8.34. If $x = \cosh t$ and $y = \sinh t$ are coordinates of a point P in the plane, where t is some real number, then identity 8.18 implies that $x^2 - y^2 = \cosh^2 t - \sinh^2 t = 1$. In other words, P is on the unit hyperbola. Furthermore, since $x = \cosh t$ is always positive, P must be on the right half of the hyperbola. Finally, because the range of $x = \cosh t$ is $x \geq 1$ in Figure 8.31(a), and the range of $y = \sinh t$ is $-\infty < y < \infty$ in Figure 8.31(b), it follows that every point on the right half of the hyperbola can be obtained from some value of t.

Note that t is *not* the angle formed by the positive x-axis and the line joining the origin to (x, y). ∎

EXAMPLE 8.17

Show that the area bounded by the positive x-axis, the line joining the origin to the point $P(x, y)$ on the curves in Figures 8.33 and 8.34, and the curves themselves is $t/2$.

SOLUTION (a) Since the area of a sector of a circle is one-half the product of the square of the radius and the angle at the centre of the circle, the shaded area in Figure 8.33 is $(1/2)(1)^2(t) = t/2$.

(b) The shaded area in Figure 8.34 is the area of triangle OQP less the area beneath the hyperbola between $x = 1$ and Q:

$$\text{Area} = \frac{1}{2}(x)(y) - \int_1^x y\, dx.$$

Since the coordinates of P can be expressed in the form $x = \cosh t$ and $y = \sinh t$ (Example 8.16), we may set

$$\text{Area} = \frac{1}{2}\cosh t \sinh t - \int_0^t \sinh t (\sinh t \, dt).$$

Now, by identity 8.21d, $\cosh t \sinh t = (1/2)\sinh 2t$, and by identity 8.21f, $\sinh^2 t = (1/2)(\cosh 2t - 1)$. Therefore

$$\text{Area} = \frac{1}{4}\sinh 2t - \frac{1}{2}\int_0^t (\cosh 2t - 1)\, dt.$$

Since the derivative of $\sinh 2t$ is $2\cosh 2t$, it follows that an antiderivative of $\cosh 2t$ is $(1/2)\sinh 2t$, and

$$\text{Area} = \frac{1}{4}\sinh 2t - \frac{1}{2}\left\{\frac{1}{2}\sinh 2t - t\right\}_0^t = \frac{t}{2}.$$

EXERCISES 8.4

In Exercises 1–10 evaluate the expression (if it has a value).

1. $3\cosh 1$
2. $\sinh(\pi/2)$
3. $\tanh\sqrt{1 - \sin 3}$
4. $\operatorname{Sin}^{-1}(\operatorname{sech} 10)$
5. $\operatorname{Cos}^{-1}(2\operatorname{csch} 1)$
6. $\coth(\sinh 5)$
7. $\sqrt{\ln|\sinh(-3)|}$
8. $\operatorname{sech}[\sec(\pi/3)]$
9. $e^{-2\cosh e}$
10. $\sinh[\operatorname{Cot}^{-1}(-3\pi/10)]$

In Exercises 11–20 y is defined as a function of x. Find dy/dx in as simplified a form as possible.

11. $y = \operatorname{csch}(2x + 3)$
12. $y = x\sinh(x/2)$
13. $y = \sqrt{1 - \operatorname{sech} x}$
14. $y = \tanh(\ln x)$
15. $\cosh(x + y) = 2x$
16. $y + \coth x = \sqrt{1 + y}$
17. $y = u\cosh u$, $u = e^x + e^{-x}$
18. $y = \tan(\cosh t)$, $t = \cos(\tanh x)$
19. $y = \operatorname{Tan}^{-1}(\sinh x)$
20. $y = \ln\sqrt{\tanh 2x}$

21. Verify the results in identities 8.21c–8.21o.

22. Verify differentiation formulas 8.22.

23. Show that $y = \sinh x$ and $y = \tanh x$ have points of inflection at $x = 0$.

24. To analyze vertical vibrations of the beam in Figure 8.35, we must solve the differential equation

$$\frac{d^4 y}{dx^4} - k^4 y = 0,$$

where $k > 0$ is a constant.

FIGURE 8.35

(a) Show that a solution is

$$y = f(x) = A\cos kx + B\sin kx + C\cosh kx + D\sinh kx$$

for any constants A, B, C, and D.

(b) If the left end ($x = 0$) is fastened horizontally and the right end ($x = L$) is pinned, then $f(x)$ must satisfy the conditions

$$f(0) = f'(0) = f(L) = f''(L) = 0.$$

Show that these restrictions imply that $C = -A$, $D = -B$, and A and B must satisfy the equations

$$A(\cos kL - \cosh kL) + B(\sin kL - \sinh kL) = 0,$$
$$A(\cos kL + \cosh kL) + B(\sin kL + \sinh kL) = 0.$$

(c) Eliminate A and B between these equations to show that k must satisfy the condition

$$\tan kL = \tanh kL.$$

25. Show that $y = \operatorname{sech} x$ has points of inflection for $x = \pm \ln(\sqrt{2} + 1)$.

26. A uniform hydro cable P metres long with mass ρ kilograms per metre of length is hung from two supports at the same level L metres apart (Figure 8.36). If the supports are b metres higher than the lowest point in the cable and H is the tension in the cable at the lowest point, it can be shown that

$$\rho g P = 2 H \sinh \frac{\rho g L}{2 H}.$$

FIGURE 8.36

(a) If $L = 70$ m, $\rho = 0.5$ kg/m, and $P = 80$ m, find H. Hint: Set $z = \rho g/(2H)$, and use Newton's iterative procedue with $z_1 = 0.01$ to solve $Pz = \sinh(Lz)$ for z.

(b) Use the result in (a) to find b. See Example 8.14 for the equation of the curve describing the cable.

27. Each hyperbolic function has associated with it an inverse hyperbolic function.

(a) Draw the inverse functions $\operatorname{Sinh}^{-1} x$, $\operatorname{Tanh}^{-1} x$, $\operatorname{Coth}^{-1} x$, and $\operatorname{Csch}^{-1} x$ for $\sinh x$, $\tanh x$, $\coth x$, and $\operatorname{csch} x$.

(b) Why do $\cosh x$ and $\operatorname{sech} x$ not have inverse functions? It is customary to associate functions $\operatorname{Cosh}^{-1} x$ and $\operatorname{Sech}^{-1} x$ with $\cosh x$ and $\operatorname{sech} x$ by restricting their domains to nonnegative numbers. Draw graphs of $\operatorname{Cosh}^{-1} x$ and $\operatorname{Sech}^{-1} x$.

(c) Obtain the following derivatives of the inverse hyperbolic functions:

$$\frac{d}{dx} \operatorname{Sinh}^{-1} x = \frac{1}{\sqrt{x^2 + 1}};$$

$$\frac{d}{dx} \operatorname{Cosh}^{-1} x = \frac{1}{\sqrt{x^2 - 1}};$$

$$\frac{d}{dx} \operatorname{Tanh}^{-1} x = \frac{1}{1 - x^2}, \quad |x| < 1;$$

$$\frac{d}{dx} \operatorname{Coth}^{-1} x = \frac{1}{1 - x^2}, \quad |x| > 1;$$

$$\frac{d}{dx} \operatorname{Sech}^{-1} x = \frac{-1}{x\sqrt{1 - x^2}};$$

$$\frac{d}{dx} \operatorname{Csch}^{-1} x = \frac{-1}{|x|\sqrt{1 + x^2}}.$$

(d) Show that

$$\operatorname{Sinh}^{-1} x = \ln\left(x + \sqrt{x^2 + 1}\right);$$

$$\operatorname{Cosh}^{-1} x = \ln\left(x + \sqrt{x^2 - 1}\right);$$

$$\operatorname{Tanh}^{-1} x = \frac{1}{2} \ln\left(\frac{1 + x}{1 - x}\right), \quad |x| < 1.$$

28. When an object of mass m fall from rest at time $t = 0$ under the influence of gravity and an air resistance proportional to the square of velocity, its velocity v as a function of time is defined by the equation

$$\frac{1}{2}\sqrt{\frac{\beta}{mg}} \ln\left(\frac{\sqrt{mg/\beta} - v}{\sqrt{mg/\beta} + v}\right) = -\frac{\beta t}{m},$$

where $\beta > 0$ is a constant and $g > 0$ is the acceleration due to gravity.

(a) Show that when this equation is solved for v in terms of t, the result is

$$v(t) = \sqrt{\frac{mg}{\beta}} \tanh\left(\sqrt{\frac{\beta g}{m}} t\right).$$

(b) What is the limit of v for large t, called the *limiting velocity*?

SUMMARY

With derivatives of trigonometric, exponential, and logarithm functions in Chapter 3, and derivatives of inverse trigonometric and hyperbolic functions in this chapter, we have completed the list of differentiation formulas.

Inverse trigonometric functions reverse the roles of trigonometric functions. A trigonometric function such as the sine function associates a value called $\sin x$ with a real number (angle) x. The corresponding inverse function, $\text{Sin}^{-1} x$, regards x as the sine of an angle, and yields the angle.

Hyperbolic functions are special combinations of exponential functions that arise sufficiently often in applications to warrant special consideration. They satisfy identities and have derivatives very similar to the trigonometric functions.

The transcendental functions that we have discussed fall into three pairs: trigonometric and inverse trigonometric, exponential and logarithm (the inverse of the exponential), and hyperbolic and inverse hyperbolic (Exercise 27 in Section 8.4). The relationship between a function and its inverse is very important. For instance, evaluation of the indeterminate forms 0^0, 1^∞, and ∞^0 in Section 4.9 was based on the fact that taking logarithms and exponentials are inverse operations. By taking logarithms, we transformed each indeterminate form into a limit to which we could apply L'Hôpital's rule. We then used exponentials to recapture the original limit. In Section 9.2, the inverse character of exponential and logarithm functions is indispensable to the solution of many differential equations. The integration technique of Section 9.5 is based on the relation of an inverse trigonometric function to the corresponding trigonometric function.

Key Terms and Formulas

In reviewing this chapter, you should be able to define or discuss the following key terms:

Algebraic function
Transcendental function
One-to-one function
Inverse function
Horizontal line test

Monotonic function
Inverse trigonometric function
Principal values
Hyperbolic function

REVIEW EXERCISES

In Exercises 1–10 find all solutions of the equation.

1. $\cos^2 x + 5 \cos x - 1 = 0$

2. $4 \sin 2x = 1$

3. $\csc^2 (x + 1) = 3$

4. $\text{Tan}^{-1} (3x + 2) = 5 - 2\pi$

5. $\cos 2x = \sin x$

6. $\ln (\sin x) + \ln (1 + \sin x) = \ln 3 - \ln 2$

7. $3 \text{Sin}^{-1} \left(e^{x+2} \right) = 2$

8. $3 \sin \left(e^{x+2} \right) = 2$

9. $\tan (x \cosh 2) = 1/4$

10. $\sinh x = 4$

In Exercises 11–20 y is defined as a function of x. Find dy/dx in as simplified a form as possible.

11. $y = \text{Sin}^{-1} (2 - 3x)$

12. $y = 3 \sinh (x^2)$

13. $y = \dfrac{\text{Cos}^{-1} x}{\text{Sin}^{-1} x}$

14. $y = \text{Tan}^{-1} \left(\dfrac{1}{x} + x \right)$

15. $y = e^{\cosh x}$

16. $\sinh y = \sin x$

17. $\text{Sec}^{-1} (x + y) = xy$

18. $y = x \text{Csc}^{-1} \left(\dfrac{1}{x^2} \right)$

19. $y = e^{2x} \cosh 2x$

20. $\ln \left[\text{Tan}^{-1} (x + y) \right] = 1$

21. Sketch graphs of the following functions:

(a) $f(x) = \tan \left(\text{Tan}^{-1} x \right)$ (b) $f(x) = \text{Tan}^{-1} (\tan x)$

CHAPTER 9

Techniques for Integration

In Chapter 5, when we defined indefinite integrals as antiderivatives, we discussed three techniques for evaluating antiderivatives. First, since every differentiation formula can be restated as an antidifferentiation formula, it follows that the more competent we are at differentiation, the more likely we are to recognize an antiderivative. Our second technique was called "guess and fix up." It is applicable to sufficiently simple integrands that we can immediately determine the antiderivative to within a multiplicative constant. It is then a matter of adjusting the constant. The third technique was to change the variable of integration. By a suitable transformation, replace a complicated antiderivative with a simpler one.

In this chapter we develop additional techniques applicable to more complex integration problems. At the same time we take the opportunity to review the applications of definite integrals in Chapter 7.

SECTION 9.1

Integration Formulas and Substitutions

Every differentiation formula can be restated as an antidifferentiation formula. A partial list of important formulas is as follows:

$$\int u^n \, du = \frac{u^{n+1}}{n+1} + C, \quad n \neq -1; \tag{9.1}$$

$$\int \frac{1}{u} \, du = \ln |u| + C; \tag{9.2}$$

$$\int e^u \, du = e^u + C; \tag{9.3}$$

$$\int a^u \, du = a^u \log_a e + C; \tag{9.4}$$

$$\int \cos u \, du = \sin u + C; \tag{9.5}$$

$$\int \sin u \, du = -\cos u + C; \tag{9.6}$$

$$\int \sec^2 u \, du = \tan u + C; \tag{9.7}$$

$$\int \csc^2 u \, du = -\cot u + C; \tag{9.8}$$

$$\int \sec u \tan u \, du = \sec u + C; \tag{9.9}$$

$$\int \csc u \cot u \, du = -\csc u + C; \tag{9.10}$$

$$\int \tan u \, du = \ln|\sec u| + C = -\ln|\cos u| + C; \tag{9.11}$$

$$\int \cot u \, du = -\ln|\csc u| + C = \ln|\sin u| + C; \tag{9.12}$$

$$\int \sec u \, du = \ln|\sec u + \tan u| + C; \tag{9.13}$$

$$\int \csc u \, du = \ln|\csc u - \cot u| + C. \tag{9.14}$$

Formulas 9.1–9.10 are restatements of differentiation formulas 5.5 and 5.6 in Chapter 5. Formulas 9.11 and 9.13 are verified in Example 9.1 below. Formulas 9.12 and 9.14 are similar.

EXAMPLE 9.1

Verify formulas 9.11 and 9.13.

SOLUTION To verify 9.11 we express $\tan u$ in terms of $\sin u$ and $\cos u$,

$$\int \tan u \, du = \int \frac{\sin u}{\cos u} \, du = -\ln|\cos u| + C = \ln|\sec u| + C.$$

If integration to $\ln|\cos u|$ is not obvious, try the substitution $v = \cos u$.

To obtain the antiderivative of $\sec u$ requires a subtle trick. We first multiply numerator and denominator by $\sec u + \tan u$,

$$\int \sec u \, du = \int \sec u \frac{\sec u + \tan u}{\sec u + \tan u} \, du = \int \frac{\sec^2 u + \sec u \tan u}{\sec u + \tan u} \, du.$$

If we now set $v = \sec u + \tan u$, then $dv = (\sec u \tan u + \sec^2 u) \, du$, and

$$\int \sec u \, du = \int \frac{1}{v} \, dv = \ln|v| + C = \ln|\sec u + \tan u| + C.$$

We could list many other integration formulas, but one of the purposes of this chapter is to develop integration techniques that will eliminate the need for excessive memorization of formulas. Our emphasis will be on the development of antiderivatives rather than the memorization of formulas, and we prefer to keep the list of formulas short.

In Chapter 5 we demonstrated how a change of variable could sometimes replace a complex integration problem with a simpler one. When the integrand contains $x^{1/n}$, where n is a positive integer, it is often useful to set $u = x^{1/n}$, or, equivalently, $x = u^n$. We illustrate in the following example.

EXAMPLE 9.2

Evaluate the following indefinite integrals:

(a) $\displaystyle\int \frac{2-\sqrt{x}}{2+\sqrt{x}}\,dx$ (b) $\displaystyle\int \frac{x^{1/3}}{1+x^{2/3}}\,dx$ (c) $\displaystyle\int \frac{\sqrt{x}}{1+x^{1/3}}\,dx$

SOLUTION (a) If we set $u=\sqrt{x}$ or $x=u^2$, then $dx=2u\,du$, and

$$\int \frac{2-\sqrt{x}}{2+\sqrt{x}}\,dx = \int \frac{2-u}{2+u}(2u\,du) = 2\int \frac{2u-u^2}{2+u}\,du.$$

Long division of $-u^2+2u$ by $u+2$ immediately gives

$$\int \frac{2-\sqrt{x}}{2+\sqrt{x}}\,dx = 2\int\left(-u+4-\frac{8}{u+2}\right)du$$

$$= 2\left(-\frac{u^2}{2}+4u-8\ln|u+2|\right)+C$$

$$= -x+8\sqrt{x}-16\ln(2+\sqrt{x})+C.$$

(b) If we set $u=x^{1/3}$ or $x=u^3$, then $dx=3u^2\,du$, and

$$\int \frac{x^{1/3}}{1+x^{2/3}}\,dx = \int \frac{u}{1+u^2}(3u^2\,du) = 3\int \frac{u^3}{1+u^2}\,du.$$

Division of u^3 by u^2+1 leads to

$$\int \frac{x^{1/3}}{1+x^{2/3}}\,dx = 3\int\left(u-\frac{u}{u^2+1}\right)du$$

$$= 3\left(\frac{u^2}{2}-\frac{1}{2}\ln|u^2+1|\right)+C$$

$$= \frac{3}{2}x^{2/3}-\frac{3}{2}\ln(x^{2/3}+1)+C.$$

(c) If we set $u=x^{1/6}$ or $x=u^6$, then $dx=6u^5\,du$, and

$$\int \frac{\sqrt{x}}{1+x^{1/3}}\,dx = \int \frac{u^3}{1+u^2}(6u^5\,du)$$

$$= 6\int\left(u^6-u^4+u^2-1+\frac{1}{1+u^2}\right)du$$

$$= 6\left(\frac{u^7}{7}-\frac{u^5}{5}+\frac{u^3}{3}-u+\operatorname{Tan}^{-1}u\right)+C$$

$$= \frac{6}{7}x^{7/6}-\frac{6}{5}x^{5/6}+2x^{1/2}-6x^{1/6}+6\operatorname{Tan}^{-1}(x^{1/6})+C.$$

∎

EXERCISES 9.1

In Exercises 1–24 evaluate the indefinite integral.

1. $\displaystyle\int \frac{x^2}{5-3x^3}\,dx$

2. $\displaystyle\int xe^{-2x^2}\,dx$

3. $\displaystyle\int \frac{x}{(x^2+2)^{1/3}}\,dx$

4. $\displaystyle\int \frac{e^x}{1+e^x}\,dx$

5. $\displaystyle\int \frac{4t+8}{t^2+4t+5}\,dt$

6. $\displaystyle\int x^2\sqrt{1-3x^3}\,dx$

7. $\displaystyle\int (x+1)(x^2+2x)^{1/3}\,dx$

8. $\displaystyle\int \frac{x^2}{(1+x^3)^3}\,dx$

9. $\displaystyle\int \frac{\sqrt{x}}{1-\sqrt{x}}\,dx$

10. $\displaystyle\int \frac{1-\sqrt{x}}{\sqrt{x}}\,dx$

11. $\displaystyle\int \frac{x+2}{x+1}\,dx$

12. $\displaystyle\int \frac{x^2+2}{x^2+1}\,dx$

13. $\displaystyle\int \frac{\sin\theta}{\cos\theta-1}\,d\theta$

14. $\displaystyle\int \frac{x+3}{\sqrt{2x+4}}\,dx$

15. $\displaystyle\int \frac{e^x}{1+e^{2x}}\,dx$

16. $\displaystyle\int \sin^3 2x\cos 2x\,dx$

17. $\displaystyle\int x^5(2x^2-5)^4\,dx$

18. $\displaystyle\int \frac{x^3}{(x+5)^2}\,dx$

19. $\displaystyle\int z^2\sqrt{3-z}\,dz$

20. $\displaystyle\int \tan 3x\,dx$

21. $\displaystyle\int \frac{(x-3)^{2/3}}{(x-3)^{2/3}+1}\,dx$

22. $\displaystyle\int \frac{\sqrt{x}}{1+x^{1/4}}\,dx$

23. $\displaystyle\int \frac{1}{x^2-4}\,dx$

24. $\displaystyle\int \frac{e^x-1}{e^x+1}\,dx$

25. Find the length of the curve $y=\ln(\cos x)$ from $(0,0)$ to $(\pi/4,-(\ln 2)/2)$.

26. Find the area bounded by the curves
$$y=\frac{x^2+1}{x+1},\qquad x+3y=7.$$

27. (a) Find the centroid of the area bounded by the curves
$$y=(3+x)^{3/2},\qquad x=1,\qquad y=0.$$

    (b) What is its second moment of area about the line $x=1$?

28. If $f(x)$ is a continuous even function, prove that
$$\int_{-a}^{a} f(x)\,dx = 2\int_{0}^{a} f(x)\,dx;$$
and if $f(x)$ is a continuous odd function,
$$\int_{-a}^{a} f(x)\,dx = 0.$$

29. (a) Show that the function $f(x)=\lambda e^{-\lambda x}$, where $\lambda > 0$ is a constant, qualifies as a probability density function on the interval $x\ge 0$. See Exercise 23 in Section 7.9 for the definition of a pdf.

    (b) If the probability that $x\ge 3$ is 0.5, what is λ?

30. A paratrooper and his parachute fall from rest with a combined mass of 100 kg. At any instant during descent, the parachute has an air resistance force acting on it that in newtons is equal to one-half its speed (in metres per second). Assuming that the paratrooper falls vertically downward and that the parachute is already open when the jump takes place, the differential equation describing his motion is
$$200\frac{dv}{dt}=1962-v,$$
where v is velocity and t is time.

    (a) By writing this equation in the form
$$\frac{1}{1962-v}\frac{dv}{dt}=\frac{1}{200},$$
and antidifferentiating with respect to t, find v as a function of t.

    (b) If x measures vertical distance fallen, then $v=dx/dt$. Use this to find the position of the paratrooper as a function of time.

31. Evaluate
$$\int_{0}^{4\pi}\sqrt{1+\cos x}\,dx.$$
Hint: $\cos x = 2\cos^2(x/2)-1$.

32. Evaluate
$$\int \frac{1}{x(3+2x^n)}\,dx$$
for any $n>0$. Hint: Find A and B in order that
$$\frac{1}{x(3+2x^n)}=\frac{A}{x}+\frac{Bx^{n-1}}{3+2x^n}.$$

33. Show that the substitution $u-x=\sqrt{x^2+bx+c}$ replaces the integral
$$\int \sqrt{x^2+bx+c}\,dx$$
with the integral of a rational function of u.

34. Use the substitution $u - x = \sqrt{x^2 + 3x + 4}$ to evaluate

$$\int \frac{1}{(x^2 + 3x + 4)^{3/2}}\, dx.$$

35. Show that when the quadratic $c + bx - x^2$ factors as $c + bx - x^2 = (p + x)(q - x)$, then either of the substitutions $(p + x)u = \sqrt{c + bx - x^2}$ or $(q - x)u = \sqrt{c + bx - x^2}$ replaces the integral

$$\int \frac{1}{\sqrt{c + bx - x^2}}\, dx$$

with the integral of a rational function of u.

36. If a, b, and n are constants, evaluate

$$\int x(a + bx)^n\, dx.$$

37. The function

$$f(x) = \frac{1}{\sqrt{2\pi}\sigma} e^{-(x-\mu)^2/(2\sigma^2)}, \quad -\infty < x < \infty,$$

where μ and σ are constants, is called the *normal probability density function*. Consequently,

$$\int_{-\infty}^{\infty} f(x)\, dx = 1.$$

If the expected value of x is defined as

$$\int_{-\infty}^{\infty} xf(x)\, dx,$$

show that μ is the expected value.

38. Evaluate, if possible,

$$\int_{-\infty}^{\infty} e^{-|a-x|} e^{-|x|}\, dx,$$

where a is a constant.

SECTION 9.2

Exponential Growth and Decay

Exponential functions play a prominent role in many areas of applied mathematics. This is often due to the fact that the derivative of the exponential function e^{kx} is ke^{kx}; that is, the derivative of e^{kx} is proportional to itself. In this section we exploit this property in a wide variety of applications.

Many quantities have functional dependence on time t which takes the form Ce^{kt} where $C > 0$ and k are constants. When $k > 0$, the quantities are said to experience exponential growth; and when $k < 0$, they experience exponential decay. We consider an example of each.

Suppose that at some initial time t, which we take as $t = 0$, the number of bacteria in a culture is 2500. Given a proper environment, the culture will grow; and by studying this growth over a limited time span, we may be able to predict the number $N(t)$ of bacteria present at any given time. If we carefully observe the culture over a small time interval, perhaps 40 seconds, starting at time $t = 0$, we could draw a graph of $N = N(t)$ as shown in Figure 9.1. Each jump (discontinuity) in this graph indicates the birth of a new bacterium. Such a "microscopic" analysis of $N(t)$, in which we attempt to predict the birth of each bacterium, is unrealistic. We are interested in a more "macroscopic" analysis, with which we might be able to predict when the number of bacteria increases by, say, 25%, or doubles. In order to do so, we take measurements of the number of bacteria at much larger time intervals than those in Figure 9.1. The result would be a sequence of points such as that in Table 9.1 which we have plotted in Figure 9.2(a).

TABLE 9.1

t (minutes)	0	5	10	15	20	25	30	35	40	45	50
N (number)	2500	2650	2810	2980	3150	3340	3540	3750	3980	4210	4470

FIGURE 9.1

FIGURE 9.2

If we join these points with a smooth curve as in Figure 9.2(b), we obtain a reasonable overview of N as a function of time. Figure 9.2(b) is, however, only a graphical representation of physical measurements. It is useful to fit a mathematical formula to this graph; that is, to find an algebraic expression for $N(t)$. To do this often requires a great deal of experimentation with different types of curves. For this particular problem, the key is in the following measurements. At any time t we measure, for a very small interval Δt, the corresponding change $N(t + \Delta t) - N(t)$ in N (Figure 9.3). It we take a number of such measurements, we find that $N(t + \Delta t) - N(t)$ is always proportional to the product of N and Δt; that is

$$N(t + \Delta t) - N(t) = kN\, \Delta t, \qquad (9.15)$$

where $k > 0$ is some constant. In Table 9.2 we have used $\Delta t = 5$ to calculate $N(t + \Delta t) - N(t)$ and $N\Delta t = 5N$. In each case, $N(t + 5) - N(t)$ is approximately 0.012 times $5N$.

TABLE 9.2

t (minutes)	0	5	10	15	20	25	30	35	40	45
$N(t+5) - N(t)$	150	160	170	170	190	200	210	230	230	260
$5N$	12500	13250	14050	14900	15750	16700	17700	18750	19900	21050
$\frac{N(t+5) - N(t)}{5N}$	0.012	0.012	0.012	0.011	0.012	0.012	0.012	0.012	0.012	0.012

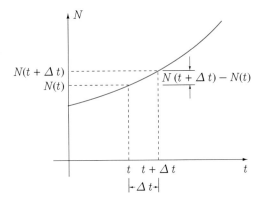

FIGURE 9.3

Experiments have shown that this is not a peculiarity of just one type of bacterium, but is characteristic of the growth pattern for many bacteria, the value of k varying from bacterium to bacterium. If we divide both sides of equation 9.15 by Δt,

$$\frac{N(t + \Delta t) - N(t)}{\Delta t} = kN,$$

and take limits as Δt approaches zero, we obtain

$$\frac{dN}{dt} = kN. \tag{9.16}$$

Experimental measurements have therefore led to mathematical law 9.16 for the rate of growth of bacteria in a culture. The time rate of change of the number of bacteria is proportional to the number of bacteria present at that time. The more bacteria present, the faster their number increases.

To find $N(t)$, we must solve this differential equation subject to the condition that $N(0) = N_0$. If we rewrite equation 9.16 in the form

$$k = \frac{1}{N}\frac{dN}{dt}$$

and take antiderivatives with respect to t, formula 9.2 gives

$$kt + C = \ln|N| = \ln N.$$

We may dispense with the absolute values since N is always positive in this case. Exponentiation of both sides to base e yields

$$e^{kt+C} = e^{\ln N} \quad \text{or} \quad N = e^C e^{kt} = De^{kt},$$

where we have set $D = e^C$. The initial condition $N(0) = N_0$ implies that $D = N_0$, and hence,

$$N = N_0 e^{kt}.$$

On the basis of experimental evidence, we have shown that the number of bacteria in a culture increases exponentially. To obtain the constant k, we must specify one additional piece of information. For example, if measurements indicate that the number of bacteria increases by 10% in 50 minutes, then $N(50) = (11/10)N_0$; that is

$$\frac{11}{10}N_0 = N_0 e^{50k}.$$

When we cancel the N_0 on each side and take natural logarithms, we have

$$\ln\left(\frac{11}{10}\right) = 50k, \quad \text{or} \quad k = \frac{1}{50}\ln\left(\frac{11}{10}\right) = 0.0019.$$

Finally, the number of bacteria in this particular culture is given by

$$N(t) = N_0 e^{(t/50)\ln(1.1)} = N_0 e^{0.0019t}.$$

An important result of this example is that *all* solutions of the differential equation

$$\frac{dN}{dt} = kN \tag{9.17a}$$

take the form

$$N = Ce^{kt}, \tag{9.17b}$$

where C is a constant. This is simply another statement of the property of the exponential function e^{kt} that its derivative is k times itself. But our derivation shows that the *only* functions that satisfy this property are constants multiplied by e^{kt}; no other function has a derivative which is proportional to itself.

As an example of a decay problem, we continue the discussion of radioactive disintegration begun in Section 2.4. Consider an ore sample that contains, along with various impurities, an amount A_0 of radioactive material, say, uranium. Disintegrations gradually reduce this amount of uranium. To describe the rate at which these disintegrations occur, we let $A(t)$ represent the amount of uranium in the sample at any given time, and choose time $t = 0$ when $A = A_0$. If we were to take measurements and perform an analysis similar to that for bacteria in a culture (see also Section 2.4), we would obtain the following *law of radioactive disintegration*: The time rate of change of the amount of radioactive material is proportional at any instant to the amount of radioactive material present. Algebraically, therefore,

$$\frac{dA}{dt} = kA, \tag{9.18}$$

where k is a constant. Since A is decreasing, dA/dt and hence k must be negative. Except for the sign of k, this is differential equation 9.17a, and its solution is therefore given by 9.17b,

$$A = Ce^{kt}.$$

The initial condition $A(0) = A_0$ requires $C = A_0$, and hence,

$$A = A_0 e^{kt}. \tag{9.19}$$

The amount of radioactive material therefore decreases exponentially, as shown in Figure 9.4. To find k we need to know A at one additional time. For example, if we know that one ten-millionth of one percent of the original amount of uranium decays in 6.5 years, then we have

$$0.999\,999\,999\,A_0 = A_0 e^{13k/2}.$$

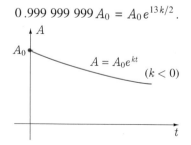

FIGURE 9.4

If we solve this equation for k, we obtain

$$k = \frac{2}{13} \ln(0.999\,999\,999) = -1.54 \times 10^{-10},$$

and therefore,
$$A = A_0 e^{-1.54 \times 10^{-10} t}.$$

The law of radioactive disintegration has an important application in the dating of once-living plants and animals. All living tissue contains two isotopes of carbon: C^{14} (carbon-14), which is radioactive, and C^{12} (carbon-12), which is stable. In living tissue, the ratio of the amount of C^{14} to that of C^{12} is $1/10\,000$ for all fragments of the tissue. When the tissue dies, however, the ratio changes due to the fact that no more carbon is produced, and the original C^{14} present decays radioactively into an element other than C^{12}. Thus, as the dead tissue ages, the ratio of C^{14} to C^{12} decreases, and by measuring this ratio, it is possible to predict how long ago the tissue was alive. Suppose, for example, the present ratio of C^{14} to C^{12} in a specimen is $1/100\,000$; that is, one-tenth that for a living tissue. Then 90% of the original amount of C^{14} in the specimen has disintegrated.

If we let $A(t)$ be the amount of C^{14} in the specimen at time t, taking $A = A_0$ to be the amount present in the living specimen at its death ($t = 0$), then $A = A_0 e^{kt}$. To determine k, we use the fact that the half-life of C^{14} is approximately 5550 years. The *half-life* of a radioactive element is the time required for one-half an original sample of the material to disintegrate. For carbon-14, this means that A is equal to $A_0/2$ when $t = 5550$,

$$\frac{A_0}{2} = A_0 e^{5550 k}.$$

When we divide by A_0 and take natural logarithms,

$$k = -\frac{1}{5550} \ln 2 = -0.000125.$$

Consequently, the amount of C^{14} in the specimen of dead tissue at any time t is given by
$$A = A_0 e^{-0.000125 t}.$$

If T is the present time, when the amount of C^{14} is known to be 10% of its original amount, then at this time
$$0.1 A_0 = A_0 e^{-0.000125 T}.$$

The solution of this equation is
$$T = \frac{\ln 10}{0.000125} = 18\,400,$$

and we conclude that the tissue died about 18 400 years ago.

EXERCISES 9.2

1. Bacteria in a culture increase at a rate proportional to the number present. If the original number increases by 25% in 2 hours, when will it double?

2. If the number of bacteria in a culture doubles in 3 hours, when will it triple?

3. If one-half of a sample of radioactive substance decays in 15 days, how long does it take for 90% of the sample to decay?

4. If 10% of a sample of radioactive material decays in 3 seconds, what is its half-life?

5. After 4 half-lives of a radioactive substance, what percentage of the original amount remains?

6. In psychology, it is sometimes assumed that if R represents the reaction of a subject to an amount S of stimulus, then dR/dS is inversely proportional to S; that is,
$$\frac{dR}{dS} = \frac{k}{S},$$
where $k > 0$ is a constant. This is called the *Weber-Fechner law*. Find R as a function of S if S_0 is the smallest stimulus detectable by the subject.

7. If the Weber-Fechner law in Exercise 6 is replaced by the Brentano-Stevens law,
$$\frac{dR}{dS} = k\frac{R}{S},$$
find $R(S)$. Can the detection threshold S_0 of Exercise 6 be used to determine the constant of integration in this case?

8. Suppose the amount of a drug injected into the body decreases at a rate proportional to the amount still present. If a dose decreases by 5% in the first hour, when will it decrease to one-half its original amount?

9. In the arctic, the snowy owl preys on the lemming, and the number of owls and lemmings in any given area fluctuate in time. If the number of lemmings increases, so does the number of owls, but with some delay. Eventually an overabundance of owls results, and a scarcity of food (lemmings) to support such a population causes a decrease in the number of owls. The lemming population consequently begins to rise and the cycle repeats. One model to describe the relationship between L, the number of lemmings, and s, the number of snowy owls, states that $s(L)$ must satisfy the differential equation
$$\frac{ds}{dL} = \frac{s(AL - B)}{L(C - Ds)},$$
where A, B, C, and D are positive constants. Solve this differential equation for an implicit definition of $s(L)$.

10. A sugar cube 1 cm on each side is dropped into a cup of coffee. If the sugar dissolves in such a way that the cube always remains a cube, compare the times for the cube to completely dissolve under the following conditions:
    (a) Dissolving occurs at a rate proportional to the surface area of the remaining cube; and
    (b) Dissolving occurs at a rate proportional to the amount of sugar remaining.

11. Solve Exercise 10 if the sugar is not in the form of a cube, but rather in free form from the sugar bowl. Assume that the sugar consists of n spherical particles each of radius r_0 centimetres.

12. An analysis of a sample of fossil remains shows that it contains only 1.51% of the original C^{14} in the living creature. When did the creature die?

13. If a fossilized creature died 100 000 years ago, what percentage of the original C^{14} remains?

14. Glucose is administered intravenously to the bloodstream at a constant rate of R units per unit time. As the glucose is added, it is converted by the body into other substances at a rate proportional to the amount of glucose in the blood at that time. Show that the amount of glucose in the blood as a function of time t is given by
$$C(t) = \frac{R}{k}(1 - e^{-kt}) + C_0 e^{-kt},$$
where k is a constant and C_0 is the amount at time $t = 0$ when the intravenous feeding is initiated. Sketch a graph of this function for $C_0 < R/k$ and $C_0 > R/k$.

15. Prove that if a quantity decreases at a rate proportional to its present amount, and if its percentage decrease in some interval of time is i%, then its percentage decrease in any interval of time of the same length is i%.

16. Water at temperature 90°C is placed in a room at constant temperature 20°C. Newton's law of cooling states that the time rate of change of the temperature T of the water is proportional to the difference between T and the temperature of the environment:
$$\frac{dT}{dt} = k(T - 20),$$
where k is a constant. If the water cools to 60°C in 40 minutes, find T as a function of t.

The exponential growth of bacteria as defined by the equation $N = N_0 e^{kt}$ is unrealistic for large t in that there is no maximum for $N(t)$. Sketch the growth curve in Exercises 17–20 and show that each has a maximum.

17. The *logistic* growth curve
$$N(t) = \frac{N_0(1 + b)}{1 + be^{-kt}}, \quad t \geq 0,$$
where b, k, and N_0 are positive constants.

18. The *Gompertz* growth curve
$$N(t) = Ae^{-be^{-kt}}, \quad t \geq 0,$$
where A, b, and k are positive constants.

19. The *monomolecular* growth curve
$$N(t) = \frac{N_0}{1 - b}(1 - be^{-kt}), \quad t \geq 0,$$
where $b < 1$, k, and N_0 are positive constants.

20. The *Von Bertalanffy* growth curve
$$N(t) = N_0(1 - be^{-kt})^3, \quad t \geq 0,$$
where $b < 1$, k, and N_0 are positive constants.

21. The differential equation

$$\frac{dN}{dt} = kN(t)\cos t, \quad t \geq 0,$$

where $k > 0$ is a constant, describes a population that undergoes periodic growths and declines. If $N(0) = N_0$, find $N(t)$ and sketch its graph.

22. Sketch a graph of the sigmoid survival curve

$$s(x) = 1 - \left(1 - e^{-kx}\right)^n, \quad x \geq 0,$$

where $k > 0$ is a constant and n is a positive integer.

23. When a deep-sea diver inhales air, his body tissues absorb extra amounts of nitrogen. Suppose the diver enters the water at time $t = 0$, drops very quickly to depth d, and remains at this depth for a very long time. The amount N of nitrogen in his body tissues increases as he remains at this depth until a maximum amount $\overline{N}$ is reached. The time rate of change of N is proportional to the difference $\overline{N} - N$. Show that if N_0 is the amount of nitrogen in his body tissues when he enters the water, then

$$N = N_0 e^{-kt} + \overline{N}(1 - e^{-kt}),$$

where $k > 0$ is a constant.

24. A dead body is found after the final calculus examination. The coroner arrives on the scene at 9:00 A.M. and immediately records the temperature of the body as $30.1\,°C$. One hour later she notes that the body's temperature has dropped to $29.2\,°C$, and that during the hour, room temperature has remained at $20\,°C$. Assuming that the body's temperature obeys Newton's law of cooling (Exercise 16), determine when death occurred. Normal body temperature is $37\,°C$.

25. (a) In Section 7.5 we proved that the rate of change of pressure P with respect to vertical distance y within a fluid with density ρ is $dP/dy = -\rho g$, where $g = 9.81$. Since air may be considered a fluid, air pressure above sea level ($y = 0$) must satisfy the equation

$$\frac{dP}{dy} = -\rho g,$$

where $\rho = \rho(y)$ is air density at altitude y. If air is assumed to be a perfect gas, the pressure of a given volume V of gas must satisfy the equation

$$P = \rho RT,$$

where $T = T(y)$ is the temperature of the air and $R > 0$ is a constant. To a reasonable approximation, temperature varies linearly with respect to y:

$$T = T_0 - \lambda y$$

where T_0 is sea level temperature and λ is a constant (about $6.10 \times 10^{-4}\,°C$ per 100 m). Show that $P(y)$ must satisfy the differential equation

$$\frac{dP}{dy} = -\frac{Pg}{R(T_0 - \lambda y)},$$

and solve the equation for

$$P(y) = P_0 \left(1 - \frac{\lambda y}{T_0}\right)^{g/(\lambda R)},$$

where P_0 is sea level pressure.

(b) Find the altitude at which air pressure is one one-millionth that of sea level pressure, if $T_0 = 293\,°K$, $g = 9.81\text{ m/s}^2$, and $R = 287\text{ m·N/kg·°K}$.

26. When light passes through a material, such as water or glass, its intensity decreases with depth of penetration. The Bouguer-Lambert law for absorption of light states that the amount of light absorbed by a small thickness of the material is proportional to the product of this thickness and the intensity of light incident on it.

(a) Show that if I is the intensity of light, then

$$I = I(x) = I_0 e^{-kx},$$

where $k > 0$ is a constant, I_0 is the intensity at the surface of the material, and x is the penetration distance from the surface of the material (Figure 9.5).

FIGURE 9.5

(b) Give an alternative statement of the Bouguer-Lambert law in terms of rates of change.

27. (a) To perform a one-hour operation on a dog a veterinarian anesthetizes the dog with sodium pentobarbital. During the operation the dog's body breaks down the drug at a rate proportional to the amount present, and only half the original dose remains after 5 hours. If the dog has mass 20 kg, and 0.02 gm of sodium pentobarbital per kilogram of body mass are required to maintain surgical anesthesia, what original dose is required?

(b) If after the hour, the operation is unfinished and the dog begins to stir, a further dose of anesthetic must be administered. What extra dose will guarantee surgical anesthesia for an additional 20 minutes?

SECTION 9.3

Integration by Parts

One of the most important integration techniques is that of **integration by parts**. It results from the product rule for differentiation:

$$\frac{d}{dx}(uv) = u\frac{dv}{dx} + v\frac{du}{dx}.$$

If we take antiderivatives of both sides of this equation with respect to x, we obtain

$$\int \frac{d}{dx}(uv)\,dx = \int u\frac{dv}{dx}\,dx + \int v\frac{du}{dx}\,dx.$$

Since the antiderivative on the left is uv, this equation may be rewritten in the form

$$\int u\,dv = uv - \int v\,du. \qquad (9.20)$$

Equation 9.20 is called the **integration-by-parts formula**. It says that if we are given an unknown integral (the left-hand side of 9.20), and we arbitrarily divide the function after the integral sign into two parts — one called u and the other called dv, then we can set the unknown integral equal to the right-hand side of 9.20. What this does is replace the integral

$$\int u\,dv \quad \text{with} \quad \int v\,du.$$

This will be useful if the new integral is easier to evaluate than the original. The problem is to choose u and dv so that the new integral is simpler than the original. For many examples, it is advantageous to let dv be the *most complicated part of the integrand (plus differential) that we can integrate mentally*. The example

$$\int xe^x\,dx$$

will clarify these ideas. To use integration by parts on this problem we must define u and dv in such a way that $u\,dv = xe^x\,dx$. There are four possibilities:

$$u = 1, \qquad dv = xe^x\,dx;$$
$$u = x, \qquad dv = e^x\,dx;$$
$$u = e^x, \qquad dv = x\,dx;$$
$$u = xe^x, \qquad dv = dx.$$

Our rule suggests the choice

$$u = x, \qquad dv = e^x\,dx,$$

in which case

$$du = dx, \qquad v = e^x.$$

Formula 9.20 then gives

$$\int xe^x\,dx = xe^x - \int e^x\,dx.$$

We have therefore replaced the integration of xe^x with the integration of e^x, a definite simplification. The final solution is therefore

$$\int xe^x\,dx = xe^x - e^x + C.$$

This example illustrates that integration by parts replaces an integral with two others: one mental integration (from dv to v) and a second integration that is hopefully simpler than the original. We should also note that in the evaluation of the mental integration we did not include a constant of integration. For example, in the integration from $dv = e^x\, dx$ to $v = e^x$, we did not write $v = e^x + D$. Had we done so, the solution would have proceeded as follows:

$$\int xe^x\, dx = x(e^x + D) - \int (e^x + D)\, dx$$
$$= x(e^x + D) - e^x - Dx + C$$
$$= xe^x - e^x + C.$$

The constant D has therefore disappeared from the eventual solution. This always occurs, and consequently, we do not include an arbitrary constant in the mental integration for v.

EXAMPLE 9.3

Evaluate the following integrals:

(a) $\int \ln x\, dx$ (b) $\int x^2 \cos x\, dx$ (c) $\int e^x \sin x\, dx$

SOLUTION (a) Integration by parts leaves no choice for u and dv in this integral; they must be

$$u = \ln x, \qquad dv = dx,$$

in which case

$$du = \frac{1}{x}\, dx, \qquad v = x.$$

Then we have

$$\int \ln x\, dx = x \ln x - \int x \frac{1}{x}\, dx = x \ln x - x + C.$$

(b) If we set

$$u = x^2, \qquad dv = \cos x\, dx,$$

then

$$du = 2x\, dx, \qquad v = \sin x, \qquad \text{and}$$

$$\int x^2 \cos x\, dx = x^2 \sin x - \int 2x \sin x\, dx.$$

Integration by parts has therefore reduced the power on x from 2 to 1. To eliminate the x in front of the trigonometric function completely, we perform integration by parts once again, this time with

$$u = x, \qquad dv = \sin x\, dx.$$

Then,

$$du = dx, \qquad v = -\cos x, \qquad \text{and}$$

$$\int x^2 \cos x\, dx = x^2 \sin x - 2\left(-x \cos x - \int -\cos x\, dx\right)$$
$$= x^2 \sin x + 2x \cos x - 2 \sin x + C.$$

(c) If we set
$$u = e^x, \quad dv = \sin x\, dx,$$
then
$$du = e^x\, dx, \quad v = -\cos x, \quad \text{and}$$
$$\int e^x \sin x\, dx = -e^x \cos x - \int -e^x \cos x\, dx.$$

Integration by parts appears to have led nowhere; the integration of $e^x \cos x$ is essentially as difficult as that of $e^x \sin x$. If, however, we persevere and integrate by parts once again with
$$u = e^x, \quad dv = \cos x\, dx,$$
then
$$du = e^x\, dx, \quad v = \sin x, \quad \text{and}$$
$$\int e^x \sin x\, dx = -e^x \cos x + e^x \sin x - \int e^x \sin x\, dx.$$

This equation can now be solved for the unknown integral. By transposing the last term on the right to the left, we find that
$$2\int e^x \sin x\, dx = e^x(\sin x - \cos x),$$
and therefore,
$$\int e^x \sin x\, dx = \frac{1}{2}e^x(\sin x - \cos x) + C,$$
to which we have added the arbitrary constant C. ∎

The technique in part (c) of this example is often called **integration by reproduction**. Two applications of integration by parts enabled us to reproduce the unknown integral and then solve the equation for it.

EXAMPLE 9.4

Evaluate
$$\int \frac{x^3}{\sqrt{9+x^2}}\, dx.$$

Solution Methods

(i) If we set $u = 9 + x^2$, then $du = 2x\, dx$, and
$$\int \frac{x^3}{\sqrt{9+x^2}}\, dx = \int \frac{u-9}{\sqrt{u}}\frac{du}{2} = \frac{1}{2}\int (u^{1/2} - 9u^{-1/2})\, du$$
$$= \frac{1}{2}\left(\frac{2}{3}u^{3/2} - 18u^{1/2}\right) + C = \frac{1}{3}(9+x^2)^{3/2} - 9\sqrt{9+x^2} + C.$$

(ii) If we set $u = \sqrt{9 + x^2}$, then $du = \left(x/\sqrt{9 + x^2}\right) dx$, and

$$\int \frac{x^3}{\sqrt{9 + x^2}} dx = \int (u^2 - 9) du = \frac{u^3}{3} - 9u + C$$
$$= \frac{1}{3}(9 + x^2)^{3/2} - 9\sqrt{9 + x^2} + C.$$

(iii) If we set
$$u = x^2, \qquad dv = \frac{x}{\sqrt{9 + x^2}} dx,$$

then
$$du = 2x\, dx, \qquad v = \sqrt{9 + x^2}, \qquad \text{and}$$

$$\int \frac{x^3}{\sqrt{9 + x^2}} dx = x^2 \sqrt{9 + x^2} - \int 2x\sqrt{9 + x^2}\, dx$$
$$= x^2 \sqrt{9 + x^2} - 2\left[\frac{1}{3}(9 + x^2)^{3/2}\right] + C$$
$$= x^2 \sqrt{9 + x^2} - \frac{2}{3}(9 + x^2)^{3/2} + C.$$

∎

Example 9.4 illustrates that there may be more than one way to evaluate an indefinite integral, and the answer may appear different depending on the technique used. According to Theorem 5.1, these solutions must be identical, up to an additive constant. To illustrate the case, note that the solution obtained by method (iii) can be written as

$$\sqrt{9 + x^2}\left[x^2 - \frac{2}{3}(9 + x^2)\right] + C = \frac{1}{3}\sqrt{9 + x^2}\left[-18 + x^2\right] + C$$
$$= \frac{1}{3}\sqrt{9 + x^2}\left[-27 + (9 + x^2)\right] + C$$
$$= -9\sqrt{9 + x^2} + \frac{1}{3}(9 + x^2)^{3/2} + C,$$

which is the solution obtained by the other two methods. It is also worth noting that the three methods work very well provided the power on x in the numerator of the integrand is odd. When it is even, they fail miserably. In such cases, trigonometric substitutions (Section 9.5) come to the rescue (see Example 9.12).

Integration-by-parts formula 9.20 applies to indefinite integrals. For definite integrals, it is replaced by

$$\int_a^b u\, dv = \{uv\}_a^b - \int_a^b v\, du. \tag{9.21}$$

EXAMPLE 9.5

Evaluate
$$\int_0^4 \frac{x}{\sqrt{x+8}}\, dx.$$

SOLUTION If we set
$$u = x, \qquad dv = \frac{1}{\sqrt{x+8}}\, dx,$$
then
$$du = dx, \qquad v = 2\sqrt{x+8}, \qquad \text{and}$$

$$\int_0^4 \frac{x}{\sqrt{x+8}}\, dx = \left\{2x\sqrt{x+8}\right\}_0^4 - \int_0^4 2\sqrt{x+8}\, dx$$
$$= 8\sqrt{12} - 2\left\{\frac{2}{3}(x+8)^{3/2}\right\}_0^4$$
$$= \frac{16}{3}\left(4\sqrt{2} - 3\sqrt{3}\right).$$

EXERCISES 9.3

In Exercises 1–18 evaluate the indefinite integral.

1. $\displaystyle\int x \sin x\, dx$
2. $\displaystyle\int x^2 e^{2x}\, dx$
3. $\displaystyle\int x^4 \ln x\, dx$
4. $\displaystyle\int \sqrt{x} \ln(2x)\, dx$
5. $\displaystyle\int z \sec^2(z/3)\, dz$
6. $\displaystyle\int x\sqrt{3-x}\, dx$
7. $\displaystyle\int \operatorname{Sin}^{-1} x\, dx$
8. $\displaystyle\int x^2\sqrt{x+5}\, dx$
9. $\displaystyle\int \frac{x}{\sqrt{2+x}}\, dx$
10. $\displaystyle\int \frac{x^2}{\sqrt{2+x}}\, dx$
11. $\displaystyle\int \frac{x}{\sqrt{2+x^2}}\, dx$
12. $\displaystyle\int (x-1)^2 \ln x\, dx$
13. $\displaystyle\int e^x \cos x\, dx$
14. $\displaystyle\int \operatorname{Tan}^{-1} x\, dx$
15. $\displaystyle\int \cos(\ln x)\, dx$
16. $\displaystyle\int e^{2x} \cos 3x\, dx$
17. $\displaystyle\int \frac{x^3}{\sqrt{5+3x^2}}\, dx$
18. $\displaystyle\int \ln(x^2+4)\, dx$

19. Consider the evaluation of $\int x^5 e^x\, dx$. Were we to use integration by parts the answer would eventually be of the form
$$\int x^5 e^x\, dx = Ax^5 e^x + Bx^4 e^x + Cx^3 e^x + Dx^2 e^x + Exe^x + Fe^x + G.$$
Differentiate this equation, and thereby obtain A, B, C, D, E, and F (with no integration).

20. The face of a dam is shown in Figure 9.6, where the curve has equation
$$y = e^{k|x|} - 1, \qquad k = \frac{1}{100}\ln 201.$$
Find the total force on the dam due to water pressure when the water level on the dam is 100 m.

FIGURE 9.6

21. If we set $u = x^{-1}$ and $dv = dx$, then $du = -x^{-2}\, dx$ and $v = x$, and
$$\int \frac{1}{x}\, dx = \left(\frac{1}{x}\right)x - \int x\left(-\frac{1}{x^2}\, dx\right) = 1 + \int \frac{1}{x}\, dx.$$
Subtraction of $\int x^{-1}\, dx$ from each side of this equation now gives $0 = 1$. What is wrong with the argument?

22. The *gamma function* $\Gamma(n)$ for $n > 0$ is defined by the improper integral

$$\Gamma(n) = \int_0^\infty x^{n-1} e^{-x} \, dx.$$

Show that when n is a positive integer, $\Gamma(n) = (n-1)!$.

23. The *Laplace transform* of a function $F(t)$ is the function $f(s)$ defined by the improper integral

$$f(s) = \int_0^\infty e^{-st} F(t) \, dt.$$

Find Laplace transforms of the following functions:

(a) e^{3t} (b) t^2 (c) $\sin t$ (d) te^{-t}

24. Show that the function

$$f(x) = \frac{x^{\alpha-1} e^{-x/\beta}}{\Gamma(\alpha) \beta^\alpha}$$

for positive constants α and β qualifies as a probability density function on the interval $x \geq 0$. See Exercise 23 in Section 7.9 for the definition of a pdf and Exercise 22 for the definition of $\Gamma(\alpha)$.

In Exercises 25–27 evaluate the indefinite integral.

25. $\int \text{Tan}^{-1} \sqrt{x} \, dx$ 26. $\int x^2 \cos^2 x \, dx$

27. $\int x e^x \sin x \, dx$

SECTION 9.4

Trigonometric Integrals

Differentiation formulas for trigonometric functions suggest three pairings of these functions: sine and cosine, tangent and secant, and cotangent and cosecant. The derivative of either function in a given pair involves at most functions of that pair. Note that integration formulas 9.5–9.14 also substantiate this pairing. Each antiderivative gives functions in the same pairing. Because of this, a first step in every integral involving trigonometric functions is to rewrite the integrand so that each term involves only one pair. For example, to evaluate

$$\int \frac{\tan x + \sec x \cot^2 x}{\sin^3 x} \, dx,$$

we would first rewrite it in the form

$$\int \left(\frac{\csc^3 x}{\cot x} + \frac{\cos x}{\sin^5 x} \right) dx.$$

The first term contains cosecants and cotangents, and the second has sines and cosines. We now consider integrals involving each of these three pairings individually.

Integrals Involving the Sine and Cosine

We frequently encounter integrals of the form

$$\int \cos^n x \sin^m x \, dx,$$

where m and n may or may not be integers. The key to evaluation of this type of integral is recognition of the fact that when either $n = 1$ or $m = 1$, the integrations become straightforward. To illustrate this, consider differentiation of $\sin^{n+1} x$ by power rule 3.16,

$$\frac{d}{dx} \sin^{n+1} x = (n+1) \sin^n x \cos x.$$

This means that differentiation of a power of $\sin x$ leads to a power of $\sin x$ one lower than the original, multiplied by $\cos x$. In order to reverse the procedure and antidifferentiate a power of $\sin x$, we need a factor of $\cos x$. It then becomes a matter of raising the power on $\sin x$ by one and dividing by the new power; that is, we can say

$$\int \sin^n x \cos x \, dx = \frac{1}{n+1} \sin^{n+1} x + C, \quad n \neq -1. \qquad (9.22\,\mathrm{a})$$

Similarly,

$$\int \cos^n x \sin x \, dx = -\frac{1}{n+1} \cos^{n+1} x + C, \quad n \neq -1. \qquad (9.22\,\mathrm{b})$$

Formulas 9.12 and 9.11 contain the $n = -1$ cases for these integrals. Of course, a simple substitution can always be made to evaluate integrals 9.22a,b. For example, if in 9.22a we set $u = \sin x$, then $du = \cos x \, dx$, and

$$\int \sin^n x \cos x \, dx = \int u^n \, du = \frac{u^{n+1}}{n+1} + C = \frac{\sin^{n+1} x}{n+1} + C.$$

Also useful are the formulas

$$\int \sin nx \, dx = -\frac{1}{n} \cos nx + C, \quad \text{and} \qquad (9.22\,\mathrm{c})$$

$$\int \cos nx \, dx = \frac{1}{n} \sin nx + C. \qquad (9.22\,\mathrm{d})$$

Trigonometric identities are used to write integrands in the forms contained in equations 9.22. Particularly helpful are

$$\sin^2 x + \cos^2 x = 1, \qquad (9.23\,\mathrm{a})$$
$$\sin 2x = 2 \sin x \cos x, \qquad (9.23\,\mathrm{b})$$
$$\cos 2x = 2 \cos^2 x - 1, \qquad (9.23\,\mathrm{c})$$
$$\cos 2x = 1 - 2 \sin^2 x. \qquad (9.23\,\mathrm{d})$$

EXAMPLE 9.6

Evaluate the following integrals:

(a) $\int \sin^3 x \, dx$ (b) $\int \sin^3 x \cos^5 x \, dx$ (c) $\int \sin 2x \cos^6 2x \, dx$

(d) $\int \sqrt{\sin x} \cos^3 x \, dx$ (e) $\int \frac{\cos^3 3x}{\sin^2 3x} \, dx$ (f) $\int \sin^2 x \, dx$

(g) $\int \cos^4 x \, dx$ (h) $\int \sin^2 x \cos^2 x \, dx$

SOLUTION (a) With identity 9.23a, we may write

$$\int \sin^3 x \, dx = \int \sin x (1 - \cos^2 x) \, dx = \int (\sin x - \cos^2 x \sin x) \, dx,$$

and the second term is in form 9.22b. Thus,

$$\int \sin^3 x \, dx = -\cos x + \frac{1}{3}\cos^3 x + C.$$

(b) Once again we use 9.23a to rewrite the integrand as two easily integrated terms,

$$\int \sin^3 x \cos^5 x \, dx = \int \sin x (1 - \cos^2 x) \cos^5 x \, dx$$
$$= \int (\cos^5 x \sin x - \cos^7 x \sin x) \, dx$$
$$= -\frac{1}{6}\cos^6 x + \frac{1}{8}\cos^8 x + C.$$

(c) This integrand is already in a convenient form for integration. Guess and fix up leads to

$$\int \sin 2x \cos^6 2x \, dx = -\frac{1}{14}\cos^7 2x + C.$$

(d) Using 9.23a, we obtain

$$\int \sqrt{\sin x} \cos^3 x \, dx = \int \sin^{1/2} x \cos x (1 - \sin^2 x) \, dx$$
$$= \int (\sin^{1/2} x \cos x - \sin^{5/2} x \cos x) \, dx$$
$$= \frac{2}{3}\sin^{3/2} x - \frac{2}{7}\sin^{7/2} x + C.$$

(e) Again we use 9.23a,

$$\int \frac{\cos^3 3x}{\sin^2 3x} \, dx = \int \frac{\cos 3x (1 - \sin^2 3x)}{\sin^2 3x} \, dx = \int \left(\frac{\cos 3x}{\sin^2 3x} - \cos 3x \right) dx$$
$$= \frac{-1}{3 \sin 3x} - \frac{1}{3}\sin 3x + C.$$

In the five examples above, at least one of the powers on the sine and cosine functions was odd, and in each case we used identity 9.23a. In the next three examples, where all powers are even, identities 9.23b,c,d are useful.

(f) Double-angle formula 9.23d can be rearranged as $\sin^2 x = (1 - \cos 2x)/2$, and therefore

$$\int \sin^2 x \, dx = \frac{1}{2} \int (1 - \cos 2x) \, dx.$$

Formula 9.22d can now be used on the second term

$$\int \sin^2 x \, dx = \frac{1}{2}\left(x - \frac{1}{2}\sin 2x \right) + C.$$

(g) With double-angle formula 9.23c rewritten in the form $\cos^2 x = (1 + \cos 2x)/2$, we obtain

$$\int \cos^4 x \, dx = \frac{1}{4} \int (1 + \cos 2x)^2 \, dx = \frac{1}{4} \int (1 + 2\cos 2x + \cos^2 2x) \, dx.$$

When x is replaced by $2x$ in 9.23c, the result is $\cos 4x = 2\cos^2 2x - 1$. This can be rearranged as $\cos^2 2x = (1 + \cos 4x)/2$, and therefore

$$\int \cos^4 x\, dx = \frac{1}{4}\int \left(1 + 2\cos 2x + \frac{1+\cos 4x}{2}\right) dx.$$

We now use 9.22d on the two cosine terms,

$$\int \cos^4 x\, dx = \frac{1}{4}\left(\frac{3x}{2} + \sin 2x + \frac{1}{8}\sin 4x\right) + C.$$

(h) In this integral we use identities 9.23b,c,

$$\int \sin^2 x \cos^2 x\, dx = \int (\sin x \cos x)^2 dx = \frac{1}{4}\int \sin^2 2x\, dx$$
$$= \frac{1}{4}\int \left(\frac{1-\cos 4x}{2}\right) dx = \frac{1}{8}\left(x - \frac{1}{4}\sin 4x\right) + C.$$

EXAMPLE 9.7

Evaluate
$$\int \sin 5x \cos 2x\, dx.$$

SOLUTION Unlike Example 9.6, in which all trigonometric functions had the same argument, this integrand is the product of trigonometric functions with different arguments. Our first step is to rewrite the integrand so that we do not have such a product. Note that a sum is obviously integrable, but not a product. The product formula

$$\sin A \cos B = \frac{1}{2}[\sin(A+B) + \sin(A-B)],$$

can be used to advantage here. We can rewrite the integrand in the form

$$\int \sin 5x \cos 2x\, dx = \int \frac{1}{2}(\sin 7x + \sin 3x)\, dx,$$

and now integration is easily handled by 9.22c,

$$\int \sin 5x \cos 2x\, dx = \frac{1}{2}\left(-\frac{1}{7}\cos 7x - \frac{1}{3}\cos 3x\right) + C = -\frac{1}{6}\cos 3x - \frac{1}{14}\cos 7x + C.$$

Two integration by parts (and reproduction) can also be used to evaluate this integral, but the above method is much simpler.

Integrals Involving Tangent and Secant

For integrals involving tangent and secant, we have two alternatives: rewrite the integrand in terms of sines and cosines, or express the integrand in terms of easily integrated combinations of tangent and secant. It will usually be more fruitful to investigate the second possibility before expressing the integrand completely in terms of sines and cosines. In this regard we note that combinations of tangent and secant that can be integrated immediately are

$$\int \tan^n x \sec^2 x \, dx = \frac{\tan^{n+1} x}{n+1} + C, \quad n \neq -1; \tag{9.24a}$$

$$\int \sec^n x \tan x \, dx = \frac{\sec^n x}{n} + C, \quad n \neq 0. \tag{9.24b}$$

Once again these results are suggested by differentiations of powers of $\tan x$ and $\sec x$. Alternatively, these integrals can be evaluated with substitutions: $u = \tan x$ for 9.24a and $u = \sec x$ for 9.24b.

To rearrange integrands into these integrable combinations, we use the identity

$$1 + \tan^2 x = \sec^2 x. \tag{9.25}$$

EXAMPLE 9.8

Evaluate the following integrals:

(a) $\int \tan^2 x \, dx$ (b) $\int \tan^3 x \, dx$ (c) $\int \sec^4 x \, dx$

(d) $\int \tan^4 3x \sec^2 3x \, dx$ (e) $\int \tan^3 x \sec^3 x \, dx$ (f) $\int \frac{\sec^4 x}{\tan^2 x} \, dx$

SOLUTION

(a) $$\int \tan^2 x \, dx = \int (\sec^2 x - 1) \, dx = \tan x - x + C$$

(b) $$\int \tan^3 x \, dx = \int \tan x (\sec^2 x - 1) \, dx$$
$$= \int (\sec^2 x \tan x - \tan x) \, dx$$
$$= \frac{1}{2} \sec^2 x + \ln|\cos x| + C$$

(c) $$\int \sec^4 x \, dx = \int \sec^2 x (1 + \tan^2 x) \, dx$$
$$= \int (\sec^2 x + \tan^2 x \sec^2 x) \, dx$$
$$= \tan x + \frac{1}{3} \tan^3 x + C$$

(d) $$\int \tan^4 3x \sec^2 3x \, dx = \frac{1}{15} \tan^5 3x + C$$

(e) $$\int \tan^3 x \sec^3 x \, dx = \int \tan x \, (\sec^2 x - 1) \sec^3 x \, dx$$
$$= \int (\sec^5 x \tan x - \sec^3 x \tan x) \, dx$$
$$= \frac{1}{5} \sec^5 x - \frac{1}{3} \sec^3 x + C$$

(f) $$\int \frac{\sec^4 x}{\tan^2 x} \, dx = \int \frac{\sec^2 x \, (1 + \tan^2 x)}{\tan^2 x} \, dx$$
$$= \int \left(\frac{\sec^2 x}{\tan^2 x} + \sec^2 x \right) dx$$
$$= -\frac{1}{\tan x} + \tan x + C$$

EXAMPLE 9.9

Evaluate
$$\int \sec^3 x \, dx.$$

SOLUTION We use integration by parts and reproduction. By setting
$$u = \sec x, \qquad dv = \sec^2 dx,$$
then
$$du = \sec x \tan x \, dx, \qquad v = \tan x,$$
and formula 9.20 gives
$$\int \sec^3 x \, dx = \sec x \tan x - \int \sec x \tan^2 x \, dx.$$

Since $\tan^2 x = \sec^2 x - 1$, we can write
$$\int \sec^3 x \, dx = \sec x \tan x - \int \sec x \, (\sec^2 x - 1) \, dx$$
$$= \sec x \tan x - \int \sec^3 x \, dx + \ln|\sec x + \tan x|.$$

We now solve for
$$\int \sec^3 x \, dx = \frac{1}{2}(\sec x \tan x + \ln|\sec x + \tan x|) + C.$$

Integrands in the following example cannot be transformed by means of identity 9.25 into forms 9.24, and we therefore express them in terms of sines and cosines.

EXAMPLE 9.10

Evaluate the following integrals:

(a) $\displaystyle\int \frac{\tan^2 2x}{\sec^3 2x}\, dx$ (b) $\displaystyle\int \frac{1}{\sec x \tan^2 x}\, dx$

SOLUTION

(a) $\displaystyle\int \frac{\tan^2 2x}{\sec^3 2x}\, dx = \int \sin^2 2x \cos 2x\, dx = \frac{1}{6} \sin^3 2x + C$

(b) $\displaystyle\int \frac{1}{\sec x \tan^2 x}\, dx = \int \frac{\cos^3 x}{\sin^2 x}\, dx = \int \frac{\cos x\,(1 - \sin^2 x)}{\sin^2 x}\, dx$
$= \displaystyle\int \left(\frac{\cos x}{\sin^2 x} - \cos x\right) dx = -\frac{1}{\sin x} - \sin x + C$

Integrals Involving Cotangent and Cosecant

Since derivatives of the cotangent and cosecant pair

$$\frac{d}{dx} \cot x = -\csc^2 x, \qquad \frac{d}{dx} \csc x = -\csc x \cot x$$

are analogous to those of the tangent and secant pair

$$\frac{d}{dx} \tan x = \sec^2 x, \qquad \frac{d}{dx} \sec x = \sec x \tan x,$$

a discussion parallel to that concerning the tangent and secant could be made here. Aside from the fact that the cotangent replaces the tangent and the cosecant replaces the secant, there are also sign changes. We leave examples of integrations of the cotangent and cosecant pair to the exercises.

EXERCISES 9.4

In Exercises 1–20 evaluate the indefinite integral.

1. $\displaystyle\int \cos^3 x \sin x\, dx$
2. $\displaystyle\int \frac{\cos x}{\sin^3 x}\, dx$
3. $\displaystyle\int \tan^5 x \sec^2 x\, dx$
4. $\displaystyle\int \csc^3 x \cot x\, dx$
5. $\displaystyle\int \cos^3 (x + 2)\, dx$
6. $\displaystyle\int \sqrt{\tan x}\, \sec^4 x\, dx$
7. $\displaystyle\int \frac{1}{\sin^4 t}\, dt$
8. $\displaystyle\int \sec^6 3x \tan 3x\, dx$
9. $\displaystyle\int \cos^2 x\, dx$
10. $\displaystyle\int \frac{\tan^3 x \sec^2 x}{\sin^2 x}\, dx$
11. $\displaystyle\int \sin^3 y \cos^2 y\, dy$
12. $\displaystyle\int \frac{\csc^2 \theta}{\cot^2 \theta}\, d\theta$
13. $\displaystyle\int \frac{\sin \theta}{1 + \cos \theta}\, d\theta$
14. $\displaystyle\int \frac{\sec^2 x}{\sqrt{1 + \tan x}}\, dx$
15. $\displaystyle\int \cos \theta \sin 2\theta\, d\theta$
16. $\displaystyle\int \frac{3 + 4 \csc^2 x}{\cot^2 x}\, dx$
17. $\displaystyle\int \sin^5 x \cos^5 x\, dx$
18. $\displaystyle\int \sin^4 x\, dx$
19. $\displaystyle\int \frac{\tan^3 x}{\sec^4 x}\, dx$
20. $\displaystyle\int \frac{\csc^4 x}{\cot^3 x}\, dx$

21. Find the area between the x-axis and $y = \sin x$ for $0 \le x \le \pi$.

22. Find the volume of the solid of revolution obtained by rotating the area bounded by $y = \tan x$, $y = 0$, and $x = \pi/4$ about the line $y = -1$.

23. Evaluate

$$\int_0^\pi \sqrt{1 - \sin^2 x}\, dx.$$

In Exercises 24–31 evaluate the indefinite integral.

24. $\int \cot^4 z\, dz$

25. $\int \dfrac{\cos^3 \theta}{3 + \sin \theta}\, d\theta$

26. $\int \dfrac{\cos^4 \theta}{1 + \sin \theta}\, d\theta$

27. $\int \sin^4 x \cos^2 x\, dx$

28. $\int \cos 6x \cos 2x\, dx$

29. $\int \cos^2 2x \sin 3x\, dx$

30. $\int \dfrac{1}{\sin x \cos^2 x}\, dx$

31. $\int \sec^5 x\, dx$

32. The alternating current in a power line is given by

$$I = f(t) = A \cos \omega t + B \sin \omega t,$$

where A, B, and ω are constants, and t is time. The root-mean-square (rms) current I_{rms} is defined by

$$(I_{\text{rms}})^2 = \dfrac{1}{T} \int_0^T I^2\, dt,$$

where the interval $0 \le t \le T$ represents any number of complete oscillations of the current. Show that I_{rms} is $1/\sqrt{2}$ times the amplitude of the current.

33. Evaluate

$$\int (1 + \sin \theta + \sin^2 \theta + \sin^3 \theta + \cdots)\, d\theta.$$

34. If n is a positive even integer, show that

$$\int \sec^n x\, dx = \sum_{r=0}^{n/2 - 1} \binom{n/2 - 1}{r} \dfrac{\tan^{2r+1} x}{2r + 1} + C.$$

35. A set of functions $f_1(x)$, $f_2(x)$, ... is said to be orthonormal over an interval $a \le x \le b$ if

$$\int_a^b f_n(x) f_m(x)\, dx = \begin{cases} 0 & m \ne n \\ 1 & m = n. \end{cases}$$

Show that the set of functions

$$\dfrac{1}{\sqrt{2\pi}},\ \dfrac{1}{\sqrt{\pi}} \sin x,\ \dfrac{1}{\sqrt{\pi}} \cos x,\ \dfrac{1}{\sqrt{\pi}} \sin 2x,$$

$$\dfrac{1}{\sqrt{\pi}} \cos 2x,\ \dfrac{1}{\sqrt{\pi}} \sin 3x,\ \ldots$$

is orthonormal over the interval $0 \le x \le 2\pi$.

SECTION 9.5

Trigonometric Substitutions

Physical and geometric problems arising from circles, ellipses and hyperbolas often give rise to integrals involving the square roots $\sqrt{a^2 - b^2 x^2}$, $\sqrt{a^2 + b^2 x^2}$, and $\sqrt{b^2 x^2 - a^2}$. *Trigonometric substitutions* replace them with integrals involving trigonometric functions.

Let us first consider two integrals that involve square roots of the form $\sqrt{a^2 - b^2 x^2}$:

(i) $\displaystyle\int \dfrac{1}{\sqrt{1 - x^2}}\, dx$ and (ii) $\displaystyle\int \dfrac{1}{\sqrt{4 - 9x^2}}\, dx$.

Integral (i) is evident from differentiation of the inverse trigonometric functions in Section 8.3,

$$\int \dfrac{1}{\sqrt{1 - x^2}}\, dx = \text{Sin}^{-1} x + C.$$

We now show that a substitution can be made on this integral which is applicable to much more difficult problems involving the square root $\sqrt{a^2 - b^2 x^2}$, including (ii). If instead of the square root $\sqrt{1 - x^2}$ in (i), we had the square root $\sqrt{1 - \sin^2 \theta}$, we could immediately simplify the latter root to $\sqrt{\cos^2 \theta} = |\cos \theta|$. This prompts us to make the substitution $x = \sin \theta$, from which we have $dx = \cos \theta\, d\theta$. Integral (i) then becomes

$$\int \dfrac{1}{\sqrt{1 - x^2}}\, dx = \int \dfrac{1}{|\cos \theta|} \cos \theta\, d\theta.$$

To eliminate the absolute values, we need to know whether $\cos\theta$ is positive or negative. But this means that we must know the possible values of θ. There is a problem. The equation $x = \sin\theta$ does not really define θ; for any given x, we do not have a unique θ, but an infinite number of possibilities. We must therefore restrict the values of θ, and we do this by specifying that $\theta = \text{Sin}^{-1} x$. In other words, although we have used the equation $x = \sin\theta$ to change variables, and will continue to do so, it is really the equation $\theta = \text{Sin}^{-1} x$ that properly defines the change.

Since θ so defined must lie in the interval $-\pi/2 \leq \theta \leq \pi/2$ (the principal values of the inverse sine function), it follows that $\cos\theta \geq 0$, and the absolute values may be dropped. Consequently,

$$\int \frac{1}{\sqrt{1-x^2}}\,dx = \int \frac{\cos\theta}{|\cos\theta|}\,d\theta = \int d\theta = \theta + C = \text{Sin}^{-1} x + C.$$

For integral (ii), the substitution $x = \sin\theta$ is of little use since the square root $\sqrt{4-9\sin^2\theta}$ does not simplify. If, however, the square root were $\sqrt{4-4\sin^2\theta}$, it would immediately reduce to $\sqrt{4-4\sin^2\theta} = \sqrt{4\cos^2\theta} = 2|\cos\theta|$. We can obtain this result if the substitution is modified to $x = (2/3)\sin\theta$ (or, more properly, $\theta = \text{Sin}^{-1}(3x/2)$). Since $dx = (2/3)\cos\theta\,d\theta$, we have

$$\int \frac{1}{\sqrt{4-9x^2}}\,dx = \int \frac{(2/3)\cos\theta}{\sqrt{4-4\sin^2\theta}}\,d\theta = \frac{2}{3}\int \frac{\cos\theta}{2\cos\theta}\,d\theta$$

$$= \frac{1}{3}\int d\theta = \frac{\theta}{3} + C = \frac{1}{3}\text{Sin}^{-1}\left(\frac{3x}{2}\right) + C.$$

For integrals containing square roots of the form $\sqrt{a^2 + b^2 x^2}$, consider the pair of integrals

(iii) $\int \frac{1}{1+x^2}\,dx$ (iv) $\int \frac{1}{(3+5x^2)^{3/2}}\,dx.$

Integral (iii) obviously has the answer $\text{Tan}^{-1} x + C$. Alternatively, if we had $1 + \tan^2\theta$ instead of $1+x^2$, the two terms in the denominator would simplify to the one term $\sec^2\theta$. We therefore substitute $x = \tan\theta$ or, more properly, $\theta = \text{Tan}^{-1} x$, from which we have $dx = \sec^2\theta\,d\theta$. With this substitution, we obtain

$$\int \frac{1}{1+x^2}\,dx = \int \frac{\sec^2\theta}{\sec^2\theta}\,d\theta = \int d\theta = \theta + C = \text{Tan}^{-1} x + C.$$

The substitution $x = \tan\theta$ in (iv) yields $(3+5x^2)^{3/2} = (3+5\tan^2\theta)^{3/2}$. For simplification, we need a 3 in front of the $\tan^2\theta$ and not a 5. This can be accomplished by modification of our substitution to

$$x = \sqrt{\frac{3}{5}}\tan\theta,$$

from which

$$dx = \sqrt{\frac{3}{5}}\sec^2\theta\,d\theta.$$

Then,

$$\int \frac{1}{(3+5x^2)^{3/2}}\,dx = \int \frac{1}{(3+3\tan^2\theta)^{3/2}}\sqrt{\frac{3}{5}}\sec^2\theta\,d\theta = \frac{1}{3\sqrt{5}}\int \frac{\sec^2\theta}{(\sec^2\theta)^{3/2}}\,d\theta$$

$$= \frac{1}{3\sqrt{5}}\int \frac{1}{\sec\theta}\,d\theta = \frac{1}{3\sqrt{5}}\int \cos\theta\,d\theta$$

$$= \frac{1}{3\sqrt{5}}\sin\theta + C.$$

FIGURE 9.7

To express $\sin\theta$ in terms of x, we draw the triangle in Figure 9.7 to fit the change of variable $\sqrt{5}\,x/\sqrt{3} = \tan\theta$. Since the hypotenuse of the triangle is $\sqrt{3 + 5x^2}$, we obtain

$$\int \frac{1}{(3+5x^2)^{3/2}}\,dx = \frac{1}{3\sqrt{5}} \frac{\sqrt{5}\,x}{\sqrt{3+5x^2}} + C = \frac{x}{3\sqrt{3+5x^2}} + C.$$

For integrals containing square roots of the form $\sqrt{b^2 x^2 - a^2}$, consider

$$\int \frac{1}{\sqrt{x^2 - 4}}\,dx.$$

If we had $\sqrt{4\sec^2\theta - 4}$ instead of $\sqrt{x^2 - 4}$, an immediate simplification would occur. We therefore substitute $x = 2\sec\theta$, from which $dx = 2\sec\theta\tan\theta\,d\theta$. Once again the real substitution is $\theta = \operatorname{Sec}^{-1}(x/2)$, in which case θ is an angle in either the first or third quadrant. With this change,

$$\sqrt{x^2 - 4} = \sqrt{4\sec^2\theta - 4} = \sqrt{4\tan^2\theta} = 2|\tan\theta| = 2\tan\theta \quad \text{and}$$

$$\int \frac{1}{\sqrt{x^2 - 4}}\,dx = \int \frac{2\sec\theta\tan\theta}{2\tan\theta}\,d\theta$$

$$= \int \sec\theta\,d\theta$$

$$= \ln|\sec\theta + \tan\theta| + C.$$

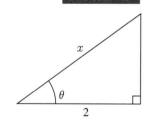

FIGURE 9.8

To express $\tan\theta$ in terms of x, we draw the triangle in Figure 9.8 to fit the change of variable $x/2 = \sec\theta$. Since the third side is $\sqrt{x^2 - 4}$, we have

$$\int \frac{1}{\sqrt{x^2 - 4}}\,dx = \ln\left|\frac{x}{2} + \frac{\sqrt{x^2 - 4}}{2}\right| + C = \ln\left|x + \sqrt{x^2 - 4}\right| + D,$$

where $D = C - \ln 2$.

We have illustrated by simple examples that trigonometric substitutions can be useful in the evaluation of integrals involving square roots of the forms $\sqrt{a^2 - b^2 x^2}$, $\sqrt{a^2 + b^2 x^2}$, and $\sqrt{b^2 x^2 - a^2}$. Essentially, the method replaces the terms under the square root by a perfect square and thereby rids the integrand of the square root. To obtain the trigonometric substitution appropriate to each square root, we suggest the following procedure. We have been working with the trigonometric identities:

$$1 - \sin^2\theta = \cos^2\theta,$$
$$1 + \tan^2\theta = \sec^2\theta,$$
$$\sec^2\theta - 1 = \tan^2\theta.$$

To determine which trigonometric substitution is appropriate to, say $\sqrt{a^2 - b^2 x^2}$, we mentally set $a = b = 1$ and obtain $\sqrt{1 - x^2}$. We note that $1 - x^2$ resembles $1 - \sin^2\theta$, and immediately set $x = \sin\theta$. To simplify the original square root $\sqrt{a^2 - b^2 x^2}$, we then modify this substitution to

$$x = \frac{a}{b}\sin\theta \qquad \left[\text{or } \theta = \operatorname{Sin}^{-1}\left(\frac{bx}{a}\right)\right].$$

Similarly, for the square root $\sqrt{a^2 + b^2 x^2}$, we initially set $x = \tan\theta$, having mentally set $a = b = 1$, and noted that $1 + x^2$ resembles $1 + \tan^2\theta$. We then modify the substitution to

$$x = \frac{a}{b}\tan\theta \qquad \left[\text{or } \theta = \text{Tan}^{-1}\left(\frac{bx}{a}\right)\right].$$

Finally, for the square root $\sqrt{b^2 x^2 - a^2}$, we initially set $x = \sec\theta$ and then modify to

$$x = \frac{a}{b}\sec\theta \qquad \left[\text{or } \theta = \text{Sec}^{-1}\left(\frac{bx}{a}\right)\right].$$

The square roots $\sqrt{a^2 - b^2 x^2}$, $\sqrt{a^2 + b^2 x^2}$, and $\sqrt{b^2 x^2 - a^2}$ are replaced respectively by

$$\begin{array}{lll}
\sqrt{a^2 - a^2 \sin^2\theta} & \sqrt{a^2 + a^2 \tan^2\theta} & \sqrt{a^2 \sec^2\theta - a^2} \\
= \sqrt{a^2 \cos^2\theta} & = \sqrt{a^2 \sec^2\theta} & = \sqrt{a^2 \tan^2\theta} \\
= a|\cos\theta| & = a|\sec\theta| & = a|\tan\theta| \\
= a\cos\theta; & = a\sec\theta; & = a\tan\theta.
\end{array}$$

In each case it is our choice of principal values for the inverse trigonometric functions that enables us to neglect the absolute value.

We now consider a number of more complex examples.

EXAMPLE 9.11

Evaluate the following integrals:

$$\text{(a)} \int \frac{1}{x^2(9x^2 + 2)}\, dx \qquad \text{(b)} \int x^2 \sqrt{3 - x^2}\, dx$$

SOLUTION (a) If we set $x = \left(\sqrt{2}/3\right)\tan\theta$, then $dx = \left(\sqrt{2}/3\right)\sec^2\theta\, d\theta$, and

$$\int \frac{1}{x^2(9x^2 + 2)}\, dx = \int \frac{1}{\left(\frac{2}{9}\tan^2\theta\right)(2\tan^2\theta + 2)}\left(\frac{\sqrt{2}}{3}\right)\sec^2\theta\, d\theta$$

$$= \frac{3}{2\sqrt{2}} \int \frac{\sec^2\theta}{\tan^2\theta \sec^2\theta}\, d\theta$$

$$= \frac{3}{2\sqrt{2}} \int \cot^2\theta\, d\theta = \frac{3}{2\sqrt{2}} \int (\csc^2\theta - 1)\, d\theta$$

$$= \frac{3}{2\sqrt{2}}(-\cot\theta - \theta) + C$$

$$= \frac{-3}{2\sqrt{2}}\left[\frac{\sqrt{2}}{3x} + \text{Tan}^{-1}\left(\frac{3x}{\sqrt{2}}\right)\right] + C \qquad \text{(Figure 9.9)}$$

$$= \frac{-1}{2x} - \frac{3}{2\sqrt{2}}\text{Tan}^{-1}\left(\frac{3x}{\sqrt{2}}\right) + C.$$

FIGURE 9.9

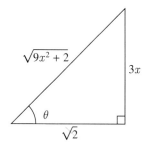

(b) If we set $x = \sqrt{3} \sin \theta$, then $dx = \sqrt{3} \cos \theta \, d\theta$, and

$$\int x^2 \sqrt{3-x^2} \, dx = \int (3 \sin^2 \theta) \sqrt{3 - 3 \sin^2 \theta} (\sqrt{3} \cos \theta \, d\theta)$$

$$= 9 \int \sin^2 \theta \sqrt{1 - \sin^2 \theta} \cos \theta \, d\theta$$

$$= 9 \int \sin^2 \theta \cos^2 \theta \, d\theta$$

$$= \frac{9}{4} \int \sin^2 2\theta \, d\theta \quad \text{(by 9.23b)}$$

$$= \frac{9}{4} \int \left(\frac{1 - \cos 4\theta}{2} \right) d\theta \quad \text{(by 9.23d)}$$

$$= \frac{9}{8} \left(\theta - \frac{1}{4} \sin 4\theta \right) + C \quad \text{(by 9.22d)}$$

$$= \frac{9}{8} \operatorname{Sin}^{-1} \left(\frac{x}{\sqrt{3}} \right) - \frac{9}{32} (2 \sin 2\theta \cos 2\theta) + C \quad \text{(by 9.23b)}$$

$$= \frac{9}{8} \operatorname{Sin}^{-1} \left(\frac{x}{\sqrt{3}} \right) - \frac{9}{16} (2 \sin \theta \cos \theta)(1 - 2 \sin^2 \theta) + C \quad \text{(by 9.23b,d)}$$

$$= \frac{9}{8} \operatorname{Sin}^{-1} \left(\frac{x}{\sqrt{3}} \right) - \frac{9}{8} \left(\frac{x}{\sqrt{3}} \right) \frac{\sqrt{3-x^2}}{\sqrt{3}} \left(1 - \frac{2x^2}{3} \right) + C \quad \text{(Figure 9.10)}$$

$$= \frac{9}{8} \operatorname{Sin}^{-1} \left(\frac{x}{\sqrt{3}} \right) - \frac{x}{8} \sqrt{3-x^2} \, (3 - 2x^2) + C.$$

FIGURE 9.10

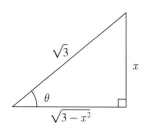

■

EXAMPLE 9.12

Use a trigonometric substitution to evaluate the integral in Example 4.

SOLUTION If we set $x = 3\tan\theta$, then $dx = 3\sec^2\theta\,d\theta$, and

$$\int \frac{x^3}{\sqrt{9+x^2}}\,dx = \int \frac{27\tan^3\theta}{\sqrt{9+9\tan^2\theta}} 3\sec^2\theta\,d\theta$$

$$= 27 \int \frac{\tan^3\theta\,\sec^2\theta}{\sec\theta}\,d\theta$$

$$= 27 \int \tan\theta\,(\sec^2\theta - 1)\sec\theta\,d\theta \qquad \text{(by 9.25)}$$

$$= 27 \int (\sec^3\theta\,\tan\theta - \sec\theta\,\tan\theta)\,d\theta$$

$$= 27 \left(\frac{1}{3}\sec^3\theta - \sec\theta\right) + C \qquad \text{(by 9.24b)}$$

$$= 9\left(\frac{\sqrt{9+x^2}}{3}\right)^3 - 27\frac{\sqrt{9+x^2}}{3} + C \qquad \text{(Figure 9.11)}$$

$$= \frac{1}{3}(9+x^2)^{3/2} - 9\sqrt{9+x^2} + C.$$

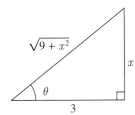

FIGURE 9.11

EXAMPLE 9.13

Verify the formula πr^2 for the area of a circle of radius r.

SOLUTION If we consider the circle $x^2 + y^2 = r^2$ in Figure 9.12, we see that its area is four times that in the first quadrant:

$$A = 4\int_0^r \sqrt{r^2 - x^2}\,dx.$$

If we set $x = r\sin\theta$, then $dx = r\cos\theta\,d\theta$, and

$$A = 4\int_0^{\pi/2} \sqrt{r^2 - r^2\sin^2\theta}\,r\cos\theta\,d\theta$$

$$= 4r^2 \int_0^{\pi/2} \cos^2\theta\,d\theta$$

$$= 4r^2 \int_0^{\pi/2} \left(\frac{1+\cos 2\theta}{2}\right) d\theta \qquad \text{(by 9.23c)}$$

$$= 2r^2 \left\{\theta + \frac{1}{2}\sin 2\theta\right\}_0^{\pi/2} \qquad \text{(by 9.22d)}$$

$$= 2r^2(\pi/2) = \pi r^2.$$

FIGURE 9.12

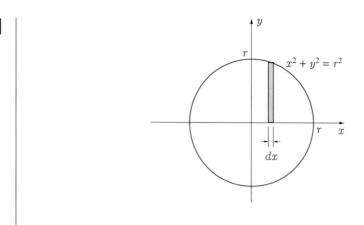

EXERCISES 9.5

In Exercises 1–20 evaluate the indefinite integral.

1. $\displaystyle\int \frac{1}{x\sqrt{2x^2-4}}\,dx$
2. $\displaystyle\int \frac{1}{\sqrt{9-5x^2}}\,dx$
3. $\displaystyle\int \frac{1}{10+x^2}\,dx$
4. $\displaystyle\int \frac{1}{x^2\sqrt{4-x^2}}\,dx$
5. $\displaystyle\int \sqrt{7-x^2}\,dx$
6. $\displaystyle\int x\sqrt{5x^2+3}\,dx$
7. $\displaystyle\int x^3\sqrt{4+x^2}\,dx$
8. $\displaystyle\int \frac{1}{1-x^2}\,dx$
9. $\displaystyle\int \frac{1}{\sqrt{x^2-5}}\,dx$
10. $\displaystyle\int \frac{x+5}{10x^2+2}\,dx$
11. $\displaystyle\int \frac{1}{x\sqrt{x^2+3}}\,dx$
12. $\displaystyle\int \frac{\sqrt{4-x^2}}{x}\,dx$
13. $\displaystyle\int \frac{x^2}{(2-9x^2)^{3/2}}\,dx$
14. $\displaystyle\int \frac{\sqrt{x^2-16}}{x^2}\,dx$
15. $\displaystyle\int \frac{1}{x^2\sqrt{2x^2+7}}\,dx$
16. $\displaystyle\int \frac{1}{x^3\sqrt{x^2-4}}\,dx$
17. $\displaystyle\int \frac{\sqrt{9-z^2}}{z^4}\,dz$
18. $\displaystyle\int \frac{y^3}{\sqrt{y^2+4}}\,dy$
19. $\displaystyle\int \frac{1}{(4x^2-9)^{3/2}}\,dx$
20. $\displaystyle\int \frac{x^2+2}{x^3+x}\,dx$

21. Show that if $a > 0$, then
$$\int \frac{1}{a^2-x^2}\,dx = \frac{1}{2a}\ln\left|\frac{a+x}{a-x}\right| + C.$$

22. Verify that the area inside the ellipse $b^2x^2 + a^2y^2 = a^2b^2$ is $\pi a b$.

23. Find the area common to the circles $x^2 + y^2 = 4$ and $x^2 + y^2 = 4x$.

24. A boy initially at O (Figure 9.13) walks along the edge of a pier (the y-axis) towing his sailboat by a string of length L.
    (a) If the boat starts at Q and the string always remains straight, show that the equation of the curved path $y = f(x)$ followed by the boat must satisfy the differential equation
    $$\frac{dy}{dx} = -\frac{\sqrt{L^2-x^2}}{x}.$$
    (b) Find $y = f(x)$.

FIGURE 9.13

25. A dog at position $(L, 0)$ in the xy-plane spots a rabbit at position $(0, 0)$ running in the positive y-direction. If the dog runs at the same speed as the rabbit and always moves directly toward the rabbit, find the equation of the path followed by the dog.

26. The parabola $x = y^2$ divides the circle $x^2 + y^2 = 4$ into two parts. Find the second moment of area of the smaller part about the x-axis.

27. Find the centroid of the area in Exercise 26.

28. Find the horizontal line that divides the ellipse $b^2 x^2 + a^2 y^2 = a^2 b^2$ into two parts so that the area of the lower part is twice the area of the upper part.

29. If a thin circular plate has radius 2 units and constant mass per unit area ρ, find its moment of inertia about any tangent line to its edge.

30. A cylindrical oil can with horizontal axis has radius r and length h. If the density of the oil is ρ, find the force on each end of the can when it is full.

In Exercises 31–37 evaluate the indefinite integral.

31. $\int \dfrac{2x^4 - x^2}{2x^2 + 1}\, dx$

32. $\int (7 - x^2)^{3/2}\, dx$

33. $\int \dfrac{1}{x - x^3}\, dx$

34. $\int \dfrac{1}{x^3 (4x^2 - 1)^{3/2}}\, dx$

35. $\int \sqrt{x^2 - 4}\, dx$

36. $\int \sqrt{1 + 3x^2}\, dx$

37. $\int \dfrac{x^2}{\sqrt{x^2 - 5}}\, dx$

38. Find the length of the curve $8y^2 = x^2(1 - x^2)$.

39. Find the length of the parabola $y = x^2$ from $(0,0)$ to $(1,1)$.

40. At what distance from the centre of a circle should a line be drawn in order that the second moment of area of the circle about that line be equal to twice the second moment of the circle about a line through its centre?

41. When a flexible cable of constant mass per unit length ρ hangs between two fixed points A and B (Figure 9.14), the shape $y = f(x)$ of the cable must satisfy the differential equation

$$\dfrac{d^2 y}{dx^2} = \dfrac{\rho g}{H}\sqrt{1 + \left(\dfrac{dy}{dx}\right)^2},$$

where $g > 0$ and $H > 0$ are constants. If we set $k = \rho g / H$ and $p = dy/dx$, then

$$\dfrac{dp}{dx} = k\sqrt{1 + p^2} \quad \text{or} \quad \dfrac{1}{\sqrt{1 + p^2}}\dfrac{dp}{dx} = k.$$

It follows that

$$\int \dfrac{1}{\sqrt{1 + p^2}}\, dp = kx + C.$$

(a) Evaluate the integral shown to find $p = dy/dx$.
(b) Integrate once more to find the shape of the cable (see also Example 8.14 in Section 8.4).

FIGURE 9.14

42. Find the area inside the loop of the strophoid $y^2(a + x) = x^2(a - x)$.

43. (a) By rationalizing the denominator, show that

$$\dfrac{1}{x + 1 + \sqrt{x^2 + 1}} = \dfrac{x + 1 - \sqrt{x^2 + 1}}{2x}.$$

(b) Use (a) to prove that

$$\int_0^1 \dfrac{1}{x + 1 + \sqrt{x^2 + 1}}\, dx = 1 - \dfrac{\sqrt{2}}{2} + \dfrac{1}{2}\ln\left(\dfrac{1 + \sqrt{2}}{2}\right).$$

44. Show that

$$\int_{-1}^{1} \sqrt{\dfrac{1 + x}{1 - x}}\, dx = \pi.$$

SECTION 9.6

Completing the Square and Trigonometric Substitutions

Trigonometric substitutions reduce integrals containing square roots of the form $\sqrt{a^2 - b^2 x^2}$, $\sqrt{a^2 + b^2 x^2}$, and $\sqrt{b^2 x^2 - a^2}$ to trigonometric integrals. For integrals containing square roots of the form $\sqrt{ax^2 + bx + c}$ we can again reduce the integral to a trigonometric integral by a trigonometric substitution, if we first complete the square.

Consider the integral

$$\int \dfrac{1}{(x^2 + 2x + 5)^{3/2}}\, dx.$$

If we complete the square in the denominator, we obtain

$$\int \frac{1}{(x^2 + 2x + 5)^{3/2}} \, dx = \int \frac{1}{[(x+1)^2 + 4]^{3/2}} \, dx.$$

Had the denominator been $(x^2 + 4)^{3/2}$, we would have set $x = 2 \tan \theta$. It is natural, then, for the denominator $[(x+1)^2 + 4]^{3/2}$ to set

$$x + 1 = 2 \tan \theta,$$

in which case $dx = 2 \sec^2 \theta \, d\theta$, and

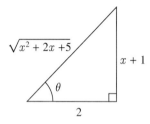

FIGURE 9.15

$$\int \frac{1}{(x^2 + 2x + 5)^{3/2}} \, dx = \int \frac{1}{(4 \tan^2 \theta + 4)^{3/2}} 2 \sec^2 \theta \, d\theta = \int \frac{2 \sec^2 \theta}{8 \sec^3 \theta} \, d\theta$$

$$= \frac{1}{4} \int \cos \theta \, d\theta = \frac{1}{4} \sin \theta + C$$

$$= \frac{x+1}{4\sqrt{x^2 + 2x + 5}} + C. \qquad \text{(Figure 9.15)}$$

Consider another example,

$$\int \sqrt{4 + x - x^2} \, dx.$$

Completion of the square leads to

$$\int \sqrt{4 + x - x^2} \, dx = \int \sqrt{-(x - 1/2)^2 + 17/4} \, dx.$$

If we now set

$$x - \frac{1}{2} = \frac{\sqrt{17}}{2} \sin \theta,$$

then

$$dx = \frac{\sqrt{17}}{2} \cos \theta \, d\theta, \qquad \text{and}$$

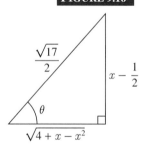

FIGURE 9.16

$$\int \sqrt{4 + x - x^2} \, dx = \int \sqrt{-\frac{17}{4} \sin^2 \theta + \frac{17}{4}} \frac{\sqrt{17}}{2} \cos \theta \, d\theta$$

$$= \int \frac{\sqrt{17}}{2} \cos \theta \frac{\sqrt{17}}{2} \cos \theta \, d\theta$$

$$= \frac{17}{4} \int \cos^2 \theta \, d\theta = \frac{17}{4} \int \left(\frac{1 + \cos 2\theta}{2} \right) d\theta$$

$$= \frac{17}{8} \left(\theta + \frac{1}{2} \sin 2\theta \right) + C = \frac{17}{8} (\theta + \sin \theta \cos \theta) + C$$

$$= \frac{17}{8} \text{Sin}^{-1} \left(\frac{2x-1}{\sqrt{17}} \right) + \frac{17}{8} \left[\frac{2}{\sqrt{17}} \left(x - \frac{1}{2} \right) \right] \frac{\sqrt{4 + x - x^2}}{\sqrt{17}/2} + C \qquad \text{(Figure 9.16)}$$

$$= \frac{17}{8} \text{Sin}^{-1} \left(\frac{2x-1}{\sqrt{17}} \right) + \frac{1}{4} (2x - 1) \sqrt{4 + x - x^2} + C.$$

As a final illustrative example, we consider

$$\int \frac{x}{\sqrt{2x^2 + 3x - 6}}\,dx = \frac{1}{\sqrt{2}} \int \frac{x}{\sqrt{x^2 + \frac{3x}{2} - 3}}\,dx = \frac{1}{\sqrt{2}} \int \frac{x}{\sqrt{\left(x + \frac{3}{4}\right)^2 - \frac{57}{16}}}\,dx.$$

If we set

$$x + \frac{3}{4} = \frac{\sqrt{57}}{4} \sec\theta,$$

then

$$dx = \frac{\sqrt{57}}{4} \sec\theta \tan\theta\,d\theta, \quad \text{and}$$

$$\int \frac{x}{\sqrt{2x^2 + 3x - 6}}\,dx = \frac{1}{\sqrt{2}} \int \frac{(\sqrt{57}/4)\sec\theta - 3/4}{\sqrt{(57/16)\sec^2\theta - 57/16}} \cdot \frac{\sqrt{57}}{4} \sec\theta \tan\theta\,d\theta$$

$$= \frac{1}{\sqrt{2}} \int \frac{(\sqrt{57}/4)\sec\theta - 3/4}{(\sqrt{57}/4)\tan\theta} \cdot \frac{\sqrt{57}}{4} \sec\theta \tan\theta\,d\theta$$

$$= \frac{1}{\sqrt{2}} \int \left(\frac{\sqrt{57}}{4}\sec^2\theta - \frac{3}{4}\sec\theta\right) d\theta$$

$$= \frac{\sqrt{57}}{4\sqrt{2}} \tan\theta - \frac{3}{4\sqrt{2}} \ln|\sec\theta + \tan\theta| + C$$

$$= \frac{\sqrt{57}}{4\sqrt{2}} \left[\frac{\sqrt{x^2 + (3/2)x - 3}}{\sqrt{57}/4}\right]$$

$$- \frac{3}{4\sqrt{2}} \ln\left|\frac{x + 3/4}{\sqrt{57}/4} + \frac{\sqrt{x^2 + (3/2)x - 3}}{\sqrt{57}/4}\right| + C \quad \text{(Figure 9.17)}$$

$$= \frac{1}{2}\sqrt{2x^2 + 3x - 6}$$

$$- \frac{3}{4\sqrt{2}} \ln\left|4x + 3 + 2\sqrt{2}\sqrt{2x^2 + 3x - 6}\right| + D,$$

FIGURE 9.17

where

$$D = C + \frac{3}{8\sqrt{2}} \ln 57.$$

We have shown that the technique of completing the square is the natural generalization of the technique of trigonometric substitutions. It reduces integrals containing square roots of the form $\sqrt{ax^2 + bx + c}$ to trigonometric integrals.

EXAMPLE 9.14

Find the volume of a donut.

SOLUTION A donut is generated (as the volume of the solid of revolution) when the circle
$$(x-a)^2 + y^2 = b^2, \quad (a > b),$$
is rotated around the y-axis (Figure 9.18). Cylindrical shells for the vertical rectangle shown yield the volume generated by the upper semicircle, which can be doubled to give
$$V = 2\int_{a-b}^{a+b} 2\pi xy\, dx.$$
If we solve the equation of the circle for $y = \pm\sqrt{b^2 - (x-a)^2}$, then
$$V = 2\int_{a-b}^{a+b} 2\pi x\sqrt{b^2 - (x-a)^2}\, dx.$$
We now set $x - a = b\sin\theta$, in which case $dx = b\cos\theta\, d\theta$, and
$$V = 4\pi \int_{-\pi/2}^{\pi/2} (a + b\sin\theta)b\cos\theta\, b\cos\theta\, d\theta$$
$$= 4\pi b^2 \int_{-\pi/2}^{\pi/2} (a\cos^2\theta + b\cos^2\theta\sin\theta)\, d\theta$$
$$= 4\pi b^2 \int_{-\pi/2}^{\pi/2} \left[a\left(\frac{1+\cos 2\theta}{2}\right) + b\cos^2\theta\sin\theta\right] d\theta$$
$$= 4\pi b^2 \left\{\frac{a}{2}\left(\theta + \frac{1}{2}\sin 2\theta\right) - \frac{b}{3}\cos^3\theta\right\}_{-\pi/2}^{\pi/2}$$
$$= 4\pi b^2 \left[\frac{a}{2}\left(\frac{\pi}{2} + \frac{\pi}{2}\right)\right] = 2\pi^2 ab^2.$$

FIGURE 9.18

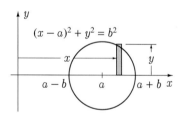

Should we use horizontal rectangles, as in Figure 9.19, the washer method yields
$$V = 2\int_0^b (\pi x_2^2 - \pi x_1^2)\, dy.$$
We can solve the equation of the circle for $x = a \pm \sqrt{b^2 - y^2}$; hence we have
$$x_1 = a - \sqrt{b^2 - y^2}, \quad x_2 = a + \sqrt{b^2 - y^2}.$$
Thus
$$V = 2\pi \int_0^b \left[\left(a + \sqrt{b^2 - y^2}\right)^2 - \left(a - \sqrt{b^2 - y^2}\right)^2\right] dy,$$

and this simplifies to

$$V = 2\pi \int_0^b 4a\sqrt{b^2 - y^2}\, dy.$$

If we set $y = b\sin\theta$, then $dy = b\cos\theta\, d\theta$, and

$$V = 8\pi a \int_0^{\pi/2} b\cos\theta\, b\cos\theta\, d\theta$$

$$= 8\pi a b^2 \int_0^{\pi/2} \left(\frac{1 + \cos 2\theta}{2}\right) d\theta$$

$$= 4\pi a b^2 \left\{\theta + \frac{1}{2}\sin 2\theta\right\}_0^{\pi/2}$$

$$= 4\pi a b^2 (\pi/2) = 2\pi^2 a b^2.$$

FIGURE 9.19

EXERCISES 9.6

In Exercises 1–10 evaluate the indefinite integral.

1. $\displaystyle\int \frac{x}{\sqrt{27 + 6x - x^2}}\, dx$
2. $\displaystyle\int \frac{1}{\sqrt{x^2 + 2x + 2}}\, dx$
3. $\displaystyle\int \frac{1}{(y^2 + 4y)^{3/2}}\, dy$
4. $\displaystyle\int \frac{1}{3x - x^2 - 4}\, dx$
5. $\displaystyle\int \frac{\sqrt{x^2 + 2x - 3}}{x + 1}\, dx$
6. $\displaystyle\int \frac{x}{(4x - x^2)^{3/2}}\, dx$
7. $\displaystyle\int \sqrt{-y^2 + 6y}\, dy$
8. $\displaystyle\int \frac{2x - 3}{x^2 + 6x + 13}\, dx$
9. $\displaystyle\int \frac{5 - 4x}{\sqrt{12x - 4x^2 - 8}}\, dx$
10. $\displaystyle\int \frac{1}{x\sqrt{6 + 4\ln x + (\ln x)^2}}\, dx$

11. Use the substitution $z = 1/x$ to evaluate

$$\int \frac{1}{x\sqrt{x^2 + 6x + 3}}\, dx.$$

12. One of the gates in a dam is circular with radius 1 m. If the gate is closed and the surface of the water is 3 m above the top of the gate, find the force due to water pressure on the gate.

13. Evaluate

$$\int \frac{1}{3x - x^2}\, dx$$

(a) by completing the square, and setting $x - 3/2 = (3/2)\sin\theta$;

(b) by multiplying numerator and denominator by -1, completing the square, and setting $x - 3/2 = (3/2)\sec\theta$.

(c) Explain the difference in the two answers.

In Exercises 14–16 evaluate the integral.

14. $\displaystyle\int \sqrt{x^2 - 2x - 3}\, dx$
15. $\displaystyle\int \frac{1}{x\sqrt{2x - x^2}}\, dx$
16. $\displaystyle\int \frac{1}{(2x + 5)\sqrt{2x - 3} + 8x - 12}\, dx$

SECTION 9.7

Partial Fractions

Partial fractions is a method that we apply to integrals of the form

$$\int \frac{N(x)}{D(x)} \, dx, \tag{9.26}$$

where $N(x)$ and $D(x)$ are polynomials in x, and the degree of $N(x)$ is less than the degree of $D(x)$. When the degree of $N(x)$ is greater than or equal to that of $D(x)$, we divide $D(x)$ into $N(x)$. For example, in the integral

$$\int \frac{x^4 + 4x^3 + 2x + 4}{x^3 + 1} \, dx,$$

the numerator has degree 4 and the denominator degree 3. By long division, we obtain

$$\int \frac{x^4 + 4x^3 + 2x + 4}{x^3 + 1} \, dx = \int \left(x + 4 + \frac{x}{x^3 + 1} \right) dx.$$

The first two terms on the right, namely, $x + 4$, can be integrated immediately, and partial fractions can be applied to the remaining term $x/(x^3 + 1)$.

A general rule is that if the degree of $N(x)$ is greater than or equal to the degree of $D(x)$, divide $D(x)$ into $N(x)$ to produce a quotient polynomial $Q(x)$ and a remainder polynomial $R(x)$ of degree less than that of $D(x)$; that is,

$$\frac{N(x)}{D(x)} = Q(x) + \frac{R(x)}{D(x)},$$

where $\deg R < \deg D$.

Integral 9.26 can then be written in the form

$$\int \frac{N(x)}{D(x)} \, dx = \int Q(x) \, dx + \int \frac{R(x)}{D(x)} \, dx.$$

The first integral on the right is trivial. This leaves integration of the rational function

$$\int \frac{R(x)}{D(x)} \, dx, \tag{9.27}$$

where $\deg R < \deg D$.

To use partial fractions on integral 9.27 we must factor the denominator $D(x)$. It is an established fact of algebra that every real polynomial, and specifically $D(x)$, can be factored into real linear factors $ax + b$ and irreducible quadratic factors $ax^2 + bx + c$. Note that a quadratic factor is irreducible if $b^2 - 4ac < 0$. In the complete factorization of $D(x)$, these factors may or may not be repeated; that is, the factorization of $D(x)$ contains terms of the form

$$(ax + b)^n \quad \text{and} \quad (ax^2 + bx + c)^n,$$

where $n \geq 1$ is an integer. When $n = 1$ the factor is nonrepeated, and when $n > 1$, it is repeated. For example, in the factorization

$$D(x) = (x-1)(2x+1)^3(3x^2+4x+5)(x^2+1)^2,$$

$x-1$ and $3x^2+4x+5$ are nonrepeated, and $2x+1$ and x^2+1 are repeated.

Having factored $D(x)$, we can break down the rational function in 9.27 into frational components. We call this the **partial fraction decomposition** of the integrand. We illustrate it with the rational function $(x^2+1)/[(x-1)(2x+1)^3(3x^2+4x+5)(x^2+1)^2]$, and then state general rules for all decompositions. The partial fraction decomposition of this particular rational function is

$$\frac{x^2+1}{(x-1)(2x+1)^3(3x^2+4x+5)(x^2+1)^2} = \frac{A}{x-1} + \frac{B}{2x+1} + \frac{C}{(2x+1)^2}$$
$$+ \frac{D}{(2x+1)^3} + \frac{Ex+F}{3x^2+4x+5}$$
$$+ \frac{Gx+H}{x^2+1} + \frac{Ix+J}{(x^2+1)^2},$$

where $A, B, \ldots, J$ are constants. The first term corresponds to the nonrepeated linear factor $x-1$, the next three corresponds to the repeated linear factor $2x+1$, the fifth term corresponds to the nonrepeated quadratic factor $3x^2+4x+5$, and the last two terms result from the repeated quadratic x^2+1.

Let us now state general rules for the partial fraction decomposition of rational functions $R(x)/D(x)$. There are three rules:

1. For each repeated or nonrepeated linear factor $(ax+b)^n$ in $D(x)$, include the following terms in the decomposition

$$\frac{A_1}{ax+b} + \frac{A_2}{(ax+b)^2} + \cdots + \frac{A_n}{(ax+b)^n}. \qquad (9.28\text{a})$$

The number of terms corresponds to the power n on $(ax+b)^n$.

2. For each repeated or nonrepeated irreducible quadratic factor $(ax^2+bx+c)^n$ in $D(x)$, include the following terms in the decomposition

$$\frac{B_1x+C_1}{ax^2+bx+c} + \frac{B_2x+C_2}{(ax^2+bx+c)^2} + \cdots + \frac{B_nx+C_n}{(ax^2+bx+c)^n}. \qquad (9.28\text{b})$$

Again the number of terms corresponds to the power n on $(ax^2+bx+c)^n$.

3. The complete decomposition is the sum of all terms in 9.28a and all terms in 9.28b. The following examples will clarify these rules.

EXAMPLE 9.15

What form do partial fraction decompositions for the following rational functions take:

(a) $\dfrac{x^2+2x+3}{3x^3-x^2-3x+1}$
(b) $\dfrac{x^2+3x-1}{x^4+x^3+x^2+x}$
(c) $\dfrac{x^3+3x^2-x}{x^5+x^4+2x^3+2x^2+x+1}$
(d) $\dfrac{3x^6-1}{(3x^2+5)(x^2+2x+3)(2x-1)^2}$

SOLUTION (a) Since $3x^3-x^2-3x+1 = (3x-1)(x-1)(x+1)$, we have nonrepeated linear factors. Rules 1. and 3. give the partial fraction decomposition

$$\frac{x^2 + 2x + 3}{3x^3 - x^2 - 3x + 1} = \frac{A}{3x - 1} + \frac{B}{x - 1} + \frac{C}{x + 1}.$$

(b) The factorization $x^4 + x^3 + x^2 + x = x(x + 1)(x^2 + 1)$ has two nonrepeated linear factors, and a nonrepeated quadratic factor. The three rules lead to

$$\frac{x^2 + 3x - 1}{x^4 + x^3 + x^2 + x} = \frac{A}{x} + \frac{B}{x + 1} + \frac{Cx + D}{x^2 + 1}.$$

(c) Since $x^5 + x^4 + 2x^3 + 2x^2 + x + 1 = (x + 1)(x^2 + 1)^2$, the partial fraction decomposition is

$$\frac{x^3 + 3x^2 - x}{x^5 + x^4 + 2x^3 + 2x^2 + x + 1} = \frac{A}{x + 1} + \frac{Cx + D}{x^2 + 1} + \frac{Ex + F}{(x^2 + 1)^2}.$$

(d) Since $x^2 + 2x + 3$ can be factored no further, the decomposition is

$$\frac{3x^6 - 1}{(3x^2 + 5)(x^2 + 2x + 3)(2x - 1)^2} = \frac{Ax + B}{3x^2 + 5} + \frac{Cx + D}{x^2 + 2x + 3} + \frac{E}{2x - 1} + \frac{F}{(2x - 1)^2}.$$
∎

To illustrate how to calculate coefficients in partial fraction decompositions, we use part (b) of Example 9.15. We bring the right side of the decomposition to a common denominator,

$$\frac{x^2 + 3x - 1}{x^4 + x^3 + x^2 + x} = \frac{A}{x} + \frac{B}{x + 1} + \frac{Cx + D}{x^2 + 1}$$
$$= \frac{A(x + 1)(x^2 + 1) + Bx(x^2 + 1) + x(x + 1)(Cx + D)}{x^4 + x^3 + x^2 + x},$$

and equate numerators

$$x^2 + 3x - 1 = A(x + 1)(x^2 + 1) + Bx(x^2 + 1) + x(x + 1)(Cx + D).$$

There are two methods for finding the constants A, B, C, and D which satisfy this equation.

Method 1 – First, we gather together terms in the various powers of x on the right side of the equation,

$$x^2 + 3x - 1 = (A + B + C)x^3 + (A + C + D)x^2 + (A + B + D)x + A.$$

Now, Exercise 35 in Section 3.7 states that two polynomials of the same degree can be equal for all values of x if and only if coefficients of corresponding powers of x are identical. Since we have equal cubic polynomials in the above equation, we equate coefficients of x^3, x^2, x, and x^0 (meaning terms with no x's):

$$\begin{aligned}
x^3 &: \quad 0 = A + B + C, \\
x^2 &: \quad 1 = A + C + D, \\
x &: \quad 3 = A + B + D, \\
x^0 &: \quad -1 = A.
\end{aligned}$$

Solutions of these four linear equations in four unknowns are $A = -1$, $B = 3/2$, $C = -1/2$, and $D = 5/2$. Hence, the partial fraction decomposition of $(x^2 + 3x - 1)/(x^4 + x^3 + x^2 + x)$ is

$$\frac{x^2 + 3x - 1}{x^4 + x^3 + x^2 + x} = -\frac{1}{x} + \frac{3/2}{x+1} + \frac{-x/2 + 5/2}{x^2 + 1}.$$

Method 2 – In this method we substitute convenient values of x into the equation

$$x^2 + 3x - 1 = A(x+1)(x^2+1) + Bx(x^2+1) + x(x+1)(Cx+D).$$

Clearly $x = 0$ is most convenient, since it yields the value of A,

$$-1 = A(1)(1),$$

that is, $A = -1$. Convenient also is $x = -1$,

$$(-1)^2 + 3(-1) - 1 = B(-1)(2);$$

it gives $B = 3/2$. The values $x = 1$ and $x = 2$ yield the equations

$$1 + 3(1) - 1 = A(2)(2) + B(1)(2) + 1(2)(C+D),$$
$$4 + 3(2) - 1 = A(3)(5) + B(2)(5) + 2(3)(2C+D).$$

When $A = -1$ and $B = 3/2$ are substituted into these, the resulting equations are

$$C + D = 2,$$
$$4C + 2D = 3.$$

Solutions of these are $C = -1/2$ and $D = 5/2$.

It is also possible to use a combination of Methods 1 and 2 for finding coefficients in a partial fraction decomposition; that is, substitute some values of x and equate some coefficients. The total number must be equal to the number of unknown coefficients in the decomposition.

Once we have completed the partial fraction decomposition of a rational function, we can integrate the function by finding antiderivatives of the component fractions. For example, with the partial fraction decomposition of $(x^2 + 3x - 1)/(x^4 + x^3 + x^2 + x)$, it is very simple to find its antiderivative,

$$\int \frac{x^2 + 3x - 1}{x^4 + x^3 + x^2 + x}\, dx = \int \left(-\frac{1}{x} + \frac{3/2}{x+1} + \frac{-x/2 + 5/2}{x^2 + 1}\right) dx$$

$$= -\int \frac{1}{x}\, dx + \frac{3}{2}\int \frac{1}{x+1}\, dx - \frac{1}{2}\int \frac{x}{x^2+1}\, dx + \frac{5}{2}\int \frac{1}{x^2+1}\, dx$$

$$= -\ln|x| + \frac{3}{2}\ln|x+1| - \frac{1}{4}\ln(x^2+1) + \frac{5}{2}\text{Tan}^{-1} x + C.$$

Other integrations by partial fraction decompositions are illustrated in the following examples.

EXAMPLE 9.16

Evaluate the following indefinite integrals:

(a) $\int \dfrac{x}{x^4 + 6x^2 + 5}\,dx$ (b) $\int \dfrac{x^2 + 1}{2x^3 - 5x^2 + 4x - 1}\,dx$

(c) $\int \dfrac{1}{x^3 + 8}\,dx$ (d) $\int \dfrac{x^2}{x^3 + 8}\,dx$

SOLUTION (a) Since $x^4 + 6x^2 + 5 = (x^2 + 1)(x^2 + 5)$, the partial fraction decomposition of $x/(x^4 + 6x^2 + 5)$ has the form

$$\frac{x}{x^4 + 6x^2 + 5} = \frac{Ax + B}{x^2 + 1} + \frac{Cx + D}{x^2 + 5}.$$

When we bring the right side to a common denominator and equate numerators, the result is

$$x = (Ax + B)(x^2 + 5) + (Cx + D)(x^2 + 1).$$

We now multiply out the right side and equate coefficients of like powers of x,

$$\begin{aligned} x^3: &\quad 0 = A + C, \\ x^2: &\quad 0 = B + D, \\ x: &\quad 1 = 5A + C, \\ x^0: &\quad 0 = 5B + D. \end{aligned}$$

Solutions of these equations are $A = 1/4$, $B = 0$, $C = -1/4$, and $D = 0$, and therefore

$$\int \frac{x}{x^4 + 6x^2 + 5}\,dx = \int \left(\frac{x/4}{x^2 + 1} + \frac{-x/4}{x^2 + 5} \right) dx$$

$$= \frac{1}{8}\ln(x^2 + 1) - \frac{1}{8}\ln(x^2 + 5) + C.$$

(b) Since $2x^3 - 5x^2 + 4x - 1 = (2x - 1)(x - 1)^2$, the partial fraction decomposition takes the form

$$\frac{x^2 + 1}{2x^3 - 5x^2 + 4x - 1} = \frac{A}{2x - 1} + \frac{B}{x - 1} + \frac{C}{(x - 1)^2}$$

$$= \frac{A(x - 1)^2 + B(x - 1)(2x - 1) + C(2x - 1)}{(2x - 1)(x - 1)^2}.$$

When we equate numerators,

$$x^2 + 1 = A(x - 1)^2 + B(x - 1)(2x - 1) + C(2x - 1).$$

We now set $x = 0$, $x = 1$, and $x = 1/2$:

$$\begin{aligned} x = 0: &\quad 1 = A + B - C, \\ x = 1: &\quad 2 = C(1), \\ x = 1/2: &\quad 5/4 = A(1/4). \end{aligned}$$

These give $A = 5$, $B = -2$, and $C = 2$, and therefore

$$\int \frac{x^2 + 1}{2x^3 - 5x^2 + 4x - 1} \, dx = \int \left(\frac{5}{2x - 1} - \frac{2}{x - 1} + \frac{2}{(x-1)^2} \right) dx$$

$$= \frac{5}{2} \ln|2x - 1| - 2 \ln|x - 1| - \frac{2}{x - 1} + C.$$

(c) Since $x^3 + 8 = (x + 2)(x^2 - 2x + 4)$, we set

$$\frac{1}{x^3 + 8} = \frac{1}{(x + 2)(x^2 - 2x + 4)} = \frac{A}{x + 2} + \frac{Bx + C}{x^2 - 2x + 4}$$

$$= \frac{A(x^2 - 2x + 4) + (Bx + C)(x + 2)}{(x + 2)(x^2 - 2x + 4)},$$

and now equate numerators,

$$1 = A(x^2 - 2x + 4) + (Bx + C)(x + 2).$$

We set $x = -2$ and equate coefficients of x^2 and 1:

$$\begin{aligned} x &= -2: & 1 &= 12A, \\ x^2 &: & 0 &= A + B, \\ 1 &: & 1 &= 4A + 2C. \end{aligned}$$

These give $A = 1/12$, $B = -1/12$, and $C = 1/3$, and therefore

$$\int \frac{1}{x^3 + 8} \, dx = \int \left(\frac{1/12}{x + 2} + \frac{-x/12 + 1/3}{x^2 - 2x + 4} \right) dx$$

$$= \frac{1}{12} \ln|x + 2| + \frac{1}{12} \int \frac{4 - x}{(x - 1)^2 + 3} \, dx.$$

For the integral on the right, we set $x - 1 = \sqrt{3} \tan \theta$, in which case $dx = \sqrt{3} \sec^2 \theta \, d\theta$, and

$$\int \frac{1}{x^3 + 8} \, dx = \frac{1}{12} \ln|x + 2| + \frac{1}{12} \int \frac{4 - (1 + \sqrt{3} \tan \theta)}{3 \sec^2 \theta} \sqrt{3} \sec^2 \theta \, d\theta$$

$$= \frac{1}{12} \ln|x + 2| + \frac{1}{12\sqrt{3}} \int (3 - \sqrt{3} \tan \theta) \, d\theta$$

$$= \frac{1}{12} \ln|x + 2| + \frac{1}{12\sqrt{3}} \left(3\theta + \sqrt{3} \ln|\cos \theta| \right) + C$$

$$= \frac{1}{12} \ln|x + 2| + \frac{\sqrt{3}}{12} \operatorname{Tan}^{-1}\left(\frac{x - 1}{\sqrt{3}} \right) + \frac{1}{12} \ln \left| \frac{\sqrt{3}}{\sqrt{x^2 - 2x + 4}} \right| + C$$

(Figure 9.20)

$$= \frac{1}{12} \ln|x + 2| + \frac{\sqrt{3}}{12} \operatorname{Tan}^{-1}\left(\frac{x - 1}{\sqrt{3}} \right) - \frac{1}{24} \ln(x^2 - 2x + 4) + D,$$

where $D = C + (1/24) \ln 3$.

FIGURE 9.20

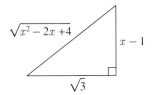

(d) Do not be misled into partial fractions in this example; the given rational function is immediately integrable,

$$\int \frac{x^2}{x^3 + 8}\, dx = \frac{1}{3} \ln |x^3 + 8| + C.$$

∎

In Section 4.4 we discussed the rate at which an infectious disease might spread through a population of susceptible individuals. If N represents the number of individuals in the population and $x(t)$ the number infected at any given time t, it is often assumed that the rate of infection is proportional to the numbers both infected and not infected:

$$\frac{dx}{dt} = kx(N - x),$$

where k is a constant. Thus, $x(t)$ must satisfy this differential equation subject to an initial condition such as, perhaps, $x(0) = 1$, where therefore one infected individual is introduced into the population at time $t = 0$. To solve this differential equation, we write

$$\frac{1}{x(N - x)} \frac{dx}{dt} = k,$$

and determine the partial fraction decomposition for the left side, namely,

$$\left(\frac{1/N}{x} + \frac{1/N}{N - x} \right) \frac{dx}{dt} = k.$$

Integration of both sides of this equation with respect to time gives

$$\frac{1}{N} \left(\ln |x| - \ln |N - x| \right) = kt + C.$$

Since x and $N - x$ are both positive, we may drop the absolute value signs and write

$$\ln \left(\frac{x}{N - x} \right) = N(kt + C).$$

To solve this equation for an explicit definition of $x(t)$, we first exponentiate both sides of the equation,

$$\frac{x}{N - x} = De^{Ft},$$

where we have substituted $D = e^{NC}$ and $F = kN$. Multiplication by $N - x$ gives

$$x = De^{Ft}(N - x) \qquad \text{or}$$

$$x\left(1 + De^{Ft}\right) = NDe^{Ft}.$$

Thus,

$$x(t) = \frac{NDe^{Ft}}{1 + De^{Ft}} = \frac{ND}{D + e^{-Ft}}.$$

Since $x(0) = 1$,

$$1 = \frac{ND}{1 + D},$$

which implies that $D = 1/(N-1)$, and

$$x(t) = \frac{\frac{N}{N-1}}{\frac{1}{N-1} + e^{-Ft}} = \frac{N}{1 + (N-1)e^{-Ft}}.$$

FIGURE 9.21

(a)
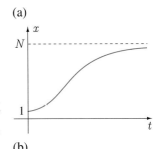

The shape of this curve is illustrated in Figure 9.21(a) (see also the discussion in Section 4.4). The time scale can be affixed to the diagram once one further piece of information is given. For instance, if it is known that one-half the population will be infected in 20 days, then $x(20) = N/2$, and the time scale is as shown in Figure 9.21(b). This added information determines F, since

$$\frac{N}{2} = \frac{N}{1 + (N-1)e^{-20F}},$$

(b)
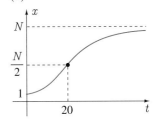

and this equation gives

$$F = \frac{1}{20} \ln(N-1).$$

EXERCISES 9.7

In Exercises 1–16 evaluate the indefinite integral.

1. $\displaystyle\int \frac{x+2}{x^2 - 2x + 1} \, dx$

2. $\displaystyle\int \frac{1}{y^3 + 3y^2 + 3y + 1} \, dy$

3. $\displaystyle\int \frac{1}{z^3 + z} \, dz$

4. $\displaystyle\int \frac{x^2 + 2x - 4}{x^2 - 2x - 8} \, dx$

5. $\displaystyle\int \frac{x}{(x-4)^2} \, dx$

6. $\displaystyle\int \frac{y+1}{y^3 + y^2 - 6y} \, dy$

7. $\displaystyle\int \frac{3x+5}{x^3 - x^2 - x + 1} \, dx$

8. $\displaystyle\int \frac{x^3}{(x^2+2)^2} \, dx$

9. $\displaystyle\int \frac{1}{x^2 - 3} \, dx$

10. $\displaystyle\int \frac{y^2}{y^2 + 3y + 2} \, dy$

11. $\displaystyle\int \frac{z^2 + 3z - 2}{z^3 + 5z} \, dz$

12. $\displaystyle\int \frac{y^2 + 6y + 4}{y^4 + 5y^2 + 4} \, dy$

13. $\displaystyle\int \frac{x}{x^4 + 7x^2 + 6} \, dx$

14. $\displaystyle\int \frac{x^2 + 3}{x^4 + x^2 - 2} \, dx$

15. $\displaystyle\int \frac{3t+4}{t^4 - 3t^3 + 3t^2 - t} \, dt$

16. $\displaystyle\int \frac{x^3 + 6}{x^4 + 2x^3 - 3x^2 - 4x + 4} \, dx$

17. Find the length of the curve $y = \ln(1 - x^2)$ from $x = 0$ to $x = 1/2$.

18. Natural factors such as space and available food supply inhibit the growth of any population. Let N represent the population of a species and M the maximum size of a stable population for which the death rate is just equal to the birth rate. If the time rate of change of N is assumed proportional to both N and $M - N$, then N must satisfy the differential equation

$$\frac{dN}{dt} = kN(M - N).$$

If $N(0) = N_0 < M$, solve this equation for $N(t)$, and sketch its graph.

19. Find the centroid of the area bounded by the curves

$$(x+2)^2 y = 4 - x, \qquad x = 0, \qquad y = 0, \qquad (x, y \geq 0).$$

20. Two substances A and B react to form a third substance C in such a way that 1 gm of A reacts with 1 gm of B to produce 2 gm of C. The rate at which C is formed is proportional to the product of the amounts of A and B present in the mixture.

    (a) Show that if 10 gm of A and 15 gm of B are brought together at time $t = 0$, the amount $x(t)$ of C present in the mixture must satisfy the differential equation

    $$\frac{dx}{dt} = k(20 - x)(30 - x), \qquad x(0) = 0,$$

    where k is a constant.

    (b) Solve this equation for $x(t)$.

In Exercises 21–26 evaluate the indefinite integral.

21. $\displaystyle\int \frac{x^3 + x + 2}{x^5 + 2x^3 + x} \, dx$

22. $\displaystyle\int \frac{1}{x^5 + x^4 + 2x^3 + 2x^2 + x + 1}\, dx$

23. $\displaystyle\int \frac{1}{(x^2 + 5)(x^2 + 2x + 3)}\, dx$

24. $\displaystyle\int \frac{1}{1 + x^3}\, dx$

25. $\displaystyle\int \frac{\sin x}{\cos x(1 + \cos^2 x)}\, dx$

26. $\displaystyle\int \frac{x^4 + 8x^3 - x^2 + 2x + 1}{x^5 + x^4 + x^2 + x}\, dx$

27. If an integrand is a rational function of $\sin x$ and $\cos x$, it can be reduced to a rational function of t by the substitution
$$t = \tan\left(\frac{x}{2}\right).$$
Show that with this substitution
$$dx = \frac{2}{1 + t^2}\, dt, \qquad \sin x = \frac{2t}{1 + t^2}, \qquad \cos x = \frac{1 - t^2}{1 + t^2}.$$

In Exercises 28–31 use the substitution of Exercise 27 to evaluate the integral.

28. $\displaystyle\int \sec x\, dx$

29. $\displaystyle\int \frac{1}{3 + 5\sin x}\, dx$

30. $\displaystyle\int \frac{1}{1 - 2\cos x}\, dx$

31. $\displaystyle\int \frac{1}{\sin x + \cos x}\, dx$

32. Show that the answer to Exercise 28 can be expressed in the usual form $\ln|\sec x + \tan x| + C$.

33. Evaluate
$$\int \frac{x^2 + x + 3}{x^4 + x^3 + 2x^2 + 11x - 5}\, dx.$$

SECTION 9.8

Using Tables and Reduction Formulas

For antiderivatives more difficult than we have considered in this chapter, it may be necessary to consult a set of integration tables; and for integrals which are amenable to our techniques, it may be more expeditious to use tables. Sets of integration tables can be found in *Standard Mathematical Tables* published by the Chemical Rubber Publishing Company, and *Handbook of Mathematical, Scientific, and Engineering Formulas, Tables, Functions, Graphs, Transforms* published by the Research and Education Association.

To illustrate the use of tables, consider the integral
$$\int \frac{1}{x^2(3 - 4x)^2}\, dx.$$

Certainly we could evaluate this integral by finding the partial fraction decomposition of the integrand. Alternatively, in tables we find the integration formula
$$\int \frac{1}{x^2(a + bx)^2}\, dx = -\frac{a + 2bx}{a^2 x(a + bx)} + \frac{2b}{a^3}\ln\left(\frac{a + bx}{x}\right).$$

Hence, if we set $a = 3$ and $b = -4$ in this formula, we obtain
$$\int \frac{1}{x^2(3 - 4x)^2}\, dx = -\frac{3 - 8x}{9x(3 - 4x)} - \frac{8}{27}\ln\left(\frac{3 - 4x}{x}\right) + C.$$

Tables do not normally include a constant of integration. Remember to add it. There is an additional difficulty with tables; they do not always place absolute value signs around arguments of logarithm functions. The above answer should read
$$\int \frac{1}{x^2(3 - 4x)^2}\, dx = \frac{8x - 3}{9x(3 - 4x)} - \frac{8}{27}\ln\left|\frac{3 - 4x}{x}\right| + C$$

in order to accommodate definite integrals wherein $(3 - 4x)/x$ is negative.

EXAMPLE 9.17

Use tables to evaluate the integral

$$\int \frac{1}{x^2 + 2x + 3}\, dx.$$

SOLUTION Tables indicate that for an integral of the form

$$\int \frac{1}{X}\, dx$$

where $X = ax^2 + bx + c$ and $q = 4ac - b^2$, there are three possibilities:

$$\int \frac{1}{X}\, dx = \frac{2}{\sqrt{q}} \text{Tan}^{-1}\left(\frac{2ax+b}{\sqrt{q}}\right), \qquad (9.29\text{a})$$

$$\int \frac{1}{X}\, dx = \frac{-2}{\sqrt{-q}} \text{Tanh}^{-1}\left(\frac{2ax+b}{\sqrt{-q}}\right), \qquad (9.29\text{b})$$

$$\int \frac{1}{X}\, dx = \frac{1}{\sqrt{-q}} \ln\left(\frac{2ax+b-\sqrt{-q}}{2ax+b+\sqrt{-q}}\right). \qquad (9.29\text{c})$$

No explanation is given as to when each should be used; it is up to the reader to make this decision. Certainly, we would use 9.29a when $q > 0$, and 9.29b,c when $q < 0$ in order to keep $\sqrt{q}$ and $\sqrt{-q}$ real. In our example, $q = 4(1)(3) - (2)^2 = 8$, and we therefore use 9.29a,

$$\int \frac{1}{x^2+2x+3}\, dx = \frac{2}{\sqrt{8}} \text{Tan}^{-1}\left(\frac{2(1)x+2}{\sqrt{8}}\right) + C = \frac{1}{\sqrt{2}} \text{Tan}^{-1}\left(\frac{x+1}{\sqrt{2}}\right) + C.$$

∎

EXAMPLE 9.18

Use tables to evaluate the integral

$$\int \frac{1}{x\sqrt{3-2x-2x^2}}\, dx.$$

SOLUTION It should be noted that none of our integration techniques seem particularly suited to this integral. Tables list the formulas

$$\int \frac{1}{x\sqrt{X}}\, dx = -\frac{1}{\sqrt{c}} \ln\left(\frac{\sqrt{X}+\sqrt{c}}{x} + \frac{b}{2\sqrt{c}}\right), \qquad \text{and} \qquad (9.30\text{a})$$

$$\int \frac{1}{x\sqrt{X}}\, dx = \frac{1}{\sqrt{-c}} \text{Sin}^{-1}\left(\frac{bx+2c}{x\sqrt{b^2-4ac}}\right), \qquad (9.30\text{b})$$

for $X = ax^2 + bx + c$. Since $c = 3$ in our example, we use 9.30a,

$$\int \frac{1}{x\sqrt{3-2x-2x^2}}\, dx = -\frac{1}{\sqrt{3}} \ln\left(\frac{\sqrt{3-2x-2x^2}+\sqrt{3}}{x} + \frac{-2}{2\sqrt{3}}\right) + C.$$

To check that 9.30a is the correct formula, and that we have used it correctly, we should calculate the derivative of this function:

$$\frac{d}{dx}\left[\frac{-1}{\sqrt{3}}\ln\left(\frac{\sqrt{3-2x-2x^2}+\sqrt{3}}{x}-\frac{1}{\sqrt{3}}\right)\right] = \frac{-1/\sqrt{3}}{\frac{\sqrt{3-2x-2x^2}+\sqrt{3}}{x}-\frac{1}{\sqrt{3}}}$$

$$\left[\frac{x(1/2)(3-2x-2x^2)^{-1/2}(-2-4x)-\left(\sqrt{3-2x-2x^2}+\sqrt{3}\right)(1)}{x^2}\right]$$

$$= \frac{-1}{\sqrt{3}}\frac{\sqrt{3}x}{\sqrt{3}\sqrt{3-2x-2x^2}+3-x}\left[\frac{-x(1+2x)-(3-2x-2x^2)-\sqrt{3}\sqrt{3-2x-2x^2}}{x^2\sqrt{3-2x-2x^2}}\right]$$

$$= \frac{-x}{\sqrt{3}\sqrt{3-2x-2x^2}+3-x}\left[\frac{-3+x-\sqrt{3}\sqrt{3-2x-2x^2}}{x^2\sqrt{3-2x-2x^2}}\right]$$

$$= \frac{1}{x\sqrt{3-2x-2x^2}}.$$

Once again, until we can be sure that the argument of the logarithm function is positive, it would be wise to insert absolute value signs,

$$\int\frac{1}{x\sqrt{3-2x-2x^2}}\,dx = -\frac{1}{\sqrt{3}}\ln\left|\frac{\sqrt{3-2x-2x^2}+\sqrt{3}}{x}-\frac{1}{\sqrt{3}}\right| + C.$$

∎

Integration tables contain a large number of entries called **reduction formulas**. To introduce these, consider the integral

$$\int x^3 e^{2x}\,dx.$$

We could use integration by parts three times to eliminate the x^3 before e^{2x}. A better procedure is to use integration by parts on

$$\int x^n e^{2x}\,dx$$

where n represents a positive integer. With

$$u = x^n, \quad dv = e^{2x}\,dx, \quad du = nx^{n-1}\,dx, \quad v = \frac{1}{2}e^{2x},$$

integration by parts formula 9.20 gives

$$\int x^n e^{2x}\,dx = \frac{1}{2}x^n e^{2x} - \int \frac{n}{2}x^{n-1}e^{2x}\,dx$$

$$= \frac{1}{2}x^n e^{2x} - \frac{n}{2}\int x^{n-1}e^{2x}\,dx. \tag{9.31}$$

We now use this formula with $n = 3$ on the given integral,

$$\int x^3 e^{2x}\, dx = \frac{1}{2} x^3 e^{2x} - \frac{3}{2} \int x^2 e^{2x}\, dx.$$

But formula 9.31 can be used again with $n = 2$ to simplify the integral on the right,

$$\int x^3 e^{2x}\, dx = \frac{1}{2} x^3 e^{2x} - \frac{3}{2}\left(\frac{1}{2} x^2 e^{2x} - \frac{2}{2} \int x e^{2x}\, dx\right)$$

$$= \frac{1}{2} x^3 e^{2x} - \frac{3}{4} x^2 e^{2x} - \frac{3}{2} \int x e^{2x}\, dx.$$

One more application of 9.31 with $n = 1$ rids the integrand of the pesky factor in front of the exponential,

$$\int x^3 e^{2x}\, dx = \frac{1}{2} x^3 e^{2x} - \frac{3}{4} x^2 e^{2x} - \frac{3}{2}\left(\frac{1}{2} x e^{2x} - \frac{1}{2} \int e^{2x}\, dx\right)$$

$$= \frac{1}{2} x^3 e^{2x} - \frac{3}{4} x^2 e^{2x} - \frac{3}{4} x e^{2x} + \frac{3}{8} e^{2x} + C.$$

Equation 9.31 is an example of a *reduction formula*. By using it on the antiderivative of $x^n e^{2x}$, we reduce the power on x from n to $n - 1$. Successive reductions eventually eliminate this term.

Many reduction formulas can be found in tables. They have the same purpose as 9.31, to reduce, and perhaps eliminate, powers of certain factors in an integrand.

EXAMPLE 9.19

Use a reduction formula to evaluate

$$\int \cos^6 x\, dx.$$

SOLUTION The following reduction formula can be found in tables,

$$\int \cos^n x\, dx = \frac{1}{n} \cos^{n-1} x \sin x + \frac{n-1}{n} \int \cos^{n-2} x\, dx. \qquad (9.32)$$

It reduces the power on $\cos x$ by two each time it is used. We begin with $n = 6$,

$$\int \cos^6 x\, dx = \frac{1}{6} \cos^5 x \sin x + \frac{5}{6} \int \cos^4 x\, dx.$$

We now set $n = 4$ in 9.32,

$$\int \cos^6 x\, dx = \frac{1}{6} \cos^5 x \sin x + \frac{5}{6}\left(\frac{1}{4} \cos^3 x \sin x + \frac{3}{4} \int \cos^2 x\, dx\right).$$

Finally, with $n = 2$ we obtain

$$\int \cos^6 x\, dx = \frac{1}{6} \cos^5 x \sin x + \frac{5}{24} \cos^3 x \sin x + \frac{5}{8}\left(\frac{1}{2} \cos x \sin x + \frac{1}{2} \int dx\right)$$

$$= \frac{1}{6} \cos^5 x \sin x + \frac{5}{24} \cos^3 x \sin x + \frac{5}{16} \cos x \sin x + \frac{5x}{16} + C.$$

Sometimes it is necessary to use more than one entry from tables on an integral, as the following example illustrates.

EXAMPLE 9.20

Evaluate
$$\int \frac{\sqrt{6x - x^2}}{x}\, dx.$$

SOLUTION We use the reduction formula

$$\int \frac{\sqrt{2ax - x^2}}{x^n}\, dx = \frac{(2ax - x^2)^{3/2}}{(3 - 2n)ax^n} + \frac{n - 3}{(2n - 3)a} \int \frac{\sqrt{2ax - x^2}}{x^{n-1}}\, dx.$$

If we set $n = 1$ and $a = 3$ in this identity,

$$\int \frac{\sqrt{6x - x^2}}{x}\, dx = \frac{(6x - x^2)^{3/2}}{(3 - 2)(3)x} + \frac{1 - 3}{(2 - 3)(3)} \int \sqrt{6x - x^2}\, dx$$

$$= \frac{(6x - x^2)^{3/2}}{3x} + \frac{2}{3} \int \sqrt{6x - x^2}\, dx.$$

We can now use the formula

$$\int \sqrt{2ax - x^2}\, dx = \frac{1}{2}\left[(x - a)\sqrt{2ax - x^2} + a^2 \operatorname{Sin}^{-1}\left(\frac{x - a}{|a|}\right)\right]$$

to finish the integration,

$$\int \frac{\sqrt{6x - x^2}}{x}\, dx = \frac{(6x - x^2)^{3/2}}{3x} + \frac{2}{3}\frac{1}{2}\left[(x - 3)\sqrt{6x - x^2} + 9\operatorname{Sin}^{-1}\left(\frac{x - 3}{3}\right)\right] + C$$

$$= \frac{(6x - x^2)^{3/2}}{3x} + \frac{1}{3}(x - 3)\sqrt{6x - x^2} + 3\operatorname{Sin}^{-1}\left(\frac{x - 3}{3}\right) + C$$

$$= \sqrt{6x - x^2} + 3\operatorname{Sin}^{-1}\left(\frac{x - 3}{3}\right) + C.$$

∎

Tables are an excellent companion to partial fractions. Partial fractions decompose complicated rational functions into simpler ones which can often be found in tables. This is illustrated in the following example.

EXAMPLE 9.21

Evaluate
$$\int \frac{3x^4 - x^3 + 18x^2 - 5x + 17}{x^5 - x^4 + 6x^3 - 6x^2 + 9x - 9}\, dx.$$

SOLUTION Since $x^5 - x^4 + 6x^3 - 6x^2 + 9x - 9 = (x - 1)(x^2 + 3)^2$, the partial fraction decomposition of the integrand is

$$\frac{3x^4 - x^3 + 18x^2 - 5x + 17}{(x - 1)(x^2 + 3)^2} = \frac{A}{x - 1} + \frac{Bx + C}{x^2 + 3} + \frac{Dx + E}{(x^2 + 3)^2}.$$

We find $A = 2$, $B = 1$, $C = 0$, $D = 3$, and $E = 1$, so that

$$\int \frac{3x^4 - x^3 + 18x^2 - 5x + 17}{x^5 - x^4 + 6x^3 - 6x^2 + 9x - 9} dx = \int \left(\frac{2}{x-1} + \frac{x}{x^2+3} + \frac{3x+1}{(x^2+3)^2} \right) dx$$

$$= 2\ln|x-1| + \frac{1}{2}\ln(x^2+3) - \frac{3}{2(x^2+3)}$$

$$+ \int \frac{1}{(x^2+3)^2} dx.$$

Using the reduction formula

$$\int \frac{1}{(a^2+x^2)^n} dx = \frac{x}{2a^2(n-1)(a^2+x^2)^{n-1}} + \frac{2n-3}{2a^2(n-1)} \int \frac{1}{(a^2+x^2)^{n-1}} dx, \quad n \geq 2,$$

(with n set equal to 2), we obtain

$$\int \frac{1}{(x^2+3)^2} dx = \frac{x}{2(3)(1)(3+x^2)} + \frac{1}{2(3)(1)} \int \frac{1}{3+x^2} dx.$$

With the integration formula

$$\int \frac{1}{a^2+x^2} dx = \frac{1}{a} \operatorname{Tan}^{-1}\left(\frac{x}{a}\right),$$

we can finish the problem,

$$\int \frac{3x^4 - x^3 + 18x^2 - 5x + 17}{x^5 - x^4 + 6x^3 - 6x^2 + 9x - 9} dx = 2\ln|x-1| + \frac{1}{2}\ln(x^2+3) - \frac{3}{2(x^2+3)}$$

$$+ \frac{x}{6(x^2+3)} + \frac{1}{6}\left[\frac{1}{\sqrt{3}}\operatorname{Tan}^{-1}\left(\frac{x}{\sqrt{3}}\right)\right] + C$$

$$= 2\ln|x-1| + \frac{1}{2}\ln(x^2+3) + \frac{x-9}{6(x^2+3)}$$

$$+ \frac{1}{6\sqrt{3}}\operatorname{Tan}^{-1}\left(\frac{x}{\sqrt{3}}\right) + C.$$

∎

EXERCISES 9.8

In Exercises 1–20 use tables to evaluate the indefinite integral.

1. $\int \frac{1}{x^2+9} dx$

2. $\int \frac{1}{x^2-16} dx$

3. $\int \frac{1}{4-3x^2} dx$

4. $\int \frac{1}{x\sqrt{x^2+5}} dx$

5. $\int \frac{x}{\sqrt{3x-2}} dx$

6. $\int \frac{\sqrt{2x+3}}{x} dx$

7. $\int \frac{1}{\sqrt{x^2+4x}} dx$

8. $\int \frac{1}{\sqrt{4x-9x^2}} dx$

9. $\int \sqrt{3+2x+4x^2}\, dx$

10. $\int \sqrt{3+2x-4x^2}\, dx$

11. $\int \frac{1}{3+\sin x} dx$

12. $\int \frac{1}{1+3\sin x} dx$

13. $\int x \operatorname{Sin}^{-1} x\, dx$

14. $\int \frac{1}{x^2} \operatorname{Sin}^{-1} x\, dx$

15. $\int e^{4x} \cos 2x\, dx$

16. $\int e^{-2x} \sin 3x\, dx$

17. $\int x^7 \ln x\, dx$

18. $\int (\ln x)^2\, dx$

19. $\displaystyle\int \frac{1}{e^x + 2e^{-2x}}\, dx$ **20.** $\displaystyle\int \frac{1}{2+x^3}\, dx$

In Exercises 21–30 use a reduction formula to aid in the evaluation of the integral.

21. $\displaystyle\int x^4 e^{3x}\, dx$ **22.** $\displaystyle\int x^3 \cos 2x\, dx$

23. $\displaystyle\int \frac{x}{\sqrt{9+x+x^2}}\, dx$ **24.** $\displaystyle\int \frac{x}{\sqrt{9+x-x^2}}\, dx$

25. $\displaystyle\int (4+x^2)^{3/2}\, dx$ **26.** $\displaystyle\int \frac{x^2}{\sqrt{2x-x^2}}\, dx$

27. $\displaystyle\int \sin^6 x\, dx$ **28.** $\displaystyle\int \frac{1}{\cos^6 x}\, dx$

29. $\displaystyle\int \cot^5 x\, dx$ **30.** $\displaystyle\int x^3 \operatorname{Sin}^{-1} x\, dx$

31. Evaluate

$$\int \sec^9 x\, dx$$

by first developing the reduction formula

$$\int \sec^n x\, dx = \frac{\tan x \sec^{n-2} x}{n-1} + \frac{n-2}{n-1}\int \sec^{n-2} x\, dx.$$

32. (a) Use a trigonometric substitution to evaluate

$$\int \frac{5x+2}{(3+x^2)^4}\, dx.$$

(b) Verify the reduction formula in Example 9.21 and use it to evaluate the integral in (a).

33. If n is a positive integer, define

$$I_n = \int_{-1}^{1} (1-x^2)^n\, dx.$$

(a) Show that

$$(2n+1) I_n = 2n I_{n-1}, \quad n \geq 2.$$

(b) Prove that

$$I_n = \frac{2^{2n+1}(n!)^2}{(2n+1)!}, \quad n \geq 1.$$

(c) Evaluate

$$\int_0^{\pi/2} \cos^{2n+1} x\, dx.$$

SECTION 9.9

Numerical Integration

To evaluate the definite integral of a continuous function $f(x)$ with respect to x from $x = a$ to $x = b$, we have used the fundamental theorem of integral calculus: Find an indefinite integral for $f(x)$, substitute the limits $x = b$ and $x = a$, and subtract. The evaluation procedure depends on our ability to produce an indefinite integral for $f(x)$. When a function $f(x)$ is complicated, it may be difficult, or even impossible to find its antiderivative. In such a case, it may be necessary to approximate the definite integral of $f(x)$ on some interval $a \leq x \leq b$, rather than evaluate it analytically. We discuss three methods for doing this, the rectangular, trapezoidal, and Simpson's rules. Each method divides the interval $a \leq x \leq b$ into subintervals and approximates $f(x)$ with an easily integrated function on each subinterval.

FIGURE 9.22

If $f(x) \geq 0$ on $a \leq x \leq b$ (Figure 9.22), the definite integral of $f(x)$ with respect to x can always be interpreted as the area bounded by $y = f(x)$, $y = 0$, $x = a$, and $x = b$. We have carefully pointed out that it is not always wise to think of a definite integral as area, but for our discussion here it is convenient to do so. Our problem is to approximate the area in Figure 9.22 when it is difficult or impossible to find an antiderivative for $f(x)$ because of the complexity of $f(x)$.

Rectangular Rule

The first method is essentially to return to equation 6.11 which defines the definite integral. This means that we subdivide the interval $a \leq x \leq b$ into n subintervals by points $a = x_0 < x_1 < \cdots < x_{n-1} < x_n = b$, and pick a point x_i^* in each subinterval. For simplicity, we choose n equal subdivisions, in which case

$$x_i = a + i\left(\frac{b-a}{n}\right),$$

and select $x_i^* = x_i$. If we denote by h the length of each subdivision, then $h = \Delta x_i = (b-a)/n$, and equation 6.11 reduces to

$$\int_a^b f(x)\,dx = \lim_{\|\Delta x_i\| \to 0} \sum_{i=1}^n f(x_i^*)\Delta x_i = \lim_{n \to \infty} h \sum_{i=1}^n f(x_i).$$

If we delete the limit from this equation, the summation becomes an approximation to the definite integral, and we write

$$\int_a^b f(x)\,dx \approx h \sum_{i=1}^n f(x_i). \tag{9.33}$$

This is the **rectangular rule** for approximating the definite integral; we have replaced the area under $y = f(x)$ with n rectangles (Figure 9.23). In other words, we have replaced the original function $f(x)$ by a function that is constant on each subinterval (Figure 9.24), but the constant value may vary from subinterval to subinterval. Such a function is called a *step function*.

Graphically, it is reasonable to expect that an increase in the number of subdivisions results in an increase in the accuracy of the approximation.

FIGURE 9.23

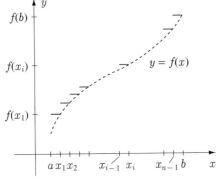

FIGURE 9.24

EXAMPLE 9.22

Use a subdivision of the interval $1 \leq x \leq 3$ into 5, 10, and 20 equal parts to approximate the definite integral

$$\int_1^3 \sin x \, dx$$

with the rectangular rule.

SOLUTION With 5 equal parts, $h = 2/5$ and $x_i = 1 + 2i/5$. Consequently,

$$\int_1^3 \sin x \, dx \approx \frac{2}{5} \sum_{i=1}^{5} \sin\left(1 + \frac{2i}{5}\right)$$

$$= \frac{2}{5}\left(\sin\frac{7}{5} + \sin\frac{9}{5} + \sin\frac{11}{5} + \sin\frac{13}{5} + \sin 3\right)$$

$$= 1.370.$$

With 10 equal parts, $h = 2/10 = 1/5$ and $x_i = 1 + i/5$ so that

$$\int_1^3 \sin x \, dx \approx \frac{1}{5} \sum_{i=1}^{10} \sin\left(1 + \frac{i}{5}\right)$$

$$= \frac{1}{5}\left(\sin\frac{6}{5} + \sin\frac{7}{5} + \sin\frac{8}{5} + \sin\frac{9}{5} + \sin 2 + \sin\frac{11}{5}\right.$$

$$\left. + \sin\frac{12}{5} + \sin\frac{13}{5} + \sin\frac{14}{5} + \sin 3\right)$$

$$= 1.455.$$

With 20 equal parts we have

$$\int_1^3 \sin x \, dx \approx \frac{1}{10} \sum_{i=1}^{20} \sin\left(1 + \frac{i}{10}\right)$$

$$= \frac{1}{10}\left(\sin\frac{11}{10} + \sin\frac{6}{5} + \sin\frac{13}{10} + \cdots + \sin\frac{29}{10} + \sin 3\right)$$

$$= 1.494.$$

When we compare these results with the correct answer

$$\int_1^3 \sin x \, dx = \{-\cos x\}_1^3 = -\cos 3 + \cos 1 = 1.530\,294\,8,$$

we see that as n increases from 5 to 10 to 20, the approximation improves, but it must be increased even further to give a reasonable approximation to this integral.

Trapezoidal Rule

Regarding area, it is clear in Figure 9.23 that were we to join successive points $(x_i, f(x_i))$ on the curve $y = f(x)$ with straight line segments as in Figure 9.25(a), the area under this broken straight line would be a much better approximation to the area under $y = f(x)$ than that provided by the rectangular rule. Effectively, we now approximate the area by n trapezoids.

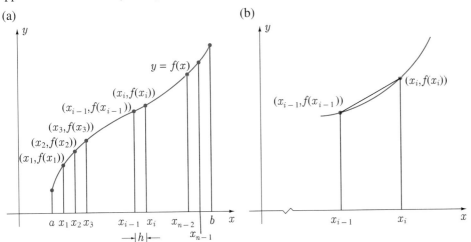

FIGURE 9.25

Since the area of a trapezoid is its width multiplied by the average of its parallel lengths, it follows that the area of the ith trapezoid in Figure 9.25(b) is given by

$$h\left(\frac{f(x_i) + f(x_{i-1})}{2}\right),$$

where again $h = x_i - x_{i-1}$. As a result, the area under $y = f(x)$ can be approximated by the sum

$$h\left[\frac{f(x_1) + f(x_0)}{2} + \frac{f(x_2) + f(x_1)}{2} + \cdots + \frac{f(x_n) + f(x_{n-1})}{2}\right]$$

$$= \frac{h}{2}[f(a) + 2f(x_1) + 2f(x_2) + \cdots + 2f(x_i) + \cdots + 2f(x_{n-1}) + f(b)]$$

$$= \frac{h}{2}\left[f(a) + 2\sum_{i=1}^{n-1} f(x_i) + f(b)\right].$$

We write therefore that

$$\int_a^b f(x)\, dx \approx \frac{h}{2}\left[f(a) + 2\sum_{i=1}^{n-1} f(x_i) + f(b)\right], \qquad (9.34\text{a})$$

where $x_i = a + ih = a + i(b-a)/n$, and call this the **trapezoidal rule** for approximating a definite integral. Note that if 9.34a is written in the form

$$\int_a^b f(x)\, dx \approx h\left[\frac{f(a) - f(b)}{2}\right] + h\sum_{i=1}^{n} f(x_i), \qquad (9.34\text{b})$$

the summation on the right, except for the first two terms, is the rectangular rule. In other words, the extra numerical calculation involved in using the trapezoidal rule rather than the rectangular rule is minimal, but it would appear that the accuracy is significantly increased. For this reason, the trapezoidal rule supplants the rectangular rule in most applications.

EXAMPLE 9.23

Use the trapezoidal rule to approximate the definite integral in Example 9.22.

SOLUTION With 5 equal partitions,

$$\int_1^3 \sin x \, dx \approx \frac{(2/5)}{2}\left[\sin 1 + 2\sum_{i=1}^{4}\sin\left(1 + \frac{2i}{5}\right) + \sin 3\right] = 1.5098.$$

With 10 equal parts,

$$\int_1^3 \sin x \, dx \approx \frac{(1/5)}{2}\left[\sin 1 + 2\sum_{i=1}^{9}\sin\left(1 + \frac{i}{5}\right) + \sin 3\right] = 1.5252.$$

Finally, with $n = 20$, we have

$$\int_1^3 \sin x \, dx \approx \frac{(1/10)}{2}\left[\sin 1 + 2\sum_{i=1}^{19}\sin\left(1 + \frac{i}{10}\right) + \sin 3\right] = 1.5290.$$

As expected, these approximations are significantly better than corresponding results using the rectangular rule. ∎

Simpson's Rule

The rectangular rule replaces a function $f(x)$ with a step function; the trapezoidal rule replaces $f(x)$ with a succession of linear functions — geometrically, a broken straight line. So far as ease of integration is concerned, the next simplest function is a quadratic function. Consider, then, replacing the curve $y = f(x)$ by a succession of parabolas on the subintervals $x_{i-1} \leq x \leq x_i$. Now the equation of a parabola with a vertical axis of symmetry is of the form $y = ax^2 + bx + c$ with three constants a, b, and c to be determined. If this parabola is to approximate $y = f(x)$ on $x_{i-1} \leq x \leq x_i$, we should have the parabola pass through the end points $(x_{i-1}, f(x_{i-1}))$ and $(x_i, f(x_i))$. But this imposes only two conditions on a, b, and c, not three. To take advantage of this flexibility, we demand that the parabola also pass through the point $(x_{i+1}, f(x_{i+1}))$ (Figure 9.26).

FIGURE 9.26

In other words, instead of replacing $y = f(x)$ with n parabolas, one on each subinterval $x_{i-1} \leq x \leq x_i$, we replace it with $n/2$ parabolas, one on each pair of subintervals $x_{i-1} \leq x \leq x_{i+1}$. Note that this requires n to be an even integer. These three conditions imply that a, b, and c must satisfy the equations

$$ax_{i-1}^2 + bx_{i-1} + c = f(x_{i-1}), \tag{9.35a}$$

$$ax_i^2 + bx_i + c = f(x_i), \tag{9.35b}$$
$$ax_{i+1}^2 + bx_{i+1} + c = f(x_{i+1}). \tag{9.35c}$$

These equations determine the values for a, b, and c, but it will not be necessary to actually solve them. Suppose for the moment that we have solved equations 9.35 for a, b, and c, and we continue our main discussion. With the parabola $y = ax^2 + bx + c$ replacing the curve $y = f(x)$ on the interval $x_{i-1} \leq x \leq x_{i+1}$, we approximate the area under $y = f(x)$ with that under the parabola, namely,

$$\int_{x_{i-1}}^{x_{i+1}} (ax^2 + bx + c)\,dx = \frac{a}{3}(x_{i+1}^3 - x_{i-1}^3) + \frac{b}{2}(x_{i+1}^2 - x_{i-1}^2) + c(x_{i+1} - x_{i-1})$$
$$= \frac{x_{i+1} - x_{i-1}}{6}[2a(x_{i+1}^2 + x_{i+1}x_{i-1} + x_{i-1}^2)$$
$$+ 3b(x_{i+1} + x_{i-1}) + 6c].$$

Now if we use $x_{i+1} - x_{i-1} = 2h$, $x_{i+1} = x_i + h$, and $x_{i-1} = x_i - h$ to write everything in terms of h and x_i, we obtain

$$\int_{x_{i-1}}^{x_{i+1}} (ax^2 + bx + c)\,dx = \frac{2h}{6}\Big\{2a\left[(x_i+h)^2 + (x_i+h)(x_i-h) + (x_i-h)^2\right]$$
$$+ 3b[(x_i+h) + (x_i-h)] + 6c\Big\}$$
$$= \frac{h}{3}\left\{a(6x_i^2 + 2h^2) + 6bx_i + 6c\right\}.$$

But if we add equation 9.35a, equation 9.35c, and four times equation 9.35b, we find that

$$f(x_{i-1}) + 4f(x_i) + f(x_{i+1}) = a(x_{i-1}^2 + 4x_i^2 + x_{i+1}^2) + b(x_{i-1} + 4x_i + x_{i+1}) + 6c$$
$$= a\left[(x_i-h)^2 + 4x_i^2 + (x_i+h)^2\right]$$
$$+ b[(x_i-h) + 4x_i + (x_i+h)] + 6c$$
$$= a(6x_i^2 + 2h^2) + 6bx_i + 6c.$$

Thus the area under the parabola may be written in the form

$$\int_{x_{i-1}}^{x_{i+1}} (ax^2 + bx + c)\,dx = \frac{h}{3}[f(x_{i-1}) + 4f(x_i) + f(x_{i+1})],$$

and the right-hand side is free of a, b, and c. The same expression would have resulted had we solved equations 9.35 for a, b, and c and then evaluated the integral of $ax^2 + bx + c$ from x_{i-1} to x_{i+1}. The above derivation, however, is much simpler.

When we add all such integrals over the $n/2$ subintervals $x_{i-1} \leq x \leq x_{i+1}$ between $x = a$ and $x = b$, we obtain

$$\frac{h}{3}[f(x_0) + 4f(x_1) + f(x_2)] + \frac{h}{3}[f(x_2) + 4f(x_3) + f(x_4)] + \cdots$$
$$+ \frac{h}{3}[f(x_{n-2}) + 4f(x_{n-1}) + f(x_n)]$$
$$= \frac{h}{3}[f(x_0) + 4f(x_1) + 2f(x_2) + 4f(x_3) + \cdots + 2f(x_{n-2}) + 4f(x_{n-1}) + f(x_n)].$$

In other words, the definite integral of $f(x)$ from $x = a$ to $x = b$ can be approximated by

$$\int_a^b f(x)\,dx \approx \frac{h}{3}\Big[f(a) + 4f(x_1) + 2f(x_2) + 4f(x_3) + \cdots$$
$$+ 2f(x_{n-2}) + 4f(x_{n-1}) + f(b)\Big], \tag{9.36}$$

where $x_i = a + ih = a + i(b-a)/n$. This result is called **Simpson's rule** for approximating a definite integral. Although the formula does not explicitly display it (except by counting terms), *do not forget that n must be an even integer.*

EXAMPLE 9.24 Approximate the definite integral in Example 9.22 using Simpson's rule with $n = 10$ and $n = 20$.

SOLUTION With $n = 10$,

$$\int_1^3 \sin x\, dx \approx \frac{1/5}{3}\left[\sin 1 + 4\sin\frac{6}{5} + 2\sin\frac{7}{5} + 4\sin\frac{8}{5} + \cdots \right.$$
$$\left. + 2\sin\frac{13}{5} + 4\sin\frac{14}{5} + \sin 3\right]$$
$$= 1.530\,308\,5.$$

With $n = 20$,

$$\int_1^3 \sin x\, dx \approx \frac{1/10}{3}\left[\sin 1 + 4\sin\frac{11}{10} + 2\sin\frac{6}{5} + 4\sin\frac{13}{10} + \cdots \right.$$
$$\left. + 2\sin\frac{14}{5} + 4\sin\frac{29}{10} + \sin 3\right]$$
$$= 1.530\,295\,7.$$

Table 9.3 lists the approximations in Examples 9.22–24. The correct answer for the integral is 1.530 294 8 (to 7 decimals). It is clear that each method gives a better approximation as the value of n increases, and that Simpson's rule is by far the most accurate.

TABLE 9.3

	Rectangular rule	Trapezoidal rule	Simpson's rule
$n = 5$	1.370	1.5098	
$n = 10$	1.455	1.5252	1.530 308 5
$n = 20$	1.494	1.5290	1.530 295 7

In practice, we use the rectangular, trapezoidal, and Simpson's rules to approximate definite integrals that cannot be handled analytically, and we will not therefore have the correct answer with which to compare the approximation. We would still like to make some statement about the accuracy of the approximation, however, since what good is the approximation otherwise? The following two theorems give error estimates for the trapezoidal rule and for Simpson's rule.

Theorem 9.1 If $f''(x)$ exists on $a \leq x \leq b$, and T_n is the error in approximating the definite integral of $f(x)$ from $x = a$ to $x = b$ using the trapezoidal rule with n equal subdivisions,

$$T_n = \int_a^b f(x)\, dx - \frac{h}{2}\left[f(a) + 2\sum_{i=1}^{n-1} f(x_i) + f(b)\right],$$

then

$$|T_n| \leq \frac{M(b-a)^3}{12\,n^2}, \qquad (9.37)$$

where M is the maximum value of $|f''(x)|$ on $a \leq x \leq b$.

> **Theorem 9.2**
>
> If $f''''(x)$ exists on $a \leq x \leq b$, and S_n is the error in approximating the definite integral of $f(x)$ from $x = a$ to $x = b$ using Simpson's rule with n equal subdivisions,
>
> $$S_n = \int_a^b f(x)\,dx - \frac{h}{3}\left[f(a) + 4\sum_{i=1}^{n/2} f(x_{2i-1}) + 2\sum_{i=1}^{n/2-1} f(x_{2i}) + f(b)\right],$$
>
> then
>
> $$|S_n| \leq \frac{M(b-a)^5}{180\,n^4}, \qquad (9.38)$$
>
> where M is the maximum value of $|f''''(x)|$ on $a \leq x \leq b$.

Proofs of these theorems can be found in books on numerical analysis. Note that because of the n^4 factor in the denominator of 9.38, the accuracy of Simpson's rule increases much more rapidly than does that of the trapezoidal rule.

For the function $f(x) = \sin x$ in Examples 9.22–24,

$$f''(x) = -\sin x \quad \text{and} \quad f''''(x) = \sin x.$$

Consequently, in both cases we can state that $M = 1$, and therefore

$$|T_n| \leq \frac{(3-1)^3}{12\,n^2} = \frac{2}{3n^2} \quad \text{and} \quad |S_n| \leq \frac{(3-1)^5}{180\,n^4} = \frac{8}{45\,n^4}.$$

For $n = 10$ and $n = 20$, we find

$$|T_{10}| \leq \frac{2}{3(10)^2} < 0.0067 \quad \text{and} \quad |S_{10}| \leq \frac{8}{45(10)^4} < 0.000018;$$

$$|T_{20}| \leq \frac{2}{3(20)^2} < 0.0017 \quad \text{and} \quad |S_{20}| \leq \frac{8}{45(20)^4} < 0.000012.$$

Differences between the correct value for the integral and the approximations listed in Table 9.3 corroborate these predictions.

> **EXAMPLE 9.25**
>
> What is the maximum possible error in using the trapezoidal rule with 100 equal subdivisions to approximate
>
> $$\int_1^3 \frac{1}{\sqrt{1+x^3}}\,dx.$$
>
> **SOLUTION** According to formula 9.37, if T_{100} is the maximum possible error, then
>
> $$|T_{100}| \leq \frac{M(3-1)^3}{12(100)^2} = \frac{2M}{3 \times 10^4},$$
>
> where M is the maximum of (the absolute value of) the second derivative of $1/\sqrt{1+x^3}$ on $1 \leq x \leq 3$. Now,

$$\frac{d^2}{dx^2}\frac{1}{\sqrt{1+x^3}} = \frac{d}{dx}\left(\frac{-3x^2}{2(1+x^3)^{3/2}}\right) = \frac{3x(5x^3-4)}{4(1+x^3)^{5/2}}.$$

Instead of maximizing the absolute value of this function on the interval $1 \le x \le 3$, which would require another derivative, we note that the maximum value of the numerator is obtained for $x = 3$, and the minimum value of the denominator occurs at $x = 1$. It follows therefore that the second derivative cannot possibly be larger than

$$\frac{3(3)[5(3)^3 - 4]}{4(1+1)^{5/2}} = \frac{1179}{16\sqrt{2}}.$$

Thus, M must be less than or equal to $1179/(16\sqrt{2})$, and we can state that

$$|T_{100}| \le \frac{2}{3 \times 10^4}\left(\frac{1179}{16\sqrt{2}}\right) \le 0.0035;$$

that is, the error in using the trapezoidal rule with 100 equal subdivisions to approximate the definite integral cannot be any larger than 0.0035. ∎

EXAMPLE 9.26

How many equal subdivisions of the interval $0 \le x \le 2$ guarantee an error of less than 10^{-5} in the approximation of the definite integral

$$\int_0^2 e^{-x^2}\, dx$$

using Simpson's rule?

SOLUTION According to Theorem 9.2, the error in using Simpson's rule with n equal subdivisions to approximate this definite integral is

$$|S_n| \le \frac{M(2-0)^5}{180\, n^4},$$

where M is the maximum of the (absolute value of the) fourth derivative of e^{-x^2} on the interval $0 \le x \le 2$. It is a short calculation to find

$$\frac{d^4}{dx^4}\left(e^{-x^2}\right) = 4(3 - 12x^2 + 4x^4)e^{-x^2}.$$

Instead of maximizing the absolute value of this function, we note that e^{-x^2} has a maximum value of 1 (when $x = 0$). Furthermore, because $|3 - 12x^2 + 4x^4| \le 3 + 12x^2 + 4x^4$, which has a maximum at $x = 2$, it follows that

$$M \le 4[3 + 12(2)^2 + 4(2)^4](1) = 460.$$

Consequently,
$$|S_n| \le \frac{460(2)^5}{180\,n^4} = \frac{3680}{45\,n^4}.$$

The error is less than 10^{-5} if n is chosen sufficiently large that
$$\frac{3680}{45\,n^4} < 10^{-5};$$

that is, if
$$n > \left(\frac{3680}{45 \times 10^{-5}}\right)^{1/4} = 53.5.$$

Since n must be an even integer, the required accuracy is guaranteed if n is chosen greater than or equal to 54. ∎

Numerical techniques are indispensable in situations where we know that y is a function of x, say $y = f(x)$, but we do not know the exact form of the function $f(x)$. What might be available is a set of experimental data listing values of y corresponding to various values of x (Table 9.4). We have plotted these points in Figure 9.27. The trapezoidal rule and Simpson's rule can be used to approximate the definite integral of $f(x)$ from $x = 1$ to $x = 3$ even though we do not have a formula for $f(x)$. Indeed, formulas 9.34 and 9.36 do not use the form of the function $f(x)$ to be integrated, only its values at the subdivision points x_i. We have these points in Table 9.4 or Figure 9.27. Since there are 11 points, and therefore 10 subdivisions of the interval $1 \le x \le 3$, the trapezoidal rule gives

$$\int_1^3 f(x)\,dx \approx \frac{1/5}{2}[f(1.0) + 2f(1.2) + 2f(1.4) + \cdots + 2f(2.6) + 2f(2.8) + f(3.0)]$$
$$= \frac{1}{10}[0.84 + 2(1.12) + 2(1.40) + \cdots + 2(1.34) + 2(0.94) + 0.42]$$
$$= 2.80.$$

Simpson's rule gives

$$\int_1^3 f(x)\,dx \approx \frac{1/5}{3}[f(1.0) + 4f(1.2) + 2f(1.4) + \cdots + 2f(2.6) + 4f(2.8) + f(3.0)]$$
$$= \frac{1}{15}[0.84 + 4(1.12) + 2(1.40) + \cdots + 2(1.34) + 4(0.94) + 0.42]$$
$$= 2.81.$$

What we lose in this type of application is the ability to predict a maximum possible error in the approximation since there is no way to estimate M in Theorems 9.1 and 9.2.

TABLE 9.4

x	1.0	1.2	1.4	1.6	1.8	2.0	2.2	2.4	2.6	2.8	3.0
y	0.84	1.12	1.40	1.60	1.75	1.82	1.79	1.62	1.34	0.94	0.42

FIGURE 9.27

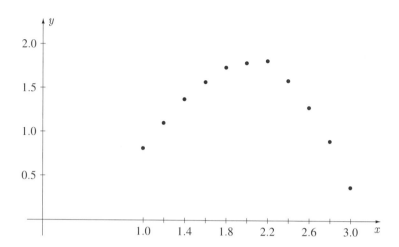

The error bounds in Theorems 9.1 and 9.2 are somewhat idealistic in the sense that they are error predictions based on the use of exact numbers. For instance, they predict that by increasing n indefinitely, any degree of accuracy is attainable. Theoretically this is true, but practically it is not. No matter how we choose to evaluate the summations in 9.34 and 9.36, be it by hand, by an electronic hand calculator, or by a high-speed computer, each calculation is rounded off to a certain number of decimals. The final sum takes into account many, many of these "approximate numbers", and must therefore be inherently inaccurate. We call this *round-off error* and it is very difficult to predict how extensive it is. It depends on both the number and nature of the operations involved in 9.34 and 9.36. In the approximation of a definite integral by the trapezoidal rule or Simpson's rule, there are two sources of error. Theorems 9.1 and 9.2 predict errors due to the methods themselves; round-off errors may also be appreciable for large n.

We should emphasize once again that although we have used area as a convenient vehicle by which to explain the approximation of definite integrals by the rectangular, trapezoidal, and Simpson's rules, it is not necessary for $f(x)$ to be nonnegative. All three methods can be used to approximate the definite integral of a function $f(x)$, be it positive, negative, or sometimes positive and sometimes negative on the interval of integration. The only condition that we have imposed is that $f(x)$ be continuous. In view of our discussion of improper integrals in Section 7.9, even this is not always necessary.

EXERCISES 9.9

 In Exercises 1–10 use the trapezoidal rule and Simpson's rule with 10 equal subdivisions to approximate the definite integral. In each case, evaluate the integral analytically to get an idea of the accuracy of the approximation.

1. $\displaystyle\int_1^2 \frac{1}{x}\,dx$

2. $\displaystyle\int_2^3 \frac{1}{\sqrt{x+2}}\,dx$

3. $\displaystyle\int_0^1 \tan x\,dx$

4. $\displaystyle\int_0^{1/2} e^x\,dx$

5. $\displaystyle\int_{-1}^1 \sqrt{x+1}\,dx$

6. $\displaystyle\int_{-3}^{-2} \frac{1}{x^3}\,dx$

7. $\displaystyle\int_{1/2}^1 \cos x\,dx$

8. $\displaystyle\int_0^1 \frac{1}{3+x^2}\,dx$

9. $\displaystyle\int_1^3 \frac{1}{x^2+x}\,dx$

10. $\displaystyle\int_0^{1/2} xe^{x^2}\,dx$

In Exercises 11–14 use the trapezoidal rule and Simpson's rule with 10 equal subdivisions to approximate the definite integral.

11. $\displaystyle\int_0^2 \frac{1}{1+x^3}\,dx$

12. $\displaystyle\int_0^1 e^{x^2}\,dx$

13. $\displaystyle\int_1^2 \sqrt{1+x^4}\,dx$

14. $\displaystyle\int_{-1}^0 \sin(x^2)\,dx$

15. Show graphically that if $y = f(x)$ is concave downward on the interval $a \le x \le b$, then the trapezoidal rule underestimates the definite integral of $f(x)$ on this interval.

16. What happens to the errors in 9.37 and 9.38 when the number of partitions is doubled?

17. The definite integral

$$\int_a^b e^{-x^2}\, dx$$

is very important in mathematical statistics. Use Simpson's rule with $n = 16$ to evaluate the integral for $a = 0$ and $b = 1$.

18. Use Simpson's rule with 10 equal intervals to approximate the definite integral for the length of the parabola $y = x^2$ between $x = 0$ and $x = 1$. Compare the answer to that of Exercise 39 in Section 9.5.

19. Use the trapezoidal rule and Simpson's rule with 10 equal subdivisions to approximate the definite integral for the length of the curve $y = \sin x$ from $x = 0$ to $x = \pi/2$.

20. The numerical techniques of this section can also be used to approximate many improper integrals. Consider

$$\int_0^4 \frac{e^x}{\sqrt{x}}\, dx.$$

   (a) Why can the trapezoidal rule and Simpson's rule not be used directly to approximate this integral?
   (b) Show that the change of variable $u = \sqrt{x}$ replaces this improper integral with an integral that is not improper and use Simpson's rule with 20 equal subdivisions to approximate its value.
   (c) Could you use the rectangular rule on the improper integral?

21. (a) Show that the definite integral

$$L = \frac{4}{3}\int_0^3 \sqrt{\frac{81 - 5x^2}{9 - x^2}}\, dx$$

represents the length L of the ellipse $4x^2 + 9y^2 = 36$.
   (b) Show that when we set $x = \sin\theta$, the θ-integral is no longer improper.
   (c) Use the trapezoidal rule and Simpson's rule with 8 equal subdivisions to approximate the θ-integral in (b).

22. To approximate the improper integral

$$\int_1^\infty \frac{1}{1 + x^4}\, dx$$

set $x = 1/t$, and use the trapezoidal rule and Simpson's rule with 10 equal subdivisions on the resulting integral.

23. Use the technique of Exercise 22 to approximate

$$\int_1^\infty \frac{x^2}{x^4 + x^2 + 1}\, dx.$$

24. Use the trapezoidal rule and Simpson's rule to approximate the definite integral from $x = -1$ to $x = 4$ for the function tabulated in Table 9.5.

25. When the truss in Figure 9.28 is subjected to a force F at its centre, point A deflects an amount y from its equilibrium position. Forces (in kilonewtons) required to produce deflections from 0 to 5 cm at intervals of 0.5 cm are listed in Table 9.6. Plot these points on a set of axes with y horizontal and F vertical. Use Simpson's rule to approximate the work done by F in deflecting A by a full 5 cm.

FIGURE 9.28

26. Show that when $f(x)$ is a cubic polynomial, evaluation of the definite integral of $f(x)$ from $x = a$ to $x = b$ by Simpson's rule always gives the *exact* answer. Illustrate with an example.

In Exercises 27–30 how many equal subdivisions of the interval of integration guarantee an error of less than 10^{-4} in the approximation of the definite integral using (a) the trapezoidal rule and (b) Simpson's rule?

27. $\int_1^4 \frac{1}{x}\, dx$

28. $\int_0^{\pi/4} \cos x\, dx$

29. $\int_0^{1/3} e^{2x}\, dx$

30. $\int_4^5 \frac{1}{\sqrt{x + 2}}\, dx$

TABLE 9.5

x	−1.0	−0.5	0	0.5	1.0	1.5	2.0	2.5	3.0	3.5	4.0
y	2.287	0.395	0	0.145	0.310	0.334	1.819	0.123	0.021	−0.037	−0.055

TABLE 9.6

y	0	0.5	1.0	1.5	2.0	2.5	3.0	3.5	4.0	4.5	5.0
F	0	1.45	2.90	4.40	5.90	7.43	9.05	10.7	13.2	15.3	18.0

SUMMARY

Antidifferentiation is a far more complicated process than differentiation; it is not possible to state a set of rules and formulas that will suffice for most functions. Certainly there are integration formulas that we must learn, but most of those given in this chapter have been differentiation formulas listed in Chapters 3 and 8 written in terms of integrals rather than derivatives. We have stressed the importance of knowing these obvious integration formulas, but beyond this it becomes an organizational problem — organizing a difficult integral into a form that utilizes these simple formulas. The three most important techniques for doing this are substitutions, integration by parts, and partial fractions.

There is a variety of substitutions for the evaluation of indefinite integrals, many suggested by the form of the integrand. For example, a term in $x^{1/n}$ (n an integer) suggests the substitution $u = x^{1/n}$. The trigonometric substitution is most important. It eliminates square roots of the form $\sqrt{a^2 \pm b^2 x^2}$ and $\sqrt{b^2 x^2 - a^2}$ and the general root $\sqrt{ax^2 + bx + c}$.

Integration by parts is a powerful integration technique. It is used to evaluate antiderivatives of transcendental functions that are multiplied by powers of x, it leads to the method of integration by reproduction, and it is used to develop reduction formulas.

The method of partial fractions decomposes complicated rational functions into simple fractions that are either immediately integrable or amenable to other methods such as trigonometric substitutions.

Tables of integrals provide formulas for simple antiderivatives as well as very complex ones. They contain reduction formulas, and are an excellent companion to partial fractions.

Even with the techniques that we have studied and tables, there are many functions that either cannot be antidifferentiated at all or can be antidifferentiated only with extreme difficulty. To approximate definite integrals of such functions, numerical techniques such as the rectangular rule, the trapezoidal rule, and Simpson's rule are essential.

Key Terms and Formulas

In reviewing this chapter, you should be able to define or discuss the following key terms:

Exponential growth
Exponential decay
Integration by parts
Integration by reproduction
Trigonometric substitution
Partial fraction decomposition
Rectangular rule
Trapezoidal rule
Simpson's rule

REVIEW EXERCISES

In Exercises 1–50 evaluate the indefinite integral.

1. $\int \sqrt{2 - x}\, dx$

2. $\int \frac{1}{(x + 3)^2}\, dx$

3. $\int \frac{x^2 + 3}{x}\, dx$

4. $\int \frac{x^2 + 3}{x + 1}\, dx$

5. $\int \frac{x^2 + 3}{x^2 + 1}\, dx$

6. $\int \frac{x}{\sqrt{x + 3}}\, dx$

7. $\int \sin^2 x \cos^3 x\, dx$

8. $\int x \sin x\, dx$

9. $\int \tan^2(2x)\, dx$

10. $\int \frac{x}{x^2 + 2x - 3}\, dx$

11. $\int \frac{1}{\sqrt{4 - 3x^2}}\, dx$

12. $\int \frac{2 - \sqrt{x}}{\sqrt{x} + 5}\, dx$

13. $\int \frac{x}{3x^2 + 4}\, dx$

14. $\int \frac{e^x}{\sqrt{1 - e^{2x}}}\, dx$

15. $\int x^2 \ln x\, dx$

16. $\int \frac{x}{(x^2 + 1)^2}\, dx$

17. $\displaystyle\int \frac{x^2}{(x^2+1)^2}\,dx$
18. $\displaystyle\int \frac{x^3}{(x^2+1)^2}\,dx$

19. $\displaystyle\int \frac{x+1}{x^3-4x}\,dx$
20. $\displaystyle\int \left(\frac{x+1}{x-1}\right)^2 dx$

21. $\displaystyle\int \frac{x^2}{(1+3x^3)^4}\,dx$
22. $\displaystyle\int \mathrm{Cos}^{-1} x\,dx$

23. $\displaystyle\int \sin x \cos 2x\,dx$
24. $\displaystyle\int \sin x \cos 5x\,dx$

25. $\displaystyle\int e^{3x}\cos 2x\,dx$
26. $\displaystyle\int \frac{1}{\sqrt{x^2+4x-5}}\,dx$

27. $\displaystyle\int \frac{1}{x^2+4x-5}\,dx$
28. $\displaystyle\int x^3\sqrt{4-x^2}\,dx$

29. $\displaystyle\int \frac{\cos 2x}{1-\sin 2x}\,dx$
30. $\displaystyle\int \frac{6x}{4-x^2}\,dx$

31. $\displaystyle\int \frac{1}{x\sqrt{\ln x}}\,dx$
32. $\displaystyle\int \frac{1}{x^2+4x-4}\,dx$

33. $\displaystyle\int \frac{\sin x}{1+\cos^2 x}\,dx$
34. $\displaystyle\int \frac{1}{x^4+x^3}\,dx$

35. $\displaystyle\int x\sec^2(3x)\,dx$
36. $\displaystyle\int \frac{1}{\sqrt{16-3x+x^2}}\,dx$

37. $\displaystyle\int \frac{\sqrt{x^2-4}}{x^2}\,dx$
38. $\displaystyle\int x^2\mathrm{Tan}^{-1} x\,dx$

39. $\displaystyle\int \frac{x^2}{x^3+3x^2+3x+1}\,dx$

40. $\displaystyle\int \frac{\ln x}{x}\,dx$

41. $\displaystyle\int \frac{x^2}{1+4x^6}\,dx$
42. $\displaystyle\int \frac{1}{x(9+x^2)^2}\,dx$

43. $\displaystyle\int \frac{x^2+2}{x^3+5x^2+4x}\,dx$
44. $\displaystyle\int \frac{x^2+2}{x^3+4x^2+4x}\,dx$

45. $\displaystyle\int \frac{x^2+2}{x^3+x^2+4x}\,dx$
46. $\displaystyle\int \frac{3x^2+2x+4}{x^3+x^2+4x}\,dx$

47. $\displaystyle\int x\mathrm{Sin}^{-1} x\,dx$
48. $\displaystyle\int \sqrt{\cot x}\,\csc^4 x\,dx$

49. $\displaystyle\int \ln(\sqrt{x}+1)\,dx$
50. $\displaystyle\int \frac{1}{(4x-x^2)^{3/2}}\,dx$

In Exercises 51–54 use the trapezoidal rule and Simpson's rule with ten equal partitions to approximate the definite integral.

51. $\displaystyle\int_1^2 \frac{\sin x}{x}\,dx$
52. $\displaystyle\int_0^1 \sqrt{\sin x}\,dx$

53. $\displaystyle\int_2^4 \frac{1}{\ln x}\,dx$
54. $\displaystyle\int_{-1}^3 \frac{1}{1+e^x}\,dx$

55. If 10% of a sample of radioactive material decays in 10 days, what is its half-life?

In Exercises 56–71 evaluate the indefinite integral.

56. $\displaystyle\int \frac{1}{x^{1/3}-\sqrt{x}}\,dx$
57. $\displaystyle\int \ln(1+x^2)\,dx$

58. $\displaystyle\int \frac{x}{x^4+16}\,dx$
59. $\displaystyle\int \csc^3 x\,dx$

60. $\displaystyle\int \frac{1}{(3x-x^2)^{3/2}}\,dx$
61. $\displaystyle\int \frac{1}{x^3\sqrt{x^2-9}}\,dx$

62. $\displaystyle\int \sin\sqrt{x}\,dx$
63. $\displaystyle\int \sin(\ln x)\,dx$

64. $\displaystyle\int x\cos x \sin 3x\,dx$
65. $\displaystyle\int \frac{x^4+3x^2+1}{x(x^2+1)^2}\,dx$

66. $\displaystyle\int \frac{1}{1+\cos 2x}\,dx$
67. $\displaystyle\int \frac{x^4+3x^2-2x+5}{x^2-3x+7}\,dx$

68. $\displaystyle\int \sin^2 x\cos 3x\,dx$
69. $\displaystyle\int \frac{1}{x^3(4-x^2)^{3/2}}\,dx$

70. $\displaystyle\int \sqrt{1-x^2}\,\mathrm{Sin}^{-1} x\,dx$
71. $\displaystyle\int \frac{1}{x+\sqrt{x^2+4}}\,dx$

72. (a) Use the substitution $z^2 = (1+x)/(1-x)$ to show that

$$\int \sqrt{\frac{1+x}{1-x}}\,dx = 2\,\mathrm{Tan}^{-1}\sqrt{\frac{1+x}{1-x}} - \sqrt{1-x^2} + C.$$

(b) Evaluate the integral in (a) by multiplying numerator and denominator of the integrand by $\sqrt{1+x}$. Verify that this answer is the same as that in (a).

CHAPTER 10

Parametric Equations and Polar Coordinates

In this chapter we introduce parametric representations for curves, and polar coordinates. Parametric representations of curves offer an alternative to the explicit and implicit forms used in Chapters 1–9. Polar coordinates provide an alternative way to identify points in a plane. They are more efficient than Cartesian coordinates in many applications. In particular, we shall see how they provide a unified approach to conic sections.

SECTION 10.1

Parametric Equations

A curve is defined explicitly by equations of the form $y = f(x)$ or $x = g(y)$, and implicitly by an equation $F(x,y) = 0$. A third method is described in the following definition.

Definition 10.1

A curve is said to be defined **parametrically** if it is given in the form

$$x = x(t), \quad y = y(t), \quad \alpha \leq t \leq \beta. \quad (10.1)$$

Variable t is called a *parameter*; it is a connecting link between x and y. Each value of t in the interval $\alpha \leq t \leq \beta$ is substituted into equations 10.1, and the pair $(x,y) = (x(t), y(t))$ represents a point on the curve. When the interval for t is unspecified, we assume that it consists of all values for which both $x(t)$ and $y(t)$ are defined.

EXAMPLE 10.1

Sketch the curve defined parametrically by $x = 3 - t$, $y = t^2 - 2$.

SOLUTION We can construct a table of values (Table 10.1) by substituting various values of t and calculating corresponding values of x and y. This leads to the points in Figure 10.1(a), which can be joined as in Figure 10.1(b). Alternatively, if we solve the first equation for t, we obtain $t = 3 - x$, and substituting this expresssion into the second equation, we have

TABLE 10.1

t	−3	−2	−1	0	1	2	3
x	6	5	4	3	2	1	0
y	7	2	−1	−2	−1	2	7

$$y = (3 - x)^2 - 2 = x^2 - 6x + 7,$$

an explicit definition of the curve. The given equations are therefore parametric equations for this parabola.

FIGURE 10.1

(a)

(b)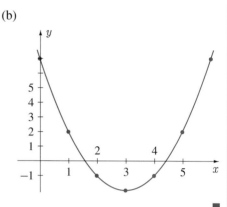

EXAMPLE 10.2 Discuss the curve defined parametrically by

$$x = 2\cos\theta, \quad y = 2\sin\theta, \quad 0 \leq \theta < 2\pi.$$

SOLUTION
If the given equations are squared and added, the result is

$$x^2 + y^2 = 4\cos^2\theta + 4\sin^2\theta = 4,$$

an implicit definition of the curve. The given equations therefore define this circle parametrically. In this case we can interpret the parameter θ as the angle in Figure 10.2. The values $0 \leq \theta < 2\pi$ guarantee that the parametric equations describe the complete circle. Additional values of θ would only duplicate points on this circle.

FIGURE 10.2

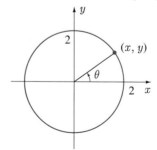

EXAMPLE 10.3

Sketch the curve defined parametrically by

$$x = t - t^3, \qquad y = t + t^3.$$

SOLUTION It is possible once again to eliminate the parameter. By adding and subtracting the given equations, we have

$$x + y = 2t, \qquad \text{and} \qquad y - x = 2t^3.$$

It follows from the first of these that $t = (x+y)/2$, and when this is substituted into the second,

$$y - x = \frac{(x+y)^3}{4}.$$

This is an implicit definition of the curve, but not a particularly simple one. An explicit definition is very cumbersome. To sketch the curve using the parametric representation, we could make a table of values as in Example 10.1. A more efficient method — one that is very helpful for complex curves — is to draw separate graphs of the functions $x = t-t^3$ and $y = t+t^3$ using the same scale on the t-axes (Figure 10.3).

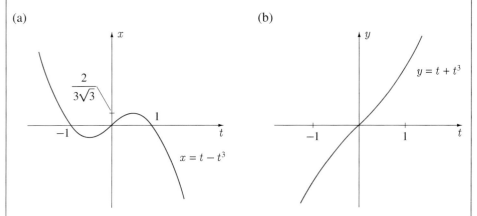

FIGURE 10.3

From these graphs we read pairs of values of x and y for the same t, and interpret them as coordinates (x, y) of a point. Beginning with $t = 0$, we obtain $x = 0$ and $y = 0$, and therefore the point $(0,0)$. As t increases from $t = 0$ to $t = 1$, values of x increase from 0 to a maximum value $2/(3\sqrt{3})$ at $t = 1/\sqrt{3}$, and then decrease to 0. Simultaneously, values of y increase steadily from 0 to 2. This is shown in Figure 10.4(a). As t increases beyond 1, values of x decrease through negative numbers while y continues to increase. This is reflected in that part of the graph in Figure 10.4(b) to the left of the y-axis. A similar procedure using negative values of t leads to the final graph in Figure 10.4(c). Alternatively, for negative t, we note that because $x = t - t^3$ and $y = t + t^3$ are odd functions, replacing t by $-t$ reverses the signs of x and y. This means that corresponding to each point (x, y) on the graph for which $t > 0$, we must have the point $(-x, -y)$.

FIGURE 10.4

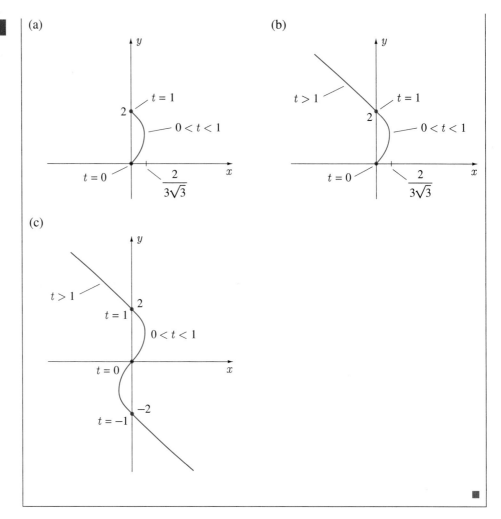

Parametric equations are frequently used in problems concerning the motion of objects. For example, suppose a stone is thrown horizontally over the edge of a cliff 100 m above a river (Figure 10.5). If the initial speed of the stone is 30 m/s, the path followed by the stone is described parametrically by

$$x = 30t, \qquad y = 100 - 4.905t^2, \qquad 0 \leq t \leq \sqrt{\frac{100}{4.905}},$$

where t is time in seconds and x and y are in metres. For these equations, t has been chosen equal to zero at the instant of projection, and $t = \sqrt{100/4.905}$ is the time at which the stone strikes the river.

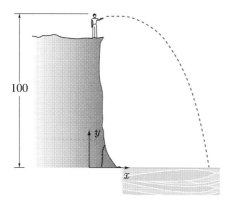

FIGURE 10.5

When parameter t is eliminated from this parametric representation of the trajectory, the resulting equation is

$$y = 100 - 4.905 \left(\frac{x}{30}\right)^2 = 100 - 0.00545\,x^2.$$

It clearly indicates that the stone follows a parabolic path. Were we to discuss the motion of the stone as regards, say, velocity and acceleration, however, we would find it necessary to return to the parametric representation which describes the motion in horizontal and vertical directions.

We say that equations 10.1 define y **parametrically as a function of** x if the curve so defined represents a function; that is, if every vertical line that intersects the curve does so exactly once. The parametric equations in Example 10.1 therefore define a function parametrically, but in Examples 10.2 and 10.3 they do not. However, each of the curves in these latter examples can be divided into subcurves that do represent functions. For instance, in Example 10.3, each of the subcurves

$$\begin{aligned} x &= t - t^3, & y &= t + t^3, & t &\leq -1/\sqrt{3}; \\ x &= t - t^3, & y &= t + t^3, & -1/\sqrt{3} &\leq t \leq 1/\sqrt{3}; \\ x &= t - t^3, & y &= t + t^3, & t &\geq 1/\sqrt{3}; \end{aligned}$$

defines a function, and does so parametrically; together all three curves make up the original curve.

When equations 10.1 do define y parametrically as a function of x, it is straightforward to find the derivative of the function. If we denote this function by $y = f(x)$, then the chain rule applied to $y = f(x)$, $x = x(t)$ gives

$$\frac{dy}{dt} = \frac{dy}{dx}\frac{dx}{dt}.$$

Provided then that $dx/dt \neq 0$, we may solve for

$$\frac{dy}{dx} = \frac{\dfrac{dy}{dt}}{\dfrac{dx}{dt}}. \qquad (10.2)$$

This formula is called the **parametric rule** for differentiation of a function defined parametrically by equations 10.1; it defines the derivative of y with respect to x in terms of derivatives of the given functions $x(t)$ and $y(t)$ with respect to t.

For example, if y is defined parametrically as a function of x by

$$x(t) = \frac{1}{t^2 + 1}, \qquad y(t) = \frac{2}{t(t^2 + 1)}, \qquad t \geq 0,$$

then

$$\frac{dy}{dx} = \frac{\dfrac{dy}{dt}}{\dfrac{dx}{dt}} = \frac{\dfrac{-2(3t^2+1)}{(t^3+t)^2}}{\dfrac{-2t}{(t^2+1)^2}} = \frac{3t^2+1}{t^3}.$$

We can also calculate d^2y/dx^2 in spite of the fact that dy/dx is in terms of t rather than x. Once again it is the chain rule that comes to the rescue:

$$\frac{d^2y}{dx^2} = \frac{d}{dx}\left(\frac{dy}{dx}\right) = \frac{d}{dt}\left(\frac{dy}{dx}\right)\frac{dt}{dx} = \frac{\dfrac{d}{dt}\left(\dfrac{dy}{dx}\right)}{\dfrac{dx}{dt}}$$

$$= \frac{\dfrac{t^3(6t)-(3t^2+1)(3t^2)}{t^6}}{\dfrac{-2t}{(t^2+1)^2}} = \frac{-3t^4-3t^2}{t^6}\cdot\frac{(t^2+1)^2}{-2t} = \frac{3(t^2+1)^3}{2t^5}.$$

EXAMPLE 10.4

Sketch the strophoid

$$x = \frac{1-t^2}{1+t^2}, \qquad y = \frac{t(1-t^2)}{1+t^2}$$

and find the points at which its tangent line is horizontal.

SOLUTION

To sketch the strophoid we first sketch graphs of the functions $x(t)$ and $y(t)$. For $x(t)$, we write $x = -1 + 2/(1+t^2)$, and first draw $2/(1+t^2)$ in Figure 10.6(a). The graph of $x(t)$ is this curve shifted downward one unit. Since $y(t) = tx(t)$, we sketch a graph of $y(t)$ by multiplication of ordinates. The result is shown in Figure 10.6(b). To sketch the strophoid, we proceed as in Example 10.3 by reading pairs of values from the graphs in Figure 10.6. The curve is shown in Figure 10.7.

FIGURE 10.6 (a) (b)

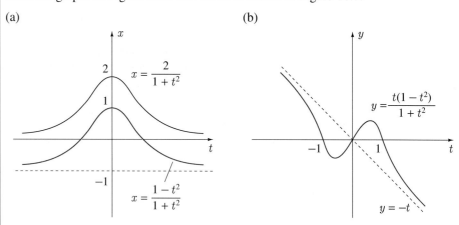

The points at which the tangent line is horizontal can be found by setting

$$0 = \frac{dy}{dx} = \frac{\frac{dy}{dt}}{\frac{dx}{dt}} = \frac{\frac{(1+t^2)(1-3t^2) - (t-t^3)(2t)}{(1+t^2)^2}}{\frac{(1+t^2)(-2t) - (1-t^2)(2t)}{(1+t^2)^2}} = \frac{1 - 4t^2 - t^4}{-4t}.$$

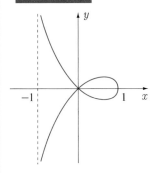

FIGURE 10.7

But this equation implies that

$$t^4 + 4t^2 - 1 = 0.$$

This is a quadratic equation in t^2, so that

$$t^2 = \frac{-4 \pm \sqrt{16 + 4}}{2} = -2 \pm \sqrt{5}.$$

Since t^2 must be nonnegative, it follows that $t^2 = \sqrt{5} - 2$, and therefore

$$t = \pm\sqrt{\sqrt{5} - 2}.$$

For these values of t,

$$x = \frac{1 - (\sqrt{5} - 2)}{1 + (\sqrt{5} - 2)} = \frac{3 - \sqrt{5}}{\sqrt{5} - 1} \cdot \frac{\sqrt{5} + 1}{\sqrt{5} + 1} = \frac{\sqrt{5} - 1}{2} = 0.62, \quad \text{and}$$

$$y = tx = \left(\pm\sqrt{\sqrt{5} - 2}\right)\left(\frac{\sqrt{5} - 1}{2}\right) = \pm 0.30.$$

The points with a horizontal tangent line are therefore $(0.62, \pm 0.30)$. ∎

Recall formula 7.17 from Section 7.3 for the length of a curve. We express it in a slightly different form,

$$L = \int_A^B \sqrt{(dx)^2 + (dy)^2},$$

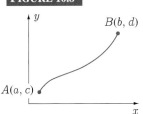

FIGURE 10.8

where A and B are the initial and final points on the curve (Figure 10.8). When the equation of the curve is $y = f(x)$, we rewrite this formula as

$$L = \int_a^b \sqrt{1 + \left(\frac{dy}{dx}\right)^2}\, dx$$

where a and b are x-coordinates of A and B, provided $a < b$. When the curve is given in the form $x = g(y)$, we write

$$L = \int_c^d \sqrt{1 + \left(\frac{dx}{dy}\right)^2}\, dy,$$

provided $c < d$.

When the curve is defined parametrically by 10.1, the above formulas are replaced by

$$L = \int_\alpha^\beta \sqrt{\left(\frac{dx}{dt}\right)^2 + \left(\frac{dy}{dt}\right)^2}\, dt. \qquad (10.3)$$

Verification of this formula is left to Exercise 42.

EXAMPLE 10.5

Find the length of the curve

$$x = e^t \cos t, \quad y = e^t \sin t, \quad 0 \le t \le \pi/2.$$

SOLUTION According to 10.3,

$$L = \int_0^{\pi/2} \sqrt{(e^t \cos t - e^t \sin t)^2 + (e^t \sin t + e^t \cos t)^2}\, dt$$

$$= \int_0^{\pi/2} \sqrt{e^{2t}(2\cos^2 t + 2\sin^2 t)}\, dt$$

$$= \int_0^{\pi/2} \sqrt{2}\, e^t\, dt = \sqrt{2}\,\{e^t\}_0^{\pi/2} = \sqrt{2}(e^{\pi/2} - 1).$$

EXERCISES 10.1

In Exercises 1–12 sketch the curve.

1. $x = 2 + t, \quad y = 3t - 1$
2. $x = t^2 + 3t + 4, \quad y = 1 - t$
3. $x = 1 + 2\cos t, \quad y = 2 + 2\sin t, \quad 0 \le t \le 2\pi$
4. $x = -2 + 4\cos t, \quad y = 3 + 4\sin t, \quad 0 \le t \le \pi$
5. $x = 1 + \cos t, \quad y = -1 - \sin t, \quad 0 \le t \le \pi/4$
6. $x = t + 1/t, \quad y = t - 1/t$
7. $x = 2\cos t, \quad y = 4\sin t, \quad 0 \le t < 2\pi$
8. $x = 1 + 3\cos t, \quad y = -2 + 2\sin t, \quad 0 \le t \le \pi$
9. $x = -1 + \sin t, \quad y = -1 - 3\cos t, \quad 0 \le t \le \pi/2$
10. $x = t - t^2, \quad y = t + t^2$
11. $x = t^2 + 1, \quad y = t^3 + 3$
12. $x = 2\cot\theta, \quad y = 2\sin^2\theta, \quad -\pi/2 \le \theta \le \pi/2, \quad \theta \ne 0$

In Exercises 13–20 assume that y is defined parametrically as a function of x, and find dy/dx.

13. $x = t^3 + 3t - 2, \quad y = t^2 - 1$
14. $x = \dfrac{u}{u-1}, \quad y = \dfrac{u^2}{u^2 - 1}$
15. $x = (v^3 + 2)\sqrt{v - 1}, \quad y = 2v^3 + 3$
16. $x = \sqrt{\dfrac{2+t}{2-t}}, \quad y = \sqrt{\dfrac{2-t}{2+t}}$
17. $x = s^{3/2} - s^{2/3}, \quad y = s^2 + 2s$
18. $x = (2t + 3)^4, \quad y = \dfrac{t}{t + 6}$
19. $x = \left(\dfrac{1+u}{1-u}\right)^{1/3}, \quad y = \left(\dfrac{1-u}{1+u}\right)^4$
20. $x = \sqrt{-t^2 + 3t + 5}, \quad y = \dfrac{1}{t^2 + 2t - 5}$

In Exercises 21–24 assume that y is defined parametrically as a function of x, and find dy/dx and d^2y/dx^2.

21. $x = t^2 + \dfrac{1}{t}, \quad y = t^2 - \dfrac{1}{t}$
22. $x = \sqrt{t - 1}, \quad y = \sqrt{t + 1}$
23. $x = 2u + 5, \quad y = 7 - 14u$
24. $x = v^2 + 2v + 3, \quad y = 2v - 4$

25. Find equations for the tangent and normal lines to the curve

$$x = t + \dfrac{1}{t}, \quad y = t - \dfrac{1}{t}$$

at the point corresponding to $t = 4$.

26. Find point(s) on the curve

$$x = \dfrac{t^3}{3} - 3t, \quad y = \dfrac{3t^2}{2} + t$$

where the slope of the tangent line to the curve is equal to 1.

27. Is there a difference between the two curves

$$y = 2x^2 - 1 \quad \text{and} \quad x = \cos t, \quad y = 2\cos^2 t - 1?$$

28. What curve is described by the parametric equations
$$x = h + a\cos\theta, \quad y = k + b\sin\theta, \quad 0 \le \theta < 2\pi?$$

29. Find parametric equations for a circle with centre (h, k) and radius r.

30. Show that the straight line through two points $P_1(x_1, y_1)$ and $P_2(x_2, y_2)$ has parametric equations
$$x = x_1 + (x_2 - x_1)t, \quad y = y_1 + (y_2 - y_1)t.$$

31. Sketch the following curves and determine whether they are related:
    (a) $x = \sec\theta, \quad y = \tan\theta, \quad -\pi/2 < \theta < \pi/2$
    (b) $x = \cosh\phi, \quad y = \sinh\phi$
    (c) $x = \dfrac{1}{2}\left(t + \dfrac{1}{t}\right), \quad y = \dfrac{1}{2}\left(t - \dfrac{1}{t}\right), \quad t \ge 0$

In Exercises 32–35 find parametric equations for the curve.

32. $y = \dfrac{x+1}{x-2}$

33. $x + y^3 + xy = 5y^2$

34. $x^2 + y^2 + 2x - 4y = 0$

35. $4 - x^2 + 2y^2 = 0$

36. Two particles move along straight lines l_1 and l_2 defined parametrically by
$$l_1: \quad x = 1 - t, \quad y = t, \quad t \ge 0;$$
$$l_2: \quad x = 4t - 5, \quad y = 2t - 1, \quad t \ge 0;$$
where t is time. When are they closest together?

37. If $x = x(t)$ and $y = y(t)$ define y as a function of x, show that
$$\dfrac{d^2 y}{dx^2} = \dfrac{\dfrac{dx}{dt}\dfrac{d^2 y}{dt^2} - \dfrac{dy}{dt}\dfrac{d^2 x}{dt^2}}{\left(\dfrac{dx}{dt}\right)^3}.$$

In Exercises 38–40 find the length of the curve.

38. $x = 3 + 4\cos t, \quad y = -2 + 4\sin t, \quad 0 \le t < 2\pi$

39. $x = e^{-t}\sin t, \quad y = e^{-t}\cos t, \quad 0 \le t \le 1$

40. $x = t + \ln t, \quad y = t - \ln t, \quad 1 \le t \le 2$

41. Set up, but do not evaluate, a definite integral representing the length of the ellipse
$$x = a\cos\theta, \quad y = b\sin\theta, \quad 0 \le \theta < 2\pi.$$

42. Verify formula 10.3 for the length of a curve.

43. The equations $x = t^2 + 2t - 1$, $y = t + 5$, $1 \le t \le 4$ define a curve parametrically. Find parametric equations that describe this curve, but have values of the parameter in the intervals (a) $0 \le t \le 3$; (b) $0 \le t \le 1$.

44. Suppose $x(t)$ and $y(t)$ in equations 10.1 are continuous on $\alpha \le t \le \beta$ and have derivatives on $\alpha < t < \beta$, and that $x'(t) \ne 0$ on $\alpha < t < \beta$. Show that Cauchy's generalized mean value theorem (Theorem 3.16) implies that there exists a number c between α and β such that
$$\dfrac{y(\beta) - y(\alpha)}{x(\beta) - x(\alpha)} = \dfrac{y'(c)}{x'(c)}.$$
Interpret this result geometrically.

45. A particle travels around the circle $x^2 + y^2 = 4$ counterclockwise at constant speed, making 2 revolutions each second. If the particle starts at point $(2, 0)$ at time $t = 0$, find parametric equations for its position in terms of t.

46. (a) A stone is embedded in the tread of a tire and the tire rolls, without slipping, along the x-axis (Figure 10.9). If the stone starts at the origin, the path that it follows is called a *cycloid*. Show that if θ is the angle through which the stone has turned, and R is the radius of the tire, parametric equations for the cycloid are
$$x = R(\theta - \sin\theta), \quad y = R(1 - \cos\theta).$$

(b) Find the area under one arch of the cycloid.

(c) Find the length of one arch of the cycloid.

FIGURE 10.9

47. If the stone in Exercise 46 is embedded in the side of the tire rather than the tread, its path is called a *trochoid* (Figure 10.10). Show that if the distance from the centre of the tire to the stone is b, parametric equations for the trochoid are
$$x = R\theta - b\sin\theta, \quad y = R - b\cos\theta.$$

FIGURE 10.10

In Exercises 48–50 sketch the curve.

48. $x = \cos\theta$, $y = \sin 3\theta$ (Curve of Lissajous)

49. $\dfrac{3t}{1+t^3}$, $y = \dfrac{3t^2}{1+t^3}$ (Folium of Descartes)

50. $x = \cos^3\theta$, $y = \sin^3\theta$ (Astroid or hypocycloid of four cusps)

SECTION 10.2

Polar Coordinates

In this section we introduce polar coordinates, an alternative coordinate system for the plane. Many problems that have complex solutions using Cartesian coordinates become much simpler to solve using polar coordinates.

Polar coordinates are defined by choosing a point O in the plane called the **pole** and a half-line originating at O called the **polar axis** (Figure 10.11). If P is a point in the plane, we join O and P. The first polar coordinate of P, denoted by r, is the length of the line segment OP. The other polar coordinate is the angle θ through which the polar axis must be rotated to coincide with line segment OP. Counterclockwise rotations are regarded as positive, and clockwise rotations as negative. In Figure 10.12, position OQ is reached through a positive rotation of $\pi/6$ radians; therefore, for point Q, $\theta = \pi/6$. But clearly we could arrive at this position in many other ways. We could, for instance, rotate the polar axis counterclockwise through any number of complete revolutions, bringing it back to its original position, and then rotate a further $\pi/6$ radians. Alternatively, we could rotate in a clockwise direction any number of complete revolutions, and then a further $-11\pi/6$ radians. In other words, polar coordinate θ for Q could be any of the values $\pi/6 + 2n\pi$, where n is an integer. Possible values of θ for point R in Figure 10.12 are $3\pi/4 + 2n\pi$. For point Q, $r = 2$ and for R, $r = 1$. Polar coordinates r and θ for a point are written in the form (r, θ) so that possible polar coordinates for Q and R are $(2, \pi/6 + 2n\pi)$ and $(1, 3\pi/4 + 2n\pi)$.

FIGURE 10.11

FIGURE 10.12

This situation is not like that for Cartesian coordinates where each point has only one set of coordinates (x, y), and every ordered pair of real numbers specifies one point. With polar coordinates, every ordered pair of real numbers (r, θ), where r must be nonnegative, represents one and only one point; but every point has an infinity of possible representations. We should point out that in some applications this is not a desirable situation. For instance, in the branch of mathematics called *tensor analysis*, it is necessary that polar coordinates assign exactly one pair of coordinates to each point. This can be accomplished in any region that does not contain the pole by demanding, for instance, that $-\pi < \theta \le \pi$. When we use polar coordinates to find areas in Section 10.4, we must also be particular about our choice of θ. For now, however, no advantage is gained by imposing restrictions on θ, and we therefore accept the fact that if (r, θ) are polar coordinates of a point, so are $(r, \theta + 2n\pi)$ for any integer n. Note also that polar coordinates for the pole are $(0, \theta)$ for any θ whatsoever.

If we introduce into a plane both a system of Cartesian coordinates (x, y) and a system of polar coordinates (r, θ), then relations will exist between the two. Suppose, for example, the pole of the polar coordinates and the origin of the Cartesian coordinates are chosen as the same point, and that the polar axis is chosen as the positive x-axis (Figure 10.13). In this case, Cartesian and polar coordinates of any point P are related by the equations

$$x = r\cos\theta, \quad y = r\sin\theta. \tag{10.4}$$

FIGURE 10.13

These equations define the Cartesian coordinates of a point in terms of its polar coordinates; that is, given its polar coordinates (r,θ), we can calculate its Cartesian coordinates (x,y) by means of 10.4. For example, if the polar coordinates of a point are $(3,2)$, then its Cartesian coordinates are

$$x = 3\cos 2 = -1.25, \quad y = 3\sin 2 = 2.73.$$

Equations 10.4 implicitly define the polar coordinates of a point in terms of its Cartesian coordinates. For instance, if the Cartesian coordinates of a point are $(1,1)$ (Figure 10.14), its polar coordinates must satisfy

$$1 = r\cos\theta, \quad 1 = r\sin\theta.$$

If we square these equations and add, we have

$$1 + 1 = r^2\cos^2\theta + r^2\sin^2\theta = r^2,$$

and therefore $r = \sqrt{2}$. It follows that $\cos\theta = \sin\theta = 1/\sqrt{2}$, from which we get $\theta = \pi/4 + 2n\pi$. Thus, polar coordinates of the point are $(\sqrt{2}, \pi/4 + 2n\pi)$.

FIGURE 10.14

Equations 10.4 define r and θ implicitly in terms of x and y, but obviously it would be preferable to have explicit definitions. There is no problem expressing r explicitly in terms of x and y,

$$r = \sqrt{x^2 + y^2}, \tag{10.5}$$

but the case for θ is not so simple. If we substitute expression 10.5 for r into equations 10.4, we obtain

$$\cos\theta = \frac{x}{\sqrt{x^2 + y^2}}, \quad \sin\theta = \frac{y}{\sqrt{x^2 + y^2}}. \tag{10.6}$$

Except for the pole, these two equations determine all possible values of θ for given x and y. What we would like to do is obtain one equation, if possible, that defines θ. If we divide the second of these equations by the first, we have

$$\tan\theta = \frac{y}{x}, \tag{10.7}$$

and this equation suggests that we set

$$\theta = \operatorname{Tan}^{-1}\left(\frac{y}{x}\right). \tag{10.8}$$

Unfortunately, neither equation 10.7 nor 10.8 is satisfactory. For instance, given the point P with Cartesian coordinates $(-1, 1)$ (Figure 10.15), equation 10.7 yields $\tan\theta = -1$, the solutions of which are $\theta = -\pi/4 + n\pi$. These are angles in the second and fourth quadrants, so only half of them are acceptable polar angles for P. Equation 10.8 gives $\theta = -\pi/4$, which is not a possible polar angle for P.

FIGURE 10.15

We suggest all angles satisfying 10.7 be found, and then a diagram be used to determine those angles that are acceptable values for θ. We illustrate in the following example.

EXAMPLE 10.6

Find all polar coordinates for points with the following Cartesian coordinates:

(a) $(1, 2)$ (b) $(-2, 3)$ (c) $(3, -1)$ (d) $(-2, -4)$

SOLUTION (a) $r = \sqrt{1^2 + 2^2} = \sqrt{5}$. Angles which satisfy $\tan\theta = 2/1$ are $1.11 + n\pi$. Since the point is in the first quadrant (Figure 10.16), acceptable values for θ are $1.11 + 2n\pi$. Polar coordinates of the point are therefore $(\sqrt{5}, 1.11 + 2n\pi)$.

(b) $r = \sqrt{(-2)^2 + 3^2} = \sqrt{13}$. Angles satisfying $\tan\theta = -3/2$ are $-0.98 + n\pi$. Since the point is in the second quadrant, possible values for θ are $(\pi - 0.98) + 2n\pi = 2.16 + 2n\pi$. Polar coordinates are therefore $(\sqrt{13}, 2.16 + 2n\pi)$.

(c) $r = \sqrt{3^2 + (-1)^2} = \sqrt{10}$. Since values of θ satisfying $\tan\theta = -1/3$ are $-0.32 + n\pi$, and the point is in the fourth quadrant, it follows that polar coordinates are $(\sqrt{10}, -0.32 + 2n\pi)$.

(d) $r = \sqrt{(-2)^2 + (-4)^2} = 2\sqrt{5}$. Since angles satisfying $\tan\theta = 2$ are $1.11 + n\pi$, and the point is in the third quadrant, polar coordinates are $(2\sqrt{5}, 4.25 + 2n\pi)$.

FIGURE 10.16

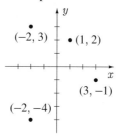

The results in equations 10.4–10.8 are valid only when the pole and origin coincide, and the polar axis and positive x-axis are identical. For a different arrangement, these relations would be changed accordingly. For example, if the pole is at the point with Cartesian coordinates (h, k) and the polar axis is as shown in Figure 10.17, equations 10.4 are replaced by

$$x = h + r\cos(\theta + \alpha), \quad y = k + r\sin(\theta + \alpha). \tag{10.9}$$

You will verify these equations in Exercise 13.

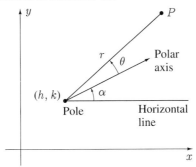

FIGURE 10.17

The usual choice of pole and polar axis is that in Figure 10.13, and unless otherwise stipulated, we assume this to be the case.

EXERCISES 10.2

In Exercises 1–8 plot the point having the given set of Cartesian coordinates, and find all possible polar coordinates.

1. $(1, -1)$
2. $(-1, \sqrt{3})$
3. $(4, 3)$
4. $(-2\sqrt{3}, 2)$
5. $(2, 6)$
6. $(-1, -4)$
7. $(7, -5)$
8. $(-5, 2)$

In Exercises 9–12 plot the point having the given set of polar coordinates, and find its Cartesian coordinates.

9. $(2, \pi/4)$
10. $(6, -\pi/6)$
11. $(7, 1)$
12. $(3, -2.4)$

13. Verify that polar and Cartesian coordinates as shown in Figure 10.17 are related by equations 10.9.

SECTION 10.3

Curves in Polar Coordinates

A curve is defined explicitly in Cartesian coordinates by equations $y = f(x)$ or $x = g(y)$, and implicitly by $F(x, y) = 0$. A point is on a curve if and only if its Cartesian coordinates (x, y) satisfy the equation of the curve.

Analogously, a curve is defined explicitly in polar coordinates if its equation is expressed in either of the forms

$$r = f(\theta) \quad \text{or} \quad \theta = g(r), \tag{10.10}$$

and is defined implicitly when its equation is given in the form

$$F(r, \theta) = 0. \tag{10.11}$$

As discussed in the preceding section, a point may be represented by an infinite number of sets of polar coordinates. A point is on a curve if at least one of these sets of polar coordinates (r, θ) satisfies the equation of the curve. It follows then, that all sets of polar coordinates for a point do not necessarily satisfy the equation. For example, the origin or pole has polar coordinates $(0, \theta)$ for any θ whatsoever. But only those coordinates of the form $(0, (2n+1)\pi)$ satisfy the equation $r = 1 + \cos\theta$. Likewise, the polar coordinates $(1, \pi)$ satisfy the equation $r = \sin(\theta/2)$, but the coordinates $(1, 3\pi)$ of the same point do not satisfy this equation.

Often we are required to transform the equation of a curve from Cartesian coordinates to polar coordinates, and vice versa. To transform from Cartesian to polar is straightforward: replace each x with $r\cos\theta$ and each y with $r\sin\theta$. For example, the equation $x^2 + y^2 = 9$ describes a circle centred at the origin with radius 3. In polar coordinates, its equation is

$$9 = (r\cos\theta)^2 + (r\sin\theta)^2 = r^2\cos^2\theta + r^2\sin^2\theta = r^2.$$

Consequently, $r = 3$ is the equation of this circle in polar coordinates.

EXAMPLE 10.7

Find equations in polar coordinates for the following curves:

(a) $2x + 3y = 3$ (b) $x^2 - 2x + y^2 = 0$ (c) $x^2 + y^2 = \sqrt{x^2 + y^2} - 4x$

SOLUTION (a) For $2x + 3y = 3$, we obtain

$$3 = 2r\cos\theta + 3r\sin\theta, \quad \text{or} \quad r = \frac{3}{2\cos\theta + 3\sin\theta}.$$

(b) For the equation $x^2 - 2x + y^2 = 0$, we have

$$0 = -2x + (x^2 + y^2) = -2r\cos\theta + r^2 = r(r - 2\cos\theta).$$

Thus,

$$r = 0 \quad \text{or} \quad r = 2\cos\theta.$$

Since $r = 0$ defines the pole, and this point also satisfies $r = 2\cos\theta$ (for $\theta = \pi/2$), it follows that we need only write $r = 2\cos\theta$.

(c) For the curve with equation $x^2 + y^2 = \sqrt{x^2 + y^2} - 4x$, we obtain

$$r^2 = r - 4r\cos\theta = r(1 - 4\cos\theta).$$

Thus,

$$0 = r^2 - r(1 - 4\cos\theta) = r(r - 1 + 4\cos\theta),$$

from which we have

$$r = 0 \quad \text{or} \quad r = 1 - 4\cos\theta.$$

Again the pole satisfies the second of these equations, and therefore the equation of the curve in polar coordinates is $r = 1 - 4\cos\theta$. ∎

To transform equations of curves from polar to Cartesian coordinates can sometimes be more difficult, principally because we have no substitution for θ. If, however, the equation involves $\cos\theta$ and/or $\sin\theta$, we use equations 10.6.

EXAMPLE 10.8

Find equations in Cartesian coordinates for the following curves:

(a) $r = 1 + \cos\theta$ (b) $r^2 \cos 2\theta = 1$ (c) $r^2 = 9\sin 2\theta$

SOLUTION (a) We use equations 10.5 and 10.6 to obtain

$$\sqrt{x^2 + y^2} = 1 + \frac{x}{\sqrt{x^2 + y^2}},$$

and this equation simplifies to

$$(x^2 + y^2 - x)^2 = x^2 + y^2.$$

(b) For $r^2 \cos 2\theta = 1$, we use double-angle formula 9.23c to write the equation in terms of $\cos\theta$ rather than $\cos 2\theta$, and then use equations 10.5 and 10.6,

$$1 = r^2(2\cos^2\theta - 1) = (x^2 + y^2)\left(2\frac{x^2}{x^2 + y^2} - 1\right) = (x^2 + y^2)\frac{x^2 - y^2}{x^2 + y^2} = x^2 - y^2.$$

(c) This time we use double-angle formula 9.23a on $\sin 2\theta$,

$$x^2 + y^2 = 18\sin\theta\cos\theta = 18\frac{y}{\sqrt{x^2 + y^2}}\frac{x}{\sqrt{x^2 + y^2}} \quad \text{or}$$

$$(x^2 + y^2)^2 = 18xy.$$

∎

Examples 10.7 and 10.8 illustrate that equations for some curves become much simpler when expressed in polar coordinates. It is also true that some curves are more easily sketched using their polar representation. For instance, consider the curve $r = 1 + \cos\theta$ of Example 10.8. To sketch this curve, we could construct a table of values (Table 10.2) and obtain the set of points in Figure 10.18(a). These points could then be joined to produce the correct curve in Figure 10.18(b). It is called a *cardioid*. The task would be much more complicated if we were to use the corresponding equation in Cartesian coordinates.

TABLE 10.2

θ	$\frac{-3\pi}{4}$	$-\frac{\pi}{2}$	$-\frac{\pi}{4}$	0	$\frac{\pi}{4}$	$\frac{\pi}{2}$	$\frac{3\pi}{4}$	π
r	0.29	1	1.71	2	1.71	1	0.29	0

FIGURE 10.18

(a)

(b)

(c)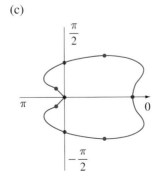

The curve shown in Figure 10.18(b) is not the only way to join the eight points in Figure 10.18(a). For instance, Figure 10.18(c) shows a curve which also passes through these eight points, but which bears little resemblance to the cardioid. This could be rectified by constructing an increasingly larger table of values, but there is an alternative method that is much more efficient.

We draw a Cartesian coordinate system consisting of a horizontal θ-axis and a vertical r-axis. On this set of axes we sketch a graph of the *function* $r = f(\theta) = 1 + \cos\theta$ (Figure 10.19). This is *not* the required curve; it is a sketch of the function $f(\theta)$, illustrating values of r for various values of θ. We might say that this sketch represents an "infinite table of values."

FIGURE 10.19

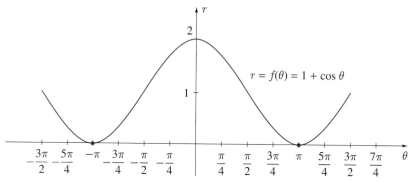

To draw the cardioid $r = 1 + \cos\theta$, we now read pairs of polar coordinates (r, θ) from this sketch, and interpret r as radial distance and θ as rotation. Suppose we begin with the two points $(2, 0)$ and $(1, \pi/2)$. Figure 10.19 indicates that as θ increases from 0 to $\pi/2$, values of r decrease from 2 to 1. This means that as we rotate from the $\theta = 0$ line to the $\theta = \pi/2$ line in the first quadrant, radial distances from the origin become smaller. This is shown in Figure 10.20(a). As rotation is increased from $\pi/2$ to π, Figure 10.19 shows that radial distances continue to decrease, eventually reaching 0 at an angle of π radians as in Figure 10.20(b). Notice that the line $\theta = \pi$ is tangent to the curve at the pole, reflecting the fact that the pole is attained for an angle of π radians. Consideration of the graph in Figure 10.19 to the left of $\theta = 0$ leads to that part of the cardioid below the $\theta = 0$ and $\theta = \pi$ lines in Figure 10.20(c). The symmetry of Figure 10.19 about the r-axis is reflected in the symmetry of Figure 10.20(c) about the $\theta = 0$ and $\theta = \pi$ lines. Since the function $r = f(\theta) = 1 + \cos\theta$ is 2π-periodic, only values of θ in the interval $-\pi < \theta \leq \pi$ need be considered. Values outside this interval retrace previous points.

FIGURE 10.20

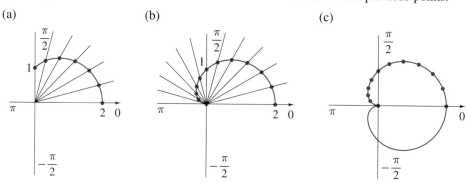

EXAMPLE 10.9

Sketch the following curves:

(a) $r^2 = 9 \sin 2\theta$ (b) $r = |1 - 4\cos\theta|$ (c) $r = \theta$, $\theta \geq 0$

SOLUTION (a) If $r^2 = 9 \sin 2\theta$, then $r = 3\sqrt{\sin 2\theta}$. A sketch of the function $\sin 2\theta$ in Cartesian coordinates is shown in Figure 10.21(a), and from this we obtain the sketch of $f(\theta) = 3\sqrt{\sin 2\theta}$ in Figure 10.21(b). The curve $r = 3\sqrt{\sin 2\theta}$ can now be obtained by reading polar coordinate pairs (r, θ) from this graph, the result being shown in Figure 10.22. It is called a *lemniscate*.

(a)

(b)

FIGURE 10.21

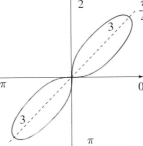

FIGURE 10.22

(b) A sketch of the function $-4\cos\theta$ is shown in Figure 10.23(a), and a vertical shift of one unit yields the sketch of $f(\theta) = 1 - 4\cos\theta$ in Figure 10.23(b). When that portion of the curve below the θ-axis is turned upside down, a sketch of the function $r = f(\theta) = |1 - 4\cos\theta|$ is obtained as in Figure 10.23(c). The values of θ at which $r = 0$ are important in sketching the curve $r = |1 - 4\cos\theta|$; they have been identified in this diagram. From Figure 10.23(c) we obtain the curve in Figure 10.24.

(a)

(b)

FIGURE 10.23

FIGURE 10.24

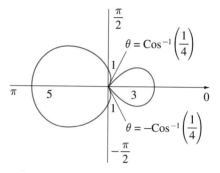

FIGURE 10.25

(c) The straight line $r = \theta$ in Figure 10.25 leads to the spiral in Figure 10.26.

FIGURE 10.26

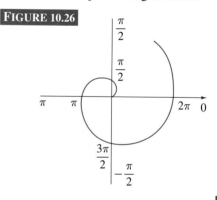

To find points of intersection of two curves whose equations are given in Cartesian coordinates, we solve the two equations simultaneously for all (real) solutions. Each solution represents a distinct point of intersection. For curves whose equations are given in polar coordinates, the situation is somewhat more complex because we have multiple names for points. To find the points of intersection, we again solve the equations simultaneously for all solutions. Each solution represents a point of intersection; but as points have many sets of polar coordinates, some of these solutions may represent the same point. In addition, it may also happen that one set of polar coordinates for a point of intersection satisfies one equation, whereas a different set satisfies the other equation. Particularly troublesome in this respect is the pole, which has so many sets of polar coordinates. The best way to handle these difficulties is to sketch the two curves.

EXAMPLE 10.10

Find points of intersection for the curves $r = \sin\theta$ and $r = 1 - \sin\theta$.

SOLUTION If we set $\sin\theta = 1 - \sin\theta$, then $\sin\theta = 1/2$. All solutions of this equation are defined by

$$\theta = \begin{cases} \dfrac{\pi}{6} + 2n\pi \\ \dfrac{5\pi}{6} + 2n\pi \end{cases}$$

where n is an integer. Sketches of the curves in Figure 10.27 indicate that these give the points of intersection $(1/2, \pi/6)$ and $(1/2, 5\pi/6)$. The figure also indicates that the origin is a point of intersection of the curves. We did not obtain this point by solving $r = \sin\theta$ and $r = 1 - \sin\theta$ because different values of θ yield $r = 0$ in $r = \sin\theta$ and $r = 1 - \sin\theta$. To obtain $r = 0$ from $r = \sin\theta$, θ must be one of the values $n\pi$, whereas to obtain $r = 0$ from $r = 1 - \sin\theta$, θ must be one of the values $\pi/2 + 2n\pi$. Thus, both curves pass through the pole, but the pole cannot be obtained by simultaneously solving the equations of the curves.

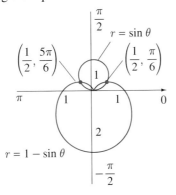

FIGURE 10.27

Slopes of Curves in Polar Coordinates

When a curve has polar equation $r = f(\theta)$, $\alpha \leq \theta \leq \beta$, substitution into equations 10.4 gives

$$x = f(\theta)\cos\theta, \quad y = f(\theta)\sin\theta, \quad \alpha \leq \theta \leq \beta. \tag{10.12}$$

These are parametric equations for the curve, where the parameter is the polar angle θ. Equation 10.2 gives

$$\frac{dy}{dx} = \frac{\dfrac{dy}{d\theta}}{\dfrac{dx}{d\theta}} = \frac{f'(\theta)\sin\theta + f(\theta)\cos\theta}{f'(\theta)\cos\theta - f(\theta)\sin\theta}. \tag{10.13}$$

This formula defines the slope of the tangent line to a curve which has polar equation $r = f(\theta)$.

SECTION 10.3 CURVES IN POLAR COORDINATES

EXAMPLE 10.11

Find the points on the cardioid $r = 1 + \sin\theta$ at which the tangent line is horizontal.

SOLUTION The cardioid is shown in Figure 10.28, and there are clearly three points at which the tangent line is horizontal. To find them we use equation 10.13 to write

$$0 = \frac{dy}{dx} = \frac{(\cos\theta)\sin\theta + (1+\sin\theta)\cos\theta}{(\cos\theta)\cos\theta - (1+\sin\theta)\sin\theta}.$$

Since the numerator must vanish, we set

$$0 = \cos\theta\sin\theta + \cos\theta + \sin\theta\cos\theta = \cos\theta(1 + 2\sin\theta).$$

From $\cos\theta = 0$, we obtain $\theta = \pi/2$ and from $1 + 2\sin\theta = 0$, we take $\theta = -\pi/6$, $-5\pi/6$. Thus, the points at which the cardioid has a horizontal tangent line have Cartesian coordinates $(0,2)$ and $(\pm\sqrt{3}/4, -1/4)$.

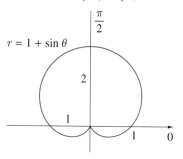

FIGURE 10.28

Lengths of Curves in Polar Coordinates

Recall that the length of a curve defined parametrically by $x = x(t)$ and $y = y(t)$ is given by formula 10.3:

$$L = \int_\alpha^\beta \sqrt{\left(\frac{dx}{dt}\right)^2 + \left(\frac{dy}{dt}\right)^2}\, dt.$$

If we substitute from equations 10.12 into this formula with t replaced by θ, we obtain

$$L = \int_\alpha^\beta \sqrt{[f'(\theta)\cos\theta - f(\theta)\sin\theta]^2 + [f'(\theta)\sin\theta + f(\theta)\cos\theta]^2}\, d\theta$$

$$= \int_\alpha^\beta \sqrt{[f'(\theta)]^2 + [f(\theta)]^2}\, d\theta.$$

Thus, we may write for the length of a curve $r = f(\theta)$, $\alpha \leq \theta \leq \beta$,

$$L = \int_\alpha^\beta \sqrt{r^2 + \left(\frac{dr}{d\theta}\right)^2}\, d\theta. \tag{10.14}$$

EXAMPLE 10.12

Find the length of the cardioid $r = 1 - \cos\theta$.

SOLUTION According to equation 10.14 (see Figure 10.29),

$$L = \int_0^{2\pi} \sqrt{(1-\cos\theta)^2 + (\sin\theta)^2}\, d\theta$$

$$= \int_0^{2\pi} \sqrt{2}\sqrt{1-\cos\theta}\, d\theta$$

$$= \sqrt{2}\int_0^{2\pi} \sqrt{1 - \left[1 - 2\sin^2\left(\frac{\theta}{2}\right)\right]}\, d\theta$$

$$= 2\int_0^{2\pi} \sin\left(\frac{\theta}{2}\right) d\theta$$

$$= 2\left\{-2\cos\left(\frac{\theta}{2}\right)\right\}_0^{2\pi}$$

$$= 8.$$

FIGURE 10.29

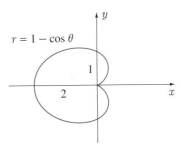

$r = 1 - \cos\theta$

EXERCISES 10.3

In Exercises 1–10 find an equation for the curve in polar coordinates, and sketch the curve.

1. $x + 2y = 5$
2. $y = -x$
3. $x^2 + y^2 = 3$
4. $x^2 - 2x + y^2 - 2y + 1 = 0$
5. $y = 4x^2$
6. $x^2 + 2y^2 = 3$
7. $x^2 + y^2 = x$
8. $x^2 + y^2 = \sqrt{x^2 + y^2} - x$
9. $(x^2 + y^2)^2 = x$
10. $y = 1/x^2$

In Exercises 11–20 find the equation of the curve in Cartesian coordinates, and sketch the curve.

11. $r = 5$
12. $\theta = 1$
13. $r = 3\sin\theta$
14. $r^2 = 4\sin 2\theta$
15. $r = 3 + 3\sin\theta$
16. $r = 2\sin 2\theta$
17. $r^2 = -4\cos 2\theta$
18. $r = 3 - 4\cos\theta$
19. $r = 5\csc\theta$
20. $r = \cot^2\theta \csc\theta$

21. Sketch the curves:

    (a) $r = 2 + 2\sin\theta$ (b) $r = 2 + 4\sin\theta$
    (c) $r = 4 + 2\sin\theta$

In Exercises 22–25 find all points of intersection for the curves.

22. $r = 2$, $r^2 = 8\cos 2\theta$
23. $r = \cos\theta$, $r = 1 + \cos\theta$
24. $r = 1 + \cos\theta$, $r = 2 - 2\cos\theta$
25. $r = 1$, $r = 2\cos 2\theta$

26. Show that if $f(\theta)$ is an even function, then the curve $r = f(\theta)$ is symmetric about the lines $\theta = 0$ and $\theta = \pi$ (or x-axis). Illustrate with two examples.

In Exercises 27–34 sketch the curve.

27. $r = \sin 3\theta$
28. $r = \cos 2\theta$
29. $r = \sin 4\theta$
30. $r^2 = \theta$
31. $r = e^\theta$
32. $r = 2\sin(\theta/2)$
33. $r = -2\cos(\theta/2)$
34. $r = 1 + \cos(\theta + \pi/6)$

In Exercises 35–38 find the slope of the curve at the given value of θ.

SECTION 10.3 CURVES IN POLAR COORDINATES

35. $r = 9\cos 2\theta$ at $\theta = \pi/6$

36. $r^2 = 9\sin 2\theta$ at $\theta = -5\pi/6$

37. $r = 3 - 5\cos\theta$ at $\theta = 3\pi/4$

38. $r = 2\cos(\theta/2)$ at $\theta = \pi/2$

39. Find the slope of the tangent line to the curve $r = 3/(1 - \sin\theta)$ at the point with polar coordinates $(6, \pi/6)$ in two ways:
    (a) by using 10.13;
    (b) by finding the equation of the curve in Cartesian coordinates, and calculating dy/dx.

40. Find the length of the cardioid $r = a(1 + \sin\theta)$. (a is a constant.)

41. (a) The electrostatic charge distribution consisting of a charge $q > 0$ at the point with polar coordinates $(s, 0)$ and a charge $-q$ at (s, π) is called a *dipole*. When s is very small, the lines of force for the dipole are defined by the equation $r = A\sin^2\theta$, where each value of the constant $A > 0$ defines a particular line of force. Sketch lines of force for $A = 1, 2,$ and 3.

    (b) The equipotential lines for the dipole are defined by $r^2 = B\cos\theta$, where B is a constant. Sketch equipotential lines for $B = \pm 1, \pm 2,$ and ± 3.

42. Sketch the *bifolium* $(x^2 + y^2)^2 = x^2 y$.

43. Curves with equations of the form $r = a(1 \pm \cos\theta)$ or $r = a(1 \pm \sin\theta)$ ($a > 0$ a constant) are called *cardioids*.
    (a) Sketch all such curves.
    (b) Find equations for the cardioids in Cartesian coordinates.

44. Curves with equations of the form $r^2 = a^2 \cos 2\theta$ or $r^2 = a^2 \sin 2\theta$ ($a > 0$ a constant) are called *lemniscates*.
    (a) Sketch all such curves.
    (b) Find equations for the lemniscates in Cartesian coordinates.

45. (a) Sketch the curves $r = b \pm a\cos\theta$ and $r = b \pm a\sin\theta$ where a and b are positive constants in the three cases $a < b$, $a = b$, and $a > b$.
    (b) Find equations for the curves in Cartesian coordinates.
    (c) Compare these curves with the cardioids of Exercise 43 when $a = b$.

46. A curve with equation of the form $r = a\sin n\theta$ or $r = a\cos n\theta$ where $a > 0$ is a constant and $n > 0$ is an integer is called a *rose*. Show that the rose has n petals.

47. Show that the roses $r = |a\sin n\theta|$ and $r = |a\cos n\theta|$, where $a > 0$ is a constant and $n > 0$ is an integer, have $2n$ petals.

SECTION 10.4

Areas in Polar Coordinates

In Section 7.1 we used definite integrals to find areas bounded by curves whose equations are conveniently expressed in Cartesian coordinates. In this section we indicate how to find areas bounded by curves whose equations are expressed in polar coordinates. We require the formula

$$\frac{1}{2}r^2(\theta_2 - \theta_1) \qquad (10.15)$$

for the area of the shaded sector of the circle in Figure 10.30. This formula results from the fact that the area of the sector is the fractional part $(\theta_2 - \theta_1)/(2\pi)$ of the area πr^2 of the circle.

FIGURE 10.30

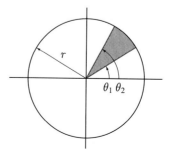

Consider finding the area in Figure 10.31(a) bounded by the radial lines $\theta = \alpha$ and $\theta = \beta$ and the curve $r = f(\theta)$. We divide the area into subareas by means of $n+1$ radial lines $\theta = \theta_i$ where

$$\alpha = \theta_0 < \theta_1 < \theta_2 < \cdots < \theta_{n-1} < \theta_n = \beta.$$

On that part of the curve $r = f(\theta)$ between $\theta = \theta_{i-1}$ and $\theta = \theta_i$, we pick any point with polar coordinates $(f(\theta_i^*), \theta_i^*)$ as in Figure 10.31(b). If between the lines $\theta = \theta_{i-1}$ and $\theta = \theta_i$ we draw the arc of a circle with centre at the pole and radius $f(\theta_i^*)$, a sector is formed with area

$$\Delta A_i = \frac{1}{2}[f(\theta_i^*)]^2 \Delta \theta_i,$$

where $\Delta \theta_i = \theta_i - \theta_{i-1}$. Since this sector approximates that part of the required area between the radial lines $\theta = \theta_{i-1}$ and $\theta = \theta_i$, we can say that an approximation to the required area is

$$\sum_{i=1}^{n} \Delta A_i = \sum_{i=1}^{n} \frac{1}{2}[f(\theta_i^*)]^2 \Delta \theta_i.$$

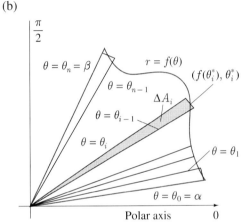

FIGURE 10.31

(a)   (b)

By increasing the number of sectors indefinitely, and at the same time requiring each of the $\Delta \theta_i$ to approach zero, we obtain a better and better approximation, and in the limit

$$\text{Area} = \lim_{||\Delta \theta_i|| \to 0} \sum_{i=1}^{n} \frac{1}{2}[f(\theta_i^*)]^2 \Delta \theta_i.$$

But this limit is the definition of the definite integral of the function $(1/2)[f(\theta)]^2$ with respect to θ from $\theta = \alpha$ to $\theta = \beta$, and we therefore write

$$\text{Area} = \int_{\alpha}^{\beta} \frac{1}{2}[f(\theta)]^2 \, d\theta. \tag{10.16}$$

In order to arrive at this integral in any given problem, without memorizing it, we use the procedure adopted for definite integrals in Cartesian coordinates discussed in Chapter 7. We draw at angle θ a representative sector of angular width $d\theta$ and radius r (Figure 10.32). The area of this sector is

$$\frac{1}{2}r^2 \, d\theta = \frac{1}{2}[f(\theta)]^2 \, d\theta.$$

If areas of all such sectors from angle α to angle β are added together, and the limit is taken as their widths approach zero, the required area is obtained. But this is the process defined by the definite integral, and we therefore write equation 10.16 for the area. Definite integral 10.16 exists when $f(\theta)$ is continuous for $\alpha \leq \theta \leq \beta$.

FIGURE 10.32

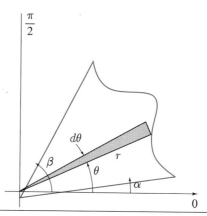

EXAMPLE 10.13

Find the area inside the cardioid $r = 1 + \sin\theta$.

SOLUTION The area of the representative sector in Figure 10.33 is

$$\frac{1}{2} r^2\, d\theta = \frac{1}{2}(1 + \sin\theta)^2\, d\theta,$$

and we must add over all sectors interior to the cardioid. Since areas on either side of the $\theta = \pi/2$ line are identical, we calculate the area to the right and double the result. To find the area to the right of the line $\theta = \pi/2$, we must identify angular positions of the first and last sectors. The first sector is at the pole, and the equation of the cardioid indicates that $r = 0$ when $\sin\theta = -1$; that is, when $\theta = -\pi/2 + 2n\pi$. But which of these values of θ shall we choose? Similarly, the last sector occurs when $r = 2$, in which case $\sin\theta = 1$, and θ could be any of the values $\pi/2 + 2n\pi$. Again, which shall we choose? If we choose $\alpha = -\pi/2$ and $\beta = \pi/2$, then all values of θ in the interval $-\pi/2 \leq \theta \leq \pi/2$ yield points on the right half of the cardioid with no duplications. Consequently,

$$\text{Area} = 2\int_{-\pi/2}^{\pi/2} \frac{1}{2}(1 + \sin\theta)^2\, d\theta = \int_{-\pi/2}^{\pi/2} (1 + 2\sin\theta + \sin^2\theta)\, d\theta$$

$$= \int_{-\pi/2}^{\pi/2} \left(1 + 2\sin\theta + \frac{1 - \cos 2\theta}{2}\right) d\theta$$

$$= \left\{\frac{3\theta}{2} - 2\cos\theta - \frac{\sin 2\theta}{4}\right\}_{-\pi/2}^{\pi/2}$$

$$= \frac{3\pi}{2}.$$

FIGURE 10.33

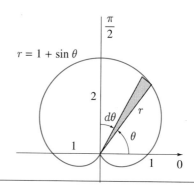

EXAMPLE 10.14

Find the area common to the circles $x^2 + y^2 = 4$ and $x^2 + y^2 = 4x$.

SOLUTION Equations for the circles in polar coordinates are $r = 2$ and $r = 4\cos\theta$, and they intersect in the points with polar coordinates $(2, \pm\pi/3)$ (Figure 10.34). If A_s is the area above the x-axis, outside $x^2 + y^2 = 4$ and inside $x^2 + y^2 = 4x$, then the area common to the circles is the area of either circle less twice A_s:

$$\text{Area} = \pi(2)^2 - 2A_s.$$

FIGURE 10.34

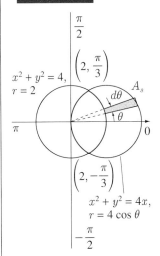

The area of the representative element is the difference in the areas of two sectors:

$$\frac{1}{2}(4\cos\theta)^2\,d\theta - \frac{1}{2}(2)^2\,d\theta = 2(4\cos^2\theta - 1)\,d\theta.$$

Since all sectors in A_s can be identified by values of θ between 0 and $\pi/3$, the required area is

$$\text{Area} = 4\pi - 2\int_0^{\pi/3} 2(4\cos^2\theta - 1)\,d\theta = 4\pi - 4\int_0^{\pi/3}[2(1 + \cos 2\theta) - 1]\,d\theta$$
$$= 4\pi - 4\{\theta + \sin 2\theta\}_0^{\pi/3} = 4\pi - 4\left(\frac{\pi}{3} + \frac{\sqrt{3}}{2}\right) = \frac{8\pi}{3} - 2\sqrt{3}.$$

EXERCISES 10.4

In Exercises 1–10 find the area enclosed by the curve.

1. $r = 3\sin\theta$
2. $r = -6\cos\theta$
3. $r = 2\sin 2\theta$
4. $r^2 = 2\sin 2\theta$
5. $r^2 = -\cos\theta$
6. $r = 2 - 2\cos\theta$
7. $r = 4 - 4\cos\theta$
8. $r = 4 - 2\cos\theta$
9. $r = \sin 3\theta$
10. $r = 2(\cos\theta + \sin\theta)$

In Exercises 11–21 find the indicated area.

11. Outside $r = 3$ but inside $r = 6\sin\theta$
12. Inside both $r = 1$ and $r = 1 - \sin\theta$
13. Inside $r = 2\sin 2\theta$ but outside $r = 1$
14. Inside both $r = 2 + 2\cos\theta$ and $r = 2 - 2\cos\theta$
15. Inside both $r = \sin\theta$ and $r = \cos\theta$
16. Inside both $r = \cos\theta$ and $r = 1 - \cos\theta$
17. Inside $r = 1 - 4\cos\theta$
18. Inside $r = 4 + 3\sin\theta$ but outside $r = 2$
19. Inside $r = |1 - 4\cos\theta|$
20. Inside the bifolium $r = \sin\theta\cos^2\theta$
21. Bounded by $\theta = 0$ and $r = \theta$, $0 \le \theta \le \pi$
22. (a) Show that in polar coordinates the strophoid

$$y^2 = x^2\frac{a-x}{a+x},$$

where $a > 0$ is a constant, takes the form $r = a\cos 2\theta \sec\theta$.

(b) Sketch the curve and find the area inside its loop.

SECTION 10.5

Conic Sections

In Section 1.4 we introduced conic sections to illustrate the algebraic-geometric interplay of plane analytic geometry. Each of the conic sections can be visualized as the curve resulting from the intersection of a plane with a pair of right-circular cones (Figure 10.35). Certainly this suggests why the conic sections are so-named, but because of the three-dimensional nature of the cone, an analysis of conic sections from this point of view is not yet possible. In this section we use plane analytic geometry to develop definitions for parabolas, ellipses, and hyperbolas.

FIGURE 10.35

(a) Circle: Plane perpendicular to axis of cones

(b) Ellipse: Plane cuts completely across one cone but not perpendicular to axis

(c) Parabola: Plane cuts only one cone but not completely across

(d) Hyperbola: Plane cuts both cones

(e) Pair of straight lines: Plane passes through vertex and cuts through both cones

(f) One straight line: Plane passes through vertex and touches both cones

The Parabola

> **Definition 10.2**
>
> A **parabola** is the curve traced out by a point that moves in a plane so that its distances from a fixed point called the *focus* and a fixed line called the *directrix* are always the same.

Suppose the focus of a parabola is the point (p, q) and the directrix is a line $y = r$ parallel to the x-axis as in Figure 10.36(a). If $P(x, y)$ is any point on the parabola, then the fact that its distance from (p, q) must be equal to its distance from $y = r$ is expressed as

$$\sqrt{(x-p)^2 + (y-q)^2} = |y - r|. \quad (10.17)$$

With the absolute values, this equation includes the case of a directrix above the focus as in Figure 10.36(b). If we square both sides of the equation and rearrange terms, we obtain

$$(x-p)^2 = (y-r)^2 - (y-q)^2 = 2y(q-r) + r^2 - q^2.$$

(a)

(b)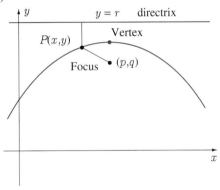

FIGURE 10.36

We can solve this equation for y in terms of x; the result is

$$y = \frac{1}{2(q-r)}[(x-p)^2 + (q^2 - r^2)]. \quad (10.18)$$

We could rewrite this equation in our accustomed form $y = ax^2 + bx + c$ for a parabola, but the present form is more informative. Firstly, the line $x = p$ through the focus and perpendicular to the directrix is the line of symmetry for the parabola. Secondly, the parabola opens upward if $q > r$, in which case the focus is above the directrix, and opens downward if $q < r$. Finally, the vertex of the parabola is found by setting $x = p$, in which case $y = (q + r)/2$, halfway between the focus and directrix.

A similar analysis shows that when the directrix is parallel to the y-axis (Figure 10.37), the equation of the parabola is of the form

$$x = \frac{1}{2(p-r)}[(y-q)^2 + (p^2 - r^2)]. \quad (10.19)$$

FIGURE 10.37

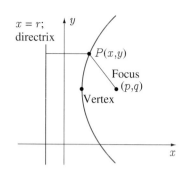

Given the focus and directrix (parallel to a coordinate axis) of a parabola, we can easily find its equation: use formulas 10.18 or 10.19, or follow the algebraic steps leading from 10.17 to 10.18. Conversely, given the equation of a parabola in the form $y = ax^2 + bx + c$ or $x = ay^2 + by + c$, we can identify its focus and directrix (see Exercises 52 and 53).

EXAMPLE 10.15 Find the equation of the parabola that has focus $(2, 4)$ and directrix $x = 6$.

SOLUTION If (x, y) is any point on the parabola (Figure 10.38), the fact that its distance from $(2, 4)$ is equal to its distance from $x = 6$ is expressed as

$$\sqrt{(x-2)^2 + (y-4)^2} = 6 - x.$$

If we square both sides and simplify, the result is $x = (16 + 8y - y^2)/8$.

FIGURE 10.38

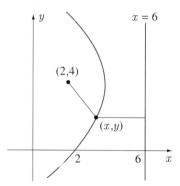

The Ellipse

Definition 10.3 An **ellipse** is the curve traced out by a point that moves in a plane so that the sum of its distances from two fixed points called *foci* remains constant.

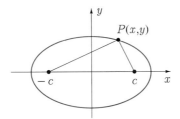

The equation of an ellipse is simplest when the foci lie on either the x-axis or the y-axis and are equidistant from the origin. Suppose the foci are $(c,0)$ and $(-c,0)$ (Figure 10.39), and the sum of the distances from these foci to a point on the ellipse is $2a$, where $a > c \geq 0$. If $P(x,y)$ is any point on the ellipse, Definition 10.3 implies that

$$\sqrt{(x+c)^2 + y^2} + \sqrt{(x-c)^2 + y^2} = 2a. \tag{10.20}$$

If we transpose the second term to the right-hand side and square both sides, we obtain

$$(x+c)^2 + y^2 = 4a^2 - 4a\sqrt{(x-c)^2 + y^2} + (x-c)^2 + y^2,$$

and this equation simplifies to

$$a^2 - cx = a\sqrt{(x-c)^2 + y^2}.$$

Squaring once again leads to

$$a^4 - 2a^2 cx + c^2 x^2 = a^2(x^2 - 2cx + c^2 + y^2) \quad \text{or}$$

$$x^2(a^2 - c^2) + a^2 y^2 = a^4 - a^2 c^2.$$

Division by $a^2(a^2 - c^2)$ gives

$$\frac{x^2}{a^2} + \frac{y^2}{a^2 - c^2} = 1. \tag{10.21}$$

It is customary to denote the y-intercepts of an ellipse by $\pm b$ ($b > 0$) (Figure 10.40), in which case $b^2 = a^2 - c^2$, and the equation of the ellipse becomes

$$\frac{x^2}{a^2} + \frac{y^2}{b^2} = 1, \quad \text{or} \quad b^2 x^2 + a^2 y^2 = a^2 b^2. \tag{10.22}$$

FIGURE 10.40

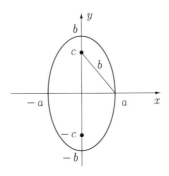

FIGURE 10.41

The line segment across the ellipse and through the foci is called the **major axis** of the ellipse; it has length $2a$ (see Figure 10.40). The midpoint of the major axis is called the **centre** of the ellipse. The line segment across the ellipse, through its centre, and perpendicular to the major axis is called the **minor axis**; it has length $2b$. Note that the line segment joining either end of the minor axis to a focus (Figure 10.40) has length a, and the triangle formed specifies the relationship among a, b, and c, namely, $a^2 = b^2 + c^2$.

A similar analysis shows that when the foci of the ellipse are on the y-axis, equidistant from the origin, the equation of the ellipse is again in form 10.22. In this case, $2b$ is the length of the major axis, $2a$ is the length of the minor axis, and $b^2 = a^2 + c^2$ (Figure 10.41).

What we should remember is that an equation of form 10.22 always specifies an ellipse. The foci are on the longer axis and can be located using $c^2 = |a^2 - b^2|$. The length of the major axis represents the sum of the distances from any point on the ellipse to the foci.

EXAMPLE 10.16

Sketch the ellipse $16x^2 + 9y^2 = 144$, indicating its foci.

FIGURE 10.42

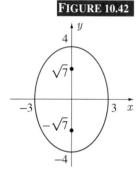

SOLUTION If we write the ellipse in the form $x^2/9 + y^2/16 = 1$, its x- and y-intercepts are ± 3 and ± 4. A sketch of the ellipse is therefore as shown in Figure 10.42. The foci must lie on the y-axis at distances $\pm c = \pm\sqrt{4^2 - 3^2} = \pm\sqrt{7}$ from the origin. ∎

If $a = b$ in equation 10.22, then $x^2 + y^2 = a^2$, and this is the equation for a circle with radius a and centre at the origin. But if $a = b$, the distance from the origin to each focus of the ellipse must be $c = 0$. In other words, a circle may be regarded as a degenerate ellipse whose foci are at one and the same point. It is also true that when c is very small compared to half the length of the major axis (Figure 10.43), the ellipse is shaped very much like a circle. On the other hand, when these lengths are almost equal (Figure 10.44), the ellipse is long and narrow.

FIGURE 10.43

FIGURE 10.44

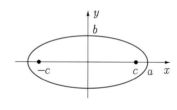

When the centre of an ellipse is at point (h, k) and the foci lie on either the line $x = h$ or $y = k$ (Figure 10.45), the equation for the ellipse is somewhat more complex that 10.22. If $2a$ and $2b$ are again the lengths of the axes of the ellipse, a calculation similar to that leading from 10.20 to 10.22 gives (see Exercise 55)

$$\frac{(x-h)^2}{a^2} + \frac{(y-k)^2}{b^2} = 1. \qquad (10.23)$$

Alternatively, the curves in Figure 10.45 are those in Figures 10.39 and 10.41 translated h units in the x-direction and k units in the y-direction. According to Section 1.5, equations for the translated ellipses can be obtained by replacing x and y in $x^2/a^2 + y^2/b^2 = 1$ by $x - h$ and $y - k$ respectively.

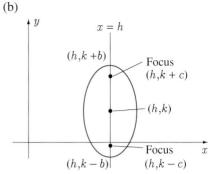

FIGURE 10.45

EXAMPLE 10.17

Sketch the ellipse $16x^2 + 25y^2 - 160x + 50y = 1175$.

SOLUTION

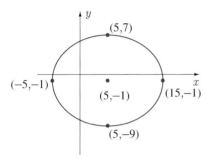

FIGURE 10.46

If we complete the squares on the x- and y-terms, we obtain

$$16(x-5)^2 + 25(y+1)^2 = 1600 \quad \text{or}$$

$$\frac{(x-5)^2}{100} + \frac{(y+1)^2}{64} = 1.$$

The centre of the ellipse is $(5, -1)$, and lengths of its major and minor axes are 20 and 16 respectively (Figure 10.46). ∎

The Hyperbola

Definition 10.4

A **hyperbola** is the path traced out by a point that moves in a plane so that the difference of its distances from two fixed points called *foci* remains constant.

Like the ellipse, the simplest hyperbolas have foci on either the x- or y-axis, equidistant from the origin. Suppose the foci are $(\pm c, 0)$ (Figure 10.47) and the difference in the distances from $P(x, y)$ to these foci is $2a$. Then Definition 10.4 implies that

$$\left| \sqrt{(x+c)^2 + y^2} - \sqrt{(x-c)^2 + y^2} \right| = 2a. \qquad (10.24)$$

This equation can be simplified by a calculation similar to that leading to 10.22; the result is

$$\frac{x^2}{a^2} - \frac{y^2}{b^2} = 1, \qquad (10.25)$$

where $b^2 = c^2 - a^2$. The hyperbola has x-intercepts equal to $\pm a$.

When the foci are on the y-axis (Figure 10.48), the equation of the hyperbola becomes

$$\frac{y^2}{b^2} - \frac{x^2}{a^2} = 1, \qquad (10.26)$$

where $2b$ is the constant difference in the distances from a point (x, y) to the foci, and $a^2 = c^2 - b^2$. This hyperbola intersects the y-axis at $\pm b$.

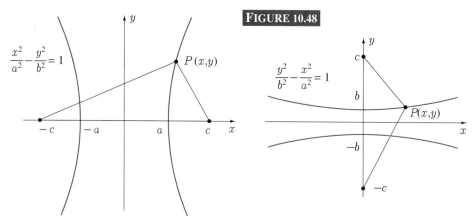

FIGURE 10.47

FIGURE 10.48

That part of the line segment joining the foci of a hyperbola which is between the two branches of the curve is called the **transverse axis** of the hyperbola; it has length $2a$ as in Figure 10.49(a) or $2b$ as in Figure 10.49(b). The midpoint of the transverse axis is called the **centre** of the hyperbola. The line segment perpendicular to the transverse axis, through its centre, and of length $2b$ as in Figure 10.49(a) or $2a$ as in Figure 10.49(b) is called the **conjugate axis**. Asymptotes of both hyperbolas are the lines $y = \pm bx/a$.

What we should remember is that an equation of form 10.25 or 10.26 specifies a hyperbola. The foci lie on the extension of the transverse axis and can be located using $c^2 = a^2 + b^2$. The length of the transverse axis represents the difference of the distances from any point on the hyperbola to the foci.

FIGURE 10.49

(a)

(b)

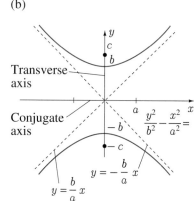

EXAMPLE 10.18

Sketch the hyperbola $16x^2 - 9y^2 = 144$, indicating its foci.

SOLUTION If we express the hyperbola in the form $x^2/9 - y^2/16 = 1$, its x-intercepts are ± 3. With asymptotes $y = \pm 4x/3$, we obtain the sketch in Figure 10.50. The foci must lie on the x-axis at distances $\pm c = \pm\sqrt{4^2 + 3^2} = \pm 5$ from the origin.

FIGURE 10.50

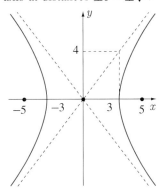

When the centre of a hyperbola is at point (h, k) and its foci are on the lines $x = h$ or $y = k$, equations 10.25 and 10.26 are modified in exactly the same way as equation 10.22 was modified for an ellipse. We replace each x by $x - h$ and each y by $y - k$ (see also Exercise 56). Consequently, equations for the hyperbolas in Figures 10.51 are

$$\frac{(x-h)^2}{a^2} - \frac{(y-k)^2}{b^2} = 1 \quad \text{and} \quad \frac{(y-k)^2}{b^2} - \frac{(x-h)^2}{a^2} = 1. \quad (10.27)$$

(a)　　　　　　　　　　　　　(b)

FIGURE 10.51

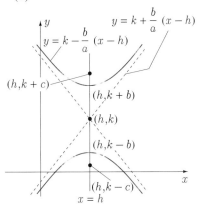

EXAMPLE 10.19

Sketch the hyperbola $x^2 - y^2 + 4x + 10y = 5$.

SOLUTION If we complete squares on the x- and y-terms, we find

$$(x+2)^2 - (y-5)^2 = -16 \quad \text{or}$$
$$\frac{(y-5)^2}{16} - \frac{(x+2)^2}{16} = 1.$$

The centre of the hyperbola is $(-2,5)$, and the length of its transverse axis (along $x = -2$) is 8. If we solve the equation for y, we obtain

$$y = 5 \pm \sqrt{(x+2)^2 + 16};$$

the asymptotes of the hyperbola are then $y = 5 \pm (x+2)$. The hyperbola can now be sketched as in Figure 10.52. Its foci are at the points $(-2, 5 + 4\sqrt{2})$ and $(-2, 5 - 4\sqrt{2})$.

FIGURE 10.52

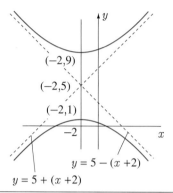

If P is a point on a conic section, the *focal radii* at P are the lines joining P to the foci (Figure 10.53). As a result, a parabola has one focal radius at each point, and an ellipse and hyperbola each have two. One of the properties of conics that makes them so useful is the fact that the normal line to the conic at any point bisects the angle between the focal radii. For the parabola, the normal bisects the angle between the focal radius and the line through P parallel to the axis of symmetry of the parabola. We will verify these facts in Exercises 59 and 60. To obtain one physical significance of these results, suppose each conic in Figure 10.53 is rotated about the x-axis to form a surface of revolution, which we regard as a mirror. It is a law of optics that when a ray of light strikes a reflecting surface, the angle between incident light and the normal to the surface is always equal to the angle between reflected light and the normal. Consequently, if a beam of light travels in the negative x-direction and strikes the parabolic mirror in Figure 10.53(a), all light is reflected toward the focus. Conversely, if F is a source of light, all light striking the mirror is reflected parallel to the x-axis. If either focus of the ellipse in Figure 10.53(b) is a light source, all light striking the elliptic mirror is reflected toward the other focus. Similarly, if either focus of the hyperbola in Figure 10.53(c) is a source, all light striking the mirror is reflected in a direction that would make it seem to originate at the other focus. Conversely, if light that is directed at one focus first strikes the mirror, it is reflected toward the other focus. This is precisely why we have parabolic reflectors in automobile headlights and searchlights, parabolic and hyperbolic reflectors in telescopes, and elliptic ceilings in whispering rooms.

FIGURE 10.53

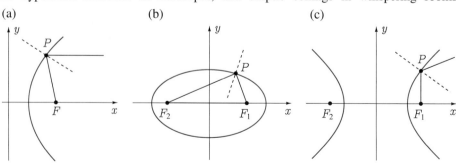

EXERCISES 10.5

In Exercises 1–14 identify the equation as representing a straight line, circle, parabola, ellipse, hyperbola, or none of these.

1. $2x + 3y = y^2$
2. $x^2 + y^2 - 3x + 2y = 25$
3. $2x - y = 3$
4. $x^2 + y^3 = 3x + 2$
5. $5x^2 = 11 - 2y^2$
6. $2x^2 - 3y^2 + 5 = 0$
7. $y^2 - x + 3y = 14 - x^2$
8. $x^2 + 2x = 3y + 4$
9. $y^2 + x^2 - 2x + 6y + 15 = 0$
10. $x^2 + 2y^2 + 24 = 0$
11. $5 + y^2 = 3x^2$
12. $x^2 + 2y^2 = 24$
13. $y^3 = 3x + 4$
14. $3 - x = 4y$

In Exercises 15–36 sketch the curve. Identify foci for each ellipse and hyperbola.

15. $y = 2x^2 - 1$
16. $\dfrac{x^2}{25} + \dfrac{y^2}{36} = 1$
17. $x^2 - \dfrac{y^2}{16} = 1$
18. $3x = 4y^2 - 1$
19. $\dfrac{y^2}{4} - \dfrac{x^2}{25} = 1$
20. $7x^2 + 3y^2 = 16$
21. $x + y^2 = 1$
22. $2y^2 + x = 3y + 5$
23. $9x^2 + 289y^2 = 2601$
24. $y^2 = 10(2 - x^2)$
25. $3x^2 - 4y^2 = 25$
26. $y^2 - x^2 = 5$
27. $y = -x^2 + 6x - 9$
28. $2x^2 - 3y^2 = 5$
29. $3x^2 + 6y^2 = 21$
30. $x^2 + 16y^2 = 2$
31. $y^2 - 3x^2 = 1$
32. $x = -(4 + y)^2$
33. $x^2 + 2x + 4y^2 - 16y + 13 = 0$
34. $x^2 - 6x - 4y^2 - 24y = 11$
35. $9x^2 + y^2 - 18x - 6y = 0$
36. $9x^2 - 16y^2 - 18x - 64y = 91$

37. Find the equation of a hyperbola that passes through the point $(1, 2)$ and has asymptotes $y = \pm 4x$.

38. Find the equation of an ellipse through the points $(-2, 4)$ and $(3, 1)$.

39. Find the width of the elliptic arch in Figure 10.54.

FIGURE 10.54

40. Find the height of the parabolic arch in Figure 10.55.

FIGURE 10.55

41. Explain how an ellipse can be drawn with a piece of string, two tacks, and a pencil.

42. Find the equation of the ellipse traced out by a point that moves so that the sum of its distances from $(\pm 4, 0)$ is always equal to 10:
    (a) by using equation 10.22 with suitable values for a and b;
    (b) by establishing and simplifying an equation similar to 10.20.

43. Find the equation of the hyperbola traced out by a point that moves so that the difference of its distances from $(0, \pm 3)$ is always equal to 1:
    (a) by using equation 10.26 with suitable values for a and b;
    (b) by establishing and simplifying an equation similar to 10.24.

44. Show that the equation of every straight line, every circle, and every conic section discussed in this section can be obtained by appropriate choices of constants A, C, D, E, and F in the equation
$$Ax^2 + Cy^2 + Dx + Ey + F = 0.$$

45. Show that the equation of the tangent line to the ellipse $b^2x^2 + a^2y^2 = a^2b^2$ at a point (x_0, y_0) is $b^2xx_0 + a^2yy_0 = a^2b^2$.

46. Show that the equation of the tangent line to the hyperbola $b^2x^2 - a^2y^2 = a^2b^2$ at a point (x_0, y_0) is $b^2xx_0 - a^2yy_0 = a^2b^2$.

47. Find the point P on that part of the ellipse $2x^2 + 3y^2 = 14$ in the first quadrant where the tangent line at P is perpendicular to the line joining P and $(2, 5)$.

48. Find the area inside the ellipse $b^2x^2 + a^2y^2 = a^2b^2$.

49. Among all rectangles that can be inscribed inside the ellipse $b^2x^2 + a^2y^2 = a^2b^2$ and have sides parallel to the axes, find the one with largest possible area.

50. A *prolate spheroid* is the solid of revolution obtained by rotating an ellipse about its major axis. An *oblate spheroid* is obtained by rotating the ellipse about its minor axis. Find volumes for the prolate and oblate spheroids generated by the ellipse $b^2x^2 + a^2y^2 = a^2b^2$ if $a > b$.

51. A sharp noise originating at one focus F_1 of an ellipse is reflected by the ellipse toward the other focus F_2. Explain why all reflected noise arrives at F_2 at exactly the same time.

52. Use equation 10.18 to show that when a parabola is written in the form $y = ax^2 + bx + c$, the following formulas identify its focus (p, q) and directrix $y = r$:

$$p = -\frac{b}{2a}, \quad q = \frac{1}{4a}(1 + 4ac - b^2),$$
$$r = \frac{1}{4a}(-1 + 4ac - b^2).$$

53. What are formulas for the focus (p, q) and directrix $x = r$ for a parabola of the type $x = ay^2 + by + c$?

54. Use the formulas in Exercises 52 and 53 to identify the focus and directrix for any parabolas in Exercises 15–36.

55. Show that when the centre of an ellipse is at point (h, k) and its foci are on the line $x = h$ or $y = k$, Definition 10.3 leads to equation 10.23.

56. Show that when the centre of a hyperbola is at point (h, k) and its foci are on the line $x = h$ or $y = k$, Definition 10.4 leads to equation 10.27.

57. Prove that the normal line to the parabola $x = ay^2$ in Figure 10.56 bisects the angle between the focal radius FP and the line PQ parallel to the x-axis. Hint: Draw the tangent line PR at P and show that $||PF|| = ||RF||$.

FIGURE 10.56

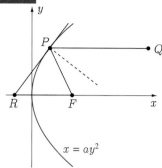

58. When a beam of light travelling in the negative x-direction (Figure 10.57) strikes a parabolic mirror with cross-section represented by $x = ay^2 + c$, all light rays are reflected to the focus F of the mirror. Show that all photons that pass simultaneously through $x = d$ arrive at F at the same time.

FIGURE 10.57

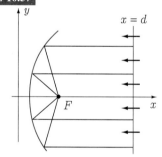

59. Prove that the normal line to an ellipse or hyperbola bisects the angle between the focal radii.

60. Prove that the normal line to the parabola $y = ax^2 + bx + c$ at any point P bisects the angle between the focal radius and the line through P parallel to the y-axis.

61. A line segment through the focus of a parabola with ends on the parabola is called a *focal chord*. It is an established result that tangents to a parabola at the ends of a focal chord are perpendicular to each other and intersect on the directrix. Prove this for the parabola $y = ax^2$.

SECTION 10.6

Conic Sections in Polar Coordinates

In Section 10.5 we defined parabolas using a focus and directrix and ellipses and hyperbolas using two foci. In this section we show that all three conics can be defined using a focus and directrix, and that in polar coordinates one equation represents all three conics.

Let F be a fixed point (the *focus*), and l be a fixed line (the *directrix*) that does not pass through F as in Figure 10.58(a). We propose to find the equation of the curve traced out by a point that moves so that its undirected distances from F and l always remain in a constant ratio ϵ called the *eccentricity*. To do this we set up polar coordinates with F as pole and polar axis directed away from l and perpendicular to l as in Figure 10.58(b).

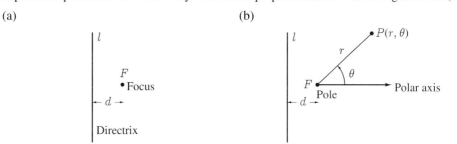

FIGURE 10.58

If (r, θ) are polar coordinates for any point P on the required curve, the fact that the ratio of the distances from F and l to P is equal to ϵ is expressed as

$$\frac{r}{d + r\cos\theta} = \epsilon. \qquad (10.28)$$

When this equation is solved for r, we have

$$r = \frac{\epsilon d}{1 - \epsilon \cos\theta}, \qquad (10.29)$$

and this is the polar equation of the curve traced out by the point. Certainly this curve should be a parabola when $\epsilon = 1$. We now verify this, and show that the curve is an ellipse when $\epsilon < 1$ and a hyperbola when $\epsilon > 1$. To do this we transform the equation into the usual Cartesian coordinates $x = r\cos\theta$ and $y = r\sin\theta$. From equations 10.5 and 10.6, we obtain

$$\sqrt{x^2 + y^2} = \frac{\epsilon d}{1 - \dfrac{\epsilon x}{\sqrt{x^2 + y^2}}},$$

and this equation simplifies to

$$\sqrt{x^2 + y^2} = \epsilon(d + x), \qquad (10.30\text{a})$$

or when squared,

$$x^2 + y^2 = \epsilon^2 (d + x)^2. \qquad (10.30\text{b})$$

When $\epsilon = 1$, the x^2-terms in 10.30b cancel, and the equation reduces to that for a parabola:

$$x = \frac{1}{2d}(y^2 - d^2), \qquad (10.31)$$

as in Figure 10.59(a). When $\epsilon \neq 1$, we write

$$x^2 + y^2 = \epsilon^2(d^2 + 2dx + x^2) \qquad \text{or}$$

$$(1 - \epsilon^2)x^2 - 2d\epsilon^2 x + y^2 = \epsilon^2 d^2.$$

When $\epsilon < 1$, we divide by $1 - \epsilon^2$,

$$x^2 - \frac{2d\epsilon^2}{1 - \epsilon^2} x + \frac{y^2}{1 - \epsilon^2} = \frac{\epsilon^2 d^2}{1 - \epsilon^2},$$

and complete the square on the x-terms,

$$\left(x - \frac{d\epsilon^2}{1 - \epsilon^2} \right)^2 + \frac{y^2}{1 - \epsilon^2} = \frac{\epsilon^2 d^2}{1 - \epsilon^2} + \frac{d^2 \epsilon^4}{(1 - \epsilon^2)^2} = \left(\frac{\epsilon d}{1 - \epsilon^2} \right)^2. \quad (10.32)$$

Comparing equation 10.32 with 10.23, we conclude that 10.32 is the equation for an ellipse with centre at position $x = d\epsilon^2/(1 - \epsilon^2)$ on the x-axis. Since one focus is at the origin, it follows that the other must also be on the x-axis at $x = 2d\epsilon^2/(1 - \epsilon^2)$ as in Figure 10.59(b).

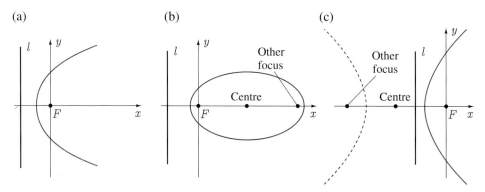

FIGURE 10.59 (a) (b) (c)

A similar calculation shows that when $\epsilon > 1$, points on the right half of the hyperbola

$$\left(x + \frac{d\epsilon^2}{\epsilon^2 - 1} \right)^2 - \frac{y^2}{\epsilon^2 - 1} = \left(\frac{\epsilon d}{\epsilon^2 - 1} \right)^2 \quad (10.33)$$

are obtained as in Figure 10.59(c). Equation 10.30a is not satisfied by points with x-coordinates less than $-d$, and therefore points on the left half of the hyperbola, shown dotted, do not satisfy 10.29.

We have shown that equation 10.29 defines an ellipse when $0 < \epsilon < 1$, a parabola when $\epsilon = 1$, and a hyperbola when $\epsilon > 1$, and this provides a unifying approach to conic sections. All three conics can be studied using a focus and a directrix.

It is clear that equation 10.29 can only yield a parabola that opens to the right and has its focus on the x-axis, and an ellipse and hyperbola with foci on the x-axis, one at the origin. To obtain parabolas that open to the left, or up, or down, and ellipses and hyperbolas with foci on the y-axis, we must change the position of the directrix. The conic sections in Figures 10.60a-c have a directrix to the right of the focus. They have equations

$$r = \frac{\epsilon d}{1 + \epsilon \cos \theta}. \quad (10.34\text{a})$$

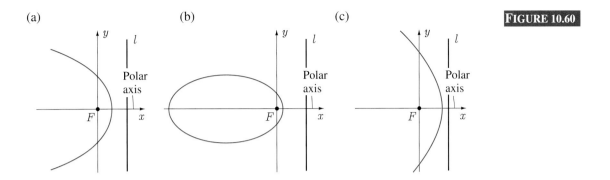

FIGURE 10.60

The conic sections in Figures 10.61 and 10.62 have equations of the form respectively,

$$r = \frac{\epsilon d}{1 - \epsilon \sin \theta} \quad \text{and} \quad r = \frac{\epsilon d}{1 + \epsilon \sin \theta}. \qquad (10.34\text{b})$$

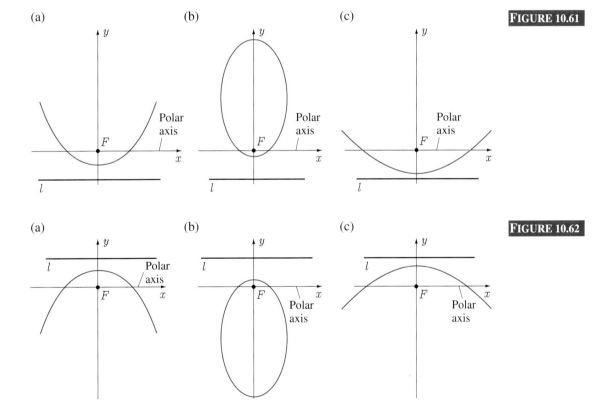

FIGURE 10.61

FIGURE 10.62

In Figures 10.59-62, one focus of the conic is chosen as the pole. In other words, the simplicity of equations 10.29 and 10.34 to describe conic sections is a direct consequence of the fact that the pole is at a focus, and the directrix is either parallel or perpendicular to the polar axis.

SECTION 10.6 CONIC SECTIONS IN POLAR COORDINATES

EXAMPLE 10.20

Sketch the curve $r = 15/(3 + 2\cos\theta)$.

SOLUTION If we write the equation in form 10.34a,

$$r = \frac{5}{1 + \left(\frac{2}{3}\right)\cos\theta},$$

the eccentricity $\epsilon = 2/3$ indicates that the curve is an ellipse. Both foci lie on the x-axis, and one is at the origin. The ends of the major axis occur when $\theta = 0$ and $\theta = \pi$, and for these values $r = 3$ and $r = 15$ (Figure 10.63). It now follows that the centre of the ellipse is at $x = -6$, and its other focus is at $x = -12$. If $b > 0$ denotes half the length of the minor axis (it is also the maximum y-value on the ellipse, occuring when $x = -6$), then $b^2 = a^2 - c^2 = 9^2 - 6^2 = 45$. Consequently, $b = 3\sqrt{5}$, and the ellipse is as shown in Figure 10.63. This information now permits us to write the equation of the ellipse in Cartesian coordinates:

$$\frac{(x+6)^2}{81} + \frac{y^2}{45} = 1.$$

FIGURE 10.63

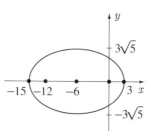

An alternative approach in Example 10.20 would be to substitute $r = \sqrt{x^2 + y^2}$ and $\cos\theta = x/\sqrt{x^2 + y^2}$ into the polar equation for the conic, simplify the equation to $(x+6)^2/81 + y^2/45 = 1$, and then sketch the ellipse from this equation. We illustrate this method in the following example.

EXAMPLE 10.21

Sketch the curve $r = 2/(3 - 4\sin\theta)$.

SOLUTION If we set $r = \sqrt{x^2 + y^2}$ and $\sin\theta = y/\sqrt{x^2 + y^2}$, then

$$\sqrt{x^2 + y^2} = \frac{2}{3 - \dfrac{4y}{\sqrt{x^2 + y^2}}} = \frac{2\sqrt{x^2 + y^2}}{3\sqrt{x^2 + y^2} - 4y}.$$

Division by $\sqrt{x^2 + y^2}$ leads to

$$3\sqrt{x^2 + y^2} - 4y = 2 \quad\text{or}\quad$$
$$3\sqrt{x^2 + y^2} = 4y + 2.$$

If we now square both sides, we obtain

$$9(x^2 + y^2) = 4 + 16y + 16y^2 \quad \text{or}$$

$$9x^2 - 7y^2 - 16y = 4.$$

If we complete the square on the y-terms, we obtain

$$9x^2 - 7\left(y + \frac{8}{7}\right)^2 = 4 - \frac{64}{7} = -\frac{36}{7}.$$

Division by $-36/7$ yields the equation

$$\frac{\left(y + \frac{8}{7}\right)^2}{\frac{36}{49}} - \frac{x^2}{\frac{4}{7}} = 1.$$

This equation describes a hyperbola with centre $(0, -8/7)$ and y-intercepts equal to $-8/7 \pm 6/7 = -2/7, -2$. Its asymptotes are

$$y = -\frac{8}{7} \pm \frac{6}{7}\sqrt{\frac{7x^2}{4}} = -\frac{8}{7} \pm \frac{3x}{\sqrt{7}}.$$

The hyperbola is shown in Figure 10.64, but only the top half is described by $r = 2/(3 - 4\sin\theta)$. The equation $3\sqrt{x^2 + y^2} = 4y + 2$ does not permit $y \leq -2$.

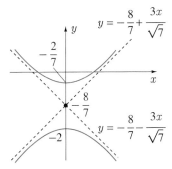

FIGURE 10.64

EXAMPLE 10.22

Find a polar representation for the ellipse

$$\frac{(x-1)^2}{4} + \frac{y^2}{9} = 1.$$

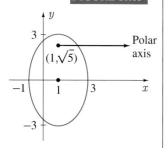

FIGURE 10.65

SOLUTION To find a polar representation for the ellipse (Figure 10.65), we could use the usual polar coordinates defined by $x = r\cos\theta$ and $y = r\sin\theta$, but the resulting equation would not be simple. Try it. We know that a simple polar representation must result if the pole is chosen at a focus of the ellipse and polar axis either parallel or perpendicular to the directrix. For this ellipse, foci are on the line $x = 1$ at distances $\pm c = \pm\sqrt{9-4} = \pm\sqrt{5}$ from the x-axis. Let us choose the pole at position $(1, \sqrt{5})$ and the polar axis parallel to the x-axis, and therefore parallel to the directrix. According to equation 10.9, polar and Cartesian coordinates are related by $x = 1 + r\cos\theta$ and $y = \sqrt{5} + r\sin\theta$. If we substitute these into the equation for the ellipse, we obtain

$$\frac{r^2 \cos^2\theta}{4} + \frac{(\sqrt{5} + r\sin\theta)^2}{9} = 1,$$

or

$$9r^2 \cos^2\theta + 4(5 + 2\sqrt{5}r\sin\theta + r^2\sin^2\theta) = 36.$$

This equation can be expressed as a quadratic equation in r:

$$r^2(9\cos^2\theta + 4\sin^2\theta) + r(8\sqrt{5}\sin\theta) - 16 = 0.$$

Solutions for r are

$$r = \frac{-8\sqrt{5}\sin\theta \pm \sqrt{320\sin^2\theta + 64(9\cos^2\theta + 4\sin^2\theta)}}{2(9\cos^2\theta + 4\sin^2\theta)}$$

$$= \frac{-8\sqrt{5}\sin\theta \pm \sqrt{576(\cos^2\theta + \sin^2\theta)}}{2(9\cos^2\theta + 4\sin^2\theta)}$$

$$= \frac{4(\pm 3 - \sqrt{5}\sin\theta)}{9\cos^2\theta + 4\sin^2\theta}.$$

Since r must be nonnegative, we must choose $+3$, and not -3, and therefore

$$r = \frac{4(3 - \sqrt{5}\sin\theta)}{9(1 - \sin^2\theta) + 4\sin^2\theta} = \frac{4(3 - \sqrt{5}\sin\theta)}{9 - 5\sin^2\theta}$$

$$= \frac{4(3 - \sqrt{5}\sin\theta)}{(3 - \sqrt{5}\sin\theta)(3 + \sqrt{5}\sin\theta)} = \frac{4}{3 + \sqrt{5}\sin\theta}.$$

EXERCISES 10.6

In Exercises 1–10 sketch the conic section.

1. $r = \dfrac{3}{1 + \cos\theta}$

2. $r = \dfrac{16}{3 + 5\cos\theta}$

3. $r = \dfrac{4}{3 - 3\sin\theta}$

4. $r = \dfrac{16}{5 + 3\cos\theta}$

5. $r = \dfrac{4}{3 - 4\sin\theta}$

6. $r = \dfrac{4}{4 - 3\sin\theta}$

7. $r = \dfrac{1}{2 - 2\cos\theta}$

8. $r = \dfrac{1}{2 + \sin\theta}$

9. $r = \dfrac{\sec\theta}{3 + 6\sec\theta}$

10. $r = \dfrac{4\csc\theta}{7\csc\theta - 2}$

In Exercises 11–16 find the Cartesian equation for the curve.

11. $r = \dfrac{3}{1 - \sin\theta}$

12. $r = \dfrac{1}{3 + \cos\theta}$

13. $r = \dfrac{1}{1 + 2\cos\theta}$ 14. $r = \dfrac{2}{1 - 3\cos\theta}$

15. $r = \dfrac{4}{6 - 3\sin\theta}$ 16. $r = \dfrac{4}{5 + 5\cos\theta}$

In Exercises 17–20 find a polar equation in one of the four forms of 10.29 or 10.34 for the conic.

17. $x^2 - 16y^2 = 16$ 18. $4x^2 + 9y^2 = 36$

19. $(x - 1)^2 + \dfrac{(y + 1)^2}{4} = 1$

20. $x^2 - 9(y - 2)^2 = 9$

21. (a) Show that the eccentricity ϵ of an ellipse or hyperbola is always equal to the distance from its centre to either focus, divided by half the length of the major or transverse axis.

    (b) Discuss the eccentricities of the ellipses in Figures 10.43 and 10.44.

22. A circle has been described as a degenerate ellipse in the sense that its foci coincide. What happens to the eccentricity of an ellipse if the distance between its foci approaches zero? Hint: See Exercise 21.

SUMMARY

In Chapter 1 we defined curves explicitly and implicitly. In this chapter we added a third description — the parametric definition

$$x = x(t), \quad y = y(t), \quad \alpha \leq t \leq \beta.$$

When such a curve defines a function, the derivative of the function can be calculated using the parametric rule:

$$\frac{dy}{dx} = \frac{\dfrac{dy}{dt}}{\dfrac{dx}{dt}};$$

second and higher-order derivatives can be calculated using the chain rule. The length of such a curve is defined by the definite integral

$$\int_\alpha^\beta \sqrt{\left(\frac{dx}{dt}\right)^2 + \left(\frac{dy}{dt}\right)^2}\, dt.$$

Polar coordinates provide an alternative way to identify the positions of points in a plane. They use distance from a point, called the pole, and rotation of a half line, called the polar axis. Many curves that have complex equations in Cartesian coordinates can be represented very simply using polar coordinates. Particularly simple are multileaved roses, cardioids, lemniscates, and some circles. Polar coordinates also provide a unified approach to parabolas, ellipses, and hyperbolas. Each curve can be described as the path traced out by a point that moves so that the ratio ϵ of its distances from a fixed point and a fixed line remains constant. The curve is an ellipse, a parabola, or a hyperbola depending on whether $0 < \epsilon < 1$, $\epsilon = 1$, or $\epsilon > 1$ respectively. Ellipses and hyperbolas can also be defined as curves traced out by a point that moves so that the sum and difference of its distances from two fixed points remains constant.

When a curve $r = f(\theta)$ in polar coordinates encloses a region R, the definite integral

$$\int_\alpha^\beta \frac{1}{2}[f(\theta)]^2\, d\theta$$

with appropriate choices of α and β can be used to find the area of R.

Parametric equations for a polar curve $r = f(\theta)$, $\alpha \leq \theta \leq \beta$ are

$$x(\theta) = f(\theta)\cos\theta, \quad y(\theta) = f(\theta)\sin\theta, \quad \alpha \leq \theta \leq \beta.$$

Its slope is given by

$$\frac{dy}{dx} = \frac{f'(\theta)\sin\theta + f(\theta)\cos\theta}{f'(\theta)\cos\theta - f(\theta)\sin\theta},$$

and its length can be calculated with the definite integral

$$\int_\alpha^\beta \sqrt{r^2 + \left(\frac{dr}{d\theta}\right)^2}\, d\theta.$$

Key Terms and Formulas

In reviewing this chapter, you should be able to define or discuss the following key terms:

Parametric equations
Parametric rule
Length of parametric curve
Pole
Polar axis
Polar coordinates
Slope of polar curve
Length of polar curve
Area in polar coordinates
Parabola

Focus
Directrix
Ellipse
Major axis
Minor axis
Hyperbola
Transverse axis
Conjugate axis
Eccentricity

REVIEW EXERCISES

In Exercises 1–30 sketch the curve.

1. $r = \cos\theta$
2. $r = -\sin\theta$
3. $r = \dfrac{3}{1 + 2\sin\theta}$
4. $r = \dfrac{3}{2 + \sin\theta}$
5. $r = \dfrac{1}{1 + \sin\theta}$
6. $r = \dfrac{3}{2 - 2\cos\theta}$
7. $r + 1 = 2\sin\theta$
8. $r^2 = 4\cos\theta$
9. $x = 2\cos t,\ y = 3\sin t,\ 0 \le t < 2\pi$
10. $x = 4 + t,\ y = 5 - 3t^2$
11. $x = \sin^2 t,\ y = \cos^2 t,\ 0 \le t \le 2\pi$
12. $x = 1 + 4\sin t,\ y = -2 + 4\cos t,\ 0 \le t \le \pi$
13. $(x^2 + y^2)^3 = x$
14. $x^2 + y^2 = 2\sqrt{x^2 + y^2}$
15. $r^2 = 4\cos^2\theta$
16. $r = 3\cos 2\theta$
17. $x + y = \sqrt{x^2 + y^2}$
18. $x^2 + y^2 = x + y$
19. $r = 4\cos 3\theta$
20. $r = \sin^2\theta - \cos^2\theta$
21. $x^2 + y^2 - 3x + 2y = 1$
22. $x^2 - 3y^2 + 4 = 0$
23. $2x^2 + 3y^2 - 6y = 0$
24. $y^2 + x + 2y = 3$
25. $r(2\cos\theta - \sin\theta) = 3$
26. $r = \cos(\theta/2)$
27. $r = \sin^2\theta$
28. $x^2 = y^2 + y$
29. $x = e^{-t},\ y = \ln t,\ t > 0$
30. $x = \sin\theta,\ y = \sin 2\theta$

In Exercises 31–35 find the indicated area.

31. Inside $r = 2 + 2\cos\theta$
32. Inside $r = 4$ but outside $r = 4\sin 2\theta$
33. Common to $r = 2$ and $r^2 = 9\cos 2\theta$
34. Inside $r = \sin^2\theta$
35. Common to $r = 1 + \sin\theta$ and $r = 2 - 2\sin\theta$

In Exercises 36 and 37 assume that y is defined as a function of x, and find dy/dx and d^2y/dx^2.

36. $x = t^3 + 2t,\ y = 3t - t^3$
37. $x = 2\sin u,\ y = 3\cos u$

In Exercises 38–40 find the length of the curve.

38. $r = 2 + 2\cos\theta$
39. $x = t^2,\ y = t^3,\ 0 \le t \le 1$
40. $x = e^t\cos t,\ y = e^t\sin t,\ 0 \le t \le \pi/2$

41. Find the equation of the tangent line to the curve $r = 2 - 2\sin\theta$ at the point with polar coordinates $(1, \pi/6)$.

CHAPTER 11

Infinite Sequences and Series

Sequences and series play an important role in many areas of applied mathematics. Sequences were first encountered in Chapter 4, although we did not use the term "sequences" at the time. In Section 4.1, Newton's iterative procedure was used to develop a set of numbers $x_1, x_2, x_3, \ldots$ to approximate a root of an equation $f(x) = 0$. The first number is chosen as some initial approximation to the solution of the equation, and subsequent numbers, defined by the formula

$$x_{n+1} = x_n - \frac{f(x_n)}{f'(x_n)},$$

are better and better approximations. This ordered set of numbers is called a sequence. Each number in the set corresponds to a positive integer, and each is calculated according to a stated formula.

A series is the sum of the numbers in a sequence. If the numbers are $x_1, x_2, x_3, \ldots$, the corresponding series is denoted symbolically by

$$\sum_{n=1}^{\infty} x_n = x_1 + x_2 + x_3 + \cdots,$$

where the three dots indicate that the addition is neverending. It is very well to write an expression like this, but it does not seem to have any meaning. No matter how fast we add, or how fast a calculator adds, or even how fast a supercomputer adds, an infinity of numbers can never be added together in a finite amount of time. We shall give meaning to such expressions, and show that they are really the only sensible way to define many of the more common transcendental functions such as trigonometric, exponential, and hyperbolic.

The first two sections of this chapter are devoted to sequences and the remaining ten to series. This is not to say that series are more important than sequences; they are not. Discussions on series invariably become discussions on sequences associated with series. We have found that difficulties with this chapter can almost always be traced back to a failure to distinguish between the two concepts. Special attention to the material in Sections 11.1 and 11.2 will be rewarded, while a cursory treatment of these sections leads to confusion in later sections.

SECTION 11.1

Infinite Sequences of Constants

Sequences are defined as follows.

Definition 11.1

An **infinite sequence of constants** is a function $f(n)$ whose domain is the set of positive integers.

For example, when $f(n) = 1/n$, the following numbers are associated with the positive integers,

$$1, \frac{1}{2}, \frac{1}{3}, \frac{1}{4}, \ldots .$$

The word "infinite" simply indicates that an infinity of numbers is defined by the sequence as there is an infinity of positive integers, but it indicates nothing about the nature of the numbers. Often we write the numbers $f(n)$ in a line separated by commas,

$$f(1), f(2), \ldots, f(n), \ldots \quad (11.1\text{a})$$

and refer to this array as the sequence rather than the rule by which it is formed. Since this notation is somewhat cumbersome, we adopt a notation similar to that used for the sequence defined by Newton's iterative procedure. We set $c_1 = f(1)$, $c_2 = f(2)$, ..., $c_n = f(n)$, ..., and write for 11.1a

$$c_1, c_2, c_3, \ldots, c_n, \ldots . \quad (11.1\text{b})$$

The first number c_1 is called the first **term** of the sequence, c_2 the second term, and for general n, c_n is called the nth term (or general term) of the sequence. For the above example, we have

$$c_1 = 1, \quad c_2 = \frac{1}{2}, \quad c_3 = \frac{1}{3}, \quad \text{etc.}$$

In some applications, it is more convenient to define a sequence as a function whose domain is the set of all integers larger than or equal to some fixed integer N, and N can be positive, negative, or zero. Indeed, in Section 11.8, we find it convenient to initiate the assignment with $N = 0$. For now we prefer to use Definition 11.1 where $N = 1$, in which case we have the natural situation where the first term of the sequence corresponds to $n = 1$, the second term to $n = 2$, etc.

EXAMPLE 11.1

The general terms of four sequences are:

(a) $\dfrac{1}{2^{n-1}}$ (b) $\dfrac{n}{n+1}$ (c) $(-1)^n |n-3|$ (d) $(-1)^{n+1}$

Write out the first six terms of each sequence.

SOLUTION The first six terms of these sequences are:

(a) $1, \dfrac{1}{2}, \dfrac{1}{4}, \dfrac{1}{8}, \dfrac{1}{16}, \dfrac{1}{32}$;

(b) $\dfrac{1}{2}, \dfrac{2}{3}, \dfrac{3}{4}, \dfrac{4}{5}, \dfrac{5}{6}, \dfrac{6}{7}$;

(c) $-2, 1, 0, 1, -2, 3$;

(d) $1, -1, 1, -1, 1, -1$.

The sequences in Example 11.1 are said to be defined **explicitly**; we have an explicit formula for the nth term of the sequence in terms of n. This allows easy determination of any term in the sequence. For instance, to find the one-hundredth term, we simply replace n by 100 and perform the resulting arithmetic. Contrast this with the sequence in the following example.

EXAMPLE 11.2

The first term of a sequence is $c_1 = 1$ and every other term is to be obtained from the formula
$$c_{n+1} = 5 + \sqrt{2 + c_n}, \quad n \geq 1.$$
Calculate c_2, c_3, c_4, and c_5.

SOLUTION To obtain c_2 we set $n = 1$ in the formula,
$$c_{1+1} = c_2 = 5 + \sqrt{2 + c_1} = 5 + \sqrt{2 + 1} = 5 + \sqrt{3} = 6.732.$$
To find c_3, we set $n = 2$,
$$c_3 = 5 + \sqrt{2 + c_2} = 5 + \sqrt{2 + \left(5 + \sqrt{3}\right)} = 5 + \sqrt{7 + \sqrt{3}} = 7.955.$$
Similarly,
$$c_4 = 5 + \sqrt{2 + c_3} = 5 + \sqrt{7 + \sqrt{7 + \sqrt{3}}} = 8.155, \quad \text{and}$$
$$c_5 = 5 + \sqrt{2 + c_4} = 5 + \sqrt{7 + \sqrt{7 + \sqrt{7 + \sqrt{3}}}} = 8.187.$$

When the terms of a sequence are defined by a formula such as the one in Example 11.2, the sequence is said to be defined **recursively**. Note that the terms for a sequence obtained from Newton's iterative procedure are so defined. To find the 100th term of a recursively defined sequence, we must know the 99th; to find the 99th, we must know the 98th; to find the 98th, we need the 97th; and so on down the line. In other words, to find a term in the sequence, we must first find every term that precedes it. Obviously it is much more convenient to have an explicit definition for c_n in terms of n, but this is not always possible. It can be very difficult to find an explicit formula for the nth term of a sequence which is defined recursively.

Sometimes it is impossible to give any algebraic formula for the terms of a sequence. This is illustrated in the following example.

EXAMPLE 11.3

The nth term of a sequence is the nth prime integer (greater than 1) when all such primes are listed in ascending order. List its first ten terms.

SOLUTION The first ten terms of the sequence are
$$2, 3, 5, 7, 11, 13, 17, 19, 23, 29.$$

No one has yet discovered a formula for the prime integers, and it is therefore impossible to express terms of the sequence explicitly or recursively. In spite of this, the sequence is well-defined; and with enough perseverence, and perhaps a high-powered computer, we could find any number of terms in this sequence. ∎

When the general term of a sequence is known explicitly, any term in the sequence is obtained by substituting the appropriate value of n. In other words, the general term specifies every term in the sequence. We therefore use the general term to abbreviate the notation for a sequence by writing the general term in braces and using this to represent the sequence. Specifically, for the sequence in Example 11.1(a), we write

$$\left\{\frac{1}{2^{n-1}}\right\}_1^\infty = 1, \frac{1}{2}, \frac{1}{4}, \frac{1}{8}, \ldots, \frac{1}{2^{n-1}}, \ldots,$$

where 1 and ∞ indicate that the first term corresponds to the integer $n = 1$, and that there is an infinite number of terms in the sequence. In general, we write

$$\{c_n\}_1^\infty = c_1, c_2, c_3, \ldots, c_n, \ldots . \qquad (11.2)$$

If, as is the case in this section, the first term of a sequence corresponds to the integer $n = 1$, we abbreviate the notation further and simply write $\{c_n\}$ in place of $\{c_n\}_1^\infty$.

Since a sequence $\{c_n\}$ is a function whose domain is the set of positive integers, we can represent $\{c_n\}$ graphically. The sequences of Example 11.1 are shown in Figure 11.1.

FIGURE 11.1

(a)

(b)

(c)

(d)
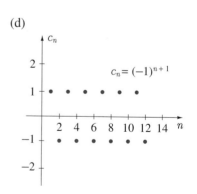

In most applications of sequences we are interested in a number called the **limit** of the sequence. Intuitively, a number L is called the limit of a sequence $\{c_n\}$ if, as we go farther and farther out in the sequence, the terms get arbitrarily close to L, and stay close to L. If such a number L exists, we write

$$L = \lim_{n \to \infty} c_n, \qquad (11.3)$$

and say that the sequence $\{c_n\}$ **converges** to L. If no such number exists, we say that the sequence does not have a limit, or that the sequence **diverges**.

For the sequences of Example 11.1, it is evident that:

(a) $\lim_{n \to \infty} \dfrac{1}{2^{n-1}} = 0$;

(b) $\lim_{n \to \infty} \dfrac{n}{n+1} = 1$;

(c) $\lim_{n \to \infty} (-1)^n |n-3|$ does not exist;

(d) $\lim_{n \to \infty} (-1)^{n+1}$ does not exist;

and in Example 11.3,

$$\lim_{n \to \infty} c_n \text{ does not exist.}$$

Note how the points on the graphs in Figures 11.1(a) and (b) cluster around the limits 0 and 1 as n gets larger and larger. No such clustering occurs in the remaining two figures.

It is usually, but not always, easy to determine whether an explicitly defined sequence has a limit, and what that limit is. It is like finding the limit of a function $f(x)$ as $x \to \infty$. Seldom is this the case, however, for recursive sequences. For instance, it is not at all clear whether the recursive sequence of Example 11.2 has a limit. In spite of the fact that differences between successive terms are approaching zero, and terms of the sequence are therefore getting closer together, the sequence might not have a limit. This is explored further in Exercise 70 at the end of this section. Likewise for recursive sequences defined by Newton's iterative procedure, we pointed out in Section 4.1 that sometimes the sequence does not converge, but gave no conditions that would ensure that it does. What we need, then, are criteria by which to determine whether a given sequence has a limit. The following two definitions lead to Theorem 11.1, which can be very useful in determining whether a recursive sequence has a limit.

Definition 11.2

A sequence $\{c_n\}$ is said to be

(i) **increasing** if $c_{n+1} > c_n$ for all $n \geq 1$; (11.4a)

(ii) **nondecreasing** if $c_{n+1} \geq c_n$ for all $n \geq 1$; (11.4b)

(iii) **decreasing** if $c_{n+1} < c_n$ for all $n \geq 1$; (11.4c)

(iv) **nonincreasing** if $c_{n+1} \leq c_n$ for all $n \geq 1$. (11.4d)

If a sequence satisfies any one of these four properties, it is said to be **monotonic**.

Definition 11.3

A sequence $\{c_n\}$ is said to have an **upper bound** U (be bounded above by U) if

$$c_n \leq U \quad (11.5)$$

for all $n \geq 1$. It has a **lower bound** V (is bounded below by V) if

$$c_n \geq V \quad (11.6)$$

for all $n \geq 1$.

Note that if U is an upper bound for a sequence, then any number greater than U is also an upper bound. If V is a lower bound, so too is any number smaller than V.

For the sequences of Example 11.1, we find that $\{1/2^{n-1}\}$ is decreasing and has an upper bound $U = 1$ and a lower bound $V = 0$; $\{n/(n+1)\}$ is increasing and has an upper bound $U = 5$ and a lower bound $V = -2$; $\{(-1)^n|n-3|\}$ is not monotonic and has neither an upper nor a lower bound; and $\{(-1)^{n+1}\}$ is not monotonic and has an upper bound $U = 1$ and a lower bound $V = -3$.

The sequence of Example 11.3 is increasing and has a lower bound $V = 2$, but has no upper bound (see Exercise 81).

The recursive sequence of Example 11.2 will be discussed in detail in Example 11.4. For a complete discussion of this sequence, however, we require the following theorem.

Theorem 11.1 *A bounded, monotonic sequence has a limit.*

To expand on this statement somewhat, consider a sequence $\{c_n\}$ whose terms are illustrated graphically in Figure 11.2(a). Suppose that the sequence is increasing and therefore monotonic, and that U is an upper bound for the sequence. We have shown the upper bound as a horizontal line in the figure; c_1 is a lower bound. Our intuition suggests that because the terms in the sequence always increase, and they never exceed U, the sequence must have a limit. Theorem 11.1 confirms this. The theorem does not suggest the value of the limit, but obviously it must be less than or equal to U.

Similarly, when a sequence is decreasing or nonincreasing and has a lower bound V (Figure 11.2(b)), it must approach a limit that is greater than or equal to V.

FIGURE 11.2

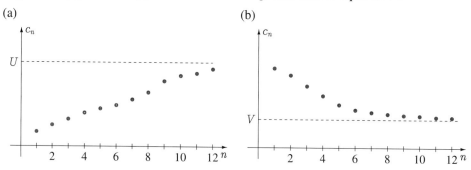

Another way of stating Theorem 11.1 is as follows.

Corollary A monotonic sequence has a limit if and only if it is bounded.

We are now prepared to give a complete and typical discussion for the recursive sequence of Example 11.2.

EXAMPLE 11.4 Find the limit of the sequence of Example 11.2, if it has one.

SOLUTION The first five terms of the sequence are

$$c_1 = 1, \quad c_2 = 6.732, \quad c_3 = 7.955, \quad c_4 = 8.155, \quad c_5 = 8.187.$$

They suggest that the sequence is increasing; that is, $c_{n+1} > c_n$. To prove this we use mathematical induction (see Appendix C). Certainly the inequaltiy is valid for $n = 1$ since $c_2 > c_1$. Suppose k is an integer for which $c_{k+1} > c_k$. Then

$$2 + c_{k+1} > 2 + c_k,$$

from which
$$\sqrt{2 + c_{k+1}} > \sqrt{2 + c_k}.$$

It follows that
$$5 + \sqrt{2 + c_{k+1}} > 5 + \sqrt{2 + c_k}.$$

The left side is c_{k+2} and the right side is c_{k+1}. Therefore we have proved that $c_{k+2} > c_{k+1}$. Hence, by mathematical induction $c_{n+1} > c_n$ for all $n \geq 1$. Since the sequence is increasing, its first term $c_1 = 1$ must be a lower bound. Certainly any upper bound, if one exists, must be at least 8.187 (c_5). We can take any number greater than 8.187 and use mathematical induction to test whether it is indeed an upper bound. It appears that $U = 10$ might be a reasonable guess for an upper bound for this sequence, and we verify this by induction as follows. Clearly $c_1 < 10$. We suppose that k is some integer for which $c_k < 10$. Then

$$2 + c_k < 12,$$

from which
$$\sqrt{2 + c_k} < \sqrt{12}.$$

Thus,
$$c_{k+1} = 5 + \sqrt{2 + c_k} < 5 + \sqrt{12} < 10.$$

By mathematical induction, then $c_n < 10$ for $n \geq 1$.

Since the sequence is monotonic and bounded, Theorem 11.1 guarantees that it has a limit, call it L. To evaluate L, we take limits on each side of the equation defining the sequence recursively:

$$\lim_{n \to \infty} c_{n+1} = \lim_{n \to \infty} \left(5 + \sqrt{2 + c_n}\right).$$

It is important to note that this cannot be done until the conditions of Theorem 11.1 have been checked. Since the terms c_n of the sequence approach L as $n \to \infty$, it follows that $5 + \sqrt{2 + c_n}$ approaches $5 + \sqrt{2 + L}$. Furthermore, as $n \to \infty$, c_{n+1} must also approach L. Do not make the mistake of saying that c_{n+1} approaches $L + 1$ as $n \to \infty$. Think about what $\lim_{n \to \infty} c_{n+1}$ means. We conclude therefore that

$$L = 5 + \sqrt{2 + L}.$$

If we transpose the 5 and square both sides of the equation, we obtain the quadratic equation
$$L^2 - 11L + 23 = 0,$$

with solutions
$$L = \frac{11 \pm \sqrt{29}}{2}.$$

Only the positive square root satisfies the original equation $L = 5 + \sqrt{2 + L}$ defining L, so that $L = (11 + \sqrt{29})/2$. The other root, $(11 - \sqrt{29})/2 \approx 2.8$ can also be eliminated on the grounds that all terms beyond the first are greater than 6. ∎

Now that we know what it means for a sequence to have a limit, and how to find limits, we can be more precise. To give a mathematical definition for the limit of a sequence, we start with our intuitive description and make a succession of paraphrases, each of which is one step closer to a precise definition:

A sequence $\{c_n\}$ has limit L if its terms get arbitrarily close to L, and stay close to L, as n gets larger and larger.

A sequence $\{c_n\}$ has limit L if its terms can be made arbitrarily close to L by choosing n sufficiently large.

A sequence $\{c_n\}$ has limit L if the difference $|c_n - L|$ can be made arbitrarily close to 0 by choosing n sufficiently large.

A sequence $\{c_n\}$ has limit L if given any real number $\epsilon > 0$, no matter how small, we can make the difference $|c_n - L|$ less than ϵ by choosing n sufficiently large.

Finally, we arrive at the following definition.

Definition 11.4

A sequence $\{c_n\}$ has **limit** L if for any given $\epsilon > 0$, there exists an integer N such that for all $n > N$
$$|c_n - L| < \epsilon.$$

This definition puts in precise terms, and in the simplest possible way, our intuitive idea of a limit. For those who have studied Section 2.5, note the similarity between Definition 11.4 and Definition 2.2. For a better understanding of Definition 11.4, it is helpful to consider its geometric interpretation. The inequality $|c_n - L| < \epsilon$, when written in the form $L - \epsilon < c_n < L + \epsilon$ is interpreted as a horizontal band of width 2ϵ around L (Figure 11.3). Definition 11.4 requires that no matter how small ϵ, we can find a stage, denoted by N, beyond which all terms in the sequence are contained in the horizontal band. For the sequence and ϵ in Figure 11.3, N must be chosen as shown. For the same sequence, but a smaller ϵ, N must be chosen correspondingly larger (Figure 11.4). Proofs of many theorems in this chapter require a working knowledge of this definition. As an example, we use it to verify that a sequence cannot have two limits, a fact that we have implicitly assumed throughout our discussions.

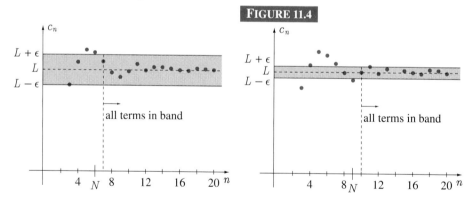

FIGURE 11.3

FIGURE 11.4

Theorem 11.2

A sequence can have at most one limit.

Proof We prove this theorem by showing that a sequence cannot have two distinct limits. Suppose to the contrary that a sequence $\{c_n\}$ has two distinct limits L_1 and L_2, where $L_2 > L_1$, and let $L_2 - L_1 = \delta$. If we set $\epsilon = \delta/3$, then according to Definition 11.4, there exists an integer N_1 such that for all $n > N_1$

$$|c_n - L_1| < \epsilon = \delta/3;$$

that is, for $n > N_1$, all terms in the sequence are within a distance $\delta/3$ of L_1.

But since L_2 is also supposed to be a limit, there exists an N_2 such that for $n > N_2$, all terms in the sequence are within a distance $\epsilon = \delta/3$ of L_2:

$$|c_n - L_2| < \epsilon = \delta/3.$$

But this is impossible (Figure 11.5), if L_1 and L_2 are a distance δ apart. This contradiction therefore implies that L_1 and L_2 are the same; that is, $\{c_n\}$ cannot have two limits.

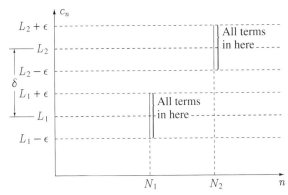

FIGURE 11.5

The following theorem, which states some of the properties of convergent sequences, can also be proved using Definition 11.4. Only the first two parts, however, are straightforward (see Exercises 71, 78, and 79).

Theorem 11.3

If sequences $\{c_n\}$ and $\{d_n\}$ have limits C and D, then:

(i) $\{kc_n\}$ has limit kC if k is a constant;
(ii) $\{c_n \pm d_n\}$ has limit $C \pm D$;
(iii) $\{c_n d_n\}$ has limit CD;
(iv) $\{c_n/d_n\}$ has limit C/D provided that $D \neq 0$, and none of the $d_n = 0$.

EXERCISES 11.1

In Exercises 1–14 state (without proof) whether the sequence is monotonic, and whether it has an upper bound, a lower bound, and a limit.

1. $\left\{\dfrac{1}{n}\right\}$
2. $\{3^n + 1\}$
3. $\{3\}$
4. $\left\{\left(\dfrac{3}{4}\right)^{n+1}\right\}$
5. $\left\{\left(\dfrac{4}{3}\right)^{n+1}\right\}$
6. $\left\{\left(-\dfrac{15}{16}\right)^{n+5}\right\}$
7. $\left\{\sin\left(\dfrac{n\pi}{2}\right)\right\}$
8. $\left\{\dfrac{n}{n^2 + n + 2}\right\}$
9. $\left\{\dfrac{(-1)^n}{n}\right\}$
10. $\{\operatorname{Tan}^{-1} n\}$
11. $\left\{\dfrac{\ln n}{n^2 + 1}\right\}$
12. $\left\{(-1)^n \sqrt{n^2 + 1}\right\}$
13. $c_1 = 2$, $c_{n+1} = \dfrac{2}{c_n - 1}$, $n \geq 1$
14. $c_1 = 4$, $c_{n+1} = -\dfrac{c_n}{n^2}$, $n \geq 1$

In Exercises 15–34 determine whether the statement is true or false. Verify any statement that is true, and give a counter example to any statement that is false.

15. A sequence can be increasing and nondecreasing.
16. A monotonic sequence must be bounded.
17. An increasing sequence must have a lower bound.

18. A decreasing sequence must have a lower bound.

19. An increasing sequence with a lower bound must have a limit.

20. An increasing sequence with an upper bound must have a limit.

21. If a sequence diverges, then either $\lim_{n\to\infty} c_n = \infty$, or, $\lim_{n\to\infty} c_n = -\infty$.

22. A sequence cannot be both increasing and decreasing.

23. A sequence can be both nonincreasing and nondecreasing.

24. If a sequence is monotonic and has a limit, it must be bounded.

25. If a sequence is bounded and has a limit, it must be monotonic.

26. If a sequence is bounded, but not monotonic, it cannot have a limit.

27. If a sequence has a limit, it must be bounded.

28. If a sequence is not monotonic, it cannot have a limit.

29. If all terms of a sequence $\{c_n\}$ are less than U, and $L = \lim_{n\to\infty} c_n$ exists, the L must be less than U.

30. A sequence $\{c_n\}$ has a limit if and only if $\{c_n^2\}$ has a limit.

31. A sequence $\{c_n\}$ of positive numbers converges if $c_{n+1} < c_n/2$.

32. A sequence $\{c_n\}$ of numbers converges if $c_{n+1} < c_n/2$.

33. If an increasing sequence has a limit L, then L must be equal to the smallest upper bound for the sequence.

34. If an infinite number of terms of a sequence all have the same value a, then either a is the limit of the sequence or the sequence has no limit.

In Exercises 35–48 discuss, with proofs, whether the sequence is monotonic, and whether it has an upper bound, a lower bound, and a limit.

35. $\left\{\dfrac{n+1}{2n+3}\right\}$

36. $\left\{\dfrac{2n+3}{n^2-5}\right\}$

37. $\left\{\dfrac{n^2+5n-4}{n^2+2n-2}\right\}$

38. $\{ne^{-n}\}$

39. $\left\{\dfrac{\ln n}{n}\right\}$

40. $\left\{\dfrac{n}{n+1}\operatorname{Tan}^{-1} n\right\}$

41. $c_1 = 1, \quad c_{n+1} = \dfrac{1}{10}(c_n^3 + 12), \quad n \geq 1$

42. $c_1 = 0, \quad c_{n+1} = \dfrac{1}{12}(c_n^4 + 5), \quad n \geq 1$

43. $c_1 = 3, \quad c_{n+1} = \sqrt{5+c_n}, \quad n \geq 1$

44. $c_1 = 1, \quad c_{n+1} = \sqrt{5+c_n}, \quad n \geq 1$

45. $c_1 = 5, \quad c_{n+1} = 1+\sqrt{6+c_n}, \quad n \geq 1$

46. $c_1 = 3, \quad c_{n+1} = 1+\sqrt{6+c_n}, \quad n \geq 1$

47. $c_1 = 1, \quad c_{n+1} = 4-\sqrt{5-c_n}, \quad n \geq 1$

48. $c_1 = 4, \quad c_{n+1} = 4-\sqrt{5-c_n}, \quad n \geq 1$

In Exercises 49–53 find an explicit formula for the general term of the sequence. In each case assume that the remaining terms follow the pattern suggested by the given terms.

49. $\dfrac{1}{2}, \dfrac{3}{4}, \dfrac{7}{8}, \dfrac{15}{16}, \dfrac{31}{32}, \ldots$

50. $4, \dfrac{7}{4}, \dfrac{10}{9}, \dfrac{13}{16}, \dfrac{16}{25}, \dfrac{19}{36}, \ldots$

51. $\dfrac{\ln 2}{\sqrt{2}}, -\dfrac{\ln 3}{\sqrt{3}}, \dfrac{\ln 4}{\sqrt{4}}, \dfrac{-\ln 5}{\sqrt{5}}, \ldots$

52. $1, 0, 1, 0, 1, 0, \ldots$

53. $1, 1, -1, -1, 1, 1, -1, -1, \ldots$

54. Verify that the sequence $\{n^{1/n}\}$ is decreasing for $n \geq 3$ by showing that the function $f(x) = x^{1/x}$ has a negative derivative for $x \geq 3$. Illustrate this result graphically. What is the limit of this sequence?

55. A sequence $\{c_n\}$ is defined by $c_n = 3n/(n^2 - 5n + 1)$.
    (a) Consider the following proof that the sequence is decreasing: $c_{n+1} < c_n$ if and only if
    $$\dfrac{3(n+1)}{(n+1)^2 - 5(n+1) + 1} < \dfrac{3n}{n^2 - 5n + 1}.$$
    But this is true if and only if
    $$3(n+1)(n^2 - 5n + 1) < 3n(n^2 - 3n - 3) \quad \text{or}$$
    $$n^3 - 4n^2 - 4n + 1 < n^3 - 3n^2 - 3n;$$
    that is, if and only if
    $$1 < n^2 + n.$$
    Since this inequality is true for all $n \geq 1$, the sequence is decreasing. Do you see any error in this proof?
    (b) Calculate the first five terms in the sequence. What do you now conclude about the argument in (a)?

In Exercises 56–59 show how L'Hôpital's rule (Theorem 4.6, Section 4.9) can be used to evaluate the limit for the sequence.

56. $\left\{\dfrac{\ln n}{\sqrt{n}}\right\}$ **57.** $\left\{\dfrac{n^3+1}{e^n}\right\}$

58. $\left\{n\sin\left(\dfrac{4}{n}\right)\right\}$ **59.** $\left\{\left(\dfrac{n+5}{n+3}\right)^n\right\}$

In Exercises 60–68 discuss, with all necessary proofs, whether the sequence is monotonic, is bounded, and has a limit.

60. $c_1 = 2$, $c_{n+1} = \dfrac{1}{3 - c_n}$, $n \geq 1$

61. $c_1 = 1$, $c_{n+1} = \dfrac{1}{4 - 2c_n}$, $n \geq 1$

62. $c_1 = 1$, $c_{n+1} = \dfrac{7}{16 - 8c_n^2}$, $n \geq 1$

63. $c_1 = 0$, $c_{n+1} = \dfrac{7}{16 - 8c_n^2}$, $n \geq 1$

64. $c_1 = 1$, $c_{n+1} = \dfrac{4c_n}{4 + c_n}$, $n \geq 1$

65. $c_1 = 4$, $c_{n+1} = \dfrac{3c_n}{2 + c_n}$, $n \geq 1$

66. $c_1 = 2$, $c_{n+1} = \dfrac{2c_n^2}{3 + c_n}$, $n \geq 1$

67. $c_1 = \dfrac{3}{2}$, $c_{n+1} = \dfrac{c_n + 2}{4 - c_n}$, $n \geq 1$

68. $c_1 = 0$, $c_{n+1} = \dfrac{3 - c_n}{5 - 2c_n}$, $n \geq 1$

69. Show that the sequence

$$c_1 = 1, \quad c_{n+1} = \dfrac{1}{4 - c_n - c_n^2}, \quad n \geq 1$$

is monotonic and bounded. Find an approximation to its limit accurate to 5 decimals.

70. (a) Prove that if a sequence converges, then differences between successive terms in the sequence must approach zero.

(b) The following example illustrates that the converse is not true. Show that differences between successive terms of the sequence $\{\ln n\}$ approach zero, but the sequence itself diverges.

71. Prove parts (i) and (ii) of Theorem 11.3.

72. Find bounds for the sequence

$$c_1 = 1, \quad c_{n+1} = \dfrac{1 + c_n}{1 + 2c_n}, \quad n \geq 1.$$

73. Determine whether the following sequence is monotonic, has bounds, and has a limit,

$$c_1 = -30, \quad c_2 = -20, \quad c_{n+1} = 5 + \dfrac{c_n}{2} + \dfrac{c_{n-1}}{3}, \quad n \geq 2.$$

74. A sequence $\{c_n\}$ has only positive terms. Prove that:

(a) if $\lim_{n \to \infty} c_n = L < 1$, there exists an integer N such that for all $n \geq N$

$$c_n < \dfrac{L + 1}{2}.$$

(b) if $\lim_{n \to \infty} c_n = L > 1$, there exists an integer N such that for all $n \geq N$,

$$c_n > \dfrac{L + 1}{2}.$$

These results are used in Theorem 11.9, Section 11.5.

75. Prove that if $\lim_{n \to \infty} c_n = L$, there exists an integer N such that for all $n \geq N$,

$$c_n < L + 1.$$

This result is used in Theorem 11.8, Section 11.4.

76. The *Fibonacci sequence* found in many areas of applied mathematics is defined by

$$c_1 = 1, \quad c_2 = 1, \quad c_{n+1} = c_n + c_{n-1}, \quad n \geq 2.$$

(a) Evaluate the first ten terms of this sequence.

(b) Is the sequence monotonic, is it bounded, and does it have a limit?

(c) Prove that

$$c_n^2 - c_{n-1}c_{n+1} = (-1)^{n+1}, \quad n \geq 2.$$

(d) Verify that an explicit formula for c_n is

$$c_n = \dfrac{1}{\sqrt{5}}\left[\left(\dfrac{1 + \sqrt{5}}{2}\right)^n - \left(\dfrac{1 - \sqrt{5}}{2}\right)^n\right].$$

(e) Define a sequence $\{b_n\}$ as the ratio of terms in the Fibonacci sequence

$$b_n = \dfrac{c_{n+1}}{c_n}.$$

Is this sequence monotonic? Does it have a limit?

77. Prove that the sequence

$$c_1 = d, \quad c_{n+1} = \sqrt{a + 2bc_n}, \quad n \geq 1$$

(a, b, and d all positive constants) is increasing if and only if $d < b + \sqrt{a+b^2}$. What happens when $d = b + \sqrt{a+b^2}$?

78. In this exercise we outline a proof of part (iii) of Theorem 11.3.

(a) Verify that

$$|c_n d_n - CD| \leq |c_n||d_n - D| + |D||c_n - C|.$$

(b) Show that given any $\epsilon > 0$, there exist positive integers N_1, N_2, and N_3 such that

$$|c_n| < |C| + 1 \qquad \text{whenever } n > N_1,$$
$$|d_n - D| < \frac{\epsilon}{2(|C|+1)} \qquad \text{whenever } n > N_2,$$
$$|c_n - C| < \frac{\epsilon}{2|D|+1} \qquad \text{whenever } n > N_3.$$

(c) Use these results to prove part (iii) of Theorem 11.3.

79. In this exercise we outline a proof for part (iv) of Theorem 11.3.

(a) Verify that when $d_n \neq 0$ and $D \neq 0$,

$$\left|\frac{c_n}{d_n} - \frac{C}{D}\right| \leq \frac{|c_n - C|}{|d_n|} + \frac{|C||d_n - D|}{|D||d_n|}.$$

(b) Show that given any $\epsilon > 0$, there exist positive integers N_1, N_2, and N_3 such that

$$|d_n| > \frac{|D|}{2} \qquad \text{whenever } n > N_1,$$
$$|c_n - C| < \frac{\epsilon|D|}{4} \qquad \text{whenever } n > N_2,$$
$$|d_n - D| < \frac{\epsilon|D|^2}{4|C|+1} \qquad \text{whenever } n > N_3.$$

(c) Now prove part (iv) of Theorem 11.3.

80. Find an explicit formula for the recursive sequence

$$c_1 = 1, \quad c_2 = 2, \quad c_{n+1} = \frac{c_n + c_{n-1}}{2}, \quad n \geq 2.$$

81. Justify the statement that the sequence of Example 11.3 has no upper bound.

SECTION 11.2

Applications of Sequences

When a sequence is monotonic and bounded, it has a limit. Verification that a recursive sequence is monotonic and bounded usually requires mathematical induction. However, when the sequence arises in an application, the derivation of the sequence may make it clear that the sequence has a limit, and it may therefore be possible to proceed directly to its evaluation without verifying monotony and bounds. Such is often the case in the field of numerical analysis, wherein recursive sequences commonly arise in the form of iterative procedures. We have already encountered Newton's iterative procedure as one example. When this method is applied to the equation

$$f(x) = x^3 - 3x + 1 = 0$$

with an initial approximation $x_1 = 0$ to find the root between 0 and 1, the sequence obtained is

$$x_1 = 0, \qquad x_{n+1} = x_n - \frac{f(x_n)}{f'(x_n)} = x_n - \frac{x_n^3 - 3x_n + 1}{3x_n^2 - 3}, \quad n \geq 1.$$

The first three terms of this sequence are illustrated in Figure 11.6, and the tangent line construction by which they are obtained makes it clear that the sequence must converge to the solution of the equation between 0 and 1. In other words, it is not necessary for us to analyze monotony and bounds for the recursive sequence algebraically; they are obvious geometrically. The sequence is increasing and has bounds $U = 1$ and $V = 0$.

To find the limit of a recursive sequence in Section 11.1, we adopted the procedure of letting n become very large on both sides of the recursive definition for the sequence (see Example 11.4). This does not work here as we are simply led back to the equation

$x^3 - 3x + 1 = 0$. Try it. Instead we evaluate terms of the sequence algebraically until they repeat

$x_1 = 0$, $x_2 = 1/3$, $x_3 = 0.347222$, $x_4 = 0.34729635$, $x_5 = 0.347296355$, $x_6 = 0.347296355$.

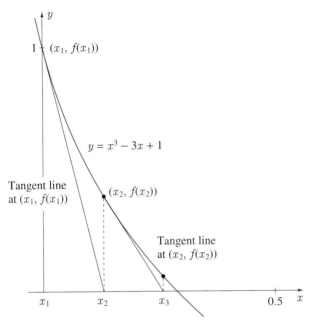

FIGURE 11.6

At this point we might be led to conclude that the required limit and solution of $x^3 - 3x + 1 = 0$ is $x = 0.347296355$. But realizing that our calculations are calculator dependent, we should at least be skeptical of the last digit on the display of our calculator. Suppose the real question were to find the solution accurate to six decimals. This could be accomplished by calculating

$$f(0.3472955) = 2.3 \times 10^{-6} \quad \text{and} \quad f(0.3472965) = -3.8 \times 10^{-7}.$$

Since one function value is positive and the other is negative, the intermediate value theorem (see Theorem 4.1) guarantees that the root is between 0.3472955 and 0.3472965. Hence, to six decimals, the solution is 0.347296.

Figure 11.6 illustrated that the sequence defined by Newton's iterative procedure must converge to a root of $f(x) = x^3 - 3x + 1 = 0$, and therefore it was unnecessary to verify existence of a limit of the sequence by proving that the sequence was monotonic and bounded. Are we proposing that a graph of the function $f(x)$ should always be drawn when using Newton's method to solve equations? Not really, although we are of the philosophy that pictures should be drawn whenever possible. Numerical analysts have proved that under very mild restrictions, the sequence defined by Newton's method always converges to a root of the equation provided the initial approximation is sufficiently close to that root. See Exercise 33 for further discussion of this point.

Another way to solve for the same root of $x^3 - 3x + 1 = 0$ is to rewrite the equation in the form

$$x = \frac{1}{3}(x^3 + 1),$$

and use this to define the following recursive sequence

$$x_1 = 0, \quad x_{n+1} = \frac{1}{3}(x_n^3 + 1), \quad n \geq 1.$$

It is straightforward to show that this sequence is bounded above by 1 and below by 0 and that it is increasing. According to Theorem 11.1, it therefore has a limit L, which must lie in the interval $0 \leq L \leq 1$. To find L, we set

$$\lim_{n \to \infty} x_{n+1} = \frac{1}{3} \lim_{n \to \infty} (x_n^3 + 1) \quad \text{or}$$

$$L = \frac{1}{3}(L^3 + 1).$$

But this implies that L is the required root of the original equation. In other words, we have defined a recursive sequence $\{x_n\}$ in such a way that its limit is the required solution of $x^3 - 3x + 1 = 0$. The first ten terms of the sequence are:

$$\begin{aligned}
x_1 &= 0, & x_6 &= 0.34729353, \\
x_2 &= 1/3, & x_7 &= 0.34729601, \\
x_3 &= 0.345679, & x_8 &= 0.34729631, \\
x_4 &= 0.34710219, & x_9 &= 0.34729635, \\
x_5 &= 0.34727295,
\end{aligned}$$

which leads us again to the conclusion that an approximation to the root of $x^3 - 3x + 1 = 0$ between 0 and 1 is $x = 0.347296$.

This method of finding the root of an equation is often called the **method of successive substitutions**. Exercises 21–32 provide more opportunities to practise this method.

In Section 3.9 and Appendix B, we essentially defined the number e as the limit of the sequence

$$e_n = \left(1 + \frac{1}{n}\right)^n, \tag{11.7}$$

but gave no proof that the sequence does indeed converge. In spite of the fact that the sequence is defined explicitly, it is far from trivial to prove that it has a limit. To do so, we first find an expanded expression for e_n by means of the binomial theorem,

$$e_n = \left(1 + \frac{1}{n}\right)^n = \sum_{r=0}^{n} \binom{n}{r}\left(\frac{1}{n}\right)^r$$

$$= 1 + \sum_{r=1}^{n} \frac{n(n-1)\cdots(n-r+1)}{r!\,n^r},$$

where we have used the expression $\binom{n}{r} = n(n-1)\cdots(n-r+1)/r!$ for binomial coefficients. Now in the product $n(n-1)\cdots(n-r+1)$, there are r factors. If each is divided by one of the n's in the denominator, e_n can be expressed in the form

$$e_n = 1 + \sum_{r=1}^{n} \frac{1}{r!}\left(\frac{n}{n}\right)\left(\frac{n-1}{n}\right)\left(\frac{n-2}{n}\right)\cdots\left(\frac{n-r+1}{n}\right)$$

$$= 1 + \sum_{r=1}^{n} \frac{1}{r!}(1)\left(1 - \frac{1}{n}\right)\left(1 - \frac{2}{n}\right)\cdots\left(1 - \frac{r-1}{n}\right). \tag{11.8}$$

To show that sequence 11.7 has a limit, we verify that it is increasing ($e_{n+1} > e_n$) and bounded ($2 < e_n < 3$). The expanded form for e_{n+1} is obtained by replacing n by $n+1$ in 11.8,

$$e_{n+1} = 1 + \sum_{r=1}^{n+1} \frac{1}{r!}(1)\left(1 - \frac{1}{n+1}\right)\left(1 - \frac{2}{n+1}\right)\cdots\left(1 - \frac{r-1}{n+1}\right).$$

If we drop the last term from this summation (when $r = n+1$), the right side becomes smaller. In addition, if the $(n+1)$'s in the denominator are all replaced by n's, each factor decreases. Thus we may write

$$e_{n+1} > 1 + \sum_{r=1}^{n} \frac{1}{r!}(1)\left(1 - \frac{1}{n}\right)\left(1 - \frac{2}{n}\right) \cdots \left(1 - \frac{r-1}{n}\right)$$

But this is e_n (see 11.8). In other words, we have verified that $e_{n+1} > e_n$, and the sequence is increasing.

Since the sequence is increasing, it is bounded below by its first term $e_1 = 2$. To verify that 3 is an upper bounded, we use expression 11.8 to write

$$e_n = 1 + \sum_{r=1}^{n} \frac{1}{r!}(1)\left(1 - \frac{1}{n}\right)\left(1 - \frac{2}{n}\right) \cdots \left(1 - \frac{r-1}{n}\right)$$

$$< 1 + \sum_{r=1}^{n} \frac{1}{r!} \quad \text{(since all omitted factors are less than one)}$$

$$= 1 + \left(1 + \frac{1}{2!} + \frac{1}{3!} + \cdots + \frac{1}{n!}\right)$$

$$< 1 + \left(1 + \frac{1}{2} + \frac{1}{2^2} + \frac{1}{2^3} + \cdots + \frac{1}{2^{n-1}}\right).$$

A little experimentation shows that the sum of the terms in the parentheses is $2 - 1/2^{n-1}$, and therefore

$$e_n < 1 + 2 - \frac{1}{2^{n-1}} = 3 - \frac{1}{2^{n-1}} < 3.$$

With an upper bound established, Theorem 11.1 now guarantees that the sequence $\{e_n\}$ has a limit. The limit can be approximated by setting n in $(1 + 1/n)^n$ equal to a very large number. For example, with $n = 10^6$, we obtain $e_{1\,000\,000} = 2.71828047$. Unfortunately, we have no idea how accurate this approximation is for e. A superior method for approximating e using infinite series is discussed in Section 11.7, and with this method we can make definitive statements about the accuracy of the approximation.

EXERCISES 11.2

In Exercises 1–12 use Newton's iterative procedure with the given initial approximation x_1 to define a sequence of approximations to a solution of the equation. Determine graphically whether the sequence is monotonic, is bounded, and has a limit. Approximate any limit that exists to seven decimals.

1. $x_1 = 1$, $x^2 + 3x + 1 = 0$
2. $x_1 = -1$, $x^2 + 3x + 1 = 0$
3. $x_1 = -1.5$, $x^2 + 3x + 1 = 0$
4. $x_1 = -3$, $x^2 + 3x + 1 = 0$
5. $x_1 = 4$, $x^3 - x^2 + x - 22 = 0$
6. $x_1 = 2$, $x^3 - x^2 + x - 22 = 0$
7. $x_1 = 2$, $x^5 - 3x + 1 = 0$
8. $x_1 = 1$, $x^5 - 3x + 1 = 0$
9. $x_1 = 0$, $x^5 - 3x + 1 = 0$
10. $x_1 = 4/5$, $x^5 - 3x + 1 = 0$
11. $x_1 = 0.85$, $x^5 - 3x + 1 = 0$
12. $x_1 = -2$, $x^5 - 3x + 1 = 0$
13. (a) Use Newton's iterative procedure with $x_1 = 1$ to approximate the root of $x^4 - 15x + 2 = 0$ between 0 and 1.

    (b) Use the method of successive substitutions with $x_1 = 1$ and $x_{n+1} = (x_n^4 + 2)/15$ to approximate the root in (a).

(c) Use Newton's method to approximate the root between 2 and 3.

(d) What happens if the sequence in (b) is used to approximate the root between 2 and 3 with $x_1 = 2$ and $x_1 = 3$?

14. A superball is dropped from the top of a building 20 m high. Each time it strikes the ground, it rebounds to 99% of the height from which it fell.

(a) If d_n denotes the distance traveled by the ball between the nth and $(n+1)^{\text{th}}$ bounces, find a formula for d_n.

(b) If t_n denotes the time between the nth and $(n+1)^{\text{th}}$ bounces, find a formula for t_n.

15. A dog sits at a farmhouse patiently watching for his master to return from the fields. When the farmer is 1 km from the farmhouse, the dog immediately takes off for the farmer. When he reaches the farmer, he turns and runs back to the farmhouse, whereupon he again turns and runs to the farmer. The dog continues this frantic action until the farmer reaches the farmhouse. If the dog runs twice as fast as the farmer, find the distance d_n run by the dog from the point when he reaches the farmer for the nth time to the point when he reaches the farmer for the $(n+1)^{\text{th}}$ time. Ignore any accelerations of the dog in the turns.

16. The equilateral triangle in Figure 11.7(a) has perimeter P. If each side of the triangle is divided into three equal parts, an equilateral triangle is drawn on the middle segment of each side, and the figure transformed into Figure 11.7(b), what is the perimeter P_1 of this figure? If each side of Figure 11.7(b) is now subdivided into three equal portions and equilateral triangles are similarly constructed to result in Figure 11.7(c), what is the perimeter P_2 of this figure? If this subdivision process is continued indefinitely, what is the perimeter P_n after the nth subdivision? What is $\lim_{n \to \infty} P_n$?

FIGURE 11.7

(a)   (b)   (c)

17. A stone of mass 100 gm is thrown vertically upward with speed 20 m/s. Air exerts a resistive force on the stone proportional to its speed, and has magnitude 1/10 N when the speed of the stone is 10 m/s. It can be shown that the height y (in metres) above the projection point attained by the stone is given by

$$y = -98.1t + 1181\left(1 - e^{-t/10}\right),$$

where t is time (measured in seconds with $t = 0$ at the instant of projection).

(a) The time taken for the stone to return to its projection point can be obtained by setting $y = 0$ and solving the equation for t. Do so (correct to 2 decimals).

(b) Find the time for the stone to return if air resistance is ignored.

18. When the beam in Figure 11.8 vibrates vertically, there are certain frequencies of vibration, called *natural frequencies of the system*. They are solutions of the equation

$$\tan x = \frac{e^x - e^{-x}}{e^x + e^{-x}}$$

divided by 20π. Find the two smallest natural frequencies.

FIGURE 11.8

19. Consider the sequence

$$c_1 = 1, \quad c_{n+1} = \frac{1}{2}\left(c_n + \frac{a}{c_n}\right), \quad n \geq 1,$$

where a is any number greater than 1.

(a) Show that $\sqrt{a} < c_n < a$ for $n \geq 2$.

(b) Prove that the sequence is decreasing for $n \geq 2$.

(c) Verify that the sequence converges to $\sqrt{a}$. This result establishes an iterative procedure for finding the square root of real numbers. The sequence converges to $\sqrt{a}$ for $0 < a < 1$ also.

(d) Show that this sequence can be obtained from Newton's iterative procedure for the solution of the equation $x^2 = a$.

20. If A_n is the area of the figure with perimeter P_n in Exercise 16, find a formula for A_n.

To solve an equation $f(x) = 0$ by the method of successive substitutions, we first rearrange the equation into the form $x = g(x)$. We then define a recursive sequence by choosing some initial approximation x_1 and setting

$$x_{n+1} = g(x_n), \quad n \geq 1.$$

Depending on the choice of $g(x)$, this sequence may or may not converge to a root of the equation. In Exercises 21–25 illustrate that the suggested rearrangement of the equation $f(x) = 0$, along with the initial approximation x_1, leads to a sequence which converges to a root of the equation. Find the root accurate to 4 decimals.

21. $f(x) = x^2 - 2x - 1;\quad x_1 = 2;\quad$ use $x = 2 + 1/x$

22. $f(x) = x^3 + 6x + 3;\quad x_1 = -1;\quad$ use $x = -(1/6)(x^3 + 3)$

23. $f(x) = x^4 - 120x + 20;\quad x_1 = 0;\quad$ use $x = (x^4 + 20)/120$

24. $f(x) = x^3 - 2x^2 - 3x + 1;\quad x_1 = 3;\quad$ use $x = (2x^2 + 3x - 1)/x^2$

25. $f(x) = 8x^3 - x^2 - 1;\quad x_1 = 0;\quad$ use $x = (1/2)(1 + x^2)^{1/3}$

In Exercises 26–31 find a rearrangement of the equation which leads, through the method of successive substitutions, to a 4 decimal approximation to the root of the equation.

26. $x^3 - 6x^2 + 11x - 7 = 0\quad$ between $x = 3$ and $x = 4$

27. $x^4 - 3x^2 - 3x + 1 = 0\quad$ between $x = 0$ and $x = 1$

28. $x^4 + 4x^3 - 50x^2 + 100x - 50 = 0\quad$ between $x = 0$ and $x = 1$

29. $\sin^2 x = 1 - x^2\quad$ between $x = 0$ and $x = 1$

30. $\sec x = 2/(1 + x^4)\quad$ between $x = 0$ and $x = 1$

31. $e^x + e^{-x} - 10x = 0\quad$ between $x = 0$ and $x = 1$

32. (a) Show that if α is the only root of the equation $x = g(x)$ and Figure 11.9(a) is a graph of $g(x)$, then Figure 11.9(b) exhibits geometrically the sequence of approximations of α determined by the method of successive substitutions.

FIGURE 11.9

(b) Illustrate graphically how the sequence defined by successive substitutions converges to the root $x = \alpha$

for the equation $x = g(x)$ if $g(x)$ is as shown in Figure 11.10.

FIGURE 11.10

(c) Illustrate graphically that the method fails for the function $g(x)$ in Figure 11.11.

FIGURE 11.11

(d) Based on the results of parts (a)–(c), what determines success or failure of the method of successive substitutions? Part (e) provides a proof of the correct answer.

(e) Prove that when $x = g(x)$ has a root $x = \alpha$, the method of successive substitutions with an initial approximation of x_1 always converges to α if $|g'(x)| \le a < 1$ on the interval $|x - \alpha| \le |x_1 - \alpha|$. In other words, when an equation $f(x) = 0$ is rearranged into the form $x = g(x)$ for successive substitutions, success or failure depends on whether the derivative of $g(x)$ is between -1 and 1 near the required root.

33. Suppose that $f''(x)$ exists on an open interval containing a root $x = \alpha$ of the equation $f(x) = 0$. Use the result of Exercises 32 to prove that if $f'(\alpha) \ne 0$, Newton's iterative sequence always converges to α provided the initial approximation x_1 is chosen sufficiently close to α. (Hint: First use Exercise 32(e) to show that Newton's sequence converges to α if on the interval $|x - \alpha| \le |x_1 - \alpha|$, $|ff''/(f')^2| \le a < 1$.) In actual fact, Newton's method usually works even when $f'(\alpha) = 0$.

SECTION 11.3

Infinite Series of Constants

If $\{c_n\}$ is an infinite sequence of constants, an expression of the form

$$c_1 + c_2 + c_3 + \cdots + c_n + \cdots \qquad (11.9)$$

is called an **infinite series**, or simply a series. We often use sigma notation to represent a series:

$$\sum_{n=1}^{\infty} c_n = c_1 + c_2 + c_3 + \cdots + c_n + \cdots. \qquad (11.10)$$

By initiating the sigma notation with $n = 1$, we once again have the convenient situation in which $n = 1$ yields the first term of the series, $n = 2$ the second, etc. We usually abbreviate the notation further by writing $\sum c_n$ in place of $\sum_{n=1}^{\infty} c_n$ whenever the lower limit of summation is chosen as unity.

From the sequences of Example 11.1 we may form the following series.

EXAMPLE 11.5

Write out the first six terms of the following series:

(a) $\sum_{n=1}^{\infty} \dfrac{1}{2^{n-1}}$ (b) $\sum_{n=1}^{\infty} \dfrac{n}{n+1}$ (c) $\sum_{n=1}^{\infty} (-1)^n |n-3|$ (d) $\sum_{n=1}^{\infty} (-1)^{n+1}$

SOLUTION   The first six terms of these series are shown below:

(a) $\sum_{n=1}^{\infty} \dfrac{1}{2^{n-1}} = 1 + \dfrac{1}{2} + \dfrac{1}{4} + \dfrac{1}{8} + \dfrac{1}{16} + \dfrac{1}{32} + \cdots$

(b) $\sum_{n=1}^{\infty} \dfrac{n}{n+1} = \dfrac{1}{2} + \dfrac{2}{3} + \dfrac{3}{4} + \dfrac{4}{5} + \dfrac{5}{6} + \dfrac{6}{7} + \cdots$

(c) $\sum_{n=1}^{\infty} (-1)^n |n-3| = -2 + 1 - 0 + 1 - 2 + 3 - \cdots$

(d) $\sum_{n=1}^{\infty} (-1)^{n+1} = 1 - 1 + 1 - 1 + 1 - 1 + \cdots$

∎

Perhaps the most frequently encountered series in applications is the **geometric series**; a series of the form

$$a + ar + ar^2 + ar^3 + \cdots.$$

Each term in a geometric series is obtained by multiplying the term that came before it by the same constant r. The number $a \neq 0$ is called the first term of the series and r is called the **common ratio**. In sigma notation,

$$\sum_{n=1}^{\infty} ar^{n-1} = a + ar + ar^2 + ar^3 + \cdots. \qquad (11.11)$$

The series in parts (a) and (d) of Example 11.5 are geometric. The first has common ratio $1/2$ the second has common ratio -1.

Another important series is

$$\sum_{n=1}^{\infty} \frac{1}{n} = 1 + \frac{1}{2} + \frac{1}{3} + \frac{1}{4} + \cdots. \qquad (11.12)$$

It is called the **harmonic series**. We shall see it many times in our discussions.

It may be quite convenient to write expressions like 11.9 and say that they are infinite series, but since we can add only finitely many numbers, 11.9 is as yet meaningless. To illustrate how we may define a meaning for such a sum, consider the geometric series of Example 11.5(a),

$$\sum_{n=1}^{\infty} \frac{1}{2^{n-1}} = 1 + \frac{1}{2} + \frac{1}{4} + \frac{1}{8} + \cdots.$$

If we nonchalantly start adding terms of this series together, a pattern soon emerges. Indeed, if we denote by S_n the sum of the first n terms of this series, we find that

$$S_1 = 1,$$
$$S_2 = 1 + \frac{1}{2} = \frac{3}{2}, \qquad \text{(sum of first two terms)}$$
$$S_3 = 1 + \frac{1}{2} + \frac{1}{4} = \frac{7}{4}, \qquad \text{(sum of first three terms)}$$
$$S_4 = 1 + \frac{1}{2} + \frac{1}{4} + \frac{1}{8} = \frac{15}{8}, \qquad \text{(sum of first four terms)}$$
$$S_5 = 1 + \frac{1}{2} + \frac{1}{4} + \frac{1}{8} + \frac{1}{16} = \frac{31}{16}, \qquad \text{(sum of first five terms)}$$

and so on. By careful examination of the pattern, we can see that the sum of the first n terms of the series is given by the formula

$$S_n = 2 - \frac{1}{2^{n-1}}.$$

As we add more and more terms of this series together, the sum S_n gets closer and closer to 2. It is always less than 2, but S_n can be made arbitrarily close to 2 by choosing n sufficiently large. If this series is to have a sum, the only reasonable sum is 2. In practice, this is precisely what we do; we define the sum of the series $\sum 1/2^{n-1}$ to be 2.

Let us now take this idea and define sums for all infinite series. We begin by defining a sequence $\{S_n\}$ as follows:

$$S_1 = c_1,$$
$$S_2 = c_1 + c_2,$$
$$S_3 = c_1 + c_2 + c_3,$$
$$\vdots \qquad \vdots$$
$$S_n = c_1 + c_2 + c_3 + \cdots + c_n,$$
$$\vdots \qquad \vdots$$

It is called the **sequence of partial sums** of the series $\sum c_n$. The nth term of the sequence $\{S_n\}$ represents the sum of the first n terms of the series. If this sequence has a limit, say, S, then the more terms of the series that we add together, the closer the sum gets to S. It seems reasonable, then, to call S the sum of the series. We therefore make the following definition.

Definition 11.5 Let $S_n = \sum_{k=1}^{n} c_k$ be the nth partial sum of a series $\sum_{n=1}^{\infty} c_n$. If the sequence of partial sums $\{S_n\}$ has limit S,
$$S = \lim_{n \to \infty} S_n,$$
we call S the sum of the series and write
$$\sum_{n=1}^{\infty} c_n = S;$$
if $\{S_n\}$ does not have a limit, we say that the series does not have a sum.

If a series has sum S, we say that the series **converges** to S, which means that its sequence of partial sums converges to S. If a series does not have a sum, we say that the series **diverges**, which means that its sequence of partial sums diverges.

According to this definition, the series $\sum 1/2^{n-1}$ of Example 11.5(a) has sum 2. Since every term of the series $\sum n/(n+1)$ in Example 11.5(b) is greater than or equal to $1/2$, it follows that the sum of the first n terms is $S_n \geq n(1/2) = n/2$. As the sequence of partial sums is therefore (increasing and) unbounded, it cannot possibly have a limit (see the corollary to Theorem 11.1). The series does not therefore have a sum; it diverges. Examination of the first few partial sums of the series $\sum (-1)^n |n - 3|$ in Example 11.5(c) leads to the result that for $n > 1$,

$$S_n = \begin{cases} \dfrac{n-4}{2} & \text{if } n \text{ is even} \\ \dfrac{1-n}{2} & \text{if } n \text{ is odd.} \end{cases}$$

Since the sequence $\{S_n\}$ does not have a limit, the series diverges. The partial sums of the series $\sum (-1)^{n+1}$ are

$$S_1 = 1, \quad S_2 = 0, \quad S_3 = 1, \quad S_4 = 0, \quad \cdots.$$

Since this sequence does not have a limit, the series does not have a sum.

Consider now geometric series 11.11. If $\{S_n\}$ is the sequence of partial sums for this series, then

$$S_n = a + ar + ar^2 + \cdots + ar^{n-1}.$$

If we multiply this equation by r,

$$rS_n = ar + ar^2 + ar^3 + \cdots + ar^n.$$

When we subtract these equations, the result is

$$S_n - rS_n = a - ar^n.$$

Hence, for $r \neq 1$, we obtain

$$S_n = \frac{a(1 - r^n)}{1 - r}.$$

Furthermore, when $r = 1$,

$$S_n = na,$$

and therefore the sum of the first n terms of a geometric series is

$$S_n = \begin{cases} \dfrac{a(1-r^n)}{1-r} & r \neq 1 \\ na & r = 1. \end{cases} \quad (11.13\text{a})$$

To determine whether a geometric series has a sum, we consider the limit of this sequence. Certainly, $\lim_{n\to\infty} na$ does not exist, unless trivially $a = 0$. In addition, $\lim_{n\to\infty} r^n = 0$ when $|r| < 1$, and does not exist when $|r| > 1$. Thus, we may state that

$$\lim_{n\to\infty} S_n = \begin{cases} \dfrac{a}{1-r} & |r| < 1 \\ \text{does not exist} & |r| \geq 1. \end{cases}$$

The geometric series therefore has sum

$$\sum_{n=1}^{\infty} ar^{n-1} = \frac{a}{1-r} \quad (11.13\text{b})$$

if $|r| < 1$, but otherwise diverges.

Next we consider harmonic series 11.12. This series does not have a sum, and we can show this by considering the following partial sums of the series:

$$S_1 = 1,$$
$$S_2 = 1 + \frac{1}{2} = \frac{3}{2},$$
$$S_4 = S_2 + \frac{1}{3} + \frac{1}{4} > \frac{3}{2} + \frac{1}{4} + \frac{1}{4} = \frac{4}{2},$$
$$S_8 = S_4 + \frac{1}{5} + \frac{1}{6} + \frac{1}{7} + \frac{1}{8} > \frac{4}{2} + \frac{1}{8} + \frac{1}{8} + \frac{1}{8} + \frac{1}{8} = \frac{5}{2},$$
$$S_{16} = S_8 + \frac{1}{9} + \frac{1}{10} + \cdots + \frac{1}{16} > \frac{5}{2} + \frac{1}{16} + \frac{1}{16} + \cdots + \frac{1}{16} = \frac{6}{2}.$$

This procedure can be continued indefinitely, and shows that the sequence of partial sums is unbounded. The harmonic series therefore diverges (see once again the corollary to Theorem 11.1).

In each of the above examples, we used Definition 11.5 for the sum of a series to determine whether the series converges or diverges; that is, we formed the sequence of partial sums $\{S_n\}$ in order to consider its limit. For most examples, it is either too difficult or impossible to evaluate S_n in a simple form, and in such cases consideration of the limit of the sequence $\{S_n\}$ is impractical. Consequently, we must develop alternative ways to decide on the convergence of a series. We do this in Sections 11.4–11.6.

We now discuss some fairly simple but important results on convergence of series. If a series $\sum c_n$ has a finite number of its terms altered in any fashion whatsoever, the new series converges if and only if the original series converges. The new series may converge to a different sum, but it converges if the original series converges. For example, if we double the first three terms of the geometric series in Example 11.5, but do not change the remaining terms, the new series is

$$2 + 1 + \frac{1}{2} + \frac{1}{8} + \frac{1}{16} + \cdots.$$

It is not geometric. But its nth partial sum, call it S_n, is very closely related to that of the geometric series, call it T_n. In fact for $n \geq 4$, we can say that $S_n = T_n + \frac{7}{4}$ ($\frac{7}{4}$ is the total change in the first three terms). Since $\lim_{n\to\infty} T_n = 2$, it follows that

$\lim_{n \to \infty} S_n = 2 + \frac{7}{4} = \frac{15}{4}$; that is, the new series converges, but it has a sum which is $\frac{7}{4}$ greater than that of the geometric series.

On the other hand, suppose we change the first 100 terms of the harmonic series, which diverges, to 0, but leave the remaining terms unaltered. The new series is

$$\overbrace{0 + 0 + \cdots + 0}^{100 \text{ terms}} + \frac{1}{101} + \frac{1}{102} + \cdots .$$

We know that the sequence of partial sums $\{T_n\}$ of the harmonic series is increasing and unbounded. The nth partial sum S_n of the new series, for $n \geq 100$, is equal to $T_n - k$, where k is the sum of the first 100 terms of the harmonic series. But because $\lim_{n \to \infty} T_n = \infty$, so also must $\lim_{n \to \infty} S_n = \infty$; that is, the new series diverges.

The following theorem indicates that convergent series can be added and subtracted, and multiplied by constants. These properties can be verified using Definition 11.5.

Theorem 11.4

If series $\sum_{n=1}^{\infty} c_n$ and $\sum_{n=1}^{\infty} d_n$ have sums C and D, then:

(i) $\quad \displaystyle\sum_{n=1}^{\infty} k c_n = kC \qquad$ (when k is a constant); $\qquad$ (11.14a)

(ii) $\quad \displaystyle\sum_{n=1}^{\infty} (c_n \pm d_n) = C \pm D.$ $\qquad$ (11.14b)

Our first convergence test is a corollary to the following theorem.

Theorem 11.5

If a series $\sum_{n=1}^{\infty} c_n$ converges, then $\lim_{n \to \infty} c_n = 0$.

Proof If series $\sum c_n$ has a sum S, its sequence of partial sums

$$S_1, S_2, \cdots, S_n, \cdots$$

has limit S. The sequence

$$0, S_1, S_2, \cdots, S_{n-1}, \cdots$$

must also have limit S; it is the sequence of partial sums with an additional term equal to 0 at the beginning. According to Theorem 11.3, if we subtract these two sequences, the resulting sequence must have limit $S - S = 0$; that is,

$$S_1 - 0, S_2 - S_1, \cdots, S_n - S_{n-1}, \cdots \to 0.$$

But $S_1 - 0 = S_1 = c_1$, $S_2 - S_1 = c_2$, etc; that is, we have shown that $\lim_{n \to \infty} c_n = 0$.

Theorem 11.5 states that a necessary condition for a series $\sum c_n$ to converge is that the sequence $\{c_n\}$ of its terms must approach zero. What we really want are sufficient conditions to guarantee convergence or divergence of a series. We can take the contrapositive of the theorem and obtain the following.

Corollary (n^{th} term test) If $\lim_{n \to \infty} c_n \neq 0$, or does not exist, then the series $\sum_{n=1}^{\infty} c_n$ diverges.

This is our first convergence test, the nth term test. It is in fact a test for divergence rather than convergence, stating that if $\lim_{n\to\infty} c_n$ exists and is equal to anything but zero, or the limit does not exist, then the series $\sum c_n$ diverges. Note well that the nth term test never indicates that a series converges. Even if $\lim_{n\to\infty} c_n = 0$, we can conclude nothing about the convergence or divergence of $\sum c_n$; it may converge or it may diverge. For example, the harmonic series $\sum 1/n$ and the geometric series $\sum 1/2^{n-1}$ both satisfy the condition $\lim_{n\to\infty} c_n = 0$, yet one series diverges and the other converges. The nth term test therefore may indicate that a series diverges, but it never indicates that a series converges.

In particular, both series $\sum (-1)^n |n-3|$ and $\sum (-1)^{n+1}$ of Example 11.5 diverge by the nth term test.

To understand series, it is crucial to distinguish clearly among three entities: the series itself, its sequence of partial sums, and its sequence of terms. For any series $\sum c_n$, we can form its sequence of terms $\{c_n\}$ and its sequence of partial sums $\{S_n\}$. Each of these sequences may give information about the sum of the series $\sum c_n$, but in very different ways.

The sum of the series is defined by its sequence of partial sums in that $\sum c_n$ has a sum only if $\{S_n\}$ has a limit. The sequence of partial sums therefore tells us definitely whether the series converges or diverges, provided we can evaluate S_n in a simple form.

The sequence of terms $\{c_n\}$, on the other hand, may or may not tell us whether the series diverges. If $\{c_n\}$ has no limit or has a limit other than zero, we know that the series does not have a sum. If $\{c_n\}$ has limit zero, we obtain no information about convergence of the series, and must continue our investigation.

EXERCISES 11.3

In Exercises 1–10 determine whether the series converges or diverges. Find the sum of each convergent series. To get a feeling for a series, it is helpful to write out its first few terms. Try it.

1. $\sum_{n=1}^{\infty} \dfrac{n+1}{2n}$

2. $\sum_{n=1}^{\infty} \dfrac{2^n}{5^{n+1}}$

3. $\sum_{n=1}^{\infty} \cos\left(\dfrac{n\pi}{2}\right)$

4. $\sum_{n=1}^{\infty} \left(\dfrac{n}{n+1}\right)^n$

5. $\sum_{n=1}^{\infty} \dfrac{7^{2n+3}}{3^{2n-2}}$

6. $\sum_{n=1}^{\infty} \dfrac{7^{n+3}}{3^{2n-2}}$

7. $\sum_{n=1}^{\infty} \sqrt{\dfrac{n^2-1}{n^2+1}}$

8. $\sum_{n=1}^{\infty} \dfrac{\cos(n\pi)}{2^n}$

9. $\sum_{n=1}^{\infty} \dfrac{4^n + 3^n}{3^n}$

10. $\sum_{n=1}^{\infty} \operatorname{Tan}^{-1} n$

In Exercises 11–14 we have given a repeating decimal. Express the decimal as a geometric series and use formula 11.13b to express it as a rational number.

11. $0.666666\ldots$
12. $0.1313131313\ldots$
13. $1.347346346346\ldots$
14. $43.020502050205\ldots$

In Exercises 15–17 complete the statement and give a short proof to substantiate your claim.

15. If $\sum c_n$ and $\sum d_n$ converge, then $\sum (c_n + d_n)\ldots$

16. If $\sum c_n$ converges and $\sum d_n$ diverges, then $\sum (c_n + d_n)\ldots$

17. If $\sum c_n$ and $\sum d_n$ diverge, then $\sum (c_n + d_n)\ldots$

In Exercises 18–21 determine whether the series converges or diverges. Find the sum of each convergent series.

18. $\sum_{n=1}^{\infty} \dfrac{2^n + 3^n}{4^n}$

19. $\sum_{n=1}^{\infty} \dfrac{3^n - 1}{2^n}$

20. $\sum_{n=1}^{\infty} \dfrac{n^2 + 2^{2n}}{4^n}$

21. $\sum_{n=1}^{\infty} \dfrac{2^n + 4^n - 8^n}{2^{3n}}$

22. Find the sum of the series

$$\sum_{n=1}^{\infty} \dfrac{1}{n(n+1)}.$$

Hint: Use partial fractions on the nth term and find the sequence of partial sums.

23. Find the total distance travelled by the superball in Exercise 14 of Section 11.2 before it comes to rest.

24. Find the time taken for the superball in Exercise 14 of Section 11.2 to come to rest.

25. What distance does the dog run from the time when it sees the farmer until the farmer reaches the farmhouse in Exercise 15 of Section 11.2?

26. Find a simplified formula for the area A_n in Exercise 20 of Section 11.2. What is $\lim_{n\to\infty} A_n$?

27. According to equation 11.13a, the nth partial sum of the geometric series $1 + r + r^2 + r^3 + \cdots$ is $S_n = (1 - r^n)/(1 - r)$. If T_n denotes the nth partial sum of the series

$$1 + 2r + 3r^2 + 4r^3 + \cdots,$$

show that
$$T_n - S_n = r(T_n - nr^{n-1}).$$

Solve this equation for T_n and take the limit as $n \to \infty$ to show that

$$\sum_{n=1}^{\infty} nr^{n-1} = \frac{1}{(1-r)^2}, \quad |r| < 1.$$

In Exercises 28–31 use the result of Exercise 27 to find the sum of the series.

28. $\dfrac{1}{2} + \dfrac{2}{2^2} + \dfrac{3}{2^3} + \dfrac{4}{2^4} + \cdots$

29. $\dfrac{2}{5} + \dfrac{4}{25} + \dfrac{6}{125} + \dfrac{8}{625} + \cdots$

30. $\dfrac{2}{3} + \dfrac{3}{27} + \dfrac{4}{243} + \dfrac{5}{2187} + \cdots$

31. $\dfrac{12}{5} + \dfrac{48}{25} + \dfrac{192}{125} + \dfrac{768}{625} + \cdots$

32. Two people flip a single coin to see who can first flip a head. Show that the probability that the first person to flip wins the game is represented by the series

$$\frac{1}{2} + \frac{1}{8} + \frac{1}{32} + \cdots + \frac{1}{2^{2n-1}} + \cdots.$$

What is the sum of this series?

33. Two people throw a die to see who can first throw a six. Find the probability that the person who first throws wins the game.

34. One of *Zeno's paradoxes* describes a race between Achilles and a tortoise. Zeno claims that if the tortoise is given a head start, then no matter how fast Achilles runs, he can never catch the tortoise. He reasons as follows: In order to catch the tortoise, Achilles must first make up the length of the head start. But while he is running this distance, the tortoise "runs" a further distance. While Achilles makes up this distance, the tortoise covers a further distance, and so on. It follows that Achilles is always making up distance covered by the tortoise, and therefore can never catch the tortoise. If the tortoise is given a head start of length L and Achilles runs $c > 1$ times as fast as the tortoise, use infinite series to show that Achilles does in fact catch the tortoise and that the distance he covers in doing so is $cL(c-1)^{-1}$.

35. Find the time between 1:05 and 1:10 when the minute and hour hands of a clock point in the same direction
    (a) by reasoning with infinite series as in Exercise 34; and
    (b) by finding expressions for the angular displacements of the hands as functions of time.

36. Repeat Exercise 35 for the instant between 10:50 and 10:55 when the hands coincide.

37. A child has a large number of identical cubical blocks, which she stacks as shown in Figure 11.12. The top block protrudes $\frac{1}{2}$ its length over the second block, which protrudes $\frac{1}{4}$ its length over the third block, which protrudes $\frac{1}{6}$ its length over the fourth block, etc. Assuming the centre of mass of each block is at its geometric centre, show that the centre of mass of the top n blocks lies directly over the edge of the $(n+1)^{\text{th}}$ block. Now deduce that if a sufficient number of blocks is piled, the top block can be made to protrude as far over the bottom block as desired without the stack falling.

FIGURE 11.12

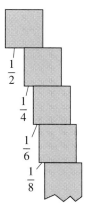

38. It is customary to assume that when a drug is administered to the human body it will be eliminated exponentially; that is, if A represents the amount of drug in the body, then

$$A = A_0 e^{-kt},$$

where $k > 0$ is a constant and A_0 is the amount injected at time $t = 0$. Suppose n successive injections of amount A_0 are administered at equally spaced time intervals T, the first injection at time $t = 0$.

(a) Show that the amount of drug in the body at time t between the nth and $(n+1)^{\text{th}}$ injection is given by

$$A_n(t) = A_0 e^{-kt}\left[\frac{1 - e^{knT}}{1 - e^{kT}}\right], \quad (n-1)T < t < nT.$$

(b) Sketch graphs of these functions on one set of axes.

(c) What is the amount of drug in the body immediately after the nth injection for very large n; that is, what is
$$\lim_{n\to\infty} A_n((n-1)T)?$$

39. Prove the following result: If $\sum_{n=1}^{\infty} c_n$ converges, then its terms can be grouped in any manner, and the resulting series is convergent with the same sum as the original series.

SECTION 11.4

Nonnegative Series

In Section 11.3 we derived the nth term test, a test for divergence of a series. In order to develop further convergence and divergence tests, we consider two classes of series:

1. Series with terms that are all nonnegative;
2. Series with both positive and negative terms.

A series with terms that are all nonpositive is the negative of a series of type 1, and therefore any test applicable to series of type 1 is easily adapted to a series with nonpositive terms.

Definition 11.6

A series $\sum_{n=1}^{\infty} c_n$ is said to be **nonnegative** if each term is nonnegative: $c_n \geq 0$.

For example, the harmonic series $\sum 1/n$ and the geometric series $\sum 1/2^{n-1}$ are both nonnegative series. We have already seen that both series have the property that $\lim_{n\to\infty} c_n = 0$, yet the harmonic series diverges and the geometric series converges. Examination of the terms of these series reveals that $1/2^{n-1}$ approaches 0 much more quickly than does $1/n$. In general, whether a nonnegative series does or does not have a sum depends on how quickly its sequence of terms $\{c_n\}$ approaches zero. The study of convergence of nonnegative series is an investigation into the question "How fast must the sequence of terms of a nonnegative series approach zero in order that the series have a sum?" This is not a simple problem; only a partial answer is provided through the tests in this section and the next. We begin with the integral test.

Theorem 11.6

(Integral test) Suppose the terms in a series $\sum_{n=1}^{\infty} c_n$ are denoted by $c_n = f(n)$ and $f(x)$ is a continuous, positive, decreasing function for $x \geq 1$. Then the series converges if and only if the improper integral $\int_1^{\infty} f(x)\,dx$ converges.

Proof Suppose first that the improper integral converges to value K. We know that this value can be interpreted as the area under the curve $y = f(x)$, above the x-axis, and to the right of the line $x = 1$ in Figure 11.13. In Figure 11.14 the area of the nth rectangle is $f(n) = c_n$, and clearly it is less than the area under the curve from $x = n-1$ to $x = n$.

FIGURE 11.13

FIGURE 11.14

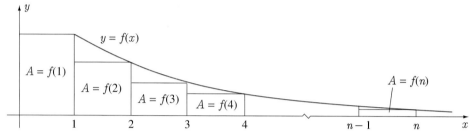

Now, the nth partial sum of the series $\sum c_n$ is

$$\begin{aligned}
S_n &= c_1 + c_2 + \cdots + c_n \\
&= f(1) + f(2) + \cdots + f(n) \\
&< f(1) + \int_1^2 f(x)\, dx + \int_2^3 f(x)\, dx + \cdots + \int_{n-1}^n f(x)\, dx \\
&= f(1) + \int_1^n f(x)\, dx \\
&< f(1) + \int_1^\infty f(x)\, dx \\
&= f(1) + K.
\end{aligned}$$

What this shows is that the sequence $\{S_n\}$ is bounded (since $f(1) + K$ is independent of n). Because the sequence is also increasing, it follows by Theorem 11.1 that it must have a limit; that is, the series $\sum c_n$ converges.

Conversely, suppose now that the improper integral diverges. This time we draw rectangles to the right of the vertical lines at $n = 1, 2, \ldots$ (Figure 11.15). Then

$$\begin{aligned}
S_n &= c_1 + c_2 + \cdots + c_n \\
&= f(1) + f(2) + \cdots + f(n) \\
&> \int_1^2 f(x)\, dx + \int_2^3 f(x)\, dx + \cdots + \int_{n-1}^n f(x)\, dx \\
&= \int_1^n f(x)\, dx.
\end{aligned}$$

Since $\lim_{n\to\infty} \int_1^n f(x)\, dx = \infty$, it follows that $\lim_{n\to\infty} S_n = \infty$, and the series therefore diverges.

FIGURE 11.15

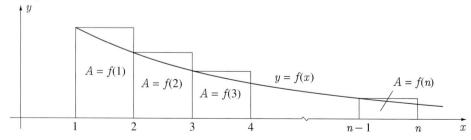

To use the integral test it is necessary to antidifferentiate the function $f(x)$, obtained by replacing n's in c_n with x's. When the antiderivative appears obvious, and other conditions of Theorem 11.6 are met, the integral test may be the easiest way to decide on convergence of the series; when the antiderivative is not obvious, it may be better to try another test.

EXAMPLE 11.6

Determine whether the following series converge or diverge:

(a) $\sum_{n=1}^{\infty} \dfrac{1}{n^2+1}$ (b) $\sum_{n=2}^{\infty} \dfrac{1}{n \ln n}$ (c) $\sum_{n=1}^{\infty} n e^{-n}$

SOLUTION (a) Since $f(x) = 1/(x^2+1)$ is continuous, positive, and decreasing for $x \geq 1$, and

$$\int_1^{\infty} \dfrac{1}{x^2+1}\,dx = \{\text{Tan}^{-1} x\}_1^{\infty} = \dfrac{\pi}{2} - \dfrac{\pi}{4} = \dfrac{\pi}{4},$$

it follows that the series $\sum 1/(n^2+1)$ converges.

(b) Since this series begins with $n=2$, we modify the integral test by considering the improper integral of $f(x) = 1/(x \ln x)$ over the interval $x \geq 2$. Since $f(x)$ is positive, continuous, and decreasing for $x \geq 2$, and

$$\int_2^{\infty} \dfrac{1}{x \ln x}\,dx = \{\ln(\ln x)\}_2^{\infty} = \infty,$$

it follows that $\sum_{n=2}^{\infty} 1/(n \ln n)$ diverges.

(c) Since xe^{-x} is continuous, positive, and decreasing for $x \geq 1$, and

$$\int_1^{\infty} xe^{-x}\,dx = \{-xe^{-x} - e^{-x}\}_1^{\infty} = \dfrac{2}{e},$$

the series converges. ∎

We mentioned before that two very important series are geometric series and the harmonic series. Convergence of geometric series was discussed in Section 11.3. The harmonic series belongs to a type of series called *p*-**series** defined as follows

$$\sum_{n=1}^{\infty} \dfrac{1}{n^p} = 1 + \dfrac{1}{2^p} + \dfrac{1}{3^p} + \cdots. \qquad (11.15)$$

When $p=1$, the *p*-series becomes the harmonic series which we know diverges. The series clearly diverges for $p < 0$ also. Consider the case when $p > 0$, but $p \neq 1$. The function $1/x^p$ is continuous, positive, and decreasing for $x \geq 1$, and

$$\int_1^{\infty} \dfrac{1}{x^p}\,dx = \left\{ \dfrac{1}{-(p-1)x^{p-1}} \right\}_1^{\infty} = \begin{cases} \infty & p < 1 \\ \dfrac{1}{p-1} & p > 1. \end{cases}$$

According to the integral test, the *p*-series converges when $p > 1$ and diverges when $p < 1$. Let us summarize results for geometric and *p*-series:

$$\text{Geometric Series} \qquad\qquad\qquad\qquad p\text{-series}$$

$$\sum_{n=1}^{\infty} ar^{n-1} = \begin{cases} \dfrac{a}{1-r} & |r| < 1 \\ \text{diverges} & |r| \geq 1 \end{cases} \qquad \sum_{n=1}^{\infty} \dfrac{1}{n^p} = \begin{cases} \text{converges} & p > 1 \\ \text{diverges} & p \leq 1 \end{cases} \quad (11.16)$$

It is unfortunate that no general formula can be given for the sum of the p-series when $p > 1$. Some interesting cases which arise frequently are cited below:

$$\sum_{n=1}^{\infty} \frac{1}{n^2} = \frac{\pi^2}{6}, \qquad \sum_{n=1}^{\infty} \frac{1}{n^3} = 1.202\ 056\ 903\ 1, \qquad \sum_{n=1}^{\infty} \frac{1}{n^4} = \frac{\pi^4}{90}. \quad (11.17)$$

Many series can be shown to converge or diverge by comparing them to known convergent and divergent series. This is the essence of the following two tests.

Theorem 11.7 (Comparison Test) If $0 \leq c_n \leq a_n$ for all n and $\sum_{n=1}^{\infty} a_n$ converges, then $\sum_{n=1}^{\infty} c_n$ converges. If $c_n \geq d_n \geq 0$ for all n and $\sum_{n=1}^{\infty} d_n$ diverges, then $\sum_{n=1}^{\infty} c_n$ diverges.

Proof Suppose first that $\sum a_n$ converges and $0 \leq c_n \leq a_n$ for all n. Since $a_n \geq 0$ the sequence of partial sums for the series $\sum a_n$ must be nondecreasing. Since the series converges, the sequence of partial sums must also be bounded (corollary to Theorem 11.1). Because $0 \leq c_n \leq a_n$ for all n, it follows that the sequence of partial sums of $\sum c_n$ is nondecreasing and bounded also. This sequence therefore has a limit and series $\sum c_n$ converges. A similar argument can be made for the divergent case.

Theorem 11.7 states that if the terms of a nonnegative series $\sum c_n$ are smaller than those of a known convergent series, then $\sum c_n$ must converge; if they are larger than a known nonnegative divergent series, then $\sum c_n$ must diverge.

In order for the comparison test to be useful, we require a catalogue of known convergent and divergent series with which we may compare other series. The geometric and p-series in equation 11.16 are extremely useful in this respect.

EXAMPLE 11.7 Determine whether the following series converge or diverge:

(a) $\displaystyle\sum_{n=2}^{\infty} \frac{\ln n}{n}$ \qquad (b) $\displaystyle\sum_{n=1}^{\infty} \frac{2n^2 - 1}{15n^4 + 14}$ \qquad (c) $\displaystyle\sum_{n=1}^{\infty} \frac{2n^2 + 1}{15n^4 - 14}$

SOLUTION (a) For this series, we note that when $n \geq 3$,

$$\frac{\ln n}{n} > \frac{1}{n}.$$

Since $\sum_{n=3}^{\infty} 1/n$ diverges (harmonic series with first two terms changed to zero), so does $\sum_{n=3}^{\infty} (\ln n)/n$ by the comparison test. Thus $\sum_{n=2}^{\infty} (\ln n)/n$ diverges. This can also be verified with the integral test.

(b) For the series $\sum (2n^2 - 1)/(15n^4 + 14)$, we note that

$$\frac{2n^2-1}{15n^4+14} < \frac{2n^2}{15n^4} = \frac{2}{15n^2}.$$

Since $\sum 2/(15n^2) = (2/15)\sum 1/n^2$ converges (p-series with $p=2$), so does the given series by the comparison test.

(c) For this series we note that the inequality

$$\frac{2n^2+1}{15n^4-14} \le \frac{3}{n^2}$$

is valid if and only if

$$45n^4 - 42 \ge 2n^4 + n^2.$$

But this inequality is valid if and only if

$$n^2(43n^2-1) \ge 42,$$

which is obviously true for $n \ge 1$. Consequently,

$$\frac{2n^2+1}{15n^4-14} \le \frac{3}{n^2},$$

and since $\sum 3/n^2 = 3\sum 1/n^2$ converges, so too does the given series. ∎

Two observations about Example 11.7 are worthwhile:

1. To use the comparison test we must first have a suspicion as to whether the given series converges or diverges in order to discuss the correct inequality; that is, if we suspect that the given series converges, we search for a convergent series for the right-hand side of the inequality $\le$, and if we suspect that the given series diverges, we search for a divergent series for the right-hand side of the opposite inequality $\ge$.

2. Recall that when the terms of a nonnegative series approach zero, whether the series has a sum depends on how fast these terms approach zero. The only difference in the nth terms of Examples 11.7(b) and (c) is the position of the negative sign. For very large n this difference becomes negligible since each nth term can, for large n, be closely approximated by $2/(15n^2)$. We might then expect similar analyses for these examples, and yet they are quite different. In addition, it is natural to ask where we obtained the factor 3 in part (c). The answer is, "by trial and error".

Each of these observations points out weaknesses in the comparison test, but these problems can be eliminated in many examples with the following test.

Theorem 11.8

(Limit Comparison Test) If $0 \le c_n$ and $0 < b_n$, and

$$\lim_{n \to \infty} \frac{c_n}{b_n} = l, \quad 0 < l < \infty, \tag{11.18}$$

then series $\sum_{n=1}^{\infty} c_n$ converges if $\sum_{n=1}^{\infty} b_n$ converges, and diverges if $\sum_{n=1}^{\infty} b_n$ diverges.

Proof Suppose that series $\sum b_n$ converges. Since sequence $\{c_n/b_n\}$ converges to l, we can use the result of Exercise 75 in Section 11.1 and say that for all n greater than or equal to some integer N,

$$\frac{c_n}{b_n} < l + 1, \quad \text{or} \quad c_n < (l+1)b_n.$$

Since the series $\sum_{n=N}^{\infty} (l+1)b_n = (l+1)\sum_{n=N}^{\infty} b_n$ converges, so also must $\sum_{n=N}^{\infty} c_n$ (by the comparison test). Hence, $\sum_{n=1}^{\infty} c_n$ converges. A similar argument can be made for the divergent case.

If $\lim_{n\to\infty} c_n/b_n = l$, then for very large n we can say that $c_n \approx lb_n$. It follows that if $\{b_n\}$ approaches zero, $\{c_n\}$ approaches zero $1/l$ times as fast as $\{b_n\}$. Theorem 11.8 implies then that if the sequence of nth terms of two nonnegative series approach zero at proportional rates, the series converge or diverge together.

To use the limit comparison test, we must find a series $\sum b_n$ so that the limit of the ratio c_n/b_n is finite and greater than zero. To obtain this series, it is sufficient in many examples to simply answer the question, "What does the given series really look like for very large n?". In both Examples 11.7(b) and (c) we see that for large n

$$c_n \approx \frac{2n^2}{15n^4} = \frac{2}{15n^2}.$$

Consequently, we calculate in Example 11.7(c) that

$$l = \lim_{n\to\infty} \frac{\frac{2n^2+1}{15n^4-14}}{\frac{2}{15n^2}} = \lim_{n\to\infty} \frac{n^2\left(2+\frac{1}{n^2}\right)}{n^4\left(15-\frac{14}{n^4}\right)} \cdot \frac{15n^2}{2} = 1.$$

Since $\sum 2/(15n^2)$ converges, so too does $\sum (2n^2+1)/(15n^4-14)$.

EXAMPLE 11.8

Determine whether the following series converge or diverge:

(a) $\sum_{n=1}^{\infty} \frac{\sqrt{n^2+2n-1}}{n^{5/2}+15n-3}$ (b) $\sum_{n=1}^{\infty} \frac{2^n+1}{3^n+5}$

SOLUTION (a) For very large n, the nth term of the series can be approximated by

$$\frac{\sqrt{n^2+2n-1}}{n^{5/2}+15n-3} \approx \frac{n}{n^{5/2}} = \frac{1}{n^{3/2}}.$$

We calculate therefore that

$$l = \lim_{n\to\infty} \frac{\frac{\sqrt{n^2+2n-1}}{n^{5/2}+15n-3}}{\frac{1}{n^{3/2}}} = \lim_{n\to\infty} \frac{n\sqrt{1+\frac{2}{n}-\frac{1}{n^2}}}{n^{5/2}\left(1+\frac{15}{n^{3/2}}-\frac{3}{n^{5/2}}\right)} \cdot n^{3/2} = 1.$$

Since $\sum 1/n^{3/2}$ converges (p-series with $p = 3/2$), so too does the given series by the limit comparison test.

(b) For large n,
$$\frac{2^n+1}{3^n+5} \approx \left(\frac{2}{3}\right)^n.$$

We calculate therefore that
$$l = \lim_{n\to\infty} \frac{\frac{2^n+1}{3^n+5}}{\left(\frac{2}{3}\right)^n} = \lim_{n\to\infty} \frac{2^n\left(1+\frac{1}{2^n}\right)}{3^n\left(1+\frac{5}{3^n}\right)} \cdot \frac{3^n}{2^n} = 1.$$

Since $\sum (2/3)^n$ converges (a geometric series with $r = 2/3$), the given series converges by the limit comparison test. ∎

EXERCISES 11.4

In Exercises 1–22 determine whether the series converges or diverges.

1. $\sum_{n=1}^{\infty} \dfrac{1}{2n+1}$

2. $\sum_{n=1}^{\infty} \dfrac{1}{4n-3}$

3. $\sum_{n=1}^{\infty} \dfrac{1}{2n^2+4}$

4. $\sum_{n=1}^{\infty} \dfrac{1}{5n^2-3n-1}$

5. $\sum_{n=2}^{\infty} \dfrac{1}{n^3-1}$

6. $\sum_{n=4}^{\infty} \dfrac{n^2}{n^4-6n^2+5}$

7. $\sum_{n=1}^{\infty} \dfrac{1}{(2n-1)(2n+1)}$

8. $\sum_{n=1}^{\infty} \dfrac{n-5}{n^2+3n-2}$

9. $\sum_{n=2}^{\infty} \dfrac{1}{\ln n}$

10. $\sum_{n=1}^{\infty} n^2 e^{-2n}$

11. $\sum_{n=2}^{\infty} \dfrac{\sqrt{n^2+2n-3}}{n^2+5}$

12. $\sum_{n=1}^{\infty} \dfrac{\sqrt{n+5}}{n^3+3}$

13. $\sum_{n=1}^{\infty} \sqrt{\dfrac{n^2+2n+3}{2n^4-n}}$

14. $\sum_{n=2}^{\infty} \dfrac{1}{n^2 \ln n}$

15. $\sum_{n=1}^{\infty} \dfrac{1}{2^n} \sin\left(\dfrac{\pi}{n}\right)$

16. $\sum_{n=1}^{\infty} \dfrac{\sqrt{n^2+1}}{n^3} \operatorname{Tan}^{-1} n$

17. $\sum_{n=1}^{\infty} \dfrac{2^n+n}{3^n+1}$

18. $\sum_{n=2}^{\infty} \dfrac{1+\ln^2 n}{n\ln^2 n}$

19. $\sum_{n=1}^{\infty} \dfrac{1+1/n}{e^n}$

20. $\sum_{n=1}^{\infty} \dfrac{\ln(n+1)}{n+1}$

21. $\sum_{n=1}^{\infty} n e^{-n^2}$

22. $\sum_{n=2}^{\infty} \dfrac{1}{n\sqrt[3]{\ln n}}$

In Exercises 23–25 find values of p for which the series converges.

23. $\sum_{n=2}^{\infty} \dfrac{1}{n^p \ln n}$

24. $\sum_{n=2}^{\infty} \dfrac{1}{n(\ln n)^p}$

25. $\sum_{n=2}^{\infty} \dfrac{1}{(\ln n)^p}$

SECTION 11.5

Further Convergence Tests for Nonnegative Series

In this section we consider two additional tests to determine whether nonnegative series converge or diverge. The first test, called the limit ratio test, indicates whether a series $\sum c_n$ resembles a geometric series for large n.

Theorem 11.9

(Limit Ratio Test) Suppose $c_n > 0$ and

$$\lim_{n \to \infty} \frac{c_{n+1}}{c_n} = L. \qquad (11.19)$$

Then,

(i) $\sum_{n=1}^{\infty} c_n$ converges if $L < 1$;
(ii) $\sum_{n=1}^{\infty} c_n$ diverges if $L > 1$ (or if $\lim_{n \to \infty} c_{n+1}/c_n = \infty$);
(iii) $\sum_{n=1}^{\infty} c_n$ may converge or diverge if $L = 1$.

FIGURE 11.16

Proof (i) By the result of part (a) in Exercise 74 of Section 11.1 (see also Figure 11.16), we can say that if $\lim_{n \to \infty} c_{n+1}/c_n = L < 1$, there exists an integer N such that for all $n \geq N$,

$$\frac{c_{n+1}}{c_n} < \frac{L+1}{2}.$$

Consequently,

$$c_{N+1} < \left(\frac{L+1}{2}\right) c_N;$$

$$c_{N+2} < \left(\frac{L+1}{2}\right) c_{N+1} < \left(\frac{L+1}{2}\right)^2 c_N;$$

$$c_{N+3} < \left(\frac{L+1}{2}\right) c_{N+2} < \left(\frac{L+1}{2}\right)^3 c_N;$$

and so on. Hence,

$$c_N + c_{N+1} + c_{N+2} + \cdots < c_N + \left(\frac{L+1}{2}\right) c_N + \left(\frac{L+1}{2}\right)^2 c_N + \cdots.$$

Since the right-hand side of this inequality is a geometric series with common ratio $(L+1)/2 < 1$, it follows by the comparison test that $\sum_{n=N}^{\infty} c_n$ converges. Therefore, $\sum_{n=1}^{\infty} c_n$ converges also.

FIGURE 11.17

(ii) By part (b) in Exercise 74 of Section 11.1 (see also Figure 11.17), if $\lim_{n \to \infty} c_{n+1}/c_n = L > 1$, there exists an integer N such that for all $n \geq N$,

$$\frac{c_{n+1}}{c_n} > \frac{L+1}{2} > 1.$$

When $\lim_{n \to \infty} c_{n+1}/c_n = \infty$, it is also true that for n greater than or equal to some N, c_{n+1}/c_n must be greater than 1. This implies that for all $n > N$, $c_n > c_N$, and therefore

$$\lim_{n \to \infty} c_n \neq 0.$$

Hence $\sum c_n$ diverges by the nth term test.

(iii) To show that the limit ratio test fails when $L = 1$, consider the two p-series $\sum 1/n$ and $\sum 1/n^2$. For each series, $L = 1$, yet the first series diverges and the second converges.

In a nonrigorous way we can justify the limit ratio test from the following standpoint. If $\lim_{n \to \infty} c_{n+1}/c_n = L$, then for large n, each term of series $\sum c_n$ is essentially L times the term before it; that is, the series resembles a geometric series with common ratio L. We would expect convergence of the series if $L < 1$ and divergence if $L > 1$. We might also anticipate some indecision about the $L = 1$ case, depending on how this limit is reached, since this case corresponds to the common ratio which separates convergent and divergent geometric series.

EXAMPLE 11.9

Determine whether the following series converge or diverge:

(a) $\sum_{n=1}^{\infty} \frac{2^n}{n^4}$ (b) $\sum_{n=1}^{\infty} \frac{n^{100}}{1 \cdot 3 \cdot 5 \cdot \cdots \cdot (2n-1)}$ (c) $\sum_{n=1}^{\infty} \frac{n^n}{n!}$

SOLUTION (a) For the series $\sum 2^n/n^4$,

$$L = \lim_{n \to \infty} \frac{\frac{2^{n+1}}{(n+1)^4}}{\frac{2^n}{n^4}} = \lim_{n \to \infty} 2\left(\frac{n}{n+1}\right)^4 = 2,$$

and the series therefore diverges by the limit ratio test.

(b) For this series,

$$L = \lim_{n \to \infty} \frac{\frac{(n+1)^{100}}{1 \cdot 3 \cdot 5 \cdot \cdots \cdot (2n+1)}}{\frac{n^{100}}{1 \cdot 3 \cdot 5 \cdot \cdots \cdot (2n-1)}} = \lim_{n \to \infty} \left(\frac{n+1}{n}\right)^{100} \frac{1}{2n+1} = 0,$$

and the series therefore converges by the limit ratio test.

(c) Since

$$L = \lim_{n \to \infty} \frac{\frac{(n+1)^{n+1}}{(n+1)!}}{\frac{n^n}{n!}} = \lim_{n \to \infty} \left(\frac{n+1}{n}\right)^n = e > 1$$

(see equation 3.29b), the series $\sum n^n/n!$ diverges. ∎

We have one last test for nonnegative series.

Theorem 11.10

(Limit Root Test) Suppose $c_n \geq 0$ and

$$\lim_{n \to \infty} \sqrt[n]{c_n} = R. \qquad (11.20)$$

Then,

(i) $\sum_{n=1}^{\infty} c_n$ converges if $R < 1$;
(ii) $\sum_{n=1}^{\infty} c_n$ diverges if $R > 1$ (or if $\lim_{n \to \infty} \sqrt[n]{c_n} = \infty$);
(iii) $\sum_{n=1}^{\infty} c_n$ may converge or diverge if $R = 1$.

Proof (i) By the result of part (a) in Exercise 74 of Section 11.1, if $\lim_{n \to \infty} \sqrt[n]{c_n} = R < 1$, there exists an integer N such that for all $n \geq N$,

$$\sqrt[n]{c_n} < \frac{R+1}{2}.$$

Consequently,

$$\sqrt[N]{c_N} < \frac{R+1}{2}, \quad \text{or} \quad c_N < \left(\frac{R+1}{2}\right)^N;$$

$$\sqrt[N+1]{c_{N+1}} < \frac{R+1}{2}, \quad \text{or} \quad c_{N+1} < \left(\frac{R+1}{2}\right)^{N+1};$$

$$\sqrt[N+2]{c_{N+2}} < \frac{R+1}{2}, \quad \text{or} \quad c_{N+2} < \left(\frac{R+1}{2}\right)^{N+2};$$

and so on. Hence,

$$c_N + c_{N+1} + c_{N+2} + \cdots < \left(\frac{R+1}{2}\right)^N + \left(\frac{R+1}{2}\right)^{N+1} + \left(\frac{R+1}{2}\right)^{N+2} + \cdots .$$

Since the right-hand side of this inequality is a convergent geometric series, the left-hand side must be a convergent series also by the comparison test. Thus, $\sum_{n=N}^{\infty} c_n$ converges, and so also must $\sum_{n=1}^{\infty} c_n$.

(ii) When $\lim_{n \to \infty} \sqrt[n]{c_n} = R > 1$, there exists an integer N such that for all $n \geq N$,

$$\sqrt[n]{c_n} > \frac{R+1}{2} > 1 \quad \text{or} \quad c_n > 1,$$

as in part (b) of Exercise 74 in Section 11.1. When $\lim_{n \to \infty} \sqrt[n]{c_n} = \infty$, it is also true that for n greater than or equal to some N, $\sqrt[n]{c_n}$ must be greater than 1. But it now follows that $\lim_{n \to \infty} c_n \neq 0$, and the series diverges by the nth term test.

(iii) To show that the test fails when $R = 1$, we show that $R = 1$ for the two p-series $\sum 1/n$ and $\sum 1/n^2$, which diverges and converges respectively. For the harmonic series, let $R = \lim_{n \to \infty} (1/n)^{1/n}$. If we take logarithms, then

$$\ln R = \ln \left[\lim_{n \to \infty} \left(\frac{1}{n}\right)^{1/n}\right] = \lim_{n \to \infty} \ln \left(\frac{1}{n}\right)^{1/n} = \lim_{n \to \infty} \frac{1}{n} \ln \left(\frac{1}{n}\right) = -\lim_{n \to \infty} \frac{1}{n} \ln n$$

$$= -\lim_{n \to \infty} \frac{\frac{1}{n}}{1} \qquad \text{(by L'Hôpital's rule)}$$

$$= 0.$$

Hence, $R = 1$ for this divergent series. A similar analysis gives $R = 1$ for the convergent series $\sum 1/n^2$.

EXAMPLE 11.10 Determine whether the following series converge or diverge:

(a) $\sum_{n=1}^{\infty} \left(\frac{n+1}{n}\right)^{n^2}$ (b) $\sum_{n=1}^{\infty} \frac{n}{(\ln n)^n}$

SOLUTION (a) Since

$$R = \lim_{n \to \infty} \left[\left(\frac{n+1}{n}\right)^{n^2}\right]^{1/n} = \lim_{n \to \infty} \left(\frac{n+1}{n}\right)^n = e > 1,$$

the series diverges by the limit root test.

(b) For this series,

$$R = \lim_{n \to \infty} \left[\frac{n}{(\ln n)^n} \right]^{1/n} = \lim_{n \to \infty} \frac{n^{1/n}}{\ln n}.$$

If we set $L = \lim_{n \to \infty} n^{1/n}$, then

$$\ln L = \ln \left(\lim_{n \to \infty} n^{1/n} \right) = \lim_{n \to \infty} \frac{1}{n} \ln n$$

$$= \lim_{n \to \infty} \frac{\frac{1}{n}}{1} \quad \text{(by L'Hôpital's rule)}$$

$$= 0.$$

Thus, $L = 1$, and it follows that

$$R = \lim_{n \to \infty} \frac{n^{1/n}}{\ln n} = 0.$$

The series therefore converges. ∎

We have developed six tests to determine whether series converge or diverge:

n^{th} term test

integral test

comparison test

limit comparison test

limit ratio test

limit root test

The form of the nth term of a series often suggests which test should be used. Keep the following ideas in mind when choosing a test:

1. The limit ratio test can be effective on factorials, products of the form $1 \cdot 3 \cdot 5 \cdot \cdots \cdot (2n-1)$, and constants raised to powers involving n (2^n, 3^{-n}, etc.)

2. The limit root test thrives on functions of n raised to powers involving n (see Example 11.10).

3. The limit comparison test is successful on rational functions of n, and fractional powers as well ($\sqrt{n}$, $\sqrt[3]{n/(n+1)}$, etc.)

4. The integral test can be effective when the nth term is easily integrated. Logarithms often require the integral test.

By definition, a series $\sum c_n$ converges if and only if its sequence of partial sums $\{S_n\}$ converges. The difficulty with using this definition to discuss convergence of a series is that S_n can seldom be evaluated in a simple form, and therefore consideration of $\lim_{n \to \infty} S_n$ is impossible. The above tests have the advantage of avoiding partial sums. On the other hand, they have one disadvantage. Although they may indicate that a series does indeed have a sum, the tests in no way suggest the value of the sum. The problem of calculating the sum often proves more difficult than showing that it exists in the first place. In Sections 11.7 and 11.11 we discuss various ways to calculate sums for known convergent series.

EXERCISES 11.5

In Exercises 1–20 determine whether the series converges or diverges.

1. $\sum_{n=1}^{\infty} \dfrac{e^n}{n^4}$

2. $\sum_{n=1}^{\infty} \dfrac{1}{n!}$

3. $\sum_{n=1}^{\infty} \dfrac{n^3}{2^n}$

4. $\sum_{n=1}^{\infty} \dfrac{1}{n^n}$

5. $\sum_{n=1}^{\infty} \dfrac{(n-1)(n-2)}{n^2 2^n}$

6. $\sum_{n=1}^{\infty} \dfrac{(2n)!}{(n!)^2}$

7. $\sum_{n=1}^{\infty} \dfrac{\sqrt{n+1}}{n^{n+1/2}}$

8. $\sum_{n=1}^{\infty} \dfrac{3^{-n} + 2^{-n}}{4^{-n} + 5^{-n}}$

9. $\sum_{n=1}^{\infty} \dfrac{e^{-n}}{\sqrt{n+\pi}}$

10. $\sum_{n=1}^{\infty} \dfrac{2 \cdot 4 \cdot \cdots \cdot (2n)}{4 \cdot 7 \cdot \cdots \cdot (3n+1)}$

11. $\sum_{n=1}^{\infty} \dfrac{n^{n-1}}{3^{n-1}(n-1)!}$

12. $\sum_{n=1}^{\infty} n\left(\dfrac{3}{4}\right)^n$

13. $\sum_{n=1}^{\infty} \dfrac{1 + 1/n}{e^n}$

14. $\sum_{n=1}^{\infty} \dfrac{2 \cdot 4 \cdot \cdots \cdot (2n)}{3 \cdot 5 \cdot \cdots \cdot (2n+1)} \left(\dfrac{1}{n^2}\right)$

15. $\sum_{n=1}^{\infty} \dfrac{n^n}{(n+1)^{n+1}}$

16. $\sum_{n=1}^{\infty} \dfrac{(n+1)^n}{n^{n+1}}$

17. $\sum_{n=1}^{\infty} \dfrac{n^4 + 3}{5^{n/2}}$

18. $\sum_{n=1}^{\infty} \dfrac{2^n + n^2 3^n}{4^n}$

19. $\sum_{n=1}^{\infty} \dfrac{n^2 2^n - n}{n^3 + 1}$

20. $\sum_{n=1}^{\infty} \dfrac{(2n)!}{(3n)!} 5^{2n}$

21. For what integer values of a is the series

$$\sum_{n=1}^{\infty} \dfrac{(n!)^2}{(an)!}$$

convergent?

SECTION 11.6

Series with Positive and Negative Terms

The convergence tests of Sections 11.4 and 11.5 are applicable to nonnegative series, series whose terms are all nonnegative. Series with infinitely many positive and negative terms are more complicated. It is fortunate, however, that all our test are still useful in discussing convergence of series with positive and negative terms. What makes this possible is the following definition and Theorem 11.11.

Definition 11.7

A series $\sum_{n=1}^{\infty} c_n$ is said to be **absolutely convergent** if the series of absolute values $\sum_{n=1}^{\infty} |c_n|$ converges.

At first glance it might seem that absolute convergence is a strange concept indeed. What possible good could it do to consider the series of absolute values, which is quite different from the original series? The fact is that when the series of absolute values converges, it automatically follows from the next theorem that the original series converges also. And since the series of absolute values has all nonnegative terms, we can use the comparison, limit comparison, limit ratio, limit root, or integral test to consider its convergence.

Theorem 11.11

If a series is absolutely convergent, then it is convergent.

Proof Let $\{S_n\}$ be the sequence of partial sums of the absolutely convergent series $\sum c_n$. Define sequences $\{P_n\}$ and $\{N_n\}$, where P_n is the sum of all positive terms in

S_n, and N_n is the sum of the absolute values of all negative terms in S_n. Then

$$S_n = P_n - N_n.$$

The sequence of partial sums for the series of absolute values $\sum |c_n|$ is

$$\{P_n + N_n\},$$

and this sequence must be nondecreasing and bounded. Since each of the sequences $\{P_n\}$ and $\{N_n\}$ is nondecreasing and a part of $\{P_n + N_n\}$, it follows that each is bounded, and therefore has a limit, say, P and N, respectively. As a result, sequence $\{P_n - N_n\} = \{S_n\}$ has limit $P - N$, and series $\sum c_n$ converges to $P - N$.

> **EXAMPLE 11.11**
>
> Show that the following series are absolutely convergent:
>
> (a) $\sum_{n=1}^{\infty} \dfrac{(-1)^n n}{(n+1)2^n}$
>
> (b)
> $$1 - \dfrac{1}{2^2} - \dfrac{1}{3^3} + \dfrac{1}{4^4} + \dfrac{1}{5^5} + \dfrac{1}{6^6} - \dfrac{1}{7^7} - \dfrac{1}{8^8} - \dfrac{1}{9^9} - \dfrac{1}{10^{10}} + \dfrac{1}{11^{11}} + \cdots + \dfrac{1}{15^{15}} - \cdots.$$
>
> **SOLUTION** (a) The series of absolute values is $\sum n/[(n+1)2^n]$. We use the limit comparison test to show that it converges. Since
>
> $$l = \lim_{n \to \infty} \dfrac{\dfrac{n}{(n+1)2^n}}{\dfrac{1}{2^n}} = 1,$$
>
> and $\sum (1/2)^n$ is convergent (a geometric series with $r = 1/2$), it follows that the series of absolute values converges. The given series therefore converges absolutely.
>
> (b) The series of absolute values is $\sum 1/n^n$. We use the comparison test to verify that this series converges. Since
>
> $$\dfrac{1}{n^n} \leq \dfrac{1}{n^2},$$
>
> and $\sum 1/n^2$ converges, so does $\sum 1/n^n$. The given series therefore converges absolutely. ∎

In Example 11.11, absolute convergence of the given series implies convergence of the series, but it is customary to omit such a statement. It is important to realize, however, that it is the given series that is being analyzed, and its convergence is guaranteed by Theorem 11.11.

We now ask whether series can converge without converging absolutely. If there are such series, and indeed there are, we must devise new convergence tests. We describe these series as follows.

> **Definition 11.8**
>
> A series that converges but does not converge absolutely is said to **converge conditionally**.

The most important type of series with both positive and negative terms is an alternating series. As the name suggests, an **alternating series** has terms that are alternately positive and negative. For example,

$$\sum_{n=1}^{\infty} \frac{(-1)^{n+1}}{n} = 1 - \frac{1}{2} + \frac{1}{3} - \frac{1}{4} + \frac{1}{5} - \cdots$$

is an alternating series called the **alternating harmonic series**.

Given an alternating series to examine for convergence, we first test for absolute convergence as in Example 11.11(a). Should this fail, we check for conditional convergence with the following test.

Theorem 11.12 (Alternating Series Test) *An alternating series $\sum_{n=1}^{\infty} c_n$ converges if the sequence of absolute values of the terms $\{|c_n|\}$ is nonincreasing and has limit zero.*

Proof Suppose $c_1 > 0$, in which case all odd terms are positive and all even terms are negative. If $\{S_n\}$ is the sequence of partial sums of $\sum c_n$, then the even partial sums S_{2n} can be expressed in two forms:

$$S_{2n} = (c_1 + c_2) + (c_3 + c_4) + \cdots + (c_{2n-1} + c_{2n}),$$
$$S_{2n} = c_1 + (c_2 + c_3) + (c_4 + c_5) + \cdots + (c_{2n-2} + c_{2n-1}) + c_{2n}.$$

Since $\{|c_n|\}$ is nonincreasing ($|c_n| \geq |c_{n+1}|$), each term in the parentheses of the first expression is nonnegative. Consequently, the subsequence $\{S_{2n}\}$ of even partial sums of $\{S_n\}$ is nondecreasing. Since each term in the parentheses in the second expression is nonpositive, as is c_{2n}, it follows that $S_{2n} \leq c_1$ for all n. By Theorem 11.1 then, the sequence $\{S_{2n}\}$ has a limit, say, S.

The subsequence of odd partial sums $\{S_{2n-1}\}$ is such that $S_{2n-1} = S_{2n} - c_{2n}$. Since $\{S_{2n}\}$ has limit S and $\{c_{2n}\}$ has limit zero, it follows that $\{S_{2n-1}\}$ has limit S also. Consequently, $\{S_n\}$ has limit S, and series $\sum c_n$ converges.

A similar proof holds when $c_1 < 0$.

EXAMPLE 11.12 Determine whether the following series converge absolutely, converge conditionally, or diverge:

(a) $\sum_{n=1}^{\infty} \frac{(-1)^{n+1}}{n}$ (b) $\sum_{n=1}^{\infty} (-1)^n \frac{\sqrt{n^2 + 5n}}{n^{3/2}}$ (c) $\sum_{n=1}^{\infty} (-1)^{n+1} \frac{4^n}{n^5 3^n}$

SOLUTION (a) The alternating harmonic series $\sum (-1)^{n+1}/n$ is not absolutely convergent because the series of absolute values $\sum 1/n$ diverges. Since the sequence of absolute values of the terms $\{1/n\}$ is decreasing with limit zero, series $\sum (-1)^{n+1}/n$ converges conditionally.

(b) For this alternating series, we first consider the series of absolute values

$$\sum_{n=1}^{\infty} \frac{\sqrt{n^2 + 5n}}{n^{3/2}}.$$

We use the limit comparison test on this series. Since

$$l = \lim_{n \to \infty} \frac{\frac{\sqrt{n^2 + 5n}}{n^{3/2}}}{\frac{1}{n^{1/2}}} = \lim_{n \to \infty} \frac{n\sqrt{1 + \frac{5}{n}}}{n^{3/2}} \cdot n^{1/2} = 1,$$

and $\sum 1/n^{1/2}$ diverges, so does $\sum \sqrt{n^2 + 5n}/n^{3/2}$. The original series does not therefore converge absolutely. We now resort to the alternating series test. The sequence $\{\sqrt{n^2 + 5n}/n^{3/2}\}$ of absolute values of the terms of the series is nonincreasing if

$$\frac{\sqrt{(n+1)^2 + 5(n+1)}}{(n+1)^{3/2}} \leq \frac{\sqrt{n^2 + 5n}}{n^{3/2}}.$$

When we square and cross multiply, the inequality becomes

$$n^3(n^2 + 7n + 6) \leq (n^2 + 5n)(n+1)^3$$
$$= n^5 + 8n^4 + 18n^3 + 16n^2 + 5n;$$

that is,

$$n^4 + 12n^3 + 16n^2 + 5n \geq 0,$$

which is obviously valid because $n \geq 1$. Since $\lim_{n \to \infty} \sqrt{n^2 + 5n}/n^{3/2} = 0$, we conclude that the alternating series $\sum (-1)^n(\sqrt{n^2 + 5n}/n^{3/2})$ converges conditionally.

(c) If we apply the limit ratio test to the series of absolute values $\sum 4^n/(n^5 3^n)$, we have

$$L = \lim_{n \to \infty} \frac{\frac{4^{n+1}}{(n+1)^5 3^{n+1}}}{\frac{4^n}{n^5 3^n}} = \lim_{n \to \infty} \frac{4}{3}\left(\frac{n}{n+1}\right)^5 = \frac{4}{3}.$$

Since $L > 1$, the series $\sum 4^n/(n^5 3^n)$ diverges. The original alternating series does not therefore converge absolutely. But $L = \frac{4}{3}$ implies that for large n, each term in the series of absolute values is approximately $\frac{4}{3}$ times the term that precedes it, and therefore

$$\lim_{n \to \infty} \frac{4^n}{n^5 3^n} = \infty.$$

Consequently,

$$\lim_{n \to \infty} (-1)^{n+1} \frac{4^n}{n^5 3^n}$$

cannot possibly exist, and the given series diverges by the nth term test. ∎

We have noted several times that the essential question for convergence of a nonnegative series is, "Do the terms approach zero quickly enough to guarantee convergence of the series?" With a series that has infinitely many positive and negative terms, this question is inappropriate. Such a series may converge because of a partial cancelling effect; for example, a negative term may offset the effect of a large positive term. This kind of process may produce a convergent series even though the series would be divergent if all terms were replaced by their absolute values. A specific example is the alternating harmonic series which converges (conditionally) because of this cancelling effect, whereas the harmonic series itself, which has no cancellations, diverges.

In Sections 11.4–6, we have obtained a number of tests for determining whether series converge or diverge. To test a series for convergence, we suggest the following procedure:

1. Try the nth term test for divergence.
2. If $\{c_n\}$ has limit zero and the series is nonnegative, try the comparison, limit comparison, limit ratio, limit root, or integral test.
3. If $\{c_n\}$ has limit zero and the series contains both positive and negative terms, test for absolute convergence using the tests in 2. If this fails and the series is alternating, test for conditional convergence with the alternating series test.

Each of the comparison, limit comparison, limit ratio, limit root, integral, and alternating series tests requires conditions to be satisfied for all terms of the series. Specifically, the comparison, limit comparison, and limit root tests require $c_n \geq 0$ for all n; the limit ratio test requires $c_n > 0$; the integral test requires $f(n)$ to be positive, continuous, and decreasing; and the alternating series test requires $\{|c_n|\}$ to be nonincreasing and $\{c_n\}$ to be alternately positive and negative. None of these requirements is essential for all n; in fact, so long as they are satisfied for all terms in the series beyond some point, say for n greater than or equal to some integer N, the particular test may be used on the series $\sum_{n=N}^{\infty} c_n$. The original series $\sum_{n=1}^{\infty} c_n$ then converges if and only if $\sum_{n=N}^{\infty} c_n$ converges.

EXERCISES 11.6

In Exercises 1–14 determine whether the series converges absolutely, converges conditionally, or diverges.

1. $\sum_{n=1}^{\infty} (-1)^n \dfrac{n}{n^3 + 1}$

2. $\sum_{n=1}^{\infty} (-1)^n \dfrac{n}{n^2 + 1}$

3. $\sum_{n=1}^{\infty} \dfrac{\cos(n\pi/2)}{2n^2}$

4. $\sum_{n=1}^{\infty} (-1)^n \dfrac{n^3}{3^n}$

5. $\sum_{n=1}^{\infty} \dfrac{(-1)^{n+1}}{\sqrt{n}}$

6. $\sum_{n=1}^{\infty} (-1)^n \dfrac{3^n}{n^3}$

7. $\sum_{n=1}^{\infty} (-1)^n \dfrac{n}{n^2 + n + 1}$

8. $\sum_{n=1}^{\infty} \dfrac{n \sin(n\pi/4)}{2^n}$

9. $\sum_{n=1}^{\infty} (-1)^{n+1} \left(\dfrac{n}{n+1} \right)$

10. $\sum_{n=1}^{\infty} (-1)^{n+1} \dfrac{\sqrt{3n-2}}{n}$

11. $\sum_{n=1}^{\infty} (-1)^n \left(\dfrac{n}{n+1} \right)^n$

12. $\sum_{n=1}^{\infty} (-1)^n \dfrac{\sqrt{n^2 + 3}}{n^2 + 5}$

13. $\sum_{n=2}^{\infty} (-1)^{n-1} \dfrac{\ln n}{n}$

14. $\sum_{n=1}^{\infty} \dfrac{\cos(n\pi/10) \cot^{-1} n}{n^3 + 5n}$

15. Discuss convergence of the series

$$\sum_{n=1}^{\infty} \dfrac{\sin(nx)}{n^2}.$$

16. Prove that if $\sum c_n$ converges absolutely, then $\sum c_n^p$ converges absolutely for all integers $p > 1$.

17. Discuss convergence of the series

$$\sum_{n=1}^{\infty} (-1)^n \dfrac{n^n}{(n+1)^{n+1}}.$$

SECTION 11.7

Approximating the Sum of a Series

We have considered only half of the convergence problem for infinite series of constants. The comparison, limit comparison, limit ratio, limit root, integral, and alternating series tests may determine whether a series converges or diverges, but they do not determine the sum of the series in the case of a convergent series. This part of the problem, as suggested before, can sometimes be more complicated.

If the convergent series is a geometric series, no problem exists; we can use formula 11.13b to find its sum. It may also happen that the nth partial sum S_n of the series can be calculated in a simple form, in which case the sum of the series is $\lim_{n\to\infty} S_n$. Cases of this latter type are very rare. These two methods exhaust our present capabilities for obtaining the sum of a convergent series, but we will find additional methods in Section 11.11.

In many applications of infinite series we only need a reasonable approximation to the sum of a series, and therefore turn our attention to the problem of estimating the sum of a convergent series. The easiest method for estimating the sum S of a series $\sum c_n$ is simply to choose the partial sum S_N for some N as an approximation; that is, truncate the series after N terms and choose

$$S \approx S_N = c_1 + c_2 + \cdots + c_N.$$

But an approximation is of value only if we can make some definitive statement about its accuracy. In truncating the series, we have neglected the infinity of terms $\sum_{n=N+1}^{\infty} c_n$, and the accuracy of the approximation is therefore determined by the size of $\sum_{n=N+1}^{\infty} c_n$; the smaller it is, the better the approximation. The problem is that we do not know the exact value of $\sum_{n=N+1}^{\infty} c_n$; if we did, there would be no need to approximate the sum of the original series in the first place. What we must do is estimate the sum $\sum_{n=N+1}^{\infty} c_n$.

When the integral test or the alternating series test are used to prove that a series converges, simple formulas give accuracy estimates on the truncated series. Let us illustrate these first.

Truncating an Alternating Series

It is very simple to obtain an estimate of the accuracy of a truncated alternating series $\sum c_n$ provided the sequence $\{|c_n|\}$ is nonincreasing with limit zero. In the proof of Theorem 11.12, we showed that when $c_1 > 0$, the subsequence $\{S_{2n}\}$ of even partial sums is nondecreasing and therefore approaches the sum of the series $\sum c_n$ from below (Figure 11.18). In a similar way, we can show that the subsequence $\{S_{2n-1}\}$ of odd partial sums is nonincreasing and approaches the sum of the series from above. It follows that the sum $\sum c_n$ must be between any two terms of the subsequences $\{S_{2n}\}$ and $\{S_{2n-1}\}$. In particular, it must be between any two successive partial sums. Thus, if $\sum c_n$ is approximated by S_N, the maximum possible error is $S_{N+1} - S_N = c_{N+1}$, the next term. *When an alternating series is truncated, the maximum possible error is the next term.*

FIGURE 11.18

EXAMPLE 11.13

Use the first twenty terms of the series $\sum_{n=1}^{\infty} (-1)^{n+1}/n^3$ to estimate its sum. Obtain an error estimate.

SOLUTION The sum of the first twenty terms of the series is $0.901\,485$. Since the series is alternating, the maximum possible error in this estimate is the twenty-first term, $1/21^3 < 0.000\,108$. Thus,

$$0.901\,485 < \sum_{n=1}^{\infty} \frac{(-1)^{n+1}}{n^3} < 0.901\,593.$$

In practical situations, we often have to decide how many terms of a series to take in order to guarantee a certain degree of accuracy. Once again this is easy for alternating series.

EXAMPLE 11.14

How many terms in the series $\sum_{n=2}^{\infty} (-1)^{n+1}/(n^3 \ln n)$ ensure a truncation error of less than 10^{-5}?

SOLUTION The maximum error in truncating this alternating series when $n = N$ is

$$\frac{(-1)^{N+2}}{(N+1)^3 \ln(N+1)}.$$

The absolute value of this error is less than 10^{-5} when

$$\frac{1}{(N+1)^3 \ln(N+1)} < 10^{-5} \quad \text{or}$$

$$(N+1)^3 \ln(N+1) > 10^5.$$

A calculator quickly reveals that the smallest integer for which this is valid is $N = 30$. Thus, the truncated series has the required accuracy after the 29th term. Note that the first term corresponds to $n = 2$ not $n = 1$.

Truncating a Series Whose Convergence was Established with the Integral Test

Suppose now that a series $\sum_{n=1}^{\infty} c_n$ has been shown to converge with the integral test; that the integral

$$\int_1^{\infty} f(x)\, dx$$

converges where $f(n) = c_n$. If the series is truncated after the Nth term, the error $c_{N+1} + c_{N+2} \cdots$ is shown as the sum of the areas of the rectangles in Figure 11.19. Clearly the sum of these areas is less than the area under $y = f(x)$ to the right of $x = N$. In other words, the error in truncating the series with the Nth term must be less than

$$\int_N^{\infty} f(x)\, dx. \qquad (11.21)$$

FIGURE 11.19

EXAMPLE 11.15

What is a maximum possible error when the series

$$\sum_{n=2}^{\infty} \frac{1}{n(\ln n)^4}$$

is truncated after 100 terms?

SOLUTION The error cannot be larger than

$$\int_{100}^{\infty} \frac{1}{x(\ln x)^4}\, dx = \left\{\frac{-1}{3(\ln x)^3}\right\}_{100}^{\infty} = \frac{1}{3(\ln 100)^3} < 0.0035.$$

When convergence of a series is established by the comparison, limit comparison, limit ratio, or limit root tests, we often estimate the truncation error $\sum_{n=N+1}^{\infty} c_n$ by comparing it to something that is summable. We illustrate this in the following two examples.

EXAMPLE 11.16

Estimate the error in truncating the series $\sum_{n=0}^{\infty} 1/n!$ after 8 terms. The first term of this series contains $0!$. This is defined as 1.

SOLUTION The sum of the first 8 terms of this series is $S_8 = 2.7182540$. The truncation error in using this approximation is

$$\sum_{n=8}^{\infty} \frac{1}{n!} = \frac{1}{8!} + \frac{1}{9!} + \frac{1}{10!} + \frac{1}{11!} + \cdots$$

$$= \frac{1}{8!}\left(1 + \frac{1}{9} + \frac{1}{9\cdot 10} + \frac{1}{9\cdot 10\cdot 11} + \cdots\right)$$

$$< \frac{1}{8!}\left(1 + \frac{1}{9} + \frac{1}{9^2} + \frac{1}{9^3} + \cdots\right)$$

$$= \frac{1}{8!}\frac{1}{1-\frac{1}{9}} \qquad \text{(using equation 11.13b)}$$

$$= \frac{9}{8\cdot 8!}$$

$$< 0.0000280.$$

We may write, therefore, that

$$2.718\,254\,0 < \sum_{n=0}^{\infty} \frac{1}{n!} < 2.718\,282\,0.$$

You might recognize that these same digits appeared in the approximation of the number e as the limit of the sequence $\{(1 + 1/n)^n\}$ in Section 11.2. In Section 11.11, we will show that $\sum_{n=0}^{\infty} 1/n!$ converges to e,

$$e = \sum_{n=0}^{\infty} \frac{1}{n!}. \qquad (11.22)$$

The advantage of the series definition is obvious; it is very easy to obtain an accuracy statement wherever the series is truncated.

EXAMPLE 11.17

How many terms in the convergent series $\sum_{n=1}^{\infty} n/[(n+1)3^n]$ ensure a truncation error of less than 10^{-5}?

SOLUTION If this series is truncated after the Nth term, the error is

$$\sum_{n=N+1}^{\infty} \frac{n}{(n+1)3^n} = \frac{N+1}{(N+2)3^{N+1}} + \frac{N+2}{(N+3)3^{N+2}} + \cdots$$

$$< \frac{1}{3^{N+1}} + \frac{1}{3^{N+2}} + \cdots$$

$$= \frac{\frac{1}{3^{N+1}}}{1 - \frac{1}{3}} \qquad \text{(using equation 11.13b)}$$

$$= \frac{1}{2 \cdot 3^N}.$$

Consequently, the error is guaranteed to be less than 10^{-5} if N satisfies the inequality

$$\frac{1}{2 \cdot 3^N} < 10^{-5} \qquad \text{or}$$

$$2 \cdot 3^N > 10^5.$$

With the help of a calculator, we find that this is true if $N \geq 10$, and therefore 10 or more terms yield the required accuracy.

EXERCISES 11.7

In Exercises 1–8 use the number of terms indicated to find an approximation to the sum of the series. In each case, obtain an estimate of the truncation error.

1. $\sum_{n=1}^{\infty} \frac{1}{n2^n}$ (10 terms)

2. $\sum_{n=1}^{\infty} \frac{(-1)^n}{n^4}$ (20 terms)

3. $\sum_{n=1}^{\infty} \frac{1}{n^n}$ (5 terms)

4. $\sum_{n=2}^{\infty} \frac{(-1)^{n+1}}{n^3 3^n}$ (3 terms)

5. $\sum_{n=1}^{\infty} \frac{1}{2^n} \sin\left(\frac{\pi}{n}\right)$ (15 terms)

6. $\sum_{n=2}^{\infty} \frac{2^n - 1}{3^n + n}$ (20 terms)

7. $\sum_{n=2}^{\infty} \frac{2^n + 1}{3^n + n}$ (20 terms)

8. $\sum_{n=1}^{\infty} \frac{(-1)^n}{n}$ (100 terms)

In Exercises 9–11 how many terms in the series guarantee an approximation to the sum with a truncation error of less than 10^{-4}?

9. $\sum_{n=1}^{\infty} \frac{(-1)^n}{n^2}$ 10. $\sum_{n=1}^{\infty} \frac{1}{n^2 4^n}$ 11. $\sum_{n=1}^{\infty} \frac{2^n}{n!}$

12. This exercise shows that we must be very careful in predicting the accuracy of a result. Consider the series

$$S = 3.125\,100\,1 - 0.000\,090\,18\left(1 + \frac{1}{10} + \frac{1}{10^2} + \frac{1}{10^3} + \cdots\right).$$

(a) Show that the sum of this series is exactly $S = 3.124\,999\,9$. To two decimals, then, the value of S is 3.12.

(b) Verify that the first four partial sums of the series are

$$S_1 = 3.125\,100\,1,$$
$$S_2 = 3.125\,009\,92,$$
$$S_3 = 3.125\,000\,902,$$
$$S_4 = 3.125\,000\,000\,2.$$

(c) If $E_n = S_n - S$ are the differences between the sum of the series and its first four partial sums, show that

$$E_1 = 0.000\,100\,2,$$
$$E_2 = 0.000\,010\,02,$$
$$E_3 = 0.000\,001\,002,$$
$$E_4 = 0.000\,000\,100\,2.$$

What can you say about the accuracy of S_1, S_2, S_3, and S_4 as approximations to S?

(d) If S is approximated by any of S_1, S_2, S_3, or S_4 to two decimals, the result is 3.13, not 3.12 as in (a). Thus, in spite of the accuracy predicted in (c), S_1, S_2, S_3, and S_4 do not predict S correctly to two decimals. Do they predict S correctly to three or four decimals?

SECTION 11.8

Power Series and Intervals of Convergence

Our work on sequences and series of constants in the first seven sections of this chapter has prepared the way for discussions on sequences and series of functions. An infinite sequence of functions is the assignment of functions to integers,

$$\{f_n(x)\}_0^{\infty} = f_0(x), f_1(x), \ldots, \qquad (11.23)$$

where we have initiated our assignment with the integer $n = 0$ rather than $n = 1$. The reason for this will become apparent shortly. From such a sequence we may form an infinite series of functions:

$$\sum_{n=0}^{\infty} f_n(x) = f_0(x) + f_1(x) + f_2(x) + \cdots. \qquad (11.24)$$

When $f_n(x)$ is a constant multiplied by x^n,

$$f_n(x) = a_n x^n, \qquad a_n = \text{a constant},$$

series 11.24 takes the form

$$\sum_{n=0}^{\infty} a_n x^n = a_0 + a_1 x + a_2 x^2 + \cdots + a_n x^n + \cdots, \qquad (11.25)$$

and is called a **power series** in x.

Unlike series of constants, the variable of summation n in 11.25 is one step out of phase with the counting of terms: $n = 0$ identifies the first term, $n = 1$ the second term, etc. Instead, the value of n now identifies the power of x. In the remainder of the chapter we often write $\sum a_n x^n$ in place of $\sum_{n=0}^{\infty} a_n x^n$ whenever the lower limit is $n = 0$.

For each value of x that we substitute into a power series, we obtain a series of constants. For example, if into the power series

$$\sum_{n=0}^{\infty} x^n = 1 + x + x^2 + x^3 + \cdots$$

we substitute $x = 1/2$, we obtain the series of constants

$$\sum_{n=0}^{\infty} \left(\frac{1}{2}\right)^n = 1 + \frac{1}{2} + \frac{1}{2^2} + \frac{1}{2^3} + \cdots,$$

a geometric series that converges to 2. When we substitute $x = -3$, we obtain the divergent geometric series

$$\sum_{n=0}^{\infty} (-3)^n = 1 - 3 + 9 - 27 + \cdots .$$

Definition 11.9

The totality of values of x for which a power series converges is called its **interval of convergence**.

Obviously the interval of convergence for a power series $\sum a_n x^n$ always includes the value $x = 0$, but what other possibilities are there? "Interval of convergence" suggests that the values of x for which a power series converges form some kind of interval. This is indeed true, as we shall soon see.

EXAMPLE 11.18

Find the interval of convergence for the power series $\sum_{n=0}^{\infty} x^n$.

SOLUTION Since the series is a geometric series with common ratio x, its interval of convergence is $-1 < x < 1$. ∎

EXAMPLE 11.19

Find intervals of convergence for the following power series:

(a) $\sum_{n=1}^{\infty} \dfrac{x^n}{n^n}$ (b) $\sum_{n=1}^{\infty} n!\, x^n$ (c) $\sum_{n=1}^{\infty} \dfrac{(-1)^{n+1}}{n} x^n$

SOLUTION (a) Consider the series of absolute values $\sum |x|^n/n^n$. We use the limit root test on this series, and calculate

$$R = \lim_{n \to \infty} \left(\dfrac{|x|^n}{n^n}\right)^{1/n} = \lim_{n \to \infty} \dfrac{|x|}{n} = 0,$$

and this is true for any x whatsoever. The power series therefore converges absolutely for all x.

(b) Consider again the series of absolute values $\sum n!\,|x|^n$. We use the limit ratio test and obtain

$$L = \lim_{n \to \infty} \dfrac{(n+1)!\,|x|^{n+1}}{n!\,|x|^n} = |x| \lim_{n \to \infty} (n+1) = \infty.$$

Since $L = \infty$ for all $x \neq 0$, the original power series does not converge absolutely for any $x \neq 0$. But this result implies much more. It shows that the terms in the sequence $\{n!\,|x|^n\}$ become indefinitely large as $n \to \infty$. But then $\lim_{n \to \infty} n!\,x^n$ cannot possibly exist, and the series diverges by the nth term test. The power series therefore converges only for $x = 0$.

(c) Once again for the series of absolute values, we obtain

$$L = \lim_{n \to \infty} \dfrac{\dfrac{|x|^{n+1}}{n+1}}{\dfrac{|x|^n}{n}} = \lim_{n \to \infty} |x| \dfrac{n}{n+1} = |x|.$$

Since $L = |x|$, it follows that the series of absolute values converges if $|x| < 1$ and diverges if $|x| > 1$. The original power series therefore converges absolutely for $-1 < x < 1$. Furthermore, when $|x| > 1$, so too is L, and it follows that the terms in the sequence $\{|x|^n/n\}$ must become indefinitely large. But then

$$\lim_{n \to \infty} \dfrac{(-1)^{n+1}}{n} x^n$$

cannot possibly exist, and the series diverges by the nth term test. If $x < -1$, terms in the sequence $\{(-1)^{n+1} x^n/n\}$ are all negative and "approach negative infinity"; if $x > 1$, the terms are alternately positive and negative with increasing absolute values. When $x = 1$, the power series reduces to the alternating harmonic series

$$\sum_{n=1}^{\infty} \dfrac{(-1)^{n+1}}{n} = 1 - \dfrac{1}{2} + \dfrac{1}{3} - \dfrac{1}{4} + \cdots,$$

which converges conditionally; and when $x = -1$, it reduces to

$$\sum_{n=1}^{\infty} \dfrac{(-1)^{n+1}}{n}(-1)^n = -1 - \dfrac{1}{2} - \dfrac{1}{3} - \dfrac{1}{4} - \cdots,$$

which diverges. The interval of convergence is therefore $-1 < x \leq 1$. ∎

Example 11.19 has illustrated that there are at least three possible types of intervals of convergence for power series $\sum a_n x^n$:

1. The power series converges only for $x = 0$;
2. The power series converges absolutely for all x;
3. There exists a number $R > 0$ such that the power series converges absolutely for $|x| < R$, diverges for $|x| > R$, and may or may not converge for $x = \pm R$.

These are in fact the only possibilities for an interval of convergence, and this is proved in Exercises 32 and 33. In (3) we call R the **radius of convergence** of the power series. It is half the length of the interval of convergence, or, the distance we may proceed in either direction along the x-axis from $x = 0$ and expect absolute convergence of the power series, with the possible exceptions of $x = \pm R$. In order to have a radius of convergence associated with every power series, we say in (1) and (2) that $R = 0$ and $R = \infty$ respectively.

Every power series $\sum a_n x^n$ now has a radius of convergence R. If $R = 0$, the power series converges only for $x = 0$; if $R = \infty$, the power series converges absolutely for all x; and if $0 < R < \infty$, the power series converges absolutely for $|x| < R$, diverges for $|x| > R$, and may or may not converge for $x = \pm R$. For many power series the radius of convergence can be calculated according to the following theorem.

Theorem 11.13

The radius of convergence of a power series $\sum_{n=0}^{\infty} a_n x^n$ is given by

$$R = \lim_{n \to \infty} \left| \frac{a_n}{a_{n+1}} \right| \quad \text{or,} \quad (11.26\text{a})$$

$$R = \lim_{n \to \infty} \frac{1}{\sqrt[n]{|a_n|}}, \quad (11.26\text{b})$$

provided either limit exists or is equal to infinity.

Proof We verify 11.26a using the limit ratio test; verification of 11.26b is similar, using the limit root test.

If the limit ratio test is applied to the series of absolute values $\sum |a_n x^n|$,

$$L = \lim_{n \to \infty} \frac{|a_{n+1} x^{n+1}|}{|a_n x^n|} = |x| \lim_{n \to \infty} \left| \frac{a_{n+1}}{a_n} \right|.$$

Assuming that limit 11.26a exists or is equal to infinity, there are three possibilities:

(i) If $\lim_{n \to \infty} |a_n / a_{n+1}| = 0$, then $\lim_{n \to \infty} |a_{n+1}/a_n| = \infty$. Therefore $L = \infty$, and the power series diverges for all $x \neq 0$. In other words,

$$R = 0 = \lim_{n \to \infty} \left| \frac{a_n}{a_{n+1}} \right|.$$

(ii) If $\lim_{n \to \infty} |a_n/a_{n+1}| = \infty$, then $\lim_{n \to \infty} |a_{n+1}/a_n| = 0$. Therefore $L = 0$, and the power series converges absolutely for all x. Consequently,

$$R = \infty = \lim_{n \to \infty} \left| \frac{a_n}{a_{n+1}} \right|.$$

(iii) If $\lim_{n\to\infty} |a_n/a_{n+1}| = R$, then $\lim_{n\to\infty} |a_{n+1}/a_n| = 1/R$. In this case, $L = |x|/R$. Since the power series converges absolutely for $L < 1$ and diverges for $L > 1$, it follows that absolute convergence occurs for $|x| < R$ and divergence for $|x| > R$. This implies that R is the radius of convergence of the power series.

With this theorem the limit ratio test and limit root test can be avoided in determining intervals of convergence for many power series. For the series in Example 11.19 we proceed as follows:

(a) For the power series $\sum_{n=1}^{\infty} x^n/n^n$, we use 11.26b to obtain

$$R = \lim_{n\to\infty} \frac{1}{\sqrt[n]{|a_n|}} = \lim_{n\to\infty} \frac{1}{(1/n^n)^{1/n}} = \lim_{n\to\infty} n = \infty.$$

The power series therefore converges absolutely for all x.

(b) For the series $\sum_{n=1}^{\infty} n!\, x^n$, we use 11.26a,

$$R = \lim_{n\to\infty} \left|\frac{a_n}{a_{n+1}}\right| = \lim_{n\to\infty} \frac{n!}{(n+1)!} = \lim_{n\to\infty} \frac{1}{n+1} = 0,$$

and the power series converges only for $x = 0$.

(c) For the power series $\sum_{n=1}^{\infty} [(-1)^{n+1}/n]\, x^n$, 11.26a gives

$$R = \lim_{n\to\infty} \left|\frac{a_n}{a_{n+1}}\right| = \lim_{n\to\infty} \left|\frac{\frac{(-1)^{n+1}}{n}}{\frac{(-1)^{n+2}}{n+1}}\right| = \lim_{n\to\infty} \left(\frac{n+1}{n}\right) = 1.$$

When $x = 1$, the power series reduces to the alternating harmonic series

$$\sum_{n=1}^{\infty} \frac{(-1)^{n+1}}{n} = 1 - \frac{1}{2} + \frac{1}{3} - \frac{1}{4} + \cdots,$$

which converges conditionally. When $x = -1$, it reduces to

$$\sum_{n=1}^{\infty} \frac{(-1)^{n+1}}{n}(-1)^n = -1 - \frac{1}{2} - \frac{1}{3} - \frac{1}{4} - \cdots,$$

which diverges. The interval of convergence is therefore $-1 < x \leq 1$.

EXAMPLE 11.20

Find the interval of convergence for the power series $\sum_{n=1}^{\infty} nx^n/5^{2n}$.

SOLUTION Since

$$R = \lim_{n\to\infty} \left|\frac{a_n}{a_{n+1}}\right| = \lim_{n\to\infty} \frac{\frac{n}{5^{2n}}}{\frac{n+1}{5^{2n+2}}} = 25,$$

the power series converges absolutely for $-25 < x < 25$. At $x = 25$, the series reduces to $\sum_{n=1}^{\infty} n$, which diverges; and at $x = -25$, it reduces to $\sum_{n=1}^{\infty} n(-1)^n$, which also diverges. The interval of convergence is therefore $-25 < x < 25$. ∎

EXAMPLE 11.21

Find the interval of convergence for the power series $\sum_{n=1}^{\infty} x^{2n+1}/(n^2 2^n)$.

SOLUTION Since coefficients of even powers of x are 0 in this power series, the sequence $\{a_n/a_{n+1}\}$ is not defined. We cannot therefore find its radius of convergence directly using Theorem 11.13. Instead we write

$$\sum_{n=1}^{\infty} \frac{1}{n^2 2^n} x^{2n+1} = x \sum_{n=1}^{\infty} \frac{1}{n^2 2^n} (x^2)^n$$

and set $y = x^2$ in the series:

$$\sum_{n=1}^{\infty} \frac{1}{n^2 2^n} (x^2)^n = \sum_{n=1}^{\infty} \frac{1}{n^2 2^n} y^n.$$

According to equation 11.26a the radius of convergence of this series in y is

$$\lim_{n \to \infty} \frac{\frac{1}{n^2 2^n}}{\frac{1}{(n+1)^2 2^{n+1}}} = \lim_{n \to \infty} 2 \left(\frac{n+1}{n} \right)^2 = 2.$$

Since $x = \pm\sqrt{y}$, it follows that the radius of convergence of the power series in x is $R = \sqrt{2}$. When $x = \sqrt{2}$, the series reduces to $\sum_{n=1}^{\infty} \sqrt{2}/n^2 = \sqrt{2} \sum_{n=1}^{\infty} 1/n^2$, which converges; and when $x = -\sqrt{2}$, it reduces to the negative of this series. The interval of convergence is therefore $-\sqrt{2} \leq x \leq \sqrt{2}$. ∎

We describe what it means for a function to be the sum of a power series in the following definition.

Definition 11.10

A function $f(x)$ is said to be the **sum** of a power series $\sum_{n=0}^{\infty} a_n x^n$ if at each point x in the interval of convergence of the power series, the sum of the series is the same as the value of $f(x)$ at that x.

We write

$$f(x) = \sum_{n=0}^{\infty} a_n x^n \tag{11.27}$$

to signify that $f(x)$ is the sum of the power series. For example, the sum of the geometric series

$$\sum_{n=0}^{\infty} x^n = 1 + x + x^2 + x^3 + \cdots$$

is $1/(1-x)$ for $|x| < 1$, and we write

$$\frac{1}{1-x} = 1 + x + x^2 + x^3 + \cdots, \quad |x| < 1.$$

We affix to the equation the interval of convergence of the power series, thus identifying for which values of x the function $1/(1-x)$ is the sum of the series. The function

$1/(1 - x)$ is defined for all x, except $x = 1$, but it represents the sum of the series only in the interval $-1 < x < 1$.

In Section 11.3 we defined the sum of an infinite series of constants as the limit of its sequence of partial sums. In the case of power series 11.27, the sequence of partial sums is a sequence of polynomials:

$$S_1(x) = a_0,$$
$$S_2(x) = a_0 + a_1 x,$$
$$S_3(x) = a_0 + a_1 x + a_2 x^2,$$
$$\vdots \qquad \vdots$$
$$S_n(x) = a_0 + a_1 x + a_2 x^2 + \cdots + a_{n-1} x^{n-1},$$
$$\vdots \qquad \vdots$$

By equation 11.27 we are saying that for each x in the interval of convergence of the power series,

$$f(x) = \lim_{n \to \infty} S_n(x).$$

For example, the nth partial sum of the geometric series $\sum_{n=0}^{\infty} x^n$ is

$$S_n(x) = 1 + x + x^2 + \cdots + x^{n-1} = \begin{cases} n & x = 1 \\ \dfrac{1 - x^n}{1 - x} & x \neq 1 \end{cases}$$

(see equation 11.13a). Algebraically, it is clear that for large n, $S_n(x)$ approaches $1/(1 - x)$, provided $|x| < 1$, and we write therefore

$$\frac{1}{1 - x} = 1 + x + x^2 + x^3 + \cdots, \quad |x| < 1.$$

In Figure 11.20 we show the first five partial sums of this geometric series together with the sum, just to illustrate how the partial sums of the series approximate the sum more closely for increasing n. Note how these figures illustrate the following facts about the geometric series and its partial sums:

1. The partial sums $S_n(x)$ are defined for all x and the sum $1/(1 - x)$ is defined for all $x \neq 1$, but the curves $y = S_n(x)$ approach $y = 1/(1 - x)$ only for $|x| < 1$.
2. For $x > 0$, the partial sums $S_n(x)$ approach $1/(1 - x)$ from below; for $x < 0$, the $S_n(x)$ oscillate about $1/(1 - x)$, but gradually approach $1/(1 - x)$.

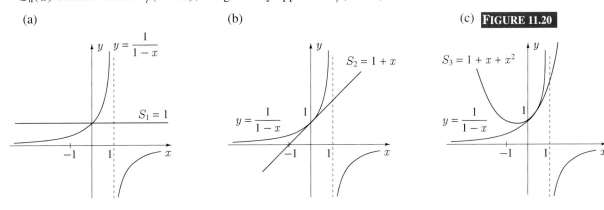

(a)   (b)   (c) **FIGURE 11.20**

(d)

(e)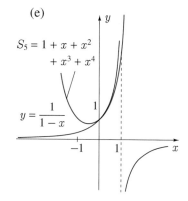

EXAMPLE 11.22

Find the sum of the series

$$\sum_{n=0}^{\infty} \frac{(-1)^n}{2^n} x^{(2n+1)/2}.$$

SOLUTION We write

$$\sum_{n=0}^{\infty} \frac{(-1)^n}{2^n} x^{(2n+1)/2} = x^{1/2} \sum_{n=0}^{\infty} \frac{(-1)^n}{2^n} x^n = x^{1/2} \sum_{n=0}^{\infty} \left(-\frac{x}{2}\right)^n$$

and note that the power series on the right is a geometric series with sum

$$\sum_{n=0}^{\infty} \left(-\frac{x}{2}\right)^n = \frac{1}{1 + \frac{x}{2}}, \quad \left|-\frac{x}{2}\right| < 1.$$

Consequently,

$$\sum_{n=0}^{\infty} \frac{(-1)^n}{2^n} x^{(2n+1)/2} = \sqrt{x} \frac{1}{1 + \frac{x}{2}} = \frac{2\sqrt{x}}{x+2}, \quad 0 \le x < 2.$$

For many power series there is no known function that is equal to the sum of the power series over its interval of convergence. In such cases we continue to write equation 11.27, but say that the power series defines the value of $f(x)$ at each x in the interval of convergence. For instance, a very important series in engineering and physics is

$$\sum_{n=0}^{\infty} \frac{(-1)^n}{2^{2n}(n!)^2} x^{2n},$$

which converges absolutely for all x. It arises so often in applications that it is given a special name, the *Bessel function of the first kind of order zero*, and is denoted by $J_0(x)$. In other words, we write

$$J_0(x) = \sum_{n=0}^{\infty} \frac{(-1)^n}{2^{2n}(n!)^2} x^{2n},$$

and say that the power series defines the value of $J_0(x)$ for each x.

We have called 11.25 a power series in x. It is also said to be a *power series about* 0 (meaning the point 0 on the x-axis), where we note that for any interval of convergence whatsoever, 0 is always at its centre. This suggests that power series about other points might be considered, and this is indeed the case. The general power series about a point c on the x-axis is

$$\sum_{n=0}^{\infty} a_n (x-c)^n = a_0 + a_1(x-c) + a_2(x-c)^2 + \cdots \qquad (11.28)$$

and is said to be a power series in $x - c$.

A power series in $x - c$ has an interval of convergence and a radius of convergence analogous to a power series in x. In particular, every power series in $x - c$ has a radius of convergence R such that if $R = 0$, the power series converges only for $x = c$; if $R = \infty$, the power series converges absolutely for all x; and if $0 < R < \infty$, the series converges absolutely for $|x - c| < R$, diverges for $|x - c| > R$, and may or may not converge for $x = c \pm R$. The radius of convergence is again given by equations 11.26, provided the limits exist or are equal to infinity. For power series in $x - c$, then, the point c is the centre of the interval of convergence.

EXAMPLE 11.23

Find the interval of convergence of the power series

$$2^2(x+2) + 2^3(x+2)^2 + 2^4(x+2)^3 + \cdots + 2^{n+1}(x+2)^n + \cdots .$$

SOLUTION By writing the power series in sigma notation,

$$\sum_{n=1}^{\infty} 2^{n+1}(x+2)^n,$$

we can use equation 11.26a (or 11.26b) to calculate its radius of convergence,

$$R = \lim_{n \to \infty} \left| \frac{a_n}{a_{n+1}} \right| = \lim_{n \to \infty} \frac{2^{n+1}}{2^{n+2}} = \frac{1}{2}.$$

Since this is a power series about -2, the ends of the interval of convergence are $x = -5/2$ and $x = -3/2$. At $x = -5/2$, the power series reduces to $\sum_{n=1}^{\infty} 2(-1)^n$, which diverges; and at $x = -3/2$, it reduces to $\sum_{n=1}^{\infty} 2$, which also diverges. The interval of convergence is therefore $-5/2 < x < -3/2$. ∎

EXERCISES 11.8

In Exercises 1–22 find the interval of convergence for the power series.

1. $\sum_{n=1}^{\infty} \frac{1}{n} x^n$

2. $\sum_{n=1}^{\infty} n^2 x^n$

3. $\sum_{n=0}^{\infty} \frac{1}{(n+1)^3} x^n$

4. $\sum_{n=0}^{\infty} n^2 3^n x^n$

5. $\sum_{n=0}^{\infty} \frac{1}{2^n} (x-1)^n$

6. $\sum_{n=4}^{\infty} (-1)^n n^3 (x+3)^n$

7. $\sum_{n=1}^{\infty} \frac{1}{\sqrt{n}} (x+2)^n$

8. $\sum_{n=2}^{\infty} 2^n \left(\frac{n-1}{n+2} \right)^2 (x-4)^n$

9. $\sum_{n=1}^{\infty} \frac{1}{n^2} x^{2n}$

10. $\sum_{n=0}^{\infty} (-1)^n x^{3n}$

11. $\sum_{n=0}^{\infty} \frac{n-1}{n+1} (2x)^n$

12. $\sum_{n=0}^{\infty} \frac{1}{\sqrt{n+1}} x^{3n+1}$

13. $\sum_{n=0}^{\infty} \frac{(-1)^n}{3^n} x^{2n+1}$ 14. $\sum_{n=1}^{\infty} \frac{(-e)^n}{n^2} x^n$

15. $\frac{x}{9} + \frac{4}{3^4} x^2 + \frac{9}{3^6} x^3 + \cdots + \frac{n^2}{3^{2n}} x^n + \cdots$

16. $x + 2^2 x^2 + 3^3 x^3 + \cdots + n^n x^n + \cdots$

17. $\frac{1}{36}(x+10)^6 + \frac{1}{49}(x+10)^7 + \frac{1}{64}(x+10)^8 + \cdots$
    $+ \frac{1}{n^2}(x+10)^n + \cdots$

18. $3x + 8(3x)^2 + 27(3x)^3 + \cdots + n^3(3x)^n + \cdots$

19. $\frac{3}{4}x^2 + x^4 + \frac{27}{16}x^6 + \cdots + \frac{3^n}{(n+1)^2} x^{2n} + \cdots$

20. $1 + \frac{1}{5}x^3 + \frac{1}{25}x^6 + \cdots + \frac{1}{5^n} x^{3n} + \cdots$

21. $\sum_{n=2}^{\infty} \frac{1}{\ln n} x^n$ 22. $\sum_{n=2}^{\infty} \frac{1}{n^2 \ln n} x^n$

In Exercises 23–25 find the radius of convergence of the power series.

23. $\sum_{n=1}^{\infty} \frac{(n!)^3}{(3n)!} x^n$

24. $\sum_{n=1}^{\infty} \frac{2 \cdot 4 \cdot 6 \cdot \cdots \cdot (2n)}{3 \cdot 5 \cdot 7 \cdot \cdots \cdot (2n+1)} x^n$

25. $\sum_{n=1}^{\infty} \frac{[1 \cdot 3 \cdot 5 \cdot \cdots \cdot (2n+1)]^2}{2^{2n}(2n)!} x^n$

In Exercises 26–29 find the sum of the power series.

26. $\sum_{n=0}^{\infty} \frac{1}{4^n} x^{3n}$ 27. $\sum_{n=1}^{\infty} (-e)^n x^n$

28. $\sum_{n=1}^{\infty} \frac{1}{3^{2n}} (x-1)^n$ 29. $\sum_{n=2}^{\infty} (x+5)^{2n}$

30. If m is a nonnegative integer, the Bessel function of order m of the first kind is defined by the power series

$$J_m(x) = \sum_{n=0}^{\infty} \frac{(-1)^n}{2^{2n+m} n! (n+m)!} x^{2n+m}.$$

(a) Write out the first five terms of $J_0(x)$, $J_1(x)$, and $J_m(x)$.

(b) Find the interval of convergence for each $J_m(x)$.

31. The hypergeometric series is

$$1 + \frac{\alpha \beta}{\gamma} + \frac{\alpha(\alpha+1)\beta(\beta+1)}{2! \, \gamma(\gamma+1)} x^2$$
$$+ \frac{\alpha(\alpha+1)(\alpha+2)\beta(\beta+1)(\beta+2)}{3! \, \gamma(\gamma+1)(\gamma+2)} x^3 + \cdots,$$

where α, β, and γ are all constants.

(a) Write this series in sigma notation.

(b) What is the radius of convergence of the hypergeometric series if γ is not zero or a negative integer?

32. Show that if a power series $\sum a_n x^n$ converges for $x_1 \neq 0$, then it converges absolutely for all x in the interval $|x| < |x_1|$.

33. Prove that the interval of convergence of a power series $\sum a_n x^n$ must be one of the three possibilities listed on page 554. Use Exercise 32 and the *completeness property* for real numbers, which states: If a nonempty set S of real numbers has an upper bound, then it has a least upper bound; that is, there exists a number l such that all numbers in S are less than or equal to l and there is no upper bound smaller than l.

SECTION 11.9

Taylor's Remainder Formula and Taylor Series

If $f(x)$ is the sum of a power series $\sum_{n=0}^{\infty} a_n(x-c)^n$, we write

$$f(x) = \sum_{n=0}^{\infty} a_n(x-c)^n. \qquad (11.29)$$

In Section 11.8 we considered this equation in situations where the power series is given and the sum is to be determined. An alternative view is to consider the function $f(x)$ as given and the power series as an expansion of the function about the point c. For example, given the function $f(x) = 1/(2x+3)$, it is quite simple to use properties of geometric series to find a power series expansion in x for $f(x)$. If we write

$$f(x) = \frac{1}{2x+3} = \frac{1}{3\left(1+\dfrac{2x}{3}\right)}$$

and interpret the right-hand side as one-third the sum of a geometric series with first term equal to 1 and common ratio $-2x/3$, then

$$f(x) = \frac{1}{3}\sum_{n=0}^{\infty}\left(-\frac{2x}{3}\right)^n, \qquad \left|-\frac{2x}{3}\right| < 1$$

$$= \sum_{n=0}^{\infty}\frac{(-1)^n 2^n}{3^{n+1}}x^n, \qquad |x| < \frac{3}{2}.$$

Naturally it is not clear at this time why we should want to expand a function in the form of an infinite series, but in Section 11.12, when we consider applications of series, we will find such expansions extremely useful.

In this section we establish two results. First we give conditions that gurarantee that a function $f(x)$ has a power series expansion about a point c and that this power series converges to $f(x)$. Second, we show that if a function has a power series expansion about a point, then it has only one such series. The first result is an immediate consequence of Theorem 11.14.

Theorem 11.14

(Taylor's Remainder Formula) If $f(x)$ and its first $n-1$ derivatives are continuous on the closed interval between c and x, and if $f(x)$ has an nth derivative on the open interval between c and x, then there exists a point z_n between c and x such that

$$f(x) = f(c) + f'(c)(x-c) + \frac{f''(c)}{2!}(x-c)^2 + \frac{f'''(c)}{3!}(x-c)^3 + \cdots$$
$$+ \frac{f^{(n-1)}(c)}{(n-1)!}(x-c)^{n-1} + \frac{f^{(n)}(z_n)}{n!}(x-c)^n. \qquad (11.30)$$

The notation $f^{(n)}(z_n)$ represents the nth derivative of $f(x)$ evaluated at $x = z_n$. This result can be proved by applying Rolle's theorem (Theorem 3.13) to the function

$$F(y) = f(x) - f(y) - f'(y)(x-y) - \frac{f''(y)}{2!}(x-y)^2 - \cdots$$
$$- \frac{f^{(n-1)}(y)}{(n-1)!}(x-y)^{n-1} - \left(\frac{x-y}{x-c}\right)^n\left[f(x) - f(c) - f'(c)(x-c)\right.$$
$$\left. - \frac{f''(c)}{2!}(x-c)^2 - \cdots - \frac{f^{(n-1)}(c)}{(n-1)!}(x-c)^{n-1}\right]$$

defined on the interval $c \leq y \leq x$, or $x \leq y \leq c$, depending on whether x is larger or smaller than c.

Note that when $n = 1$, Taylor's remainder formula reduces to the mean value theorem of Section 3.9. In other words, the mean value theorem is a special case of Taylor's remainder formula.

Consider now a function $f(x)$ that has derivatives of all orders on the closed interval between c and x. Taylor's remainder formula can therefore be written down for all values of n. Let us write it out for $n = 1, 2, 3$, and 4:

$$f(x) = f(c) + f'(z_1)(x - c), \tag{11.31a}$$

$$f(x) = f(c) + f'(c)(x - c) + \frac{f''(z_2)}{2!}(x - c)^2, \tag{11.31b}$$

$$f(x) = f(c) + f'(c)(x - c) + \frac{f''(c)}{2!}(x - c)^2 + \frac{f'''(z_3)}{3!}(x - c)^3, \tag{11.31c}$$

$$f(x) = f(c) + f'(c)(x - c) + \frac{f''(c)}{2!}(x - c)^2$$
$$+ \frac{f'''(c)}{3!}(x - c)^3 + \frac{f''''(z_4)}{4!}(x - c)^4. \tag{11.31d}$$

The last terms in these equations are called **remainders**, and when they are denoted by R_1, R_2, R_3, and R_4 respectively,

$$f(x) = f(c) + R_1, \tag{11.32a}$$

$$f(x) = f(c) + f'(c)(x - c) + R_2, \tag{11.32b}$$

$$f(x) = f(c) + f'(c)(x - c) + \frac{f''(c)}{2!}(x - c)^2 + R_3, \tag{11.32c}$$

$$f(x) = f(c) + f'(c)(x - c) + \frac{f''(c)}{2!}(x - c)^2$$
$$+ \frac{f'''(c)}{3!}(x - c)^3 + R_4, \tag{11.32d}$$

and in general,

$$f(x) = f(c) + f'(c)(x - c) + \frac{f''(c)}{2!}(x - c)^2 + \cdots$$
$$+ \frac{f^{(n-1)}(c)}{(n-1)!}(x - c)^{n-1} + R_n. \tag{11.32e}$$

We call R_n the remainder for the simple reason that if we were to drop the remainders from equations (11.32), then what remains is a sequence of polynomial approximations to $f(x)$:

$$f(x) \approx f(c), \tag{11.33a}$$

$$f(x) \approx f(c) + f'(c)(x - c), \tag{11.33b}$$

$$f(x) \approx f(c) + f'(c)(x - c) + \frac{f''(c)}{2!}(x - c)^2, \tag{11.33c}$$

$$f(x) \approx f(c) + f'(c)(x - c) + \frac{f''(c)}{2!}(x - c)^2 + \frac{f'''(c)}{3!}(x - c)^3. \tag{11.33d}$$

We give a detailed discussion on the accuracy of these approximations in Section 11.12, but for now we can at least appreciate why Theorem 11.14 is called Taylor's remainder formula.

The nth remainder is

$$R_n = R_n(c, x) = \frac{f^{(n)}(z_n)}{n!}(x - c)^n, \quad (11.34)$$

where z_n (which, as the notation suggests, varies with n) is always between c and x. We have written $R_n(c, x)$ to emphasize the fact that the remainders depend on the point of expansion c and the value of x being considered.

Suppose the sequence of remainders has limit zero,

$$\lim_{n \to \infty} R_n = 0. \quad (11.35)$$

What this means is that the sequence of polynomials in 11.33 approximates $f(x)$ more and more closely as n gets larger and larger. In fact, these polynomials are the partial sums for the following infinite series,

$$f(c) + f'(c)(x - c) + \frac{f''(c)}{2!}(x - c)^2 + \frac{f'''(c)}{3!}(x - c)^3 + \cdots . \quad (11.36)$$

Thus, if the sequence of remainders $\{R_n\}$ approaches zero, then power series 11.36 converges to $f(x)$. We may write therefore that

$$f(x) = f(c) + f'(c)(x - c) + \frac{f''(c)}{2!}(x - c)^2 + \frac{f'''(c)}{3!}(x - c)^3 + \cdots, \quad (11.37)$$

and this equation is valid for all those x's for which $\lim_{n \to \infty} R_n(c, x) = 0$. This power series in $x - c$ is called the **Taylor series** of $f(x)$ about the point c. We have shown that *if the sequence of remainders $\{R_n\}$ for a function $f(x)$ exists and has limit zero, the Taylor series converges to $f(x)$ for that x.*

When $c = 0$, the Taylor series takes the form

$$f(x) = f(0) + f'(0)x + \frac{f''(0)}{2!}x^2 + \frac{f'''(0)}{3!}x^3 + \cdots, \quad (11.38)$$

a power series in x called the **Maclaurin series** for $f(x)$.

To obtain the Taylor or Maclaurin series for a function $f(x)$ is quite simple; evaluate all derivatives of $f(x)$ at the point of expansion and use formula 11.37 or 11.38. To show that the series converges to $f(x)$ is a different matter. We must show that the sequence $\{R_n\}$ has limit zero, and this can be very difficult because R_n is defined in terms of an unknown point z_n. This problem is usually circumvented by finding a maximum value for R_n and showing that this maximum value approaches zero. We illustrate this procedure in the next three examples. To do so, however, we require the result of the following theorem. Apart from the fact that Theorem 11.15 is needed for Examples 11.24 and 11.25, its proof is interesting in that it interchanges the usual role of sequences and series. Heretofore, we have used sequences to determine convergence of series. This theorem uses series to verify convergence of a certain sequence.

Theorem 11.15

The sequence $\{|x|^n/n!\}$ has limit 0 for any x whatsoever,

$$\lim_{n \to \infty} \frac{|x|^n}{n!} = 0. \quad (11.39)$$

Proof Consider the series $\sum_{n=1}^{\infty} |x|^n/n!$. Since the ratio test gives

$$L = \lim_{n\to\infty} \frac{\frac{|x|^{n+1}}{(n+1)!}}{\frac{|x|^n}{n!}} = \lim_{n\to\infty} \frac{|x|}{n+1} = 0,$$

the series converges for any x. But then the nth term test implies that its nth term must have limit 0.

EXAMPLE 11.24 Find the Maclaurin series for $\sin x$ and show that it converges to $\sin x$ for all x.

SOLUTION If $f(x) = \sin x$, then:

$$f(0) = \sin x_{|x=0} = 0;$$
$$f'(0) = \cos x_{|x=0} = 1;$$
$$f''(0) = -\sin x_{|x=0} = 0;$$
$$f'''(0) = -\cos x_{|x=0} = -1;$$
$$f''''(0) = \sin x_{|x=0} = 0; \quad \text{etc.}$$

Taylor's remainder formula for $\sin x$ and $c = 0$ yields

$$\sin x = x - \frac{x^3}{3!} + \frac{x^5}{5!} - \frac{x^7}{7!} + \cdots + (n^{\text{th}} \text{ term}) + R_n,$$

where

$$R_n = \frac{d^n}{dx^n}(\sin x)_{|x=z_n} \frac{x^n}{n!}.$$

But the nth derivative of $\sin x$ is $\pm \sin x$ or $\pm \cos x$, so that

$$\left| \frac{d^n}{dx^n}(\sin x)_{|x=z_n} \right| \leq 1.$$

Hence,

$$|R_n| \leq \frac{|x|^n}{n!}.$$

But according to equation 11.39, $\lim_{n\to\infty} |x|^n/n! = 0$ for any x whatsoever, and therefore

$$\lim_{n\to\infty} R_n = 0.$$

The Maclaurin series for $\sin x$ therefore converges to $\sin x$ for all x, and we may write

$$\sin x = x - \frac{x^3}{3!} + \frac{x^5}{5!} - \frac{x^7}{7!} + \cdots$$
$$= \sum_{n=0}^{\infty} \frac{(-1)^n}{(2n+1)!} x^{2n+1}, \quad -\infty < x < \infty.$$

Often it is necessary to consider separately points on either side of the point of expansion. This is illustrated in the next example.

EXAMPLE 11.25

Find the Maclaurin series for e^x and show that it converges to e^x for all x.

SOLUTION Since
$$\frac{d^n}{dx^n}(e^x)|_{x=0} = e^x|_{x=0} = 1,$$

Taylor's remainder formula for e^x and $c = 0$ gives

$$e^x = 1 + x + \frac{x^2}{2!} + \frac{x^3}{3!} + \cdots + \frac{x^{n-1}}{(n-1)!} + R_n,$$

where
$$R_n = \frac{d^n}{dx^n}(e^x)|_{x=z_n}\frac{x^n}{n!} = e^{z_n}\frac{x^n}{n!}.$$

Now, if $x < 0$, then $x < z_n < 0$, and

$$|R_n| < e^0 \frac{|x|^n}{n!},$$

which approaches zero as n becomes infinite (see equation 11.39). If $x > 0$, then $0 < z_n < x$, and

$$|R_n| < e^x \frac{|x|^n}{n!},$$

which again has limit zero as n approaches infinity. Thus, for any x whatsoever, $\lim_{n\to\infty} R_n = 0$, and the Maclaurin series for e^x converges to e^x:

$$e^x = 1 + x + \frac{x^2}{2!} + \frac{x^3}{3!} + \cdots$$
$$= \sum_{n=0}^{\infty} \frac{1}{n!}x^n, \quad -\infty < x < \infty.$$

∎

The following example is more difficult. The remainders do not apporach zero for all x so that the Taylor series does not converge to the function for all x. The problem would have been even more difficult had we not suggested the values of x to consider.

EXAMPLE 11.26

Find the Taylor series for $\ln x$ about the point 1 and show that it converges to $\ln x$ for $1/2 \leq x \leq 2$.

SOLUTION If $f(x) = \ln x$, then:

$$f(1) = \ln x|_{x=1} = 0;$$
$$f'(1) = \frac{1}{x}\Big|_{x=1} = 1;$$
$$f''(1) = \frac{-1}{x^2}\Big|_{x=1} = -1;$$

$$f'''(1) = \frac{2}{x^3}\Big|_{x=1} = 2;$$

$$f''''(1) = \frac{-3!}{x^4}\Big|_{x=1} = -3!;\quad \text{etc.}$$

Taylor's remainder formula for $\ln x$ and $c = 1$ yields

$$\ln x = (x-1) - \frac{1}{2!}(x-1)^2 + \frac{2!}{3!}(x-1)^3 + \cdots + \frac{(-1)^n(n-2)!}{(n-1)!}(x-1)^{n-1} + R_n,$$

where

$$R_n = \frac{d^n}{dx^n}(\ln x)\Big|_{x=z_n} \frac{(x-1)^n}{n!} = \frac{(-1)^{n+1}(n-1)!}{(z_n)^n}\frac{(x-1)^n}{n!} = \frac{(-1)^{n+1}}{n}\left(\frac{x-1}{z_n}\right)^n,$$

and z_n is between 1 and x.

If $1 < x \le 2$, then the largest value of $x - 1$ is 1. Furthermore, z_n must be larger than 1. It follows that

$$|R_n| < \frac{1}{n}\left(\frac{1}{1}\right)^n = \frac{1}{n}$$

and therefore

$$\lim_{n\to\infty} R_n = 0.$$

If $1/2 \le x < 1$, then $-1/2 \le x - 1 < 0$. Combine this with $x < z_n < 1$, and we can state that $-1 < (x-1)/z_n < 0$. Then,

$$|R_n| < \frac{1}{n} \quad \text{and} \quad \lim_{n\to\infty} R_n = 0.$$

Thus, for $1/2 \le x \le 2$, the sequence of remainders $\{R_n\}$ approaches zero, and the Taylor series converges to $\ln x$ for those values of x:

$$\ln x = (x-1) - \frac{1}{2}(x-1)^2 + \frac{1}{3}(x-1)^3 + \cdots$$

$$= \sum_{n=1}^{\infty} \frac{(-1)^{n+1}}{n}(x-1)^n, \quad \frac{1}{2} \le x \le 2.$$

This series actually converges to $\ln x$ on the larger interval $0 < x \le 2$. We will prove this in Example 11.30. ∎

We have shown that if the sequence of Taylor's remainders $\{R_n\}$ for a function $f(x)$ and a point c has limit zero, then $f(x)$ has at least one power series expansion about c, namely its Taylor series,

$$f(x) = \sum_{n=0}^{\infty} \frac{f^{(n)}(c)}{n!}(x-c)^n, \quad f^{(0)}(c) = f(c), \tag{11.40}$$

and this series converges to $f(x)$. We now show that this is the *only* power series expansion of $f(x)$ about c. To do this we require the following theorem.

> **Theorem 11.16**
>
> If a function $f(x)$ has a power series expansion $f(x) = \sum_{n=0}^{\infty} a_n(x-c)^n$ with positive radius of convergence R, then each of the following series has radius of convergence R:
>
> $$f'(x) = \sum_{n=0}^{\infty} na_n(x-c)^{n-1}, \qquad (11.41)$$
>
> $$\int f(x)\,dx = \sum_{n=0}^{\infty} \frac{a_n}{n+1}(x-c)^{n+1} + C. \qquad (11.42)$$
>
> In other words, power series expansions for $f'(x)$ and $\int f(x)\,dx$ can be obtained by term-by-term differentiation and integration of the power series of $f(x)$, and all three series have the same radius of convergence.

Due to the difficulty in proving this theorem, and in order to preserve the continuity of our discussion, we omit a proof. Note that the theorem is stated in terms of radii of convergence rather than intervals of convergence. This is due to the fact that in differentiating a power series we may lose the end points of the original interval of convergence, and in integrating we may pick them up.

The next theorem implies that there is at most one power series expansion of a function about a given point.

> **Theorem 11.17**
>
> If a power series expansion of a function has a positive radius of convergence, then it is the Taylor series.

Proof Suppose the power series expansion

$$f(x) = \sum_{n=0}^{\infty} a_n(x-c)^n = a_0 + a_1(x-c) + a_2(x-c)^2 + \cdots$$

has a positive radius of convergence. Substitution of $x = c$ gives

$$f(c) = a_0.$$

If we differentiate the power series according to Theorem 11.16, we obtain

$$f'(x) = a_1 + 2a_2(x-c) + 3a_3(x-c)^2 + \cdots,$$

and if we substitute $x = c$, the result is

$$f'(c) = a_1.$$

If we differentiate the power series for $f'(x)$, we obtain

$$f''(x) = 2a_2 + 3 \cdot 2a_3(x-c) + 4 \cdot 3a_4(x-c)^2 + \cdots,$$

and substitute $x = c$,

$$f''(c) = 2a_2 \quad \text{or} \quad a_2 = \frac{f''(c)}{2!}.$$

Continued differentiation and substitution leads to the result that for all n,

$$a_n = \frac{f^{(n)}(c)}{n!}.$$

The power series $f(x) = \sum_{n=0}^{\infty} a_n(x-c)^n$ is therefore the Taylor series.

The following corollary is an immediate consequence of this theorem.

Corollary *If two power series $\sum_{n=0}^{\infty} a_n(x-c)^n$ and $\sum_{n=0}^{\infty} b_n(x-c)^n$ with positive radii of convergence have identical sums,*

$$\sum_{n=0}^{\infty} a_n(x-c)^n = \sum_{n=0}^{\infty} b_n(x-c)^n,$$

then $a_n = b_n$ for all n.

Our theory of power series expansions of functions is now complete. Given a function $f(x)$ and a point c, there can only be one power series expansion for $f(x)$ about c, its Taylor series. To show that the Taylor series does indeed have sum $f(x)$, it is sufficient to show that the sequence of Taylor's remainders $\{R_n\}$ has limit zero. When this is the case we write

$$f(x) = \sum_{n=0}^{\infty} \frac{f^{(n)}(c)}{n!}(x-c)^n. \quad (11.43)$$

Equation 11.43 says, then, that the right-hand side is the Taylor series of $f(x)$ and that the Taylor series converges to $f(x)$. This point is most important because there do exist functions $f(x)$ whose Taylor series exist but do not converge to $f(x)$ (see Exercise 12). We therefore write 11.43 only when the Taylor series of $f(x)$ exists and that it converges to $f(x)$.

Only one point of consideration remains. Although we have a formula for calculation of the Taylor series of a function $f(x)$, determination of $f^{(n)}(c)$ could be very complicated. In addition, if a reasonable formula for $f^{(n)}(c)$ cannot be found, how will we ever prove that the sequence of Taylor's remainders has limit zero? Fortunately, in more common situations, special devices exist that enable us to find Taylor series without actually calculating $f^{(n)}(c)$. Indeed, if by any method, we can obtain a power series representation of $f(x)$, then we can be assured by Theorem 11.17 that it is the Taylor series. In Section 11.10 we illustrate some of these techniques.

EXERCISES 11.9

In Exercises 1–6 use Taylor's remainder formula to find Taylor series for the function $f(x)$ about the point indicated. In each case, find the interval on which the Taylor series converges to $f(x)$.

1. $f(x) = \cos x$ about $x = 0$
2. $f(x) = e^{5x}$ about $x = 0$
3. $f(x) = \sin(10x)$ about $x = 0$
4. $f(x) = \sin x$ about $x = \pi/4$
5. $f(x) = e^{2x}$ about $x = 1$
6. $f(x) = 1/(3x+2)$ about $x = 2$

7. Prove the corollary to Theorem 11.17.

8. Prove that if a power series with positive radius of convergence has sum zero, $\sum_{n=0}^{\infty} a_n(x-c)^n = 0$, then $a_n = 0$ for all n.

9. In Section 4.4 we stated the second-derivative test for determining whether a critical point x_0 at which $f'(x_0) = 0$ yields a relative maximum or a relative minimum. Use Taylor's remainder formula to verify this result when $f'(x)$ and $f''(x)$ are continuous on an open interval containing x_0.

10. Extend the result of Exercise 9 to verify the extrema test of Exercise 34 in Section 4.4.

11. There is an integral form for the remainder $R_n(c, x)$ in Taylor's remainder formula that is sometimes more useful than the derivative form in Theorem 11.15: If $f(x)$ and its first n derivatives are continuous on the closed interval between c and x, then

$$f(x) = f(c) + f'(c)(x - c) + \frac{f''(c)}{2!}(x - c)^2 + \cdots$$
$$+ \frac{f^{(n-1)}(c)}{(n-1)!}(x - c)^{n-1} + R_n(c, x),$$

where

$$R_n(c, x) = \frac{1}{(n-1)!} \int_c^x (x - t)^{n-1} f^{(n)}(t) \, dt.$$

Use the following outline of steps to prove this result.

(a) Show that

$$f(x) = f(c) + \int_c^x f'(t) \, dt.$$

(b) Use integration by parts with $u = f'(t)$, $du = f''(t) \, dt$, $dv = dt$, and $v = t - x$ on the integral in (a) to obtain

$$f(x) = f(c) + f'(c)(x - c) + \int_c^x (x - t) f''(t) \, dt.$$

(c) Use integration by parts with $u = f''(t)$, $du = f'''(t) \, dt$, $dv = (x - t) \, dt$, and $v = -(1/2)(x - t)^2$ to obtain

$$f(x) = f(c) + f'(c)(x - c) + \frac{f''(c)}{2!}(x - c)^2$$
$$+ \frac{1}{2!} \int_c^x (x - t)^2 f'''(t) \, dt.$$

(d) Continue this process to obtain the integral form for Taylor's remainder formula.

12. (a) Sketch a graph of the function

$$f(x) = \begin{cases} e^{-1/x^2} & x \neq 0 \\ 0 & x = 0. \end{cases}$$

(b) Use L'Hôpital's rule to show that for every positive integer n,

$$\lim_{x \to 0} \frac{e^{-1/x^2}}{x^n} = 0.$$

(c) Prove, by mathematical induction, that $f^{(n)}(0) = 0$ for $n \geq 1$.

(d) What is the Maclaurin series for $f(x)$?

(e) For what values of x does the Maclaurin series of $f(x)$ converge to $f(x)$?

SECTION 11.10

Taylor Series Expansions of Functions

In Section 11.9 we pointed out that if by any means we can find a power series representation for a function $f(x)$, then we can be assured by Theorem 11.17 that it is the Taylor series for $f(x)$. In this section we illustrate various techniques for producing power series. This has the advantage of avoiding Taylor remainders in establishing convergence of the series to $f(x)$. Our first example uses geometric series.

EXAMPLE 11.27

Find (a) the Maclaurin series for $1/(4 + 5x)$; (b) the Taylor series about 5 for $1/(13 - 2x)$.

SOLUTION (a) If we write

$$\frac{1}{4 + 5x} = \frac{1}{4\left(1 + \frac{5x}{4}\right)}$$

and interpret the right-hand side as one-quarter of the sum of a geometric series with first term equal to 1 and common ratio $-5x/4$, then

$$\frac{1}{4+5x} = \frac{1}{4}\left[1 + \left(-\frac{5x}{4}\right) + \left(-\frac{5x}{4}\right)^2 + \cdots\right], \qquad \left|-\frac{5x}{4}\right| < 1,$$

$$= \sum_{n=0}^{\infty} \frac{(-1)^n 5^n}{4^{n+1}} x^n, \qquad |x| < \frac{4}{5}.$$

(b) By a similar procedure, we have

$$\frac{1}{13-2x} = \frac{1}{3-2(x-5)}$$

$$= \frac{1}{3\left[1 - \frac{2}{3}(x-5)\right]}$$

$$= \frac{1}{3}\left[1 + \frac{2}{3}(x-5) + \left(\frac{2}{3}\right)^2 (x-5)^2 + \cdots\right], \qquad \left|\frac{2}{3}(x-5)\right| < 1,$$

$$= \sum_{n=0}^{\infty} \frac{2^n}{3^{n+1}}(x-5)^n, \qquad |x-5| < \frac{3}{2}.$$

In both examples, properties of geometric series gave not only the required series, but also their intervals of convergence. To appreciate the simplicity of these solutions, we suggest using Taylor remainders in an attempt to obtain the series with the same intervals of convergence. You will quickly abort.

Addition and Subtraction of Power Series

In Section 11.3 we noted that convergent series of constants can be added or subtracted to give convergent series. It follows, therefore, that if $f(x) = \sum a_n(x-c)^n$ and $g(x) = \sum b_n(x-c)^n$, then

$$f(x) \pm g(x) = \sum_{n=0}^{\infty} (a_n \pm b_n)(x-c)^n, \qquad (11.44)$$

and these results are valid for every x that is common to the intervals of convergence of the two series. We use this result in the following example.

EXAMPLE 11.28

Find the Maclaurin series for $f(x) = 5x/(x^2 - 3x - 4)$.

SOLUTION We decompose $f(x)$ into its partial fractions,

$$f(x) = \frac{5x}{x^2 - 3x - 4} = \frac{4}{x-4} + \frac{1}{x+1},$$

and expand each of these terms in a Maclaurin series,

$$\frac{4}{x-4} = \frac{-1}{1-\frac{x}{4}} = -\left(1 + \frac{x}{4} + \frac{x^2}{4^2} + \cdots\right), \qquad |x| < 4, \qquad \text{and}$$

$$\frac{1}{1+x} = 1 - x + x^2 - x^3 + \cdots, \qquad |x| < 1.$$

Addition of these series within their common interval of convergence gives the Maclaurin series for $f(x)$:

$$\frac{5x}{x^2 - 3x - 4} = \left(-1 - \frac{x}{4} - \frac{x^2}{4^2} - \frac{x^3}{4^3} - \cdots\right) + \left(1 - x + x^2 - x^3 + \cdots\right)$$

$$= \left(-1 - \frac{1}{4}\right)x + \left(1 - \frac{1}{4^2}\right)x^2 + \left(-1 - \frac{1}{4^3}\right)x^3 + \cdots$$

$$= \sum_{n=1}^{\infty} \left[(-1)^n - \frac{1}{4^n}\right] x^n, \qquad |x| < 1.$$

∎

Differentiation and Integration of Power Series

Perhaps the most useful technique for generating power series expansions is to differentiate or integrate known expansions according to Theorem 11.16.

EXAMPLE 11.29

Find Maclaurin series for the following functions:

(a) $\cos x$ (b) $\dfrac{1}{(1-x)^3}$

SOLUTION (a) Recall from Example 11.24 that

$$\sin x = x - \frac{x^3}{3!} + \frac{x^5}{5!} - \frac{x^7}{7!} + \cdots, \qquad -\infty < x < \infty.$$

Term-by-term differentiation gives

$$\cos x = 1 - \frac{x^2}{2!} + \frac{x^4}{4!} - \frac{x^6}{6!} + \cdots$$

$$= \sum_{n=0}^{\infty} \frac{(-1)^n}{(2n)!} x^{2n}, \qquad -\infty < x < \infty.$$

(b) Term-by-term differentiation of the series

$$\frac{1}{1-x} = 1 + x + x^2 + x^3 + \cdots + x^n + \cdots, \qquad |x| < 1$$

gives

$$\frac{1}{(1-x)^2} = 1 + 2x + 3x^2 + \cdots + nx^{n-1} + \cdots, \qquad |x| < 1.$$

Another differentiation yields

$$\frac{2}{(1-x)^3} = 2 + 3 \cdot 2x + 4 \cdot 3x^2 + \cdots + n(n-1)x^{n-2} + \cdots, \qquad |x| < 1.$$

Division by 2 now gives

$$\frac{1}{(1-x)^3} = 1 + 3x + 6x^2 + \cdots + \frac{n(n-1)}{2}x^{n-2} + \cdots$$

$$= \sum_{n=2}^{\infty} \frac{n(n-1)}{2} x^{n-2}$$

$$= \frac{1}{2} \sum_{n=0}^{\infty} (n+2)(n+1) x^n, \qquad |x| < 1.$$

∎

EXAMPLE 11.30 Find the Taylor series about 1 for $\ln x$.

SOLUTION Noting that $\ln x$ is an antiderivative of $1/x$, we first expand $1/x$ in a Taylor series about 1:

$$\frac{1}{x} = \frac{1}{(x-1)+1} = 1 - (x-1) + (x-1)^2 - (x-1)^3 + \cdots, \qquad |x-1| < 1.$$

If we integrate this series term-by-term, we have

$$\ln |x| = \left[x - \frac{1}{2}(x-1)^2 + \frac{1}{3}(x-1)^3 - \frac{1}{4}(x-1)^4 + \cdots \right] + C.$$

Substitution of $x = 1$ implies that $0 = 1 + C$; that is, $C = -1$, and hence,

$$\ln |x| = (x-1) - \frac{1}{2}(x-1)^2 + \frac{1}{3}(x-1)^3 - \cdots$$

$$= \sum_{n=1}^{\infty} \frac{(-1)^{n+1}}{n} (x-1)^n.$$

According to Theorem 11.16, the radius of convergence of this series is also $R = 1$. So the series certainly converges to $\ln |x|$ for $0 < x < 2$. But when $x = 2$, the series reduces to the alternating harmonic series, which converges conditionally. Consequently, we can extend the interval of convergence to include $x = 2$ and write

$$\ln x = \sum_{n=1}^{\infty} \frac{(-1)^{n+1}}{n} (x-1)^n, \qquad 0 < x \leq 2.$$

Convergence of this series at $x = 2$ does not, by itself, imply convergence to $\ln 2$, as this equation suggests. It is, however, true, and this is a direct application of the following theorem.

∎

> **Theorem 11.18**
>
> If the Taylor series $\sum_{n=0}^{\infty} a_n(x-c)^n$ of a function $f(x)$ converges at the endpoint $x = c + R$ of its interval of convergence, and if $f(x)$ is continuous at $x = c + R$, then the Taylor series evaluated at $c + R$ converges to $f(c + R)$. The same result is valid at the other endpoint $x = c - R$.

Comparison of the solutions in Examples 11.26 and 11.30 indicates once again the advantage of avoiding the use of Taylor's remainder formula.

Multiplication of Power Series

In Example 11.28 we added the Maclaurin series for $4/(x-4)$ and $1/(1+x)$ to obtain the Maclaurin series for $5x/(x^2 - 3x - 4)$. An alternative procedure might be to multiply the two series since

$$\frac{5x}{x^2 - 3x - 4} = 5x \left(\frac{1}{x-4}\right)\left(\frac{1}{x+1}\right)$$

$$= \frac{5x}{-4}\left(1 + \frac{x}{4} + \frac{x^2}{4^2} + \frac{x^3}{4^3} + \cdots\right)(1 - x + x^2 - x^3 + \cdots).$$

The rules of algebra would demand that we multiply every term of the first series by every term of the second. If we do this and group all products with like powers of x, we obtain

$$\frac{5x}{x^2 - 3x - 4} = \frac{5x}{-4}\left[1 + \left(-1 + \frac{1}{4}\right)x + \left(1 - \frac{1}{4} + \frac{1}{4^2}\right)x^2\right.$$
$$\left. + \left(-1 + \frac{1}{4} - \frac{1}{4^2} + \frac{1}{4^3}\right)x^3 + \cdots\right].$$

It is clear that the coefficient of x^n is a finite geometric series to which we can apply formula 11.13a:

$$(-1)^n\left[1 - \frac{1}{4} + \frac{1}{4^2} - \cdots + \frac{(-1)^n}{4^n}\right] = (-1)^n\left[\frac{1 - \left(-\frac{1}{4}\right)^{n+1}}{1 + \frac{1}{4}}\right]$$

$$= (-1)^n \frac{4}{5}\left[1 - \left(-\frac{1}{4}\right)^{n+1}\right].$$

Consequently,

$$\frac{5x}{x^2 - 3x - 4} = \frac{5x}{-4}\sum_{n=0}^{\infty}(-1)^n \frac{4}{5}\left[1 - \left(-\frac{1}{4}\right)^{n+1}\right]x^n$$

$$= \sum_{n=0}^{\infty}(-1)^{n+1}\left[1 - \frac{(-1)^{n+1}}{4^{n+1}}\right]x^{n+1}$$

$$= \sum_{n=0}^{\infty}\left[(-1)^{n+1} - \frac{1}{4^{n+1}}\right]x^{n+1}$$

$$= \sum_{n=1}^{\infty}\left[(-1)^n - \frac{1}{4^n}\right]x^n.$$

For this example, then, multiplication as well as addition of power series leads to the Maclaurin series. Clearly, addition of power series is much simpler for this example, but we have at least demonstrated that power series can be multiplied together. That this is generally possible is stated in the following theorem.

Theorem 11.19

If $f(x) = \sum_{n=0}^{\infty} a_n(x-c)^n$ and $g(x) = \sum_{n=0}^{\infty} b_n(x-c)^n$, then

$$f(x)g(x) = \sum_{n=0}^{\infty} d_n(x-c)^n, \qquad (11.45\text{a})$$

where

$$d_n = \sum_{i=0}^{n} a_i b_{n-i} = a_0 b_n + a_1 b_{n-1} + \cdots + a_{n-1} b_1 + a_n b_0, \qquad (11.45\text{b})$$

and this series converges absolutely at every point at which the series for $f(x)$ and $g(x)$ converge absolutely.

EXAMPLE 11.31

Find the Maclaurin series for $f(x) = [1/(x-1)] \ln(1-x)$.

SOLUTION If we integrate the Maclaurin series

$$\frac{1}{1-x} = 1 + x + x^2 + x^3 + \cdots, \qquad |x| < 1,$$

we find

$$-\ln|1-x| = \left(x + \frac{x^2}{2} + \frac{x^3}{3} + \frac{x^4}{4} + \cdots \right) + C.$$

By setting $x = 0$, we obtain $C = 0$, and

$$\ln(1-x) = -x - \frac{x^2}{2} - \frac{x^3}{3} - \frac{x^4}{4} - \cdots .$$

Note that we have dropped the absolute value signs since the radius of convergence of the series is 1. We now form the Maclaurin series for $f(x)$:

$$\frac{1}{x-1} \ln(1-x) = \frac{-1}{1-x} \ln(1-x)$$

$$= (1 + x + x^2 + x^3 + \cdots)\left(x + \frac{x^2}{2} + \frac{x^3}{3} + \cdots \right)$$

$$= x + \left(1 + \frac{1}{2}\right) x^2 + \left(1 + \frac{1}{2} + \frac{1}{3}\right) x^3 + \cdots$$

$$= \sum_{n=1}^{\infty} \left(1 + \frac{1}{2} + \frac{1}{3} + \cdots + \frac{1}{n}\right) x^n.$$

Since both of the multiplied series converge absolutely on the interval $|x| < 1$, the Maclaurin series for $(x-1)^{-1} \ln(1-x)$ is also valid on this interval.

EXAMPLE 11.32

Find the first three nonzero terms in the Maclaurin series for $\tan x$.

SOLUTION If $\tan x = \sum a_n x^n$, then by setting $\tan x = \sin x / \cos x$, we have

$$\sin x = \cos x \sum_{n=0}^{\infty} a_n x^n.$$

We now substitute the Maclaurin series for $\sin x$ and $\cos x$:

$$x - \frac{x^3}{3!} + \frac{x^5}{5!} - \frac{x^7}{7!} + \cdots = \left(1 - \frac{x^2}{2!} + \frac{x^4}{4!} - \frac{x^6}{6!} + \cdots\right)\left(a_0 + a_1 x + a_2 x^2 + \cdots\right).$$

According to the corollary to Theorem 11.17, two power series can be identical only if their corresponding coefficients are equal. We therefore multiply the right-hand side and equate coefficients of like powers of x:

$x^0:\quad 0 = a_0;$

$x:\quad 1 = a_1;$

$x^2:\quad 0 = a_2 - \dfrac{a_0}{2!},\quad$ which implies $a_2 = 0;$

$x^3:\quad -\dfrac{1}{3!} = a_3 - \dfrac{a_1}{2!},\quad$ which implies $a_3 = \dfrac{1}{2!} - \dfrac{1}{3!} = \dfrac{1}{3};$

$x^4:\quad 0 = a_4 - \dfrac{a_2}{2!} + \dfrac{a_0}{4!},\quad$ from which $a_4 = 0;$

$x^5:\quad \dfrac{1}{5!} = a_5 - \dfrac{a_3}{2!} + \dfrac{a_1}{4!},\quad$ from which $a_5 = \dfrac{1}{5!} + \dfrac{1}{6} - \dfrac{1}{4!} = \dfrac{2}{15}.$

The first three nonzero terms in the Maclaurin series for $\tan x$ are therefore

$$\tan x = x + \frac{1}{3}x^3 + \frac{2}{15}x^5 + \cdots.$$

∎

Binomial Expansion

One of the most widely used power series is the binomial expansion. We are well acquainted with the binomial theorem, which predicts the product $(a+b)^m$ for any positive integer m:

$$(a+b)^m = \sum_{n=0}^{m} \binom{m}{n} a^n b^{m-n}. \qquad (11.46)$$

With the usual definition of the binomial coefficients,

$$\binom{m}{n} = \frac{m!}{(m-n)!\,n!} = \frac{m(m-1)(m-2)\cdots(m-n+1)}{n!},$$

the binomial theorem becomes

$$(a+b)^m = a^m + ma^{m-1}b + \frac{m(m-1)}{2!}a^{m-2}b^2 + \cdots + mab^{m-1} + b^m. \qquad (11.47)$$

Even when m is not a positive integer, this form for the binomial theorem remains almost intact. To show this, we consider the power series

$$1 + \sum_{n=1}^{\infty} \frac{m(m-1)(m-2)\cdots(m-n+1)}{n!} x^n$$

for any real number m except a nonnegative integer. The radius of convergence of this power series is

$$R = \lim_{n \to \infty} \left| \frac{m(m-1)(m-2)\cdots(m-n+1)}{n!} \cdot \frac{(n+1)!}{m(m-1)(m-2)\cdots(m-n)} \right|$$

$$= \lim_{n \to \infty} \left| \frac{n+1}{m-n} \right| = 1.$$

The power series therefore converges absolutely for $|x| < 1$. Whether the series converges at the end points $x = \pm 1$ depends on the value of m. For the time being, we will work on the interval $|x| < 1$, and at the end of the discussion, we will state the complete result. Let us denote the sum of the series by

$$f(x) = 1 + \sum_{n=1}^{\infty} \frac{m(m-1)(m-2)\cdots(m-n+1)}{n!} x^n, \qquad |x| < 1.$$

If we differentiate this series term-by-term according to Theorem 11.16,

$$f'(x) = \sum_{n=1}^{\infty} \frac{m(m-1)\cdots(m-n+1)}{(n-1)!} x^{n-1}, \qquad |x| < 1,$$

and then multiply both sides by x, we have

$$xf'(x) = \sum_{n=1}^{\infty} \frac{m(m-1)\cdots(m-n+1)}{(n-1)!} x^n, \qquad |x| < 1.$$

If we add these results, we obtain

$$f'(x) + xf'(x) = \sum_{n=1}^{\infty} \frac{m(m-1)\cdots(m-n+1)}{(n-1)!} x^{n-1} + \sum_{n=1}^{\infty} \frac{m(m-1)\cdots(m-n+1)}{(n-1)!} x^n.$$

We now change the variable of summation in the first sum,

$$(1+x)f'(x) = \sum_{n=0}^{\infty} \frac{m(m-1)\cdots(m-n)}{n!} x^n + \sum_{n=1}^{\infty} \frac{m(m-1)\cdots(m-n+1)}{(n-1)!} x^n.$$

When these summations are added over their common range, beginning at $n = 1$, and the $n = 0$ term in the first summation is written out separately, the result is

$$(1+x)f'(x) = m + \sum_{n=1}^{\infty} \frac{m(m-1)\cdots(m-n+1)}{(n-1)!} \left(\frac{m-n}{n} + 1 \right) x^n$$

$$= m \left[1 + \sum_{n=1}^{\infty} \frac{m(m-1)\cdots(m-n+1)}{n!} x^n \right]$$

$$= mf(x).$$

Consequently, the function $f(x)$ must satisfy the differential equation

$$\frac{f'(x)}{f(x)} = \frac{m}{1+x}.$$

Integration immediately gives

$$\ln|f(x)| = m \ln|1+x| + C, \quad \text{or}$$

$$f(x) = D(1+x)^m.$$

To evaluate D, we note that from the original definition of $f(x)$ as the sum of the power series, $f(0) = 1$, and this implies that $D = 1$. Thus,

$$f(x) = (1+x)^m,$$

and we may write finally that

$$(1+x)^m = 1 + \sum_{n=1}^{\infty} \frac{m(m-1)(m-2)\cdots(m-n+1)}{n!} x^n, \quad |x| < 1 \quad (11.48\text{a})$$

$$= 1 + mx + \frac{m(m-1)}{2!} x^2 + \frac{m(m-1)(m-2)}{3!} x^3 + \cdots, \quad |x| < 1. \,(11.48\text{b})$$

This is called the binomial expansion of $(1+x)^m$. We have verified the binomial expansion for m any real number except a nonnegative integer, but in the case of a nonnegative integer, the series terminates after $m+1$ terms and is therefore valid for these values of m also. We mentioned earlier that the binomial expansion may also converge at the end points $x = \pm 1$ depending on the value of m. The complete result states that 11.48 is valid for

$$\begin{array}{ll} -\infty < x < \infty & \text{if } m \text{ is a nonnegative integer,} \\ -1 < x < 1 & \text{if } m \leq -1, \\ -1 < x \leq 1 & \text{if } -1 < m < 0, \\ -1 \leq x \leq 1 & \text{if } m > 0 \text{ but not an integer.} \end{array}$$

It is not difficult to generalize this result to expand $(a+b)^m$ for real m. If $|b| < |a|$, we write

$$(a+b)^m = a^m \left(1 + \frac{b}{a}\right)^m$$

and now expand the bracketed term by means of 11.48:

$$(a+b)^m = a^m \left[1 + m\left(\frac{b}{a}\right) + \frac{m(m-1)}{2!}\left(\frac{b}{a}\right)^2 + \cdots\right] \quad |b| < |a|,$$

$$= a^m + ma^{m-1}b + \frac{m(m-1)}{2!} a^{m-2} b^2 + \cdots \quad |b| < |a|, \quad (11.49)$$

which, as we predicted, is equation 11.47 except that the series does not terminate.

EXAMPLE 11.33

Use the binomial expansion to find the Maclaurin series for $1/(1-x)^3$.

SOLUTION By 11.48b, we have

$$\frac{1}{(1-x)^3} = (1-x)^{-3}$$
$$= 1 + (-3)(-x) + \frac{(-3)(-4)}{2!}(-x)^2 + \frac{(-3)(-4)(-5)}{3!}(-x)^3 + \cdots$$
$$= 1 + 3x + \frac{3 \cdot 4}{2}x^2 + \frac{4 \cdot 5}{2}x^3 + \frac{5 \cdot 6}{2}x^4 + \cdots$$
$$= \frac{1}{2}\sum_{n=0}^{\infty}(n+1)(n+2)x^n, \qquad |x| < 1.$$

This result was also obtained in Example 11.29 by differentiation of the Maclaurin series for $1/(1-x)$.

EXAMPLE 11.34

Find the Maclaurin series for $\text{Sin}^{-1} x$.

SOLUTION By the binomial expansion, we have

$$\frac{1}{\sqrt{1-x^2}} = 1 + \left(-\frac{1}{2}\right)(-x^2) + \frac{\left(-\frac{1}{2}\right)\left(-\frac{3}{2}\right)}{2!}(-x^2)^2 + \frac{\left(-\frac{1}{2}\right)\left(-\frac{3}{2}\right)\left(-\frac{5}{2}\right)}{3!}(-x^2)^3 + \cdots$$
$$= 1 + \frac{1}{2}x^2 + \frac{3}{2^2 2!}x^4 + \frac{3 \cdot 5}{2^3 3!}x^6 + \frac{3 \cdot 5 \cdot 7}{2^4 4!}x^8 + \cdots, \qquad |x| < 1.$$

Integration of this series gives

$$\text{Sin}^{-1} x = \left(x + \frac{1}{2 \cdot 3}x^3 + \frac{3}{2^2 2! 5}x^5 + \frac{3 \cdot 5}{2^3 3! 7}x^7 + \frac{3 \cdot 5 \cdot 7}{2^4 4! 9}x^9 + \cdots\right) + C.$$

Evaluation of both sides of this equation at $x = 0$ gives us $C = 0$. According to Theorem 11.16, the radius of convergence of this series must be 1, and we can write

$$\text{Sin}^{-1} x = x + \sum_{n=1}^{\infty} \frac{1 \cdot 3 \cdot 5 \cdot \cdots \cdot (2n-1)}{2^n n!(2n+1)} x^{2n+1}$$
$$= x + \sum_{n=1}^{\infty} \frac{1 \cdot 2 \cdot 3 \cdot 4 \cdot 5 \cdot \cdots \cdot (2n-2)(2n-1)(2n)}{2 \cdot 4 \cdot \cdots \cdot (2n) 2^n n!(2n+1)} x^{2n+1}$$
$$= x + \sum_{n=1}^{\infty} \frac{(2n)!}{(2n+1) 2^{2n}(n!)^2} x^{2n+1}, \qquad |x| < 1.$$

In this case, integration of the series for $(1-x^2)^{-1/2}$ does not pick up the end points $x = \pm 1$, although this is difficult to prove.

EXERCISES 11.10

In Exercises 1–20 find the power series expansion of the function about the indicated point.

1. $f(x) = \dfrac{1}{3x+2}$ about $x = 0$

2. $f(x) = \dfrac{1}{4+x^2}$ about $x = 0$

3. $f(x) = \dfrac{1}{x+3}$ about $x = 2$

4. $f(x) = \cos(x^2)$ about $x = 0$

5. $f(x) = \dfrac{1}{\sqrt{1+x}}$ about $x = 0$

6. $f(x) = e^{5x}$ about $x = 0$

7. $f(x) = \cosh x$ about $x = 0$

8. $f(x) = \sinh x$ about $x = 0$

9. $f(x) = \ln(1+2x)$ about $x = 0$

10. $f(x) = (1+3x)^{3/2}$ about $x = 0$

11. $f(x) = 1/x$ about $x = 4$

12. $f(x) = x^4 + 3x^2 - 2x + 1$ about $x = 0$

13. $f(x) = \dfrac{1}{(x+2)^3}$ about $x = 0$

14. $f(x) = x^4 + 3x^2 - 2x + 1$ about $x = -2$

15. $f(x) = \dfrac{1}{x^2 + 8x + 15}$ about $x = 0$

16. $f(x) = e^x$ about $x = 3$

17. $f(x) = \mathrm{Tan}^{-1} x$ about $x = 0$

18. $f(x) = \sqrt{x+3}$ about $x = 0$

19. $f(x) = \dfrac{x^2}{(1+x^2)^2}$ about $x = 0$

20. $f(x) = x(1-x)^{1/3}$ about $x = 0$

In Exercises 21–23 find the first four nonzero terms in the Maclaurin series for the function.

21. $f(x) = \tan 2x$

22. $f(x) = \sec x$

23. $f(x) = e^x \sin x$

24. Find the Maclaurin series for $\cos^2 x$.

In Exercises 25–28 find the Maclaurin series for the function.

25. $f(x) = \dfrac{1}{x^6 - 3x^3 - 4}$

26. $f(x) = \mathrm{Sin}^{-1}(x^2)$

27. $f(x) = \dfrac{2x^2 + 4}{x^2 + 4x + 3}$

28. $f(x) = \ln\left[\dfrac{1 + x/\sqrt{2}}{1 - x/\sqrt{2}}\right]$

29. If during a working day, one person drinks from a fountain every 30 seconds (on the average), then the probability that exactly n people drink in a time interval of length t seconds is given by the *Poisson distribution*:

$$P_n(t) = \dfrac{1}{n!}\left(\dfrac{t}{30}\right)^n e^{-t/30}.$$

Calculate $\sum_{n=0}^{\infty} P_n(t)$ and interpret the result.

30. A certain experiment is to be performed until it is successful. The probability that it will be successful in any given attempt is p ($0 < p < 1$), and therefore the probability that it will fail is $q = 1 - p$. The expected number of times that the experiment must be performed in order to be successful can be shown to be represented by the infinite series

$$\sum_{n=1}^{\infty} npq^{n-1} = \sum_{n=1}^{\infty} np(1-p)^{n-1}.$$

(a) What is the sum of this series?

(b) If p is the probability that a single die will come up 6, is the answer in (a) what you would expect?

31. Find the Maclaurin series for the *error function* $\mathrm{erf}(x)$ defined by

$$\mathrm{erf}(x) = \dfrac{2}{\sqrt{\pi}} \int_0^x e^{-t^2} dt.$$

32. Find Maclaurin series for the *Fresnel integrals* $C(x)$ and $S(x)$ defined by

$$C(x) = \int_0^x \cos(\pi t^2/2)\, dt, \quad S(x) = \int_0^x \sin(\pi t^2/2)\, dt.$$

Show that Bessel functions of the first kind (defined in Exercise 30 of Section 11.8) satisfy the properties in Exercises 33 and 34.

33. $2m J_m(x) - x J_{m-1}(x) = x J_{m+1}(x)$

34. $J_{m-1}(x) - J_{m+1}(x) = 2 J_m'(x)$

35. If the function $(1 - 2\mu x + x^2)^{-1/2}$ is expanded in a Maclaurin series in x,

$$\frac{1}{\sqrt{1-2\mu x + x^2}} = \sum_{n=0}^{\infty} P_n(\mu) x^n,$$

the coefficients $P_n(\mu)$ are called the *Legendre polynomials*. Find $P_0(\mu)$, $P_1(\mu)$, $P_2(\mu)$, and $P_3(\mu)$.

36. (a) If we define $f(x) = x/(e^x - 1)$ at $x = 0$ as $f(0) = 1$, it turns out that $f(x)$ has a Maclaurin series expansion with positive radius of convergence. When this expansion is expressed in the form

$$\frac{x}{e^x - 1} = 1 + B_1 x + \frac{B_2}{2!} x^2 + \frac{B_3}{3!} x^3 + \cdots,$$

the coefficients $B_1, B_2, B_3, \ldots$ are called the *Bernoulli numbers*. Write this equation in the form

$$x = (e^x - 1) \left(1 + B_1 x + \frac{B_2}{2!} x^2 + \cdots \right),$$

and substitute the Maclaurin series for e^x to find the first five Bernoulli numbers.

(b) Show that the odd Bernoulli numbers all vanish for $n \geq 3$.

37. By substituting power series expansions for $e^{xt/2}$ and $e^{-x/(2t)}$ in terms of powers of t and $1/t$ respectively, show that

$$e^{x(t-1/t)/2} = \sum_{n=0}^{\infty} J_n(x) t^n.$$

For a definition of $J_n(x)$, see Exercises 30 in Section 11.8.

SECTION 11.11

Sums of Power Series and Series of Constants

Theorem 11.16 provides an important technique for finding sums of power series. If an unknown series can be reduced to a known series by differentiations or integrations, then its sum can be obtained when these operations are reversed.

EXAMPLE 11.35

Find the sum of the series $\sum_{n=0}^{\infty} (n+1) x^n$.

SOLUTION The series converges absolutely for $|x| < 1$. If we denote the sum of the series by $S(x)$,

$$S(x) = \sum_{n=0}^{\infty} (n+1) x^n,$$

and use Theorem 11.16 to integrate the series term-by-term,

$$\int S(x) \, dx = \sum_{n=0}^{\infty} x^{n+1} + C.$$

But the series on the right is a geometric series with sum $x/(1-x)$, provided $|x| < 1$, and we may write therefore

$$\int S(x) \, dx = \frac{x}{1-x} + C, \qquad |x| < 1.$$

If we now differentiate this equation, we obtain

$$S(x) = \frac{(1-x)(1) - x(-1)}{(1-x)^2} = \frac{1}{(1-x)^2},$$

and therefore

$$\sum_{n=0}^{\infty} (n+1) x^n = \frac{1}{(1-x)^2}, \qquad |x| < 1.$$

> **EXAMPLE 11.36**
>
> Find the sum of the series $\sum_{n=1}^{\infty} x^n/n$.
>
> **SOLUTION** The series converges for $-1 \leq x < 1$. If we set
> $$S(x) = \sum_{n=1}^{\infty} \frac{1}{n} x^n,$$
> and differentiate the series term-by-term,
> $$S'(x) = \sum_{n=1}^{\infty} x^{n-1}.$$
> This is a geometric series with sum $1/(1-x)$, so that
> $$S'(x) = \frac{1}{1-x}.$$
> Antidifferentiation now gives
> $$S(x) = -\ln(1-x) + C.$$
> Since $S(0) = 0$, it follows that
> $$0 = -\ln(1) + C.$$
> Hence, $C = 0$, and $S(x) = -\ln(1-x)$. We have shown therefore that
> $$\sum_{n=1}^{\infty} \frac{1}{n} x^n = -\ln(1-x), \qquad -1 \leq x < 1.$$

By substituting values of x into known power series we obtain formulas for sums of series of constants. For instance, in Example 11.25, we verified that the Maclaurin series for e^x is

$$e^x = \sum_{n=0}^{\infty} \frac{1}{n!} x^n, \qquad -\infty < x < \infty.$$

By substituting $x = 1$, we obtain a series which converges to e,

$$e = \sum_{n=0}^{\infty} \frac{1}{n!} = 1 + \frac{1}{1!} + \frac{1}{2!} + \frac{1}{3!} + \cdots.$$

This was the series that we truncated in Example 11.16.

The following example is another illustration of this idea.

> **EXAMPLE 11.37**
>
> Find the sum of the series
> $$\sum_{n=0}^{\infty} \frac{(-1)^n}{(2n+1)2^n}.$$
>
> **SOLUTION** There are many power series which reduce to the given series upon substitution of a specific value of x. For instance, substitution of $-1/2$, 1, and $1/\sqrt{2}$ into the following power series, respectively, lead to the given series:
>
> $$\sum_{n=0}^{\infty} \frac{1}{2n+1} x^n, \quad \sum_{n=0}^{\infty} \frac{(-1)^n}{(2n+1)2^n} x^n, \quad \sum_{n=0}^{\infty} \frac{\sqrt{2}(-1)^n}{2n+1} x^{2n+1}.$$

Which should we consider? Although it is not the simplest, the third series looks most promising; the fact that the power on x corresponds to the coefficient in the denominator suggests that we can find the sum of this series. We therefore set

$$S(x) = \sum_{n=0}^{\infty} \frac{\sqrt{2}(-1)^n}{2n+1} x^{2n+1},$$

and this series converges for $-1 \leq x \leq 1$. If we differentiate the series with respect to x, then

$$S'(x) = \sum_{n=0}^{\infty} \sqrt{2}(-1)^n x^{2n} = \sum_{n=0}^{\infty} \sqrt{2}(-x^2)^n = \frac{\sqrt{2}}{1-(-x^2)} = \frac{\sqrt{2}}{1+x^2}.$$

Antidifferentiation now gives

$$S(x) = \int \frac{\sqrt{2}}{1+x^2} dx = \sqrt{2} \operatorname{Tan}^{-1} x + C.$$

Since $S(0) = 0$, it follows that $C = 0$, and

$$\sum_{n=0}^{\infty} \frac{\sqrt{2}(-1)^n}{2n+1} x^{2n+1} = \sqrt{2} \operatorname{Tan}^{-1} x, \qquad |x| \leq 1.$$

If we now set $x = 1/\sqrt{2}$,

$$\sum_{n=0}^{\infty} \frac{\sqrt{2}(-1)^n}{2n+1} \left(\frac{1}{\sqrt{2}}\right)^{2n+1} = \sqrt{2} \operatorname{Tan}^{-1} \left(\frac{1}{\sqrt{2}}\right).$$

Consequently,

$$\sum_{n=0}^{\infty} \frac{(-1)^n}{(2n+1)2^n} = \sqrt{2} \operatorname{Tan}^{-1} \left(\frac{1}{\sqrt{2}}\right). \blacksquare$$

EXERCISES 11.11

In Exercises 1–10 find the sum of the power series.

1. $\sum_{n=1}^{\infty} nx^{n-1}$

2. $\sum_{n=2}^{\infty} n(n-1)x^{n-2}$

3. $\sum_{n=1}^{\infty} (n+1)x^{n-1}$

4. $\sum_{n=1}^{\infty} n^2 x^{n-1}$

5. $\sum_{n=1}^{\infty} (n^2 + 2n)x^n$

6. $\sum_{n=0}^{\infty} \frac{1}{n+1} x^n$

7. $\sum_{n=0}^{\infty} \frac{(-1)^n}{2n+1} x^{2n+1}$

8. $\sum_{n=1}^{\infty} \frac{(-1)^n}{n} x^{2n}$

9. $\sum_{n=2}^{\infty} n3^n x^{2n}$

10. $\sum_{n=0}^{\infty} \left(\frac{n+1}{n+2}\right) x^n$

In Exercises 11–20 verify that the sum of the series is as indicated.

11. $\sum_{n=0}^{\infty} \frac{2^n}{n!} = e^2$

12. $\sum_{n=0}^{\infty} \frac{(-1)^n}{(2n+1)!} = \sin 1$

13. $\sum_{n=0}^{\infty} \frac{(-1)^n 3^{2n}}{(2n)!} = \cos 3$

14. $\sum_{n=1}^{\infty} \frac{(-1)^n}{n!} = \frac{1}{e} - 1$

15. $\sum_{n=1}^{\infty} \frac{(-1)^n}{3^{2n}(2n+1)!} = 3\sin\left(\frac{1}{3}\right) - 1$

16. $\sum_{n=2}^{\infty} \frac{(-1)^{n+1} 2^{2n+3}}{(2n)!} = -8(1 + \cos 2)$

17. $\sum_{n=1}^{\infty} \frac{2^n}{n 3^n} = \ln 3$ **18.** $\sum_{n=1}^{\infty} \frac{1}{n 2^n} = \ln 2$

19. $\sum_{n=1}^{\infty} \frac{(-1)^n}{2^{2n}} = -\frac{1}{5}$ **20.** $\sum_{n=1}^{\infty} \frac{n}{2^n} = 2$

21. Find the Maclaurin series for $\mathrm{Tan}^{-1} x$ and use it to evaluate
$$\sum_{n=1}^{\infty} \frac{(-1)^n}{2n+1}.$$

22. Find the Maclaurin series for $x/(1+x^2)^2$ and use it to evaluate
$$\sum_{n=1}^{\infty} \frac{n(-1)^n}{3^{2n}}.$$

SECTION 11.12

Applications of Taylor Series and Taylor's Remainder Formula

If Taylor's remainder $R_n(c, x)$ is truncated from remainder formula 11.32e, a polynomial approximation to $f(x)$ is obtained

$$f(x) \approx f(c) + f'(c)(x - c) + \frac{f''(c)}{2!}(x - c)^2 + \cdots + \frac{f^{(n-1)}(c)}{(n-1)!}(x - c)^{n-1}.$$

The remainder

$$R_n(c, x) = \frac{f^{(n)}(z_n)}{n!}(x - c)^n,$$

where z_n is between c and x represents the error of the approximation; the smaller R_n, the better the approximation. In this section we use Taylor series and Taylor's remainder formula in a number of situations that require approximations.

Approximations of Functions

Consider using the first three terms of the Maclaurin series for e^x to approximate e^x on the interval $0 \leq x \leq 1/2$. Taylor's remainder formula, with $c = 0$, states that

$$e^x = 1 + x + \frac{x^2}{2} + R_3$$

where

$$R_3 = \frac{d^3}{dx^3}(e^x)|_{x=z} \frac{x^3}{3!} = \frac{e^z x^3}{6},$$

and z is between 0 and x. Although z is unknown — except that it is between 0 and x, we can say that because only the values $0 \leq x \leq \frac{1}{2}$ are under consideration, z must be less than $\frac{1}{2}$. It follows that

$$R_3 < \frac{e^{1/2} x^3}{6} \leq \frac{\sqrt{e}(1/2)^3}{6} < 0.035.$$

Thus the quadratic function $1 + x + x^2/2$ approximates e^x on the interval $0 \leq x \leq \frac{1}{2}$ with an error no greater than 0.035.

In the following example, we determine the number of terms of a Maclaurin series required to guarantee a certain accuracy.

EXAMPLE 11.38

How many terms in the Maclaurin series of $\ln(1+x)$ guarantee a truncation error of less than 10^{-6} for any x in the interval $0 \leq x \leq 1/2$?

SOLUTION The nth derivative of $\ln(1+x)$ is

$$\frac{d^n}{dx^n}\ln(1+x) = \frac{(-1)^{n+1}(n-1)!}{(x+1)^n}, \quad n \geq 1,$$

and therefore Taylor's remainder formula with $c = 0$ states that

$$\ln(1+x) = x - \frac{x^2}{2} + \frac{x^3}{3} - \frac{x^4}{4} + \cdots + \frac{(-1)^n}{n-1}x^{n-1} + R_n.$$

Now that we see terms in the Maclaurin series, for $0 \leq x \leq \frac{1}{2}$, the series is alternating. We could therefore discuss accuracy from an alternating series point of view and avoid Taylor remainders totally. Recall from Section 11.7 that when an alternating series is truncated, the maximum possible error is the next term. Consequently, if the Maclaurin series for $\ln(1+x)$ is truncated after the term $(-1)^n x^{n-1}/(n-1)$, the maximum error is

$$\frac{(-1)^{n+1}}{n}x^n.$$

Considering values of x in the interval $0 \leq x \leq \frac{1}{2}$, this error is a maximum at $x = \frac{1}{2}$. Thus, the error in truncating the series after $(-1)^n x^{n-1}/(n-1)$ is no greater in absolute value than

$$\frac{(1/2)^n}{n}.$$

This is less than 10^{-6} if

$$\frac{(1/2)^n}{n} < 10^{-6} \quad \text{or}$$

$$n2^n > 10^6.$$

A calculator quickly indicates that the smallest value of n for which this is true is $n = 16$. Consequently, if $\ln(1+x)$ is approximated by the 15th degree polynomial

$$\ln(1+x) \approx x - \frac{x^2}{2} + \frac{x^3}{3} - \frac{x^4}{4} + \cdots + \frac{x^{15}}{15}$$

on the interval $0 \leq x \leq 1/2$, the truncation error is less than 10^{-6}.

It is interesting to note that Taylor remainders do no better. To see this we calculate that

$$R_n = \frac{d^n}{dx^n}\ln(1+x)\Big|_{x=z}\frac{x^n}{n!} = \frac{(-1)^{n+1}(n-1)!}{(z+1)^n}\frac{x^n}{n!} = \frac{(-1)^{n+1}}{n(z+1)^n}x^n.$$

Since z is between 0 and x and $0 \leq x \leq \frac{1}{2}$, we can state that x must be less than or equal to $\frac{1}{2}$, and z must be greater than 0. Hence,

$$|R_n| < \frac{1}{n(1)^n}(1/2)^n = \frac{1}{n2^n},$$

and this is the same error expression previously obtained when analysis was performed from an alternating series point of view. ∎

Evaluation of Definite Integrals

In Section 9.9 we developed three numerical techniques for approximating definite integrals of functions $f(x)$ that have no obvious antiderivatives: the rectangular rule, the trapezoidal rule, and Simpson's rule. Each method divides the interval of integration into a number of subintervals and approximates $f(x)$ by a more elementary function on each subinterval. The rectangular rule replaces $f(x)$ by a step function, the trapezoidal rule replaces $f(x)$ by a succession of linear functions, and Simpson's rule uses a sequence of quadratic functions.

Another possibility is to replace $f(x)$ by a truncated power series (a polynomial) over the entire interval of integration. For instance, consider the definite integral

$$\int_0^{1/2} \frac{\sin x}{x}\, dx,$$

where $(\sin x)/x$ is defined as 1 at $x = 0$. Suppose we expand the integrand in a Maclaurin series:

$$\frac{1}{x}\sin x = \frac{1}{x}\left(x - \frac{x^3}{3!} + \frac{x^5}{5!} - \cdots\right) = 1 - \frac{x^2}{3!} + \frac{x^4}{5!} - \cdots, \qquad -\infty < x < \infty.$$

According to Theorem 11.16,

$$\int \frac{\sin x}{x}\, dx = \left(x - \frac{x^3}{3 \cdot 3!} + \frac{x^5}{5 \cdot 5!} - \cdots\right) + C, \qquad -\infty < x < \infty,$$

and therefore

$$\int_0^{1/2} \frac{\sin x}{x}\, dx = \frac{1}{2} - \frac{\left(\frac{1}{2}\right)^3}{3 \cdot 3!} + \frac{\left(\frac{1}{2}\right)^5}{5 \cdot 5!} - \cdots.$$

If we approximate this alternating series with its first three terms, the truncation error is the fourth term. Since

$$\frac{1}{2} - \frac{\left(\frac{1}{2}\right)^3}{3 \cdot 3!} + \frac{\left(\frac{1}{2}\right)^5}{5 \cdot 5!} = 0.493\,107\,6 \qquad \text{and}$$

$$\frac{\left(\frac{1}{2}\right)^7}{7 \cdot 7!} < 0.000\,000\,222,$$

it follows that

$$0.493\,107\,4 < \int_0^{1/2} \frac{\sin x}{x}\, dx < 0.493\,107\,6.$$

Consequently, using only three terms of the Maclaurin series for $(\sin x)/x$, we can say that to five decimals

$$\int_0^{1/2} \frac{\sin x}{x}\, dx = 0.49311.$$

EXAMPLE 11.39

A very important function in engineering and physics is the error function erf(x) defined by

$$\text{erf}(x) = \frac{2}{\sqrt{\pi}} \int_0^x e^{-t^2}\, dt.$$

Calculate erf(1) correct to three decimals.

SOLUTION If we replace the integrand by its Maclaurin series, we have

$$\frac{\sqrt{\pi}}{2}\text{erf}(1) = \int_0^1 \left[1 - t^2 + \frac{(-t^2)^2}{2!} + \frac{(-t^2)^3}{3!} + \cdots\right] dt$$

$$= \left\{t - \frac{t^3}{3} + \frac{t^5}{5 \cdot 2!} - \frac{t^7}{7 \cdot 3!} + \cdots\right\}_0^1$$

$$= 1 - \frac{1}{3} + \frac{1}{5 \cdot 2!} - \frac{1}{7 \cdot 3!} + \frac{1}{9 \cdot 4!} - \cdots.$$

Since

$$\frac{2}{\sqrt{\pi}}\left(1 - \frac{1}{3} + \frac{1}{5 \cdot 2!} - \frac{1}{7 \cdot 3!} + \frac{1}{9 \cdot 4!} - \frac{1}{11 \cdot 5!}\right) = 0.842\,593\,67$$

and the seventh nonzero term is

$$\frac{2}{\sqrt{\pi}}\frac{1}{13 \cdot 6!} < 0.000\,120\,56,$$

it follows that

$$0.842\,593\,67 < \text{erf}(1) < 0.842\,714\,23,$$

and to three decimals

$$\text{erf}(1) = 0.843.$$

Limits

We have customarily used L'Hôpital's rule to evaluate limits of the indeterminate form $0/0$. Maclaurin and Taylor series can sometimes be used to advantage. Consider

$$\lim_{x \to 0} \frac{x - \sin x}{x^3}.$$

Three applications of L'Hôpital's rule give a limit of $\frac{1}{6}$. Alternatively, if we substitute the Maclaurin series for $\sin x$,

$$\lim_{x \to 0} \frac{x - \sin x}{x^3} = \lim_{x \to 0} \frac{1}{x^3}\left[x - \left(x - \frac{x^3}{3!} + \frac{x^5}{5!} - \cdots\right)\right]$$

$$= \lim_{x \to 0}\left[\frac{1}{6} - \frac{x^2}{5!} + \cdots\right]$$

$$= \frac{1}{6}.$$

Here is another example.

EXAMPLE 11.40

Evaluate
$$\lim_{\lambda \to 0^+} \frac{\lambda^{-5}}{e^{c/\lambda} - 1}$$
where $c > 0$ is a constant (see also Exercise 58 in Section 4.9).

SOLUTION We begin by making the change of variable $v = 1/\lambda$ in the limit,
$$\lim_{\lambda \to 0^+} \frac{\lambda^{-5}}{e^{c/\lambda} - 1} = \lim_{v \to \infty} \frac{v^5}{e^{cv} - 1}.$$

We now expand e^{cv} into its Maclaurin series,
$$\lim_{\lambda \to 0^+} \frac{\lambda^{-5}}{e^{c/\lambda} - 1} = \lim_{v \to \infty} \frac{v^5}{\left(1 + cv + \frac{c^2 v^2}{2!} + \cdots\right) - 1}$$
$$= \lim_{v \to \infty} \frac{v^5}{cv + \frac{c^2 v^2}{2!} + \cdots}.$$

If we now divide numerator and denominator by v^5,
$$\lim_{\lambda \to 0^+} \frac{\lambda^{-5}}{e^{c/\lambda} - 1} = \lim_{v \to \infty} \frac{1}{\frac{c}{v^4} + \frac{c^2}{2 v^3} + \frac{c^3}{3! v^2} + \frac{c^4}{4! v} + \frac{c^5}{5!} + \frac{c^6 v}{6!} + \cdots} = 0.$$

Differential Equations

Many differential equations arising in physics and engineering have solutions that can be expressed only in terms of infinite series. One such equation is Bessel's differential equation of order zero for a function $y = f(x)$:
$$xy'' + y' + xy = 0.$$

Before considering this rather difficult differential equation, we introduce the ideas through an easier example.

EXAMPLE 11.41

Determine whether the differential equation
$$\frac{dy}{dx} - 2y = x$$
has a solution which can be expressed as a power series $y = \sum_{n=0}^{\infty} a_n x^n$ with positive radius of convergence.

SOLUTION If
$$y = f(x) = \sum_{n=0}^{\infty} a_n x^n = a_0 + a_1 x + a_2 x^2 + \cdots$$

is to be a solution of the differential equation, we may substitute the power series into the differential equation,

$$[a_1 + 2a_2x + 3a_3x^2 + 4a_4x^3 + \cdots] - 2[a_0 + a_1x + a_2x^2 + \cdots] = x.$$

We now gather together like terms in the various powers of x,

$$0 = (a_1 - 2a_0) + (2a_2 - 2a_1 - 1)x + (3a_3 - 2a_2)x^2 + (4a_4 - 2a_3)x^3 + \cdots.$$

Since the power series on the right has sum zero, its coefficients must all vanish (see Exercise 8 in Section 11.9), and therefore we must set

$$a_1 - 2a_0 = 0,$$
$$2a_2 - 2a_1 - 1 = 0,$$
$$3a_3 - 2a_2 = 0,$$
$$4a_4 - 2a_3 = 0,$$

and so on. These equations imply that

$$a_1 = 2a_0;$$
$$a_2 = \frac{1}{2}(1 + 2a_1) = \frac{1}{2}(1 + 4a_0);$$
$$a_3 = \frac{2}{3}a_2 = \frac{2}{3!}(1 + 4a_0);$$
$$a_4 = \frac{2}{4}a_3 = \frac{2^2}{4!}(1 + 4a_0).$$

The pattern emerging is

$$a_n = \frac{2^{n-2}}{n!}(1 + 4a_0), \qquad n \geq 2.$$

Thus,

$$f(x) = a_0 + 2a_0x + \frac{1}{2}(1 + 4a_0)x^2 + \cdots + \frac{2^{n-2}}{n!}(1 + 4a_0)x^n + \cdots$$
$$= a_0 + 2a_0x + \frac{1}{4}(1 + 4a_0)\left(\frac{2^2}{2!}x^2 + \frac{2^3}{3!}x^3 + \cdots + \frac{2^n}{n!}x^n + \cdots\right).$$

We can find the sum of the series in parentheses by noting that the Maclaurin series for e^{2x} is

$$e^{2x} = 1 + (2x) + \frac{(2x)^2}{2!} + \frac{(2x)^3}{3!} + \cdots.$$

Therefore the solution of the differential equation is

$$y = f(x) = a_0 + 2a_0x + \frac{1}{4}(1 + 4a_0)\left[e^{2x} - 1 - 2x\right]$$
$$= -\frac{1}{4} - \frac{x}{2} + \frac{1}{4}(1 + 4a_0)e^{2x}$$
$$= Ce^{2x} - \frac{1}{4} - \frac{x}{2}.$$

Using power series to solve the differential equation in Example 11.41 is certainly not the most expedient method. You will learn better methods if and when you undertake an indepth study of differential equations. But the example clearly illustrated the procedure by which power series are used to solve differential equations. We now apply the procedure to Bessel's differential equation of order zero.

EXAMPLE 11.42

Find a power series solution $y = \sum_{n=0}^{\infty} a_n x^n$, with positive radius of convergence, for Bessel's differential equation of order zero,

$$xy'' + y' + xy = 0.$$

SOLUTION In this example we abandon the $\cdots$ notation used in Example 11.41, and maintain sigma notation throughout. When we substitute $y = \sum_{n=0}^{\infty} a_n x^n$ into the differential equation, we obtain

$$0 = x \sum_{n=2}^{\infty} n(n-1) a_n x^{n-2} + \sum_{n=1}^{\infty} n a_n x^{n-1} + x \sum_{n=0}^{\infty} a_n x^n$$

$$= \sum_{n=2}^{\infty} n(n-1) a_n x^{n-1} + \sum_{n=1}^{\infty} n a_n x^{n-1} + \sum_{n=0}^{\infty} a_n x^{n+1}.$$

In order to bring these three summations together as one, and combine terms in like powers of x, we lower the index of summation in the last term by 2,

$$0 = \sum_{n=2}^{\infty} n(n-1) a_n x^{n-1} + \sum_{n=1}^{\infty} n a_n x^{n-1} + \sum_{n=2}^{\infty} a_{n-2} x^{n-1}.$$

We now combine the three summations over their common interval, beginning at $n=2$, and write separately the $n=1$ term in the second summation,

$$0 = a_1 + \sum_{n=2}^{\infty} [n(n-1) a_n + n a_n + a_{n-2}] x^{n-1}.$$

But the only way a power series can be equal to zero is for all of its coefficients to be equal to zero; that is,

$$a_1 = 0; \qquad n(n-1) a_n + n a_n + a_{n-2} = 0, \quad n \geq 2.$$

Thus,

$$a_n = -\frac{a_{n-2}}{n^2}, \qquad n \geq 2,$$

a recursive relation defining the unknown coefficient a_n of x^n in terms of the coefficient a_{n-2} of x^{n-2}. Since $a_1 = 0$, it follows that

$$0 = a_1 = a_3 = a_5 = \cdots.$$

For $n = 2$,

$$a_2 = -\frac{a_0}{2^2}.$$

For $n = 4$,
$$a_4 = -\frac{a_2}{4^2} = \frac{a_0}{2^2 4^2} = \frac{a_0}{2^4 (2!)^2}.$$

For $n = 6$,
$$a_6 = -\frac{a_4}{6^2} = \frac{-a_0}{2^4 (2!)^2 6^2} = -\frac{a_0}{2^6 (3!)^2}.$$

The solution is therefore
$$y = a_0 - \frac{a_0}{2^2}x^2 + \frac{a_0}{2^4(2!)^2}x^4 - \frac{a_0}{2^6(3!)^2}x^6 + \cdots$$
$$= a_0 \sum_{n=0}^{\infty} \frac{(-1)^n}{2^{2n}(n!)^2} x^{2n}.$$

The function defined by the infinite series
$$J_0(x) = \sum_{n=0}^{\infty} \frac{(-1)^n}{2^{2n}(n!)^2} x^{2n}, \qquad -\infty < x < \infty$$

is called the zero-order Bessel function of the first kind. ∎

EXERCISES 11.12

In Exercises 1–10 find a maximum possible error in using the given terms of the Taylor series to approximate the function on the interval specified.

1. $e^x \approx 1 + x + \frac{x^2}{2} + \frac{x^3}{6}$ for $0 \leq x \leq 0.01$

2. $e^x \approx 1 + x + \frac{x^2}{2} + \frac{x^3}{6}$ for $0 \leq x < 0.01$

3. $e^x \approx 1 + x + \frac{x^2}{2} + \frac{x^3}{6}$ for $-0.01 \leq x \leq 0$

4. $e^x \approx 1 + x + \frac{x^2}{2} + \frac{x^3}{6}$ for $|x| \leq 0.01$

5. $\sin x \approx x - \frac{x^3}{3!}$ for $0 \leq x \leq 1$

6. $\cos x \approx 1 - \frac{x^2}{2!} + \frac{x^4}{4!}$ for $|x| \leq 0.1$

7. $\ln(1-x) \approx -x - \frac{x^2}{2} - \frac{x^3}{3}$ $0 \leq x \leq 0.01$

8. $\frac{1}{(1-x)^3} \approx 1 + 3x + 6x^2 + 10x^3$ for $|x| < 0.2$

9. $\text{Tan}^{-1} x \approx x - \frac{x^3}{3} + \frac{x^5}{5} - \cdots - \frac{x^{99}}{99}$ for $|x| < 1$

10. $\ln x \approx (x-1) - \frac{1}{2}(x-1)^2 + \frac{1}{3}(x-1)^3 - \frac{1}{4}(x-1)^4$ for $|x-1| \leq 1/2$

In Exercises 11–18 evaluate the integral correct to three decimals.

11. $\int_0^1 \frac{\sin x}{x} \, dx$

12. $\int_0^{1/2} \cos(x^2) \, dx$

13. $\int_0^{2/3} \frac{1}{x^4 + 1} \, dx$

14. $\int_{-1}^{1} x^{11} \sin x \, dx$

15. $\int_0^{1/2} \frac{1}{\sqrt{1+x^3}} \, dx$

16. $\int_0^{0.3} e^{-x^2} \, dx$

17. $\int_{-0.1}^{0} \frac{1}{x-1} \ln(1-x) \, dx$

18. $\int_0^{1/2} \frac{1}{x^6 - 3x^3 - 4} \, dx$

In Exercises 19–24 use series to evaluate the limit.

19. $\lim_{x \to 0} \frac{\tan x}{x}$

20. $\lim_{x \to 0} \frac{1 - \cos x}{x^2}$

21. $\lim_{x \to 0} \frac{(1 - \cos x)^2}{3x^4}$

22. $\lim_{x \to 0} \frac{\sqrt{1+x} - 1}{x}$

23. $\lim_{x \to \infty} x \sin\left(\frac{1}{x}\right)$

24. $\lim_{x \to 0} \left(\frac{e^x + e^{-x}}{e^x - e^{-x}} - \frac{1}{x} \right)$

In Exercises 25–28 determine where the Maclaurin series for the function may be truncated in order to guarantee the accuracy indicated.

25. $\sin(x/3)$ on $|x| \leq 4$ with error less than 10^{-3}

26. $1/\sqrt{1+x^3}$ on $0 < x < 1/2$ with error less than 10^{-4}

27. $\ln(1-x)$ on $|x| < 1/3$ with error less than 10^{-2}

28. $\cos^2 x$ on $|x| < 0.1$ with error less than 10^{-3}

In Exercises 29–34 find a series solution in powers of x for the differential equation.

29. $y' + 3y = 4$
30. $y'' + y' = 0$
31. $xy' - 4y = 3x$
32. $4xy'' + 2y' + y = 0$
33. $y'' + y = 0$
34. $xy'' + y = 0$

35. Find the natural logarithm of $0.999\,999\,999\,9$ accurate to 15 decimal places.

36. In special relativity theory, the kinetic energy K of an object moving with speed v is defined by

$$K = c^2(m - m_0),$$

where c is a constant (the speed of light), m_0 is the rest mass of the object, and m is its mass when moving with speed v. The masses m and m_0 are related by

$$m = \frac{m_0}{\sqrt{1 - v^2/c^2}}.$$

Use the binomial expansion to show that

$$K = \frac{1}{2}m_0 v^2 + m_0 c^2 \left(\frac{3}{8}\frac{v^4}{c^4} + \frac{5}{16}\frac{v^6}{c^6} + \cdots \right),$$

and hence, to a first approximation, kinetic energy is defined by the classical expression $m_0 v^2/2$.

37. The ellipse $b^2 x^2 + a^2 y^2 = a^2 b^2$ can be represented parametrically by

$$x = a\cos t, \quad y = b\sin t, \quad 0 \leq t < 2\pi.$$

(a) Show that the length of the circumference of the ellipse is defined by the definite integral

$$L = 4b \int_0^{\pi/2} \sqrt{1 - k^2 \sin^2 t}\, dt, \quad k^2 = 1 - \frac{a^2}{b^2}.$$

(b) Use the binomial expansion to show that

$$L = 2\pi b \left(1 - \frac{k^2}{4} - \frac{3k^4}{64} - \cdots \right)$$

so that to a first approximation, L is the circumference of a circle of radius b.

38. In determining the radiated power from a half-wave antenna, it is necessary to evaluate

$$\int_0^{2\pi} \frac{1 - \cos\theta}{\theta}\, d\theta.$$

Find a two-decimal approximation for this integral.

39. Planck's law for the energy density Ψ of blackbody radiation of wavelength λ states that

$$\Psi(\lambda) = \frac{8\pi ch \lambda^{-5}}{e^{ch/(\lambda kT)} - 1},$$

where $h > 0$ is Planck's constant, c is the (constant) speed of light, and T is temperature, also assumed constant. Show that for long wavelengths, Planck's law reduces to the Rayleigh-Jeans law:

$$\Psi(\lambda) = \frac{8\pi kT}{\lambda^4}.$$

SUMMARY

An infinite sequence of constants is the assignment of numbers to positive integers. In most applications of sequences, the prime consideration is whether the sequence has a limit. If the sequence has its terms defined explicitly, then our ability to take limits of continuous functions ("limits at infinity" in Chapter 2 and L'Hôpital's rule in Chapter 4) can be very helpful. If the sequence is defined recursively, existence of the limit can sometimes be established by showing that the sequence is monotonic and bounded.

An expression of the form

$$\sum_{n=1}^{\infty} c_n = c_1 + c_2 + \cdots + c_n + \cdots$$

is called an infinite series. We define the sum of this series as the limit of its sequence of partial sums $\{S_n\}$, provided the sequence has a limit. Unfortunately, for most series we cannot find a simple formula for S_n, and therefore analysis of the limit of the sequence $\{S_n\}$ is impossible. To remedy this, we developed various convergence tests that avoided the sequence $\{S_n\}$: n^{th} term, comparison, limit comparison, limit ratio, limit root, integral, and alternating series tests. Note the sequences that are associated with a series $\sum c_n$:

$\{S_n\}$	sequence of partial sums for the definition of a sum;
$\{c_n\}$	sequence of terms for the nth term test;
$\{c_n/b_n\}$	sequence for limit comparison test;
$\{c_{n+1}/c_n\}$	sequence for the limit ratio test;
$\{\sqrt[n]{c_n}\}$	sequence for the limit root test;
$\{|c_n|\}$	sequence for the alternating series test.

Depending on the limits of these sequences — if they exist, we may be able to infer something about convergence of the series.

From infinite sequences and series of constants we proceeded to infinite sequences and series of functions — in particular, power series. We first considered situations where a power series was given, and the sum was to be determined. We saw that every power series $\sum a_n(x-c)^n$ has a radius of convergence R and an associated interval of convergence. If $R = 0$, the interval of convergence consists of only one point $x = c$; if $R = \infty$, the power series converges absolutely for all x; and if $0 < R < \infty$, the interval of convergence must be one of four possibilities: $c - R < x < c + R$, $c - R \leq x < c + R$, $c - R < x \leq c + R$, or $c - R \leq x \leq c + R$. The radius of convergence is given by $\lim_{n \to \infty} |a_n/a_{n+1}|$ or $\lim_{n \to \infty} |a_n|^{-1/n}$ provided the limits exist or are equal to infinity. If at each point in the interval of convergence of the power series the value of a function $f(x)$ is the same as the sum of the series, we write $f(x) = \sum a_n(x-c)^n$ and call $f(x)$ the sum of the series.

We also considered situations where a function $f(x)$ and a point c are given, and ask whether $f(x)$ has a power series expansion about c. We saw that there can be at most one power series expansion of $f(x)$ about c with a positive radius of convergence, and this series must be its Taylor series. One way to verify that $f(x)$ does indeed have a Taylor series about c and that this series converges to $f(x)$ is to show that the sequence of Taylor's remainders $\{R_n(c,x)\}$ exists and has limit zero. Often, however, it is much easier to find Taylor series by adding, multiplying, differentiating, and integrating known series.

When a Taylor series is truncated, Taylor's remainder $R_n(c,x)$ represents the truncation error and, in spite of the fact that R_n is expressed in terms of some unknown point z_n, it is often possible to calculate a maximum value for the error. Sometimes $R_n(c,x)$ can be avoided altogether. For instance, if the Taylor series is an alternating series, then the maximum possible truncation error is the value of the next term.

Power series are often used in situations that require approximations. Taylor series provide polynomial approximations to complicated functions, and they offer an alternative to the numerical techniques of Section 9.9 in the evaluation of definite integrals. Power series are also useful in situations that do not require approximations. They are sometimes helpful in evaluating limits, and they are the only way to solve many differential equations.

Key Terms and Formulas

In reviewing this chapter, you should be able to define or discuss the following key terms:

Sequence of constants
Explicit sequence
Recursive sequence
Limit of a sequence
Convergent sequence
Divergent sequence
Increasing sequence
Decreasing sequence
Nonincreasing sequence
Nondecreasing sequence
Monotonic sequence
Upper bound
Lower bound
Successive substitutions
Series of constants
Geometric series
Harmonic series
Sequence of partial sums
Convergent series
Divergent series

n^{th} term test
Nonnegative series
Integral test
p-series
Comparison test
Limit comparison test
Limit ratio test
Limit root test
Absolutely convergent series
Conditionally convergent series
Alternating harmonic series
Truncation error
Power series
Interval of convergence
Radius of convergence
Sum of a power series
Taylor remainder formula
Taylor series
Maclaurin series
Binomial expansion

REVIEW EXERCISES

In Exercises 1–4 discuss, with all necessary proofs, whether the sequence is monotonic and has an upper bound, a lower bound, and a limit.

1. $\left\{ \dfrac{n^2 - 5n + 3}{n^2 + 5n + 4} \right\}$

2. $c_1 = 1$, $c_{n+1} = (1/2)\sqrt{c_n^2 + 1}$, $n \geq 1$

3. $\left\{ \dfrac{\operatorname{Tan}^{-1}(1/n)}{n^2 + 1} \right\}$

4. $c_1 = 7$, $c_{n+1} = 15 + \sqrt{c_n - 2}$, $n \geq 1$

5. Use Newton's iterative procedure and the method of successive substitutions to approximate the root of the equation

$$x = \left(\frac{x+5}{x+4} \right)^2$$

between $x = 1$ and $x = 2$.

6. For what values of k does the sequence

$$c_1 = k, \quad c_{n+1} = c_n^2, \quad n \geq 1$$

converge?

7. Find an explicit definition for the sequence

$$c_1 = 1, \quad c_{n+1} = \sqrt{1 + c_n^2}, \quad n \geq 1.$$

8. Use the derivative of the function $f(x) = (\ln x)/x$ to prove that the sequence $\{\ln n/n\}$ is decreasing for $n \geq 3$.

In Exercises 9–28 determine whether the series converges or diverges. In the case of a convergent series that has both positive and negative terms, indicate whether it converges absolutely or conditionally.

9. $\displaystyle\sum_{n=1}^{\infty} \dfrac{n^2 - 3n + 2}{n^3 + 4n}$

10. $\displaystyle\sum_{n=1}^{\infty} \dfrac{n^2 + 5n + 3}{n^4 - 2n + 5}$

11. $\displaystyle\sum_{n=1}^{\infty} \dfrac{5^{2n}}{n!}$

12. $\displaystyle\sum_{n=1}^{\infty} \dfrac{n^2 + 3}{n3^n}$

13. $\displaystyle\sum_{n=1}^{\infty} \dfrac{(\ln n)^2}{\sqrt{n}}$

14. $\displaystyle\sum_{n=1}^{\infty} (-1)^n \left(\dfrac{n+1}{n^2} \right)$

15. $\displaystyle\sum_{n=1}^{\infty} (-1)^n \left(\dfrac{n+1}{n^3} \right)$

16. $\displaystyle\sum_{n=1}^{\infty} \operatorname{Cos}^{-1}\left(\dfrac{1}{n} \right)$

17. $\displaystyle\sum_{n=1}^{\infty} \dfrac{1}{n}\operatorname{Cos}^{-1}\left(\dfrac{1}{n} \right)$

18. $\displaystyle\sum_{n=1}^{\infty} \dfrac{1}{n^2}\operatorname{Cos}^{-1}\left(\dfrac{1}{n} \right)$

19. $\displaystyle\sum_{n=1}^{\infty} \dfrac{2 \cdot 4 \cdot 6 \cdot \,\cdots\, \cdot (2n)}{n!}$

20. $\displaystyle\sum_{n=1}^{\infty} \dfrac{3 \cdot 6 \cdot 9 \cdot \,\cdots\, \cdot (3n)}{(2n)!}$

21. $\sum_{n=1}^{\infty} \sqrt{\dfrac{n^2+1}{n^2+5}}$
22. $\sum_{n=1}^{\infty} (-1)^{n+1} \left(1 + \dfrac{1}{n}\right)^3$
23. $\sum_{n=1}^{\infty} \dfrac{1}{n^2} \sin n$
24. $\sum_{n=1}^{\infty} \dfrac{10^n}{5^{3n+2}}$
25. $\sum_{n=1}^{\infty} (-1)^n \dfrac{\ln n}{n}$
26. $\sum_{n=1}^{\infty} \dfrac{1}{e^{n\pi}}$
27. $\sum_{n=1}^{\infty} \dfrac{2^n + 2^{-n}}{3^n}$
28. $\sum_{n=1}^{\infty} \dfrac{1}{\sqrt{n}} \cos(n\pi)$

In Exercises 29–36 find the interval of convergence for the power series.

29. $\sum_{n=0}^{\infty} \dfrac{n+1}{n^2+1} x^n$
30. $\sum_{n=1}^{\infty} \dfrac{1}{n^2 2^n} x^n$
31. $\sum_{n=0}^{\infty} (n+1)^3 x^n$
32. $\sum_{n=1}^{\infty} \dfrac{1}{n^n} x^n$
33. $\sum_{n=0}^{\infty} \dfrac{1}{4^n} (x-2)^n$
34. $\sum_{n=2}^{\infty} \sqrt{\dfrac{n+1}{n-1}} (x+3)^n$
35. $\sum_{n=1}^{\infty} n3^n x^{2n}$
36. $\sum_{n=1}^{\infty} \dfrac{2^n}{n} x^{3n}$

In Exercises 37–45 find the power series expansion of the function about the indicated point.

37. $f(x) = \sqrt{1 + x^2}$, about $x = 0$
38. $f(x) = e^{x+5}$, about $x = 0$
39. $f(x) = \cos(x + \pi/4)$, about $x = 0$
40. $f(x) = x \ln(2x + 1)$, about $x = 0$
41. $f(x) = \sin x$, about $x = \pi/4$
42. $f(x) = x/(x^2 + 4x + 3)$, about $x = 0$
43. $f(x) = e^x$, about $x = 3$
44. $f(x) = (x+1)\ln(x+1)$, about $x = 0$
45. $f(x) = x^3 e^{x^2}$, about $x = 0$

46. How many terms in the Maclaurin series for $f(x) = e^{-x^2}$ guarantee a truncation error of less than 10^{-5} for all x in the interval $0 \leq x \leq 2$?

47. Find a power series solution in powers of x for the differential equation
$$y'' - 4y = 0.$$

48. Find the Maclaurin series for $f(x) = \sqrt{1 + \sin x}$ valid for $-\pi/2 \leq x \leq \pi/2$ by first showing that $f(x)$ can be written in the form
$$f(x) = \sin(x/2) + \cos(x/2).$$
Why is the restriction $-\pi/2 \leq x \leq \pi/2$ necessary?

49. On a calculator take the cosine of 1 (radian). Take the cosine of this result, and again, and again, and again, ... What happens? Interpret what is going on.

APPENDIX A

Trigonometry Review

Many physical systems exhibit an oscillatory nature: vibrations of a plucked guitar string, motion of a pendulum, alternating electric currents, fluctuations in room temperature as the thermostat continually engages and disengages the furnace on a cold day, and the rise and fall of tides and waves. Magnitudes of these oscillations are best described by the sine and cosine functions of trigonometry. In addition, rates at which these oscillations occur can be represented by the *derivatives* of these functions. The study of oscillatory systems requires a knowledge of the sine and cosine functions and their derivatives. In this appendix, we briefly review the trigonometric functions and their properties, placing special emphasis on those aspects that are most useful to the study of calculus.

In trigonometry, angles are measured in either degrees or radians; in calculus angles are always measured in radians. In preparation, then, for the calculus of trigonometric functions, we will work completely in radian measure. By definition, a **radian** is that angle subtended at the centre of a circle of radius r by an arc of equal length r (Figure A.1). When an arc has length s (Figure A.2), the number of "units of radius" in this arc is s/r; therefore the angle θ subtended at the centre of the circle is

$$\theta = \frac{s}{r}. \qquad (A.1)$$

In particular, if s contains π units of radius ($s = \pi r$), then s represents one-half the circumference of the complete circle, and $\theta = \pi r/r = \pi$ radians. In degree measure, this angle is $180°$; hence we can state that π radians is equivalent to $180°$. This statement enables us to convert angles expressed in degrees to radian measure and vice versa. If the degreee measure of an angle is ϕ, then it is $\pi\phi/180$ radians; conversely if an angle measures θ radians, then its degree measure is $180\theta/\pi$. For example, the radian measure for $\phi = 45°$ is $45\pi/180 = \pi/4$ radians.

FIGURE A.1

FIGURE A.2

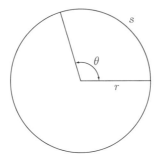

Elementary trigonometry is concerned with relationships among angles and lengths, and, in particular, angles and sides of triangles. This naturally restricts angles to the range

$0 \leq \theta \leq \pi$. For the purposes of calculus, however, we need to talk about angles θ, where θ is any real number — positive, negative, or zero. To do this we first define what we mean by the **standard position of an angle**. If $\theta > 0$, we draw a line segment OP through the origin O of the xy-plane (Figure A.3) in such a way that the positive x-axis must rotate counterclockwise through an angle θ to coincide with OP. In other words, we now regard an angle as rotation, rotation of the positive x-axis to some terminal position. If $0 < \theta < \pi/2$, then OP lies in the first quadrant; and if $\pi/2 < \theta < \pi$, then OP is in the second quadrant. But now if $\theta > \pi$, we have a geometric representation of θ also. For example, angles $7\pi/4$ and $9\pi/4$ are shown in Figure A.4. When $\theta < 0$, we regard θ as a clockwise rotation (Figure A.5).

Note that for any angle θ, the angles $\theta + 2n\pi$ for n an integer have the same terminal position of OP as θ. They are different angles, however, because the positive x-axis must encircle the origin one or more times before reaching the terminal position in the case of $\theta + 2n\pi$.

FIGURE A.3

FIGURE A.4

FIGURE A.5

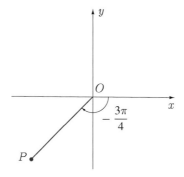

With this representation of angles as rotations, it is easy to define the six trigonometric functions. If θ is an angle in standard position (Figure A.6) and (x, y) are the coordinates of P, we define

$$\sin \theta = \frac{y}{r}, \quad \cos \theta = \frac{x}{r}, \quad \tan \theta = \frac{y}{x},$$
$$\csc \theta = \frac{r}{y}, \quad \sec \theta = \frac{r}{x}, \quad \cot \theta = \frac{x}{y}, \quad (A.2)$$

wherever these ratios are defined, and where $r = \sqrt{x^2 + y^2}$ is always assumed positive. Since x is positive in the first and fourth quadrants, and y is positive in the first and second, signs of the trigonometric functions in the various quadrants are as shown in Figure A.7. Furthermore, since $r \neq 0$, $\sin \theta$ and $\cos \theta$ are defined for all θ, whereas $\tan \theta$ and $\sec \theta$ are not defined for $x = 0$, and $\csc \theta$ and $\cot \theta$ do not exist when $y = 0$.

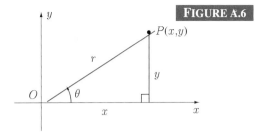

FIGURE A.6

FIGURE A.7

Sine and cosecant positive; all others negative	All trigonometric functions positive
Tangent and cotangent positive; all others negative	Cosine and secant positive; all others negative

Definitions (A.2) indicate that

$$\csc\theta = \frac{1}{\sin\theta}, \qquad \sec\theta = \frac{1}{\cos\theta}, \qquad \cot\theta = \frac{1}{\tan\theta}. \qquad (A.3)$$

In addition, the fact that $r^2 = x^2 + y^2$ leads to the identity

$$\sin^2\theta + \cos^2\theta = 1, \qquad (A.4)$$

which in turn implies that

$$1 + \tan^2\theta = \sec^2\theta, \qquad (A.5a)$$
$$1 + \cot^2\theta = \csc^2\theta. \qquad (A.5b)$$

Identity (A.4) is simply a restatement of the Pythagorean relation ($r^2 = x^2 + y^2$), which allows us to express the hypotenuse of a right-angled triangle in terms of the other two sides. If the triangle is not right-angled (Figure A.8), it is not possible to express one side, c, in terms of the other two sides, a and b, alone, but it is possible to express c in terms of a and b and the angle θ between them. The coordinates of P and Q in Figure A.8 are $(b\cos\theta, b\sin\theta)$ and $(a, 0)$, and therefore by distance formula 1.7 in Section 1.2,

$$c^2 = ||PQ||^2 = (b\cos\theta - a)^2 + (b\sin\theta)^2$$
$$= b^2(\cos^2\theta + \sin^2\theta) + a^2 - 2ab\cos\theta.$$

Since $\sin^2\theta + \cos^2\theta = 1$, we have

$$c^2 = a^2 + b^2 - 2ab\cos\theta. \qquad (A.6)$$

This result, called the **cosine law**, generalizes the Pythagorean relation to triangles that are not right-angled. It reduces to the Pythagorean relation $c^2 = a^2 + b^2$ when $\theta = \pi/2$.

FIGURE A.8

FIGURE A.9

If A, B, and C are the angles in the triangle of Figure A.9, and a, b, and c are the lengths of the opposite sides, then by drawing altitudes of the triangle we can show that

$$\frac{\sin A}{a} = \frac{\sin B}{b} = \frac{\sin C}{c}. \qquad (A.7)$$

This result, known as the **sine law**, is also useful in many problems.

FIGURE A.10

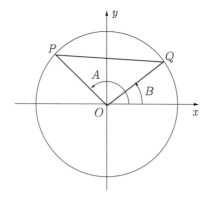

There is a large number of identities satisfied by the trigonometric functions. They can all be derived from the following **compound-angle formulas** for sines and cosines:

$$\sin(A+B) = \sin A \cos B + \cos A \sin B, \tag{A.8a}$$
$$\sin(A-B) = \sin A \cos B - \cos A \sin B, \tag{A.8b}$$
$$\cos(A+B) = \cos A \cos B - \sin A \sin B, \tag{A.8c}$$
$$\cos(A-B) = \cos A \cos B + \sin A \sin B. \tag{A.8d}$$

Let us prove one of these, say, identity A.8d, where A and B are the angles in Figure A.10. If P and Q lie on the circle $x^2 + y^2 = r^2$, then their coordinates are $(r\cos A, r\sin A)$ and $(r\cos B, r\sin B)$ respectively. According to formula 1.7, the length of PQ is

$$\begin{aligned}\|PQ\|^2 &= (r\cos A - r\cos B)^2 + (r\sin A - r\sin B)^2 \\ &= r^2(\cos^2 A + \sin^2 A) + r^2(\cos^2 B + \sin^2 B) \\ &\quad - 2r^2(\cos A \cos B + \sin A \sin B) \\ &= 2r^2 - 2r^2(\cos A \cos B + \sin A \sin B).\end{aligned}$$

But according to the cosine law A.6,

$$\begin{aligned}\|PQ\|^2 &= \|OP\|^2 + \|OQ\|^2 - 2\|OP\|\|OQ\|\cos(A-B) \\ &= r^2 + r^2 - 2r^2\cos(A-B) \\ &= 2r^2 - 2r^2\cos(A-B).\end{aligned}$$

Comparison of these two expressions for $\|PQ\|^2$ immediately implies identity A.8d.

With equations A.8, it is easy to derive the following identities:

$$\sin(-\theta) = -\sin\theta, \tag{A.9a}$$
$$\cos(-\theta) = \cos\theta, \tag{A.9b}$$
$$\tan(-\theta) = -\tan\theta, \tag{A.9c}$$
$$\csc(-\theta) = -\csc\theta, \tag{A.9d}$$
$$\sec(-\theta) = \sec\theta, \tag{A.9e}$$
$$\cot(-\theta) = -\cot\theta; \tag{A.9f}$$
$$\sin(\pi/2 - \theta) = \cos\theta, \tag{A.10a}$$
$$\cos(\pi/2 - \theta) = \sin\theta, \tag{A.10b}$$
$$\tan(\pi/2 - \theta) = \cot\theta, \tag{A.10c}$$
$$\csc(\pi/2 - \theta) = \sec\theta, \tag{A.10d}$$

$$\sec(\pi/2 - \theta) = \csc\theta, \qquad (A.10\,e)$$
$$\cot(\pi/2 - \theta) = \tan\theta; \qquad (A.10\,f)$$
$$\sin(\pi - \theta) = \sin\theta, \qquad (A.11\,a)$$
$$\cos(\pi - \theta) = -\cos\theta, \qquad (A.11\,b)$$
$$\tan(\pi - \theta) = -\tan\theta, \qquad (A.11\,c)$$
$$\csc(\pi - \theta) = \csc\theta, \qquad (A.11\,d)$$
$$\sec(\pi - \theta) = -\sec\theta, \qquad (A.11\,e)$$
$$\cot(\pi - \theta) = -\cot\theta. \qquad (A.11\,f)$$

By expressing $\tan(A + B)$ as $\sin(A + B)/\cos(A + B)$ and using identities A.8a and A.8c, we find a compound-angle formula for the tangent function,

$$\tan(A + B) = \frac{\tan A + \tan B}{1 - \tan A \tan B}, \qquad (A.12\,a)$$

and similarly,

$$\tan(A - B) = \frac{\tan A - \tan B}{1 + \tan A \tan B}. \qquad (A.12\,b)$$

By setting $A = B$ in A.8a,c and A.12a, we obtain the **double-angle formulas**,

$$\sin 2A = 2\sin A \cos A, \qquad (A.13\,a),$$
$$\cos 2A = \cos^2 A - \sin^2 A, \qquad (A.13\,b)$$
$$= 2\cos^2 A - 1, \qquad (A.13\,c)$$
$$= 1 - 2\sin^2 A, \qquad (A.13\,d)$$
$$\tan 2A = \frac{2\tan A}{1 - \tan^2 A}. \qquad (A.13\,e)$$

When pairs of compound-angle formulas A.8 are added or subtracted, the **product formulas** result. For example, subtracting A.8c from A.8d gives

$$\sin A \sin B = \frac{1}{2}[-\cos(A + B) + \cos(A - B)]. \qquad (A.14\,a)$$

The other product formulas are

$$\sin A \cos B = \frac{1}{2}[\sin(A + B) + \sin(A - B)], \qquad (A.14\,b)$$
$$\cos A \cos B = \frac{1}{2}[\cos(A + B) + \cos(A - B)]. \qquad (A.14\,c)$$

By setting $X = A + B$ and $Y = A - B$ in A.14, we obtain the **sum and difference formulas**,

$$\sin X + \sin Y = 2\sin\left(\frac{X + Y}{2}\right)\cos\left(\frac{X - Y}{2}\right), \qquad (A.15\,a)$$
$$\sin X - \sin Y = 2\cos\left(\frac{X + Y}{2}\right)\sin\left(\frac{X - Y}{2}\right), \qquad (A.15\,b)$$
$$\cos X + \cos Y = 2\cos\left(\frac{X + Y}{2}\right)\cos\left(\frac{X - Y}{2}\right), \qquad (A.15\,c)$$
$$\cos X - \cos Y = -2\sin\left(\frac{X + Y}{2}\right)\sin\left(\frac{X - Y}{2}\right). \qquad (A.15\,d)$$

The following two examples are typical of problems in the calculus of trigonometric functions.

EXAMPLE A.1

Find $R > 0$ and $0 < \phi < \pi$ such that $3\cos\theta - 4\sin\theta$ can be expressed in the form

$$R\sin(\theta + \phi) = 3\cos\theta - 4\sin\theta$$

for all real θ.

SOLUTION If we expand $R\sin(\theta + \phi)$ according to A.8a, we get

$$R\sin\theta\cos\phi + R\cos\theta\sin\phi = 3\cos\theta - 4\sin\theta.$$

Since this equation is to be valid for all θ, it must therefore be valid for $\theta = 0$ and $\theta = \pi/2$. Substitution of these values gives the two equations

$$R\sin\phi = 3 \quad \text{and} \quad R\cos\phi = -4.$$

If we square and add these equations, we find

$$R^2\sin^2\phi + R^2\cos^2\phi = R^2 = 3^2 + (-4)^2 = 25.$$

Consequently, $R = 5$, and

$$\sin\phi = \frac{3}{5}, \qquad \cos\phi = -\frac{4}{5}.$$

The only angle in the range $0 < \phi < \pi$ satisfying these equations is $\phi = 2.50$ radians. Thus, $3\cos\theta - 4\sin\theta$ can be expressed in the form

$$3\cos\theta - 4\sin\theta = 5\sin(\theta + 2.50).$$

EXAMPLE A.2

Write the expression $\cos^4\theta$ in terms of $\cos 2\theta$ and $\cos 4\theta$.

SOLUTION If we replace A by θ in double-angle formula A.13c,

$$\cos 2\theta = 2\cos^2\theta - 1.$$

It follows that

$$\cos^2\theta = \frac{1 + \cos 2\theta}{2}.$$

Consequently,

$$\cos^4\theta = (\cos^2\theta)^2 = \frac{1}{4}(1 + \cos 2\theta)^2 = \frac{1}{4}(1 + 2\cos 2\theta + \cos^2 2\theta).$$

But if we now replace θ by 2θ in the identity $\cos^2\theta = (1 + \cos 2\theta)/2$, we obtain

$$\cos^2 2\theta = \frac{1 + \cos 4\theta}{2}.$$

Thus,
$$\cos^4\theta = \frac{1}{4}\left(1 + 2\cos 2\theta + \frac{1 + \cos 4\theta}{2}\right) = \frac{1}{8}(3 + 4\cos 2\theta + \cos 4\theta).$$

∎

By regarding arguments of the trigonometric functions as angles, we have been stressing geometric properties of these functions. In particular, identities A.8–A.15 have all been based on definitions of the trigonometric functions as functions of angles. What is important about a function — be it a trigonometric function or any other kind of function — is that there is a number associated with each value of the independent variable. How we arrive at this number is irrelevant. So far as properties of the function are concerned, only its values are taken into account. Thus, when we discuss properties of a trigonometric function, what is important is not that its argument can be regarded as an angle or that its values can be defined as ratios of sides of a triangle, but that we know its values. Indeed, it is sometimes unwise to regard arguments of trigonometric functions as angles. Consider, for example, the motion of a mass m suspended from a spring with spring constant k (Figure A.11). If $x = 0$ is the position at which the mass would hang motionless, then when vertical oscillations are initiated, the position of m as a function of time t is always of the form

$$x = A\cos\left(\sqrt{\frac{k}{m}}t\right) + B\sin\left(\sqrt{\frac{k}{m}}t\right),$$

where A and B are constants. Clearly there are no angles associated with the motion of m, and it is therefore unnatural to attempt to interpret the argument $(\sqrt{k/m})t$ as a physical angle.

FIGURE A.11

We will now consider the trigonometric functions as those of a real variable. If it is convenient to regard the argument as an angle, then we will do so, but we are not compelled to. To emphasize this, we replace θ with our usual generic label for the independent variable of a function, namely, x. With this change, the trigonometric functions are $\sin x$, $\cos x$, $\tan x$, $\csc x$, $\sec x$, and $\cot x$. Their graphs are shown in Figures A.12.

(a)

(b)

(c)

FIGURE A.12

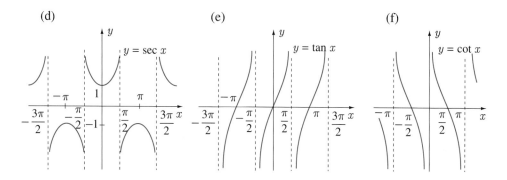

Trigonometric functions are *periodic*. A function $f(x)$ is said to be periodic if there exists a number T such that for all x,

$$f(x + T) = f(x). \tag{A.16}$$

The smallest such positive number T is called the **period** of $f(x)$. Clearly, then, $\sin x$, $\cos x$, $\csc x$, and $\sec x$ are periodic with period 2π, whereas $\tan x$ and $\cot x$ have period π.

When a periodic function is added to a periodic function, the resulting function is periodic. For example, the function $f(x) = 3\cos x - 4\sin x$ is the addition of two 2π-periodic functions $3\cos x$ and $-4\sin x$. In Example A.1, we showed that $f(x)$ can be expressed in the form $f(x) = 5\sin(x + 2.50)$, and using this representation, it is easy to produce the graph in Figure A.13. The function again has period 2π. Note that the 5 in $f(x)$ indicates that oscillations occur between the limits ± 5, and the 2.50 represents a shift of the usual sine curve 2.50 units to the left.

FIGURE A.13

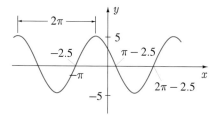

The curve in Figure A.13 is an example of a **general sine function**,

$$f(x) = A\sin(\omega x + \phi), \tag{A.17}$$

where A and ω are positive constants and ϕ is also constant. The graph of the general sine function is shown in Figure A.14. The number A, which represents half the range of the function, is called the **amplitude** of the oscillations. The period is $2\pi/\omega$, and $-\phi/\omega$ is called the **phase shift**.

FIGURE A.14

EXERCISES

In Exercises 1–10, express the angle in radians.
1. $30°$ 2. $60°$ 3. $135°$ 4. $-90°$
5. $-300°$ 6. $765°$ 7. $72°$ 8. $-128°$
9. $321°$ 10. $-213°$

In Exercises 11–20, express the angle in degrees.
11. $\pi/3$ 12. $-5\pi/4$ 13. $3\pi/2$ 14. 8π
15. $-5\pi/6$ 16. 1 17. -3 18. 2.5
19. -3.6 20. 11

21. What angle is subtended at the centre of a circle of radius 4 by an arc of length (a) 2; (b) 7; (c) 3.2?

22. Fill in entries for the following table.

	$\sin x$	$\cos x$	$\tan x$	$\csc x$	$\sec x$	$\cot x$
0						
$\pi/6$						
$\pi/4$						
$\pi/3$						
$\pi/2$						

For each of the triangles in Exercises 23–26 use the cosine law and/or the sine law to find the lengths of all sides and all interior angles.

23.

24.

25.

26.
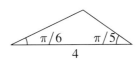

27. Use compound-angle formulas A.8 to prove identities A.12.

28. Verify double-angle formulas A.13.

29. Use compound-angle formulas A.8 to prove the product formulas A.14.

30. Show that the sum and difference formulas A.15 can be obtained from A.14.

In Exercises 31–40 sketch a graph of the function.
31. $f(x) = 3 \sin x$
32. $f(x) = \sin 2x$
33. $f(x) = 3 \sin 2x$
34. $f(x) = \sin(x + \pi/4)$
35. $f(x) = 3 \sin(x + \pi/4)$
36. $f(x) = \sin(2x + \pi/4)$
37. $f(x) = 3 \sin(2x + \pi/4)$
38. $f(x) = 4 \cos(x/3)$
39. $f(x) = 2 \sin(x/2 - \pi)$
40. $f(x) = 5 \cos(\pi/2 - 3x)$

41. Show that compound-angle formulas A.8a,b,c can be derived from A.8d and A.4.

In Exercises 42–49 find all solutions of the equation.
42. $\sin x = \sqrt{3}/2$
43. $\cos x = 2$
44. $3 \sin x = \cos x$
45. $3 \cos^2 x = 1$
46. $\sin 2x = \sin x$
47. $\sin^2 x + 3 \sin x + 1 = 0$
48. $\sin^2 x + \sin x - 1 = 0$
49. $\sin x + \cos x = 1$

In Exercises 50–53 express each function as a general sine function, identifying its amplitude, period, and phase shift. Sketch a graph of each function.
50. $f(x) = \sin x - 2 \cos x$
51. $f(x) = 4 \sin 2x + \cos 2x$
52. $f(x) = -2 \sin 3x - 4 \cos 3x$
53. $f(x) = \sin x \cos x$

54. Which of the trigonometric functions in Figures A.12 are even and which are odd?

In Exercises 55–59 verify the identity.
55. $\cos 3x = 4 \cos^3 x - 3 \cos x$
56. $\sin 4x = 8 \cos^3 x \sin x - 4 \cos x \sin x$
57. $\tan 3x = \dfrac{3 \tan x - \tan^3 x}{1 - 3 \tan^2 x}$
58. $\tan\left(\dfrac{x}{2}\right) = \dfrac{\sin x}{1 + \cos x}$
59. $\dfrac{1 + \tan x}{1 - \tan x} = \tan\left(x + \dfrac{\pi}{4}\right)$

60. Show that a function $f(x) = A \cos \omega x + B \sin \omega x$ can always be written in the form
$$f(x) = \sqrt{A^2 + B^2} \sin(\omega x + \phi),$$

where ϕ is defined by the equations

$$\sin\phi = \frac{A}{\sqrt{A^2 + B^2}} \quad \text{and} \quad \cos\phi = \frac{B}{\sqrt{A^2 + B^2}}.$$

61. In Exercise 60 can we replace the two equations defining ϕ with the single equation $\tan\phi = A/B$?

In Exercises 62–65 find all solutions of the equation in the interval $0 \leq x < 2$.

62. $\sin 4x = \cos 2x$

63. $\cos x + \cos 3x = 0$

64. $\sin 2x + \cos 3x = \sin 4x$

65. $\sin x + \cos x = \sqrt{3} \sin x \cos x$

66. Verify that if A, B, and C are the angles of a triangle, then

$$\tan A + \tan B + \tan C = \tan A \tan B \tan C.$$

Hint: Expand $\tan(A + B + C)$ in terms of $\tan A$, $\tan B$, and $\tan C$.

APPENDIX B

The Exponential and Logarithm Functions

In this appendix we give a brief introduction to the exponential and logarithm functions and their properties.

In elementary algebra we learned the basic rules for products and quotients of powers:

$$a^b a^c = a^{b+c}; \tag{B.1a}$$

$$\frac{a^b}{a^c} = a^{b-c}; \tag{B.1b}$$

$$\left(a^b\right)^c = a^{bc}. \tag{B.1c}$$

These rules are used to develop the **exponential function**

$$f(x) = a^x; \tag{B.2}$$

that is, a raised to the power of x for variable x.

Consider first the case when $a > 1$. The meaning of a^x when x is an integer is clear, and when $x = 1/n$, where $n > 0$ is an integer, $a^x = a^{1/n}$ is the nth root of a. When x is a positive rational number n/m (n and m positive integers), B.1c implies that

$$a^x = a^{n/m} = (a^n)^{1/m} \quad \text{or}$$
$$a^x = a^{n/m} = \left(a^{1/m}\right)^n.$$

When x is a negative rational, we write $a^x = 1/a^{-x}$, where $-x$ is positive. These results lead to the points and the graph of $y = a^x$ in Figure B.1.

FIGURE B.1

FIGURE B.2

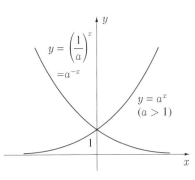

THE EXPONENTIAL AND LOGARITHM FUNCTIONS 605

There is a difficulty with this graph and the definition of a^x when x is not a rational number, a fact we should at least mention here. For instance, if $x = \sqrt{2}$, an irrational number, what is the value of $a^{\sqrt{2}}$? Calculators with a y^x button suggest a possible definition. If, for example, a is equal to 3, the calculator has no difficulty evaluating $3^{\sqrt{2}}$. But what does it actually calculate? It uses a decimal approximation for $\sqrt{2}$. Since a decimal approximation represents a rational number, the calculator approximates $3^{\sqrt{2}}$ with 3 raised to a rational number approximating $\sqrt{2}$. The more sophisticated the calculator, the better the approximation for $\sqrt{2}$. This suggests that we define $a^{\sqrt{2}}$ as follows. Find a set of rational numbers which approximates $\sqrt{2}$ more and more closely, say,

$$c_1 = 1.4, \quad c_2 = 1.41, \quad c_3 = 1.414, \quad \cdots$$

where each number picks up another decimal place in the nonrecurring, nonterminating decimal expansion of $\sqrt{2}$. Define $a^{\sqrt{2}}$ as the limit

$$a^{\sqrt{2}} = \lim_{n \to \infty} a^{c_n}.$$

In other words, define $a^{\sqrt{2}}$ as the limit $\lim_{n \to \infty} a^{c_n}$ where $c_1, c_2, \cdots$ is any set of rational numbers which approximates $\sqrt{2}$ more closely. With this definition, it is possible to prove that a^x is a continuous function, and its graph is indeed as shown in Figure B.1.

When one curve is obtained from another by replacing each x by $-x$, the two curves are mirror images of each other in the y-axis. The graph of $f(x) = a^{-x}$ is therefore the reflection of $f(x) = a^x$ in the y-axis (Figure B.2). But

$$a^{-x} = \frac{1}{a^x} = \left(\frac{1}{a}\right)^x,$$

where $0 < 1/a < 1$. In other words, the graph of the exponential function for $0 < a < 1$ is as shown in Figure B.2.

For the case in which $a = 1$, the graph of $f(x) = a^x = 1^x = 1$ is a horizontal straight line. We do not consider $f(x) = a^x$ for $a < 0$. Can you see why?

EXAMPLE B.1

Sketch graphs of the exponential functions 2^x and 3^x on the same axes.

SOLUTION Graphs of these functions are shown in Figure B.3. Things to notice are that both curves pass through the point $(0, 1)$. More generally, $a^0 = 1$ for any a. When $x > 0$, the graph of 3^x is higher than that of 2^x, whereas the opposite is true for $x < 0$.

FIGURE B.3

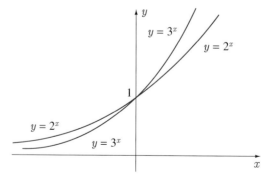

EXAMPLE B.2

Sketch graphs of the functions x^4 (a power function) and 2^x (an exponential function) on the same axes.

SOLUTION Sketches of the graphs of these functions are shown in Figure B.4, but no attempt has been made to use a scale on either the x- or y-axis. Notice here that for $x > 16$, $2^x > x^4$. This is always the situation for power and exponential functions. Given any exponential function a^x ($a > 1$), and any power function x^n ($n > 1$), there always exists a value of x, say, X, such that when $x > X$, we have $a^x > x^n$. In short, exponential functions grow more rapidly than power functions for large values of x.

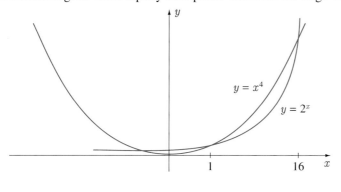

FIGURE B.4

In terms of the exponential function, rules B.1 take the form:

$$a^{x_1} a^{x_2} = a^{x_1 + x_2}, \tag{B.3a}$$

$$\frac{a^{x_1}}{a^{x_2}} = a^{x_1 - x_2}, \tag{B.3b}$$

$$(a^{x_1})^{x_2} = a^{x_1 x_2}. \tag{B.3c}$$

Since the exponential function $f(x) = a^x$ ($a > 1$) is increasing, it has an inverse function. See Section 8.1 if you have not studied inverse functions. Its graph can be obtained by reflecting the curve $y = a^x$ in the line $y = x$ (Figure B.5). This inverse function is called the **logarithm function** to base a, and is denoted by

$$y = \log_a x. \tag{B.4}$$

Note that $\log_a x$ is defined only for $x > 0$; that is, we cannot take the logarithm of zero or a negative number, and the logarithm of 1 for any $a > 1$ is equal to 0.

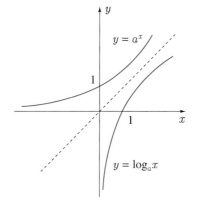

FIGURE B.5

We can also define $\log_a x$ for $0 < a < 1$, but such a logarithm is simply the negative of $\log_{1/a} x$, where $1/a > 1$ (see Exercise 22). Consequently, unless otherwise specified, we shall always assume the base of logarithms to be greater than unity.

Figure B.5 gives a graphical representation of the logarithm function, and calculators can be used to find decimal approximations for logarithms of numbers. But what really is a logarithm? To find out, we suppose (x, y) is a point on $y = \log_a x$. Then (y, x) is a point on the exponential curve; that is,

$$y = \log_a x \quad \text{only if} \quad x = a^y. \tag{B.5}$$

These equations contain the real meaning of logarithm. They state that y is the logarithm of x to base a only if x is a to the power y. In other words, the *logarithm of x to base a is the power to which a must be raised in order to produce x*. Logarithms then are powers. For example, to find $\log_{10} 100$, we search for a number to fill the box in

$$100 = 10^{[\ ]}.$$

Clearly the number is 2, and therefore $\log_{10} 100 = 2$. Similarly, to calculate $\log_2 (1/16)$, we consider

$$\frac{1}{16} = 2^{[\ ]}.$$

Since the missing number is -4, we have that $\log_2 (1/16) = -4$.

When the first of equations B.5 is substituted into the second, the result is

$$x = a^{\log_a x} \quad (x > 0). \tag{B.6a}$$

When the second is substituted into the first,

$$y = \log_a (a^y),$$

or, since y is arbitrary, we may replace it with x and write

$$x = \log_a (a^x). \tag{B.6b}$$

These equations are statements of the fact that the exponential and logarithm functions are inverses. Substitute a value of x into one function, substitute the result into the other, and the final result is the original value of x.

Corresponding to rules B.3 for the exponential function are the following rules for logarithms:

$$\log_a(x_1 x_2) = \log_a x_1 + \log_a x_2, \tag{B.7a}$$

$$\log_a \left(\frac{x_1}{x_2}\right) = \log_a x_1 - \log_a x_2, \tag{B.7b}$$

$$\log_a \left(x_1^{x_2}\right) = x_2 \log_a x_1. \tag{B.7c}$$

To prove B.7a, say, we set $z = \log_a(x_1 x_2)$, in which case

$$a^z = x_1 x_2$$
$$= \left(a^{\log_a x_1}\right)\left(a^{\log_a x_2}\right) \quad \text{(using B.6a)}$$
$$= a^{\log_a x_1 + \log_a x_2}. \quad \text{(using B.3a)}$$

Thus,

$$\log_a x_1 + \log_a x_2 = z = \log_a (x_1 x_2).$$

We leave proofs of B.7b and B.7c to the exercises.

EXAMPLE B.3

Simplify the following expressions:

(a) $3^{\log_3(x^2)}$ (b) $10^{-4\log_{10} x}$ (c) $\log_a(a^{-x+3})$ (d) $\log_2 8 + \log_3(1/27)$

SOLUTION (a) Identity B.6a implies that
$$3^{\log_3(x^2)} = x^2.$$

(b) Since -4 intervenes between the logarithm and exponential operations, we cannot use B.6a immediately. The -4 can be relocated, however, with B.7c:
$$10^{-4\log_{10} x} = 10^{\log_{10}(x^{-4})} = \frac{1}{x^4} \quad (\text{if } x > 0).$$

(c) Identity B.6b gives
$$\log_a(a^{-x+3}) = -x + 3.$$

(d) Since $\log_2 8 = 3$ and $\log_3(1/27) = -3$,
$$\log_2 8 + \log_3(1/27) = 3 - 3 = 0.$$

■

EXAMPLE B.4

Solve the following equations:

(a) $\log_5 x = -3$ (b) $\log_{10} x + \log_{10}(x + 1) = 0$ (c) $10^x - 12 + 10^{-x} = 0$

SOLUTION (a) By means of B.5,
$$x = 5^{-3} = \frac{1}{125}.$$

(b) Since $\log_{10} x + \log_{10}(x + 1) = \log_{10}[x(x + 1)]$, we can write
$$0 = \log_{10}[x(x + 1)].$$

If we now take exponentials to base 10,
$$10^0 = 10^{\log_{10}[x(x+1)]} \quad \text{or}$$
$$1 = x(x + 1).$$

This quadratic equation has solutions
$$x = \frac{-1 \pm \sqrt{1 + 4}}{2} = \frac{-1 \pm \sqrt{5}}{2}.$$

Since x must be positive (the original equation demands this), the only solution is $x = (\sqrt{5} - 1)/2$.

(c) If we multiply the equation by 10^x, the result is

$$0 = 10^{2x} - 12(10^x) + 1 = (10^x)^2 - 12(10^x) + 1.$$

But this is a quadratic equation in 10^x so that

$$10^x = \frac{12 \pm \sqrt{144 - 4}}{2} = 6 \pm \sqrt{35}.$$

Finally, we have

$$x = \log_{10}\left(6 \pm \sqrt{35}\right).$$

EXAMPLE B.5

In acoustics, the minimum sound level detectable by the normal human ear is taken as the "tick of a watch at 20 feet under otherwise quiet conditions." This sound level, called the audible sound threshold, is said to have a decibel reading of 0. All other sounds are given decibel readings relative to this audible sound threshold. For example, a normal voice is approximately 70 decibels, the sound of a car is 100 decibels, and a jet engine 160 decibels. Mathematically, the loudness of a sound is said to be L decibels if

$$L = 10 \log_{10}\left(\frac{I}{I_0}\right),$$

where I is the intensity of the sound and I_0 is the intensity of sound at the audible threshold. Use this definition to answer the following questions.

(a) Express the intensity I of a sound in terms of I_0 and its decibel reading L.

(b) Intensity is a measure of energy per unit area per unit time transmitted by sound waves. Find the intensity of the normal voice, a car, and a jet engine relative to the intensity I_0.

(c) If the pain threshold for sound has an intensity 10^{14} times I_0, what is its decibel reading?

(d) If the intensity I_1 of one sound is 10 times the intensity I_2 of a second sound, how do their decibel readings compare?

SOLUTION (a) When we take both sides as exponents of powers of 10, and use properties B.7c and B.6a,

$$10^L = 10^{10 \log_{10}(I/I_0)} = 10^{\log_{10}(I/I_0)^{10}} = \left(\frac{I}{I_0}\right)^{10}.$$

If we take 10th roots of both sides,

$$\left(10^L\right)^{1/10} = \frac{I}{I_0} \quad \text{or} \quad I = I_0 \, 10^{L/10}.$$

(b) Since the decibel level of the normal voice is 70, its intensity is

$$I = I_0 \, 10^{70/10} = 10^7 I_0.$$

Similarly, the intensities of a car and a jet are 10^{10} and 10^{16} times I_0.

(c) For an intensity of $10^{14} I_0$, the decibel reading is

$$L = 10 \, \log_{10}(10^{14}) = 10(14) = 140.$$

(d) If L_1 and L_2 are the decibel readings for sounds with intensities I_1 and I_2, then

$$L_1 = 10 \, \log_{10}(I_1/I_0) \quad \text{and} \quad L_2 = 10 \, \log_{10}(I_2/I_0).$$

When we subtract these readings,

$$\begin{aligned}
L_1 - L_2 &= 10 \, \log_{10}(I_1/I_0) - 10 \, \log_{10}(I_2/I_0) \\
&= 10[\log_{10}(I_1/I_0) - \log_{10}(I_2/I_0)] \\
&= 10 \, \log_{10}\left(\frac{I_1/I_0}{I_2/I_0}\right) \quad \text{(using (B.7b))} \\
&= 10 \, \log_{10}\left(\frac{I_1}{I_2}\right) \\
&= 10 \, \log_{10}\left(\frac{10\,I_2}{I_2}\right) \quad \text{(since } I_1 = 10\,I_2\text{)} \\
&= 10 \, \log_{10} 10 \\
&= 10.
\end{aligned}$$

Thus, when the intensity of one sound is 10 times that of another, their decibel readings differ by 10. ∎

It is sometimes necessary to change from one base of logarithms to another. If we take logarithms to base b on both sides of identity B.6a, we immediately obtain

$$\log_b x = (\log_a x)(\log_b a). \tag{B.8}$$

This equation defines $\log_b a$ as the conversion factor from logarithms to base a to logarithms to base b.

Before the discovery of calculus, the base of logarithms was invariably chosen to be 10. Such logarithms are called *common logarithms*; they correspond to the exponential function 10^x. Another base for exponentials and logarithms, however, that is much more convenient in most applications is a number, denoted by the letter e, and defined in a variety of ways, one of which is

$$e = \lim_{z \to 0} (1 + z)^{1/z}. \tag{B.9a}$$

An alternative is obtained by setting $v = 1/z$,

$$e = \lim_{v \to \infty} \left(1 + \frac{1}{v}\right)^v. \tag{B.9b}$$

To prove the existence of either limit is extremely difficult. Substitution of various values of v into $(1+1/v)^v$ may suggest, however, whether a limit exists. The values in Table B.1

are steadily increasing but as they rapidly get closer together, they are indeed approaching a limit — to twelve decimals

$$e = 2.718281828459.$$

TABLE B.1

v	$\left(1 + \dfrac{1}{v}\right)^v$
1	2.000 000
3	2.370 370
5	2.488 320
10	2.593 742
100	2.704 814
1 000	2.716 924
10 000	2.718 146
100 000	2.718 255
1 000 000	2.718 282

It is an irrational number with a nonterminating, nonrepeating decimal expansion. Why this number is so convenient as a base for logarithms and exponentials is shown in Section 3.9. For now, let us rewrite some of the more important formulas of this appendix with set equal to e. The exponential function to base e is e^x, and equations B.3 in terms of e^x read

$$e^{x_1} e^{x_2} = e^{x_1 + x_2}, \tag{B.10a}$$

$$\frac{e^{x_1}}{e^{x_2}} = e^{x_1 - x_2}, \tag{B.10b}$$

$$(e^{x_1})^{x_2} = e^{x_1 x_2}. \tag{B.10c}$$

Logarithms to base e are usually given the notation $\ln x$ rather than $\log_e x$, and are called **natural**, or **Naperian logarithms** after the Scottish mathematician John Napier (1550–1617), who invented logarithms,

$$\ln x = \log_e x. \tag{B.11}$$

In terms of $\ln x$, rules B.7 are

$$\ln(x_1 x_2) = \ln x_1 + \ln x_2, \tag{B.12a}$$

$$\ln\left(\frac{x_1}{x_2}\right) = \ln x_1 - \ln x_2, \tag{B.12b}$$

$$\ln\left(x_1^{x_2}\right) = x_2 \ln x_1. \tag{B.12c}$$

Identities B.6, which express the inverse character of exponential and logarithm functions, become

$$x = e^{\ln x}, \quad x > 0, \tag{B.13a}$$

$$x = \ln(e^x). \tag{B.13b}$$

EXERCISES

In Exercises 1–13 find all values of x satisfying the equation.

1. $\log_{10}(2+x) = -1$
2. $10^{3x} = 5$
3. $\log_{10}(x^2 + 2x + 1) = 1$
4. $\ln(x^2 + 2x + 10) = 1$
5. $10^{5-x^2} = 100$
6. $10^{1-x^2} = 100$
7. $\log_{10}(x-3) + \log_{10} x = 1$
8. $\log_{10}(3-x) + \log_{10} x = 1$
9. $\log_{10}[x(x-3)] = 1$
10. $2\ln x + \ln(x-1) = 2$
11. $\log_a x + \log_a(x+2) = 2$
12. $\log_a[x(x+2)] = 2$
13. $\log_{10}\left[\log_{10}\left(\dfrac{x+3}{200x}\right) + 4\right] = -1$

In Exercises 14–17 sketch a graph of the function.

14. $f(x) = e^{-x^2}$
15. $f(x) = \ln(\cos x)$
16. $f(x) = \log_a |x^2 - 1|$
17. $f(x) = a^{\log_a(2x+1)}$

18. Is there a difference between the graphs of the functions $f(x) = \log_a(x^2)$ and $g(x) = 2\log_a x$?

19. (a) In the early period of reforestation, the percentage increase of timber each year is almost constant. If the original amount A_0 of a certain timber increases 3.5% the first year and 3.5% each year thereafter, find an expression for the amount of timber after t years.

    (b) How long does it take for timber of this type to double?

20. A new car costs $20 000. In any year it depreciates to 75% of its value at the beginning of that year. What is the value of the car after t years?

21. If the effective height of the earth's atmosphere (in metres) is the solution of the equation

    $$10^{-6} = (1 - 2.08 \times 10^{-6} y)^{56},$$

    find y.

22. Show that if y is the logarithm of x to base a, then $-y$ is the logarithm of x to base $1/a$.

23. Prove B.7b and B.7c.

24. Is identity B.7a valid for any x_1 and x_2?

25. The magnitude of an earthquake is measured in much the same way as noise level. An earthquake of minimal size is taken as having value 0 on the Richter scale. Any other earthquake of intensity I is said to have magnitude R on the Richter scale if

    $$R = \log_{10}\left(\dfrac{I}{I_0}\right),$$

    where I_0 is the intensity of the minimal earthquake being used as reference.

    (a) Express the intensity of an earthquake in terms of I_0 and its reading on the Richter scale.

    (b) What are readings on the Richter scale of earthquakes that have intensities 1.20×10^6 and 6.20×10^4 times I_0?

26. (a) If P dollars is invested at $i\%$ compounded n times per year, show that the accumulated value after t years is

    $$A = P\left(1 + \dfrac{i}{100n}\right)^{nt}.$$

    (b) How long does it take to double an investment if interest is 8% compounded semiannually?

    (c) Calculate the maximum possible value of A if i is fixed but the number of times that interest is compounded is unlimited; that is, calculate $\lim_{n\to\infty} A$. This method of calculating interest is called *continuously compounded interest*.

    (d) What is the accumulated value of a $1000 investment after 10 years at 6% compounded continuously? Compare this to the accumulated value if interest is calculated only once each year.

In Exercises 27–30 find all values of x satisfying the equation.

27. $3a^{2x} + 3a^{-2x} = 10$
28. $2^x + 4^x = 8^x$
29. $3^{x+4} = 7^{x-1}$
30. $\log_x 2 = \log_{2x} 8$

31. Sketch a graph of the function $f(x) = \ln\left(x + \sqrt{x^2 + 1}\right)$.

32. Repair costs on the car in Exercise 20 are estimated at $50 the first year, increasing by 20% each year thereafter. Set up a function $C(t)$ that represents the average yearly cost associated with owning the car for t years. Hint: You may need the result of Exercise 27 in Section 6.1.

33. A straight wire conductor has length $2L$ and circular cross-section of radius R. If the wire carries current $I > 0$, then the magnitude of the "vector" potential at a distance r from the centre of the wire is

$$f(r) = \begin{cases} \dfrac{\mu_0 I}{4\pi}\left[\ln\left(1 + \dfrac{4L^2}{R^2}\right) - 1 + \dfrac{r^2}{R^2}\right] & 0 \le r \le R \\ \dfrac{\mu_0 I}{4\pi} \ln\left(1 + \dfrac{4L^2}{r^2}\right) & r > R \end{cases}$$

where μ_0 is a positive constant.

(a) Sketch a graph of this function.
(b) Find the radius $r > R$ for which $f(r) = f(0)$.

In Exercises 34–36 solve the given equation for x in terms of y.

34. $y = \dfrac{e^{2x} - e^{-2x}}{2}$

35. $y = \dfrac{e^x + e^{-x}}{2}$

36. $y = \dfrac{e^x - e^{-x}}{e^x + e^{-x}}$

APPENDIX C

Mathematical Induction

Mathematical induction is a type of proof used to verify propositions that involve integers. It is best introduced through a simple example.

Suppose we are required to verify that the sum of the first n positive integers is

$$1 + 2 + 3 + \cdots + n = \frac{n(n+1)}{2}. \tag{C.1}$$

It is a simple matter to check that this result is valid for the first few integers:

For $n = 1$, the sum is 1, and the formula gives $1(1 + 1)/2 = 1$.
For $n = 2$, the sum is $1 + 2 = 3$; the formula gives $2(2 + 1)/2 = 3$.
For $n = 3$, the sum is $1 + 2 + 3 = 6$; the formula gives $3(3 + 1)/2 = 6$.

We shall see later that to use induction, it is necessary to verify the result at this stage only for $n = 1$.

We now suppose that k is some integer for which the result is valid; that is, we suppose that

$$1 + 2 + 3 + \cdots + k = \frac{k(k+1)}{2}.$$

On the basis of this supposition, the sum of the first $k + 1$ integers is

$$\begin{aligned}
1 + 2 + \cdots + (k+1) &= (1 + 2 + \cdots + k) + (k+1) \\
&= \frac{k(k+1)}{2} + (k+1) \\
&= \frac{1}{2}[k(k+1) + 2(k+1)] \\
&= \frac{(k+1)(k+2)}{2}.
\end{aligned}$$

But this is precisely equation C.1 with n replaced by $k + 1$. We have shown that if C.1 is valid for some integer k, then it must also be valid for the next integer $k + 1$.

We now put the above facts together to verify that C.1 is valid for all $n \geq 1$. We have proved that it is valid for $n = 1, 2,$ and 3. But if it is true for 3, then it must be true for 4. If it is true for 4, then it is true for 5. If it is true for 5, it is true for 6; and so on, and so on. Consequently, C.1 must be valid for every positive integer n.

This example contains the two essential parts of every inductive proof:

1. Verification of the result for the smallest integer in question;
2. Verification of the fact that if the result is valid for some integer k, then it must be valid for the next integer $k + 1$.

Once these two facts are established, it follows by the same argument as above that the result is valid for the given set of integers. This is the principle of mathematical induction, which we state formally as follows.

Principle of Mathematical Induction

Suppose with each integer n greater than or equal to some fixed integer N, there is associated a proposition P. Then P is true for all $n \geq N$ provided:

(a) P_N is valid;

(b) the validity of P_k implies the validity of P_{k+1}.

All inductive arguments must therefore contain these two essential parts. Part (a) is usually quite simple to establish; but part (b) can sometimes be quite difficult. Be sure you understand what we are doing in part (b). We are proving that under the assumption that P_k is true, then P_{k+1} must also be true. It is the principle of mathematical induction that puts (a) and (b) together to state that P is true for all $n \geq N$.

EXAMPLE C.1

Verify that

$$1^2 + 2^2 + 3^2 + \cdots + n^2 = \frac{n(n+1)(2n+1)}{6}, \quad n \geq 1.$$

SOLUTION When $n = 1$ the left-hand side of this equation has value 1, and the right-hand side is $1(2)(3)/6 = 1$. The required result is therefore true for $n = 1$. Next we suppose k is some integer for which the result is valid; that is, we suppose that

$$1^2 + 2^2 + 3^2 + \cdots + k^2 = \frac{k(k+1)(2k+1)}{6},$$

and we must write this down because we are going to need it later on. Our objective now is to verify that the result is valid for $k + 1$; that is, we must verify that

$$1^2 + 2^2 + 3^2 + \cdots + (k+1)^2 = \frac{(k+1)(k+2)[2(k+1)+1]}{6}.$$

If we begin on the left, we have

$$1^2 + 2^2 + \cdots + (k+1)^2 = (1^2 + 2^2 + \cdots + k^2) + (k+1)^2$$

$$= \frac{k(k+1)(2k+1)}{6} + (k+1)^2$$

$$= \frac{(k+1)}{6}[k(2k+1) + 6(k+1)]$$

$$= \frac{(k+1)}{6}(2k^2 + 7k + 6)$$

$$= \frac{(k+1)(k+2)(2k+3)}{6}$$

$$= \frac{(k+1)(k+2)[2(k+1)+1]}{6},$$

which is the right-hand side of the required result. We have therefore verified that the result is valid for $k + 1$. Consequently, by mathematical induction the formula is correct for all $n \geq 1$. ∎

EXAMPLE C.2

Prove that 7 divides $23^{3n} - 1$ for all positive integers n.

SOLUTION When $n = 1$, $23^{3n} - 1 = 23^3 - 1 = 12\,166$, and this is divisible by 7. The result is therefore true for $n = 1$. Next we suppose k is some integer for which the result is valid; that is, we suppose 7 divides $23^{3k} - 1$. We must now verify that 7 divides $23^{3(k+1)} - 1$. Since

$$23^{3(k+1)} - 1 = 23^{3k+3} - 1 = (23^{3k})(23^3) - 1$$
$$= [(23^{3k})(23^3) - 23^3] + (23^3 - 1)$$
$$= 23^3(23^{3k} - 1) + (23^3 - 1),$$

and both $23^{3k} - 1$ and $23^3 - 1$ are divisible by 7, it follows that $23^{3(k+1)} - 1$ must also be divisible by 7. The result is therefore true for $k + 1$, and by mathematical induction, it is true for all $n \geq 1$. ∎

EXAMPLE C.3

Verify that if n lines are drawn in the plane, no two of which are identical, then the maximum number of points of intersection of these lines is $n(n-1)/2$.

SOLUTION If we draw two lines in the plane, there can be at most one point of intersection. This is precisely what is predicted by $n(n-1)/2$ when $n = 2$. The result is therefore correct for $n = 2$. Suppose the proposition is true for some integer k; that is, suppose when k lines are drawn in the plane, the maximum number of points of intersection is $k(k-1)/2$. We must now verify that when any $k + 1$ lines are drawn, the maximum number of points of intersection is $(k+1)k/2$. To do this, we remove one of the $k + 1$ lines leaving k lines. But for these k lines the maximum number of points of intersection is $k(k-1)/2$. If the line that was removed is now replaced, it can add at most k more points of intersection, one with each of the k lines. The maximum number of points of intersection of the $k + 1$ lines is therefore

$$\frac{k(k-1)}{2} + k = \frac{k}{2}(k - 1 + 2) = \frac{k(k+1)}{2},$$

and this is the result for $n = k + 1$. By mathematical induction, then, the result is valid for all $n \geq 2$. ∎

EXERCISES

In Exercises 1–5 use mathematical induction to establish the formula for positive integer n.

1. $1^3 + 2^3 + 3^3 + \cdots + n^3 = \dfrac{n^2(n+1)^2}{4}$

2. $1 + 3 + 6 + 10 + \cdots + \dfrac{n(n+1)}{2} = \dfrac{n(n+1)(n+2)}{6}$

3. $\dfrac{1}{1 \cdot 2} + \dfrac{1}{2 \cdot 3} + \dfrac{1}{3 \cdot 4} + \cdots + \dfrac{1}{n(n+1)} = \dfrac{n}{n+1}$

4. $\dfrac{1}{1 \cdot 2 \cdot 3} + \dfrac{1}{2 \cdot 3 \cdot 4} + \dfrac{1}{3 \cdot 4 \cdot 5} + \cdots + \dfrac{1}{n(n+1)(n+2)}$
   $= \dfrac{n(n+3)}{4(n+1)(n+2)}$

5. $a + ar + ar^2 + ar^3 + \cdots + ar^{n-1} = \dfrac{a(1 - r^n)}{1 - r}$, $(r \neq 1)$

6. Prove that $x - y$ divides $x^n - y^n$ for $n \geq 1$.

7. Verify that $x + y$ divides $x^{2n+1} + y^{2n+1}$ for $n \geq 0$.

8. Prove that 576 divides $5^{2n+2} - 24n - 25$ for $n \geq 1$.

9. Prove that if the proposition

$$1 + 3 + 5 + \cdots + (2n-1) = n^2 + 4$$

is valid for some integer k, then it is also valid for $k+1$. Is the result valid for all $n \geq 1$?

10. Verify that the sum of the interior angles of a polygon with n sides is $(n-2)\pi$ radians.

11. Show that if n is a positive integer so too is $(n^3 + 6n^2 + 2n)/3$.

12. Prove that:
    (a) $1 + \cos\theta + \cos 2\theta + \cdots + \cos n\theta = \dfrac{1 - \cos\theta + \cos n\theta - \cos(n+1)\theta}{2 - 2\cos\theta}$;

    (b) $\sin\theta + \sin 3\theta + \cdots + \sin(2n-1)\theta = \dfrac{\sin^2 n\theta}{\sin\theta}$.

13. Verify that

$$\left(1 - \frac{1}{4}\right)\left(1 - \frac{1}{9}\right)\left(1 - \frac{1}{16}\right) \cdots \left(1 - \frac{1}{n^2}\right) = \frac{n+1}{2n}.$$

14. Use mathematical induction to show that for $n \geq 2$, there exist constants a_r and b_r, $r = 0, 1, 2, \ldots, n$ such that

$$\sin^n x = \sum_{r=0}^{n} (a_r \cos rx + b_r \sin rx).$$

Answers to Even-numbered Exercises

CHAPTER 1

Exercises 1.1

2. $-3 < x < 4$ **4.** $-1 < x \leq 6$ **6.** $x \geq 3$
8. $-\infty < x < \infty$ **10.** No **12.** $[3, 10]$
14. $(4, 22]$ **16.** $(-\infty, -2]$ **18.** $(-\infty, -2]$
20. $(-\infty, \infty)$ **22.** $x < 3$ **24.** $0 < x < 3/5$
26. $-3 \leq x \leq 3$ **28.** $-5/3 < x < 3$
30. $-1/\sqrt{2} \leq x \leq 1/\sqrt{2}$ **32.** $-11/2 \leq x \leq -1/2$
34. $x \leq -3, x \geq 3$ **36.** $x \leq -1, x \geq 1$
38. $x < 1/2, x > 11/9$
40. $-2 - \sqrt{10} < x < -2 + \sqrt{10}$
42. $x < -1 - \sqrt{7/2}, x > -1 + \sqrt{7/2}$
44. All points
48. $x < -2, -4/\sqrt{5} < x < 4/\sqrt{5}, x > 2$
50. $\sqrt{3} - 1 \leq x \leq \sqrt{5} - 1, -\sqrt{5} - 1 \leq x \leq -\sqrt{3} - 1$
52. $x < -\sqrt{37}, -3 < x < 3, x > \sqrt{37}$
54. $x < -11/7, -1 < x < 3$ **56.** $x \geq 5$
58. $(1 - \sqrt{5})/2 \leq x \leq (1 + \sqrt{5})/2$
60. $x < (3 - \sqrt{101})/2, 5 < x < (3 + \sqrt{101})/2$
62. $|x + 1| + |x - 2| > 4, x < -3/2, x > 5/2$
64. $||x - 2| - |x + 6|| < 3, -7/2 < x < -1/2$
66. **(a)** $-11 < x < -1$ **(b)** $x < -11, -2 < x < -1$
  (c) $x < -11, -5 < x < -1$ **(d)** $-11 < x < -5, -2 < x < -1$

Exercises 1.2

4. $3\sqrt{5}$ **6.** $\sqrt{58}$ **8.** $(1, -1/2)$
10. $(-1/2, 1/2)$

12.

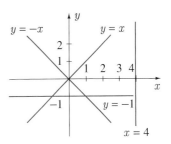

14. $(15/2, 15\sqrt{3}/2)$
16. $(\sqrt{2}, -\sqrt{7}), (0, -\sqrt{7} - \sqrt{2}), (-\sqrt{2}, -\sqrt{7})$
18. $20(\sqrt{2} + 1)$ cm **20.** $(2, 3/2)$ **22.** $(10/3, 40/3)$

Exercises 1.3

2.   **4.**

6.   **8.**

10.

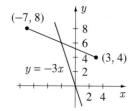

12. 2.68   **14.** 1.25   **16.** 0
18. parallel   **20.** perpendicular
22. 1.37   **24.** 1.11   **26.** $(1,2)$
28. $(-10,-14)$ **30.** $(37/73, 153/146)$
32. $x+y=2$ **34.** $3y=2x$ **36.** $5x+3y=30$
38. $y=-1$ **42.** $(-11/10, 19/10)$

Exercises 1.4A

2.   **4.**

6.   **8.**

10.   **12.**

 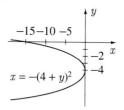

14. $y = x^2/2 + 1$ **16.** $y = (x^2 - 2x - 3)/3$
18. $(0,1), (-1,0)$ **20.** No points   **22.** $(0,1), (-3,-2)$
24. (a) $(v^2/4.905)\sin\theta\cos\theta$ **(b)** $(v^2/19.62)\sin^2\theta$
26. $(1,1), (4,4)$ **28.** $y = (x^2 - 2x + 5)/2$

Exercises 1.4B

2.   **4.**

6.   **8.**

 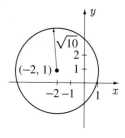

10. No points satisfy the equation
12. $(x-1)^2 + y^2 = 1$
14. $(x-3/2)^2 + (y+3/2)^2 = 9/2$
16. $3x^2 + 3y^2 - 14x - 4y = 4$
18. $(0,2), (-7/5, -11/5)$ **20.** No points
24. $7(x^2+y^2) + 6x - 44y + 24 = 0$

Exercises 1.4C

2.   **4.**

6.

8.

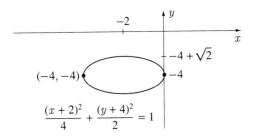

10. $32/\sqrt{15}$ **12.** $(-3,0), (21/25, 96/25)$
14. $(2, 3\sqrt{3}/2)$ **16.** $(\pm 2, 0), (\pm\sqrt{15}/2, -1/4)$

Exercises 1.4D

2.

4.

6.

8.

10.
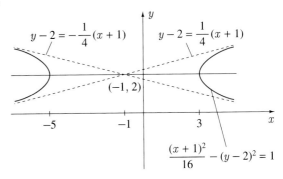

12. $\pm(\sqrt{2}, 1/\sqrt{2})$ **14.** No points   **16.** $(0,0), (3, \pm\sqrt{3})$
18. $(5, 1 \pm 3\sqrt{3})$

Exercises 1.5

2. $x \ne 2$ **4.** $|x| > 2$ **6.** $2 \le |x| < 3$

8.

10.

12.

14.

16.

18.

20.

22.

24.

26.

28.

30.

32.

34.

36. (a)   **(b)**

(c) $40[\![1 + x/50]\!]$, $x/50 \neq$ integer; $4x/5$, $x/50$ an integer; $0 < x \leq 500$

38.

42.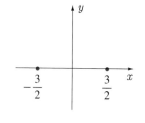

40. Not a function

44.

46.

50.

22 or 23

52.

54.

56.

58.

60.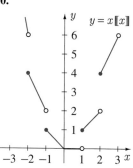

62. No

Exercises 1.6

2. odd **4.** odd **6.** neither
8. odd **10.** odd

14.

$y = (x+1)^2 - 2(x+1)$
(Function)

16.

$y = \sqrt{4 - (x-2)^2}$
(Function)

18.
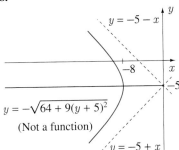
$y = -\sqrt{64 + 9(y+5)^2}$
(Not a function)

20.

$|x| + |y+2| = 2$
(Not a function)

22.

$|x| - |y| = 1$
(Not a function)

24.

$(y-1)^2 = (x-1)^2[1-(x-1)^2]$
(Not a function)

26.
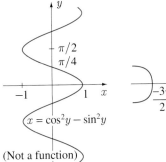
$x = \cos^2 y - \sin^2 y$
(Not a function)

28.
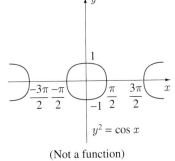
$y^2 = \cos x$
(Not a function)

30.

$y = \sin 2x$

32.
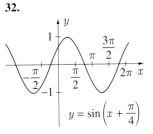
$y = \sin\left(x + \dfrac{\pi}{4}\right)$

34.

$y = \sin\left(2x + \dfrac{\pi}{4}\right)$

36.

$y = 4\cos(x/3)$

38.
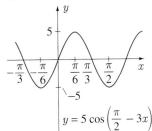
$y = 5\cos\left(\dfrac{\pi}{2} - 3x\right)$

40.

$y = \tan(3x)$

42.

$y = \cot\left(x + \dfrac{\pi}{4}\right)$

44.

$y = \sqrt{1 - \cos^2 x}$

46.
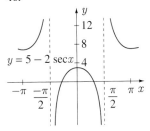
$y = 5 - 2\sec x$

48.
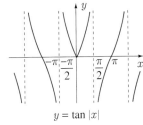
$y = \tan |x|$

50.

$y = 3\csc\left(\dfrac{x}{2}\right)$

52.
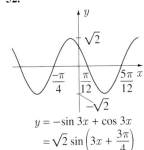
$y = -\sin 3x + \cos 3x$
$= \sqrt{2}\sin\left(3x + \dfrac{3\pi}{4}\right)$

54.
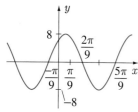
$y = 4 \sin 3x + 4\sqrt{3} \cos 3x$
$= 8 \sin\left(3x + \dfrac{\pi}{3}\right)$

56.

$y = -2 \sin 2x + 4 \cos 2x$
$= 2\sqrt{5} \sin (2x + 2.034)$

58.

60.

62.

64.

66.

68. (a)

(b)

(c) $f(x - a) H(x - a)$
(d) eliminates that part of graph to left of $x = a$

70. They are identical for $x > 0$. $y = f(x)$ is symmetric about $x = 0$, whereas $g(x)$ is not defined for $x \leq 0$.

72.
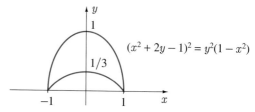

Review Exercises

2. $-2 \leq x \leq 2$
4. $x < -2, 3 < x < 4$
6. $x < -3, -1 < x < 1, x > 3$
8. $5\sqrt{2}, (-1/2, -3/2)$
10. $y = 2x + 4$
12. $x + y = 2$
14. $|x| \geq \sqrt{5}$
16. $x \neq 0, -1$
18. $x \geq 0$
20. $|x + 1| > \sqrt{14}/2$
22. $-1 \leq x < 0, x \geq 1$
24. parabola
26. ellipse
28. none of these
30. parabola
32. circle
34. hyperbola

36.

38.

40.

42.

44.

46.

48.

50.

52.

54.

56.

58.

60.

62. $\sqrt{(x+1)(2-x)}$ **64.** $\sqrt{-x}+1$

68.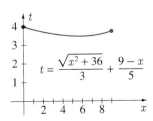

CHAPTER 2

Exercises 2.1

2. 0	**4.** $-3/2$	**6.** 0
8. -12	**10.** 6	**12.** -4
14. 0	**16.** -1	**18.** 0
20. $-7/4$	**22.** does not exist	**24.** 2
26. $1/\sqrt{2}$	**28.** $2\sqrt{2}$	**30.** 1
32. does not exist	**34.** $-1/4$	**36.** 1
38. 4	**40.** -2	**42.** False
44. $1/(2\sqrt{x})$	**48.** L	**50.** Cannot find
52. $-L$	**54.** Cannot find	

Exercises 2.2

2. $-\infty$	**4.** ∞	**6.** ∞
8. $13/2$	**10.** does not exist	**12.** ∞
14. 1	**16.** ∞	**18.** does not exist
20. ∞	**22.** 0	**24.** ∞
26. ∞		

Exercises 2.3

2. $-1/2$	**4.** ∞	**6.** $1/3$
8. $-\infty$	**10.** $3/2$	**12.** 0
14. 1	**16.** ∞	**18.** ∞
20. $3/4^{1/3}$	**22.** 0	**24.** 0
26. does not exist	**28.** 0	**30.** ∞
32. 1	**34.** -1	**36.** 0
38. ∞	**40.** $a=d$, $(b-e)/(2\sqrt{a})$	

Exercises 2.4

2.

4.

6.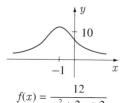
$f(x) = \dfrac{12}{x^2 + 2x + 2}$

8.
Discontinuous at $x = 0, 1$
$f(x) = \dfrac{x^3 + x^2 - 2x}{x^2 - x}$

10.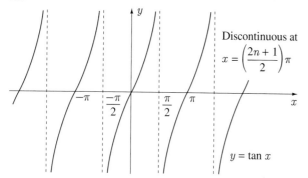
Discontinuous at $x = \left(\dfrac{2n+1}{2}\right)\pi$
$y = \tan x$

12. See Figure 2.5(**b**). Discontinuous at $x = 0$.

14.
Discontinuous at $x = \pm 3$
$f(x) = \dfrac{x+12}{x^2-9},\ -5 \le x \le 4$

16.
Discontinuous at $x = 3$
$f(x) = \dfrac{x^3 - 27}{|x - 3|}$

18. and **20.**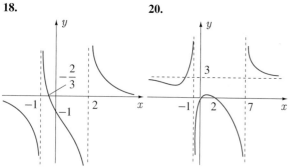
Discontinuous at $x = -1, 2$
$f(x) = \dfrac{3x+2}{x^2 - x - 2}$

Discontinuous at $x = -1, 7$
$f(x) = \dfrac{3x^2 - 6x}{x^2 - 6x - 7}$

22.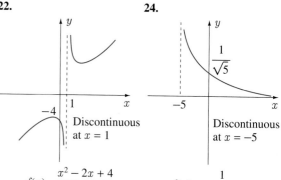
Discontinuous at $x = 1$
$f(x) = \dfrac{x^2 - 2x + 4}{x - 1}$

24. Discontinuous at $x = -5$
$f(x) = \dfrac{1}{\sqrt{5+x}}$

26.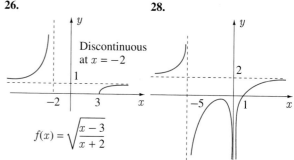
Discontinuous at $x = -2$
$f(x) = \sqrt{\dfrac{x-3}{x+2}}$

28. $f(x) = \dfrac{2x^3 - 2}{x^3 + 5x^2}$ Discontinuous at $x = -5, 0$

30.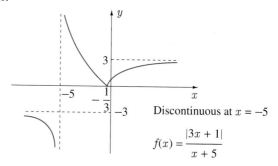
Discontinuous at $x = -5$
$f(x) = \dfrac{|3x+1|}{x+5}$

32. $x = -4$ in 2.; $x = -1$ in 7.; $x = 0$ and $x = 1$ in 8.; $x = 2$ in 9.; $x = 0$ in 13.; $x = 1$ in 19.

36.
Discontinuous at $x = 0$

38.

Discontinuous at $x = a$

Discontinuous at $x = a, a + \epsilon$

Discontinuous at $x = a$

40. (a)

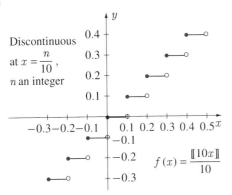

Discontinuous at $x = \dfrac{n}{10}$, n an integer

$f(x) = \dfrac{[\![10x]\!]}{10}$

42. (a)

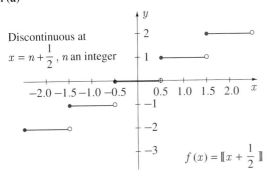

Discontinuous at $x = n + \dfrac{1}{2}$, n an integer

$f(x) = [\![x + \dfrac{1}{2}]\!]$

44.

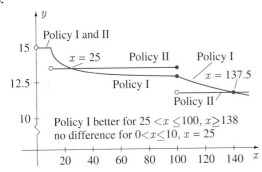

Policy I better for $25 < x \leq 100$, $x \geq 138$
no difference for $0 < x \leq 10$, $x = 25$

46. None

Exercises 2.5

10. $\lim_{x \to a^+} f(x) = L$ if given any $\epsilon > 0$, there exists a $\delta > 0$ such that $|f(x) - L| < \epsilon$ whenever $0 < x - a < \delta$.

12. $\lim_{x \to \infty} f(x) = L$ if given any $\epsilon > 0$, there exists an X such that $|f(x) - L| < \epsilon$ whenever $x > X$.

14. $\lim_{x \to a} f(x) = \infty$ if given any $M > 0$, there exists a $\delta > 0$ such that $f(x) > M$ whenever $0 < |x - a| < \delta$.

16. $\lim_{x \to \infty} f(x) = \infty$ if given any $M > 0$, there exists an X such that $f(x) > M$ whenever $x > X$.

18. $\lim_{x \to -\infty} f(x) = \infty$ if given any $M > 0$, there exists an X such that $f(x) > M$ whenever $x < X$.

36. No

Review Exercises

2. -2 **4.** 1 **6.** $-1/4$
8. 0 **10.** 0 **12.** ∞
14. $1/2$ **16.** 0 **18.** $-\sqrt{3}/2$
20. $-\infty$

22.   **24.**

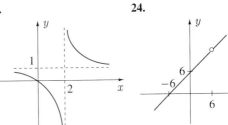

$f(x) = \dfrac{x}{x-2}$ Discontinuous at $x = 2$

$f(x) = \dfrac{x^2 - 36}{x - 6}$ Discontinuous at $x = 6$

26.   **28.**

$f(x) = \left|\dfrac{x+1}{x-1}\right|$ Discontinuous at $x = 1$

$f(x) = \dfrac{x+1}{|x-1|}$ Discontinuous at $x = 1$

30.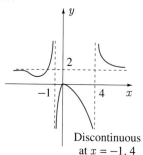
Discontinuous at $x = -1, 4$
$f(x) = \dfrac{2x^2}{x^2 - 3x - 4}$

32.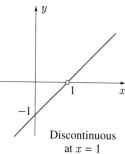
Discontinuous at $x = 1$
$f(x) = \dfrac{x^3 - 3x^2 + 3x - 1}{x^2 - 2x + 1}$

14. **16.**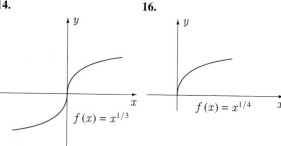

18. Yes
20. Not necessarily
22. Yes; derivative is zero
24. Yes; derivative is zero

CHAPTER 3

Exercises 3.1

2. $6x$
4. $3x^2 + 4x$
6. $-9/(x-5)^2$
8. $3x^2 + 4x$
10. $(2x^2 - 2)/(x^2 + x + 1)^2$
12. $2\pi r$
14. $4\pi r^2$
16. m
18. 0
20. $y = 2x + 2$
22. $x + 4y = 3$
24. 1.107
26. 2.897
28. $8x^7$
30. $-4/(x-2)^5$
32. $(3x+2)/(2\sqrt{x+1})$
34. $\sqrt{A}/(4\sqrt{\pi})$
36. $|x|/x$

Exercises 3.2

2. $9x^2 + 4$
4. $20x^4 - 30x^2 + 3$
6. $-6/x^4$
8. $1/x^3 - 12/x^5$
10. $20x^3 - 5/(4x^6)$
12. $1/(2\sqrt{x})$
14. $-(3/2)/x^{5/2} + (3/2)\sqrt{x}$
16. $\pi^2 x^{\pi-1}$
18. $4 + 3/x^4$
20. $6(2x + 5)^2$
22. $4y = x + 24$, $4x + y = 23$
24. $y = 6x - 10$, $x + 6y = 51$
26. $75 - 7x/50$, $50 - x/15 - x^2/20\,000$, $25 - 11x/150 + x^2/20\,000$
28. $(2, -12)$
32. $n|x|^{n-1}\,\text{sgn}(x)$

Exercises 3.3

2. right; no left; no derivative
4. no right; no left; no derivative
6. no right; no left; no derivative
8. right; no left; no derivative
10. True
12. True

Exercises 3.4

2. $4(2 - 3x^2 - x^3)$
4. $-10x/(4x^2 - 5)^2$
6. $x^2(4x^2 + 3)/(4x^2 + 1)^2$
8. $(3x + 2)/[2\sqrt{x+1}]$
10. $-(2x^2 + 20x + 1)/(2x^2 - 1)^2$
12. $(17 - 4x - 7x^2)/(x^2 - 5x + 1)^2$
14. $(x - 8\sqrt{x} - 2)/[\sqrt{x}(\sqrt{x} - 4)^2]$
16. $-27/(x+1)^4$
18. (b) $x(x-4)/(x^2 + x - 2)^2$
20. $f'(x)g(x)h(x) + f(x)g'(x)h(x) + f(x)g(x)h'(x)$
22. 1.23 at $(\pm 1/\sqrt{2}, 1/2)$
24. 0.668 at $(\sqrt{2}, 2\sqrt{2} + 2)$; 0.415 at $(-\sqrt{2}, 2 - 2\sqrt{2})$
30. $2/3$
32. $(5 + \sqrt{55}, -\sqrt{55}/(11 + \sqrt{55}))$, $(5 - \sqrt{55}, \sqrt{55}/(11 - \sqrt{55}))$

Exercises 3.5

2. 6
4. 18
6. $6 + 60/t^6$
8. $(3u^2 - 6u - 1)/[4u^{3/2}(u+1)^3]$
10. $-(\sqrt{x} + 3)/[4\sqrt{x}(\sqrt{x} + 1)^3]$
12. (b) $(bT_b - aT_a)/(b - a)$, $ab(T_a - T_b)/(b - a)$
14. $5/4, -5, 41/4, -13/2$
16. $n!/(2-x)^{n+1} + (-1)^{n+1} n!/(2+x)^{n+1}$

Exercises 3.6

2. $1/[2\sqrt{x}(u+1)^2]$
4. $-2(x-1)(s^2 + 2)/(s^2 - 2)^2$
6. $-21/[(t-4)^2(x+1)^2]$
8. $(4u + 5u^{3/2})(x^2 + 2x - 1)/[2(x - x^2)^2]$
10. $(3x+2)/[2\sqrt{x+1}]$

12. $(x+1)/(2x+1)^{3/2}$
14. $(2x-1)(6x+23)/(3x+5)^2$
16. $x^2(18-55x^2)/[3(2-5x^2)^{2/3}]$
18. $(x+1)(15x+11)(3x+1)^2$
20. $(9x-8)/[2x^3\sqrt{2-3x}]$
22. $-1/[(2+x)^{5/4}(2-x)^{3/4}]$
24. $(x+5)^3(11x^3+15x^2+8)/[2\sqrt{1+x^3}]$
26. $(x+5)^3(13x^4+25x^3+10x+10)/[2\sqrt{1+x^3}]$
28. $(7x^2+6x+4\sqrt{1+x})/[4\sqrt{1+x}\sqrt{1+x\sqrt{1+x}}]$
30. $(4/3)x(s-1)/[(x^2+5)^2(2s-s^2)^{2/3}]$
32. $(5/2)(2-x)/[x^2\sqrt{x-1}(u+5)^2]$
34. $[1+(2\sqrt{t}+1)/(4\sqrt{t}\sqrt{t+\sqrt{t}})][-4x/(x^2-1)^2]$
36. $(1-k-3k^2)(x^2+5)^4(11x^2+5)/[2\sqrt{k}(1+k+k^2)^2]$
38. -0.067 N/s
44. $(6u^4+6u^3-2)(2\sqrt{x+1}+1)^2/[4u^3(x+1)]-[3(u^2+u)^2+1]/[4u^2(x+1)^{3/2}]$
46. $-3[2\sqrt{x}+(s+6)(1+3x+4\sqrt{x})]/[2x^{3/2}(1+\sqrt{x})^4(s+6)^3]$
48. $-8f(3-4x)f'(3-4x)$; $8(4x-3)(16x^2-24x+7)(48x^2-72x+25)$
50. $(1-1/x^2)f'(x+1/x)$; $(x^2-1)(3x^4+4x^2+3)/x^4$
52. $12f(1-3x)f'(1-3x)/\sqrt{3-4[f(1-3x)]^2}$;
$12(1-3x)(9x^2-6x-1)(27x^2-18x+1)/\sqrt{3-4(1-3x)^2(9x^2-6x-1)^2}$
54. $f'(x-f(x))[1-f'(x)]$; $3(1-x^2)(3x^6-18x^4+27x^2-2)$

56. **58.**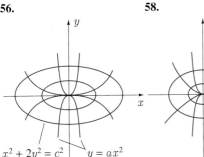

60. 444
62. $-x(x+1)^2/\sqrt{1-x^2}$
64. $(4.05,-0.164),(-4.51,-0.521)$
66. $(-4^{1/3},\pm(4^{1/3}-1)^{3/2})$

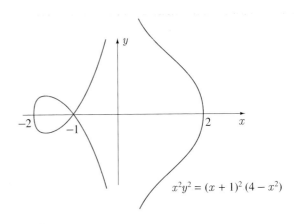

Exercises 3.7

2. $-4x^3/(2y+3y^2)$
4. $(6x^2-3y^4+5y)/(12xy^3-5x)$
6. $(1-x-y)/(x+y)$
8. $(48x\sqrt{x+y}-1)/[1+(4y-2)\sqrt{x+y}]$
10. $-(6x+5y)/(5x+2y)$
12. $-[2(3y^2+1)^2+24x^2y]/(3y^2+1)^3$
14. $-(25/4)/(y+1)^3$ **16.** -1
18. $-(2xy^3+2)/(3x^2y^2+4)$, $[-2y^3(3x^2y^2+4)^2+12xy^2(2xy^3+2)(3x^2y^2+4)-6x^2y(2xy^3+2)^2]/(3x^2y^2+4)^3$
20. $\sqrt{1-y^2}/(1-2y^2)$ **22.** $-3/4$
24. $0,-1/2$ **26.** $(1,\pm 1),(-1,\pm 1)$
28. (a) $(x^2+4x+2)/(x^2+3x+2)$ (b) $(2y+1)(3-y)/(7y)$
30. $(1,2)$
34. $(3r^3-6ar^2+x)/(3r^2-6ar)$
36. (a) -1 **42.** (a) $-\sqrt{y/x}$
44. $2xu/[(u^2-1)(1+2x^2\sqrt{u^2-1})]$

Exercises 3.8

2. $-\sin x-20\cos 5x$ **4.** $-9\sec^2 3x/\tan^4 3x$
6. $2\csc(4-2x)\cot(4-2x)$ **8.** $\cot x^2-2x^2\csc^2 x^2$
10. $[(x^2+x)\cos x+\sin x]/(x+1)^2$
12. $(3/4)\tan^2 x\sec^2 x/(1+\tan^3 x)^{3/4}$
14. $2\cos 4x$
16. $(\cos y+y\sin x)/(\cos x+x\sin y)$
18. $\sec^2(x+y)/[1-\sec^2(x+y)]$
20. $3(1-x^2y)/(x^3+2\tan y\sec^2 y)$
22. $-\sec^2 x\sin(\tan x)$ **24.** $4x\sin(2x^2)$
26. $-\sec v\tan v\sec^2\sqrt{x}/[4\sqrt{x}\sqrt{3-\sec v}]$
28. $-x\sin x^2\sec^3 u\tan u/[\sqrt{1+\cos x^2}(1+\sec^3 u)^{2/3}]$

30. $[18x^2 \tan^2(3x^2-4)\sec^2(3x^2-4) - (2+x\cot x)(1+\tan^3(3x^2-4))]/(x^3 \sin x)$
32. $2(y+1)[\sec^2 y - x - (y+1)\sec^2 y \tan y]/(\sec^2 y - x)^3$
34. 0   **36.** 2   **38.** Does not exist
40. No   **44.** $\cos x |\sin x|/\sin x,\ x \neq n\pi$

Exercises 3.9

2. $6x/(3x^2+1)$ **4.** $-2e^{1-2x}$
6. $\ln x + 1$ **8.** $-4\log_{10} e/(3-4x)$
10. $-\tan x$ **12.** $2x + (4x^3 + 3x^2)e^{4x}$
14. $2e^{2x}\cos(e^{2x})$ **16.** $e^{-2x}(3\cos 3x - 2\sin 3x)$
18. $4(e^x + e^{-x})^{-2}$ **20.** $2\sin(\ln x)$
22. $-\tan v \sin 2x$
24. $-(y\cos x + 2x \ln y + e^y)/(xe^y + \sin x + x^2/y)$
26. $2x^2/\sqrt{x^2+1}$ **28.** $(1 + 5e^{4x})^{-1}$
30. $(y^2/x^2)(e^{1/x} - 1)/(1 - e^{1/y})$

Exercises 3.10

2. $4x^{4\cos x}(\cos x/x - \sin x \ln x)$
4. $(\sin x)^x[x \cot x + \ln(\sin x)]$
6. $x(1+1/x)^{x^2}[2\ln(1+1/x) - 1/(1+x)]$
8. $(3/x^2)(2/x)^{3/x}(\ln x - \ln 2 - 1)$
10. $(1/x)(\ln x)^{\ln x}[\ln(\ln x) + 1]$
12. $(1 + 9x^2 + 6x^4)/2\sqrt{x}(1+x^2)^{3/2}]$
14. $(x^2 + 3x)^2(x^2+5)^3(14x^3 + 33x^2 + 30x + 45)$
16. $\sqrt{x}e^{-2x}(3-4x)/2$
18. $e^x[(x-1)\ln(x-1) - 1]/\{(x-1)[\ln(x-1)]^2\}$
20. $(1 - 6x^2 - 3x^4)/[2\sqrt{x}(1+x^2)^{3/2}]$
22. $x^3(13x^4 - 44x^3 + 10x - 32)/(2\sqrt{1+x^3})$
24. $\sin 2x \sec 5x[2\cot 2x + 5\tan 5x - 6\csc^2 x/(1 - 2\cot x)]/(1 - 2\cot x)^3$
26. (b) $(a - br)^{-1}$

Exercises 3.11

2. 2
4. Cannot be applied since $f'(0)$ does not exist
6. $(-2 \pm \sqrt{19})/3$
8. $1 + \sqrt{3}$
10. $-3 + \sqrt{6}$
12. $(4 - \ln 5)/(2\ln 5)$
14. Cannot be applied since $f(\pi/2)$ is not defined
16. Cannot be applied since $g'(0) = 0$
18. $(1 + \sqrt{6})/(1 - \sqrt{6})$

Review Exercises

2. $6x + 2 - 1/x^2$
4. $(1/3)x^{-2/3} - (10/9)x^{2/3}$
6. $x(x^2+2)(x^3-3)^2(13x^3 + 18x - 12)$
8. $17/(x+5)^2$
10. $-4(x^2+2)/(x^2+5x-2)^2$
12. $2x(y-1)/(2y-x^2)$
14. $-(4xy\sqrt{1+x} + y)/[2\sqrt{1+x}(x^2 + \sqrt{1+x})]$
16. $-8\sec^2(1-4x)\tan(1-4x)$
18. $2\sec^2 2x \sec(\tan 2x)\tan(\tan 2x)$
20. $(1/2)\sin 4x$
22. $(1 - 2\sin 2t)(1 + 2\sin 2x)$
24. $(-x/v)(\cos^2 v - v\sin 2v)$
26. $(1/2)e^{-2y}$
28. $2x(x^2+1)/(x^2-1) + 2x\ln(x^2-1)$
30. 1
32. $16x/(x^2+1)^2$
34. $x(20 - 23x - 3x^2)/[2\sqrt{1-x}(x+5)^2]$
36. $(x-y)/(x+y)$
38. $(y^3 + 4y + 6)^2/(-y^4 + 4y^3 + 4y^2 + 12y - 12)$
40. $1/(2\sqrt{x})$
42. $(\cos x)^x[\ln(\cos x) - x\tan x]$
44. $(\log_{10} e)^2/(x\log_{10} x)$
46. $1/(e^y - e^{-y})$
48. $[(x+y)(1-2xy) - 1]/[x^2(x+y) + 1]$
50. $x + 25y = 5,\ 125x - 5y + 1 = 0$
52. $23x + 9y = 11,\ 27x - 69y = 119$
54. $-42(x^2 - 5xy + y^2)/(5x - 2y)^3$
56. No
58. $(-3, 1)$; finding the shortest distance from $(-6, 7)$ to the curve
60. $\sqrt{3}l/2$
62. $(\sqrt{6}/4, \pm\sqrt{2}/4),\ (-\sqrt{6}/4, \pm\sqrt{2}/4)$
64. $(\sqrt{93} - 9)/6$

CHAPTER 4

Exercises 4.1

2. $-1.561553,\ 2.561553$ **4.** 3.044723
6. 0.754878 **8.** 2.485584
10. ± 0.795324
12. $-2.931137,\ -2.467518,\ -1.555365,\ -0.787653,$
    $0.056258,\ 0.642851$

14. 4.188760   **16.** 0.852606
18. 0, −1.2483   **20.** −1.892, −0.173, 3.064
22. ±1.0986   **24.** 0.3212
26. (−0.1725, 0.6848), (−1.8920, 0.7956), (3.0644, 16.5196)
28. (−1.3532, 3.8312)   **30.** (a) 3.83 s (b) 4.08 s
32. 0.0002897

Exercises 4.2

2. decreasing for all x
4. increasing for $x \leq 5/4$; decreasing for $x \geq 5/4$
6. increasing for $x \leq 1/4$; decreasing for $x \geq 1/4$
8. increasing for all x
10. decreasing for all x
12. decreasing for $x \leq 1$; increasing for $x \geq 1$
14. decreasing for $-1 \leq x \leq 1$; increasing for $x \leq -1$ and $x \geq 1$
16. decreasing for $x \leq -1$ and $0 < x \leq 1$; increasing for $-1 \leq x < 0$ and $x \geq 1$
18. decreasing for $0 \leq x < 1$ and $x > 1$; increasing for $x < -1$ and $-1 < x \leq 0$
20. decreasing for $x \leq -1$ and $0 \leq x \leq 1$; increasing for $-1 \leq x \leq 0$ and $x \geq 1$
22. decreasing for $x \leq 0$ and $x \geq 2$; increasing for $0 \leq x \leq 2$
24. decreasing for $0 < x \leq 1/e$; increasing for $x \geq 1/e$

26.

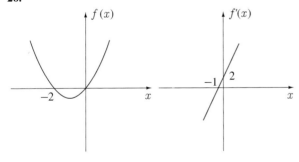

$f(x) = x^2 + 2x$ $f'(x) = 2x + 2$

28.

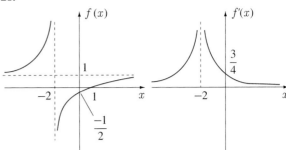

$f(x) = \dfrac{x-1}{x+2}$ $f'(x) = \dfrac{3}{(x+2)^2}$

30.

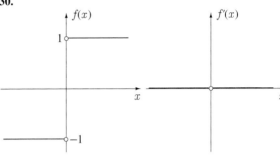

$f(x) = \dfrac{|x|}{x}$ $f'(x) = 0,\ x \neq 0$

32.

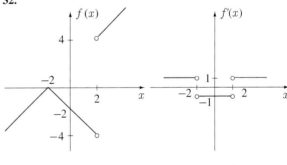

$f(x) = \dfrac{|x^2 - 4|}{x - 2}$ $f'(x) = \dfrac{|x^2 - 4|}{x^2 - 4}$

34. decreasing for $x \leq 0.8612$; increasing for $x \geq 0.8612$
36. decreasing for $-\pi \leq x \leq -2.2889$ and $2.2889 \leq x \leq \pi$; increasing for $-2.2889 \leq x \leq 2.2889$
44. Yes

Exercises 4.3

2. relative minimum at $x = -1$; relative maximum at $x = -4$

4. $x = 1$ critical, but does not give a relative extremum
6. relative maximum at $x = 1 - \sqrt{2}$; relative minimum at $x = 1 + \sqrt{2}$
8. relative minimum at $x = -3^{1/4}$; relative maximum at $x = 3^{1/4}$; $x = 0$ critical, but does not give a relative extremum
10. $x = 0$ critical, but does not give a relative extremum
12. $x = 2$ critical, but does not give a relative extremum
14. relative maximum at $x = -1$; relative minimum at $x = 1$
16. $x = 0$ critical, but does not give a relative extremum
18. relative maximum at $x = 7$; $x = 1$ critical, but does not give a relative extremum
20. relative minima at $x = 2(3n+2)\pi/3$; relative maxima at $x = 2(3n+1)\pi/3$
22. relative minima at $x = 2 + 50^{1/3}$ and $x = 0$
24. relative maximum at $x = 1$
26. relative maximum at $x = 0$
28. relative maximum at $x = -2$; relative minimum at $x = 2$
30. relative maxima at $x = (8n+1)\pi/4$; relative minima at $x = (8n+5)\pi/4$
32. relative minimum at $x = 0.464$
34. $x = 3\pi/4$ and $7\pi/4$ are critical, but do not give relative extrema
36. relative minimum at $x = 1/e$
38. relative maximum at $x = 1/2$
40. relative maximum at $x = -2/3$; relative minimum at $x = 0$
42. $x = 0$ is critical **44.** False **46.** True
48. False **50.** False
52. (a)

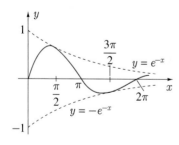

(b) $(4n+1)\pi/4$
54. relative minima at $x = 2.3301$ and $x = -2.1284$; relative maximum at $x = -0.2016$
56. relative maximum at $x = -4.1072$
60. $1/\sqrt{LC}$, E_0/R

62. $3\sqrt{3}/4$

Exercises 4.4

2. concave upward on $x \leq -2/3$ and $x \geq 0$; concave downward on $-2/3 \leq x \leq 0$; points of inflection at $(0, 2)$ and $(-2/3, 470/27)$
4. concave upward on $x < -1$ and $x > 1$; concave downward on $-1 < x < 1$
6. concave upward for all x
8. concave upward on $-2\pi < x \leq -5\pi/6$, $-\pi/6 \leq x \leq 7\pi/6$, $11\pi/6 \leq x < 2\pi$; concave downward on $-5\pi/6 \leq x \leq -\pi/6$, $7\pi/6 \leq x \leq 11\pi/6$; points of inflection at $(-5\pi/6, 2+25\pi^2/36)$, $(-\pi/6, 2+\pi^2/36)$, $(7\pi/6, 2+49\pi^2/36)$, $(11\pi/6, 2+121\pi^2/36)$
10. concave downward on $0 < x \leq e^{-3/2}$; concave upward on $x \geq e^{-3/2}$; point of inflection at $(e^{-3/2}, -3/(2e^3))$
12. concave downward on $x \leq 1$; concave upward on $x \geq 1$; point of inflection at $(1, e^{-2})$
14. concave downward on $x \leq -\ln 2$; concave upward on $x \geq -\ln 2$; point of inflection at $(-\ln 2, (\ln 2)^2 - 2)$
16. relative minimum at $x = 1$; relative maximum at $x = -1$
18. relative minumum at $x = 1/5$
20. relative minimum at $x = 1/\sqrt{e}$
22. relative minimum at $x = 0$; relative maximum at $x = 1$

26. **28.**

30.

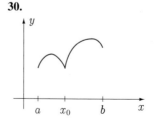

32. Not necessarily
34.(a) relative minimum at $x = 0$; horizontal points of inflection at $x = \pm 1$ **(b)** relative minimum at $x = 0$; relative maximum at $x = 4/5$

Exercises 4.5

2.

$y = x^3 - 6x^2 + 12x + 9$

4.

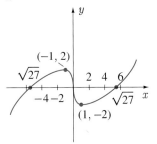

$y = x - 3x^{1/3}$

6.

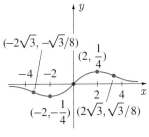

$y = \dfrac{x}{x^2 + 4}$

8.

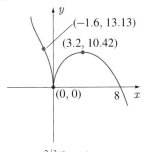

$y = x^{2/3}(8 - x)$

10.

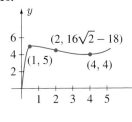

$y = 2x^{3/2} - 9x + 12x^{1/2}$

12.

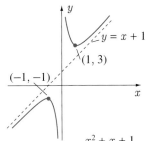

$y = \dfrac{x^2 + x + 1}{x}$

14.

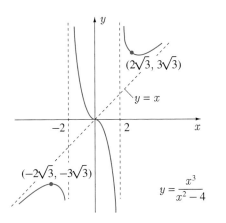

$y = \dfrac{x^3}{x^2 - 4}$

16.

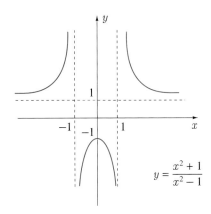

$y = \dfrac{x^2 + 1}{x^2 - 1}$

18.

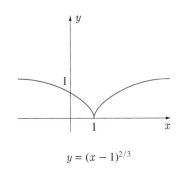

$y = (x - 1)^{2/3}$

20.

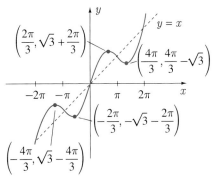

$y = x + 2 \sin x$

22.

$$P = \frac{MI}{I+K} - R$$

24.

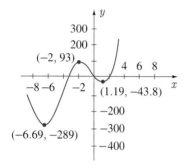

$$y = x^4 + 10x^3 + 6x^2 - 64x + 5$$

26.

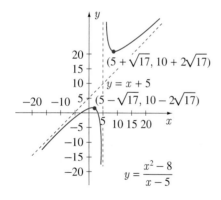

$$y = \frac{x^2 - 8}{x - 5}$$

28.

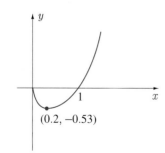

$$y = x^{5/4} - x^{1/4}$$

30.

32.

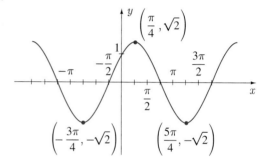

$$y = \sin x + \cos x$$

34.

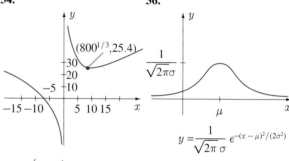

$$y = \left(\frac{x+8}{x}\right)\sqrt{x^2 + 100}$$

36.

$$y = \frac{1}{\sqrt{2\pi}\,\sigma} e^{-(x-\mu)^2/(2\sigma^2)}$$

38.

$$y = \frac{1 + x + x^2 + x^3}{1 + x^3}$$

Exercises 4.6

2. $-4, 6/11$ 4. $\pi/3 - \sqrt{3}, 11\pi/3 + \sqrt{3}$
6. no absolute minimum, 12
8. no absolute minimum, $\sqrt{3}/6$
12. $720\sqrt{10}$ m   14. $12^{1/3} \times 12^{1/3} \times 12^{1/3}/2$ m
16. $90 \times 180/\pi$ m
18. $ab/4$ 20. $a/\sqrt{3}$
22. 15 cm $\times$ 22.5 cm
24. $(-1 - \sqrt{6}/2, 5/2 + \sqrt{6})$
26. smallest area is 20
28. width $= 50/\sqrt{3}$ cm, depth $= 50\sqrt{2/3}$ cm
30. (a) $7/4$ km from P (b) directly to Q
32. lengths in x- and y-directions are $\sqrt{2}\,a$ and $\sqrt{2}\,b$
34. $\sqrt{2/e}$ 36. $4\pi h r^2/27$ 38. 11.8 m
40. $[p\beta/(r\alpha - r)]^{1/\alpha}$
42. $dI_1^{1/3}/(I_1^{1/3} + I_2^{1/3})$ from I_1 source
44. (a) width $= 5/3^{1/3}$ m, height $= 4(3^{2/3})$ m
    (b) width $= 4.19$ m, height $= 5.70$ m
48. $l = 1/3$ m, $w = h = 2/3$ m; $1/3$ m apart
50. $2\pi(1 - \sqrt{6}/3)$ 54. $(2, 2\sqrt{5}/3)$
56. $x = -(A + bB)/(2aB)$ 60. $3a/4$

Exercises 4.7

2. (a) right, left (b) left (c) 5 (d) 4 (e) right, left (f) twice
   (g) $9/2$ (h) $9/2$ (i) negative, positive (j) once
4. $2t - 7$ m/s, 2 m/s^2
6. $-6t^2 + 4t + 16$ m/s, $-12t + 4$ m/s^2
8. $-12\sin 4t$ m/s, $-48\cos 4t$ m/s^2
10. $1 - 4/t^2$ m/s, $8/t^3$ m/s^2
12. $(4t - 5t^2)/(2\sqrt{1-t})$ m/s, $(1/4)(15t^2 - 24t + 8)/(1-t)^{3/2}$ m/s^2
14. (a) $3t^2 - 18t + 15$ m/s, $6t - 18$ m/s^2 (b) $t = 1, 5$ (c) slowing down (d) 2.18 s, 3.82 s (e) increasing
16. No
18.

20.

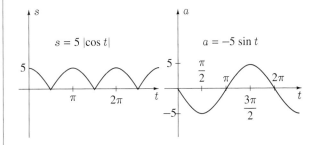

22. Not always   26. Not always   28. Not always

Exercises 4.8

2. $2\sqrt{13}/3$ m/s   4. $13/(5\pi)$ cm/s
6. decreasing at 20 m^2/s   8. -2.5 N/m^2/s
10. $7/\sqrt{5}$ m/s
12. (a) $2/(25\pi)$ m/min (b) $2/(25\pi)$ m/min
14. (a) $10(50 - x)/y$ m/s (b) $10/\sqrt{3}$ m/s (c) Car 1
16. -27.14 cm/s   18. 115.8 cm/s
20. $-31\sqrt{15}/10$ 22. 102 m/s
24. $6/\sqrt{13}$ m/s
26. (a) $1/(50\pi)$ m/min (b) $\sqrt{13}/(300\pi)$ m/min
28. $4\pi x l(k - x)\sin\theta/[y(l\cos\theta - x)]$ m/s
30. (a) $\pm 11\pi/360$ rad/min (b) $-11\pi/60$ cm/min (c) 0.23 cm/min

Exercises 4.9

2. 6   4. 0   6. ∞
8. 0   10. Does not exist   12. ∞

14. $1/\sqrt{5}$ **16.** na^{n-1} **18.** 1
20. 0   **22.** $1/12$ **24.** 2^{15}
26. 0   **28.** 4   **30.** 0
32. 1   **34.** e **36.** 1
38. Does not exist   **40.** 2   **42.** $-1/3$

44.

$y = x^2 e^{3x}$

46.

$y = e^{1/x}$

48.

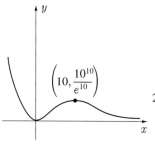

$y = x^2 \ln x$

50.

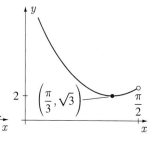

$y = \dfrac{x^2}{\ln x}$

52.

$y = x^{10} e^{-x}$

54.

$y = 2\csc x - \cot x$

56. ∞ **58.** (b) 0.0002897
60. $a = b = \pm 7$, c arbitrary

Exercises 4.10

2. $-2\,dx/(x-1)^2$
4. $[2x\cos(x^2+2) + \sin x]\,dx$

6. $x^2(9 - 16x^2)\,dx/\sqrt{3-4x^2}$
8. $[1 - 12/(x-1)^3]\,dx$
10. $2(x^2 + 5x - 5)\,dx/(x^2+5)^2$
12. -4% **14.** 3250 m

Review Exercises

2. (a)   (b)

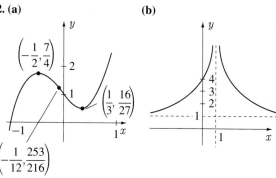

$y = 4x^3 + x^2 - 2x + 1$ $y = \dfrac{x^2 - 2x + 4}{x^2 - 2x + 1}$

4. No solution   **6.** No solution   **8.** $-3/2$
10. 8   **12.** 0   **14.** 0
16. 0   **18.** e^2
20. (a) 0.312908   (b) 1.051888   **24.** $2l^3/27$

26. (a)   (b)

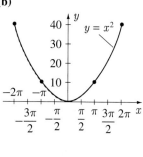

$y = \dfrac{x^3}{x^2 - 1}$ $y = x^2 + \sin^2 x$

28. $67\,t$ **30.** $4\sqrt{5}$ m/s
32. $x = 100bq/(ap+bq)$ hectares, $y = 100ap/(ap+bq)$ hectares

CHAPTER 5

Exercises 5.1

2. $x^5/5 + x^3 + 5x^2/2 + C$
4. $-\cos x + C$
6. $(2/3)x^{3/2} + C$
8. $-1/x + 2/(3x^3) + C$
10. $-1/x + \sqrt{x} + C$
12. $1/(2x) + (3/4)x^4 + C$
14. $(4/3)x^{3/2} + (6/5)x^{5/2} - (10/7)x^{7/2} + C$
16. $(2/5)x^{5/2} + (2/3)x^{3/2} + C$
18. $x^3/3 + 2x^5/5 + x^7/7 + C$
20. $(2/5)x^{5/2} - (4/3)x^{3/2} + 2\sqrt{x} + C$
22. $y = x^4/2 + 2x^2 + 5$
24. $y = 2x - 2x^2 + x^8$
26. $f(x) = -5x^3/6 + 10x - 31/3$
28. $(2/3)(x + 2)^{3/2} + C$
30. $-(2/3)(2 - x)^{3/2} + C$
32. $(1/5)(2x - 3)^{5/2} + C$
34. $-(1/16)(1 - 2x)^8 + C$
36. $-(1/15)/(1 + 3x)^5 + C$
38. $(1/72)(2 + 3x^3)^8 + C$
40. $(1/2) \sin 2x + C$
42. $(3/4) \sin^2 2x + C$
44. $-(1/4) \cot 4x + C$
46. $-(1/2)e^{-x^2} + C$
48. $(1/4)e^{4x-3} + C$
50. $-(2/5) \ln |7 - 5x| + C$
52. $-(1/4) \ln |1 - 4x^3| + C$
54. $(1/2) 3^{2x} \log_3 e + C$
56. $-(1/5)(1 + \cos x)^5 + C$
58. $(1/8)(1 + e^{2x})^4 + C$
60. $y = x^4/4 + 1/x + C$
62. $-(2/3)/(3x + 5)^{1/2} + C$
64. $-(1/12)/(2 + 3x^4) + C$
66. $y = -3 - 1/x, x < 0; y = 2 - 1/x, x > 0$
68. $C, x < a; x + D, x > a$

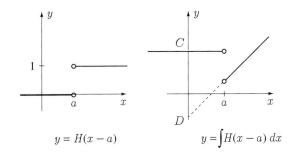

$y = H(x - a)$ $y = \int H(x - a)\,dx$

Exercises 5.2

2. $6t - t^2 + 5,\ 3t^2 - t^3/3 + 5t$
4. $60t^2 - 4t^3,\ 20t^3 - t^4 + 4$
6. $t^3/3 + 5t^2/2 + 4t - 2,\ t^4/12 + 5t^3/6 + 2t^2 - 2t - 3$
8. $4 - 3 \cos t,\ 4 + 4t - 3 \sin t$
10. (a) $t^3 - t^2 - 3t + 1$ (b) $(\sqrt{10} + 1)/3$
12. (a) $350/3$ m (b) $650/3$ m, 20 s
14. 5.1 m
16. $|v| \geq 19.8$ m/s

18.

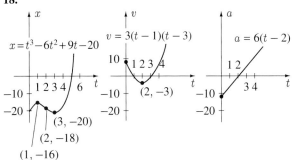

20. Less than or equal to 47.8 km/hr; defense
22.(a) Yes (b) No

Exercises 5.3

2. $-(1/3)(1 - 2x)^{3/2} + C$
4. $-(10/63)(5 - 42x)^{3/4} + C$
6. $-(1/2)/(x^2 + 4) + C$
8. $-1/(x - 2) - 2/(x - 2)^2 - (4/3)/(x - 2)^3 + C$
10. $(1/6)(2x + 3)^{3/2} - (3/2)\sqrt{2x + 3} + C$
12. $(1/5)(s^2 + 5)^{5/2} - (5/3)(s^2 + 5)^{3/2} + C$
14. $(2/3)(1 - \cos x)^{3/2} + C$
16. $(2/7)(y - 4)^{7/2} + (16/5)(y - 4)^{5/2}$
    $+ (32/3)(y - 4)^{3/2} + C$
18. $(1/729)(3x^3 - 5)^9 + (5/324)(3x^3 - 5)^8 +$
    $(25/567)(3x^3 - 5)^7 + C$
20. $(3/4)(x^2 + 2x + 2)^{2/3} + C$
22. $(7/192)/(3 - 4 \sin x)^3 + (3/64)/(3 - 4 \sin x)^2 -$
    $(1/64)/(3 - 4 \sin x) + C$
24. $(4/5)(1 + \sqrt{x})^{5/2} - (4/3)(1 + \sqrt{x})^{3/2} + C$
26. $(1/2) \tan^2 x + C$
28. $(\ln x)^2/2 + C$
30. $-(1/2)/\ln(x^2 + 1) + C$
32. $(2/3)(1 + x)^{3/2} - 2\sqrt{1 + x} + C$ if $x \geq 0$;
    $-(2/3)(1 + x)^{3/2} + 2\sqrt{1 + x} + C$ if $-1 < x \leq 0$
34. $-(2/5)(1/x - 1)^{5/2} - (2/3)(1/x - 1)^{3/2} + C$
36. $-2/[\sqrt{(1 - x)/(1 + x)} - 3] + C$

Exercises 5.4

2. (a) $(24EI)^{-1}gx^2[4M(3L - x) + m(6L^2 - 4Lx + x^2)]$
   (b) $(24EI)^{-1}gL^3(8M + 3mL)$
4. $x^2 + (y - 2)^2 = 13$
6. $\pi R^2 \sqrt{2H}/(5cA\sqrt{g})$
8. $-\sqrt{2GMh/(R^2 + Rh)},\ \sqrt{2GM/R}$
10. $y = Cx^k$

ANSWERS TO EVEN-NUMBERED EXERCISES

12. (a) $(192\,EI)^{-1}[8mgx^4 + 32Ax^3 - (8AL^2 + mgL^3)x]$
(b) $-3mgL/16$
14. $(1 + e^{-t/20})/20$ m^3, $1/20$ m^3

Review Exercises

2. $-(1/4)/x^4 + x^2 + (1/2)/x^2 + C$
4. $-1/x - (4/3)x^{3/2} + C$
6. $(1/30)(1 + 3x^2)^5 + C$
8. $(2/5)x^{5/2} + 10\sqrt{x} + C$
10. $-2/\sqrt{x} - 30\sqrt{x} + C$
12. $-(1/3)(1 - x^2)^{3/2} + C$
14. $x^3/3 - (4/5)x^5 + (4/7)x^7 + C$
16. $(2/3)(2 - x)^{3/2} - 4\sqrt{2 - x} + C$
18. $4x + (8/3)x^{3/2} + x^2/2 + C$
20. $(1/5)\sin^5 x + C$
22. $-(1/8)e^{-4x^2} + C$
24. $(1/5)\ln|\ln x| + C$
26. $y = x^4 + 7x/2 - 1/2$
28. $[(\sqrt{3} - \sqrt{6})t + \sqrt{6}]^2$ km, $\sqrt{6}/(\sqrt{6} - \sqrt{3})$ hours
30. 11.9 m/s
32. $(2/3)(1 + x)^{3/2} - x + C$
34. $-(3/56)(3 - 2x^3)^7 + (1/32)(3 - 2x^3)^8 - (1/216)(3 - 2x^3)^9 + C$
36. $(1/4)\sin^4 x - (1/6)\sin^6 x + C$
38. $-\ln|\cos x| + C$
40. $y = 2x/(1 + x)$

CHAPTER 6

Exercises 6.1

2. $\sum_{k=1}^{10} k/2^k$
4. $\sum_{k=1}^{14\,641} \sqrt{k}$
6. $\sum_{k=1}^{1019} (-1)^k(k + 1)$
8. $\sum_{k=1}^{225} (\tan k)/(1 + k^2)$
10. $\sum_{k=1}^{9} (10^k - 1)/10^k$
16. 258
18. $4n(4n^2 - 1)/3$
20. 5350
24. $f(n) - f(0)$
26. No
28. $(1 - 2^{-18})/4$
30. $3960[1 - (0.99)^{15}]$

Exercises 6.3

2. 6
4. 4
6. 3/2
8. 0
10. (c) $\log_2 e$
12. 2

Exercises 6.4

2. 20/3
4. 2/3
6. 1
8. $-7/2$
10. 3/4
12. 0
14. $5/16 - 1/(\pi + 1)$
16. 125/12
18. $-65/4$
20. $3/\sqrt{2}$
22. $-1/2$
24. 0
26. $(e^3 - 1)/e$
28. $\ln(2/3)$
30. $20\log_3 e$
32. 88/3
34. $-1/6$
36. $\sqrt{2}\pi/8$, 0
38. $\pi/2$, 0
40. 5, $\cos 1$

Exercises 6.5

2. $1/\sqrt{x^2 + 1}$
4. $-x^3 \cos x$
6. $2\sqrt{2x + 1}$
8. $5\sqrt{(5x + 4)^3 + 1}$
10. $256(3x^2 + 3x + 1) + 2/\sqrt{x} - 2/\sqrt{x + 1}$
12. $2x[-2\sec(1 + 2x^2) + \sec(1 + x^2)]$
14. $\cos x/\sqrt{\sin x + 1} + \sin x/\sqrt{\cos x + 1}$
16. $(2\sqrt{2} - 1)/(2x^{1/4})$
18. $-\ln(x^2 + 1)$
20. $3e^{-36x^2} + 2e^{-16x^2}$

Exercises 6.6

2. 0
4. 31/5
6. $2\sqrt{2}/3$
8. 11/5
10. $2/\pi$
12. 1
14. 23/9
16. 1/2
18. 0
20. 9/7
22. $2cR^2/3$
24. 0
26. $\pi/2$
28. 0.6032
30. 1.400

Exercises 6.7

2. 4/15
4. 7/384
6. Does not exist
8. $\sqrt{3/2} - 2\sqrt{2}/3$
10. $3(5^{2/3} - 2^{2/3})/4$
12. $\sqrt{2}(7\sqrt{7} - 8)/18$
14. 52/5
16. $(16 - 9\sqrt{2})/6$
18. 5/18
20. $(1/3)(\ln 2)^3$
22. $2(\sqrt{7} - 1)/3$
24. $1/\sqrt{3}$
26. $-8\sqrt{5}/45$

Review Exercises

2. 4/15
4. $-4/3$
6. 1/6
8. 0
10. 66/5
12. $16\sqrt{6}$
14. $-4/15$
16. $(2112\sqrt{6} - 704)/105$
18. 120 852.5
20. 20
24. 2
26. $4/(15\pi)$

28. $-x^2(x+1)^3$
30. $-4x\cos 2x$
32. $4x\sin^2 x^2$
34. $(33\sqrt{6} - 56)/60$
36. $(24\sqrt{3} - 14\sqrt{2} - 8\sqrt{5})/3$
38. $24\sqrt{5} - 8$
40. $(3\sqrt{3} - 2\sqrt{2} - 1)/3$

CHAPTER 7

Exercises 7.1

2. 8
4. $343/6$
6. $(e^6 - e^3)/3 + 3/2$
8. $8/3$
10. 2
12. $20/3$
14. $4\sqrt{2}/3 - e\ln 2$
16. $10/3$
18.(a) 22 **(b)** 37, 15 **(c)** 29, 7 **(d)** $203/8, 27/8$
20. $4\int_1^2 \sqrt{4-x^2}\,dx$
22. $2\int_0^{\sqrt{(\sqrt{65}-1)/2}} (\sqrt{16-y^2} - y^2)\,dy$
24. $235/3$
26. $7/6$
28. $1 - \ln 2$
30. $4\sqrt{e} - 16/3$
32. $32/3$
34. $14/\sqrt{21}$
36. 2.182
38. 8.436
40. 2.067
42. 7.177
44. $0 < m < 1, (m - 1 - \ln m)/3$

Exercises 7.2

2. $80\sqrt{5}\pi/3$
4. $5888\pi/15$
6. π
8. $1088\pi/15$
10. $32\pi/3$
12. $\pi(e^2 - 4e + 5)/2$
14. 8π
16. $14\pi/3$
18. $40\pi/3$
20. $\pi/3$
22. $344\pi/3$
24. $1472\pi/15$
26. $775\pi/6$
28. $272\pi/15$
30. $68\pi/9$
32. $16\pi(13 + 2^{17/4} - 3^{9/4})/45$
34. 111.303
36. 21.186
38. $4\pi/3$
40.(a) $x^2(6-x)/16$ **(b)** $1161\pi/10$ m^3
44. $2\pi R^3(1 - H/\sqrt{r^2 + H^2})/3$

Exercises 7.3

2. $17/3$
4. $(85^{3/2} - 13^{3/2})/27$
6. $(e - e^{-1})/2$
8. $23/18$
10. $3011/480$
12. $\int_1^2 \sqrt{36x^2 - 48x + 17}\,dx$
14. $\int_{-\sqrt{3}}^{2\sqrt{2}} \sqrt{(1+2y^2)/(1+y^2)}\,dy$
16. $\int_0^{\pi/4} \sec x\,dx$
18. $\sqrt{2}\int_0^1 (3-2x^2)/\sqrt{1-x^2}\,dx$
20. $2\int_0^2 \sqrt{(16+5x^2)/(4-x^2)}\,dx$
22. 6
24. $(b^{2n+1} - a^{2n+1})/[4(2n-1)] + (1/a^{2n-1} - 1/b^{2n-1})/(2n+1)$

Exercises 7.4

2. $8/15$ J
4. 8829 J
6. 49.05 J
8. 184 J
10. 1.78×10^7 J
12. 114 450 J
14. 7.60×10^6 J
16. $q_3(q_1 - 8q_2)/(48\pi\epsilon_0)$
18. kb^2 J
20. 3.22×10^5 J
22. (b) 9.8087×10^5 J **(c)** 9.82×10^5 J
24. $C\ln 2$

Exercises 7.5

2. 6.78×10^{10} N
4. 9.25×10^3 N
6. 7.60×10^5 N, 4.41×10^5 N, 4.91×10^4 N
12. 2.943×10^6 N, 2.969×10^6 N
14. 1.60 m × 1.60 m
20.(b) 21.65 cm
22. 0.52

Exercises 7.6

2. $(3a/8, 3h/5)$
4. $(4r/(3\pi), 4r/(3\pi))$
6. $(-1/2, -1/2)$
8. $(0, 0)$
10. $(33/5, 1/2)$
12. $(177/85, 9/17)$
14. $25\rho/3$
16. $(M + 2\rho L)^{-1}(\sum_{i=1}^{6} m_i x_i + 2\rho L^2)$
18. $(0, 192/205)$
20. On the axis of symmetry, $1637/350$ from smaller end
24. $(12/5, -3/5)$
26. $\bar{x} = -2/(8\sqrt{3} - 1), \bar{y} = 0$
28. $(0, 1.944)$
30. $(0.684, 0.319)$
32. $2\pi^2 ab^2$

Exercises 7.7

2. $32\rho/3$
4. $603\rho/10$
6. $2\rho/15$
8. $344\rho/105$
10. $1143\rho/20$
12. $\rho h(y_2^3 + y_2^2 y_1 - y_1^2 y_2 - y_1^3)/4$; no
16. $\tilde{x} = \bar{x}$
18. 0.680
20. 0.519

Exercises 7.8

2. $\pi r\sqrt{r^2 + h^2}$
4. $b^2 h/3$
6. (a) 20 000 **(b)** 7000
8. $31\,250/\sqrt{3}$ cm^3
10. $\pi/2$
12. $r^3(3\pi - 4)/6$

14. 15   **16.** 16.5
18. 140 m^3 **20.** 50 days

Exercises 7.9

2. Diverges   **4.** $-1/16\,464$ **6.** Diverges
8. 2   **10.** $\sqrt{21}$ **12.** Diverges
14. Diverges   **16.** Diverges   **18.** $17\sqrt{5}/3$
20. (a) No (b) No   **22.** (a) 1/2 (b) $\pi/5$
28. No   **30.** Diverges   **32.** Converges

Review Exercises

2. (a) 5/4 (b) $46\pi/21$, $14\pi/15$ (c) $(28/75, 92/105)$
 (d) 73/60, 1/4
4. (a) 1/2 (b) $\pi/3$, π (c) $(1, 1/3)$ (d) 1/12, 7/12
6. $10\pi/3$ **8.** $16\pi/105$ **10.** $43\pi/12$
12. $-5/6$, 3/5   **14.** $-28/3$, 14   **16.** 1.04×10^5 J
18. $4\sqrt{5}\pi$ **22.** $2\sqrt{3}$ **24.** 0
26. Diverges   **28.** Diverges   **30.** 6.58×10^4 J

CHAPTER 8

Exercises 8.1

2.

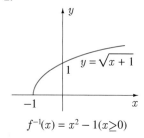

$f^{-1}(x) = x^2 - 1 \,(x \geq 0)$

4.

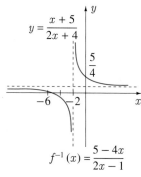

$f^{-1}(x) = \dfrac{5 - 4x}{2x - 1}$

6.

$f^{-1}(x) = \left(\dfrac{x-2}{3}\right)^{1/3}$

8.

$f^{-1}(x) = \begin{cases} x, & x<0 \\ \dfrac{x}{3}, & x\geq 0 \end{cases}$

10.

No inverse function $f^{-1}(x) = 1 + \sqrt{x - 3}$

14.

No inverse function

16. $-x^{-1/4}$, $x < 0$; $x^{-1/4}$, $x > 0$
18. $-\sqrt{\sqrt{2+x} - 2}$, $x < 0$; $\sqrt{\sqrt{2+x} - 2}$, $x \geq 0$
20. $-\sqrt{(x + \sqrt{x^2 + 16x})/2}$, $x < 0$;
 $\sqrt{(x + \sqrt{x^2 + 16x})/2}$, $x \geq 0$

22.

Exercises 8.2

2. 0.253   **4.** $-2\pi/3$ **6.** Does not exist
8. $-\pi/4$ **10.** Does not exist   **12.** 0.164
14. $\pi/4$ **16.** $-\pi/4$ **18.** $-0.876 + n\pi$
20. No solutions   **22.** No solutions

24.   **26.**

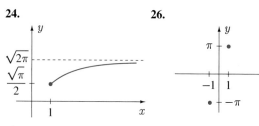

$y = \sqrt{\text{Tan}^{-1} x} + \sqrt{\text{Sec}^{-1} x}$ $y = \text{Sin}^{-1} x + \text{Csc}^{-1} x$

28. $(3n \pm 1)\pi/3$ **30.** No solutions
32. $\text{Tan}^{-1}(2n\pi \pm 3\pi/4) + m\pi$

34. (a)   **(b)**

$y = \sin(\text{Sin}^{-1} x)$ $y = \text{Sin}^{-1}(\sin x)$

36. $\sqrt{17}\sin(2x + 0.24)$ **38.** $2\sqrt{5}\sin(2x + 2.03)$
44. $E_0/\sqrt{R^2 + [\omega L - 1/(\omega C)^2]^2}\cos(\omega t + \phi)$ where
$\phi = \text{Tan}^{-1}\{[(\omega L - 1/(\omega C)]/R\}$

Exercises 8.3

2. $-2x/[1 + (x^2 + 2)^2]$ **4.** $-2x/[1 + (2 - x^2)^2]$
6. $\text{Csc}^{-1}(x^2 + 5) - 2x^2/[(x^2 + 5)\sqrt{(x^2 + 5)^2 - 1}]$
8. $1/[2(x + 3)\sqrt{x + 2}]$ **10.** $-1/(x\sqrt{x^2 - 1})$
12. $2x\text{Sec}^{-1} x + x/\sqrt{x^2 - 1}$
14. $-1/(1 + x^2)$ **16.** $-1/[\sqrt{x}(x + 1)]$
18. $x(t - \sqrt{1 - t^2}\text{Cos}^{-1} t)/[\sqrt{1 - x^2}\sqrt{1 - t^2}]$
20. $(5\sqrt{1 - x^2 y^2} - y)/(x - 2\sqrt{1 - x^2 y^2})$
22. $-18/(x^3\sqrt{x^2 - 9})$
24. $-(1/x^2)\text{Csc}^{-1}(3x) - 2/(x^2\sqrt{9x^2 - 1})$
26. $(\text{Cos}^{-1} x)^2 - 2 + 2x(1 - \text{Cos}^{-1} x)/\sqrt{1 - x^2}$
28. $2\sqrt{4x - x^2}$
30. $\sqrt{2}(1 - x^2)/[(1 + x^2)\sqrt{1 + x^4}]$

32.

$y = x\text{Sin}^{-1} x + \sqrt{1 - x^2}$

34. 1.23   **36.** $\text{Tan}^{-1}\mu$
38.(a) 2/15 rad/s   **(b)** 0.0369 rad/s

Exercises 8.4

2. 2.30   **4.** 9.08×10^{-5} **6.** 1.00
8. 0.266   **10.** 5.07

12. $\sinh(x/2) + (x/2)\cosh(x/2)$
14. $(1/x)\text{sech}^2(\ln x)$
16. $2\sqrt{1 + y}\text{csch}^2 x/(2\sqrt{1 + y} - 1)$
18. $-\sinh t\,\text{sech}^2 x\,\sec^2(\cosh t)\sin(\tanh x)$
20. $2\,\text{csch}\,4x$ **26.(a)** 189.3 N **(b)** 17.0 m
28.(b) $\sqrt{mg/\beta}$

Review Exercises

2. $(4n + 1)\pi/4 \pm 0.659$ **4.** -1.79
6. $(4n + 1)\pi/2 \pm 0.604$
8. $\ln[(4n + 1)\pi/2 \pm 0.84] - 2$, $n \geq 0$
10. 2.09   **12.** $6x\cosh x^2$
14. $(x^2 - 1)/(x^4 + 3x^2 + 1)$ **16.** $\cos x\,\text{sech}\,y$
18. $\text{Csc}^{-1}(1/x^2) + 2x^2/\sqrt{1 - x^4}$
20. -1

CHAPTER 9

Exercises 9.1

2. $-(1/4)e^{-2x^2} + C$ **4.** $\ln(1 + e^x) + C$
6. $-(2/27)(1 - 3x^3)^{3/2} + C$
8. $-(1/6)(1 + x^3)^{-2} + C$
10. $2\sqrt{x} - x + C$ **12.** $\text{Tan}^{-1} x + x + C$
14. $(1/6)(2x + 4)^{3/2} + \sqrt{2x + 4} + C$
16. $(1/8)\sin^4 2x + C$
18. $(1/2)(x + 5)^2 - 15x + 75\ln|x + 5| + 125/(x + 5) + C$
20. $(1/3)\ln|\sec 3x| + C$
22. $(4/5)x^{5/4} - x + (4/3)x^{3/4} - 2\sqrt{x} + 4x^{1/4} -$
$4\ln(x^{1/4} + 1) + C$
24. $2\ln(e^x + 1) - x + C$ **26.** $(35 - 12\ln 6)/6$
30.(a) $1962(1 - e^{-t/200})$ m/s
(b) $1962t + 392\,400(e^{-t/200} - 1)$ m
32. $(1/(3n))\ln|x^n/(3 + 2x^n)| + C$
34. $(2/7)(2x + 3)/\sqrt{x^2 + 3x + 4} + C$
36. $(1/b^2)[bx - a\ln|a + bx|] + C$, $n = -1$;
$(1/b^2)[\ln|a + bx| + a/(a + bx)] + C$, $n = -2$;
$(1/b^2)[(a + bx)^{n+2}/(n + 2) - a(a + bx)^{n+1}/(n + 1)] + C$,
$n \neq -2, -1$
38. $e^{-|a|}(1 + |a|)$

Exercises 9.2

2. 4.75 hr   **4.** 19.74 s   **6.** $k\ln(S/S_0)$
8. 13.51 hr

10. (a) Finitely long (b) Infinitely long

12. 33 574 years ago

14.
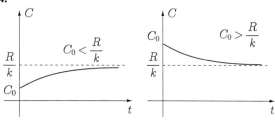

$$C = \frac{R}{k}(1 - e^{-kt}) + C_0 e^{-kt}$$

16. $20 + 70 e^{-0.01399 t}$

18.

$N = Ae^{-be^{-kt}}$

20.

$N = N_0 (1 - be^{-kt})^3$

22.
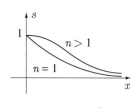

$s = 1 - (1 - e^{-kx})^n$

24. 3:25 A.M.

Exercises 9.3

2. $(x^2/2 - x/2 + 1/4) e^{2x} + C$
4. $(2/3) x^{3/2} \ln(2x) - (4/9) x^{3/2} + C$
6. $-(2/5)(2 + x)(3 - x)^{3/2} + C$
8. $(2x^2/3)(x + 5)^{3/2} - (8x/15)(x + 5)^{5/2} + (16/105)(x + 5)^{7/2} + C$
10. $2x^2 \sqrt{x + 2} - (8x/3)(x + 2)^{3/2} + (16/15)(x + 2)^{5/2} + C$
12. $(1/3)(x - 1)^3 \ln x - x^3/9 + x^2/2 - x + (1/3) \ln x + C$
14. $x \tan^{-1} x - (1/2) \ln(1 + x^2) + C$
16. $e^{2x}(2 \cos 3x + 3 \sin 3x)/13 + C$
18. $x \ln(x^2 + 4) - 2x + 4 \tan^{-1}(x/2) + C$

20. 5.92×10^9 N
26. $x^3/6 + (x^2/4 - 1/8) \sin 2x + (x/4) \cos 2x + C$

Exercises 9.4

2. $-(1/2) \sin^{-2} x + C$ 4. $-(1/3) \csc^3 x + C$
6. $(2/3) \tan^{3/2} x + (2/7) \tan^{7/2} x + C$
8. $(1/18) \sec^6 3x + C$ 10. $(1/4) \sec^4 x + C$
12. $\tan \theta + C$ 14. $2 \sqrt{1 + \tan x} + C$
16. $7 \tan x - 3x + C$
18. $(1/32)(12 x - 8 \sin 2x + \sin 4x) + C$
20. $(1/2) \tan^2 x + \ln |\tan x| + C$
22. $\pi(1 - \pi/4 + \ln 2)$
24. $-(1/3) \cot^3 z + \cot z + z + C$
26. $\theta/2 + (1/4) \sin 2\theta + (1/3) \cos^3 \theta + C$
28. $(1/8) \sin 4x + (1/16) \sin 8x + C$
30. $\sec x + \ln |\csc x - \cot x| + C$

Exercises 9.5

2. $(1/\sqrt{5}) \operatorname{Sin}^{-1}(\sqrt{5} x/3) + C$
4. $-\sqrt{4 - x^2}/(4x) + C$
6. $(1/15)(5 x^2 + 3)^{3/2} + C$
8. $(1/2) \ln |(1 + x)/(1 - x)| + C$
10. $(1/20) \ln(5 x^2 + 1) + (\sqrt{5}/2) \operatorname{Tan}^{-1}(\sqrt{5} x) + C$
12. $\sqrt{4 - x^2} + 2 \ln |(2 - \sqrt{4 - x^2})/x| + C$
14. $\ln |x + \sqrt{x^2 - 16}| - \sqrt{x^2 - 16}/x + C$
16. $(1/16) \operatorname{Sec}^{-1}(x/2) + \sqrt{x^2 - 4}/(8 x^2) + C$
18. $(1/3)(y^2 + 4)^{3/2} - 4\sqrt{y^2 + 4} + C$
20. $2 \ln |x| - (1/2) \ln(1 + x^2) + C$
24. $L \ln[(L + \sqrt{L^2 - x^2})/x] - \sqrt{L^2 - x^2}$
26. 1.053   28. $0.404 b$ 30. $9.81 \pi \rho r^3$
32. $(147/8) \operatorname{Sin}^{-1}(x/\sqrt{7}) + (x/8)(35 - 2 x^2) \sqrt{7 - x^2} + C$
34. $-6 \operatorname{Sec}^{-1}(2 x) - (12 x^2 - 1)/(2 x^2 \sqrt{4 x^2 - 1}) + C$
36. $(1/2)[x\sqrt{1 + 3 x^2} + (1/\sqrt{3}) \ln |\sqrt{3} x + \sqrt{1 + 3 x^2}|] + C$
38. $\sqrt{2} \pi$ 40. $r/2$ 42. $a^2 (4 - \pi)/2$

Exercises 9.6

2. $\ln |\sqrt{x^2 + 2 x + 2} + x + 1| + C$
4. $-(2/\sqrt{7}) \operatorname{Tan}^{-1}[(2 x - 3)/\sqrt{7}] + C$
6. $(x/2)/\sqrt{4x - x^2} + C$
8. $\ln(x^2 + 6x + 13) - (9/2) \operatorname{Tan}^{-1}[(x + 3)/2] + C$
10. $\ln |\sqrt{6 + 4 \ln x + (\ln x)^2} + \ln x + 2| + C$
12. $39\,240 \pi$ N
14. $(1/2)(x - 1) \sqrt{x^2 - 2x - 3} - 2 \ln |x - 1 + \sqrt{x^2 - 2x - 3}| + C$

16. $(1/2)\text{Tan}^{-1}[(\sqrt{2x-3}+2)/2]+C$

Exercises 9.7

2. $-(1/2)(y+1)^{-2}+C$
4. $x+(10/3)\ln|x-4|+(2/3)\ln|x+2|+C$
6. $-(1/6)\ln|y|-(2/15)\ln|y+3|+(3/10)\ln|y-2|+C$
8. $(1/2)\ln(x^2+2)+(x^2+2)^{-1}+C$
10. $y-4\ln|y+2|+\ln|y+1|+C$
12. $\ln(y^2+1)+\text{Tan}^{-1}y-\ln(y^2+4)+C$
14. $-[1/(3\sqrt{2})]\text{Tan}^{-1}(x/\sqrt{2})+(2/3)\ln|(x-1)/(x+1)|+C$
16. $-(5/27)\ln|x-1|-(7/9)(x-1)^{-1}+(32/27)\ln|x+2|+(2/9)(x+2)^{-1}+C$

18.

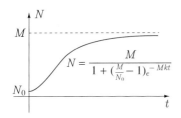

20.(b) $60(1-e^{10kt})/(2-3e^{10kt})$ gm
22. $(1/4)\ln|x+1|-(1/8)\ln(x^2+1)+(1/2)\text{Tan}^{-1}x+(1/4)(x+1)/(x^2+1)+C$
24. $(1/3)\ln|x+1|+(1/\sqrt{3})\text{Tan}^{-1}[(2x-1)/\sqrt{3}]-(1/6)\ln(x^2-x+1)+C$
26. $\ln|x|+\ln(x^2-x+1)-2\ln|x+1|-3/(x+1)+(2/\sqrt{3})\text{Tan}^{-1}[(2x-1)/\sqrt{3}]+C$
28. $\ln|[1+\tan(x/2)]/[1-\tan(x/2)]|+C$
30. $(1/\sqrt{3})\ln|[\sqrt{3}\tan(x/2)-1]/[\sqrt{3}\tan(x/2)+1]|+C$

Exercises 9.8

2. $(1/8)\ln|(x-4)/(x+4)|+C$
4. $-(1/\sqrt{5})\ln|(\sqrt{5}+\sqrt{x^2+5})/x|+C$
6. $2\sqrt{2x+3}+\sqrt{3}\ln|(\sqrt{2x+3}-\sqrt{3})/(\sqrt{2x+3}+\sqrt{3})|+C$
8. $(1/3)\text{Sin}^{-1}[(9x-2)/2]+C$
10. $(1/8)(4x-1)\sqrt{3+2x-4x^2}+(13/16)\text{Sin}^{-1}[(4x-1)/\sqrt{13}]+C$
12. $[1/(2\sqrt{2})]\ln|[\tan(x/2)+3-2\sqrt{2}]/[\tan(x/2)+3+2\sqrt{2}]|+C$
14. $\ln|(1-\sqrt{1-x^2})/x|-(1/x)\text{Sin}^{-1}x+C$
16. $-(1/13)e^{-2x}(2\sin 3x+3\cos 3x)+C$
18. $x(\ln x)^2-2x\ln x+2x+C$

20. $[1/(3\cdot 2^{5/3})]\ln|(x+2^{1/3})^2/(2^{2/3}-2^{1/3}x+x^2)|+[1/(\sqrt{3}\cdot 2^{2/3})]\text{Tan}^{-1}[(2^{2/3}x-1)/\sqrt{3}]+C$
22. $(x^3/2-3x/4)\sin 2x+(3x^2/4-3/8)\cos 2x+C$
24. $-\sqrt{9+x-x^2}+(1/2)\text{Sin}^{-1}[(2x-1)/\sqrt{37}]+C$
26. $-(1/2)(x+3)\sqrt{2x-x^2}+3\text{Sin}^{-1}\sqrt{x/2}+C$
28. $(1/15)\tan x(3\sec^4 x+4\sec^2 x+8)+C$
30. $(x^4/4-3/32)\text{Sin}^{-1}x+(3x/16)(1-x^2)^{3/2}+[(8x^3-3x)/32]\sqrt{1-x^2}+C$
32. $[5/(216\sqrt{3})]\text{Tan}^{-1}(x/\sqrt{3})+(1/216)(5x^5+40x^3+99x-180)/(3+x^2)^3+C$

Exercises 9.9

2. 0.47215, 0.47214
4. 0.64886, 0.64872
6. $-0.069570, -0.069445$
8. 0.30220, 0.30230
10. 0.14221, 0.14201
12. 1.4672, 1.4627
14. 0.31117, 0.31026
16. 1/4 previous; 1/16 previous
18. 1.4789
20. 32.91
22. 0.2437, 0.2438
24. 2.113, 1.729
28.(a) 21 (b) 4
30.(a) 3 (b) 2

Review Exercises

2. $-1/(x+3)+C$
4. $x^2/2-x+4\ln|x+1|+C$
6. $(2/3)(x+3)^{3/2}-6\sqrt{x+3}+C$
8. $-x\cos x+\sin x+C$
10. $(3/4)\ln|x+3|+(1/4)\ln|x-1|+C$
12. $14\sqrt{x}-x-70\ln(5+\sqrt{x})+C$
14. $\text{Sin}^{-1}(e^x)+C$
16. $-(1/2)/(x^2+1)+C$
18. $(1/2)\ln(x^2+1)+(1/2)/(x^2+1)+C$
20. $x+4\ln|x-1|-4/(x-1)+C$
22. $x\text{Cos}^{-1}x-\sqrt{1-x^2}+C$
24. $-(1/12)\cos 6x+(1/8)\cos 4x+C$
26. $\ln|x+2+\sqrt{x^2+4x-5}|+C$
28. $(1/5)(4-x^2)^{5/2}-(4/3)(4-x^2)^{3/2}+C$
30. $-3\ln|4-x^2|+C$
32. $[1/(2\sqrt{2})]\ln|(x+2-2\sqrt{2})/\sqrt{x^2+4x-4}|+C$
34. $\ln|x|+1/x-1/(2x^2)-\ln|x+1|+C$
36. $\ln|2\sqrt{x^2-3x-16}+2x-3|+C$
38. $-x^2/6+(x^3/3)\text{Tan}^{-1}x+(1/6)\ln(x^2+1)+C$
40. $(\ln x)^2/2+C$
42. $(1/81)\ln|x|-(1/162)\ln|x^2+9|+(1/18)/(x^2+9)+C$
44. $(1/2)\ln|x|+(1/2)\ln|x+2|+3/(x+2)+C$

46. $\ln|x| + \ln(x^2 + x + 4) + C$
48. $-(2/3)\cot^{3/2} x - (2/7)\cot^{7/2} x + C$
50. $(x-2)/(4\sqrt{4x-x^2}) + C$
52. 0.63665, 0.64041
54. 1.2667, 1.2647
56. $-2\sqrt{x} - 3x^{1/3} - 6x^{1/6} - 6\ln|1 - x^{1/6}| + C$
58. $(1/8)\text{Tan}^{-1}(x^2/4) + C$
60. $(4x-6)/(9\sqrt{3x-x^2}) + C$
62. $2\sin\sqrt{x} - 2\sqrt{x}\cos\sqrt{x} + C$
64. $(1/32)\sin 4x + (1/8)\sin 2x - (x/8)\cos 4x - (x/4)\cos 2x + C$
66. $(1/2)\tan x + C$
68. $(1/6)\sin 3x - (1/20)\sin 5x - (1/4)\sin x + C$
70. $(1/4)(\text{Sin}^{-1} x)^2 + (x/2)\sqrt{1-x^2}\,\text{Sin}^{-1} x - x^2/4 + C$
72.(b) $\text{Sin}^{-1} x - \sqrt{1-x^2} + C$

CHAPTER 10

Exercises 10.1

2.
$x = t^2 + 3t + 4$
$y = 1 - t$

4.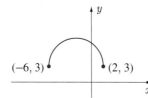
$x = -2 + 4\cos t$, $0 \le t \le \pi$
$y = 3 + 4\sin t$

6.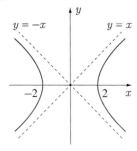
$x = t + 1/t$
$y = t - 1/t$

8.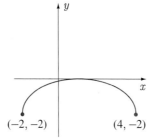
$x = 1 + 3\cos t$, $0 \le t \le \pi$
$y = -2 + 2\sin t$

10.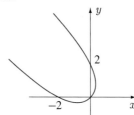
$x = t - t^2$
$y = t + t^2$

12.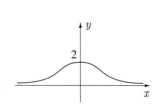
$x = 2\cot\theta$, $-\dfrac{\pi}{2} \le \theta \le \dfrac{\pi}{2}$
$y = 2\sin^2\theta$

14. $2u/(u+1)^2$ **16.** $-1/x^2$
18. $3/[4(t+6)^2(2t+3)^3]$
20. $-4(t+1)\sqrt{-t^2+3t+5}/[(3-2t)(t^2+2t-5)^2]$
22. x/y, $2/y^3$
24. $1/(v+1)$, $-(1/2)/(v+1)^3$
26. $(8/3, 1/2)$, $(28/3, 28)$
28. Ellipse $b^2(x-h)^2 + a^2(y-k)^2 = a^2 b^2$
32. $x = t$, $y = (t+1)/(t-2)$
34. $x = -1 + \sqrt{5}\cos t$, $y = 2 + \sqrt{5}\sin t$, $0 \le t < 2\pi$
36. $31/26$ **38.** 8π **40.** 1.73
46.(b) $3\pi R^2$ **(c)** $8R$

48.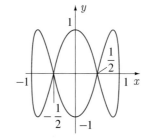
$x = \cos\theta$, $y = \sin 3\theta$

50.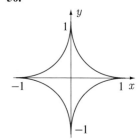
$x = \cos^3\theta$, $y = \sin^3\theta$

Exercises 10.2

2. $(2, 2\pi/3 + 2n\pi)$ **4.** $(4, 5\pi/6 + 2n\pi)$
6. $(\sqrt{17}, -1.82 + 2n\pi)$ **8.** $(\sqrt{29}, 2.76 + 2n\pi)$
10. $(3\sqrt{3}, -3)$ **12.** $(-2.21, -2.03)$

Exercises 10.3

2.
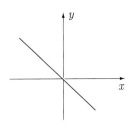

$\theta = \dfrac{3\pi}{4},\ \theta = \dfrac{-\pi}{4}$

4.
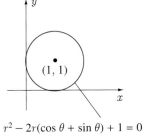

$r^2 - 2r(\cos\theta + \sin\theta) + 1 = 0$

6.
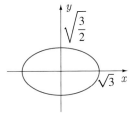

$r^2 = \dfrac{3}{1 + \sin^2\theta}$

8.
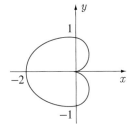

$r = 1 - \cos\theta$

10.

$r^3 = \sec^2\theta \csc\theta$

12.

$y = (\tan 1)\, x$

14.
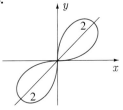

$(x^2 + y^2)^2 = 8xy$

16.
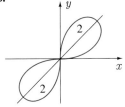

$(x^2 + y^2)^{3/2} = 4xy$

18.
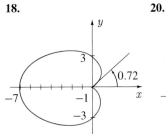

$x^2 + y^2 = 3\sqrt{x^2 + y^2} - 4x$

20.

$y^3 = x^2$

22. $(2, \pm\pi/6),\ (2, \pm 5\pi/6)$

24. $(0, \theta),\ (4/3, \pm 1.23)$

28.
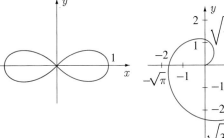

$r = \cos 2\theta$

30.

$r^2 = \theta$

32.

$r = 2\sin(\theta/2)$

34.
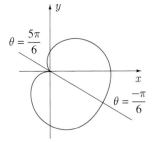

$r = 1 + \cos\left(\theta + \dfrac{\pi}{6}\right)$

36. does not exist   **38.** $1/2$ **40.** $8a$

42.

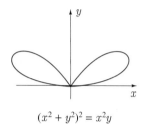

$(x^2 + y^2)^2 = x^2 y$

44.

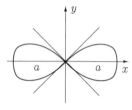

$(x^2 + y^2)^2 = a^2(x^2 - y^2)$

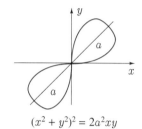

$(x^2 + y^2)^2 = 2a^2 xy$

Exercises 10.4

2. 9π **4.** 2 **6.** 6π
8. 18π **10.** 2π **12.** $(5\pi - 8)/4$
14. $6\pi - 16$ **16.** $(7\pi - 12\sqrt{3})/12$
18. $33[\pi + 2\,\mathrm{Sin}^{-1}(2/3)]/4 + 7\sqrt{5}$
20. $\pi/32$ **22.(b)** $a^2(4-\pi)/2$

Exercises 10.5

2. Circle **4.** None of these
6. Hyperbola **8.** Parabola
10. None of these **12.** Ellipse
14. Straight line

16.

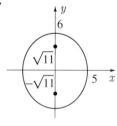

$\dfrac{x^2}{25} + \dfrac{y^2}{36} = 1$

18.

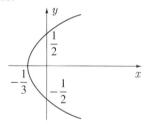

$3x = 4y^2 - 1$

20.

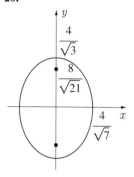

$7x^2 + 3y^2 = 16$

22.

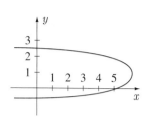

$2y^2 + x = 3y + 5$

24.

$y^2 = 10(2 - x^2)$

26.

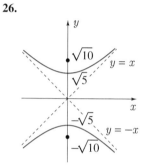

$y^2 - x^2 = 5$

28.

$2x^2 - 3y^2 = 5$

30.

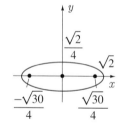

$x^2 + 16y^2 = 2$

32.

$x = -(4+y)^2$

34.

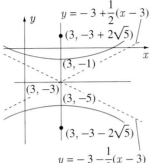

$x^2 - 6x - 4y^2 - 24y = 11$

36.

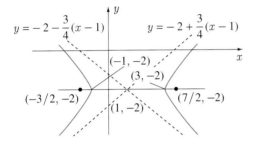

$9x^2 - 16y^2 - 18x - 64y = 91$

38. $3x^2 + y^2 = 28$ **40.** $25/4$

42. $9x^2 + 25y^2 = 225$ **48.** πab

50. $4\pi ab^2/3$, $4\pi a^2 b/3$

54. In Exercise 15, focus is $(0, -7/8)$, directrix is $y = -9/8$; in Exercise 18, focus is $(-7/48, 0)$, directrix is $x = -25/48$; in Exercise 21, focus is $(3/4, 0)$, directrix is $x = 5/4$; in Exercise 22, focus is $(6, 3/4)$, directrix is $x = 25/4$; in Exercise 27, focus is $(3, -1/4)$, directrix is $y = 1/4$; in Exercise 32, focus is $(-1/4, -4)$, directrix is $x = 1/4$.

Exercises 10.6

2.

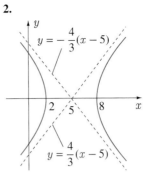

$r = \dfrac{16}{3 + 5\cos\theta}$

4.

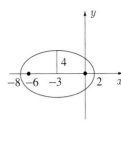

$r = \dfrac{16}{5 + 3\cos\theta}$

6.

$r = \dfrac{4}{4 - 3\sin\theta}$

8.

$r = \dfrac{1}{2 + \sin\theta}$

10.

$r = \dfrac{4\csc\theta}{7\csc\theta - 2}$

12. $8x^2 + 2x + 9y^2 = 1$

14. $8x^2 + 12x - y^2 + 4 = 0$

16. $40x + 25y^2 = 16$

18. With $x = \sqrt{5} + r\cos\theta$, $y = r\sin\theta$, the equation becomes $r = 4/(3 + \sqrt{5}\cos\theta)$.

20. With $x = \sqrt{10} + r\cos\theta$, $y = 2 + r\sin\theta$, the equation becomes $r = 1/(3 - \sqrt{10}\cos\theta)$.

22. $\epsilon \to 0$

Review Exercises

2.

$r = -\sin\theta$

4.

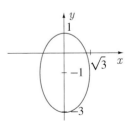

$r = \dfrac{3}{2 + \sin\theta}$

6.

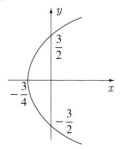

$r = \dfrac{3}{2 - 2\cos\theta}$

8.

$r^2 = 4\cos\theta$

10.

$x = 4 + t$
$y = 5 - 3t^2$

12.

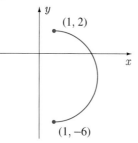

$x = 1 + 4\sin t$, $0 \le t \le \pi$
$y = -2 + 4\cos t$

14.

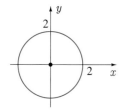

$x^2 + y^2 = 2\sqrt{x^2 + y^2}$

16.

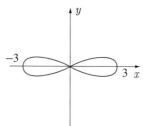

$r = 3\cos 2\theta$

18.

$x^2 + y^2 = x + y$

20.

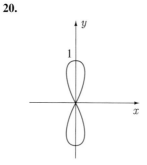

$r = \sin^2\theta - \cos^2\theta$

22.

$x^2 - 3y^2 + 4 = 0$

24.

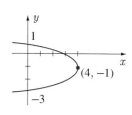

$y^2 + x + 2y = 3$

26.

$y = \cos(\theta/2)$

28.

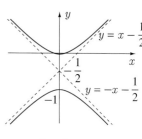

$x^2 = y^2 + y$

30.

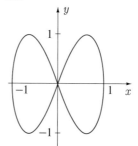

$x = \sin\theta$
$y = \sin 2\theta$

32. 12π **34.** $3\pi/8$
36. $(3 - 3t^2)/(2 + 3t^2)$, $-30t/(2 + 3t^2)^3$
38. 16 **40.** $\sqrt{2}(e^{\pi/2} - 1)$

CHAPTER 11

Exercises 11.1

2. Increasing, $V = 4$
4. Decreasing, $U = 9/16$, $V = 0$, $L = 0$
6. Not monotonic, $U = 1$, $V = -1$, $L = 0$

8. Decreasing, $U = 1/4$, $V = 0$, $L = 0$
10. Increasing, $U = \pi/2$, $V = \pi/4$, $L = \pi/2$
12. Not monotonic
14. Not monotonic, $U = 4$, $V = -4$, $L = 0$
16. False $\{n\}$ **18.** False $\{-n\}$
20. True **22.** True
24. True **26.** False $\{(-1)^n/n\}$
28. False $\{(-1)^n/n\}$ **30.** False $\{(-1)^n\}$
32. False $\{-3^n\}$ **34.** True
36. Not monotonic, $U = 9/4$, $V = -7$, $L = 0$
38. Decreasing, $U = 1/e$, $V = 0$, $L = 0$
40. Increasing, $U = \pi/2$, $V = 0$, $L = \pi/2$
42. Increasing, $U = 1$, $V = 0$, $L = 0.419241$
44. Increasing, $U = 5$, $V = 0$, $L = (\sqrt{21} + 1)/2$
46. Increasing, $U = 10$, $V = 3$, $L = (3 + \sqrt{29})/2$
48. Decreasing, $U = 4$, $V = 0$, $L = (7 - \sqrt{5})/2$
50. $(3n + 1)/n^2$ **52.** $[1 + (-1)^{n+1}]/2$
54. 1 **56.** 0 **58.** 4
60. Decreasing, $U = 2$, $V = 0$, $L = (3 - \sqrt{5})/2$
62. Decreasing, $U = 1$, $V = 0$, $L = 1/2$
64. Decreasing, $U = 1$, $V = 0$, $L = 0$
66. Decreasing, $U = 2$, $V = 0$, $L = 0$
68. Increasing, $U = 1$, $V = 0$, $L = (3 - \sqrt{3})/2$
72. $V = 1/2$, $U = 1$
76.(a) 1, 1, 2, 3, 5, 8, 13, 21, 34, 55 (b) Increasing, $V = 1$, no upper bound, no limit (e) $(\sqrt{5} + 1)/2$
80. $[5 + (-1)^n/2^{n-2}]/3$

Exercises 11.2

2. -0.3819660 **4.** -2.6180340
6. 3.0447231 **8.** 1.2146480
10. 0.3347341 **12.** -1.3887920
14. (a) $40(0.99)^n$ m (b) $4(0.981)^{-1/2}(0.99)^{n/2}$ s
16. $4P/3$, $16P/9$, $(4/3)^n P$, does not exist
18. 0.0625, 0.1125
20. $(\sqrt{3}P^2/36)[1 + (1/3) + (4/3^3) + (4^2/3^5) + \cdots + (4^{n-1}/3^{2n-1})]$
22. -0.4814 **24.** 2.9122 **26.** 3.3247
28. 0.7849 **30.** 0.7953

Exercises 11.3

2. $2/15$ **4.** Diverges
6. 10 804.5 **8.** $-1/3$
10. Diverges **12.** $13/99$
14. 430 162/9999 **16.** Diverges
18. 4 **20.** Diverges
22. 1 **24.** 804 s
26. $(\sqrt{3}P^2/180)[8 - 3(4/9)^n]$, $2\sqrt{3}P^2/45$
28. 2 **30.** $51/64$
32. $2/3$
36. $600/11$ minutes after 10:00
38.(c) $A_0 e^{kT}/(e^{kT} - 1)$

Exercises 11.4

2. Diverges **4.** Converges
6. Converges **8.** Diverges
10. Converges **12.** Converges
14. Converges **16.** Converges
18. Diverges **20.** Diverges
22. Diverges
24. Converges for $p > 1$, diverges for $p \leq 1$

Exercises 11.5

2. Converges **4.** Converges
6. Diverges **8.** Diverges
10. Converges **12.** Converges
14. Converges **16.** Diverges
18. Converges **20.** Converges

Exercises 11.6

2. Converges conditionally **4.** Converges absolutely
6. Diverges **8.** Converges absolutely
10. Converges conditionally **12.** Converges conditionally
14. Converges absolutely

Exercises 11.7

2. -0.9470 **4.** -0.0127 **6.** 1.07
8. -0.7 **10.** 4

Exercises 11.8

2. $-1 < x < 1$ **4.** $-1/3 < x < 1/3$
6. $-4 < x < -2$ **8.** $7/2 < x < 9/2$
10. $-1 < x < 1$ **12.** $-1 \leq x < 1$
14. $-1/e \leq x \leq 1/e$ **16.** $x = 0$
18. $-1/3 < x < 1/3$ **20.** $-5^{1/3} < x < 5^{1/3}$
22. $-1 \leq x \leq 1$ **24.** 1
26. $4/(4 - x^3)$, $|x| < 4^{1/3}$

28. $(x-1)/(10-x)$, $-8 < x < 10$
30.(b) $-\infty < x < \infty$

Exercises 11.9

2. $\sum_{n=0}^{\infty} (5^n/n!) x^n$, $-\infty < x < \infty$
4. $(1/\sqrt{2})[1 + (x-\pi/4) - (x-\pi/4)^2/2! - (x-\pi/4)^3/3! + (x-\pi/4)^4/4! + \cdots]$, $-\infty < x < \infty$
6. $1/8 - (3/8^2)(x-2) + (3^2/8^3)(x-2)^3 - (3^3/8^4)(x-2)^4 + \cdots$, $-2/3 < x < 14/3$
12.(d) 0 (e) $x = 0$

Exercises 11.10

2. $\sum_{n=0}^{\infty} [(-1)^n / 4^{n+1}] x^{2n}$, $|x| < 2$
4. $\sum_{n=0}^{\infty} [(-1)^n/(2n)!] x^{4n}$, $-\infty < x < \infty$
6. $\sum_{n=0}^{\infty} (5^n/n!) x^n$, $-\infty < x < \infty$
8. $\sum_{n=0}^{\infty} [1/(2n+1)!] x^{2n+1}$, $-\infty < x < \infty$
10. $1 + 9x/2 + 27x^2/8 + \sum_{n=3}^{\infty} \{(-1)^n (2n-5)! 3^{n+1} / [2^{2n-3} n! (n-3)!]\} x^n$, $-1/2 \le x \le 1/3$
12. $x^4 + 3x^2 - 2x + 1$
14. $33 - 46(x+2) + 27(x+2)^2 - 8(x+2)^3 + (x+2)^4$
16. $\sum_{n=0}^{\infty} (e^n/n!)(x-3)^n$, $-\infty < x < \infty$
18. $\sqrt{3}(1 + x/6) + \sum_{n=2}^{\infty} \{4\sqrt{3}(-1)^{n-1}(2n-3)! / [12^n n! (n-2)!]\} x^n$, $|x| \le 3$
20. $x - x^2/3 - \sum_{n=3}^{\infty} \{(2)(5) \cdots (3n-7)/[3^{n-1}(n-1)!]\} x^n$, $|x| \le 1$
22. $1 + x^2/2 + 5x^4/24 + 61 x^6/720$
24. $1 + \sum_{n=1}^{\infty} [(-1)^n 2^{2n-1}/(2n)!] x^{2n}$, $-\infty < x < \infty$
26. $x^2 + \sum_{n=1}^{\infty} \{(2n)! / [(2n+1) 2^{2n} (n!)^2]\} x^{4n+2}$, $|x| < 1$
28. $\sqrt{2} \sum_{n=0}^{\infty} \{1/[2^n (2n+1)]\} x^{2n+1}$, $|x| < \sqrt{2}$
30.(a) $1/p$ (b) 6
32. $\sum_{n=0}^{\infty} \{(-1)^n \pi^{2n}/[(4n+1) 2^{2n} (2n)!]\} x^{4n+1}$, $-\infty < x < \infty$;
$\sum_{n=0}^{\infty} \{(-1)^n \pi^{2n+1}/[(4n+3) 2^{2n+1} (2n+1)!]\} x^{4n+3}$, $-\infty < x < \infty$
36.(a) $-1/2$, $1/6$, 0, $-1/30$, 0

Exercises 11.11

2. $2/(1-x)^3$, $|x| < 1$
4. $(x+1)/(1-x)^3$, $|x| < 1$
6. $-(1/x) \ln(1-x)$, $-1 \le x < 1$
8. $-\ln(1 + x^2)$, $-1 \le x \le 1$
10. $1/(x - x^2) - (1/x^2) \ln(1-x)$, $|x| < 1$
22. $-9/100$

Exercises 11.12

2. 4.2×10^{-10} 4. 4.2×10^{-10} 6. 1.4×10^{-9}
8. 0.115 10. 0.00625 12. 0.497
14. 0.133 16. 0.291 18. -0.122
20. $1/2$ 22. $1/2$ 24. 0
26. $1 - x^3/2 + 3x^6/8 - 5x^9/16$ 28. $1 - x^2$
30. $a_0 + a_1 \sum_{n=1}^{\infty} [(-1)^{n+1}/n!] x^n = C + De^{-x}$
32. $a_0 \sum_{n=0}^{\infty} [(-1)^n/(2n)!] x^n$
34. $a_1 \sum_{n=1}^{\infty} \{(-1)^{n+1}/[n!(n-1)!]\} x^n$
38. 2.44

Review Exercises

2. Decreasing, $U = 1$, $V = 0$, $L = 1/\sqrt{3}$
4. Increasing, $U = 100$, $V = 7$, $L = (31 + \sqrt{53})/2$
6. $|k| \le 1$ 10. Converges
12. Converges 14. Converges conditionally
16. Diverges 18. Converges
20. Converges 22. Diverges
24. Converges 26. Converges
28. Converges conditionally 30. $-2 \le x \le 2$
32. $-\infty < x < \infty$ 34. $-4 < x < -2$
36. $-2^{-1/3} \le x < 2^{-1/3}$
38. $e^5 \sum_{n=0}^{\infty} (1/n!) x^n$, $-\infty < x < \infty$
40. $\sum_{n=2}^{\infty} [(-1)^n 2^{n-1}/(n-1)] x^n$, $-1/2 < x \le 1/2$
42. $(1/2) \sum_{n=0}^{\infty} (-1)^{n+1} (1 - 1/3^n) x^n$, $|x| < 1$
44. $x + \sum_{n=2}^{\infty} \{(-1)^n/[n(n-1)]\} x^n$, $-1 < x \le 1$
46. $\sum_{n=0}^{18} [(-1)^n/n!] x^{2n}$
48. $\sum_{n=0}^{\infty} \{(-1)^n/[2^{2n}(2n)!]\} x^{2n} + \sum_{n=0}^{\infty} \{(-1)^n/[2^{2n+1}(2n+1)!]\} x^{2n+1}$

APPENDIX E

Additional Exercises

When you solve the exercises in a particular section of this text, you already know the nature of the problem and the techniques that should be used. For example, a problem in Exercises 3.7 will use implicit differentiation, a problem in Exercises 9.5 will use a trigonometric substitution, etc. To ensure that you understand all parts of a chapter, we recommended the review exercises at the end of each chapter. They require you to first identify the particular section that is useful in solving a problem, and then to follow through with its application. Now that you have studied all eleven chapters in the book, you should test your understanding of calculus as a whole. The following exercises, which review all aspects of the course, will do this. They test your ability to recognize the essential nature of a problem — be it a maximum-minimum problem, a related rate problem, an application of definite integrals, a problem in infinite series, etc. — and they tax your organizational, analytical, and interpretive skills.

The first 200 or so exercises are a reasonably straightforward review of the many types of problems that you have already encountered. The next 100 are more difficult; they may require some originality on your part or some extra calculations. The next 50 are substantially more involved, and the last 15 or so are delightfully challenging. Set your sights high; attempt as many problems as you can. Answers to selected exercises are found at the end of the appendix.

ADDITIONAL EXERCISES

1. Find all points (x, y) satisfying $y + |y| = x + |x|$.

2. (a) Sketch the graph of an even, 2π-periodic function $f(x)$ for which
$$f(x) = x^3 - \pi x, \quad 0 < x < \pi.$$
Can $f(x)$ be made continuous?

   (b) Sketch the graph of an odd, 2π-periodic function $f(x)$ for which
$$f(x) = x^3 - \pi x, \quad 0 < x < \pi.$$
Can it be made continuous?

In Exercises 3–18 evaluate the limit if it exists. If a limit does not exist, explain why not.

3. $\lim\limits_{x \to 0} \dfrac{x^2 - x}{2x + x^2}$

4. $\lim\limits_{x \to -4^-} \dfrac{x^2 - 16}{x + 4}$

5. $\lim\limits_{x \to 0} \dfrac{\sqrt{2 + x} - \sqrt{2 - x}}{x}$

6. $\lim\limits_{x \to 5} \dfrac{|x^3 - 125|}{x - 5}$

7. $\lim\limits_{x \to 3} \dfrac{x - 3}{x^2 - 6x + 9}$

8. $\lim\limits_{x \to 0^-} \dfrac{3x}{2 - \sqrt{x^2 + 4}}$

9. $\lim\limits_{x \to \infty} \dfrac{2 - x^5}{3 + 4x^5}$

10. $\lim\limits_{x \to -\infty} \tan x$

11. $\lim\limits_{x \to 1^+} \left[\left(1 - \dfrac{1}{x}\right)^3 + x - \dfrac{2}{x^2} \right]$

12. $\lim\limits_{x \to 0} \dfrac{\sin 4x^{12}}{3x^{12}}$

13. $\lim\limits_{x \to \infty} \left(\sqrt{x^2 + 3} - \sqrt{x^2 + 1} \right)$

14. $\lim\limits_{x \to -\infty} \left(\sqrt{x^2 + x} - \sqrt{x^2 + 2x} \right)$

ADDITIONAL EXERCISES 651

15. $\lim_{x \to -\pi/2^+} \tan x$

16. $\lim_{x \to 1} e^{\frac{1}{x-1}}$

17. $\lim_{x \to 2} \frac{\sqrt{x^2 - 4x + 4}}{x - 2}$

18. $\lim_{x \to 1} \frac{x + 2\sqrt{x} - 3}{x^2 - 1}$

In Exercises 19–69 assume that y is defined as a function of x by the given equation(s), and find dy/dx.

19. $y = x^2(1 - 3x)^4$

20. $y = \frac{x^2}{\sqrt{1 - 3x^2}}$

21. $y = \frac{\sin 2x}{\cos 3x}$

22. $y = (2x + 1)^3 \ln(x + 1)$

23. $y = (x^2 + 2x - 3)e^{3x}$

24. $y = \sqrt{\ln(e^{2x} + 1)}$

25. $y = \text{Sin}^{-1}(\sqrt{1 - 4x})$

26. $y = x^2 \csc^2(x^2 - 1)$

27. $y = \frac{x}{1 + \sec(1 - 2x^3)}$

28. $y = \sqrt[3]{1 + \sqrt{2 - x}}$

29. $x^2 y + y^3 x - 2x = 5$

30. $\sin(x + y) = xy$

31. $x^3 e^y + x - 2y = 3x^2$

32. $x \text{Tan}^{-1}(xy) = y + 3$

33. $y = \cos(v^3 + 2), \quad v = \frac{x}{x+1}$

34. $y = \text{Sec}^{-1}(1 - 2u), \quad u = \tan x + 1$

35. $y = \sqrt{v - \sqrt{v}}, \quad v = \sqrt{x + \sqrt{x}}$

36. $y = t^2 e^{-3t}, \quad t = x + \ln x$

37. $x = t + \frac{1}{t}, \quad y = 2t - \frac{3}{t^2}$

38. $x = 4 + 3\cos t, \quad y = 2 - 4\sin t$

39. $x = \frac{\theta}{1 + \theta^2}, \quad y = \frac{\theta^2}{1 + \theta^2}$

40. $x = e^{3v} - v, \quad y = e^{-3v} - v^2$

41. $y = 3 + 2x - 6\ln(3 + 2x) - \frac{9}{3 + 2x}$

42. $y = \ln\left(\frac{4 - x}{x}\right)$

43. $y = \ln\left[\frac{(2 + x)^3}{8 + x^3}\right] + 2\sqrt{3}\,\text{Tan}^{-1}\left(\frac{x - 1}{\sqrt{3}}\right)$

44. $y = \ln\left(\frac{2x^2 - 2x + 1}{2x^2 + 2x + 1}\right) + 2\,\text{Tan}^{-1}\left(\frac{2x}{1 - 2x^2}\right)$

45. $y = (8 - 12x + 27x^2)\sqrt{1 + 3x}$

46. $y = \ln\left(x + \sqrt{x^2 + 9}\right)$

47. $y = \sqrt{x^2 - 4} - 2\,\text{Sec}^{-1}(x/2)$

48. $y = -\frac{\sqrt{7x^2 + 49}}{x^2} + \ln\left(\frac{\sqrt{7} + \sqrt{x^2 + 7}}{x}\right)$

49. $y = 2x(3 - x^2)^{3/2} + 9x\sqrt{3 - x^2} + 27\,\text{Sin}^{-1}\left(\frac{x}{\sqrt{3}}\right)$

50. $y = (3x^2 + 2)(1 - x^2)^{3/2}$

51. $y = x\sqrt{16 - x^2} + 16\cos(x/4)$

52. $y = \frac{5}{\sqrt{25 - x^2}} - \ln\left(\frac{5 + \sqrt{25 - x^2}}{x}\right)$

53. $y = \ln\left(\frac{3\sqrt{16 - x^2} + \sqrt{7}x}{\sqrt{9 - x^2}}\right)$

54. $y = \frac{3x + 12}{\sqrt{6 + 3x + 4x^2}}$

55. $y = \text{Cos}^{-1}\left(1 - \frac{x}{4}\right)$

56. $y = \frac{x - 7}{\sqrt{14x - x^2}}$

57. $y = \ln\left(x + 3 + \sqrt{6x + x^2}\right)$

58. $y = \text{Csc}^{-1}\sqrt{\frac{3x^5}{2}}$

59. $y = \text{Tan}^{-1}\left(\frac{4\tan(x/2) + 3}{\sqrt{7}}\right)$

60. $y = \text{Tan}^{-1}\left(\frac{3\tan x}{2}\right)$

61. $y = 3\cot\left(\frac{3x}{2}\right) - \cot^3\left(\frac{3x}{2}\right)$

62. $y = (3x^2 - 6)\cos x + (x^3 - 6x)\sin x$

63. $y = \text{Sin}^{-1}\left(\frac{3\cos 2x}{\sqrt{10}}\right)$

64. $y = \ln\left(7\sin 3x + \sqrt{1 + 49\sin^2 3x}\right)$

65. $y = (1 + x^2)\text{Cot}^{-1}x + x$

66. $y = \frac{\sqrt{4x^2 - 1}}{x} - \frac{\text{Sec}^{-1}(2x)}{x}$

67. $y = e^{3x}(3\sin 2x - 2\cos 2x)$

68. $y = 2x - \ln(5 + 3e^{2x})$

69. $y = 13xe^{2x}(2\sin 3x - 3\cos 3x) + e^{2x}(5\sin 3x + 12\cos 3x)$

70. For what value of k are the curves $y = 1 - x^2$ and $y = x^2 - 4x + k + 4$ tangent? What is the point of tangency?

71. Find the equation of the tangent line to the curve $y = (x + 9)/(x + 5)$ that passes through the origin.

72. Find the first and second derivatives of the function $y = f(x)$ defined by the equation $y^3 + (x - 1)y + 16 = 3x$ when $x = 2$.

73. Verify that when n is a positive integer
$$\frac{d}{dx}(\sin^n x \cos nx) = n \sin^{n-1} x \cos(n+1)x.$$

74. Show that $f(x) = \ln(x + \sqrt{x^2 + 1})$ is an odd function and sketch its graph.

75. A chord of the parabola $y = x^2 - 2x + 5$ joins the points with abscissas $x = 1$ and $x = 3$. Find the equation of the tangent line to the parabola parallel to the chord.

76. Find the second derivative of the function $y = f(x)$ defined by the equation $xy^2 + x^2y = 16$ at the point where its first derivative is equal to 0.

77. Show that the function defined by $y^3 + x^4y = 3$ is decreasing when $x > 0$.

In Exercises 78–103 evaluate the indefinite integral.

78. $\int \dfrac{1}{3 + 4e^{2x}} dx$

79. $\int \dfrac{1}{4e^{3x} + 5e^{-3x}} dx$

80. $\int (x-1)^2 \ln x \, dx$

81. $\int \ln(x^2 + 9) \, dx$

82. $\int \sin^2 3x \, dx$

83. $\int \cos^3(5x+1) \, dx$

84. $\int \sin 3x \cos^2 4x \, dx$

85. $\int \dfrac{1}{\sqrt{x^2 + 5}} dx$

86. $\int \dfrac{x+5}{\sqrt{x^2 - 3}} dx$

87. $\int \dfrac{\sqrt{x^2 + 1}}{x^4} dx$

88. $\int \dfrac{1}{x^2 \sqrt{7 - x^2}} dx$

89. $\int \dfrac{1}{(x^2 + 2x + 5)^{5/2}} dx$

90. $\int \dfrac{x}{\sqrt{2x - x^2}} dx$

91. $\int \dfrac{2 - 5x - x^2}{x^3 + x^2 + 5x + 5} dx$

92. $\int \dfrac{x^3 + 5x^2 + 12x + 12}{x^3 + 2x^2 + 4x} dx$

93. $\int \csc^3 2x \, dx$

94. $\int \csc^4 x \, dx$

95. $\int \dfrac{7x^4 - 6x^3 + 16x^2 - 6x + 9}{2x^5 - 4x^4 + 5x^3 - 6x^2 + 3x} dx$

96. $\int \dfrac{e^{2x} + 1}{(2e^x + 5)^3} dx$

97. $\int \dfrac{x+2}{\sqrt{1 - x^2}} dx$

98. $\int \dfrac{x^2}{(5 - x^2)^{3/2}} dx$

99. $\int \dfrac{1}{\sqrt{-2x^2 + 4x + 3}} dx$

100. $\int (6x - 3) \ln(x^2 - x + 5) \, dx$

101. $\int \cot^4\left(\dfrac{x}{2}\right) dx$

102. $\int \dfrac{\tan^3 x}{\sec x} dx$

103. $\int \dfrac{1}{\sec x - \tan x} dx$

In Exercises 104–123 determine whether the series converges or diverges. In the case of a series with both positive and negative terms, determine whether convergence is absolute or conditional.

104. $\sum_{n=1}^{\infty} \dfrac{n^2 + 1}{3n^3 + 5}$

105. $\sum_{n=1}^{\infty} \dfrac{(n^3 + 4n)^{2/3}}{n^4 - 3n^3 + 1}$

106. $\sum_{n=1}^{\infty} \dfrac{(-1)^{n+1} n^2}{2^n}$

107. $\sum_{n=1}^{\infty} \dfrac{(-1)^n}{\sqrt{n+1}}$

108. $\sum_{n=2}^{\infty} \dfrac{1}{n \ln n}$

109. $\sum_{n=1}^{\infty} n^{1/n}$

110. $\sum_{n=1}^{\infty} \dfrac{\sin(n+1)}{n^2 + 4}$

111. $\sum_{n=1}^{\infty} \dfrac{1 + 2^{-n}}{1 + 3^{-n}}$

112. $\sum_{n=1}^{\infty} \sqrt{\dfrac{n^2 + n - 1}{2n^4 + n^3 + 5}}$

113. $\sum_{n=1}^{\infty} \left(\dfrac{n^2 - 2}{2n^2 + 5}\right)^n$

114. $\sum_{n=1}^{\infty} \dfrac{n^2}{n^{3n}}$

115. $\sum_{n=1}^{\infty} \dfrac{\operatorname{Tan}^{-1}(n^2 + 5)}{2n^2 + 4}$

116. $\sum_{n=1}^{\infty} \dfrac{1 \cdot 3 \cdot 5 \cdots (2n+1)}{2 \cdot 4 \cdot 6 \cdots (2n)} \left(\dfrac{1}{n}\right)$

117. $\sum_{n=1}^{\infty} \dfrac{(n!)^2}{(2n+1)!}$

118. $\sum_{n=1}^{\infty} \dfrac{\sqrt{n + \sqrt{1 + n^2}}}{n^2}$

119. $\sum_{n=1}^{\infty} \left(\dfrac{1}{2} - \dfrac{1}{n+1}\right)^{10}$

120. $\sum_{n=1}^{\infty} \dfrac{3}{n(1 + \ln n)^3}$

121. $\sum_{n=1}^{\infty} \dfrac{3^{-2n} + 2^{-2n}}{4^{-n} + 5^{-n}}$

122. $\sum_{n=1}^{\infty} \dfrac{(n-1)^n}{n^{n-1}}$

123. $\sum_{n=1}^{\infty} \operatorname{Sin}^{-1}\left(\dfrac{1}{n^2}\right)$

124. Find point(s) on the curve $y = 3x^2/(x + 1)$ where the inclination of the tangent line is equal to 1.

125. Can a continuously differentiable function defined by the equation $\sin(x + y) = 2x + 3$ have a critical point?

126. When a projectile (such as an arrow) is fired up an inclined plane (Figure E.1), the distance that it travels before striking the plane is called its range R. It can be shown that

$$R = \dfrac{v^2 \cos\theta \sin(\theta - \alpha)}{4.905 \cos^2\alpha},$$

where v is the initial speed of the projectile, α is the angle of inclination of the plane, and θ is the angle of projection with the horizontal. Find the angle θ for which R is greatest.

FIGURE E.1

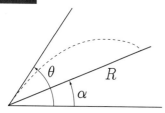

127. Evaluate

$$\lim_{x \to 0} \frac{\sqrt{1+\sin x} - \sqrt{1-\sin x}}{\tan x}.$$

128. A differential equation in $y = f(x)$ is said to be homogeneous, linear, second order if it is of the form

$$a(x)\frac{d^2y}{dx^2} + b(x)\frac{dy}{dx} + c(x)y = 0,$$

where $a(x)$, $b(x)$, and $c(x)$ are functions of x. Show that if $f_1(x)$ and $f_2(x)$ are two solutions of this equation, so is $y = p_1 f_1(x) + p_2 f_2(x)$ for any constants p_1 and p_2.

In Exercises 129–132 find the area bounded by the curves.

129. $y = 2/\sqrt{x-1}$, $x + 3y = 8$

130. $y = 8 - 3x^2$, $y = 3x^2/(x^2+1)$

131. $x + y = 2e$, $x = y \ln y$, $x = 0$

132. $x = (2/\pi)\operatorname{Sin}^{-1} y$, $y = 2 - x^2$, $y = 0$

In Exercises 133–135 find the volume of the solid of revolution obtained by rotating the area bounded by the curves about the line indicated.

133. $x = y(y-1)$, $x = 4y(y-1)$ about $x = 0$

134. $y = (x^2 + x)/(x-2)$, $y = 0$ about $x = 1$

135. $x = e^{2y}$, $x = 2$, $y = 0$ about $y = 2$

136. Find the length of $y = \ln(1 - x^2)$ between the lines $x = 0$ and $x = 1/2$.

137. A thin uniform rod of length L and mass M spins in a plane about one of its ends with angular velocity ω. What is its kinetic energy?

138. The length of the hypotenuse of a right-angled triangle is 1 metre, and the lengths of the other two sides are a and b. When the triangle is submerged vertically in a fluid so that the edge with length b is in the surface of the fluid, the fluid force on each side of the triangle is twice that when the side of length a is in the surface. What are a and b?

139. Find the x-coordinate of the centroid of the area bounded by the curves

$$y = \frac{x^2 + 2x - 3}{x^2 + 1}, \qquad y = 0.$$

140. Find moments of inertia of an elliptic plate of uniform mass per unit area about its major and minor axes.

141. If $f(x)$ satisfies the condition $f(f(x)) = x$ for all x in its domain, what is $f^{-1}(x)$? Give an example.

142. In Kintsch's model of choice behaviour, he states that

$$\sum_{n=2}^{\infty} (1-b)^{n-2} bs^n = \frac{bs^2}{1-(1-b)s}.$$

Is this true, and would you place any restrictions on b and s?

143. Let P be a point on the circumference of a circle. Where should a chord QR be drawn parallel to the tangent at P in order that triangle PQR have maximum area?

144. In Figure E.2, a 5 m rod CD is such that its end C can only move horizontally along the line AGB and end D can only move horizontally along EF. When C is 3 m to the left of G, it is moving toward A at 4 m/s. In what direction and how fast is D moving at this instant?

FIGURE E.2

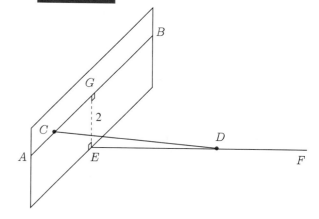

145. Evaluate

$$\lim_{n \to \infty} \frac{1}{n^2}(1 + 2 + 3 + \cdots + n).$$

146. Find d^2y/dx^2 if
$$y = \int_0^{x^2} t \cos t \, dt.$$

147. Are either of the following valid?
$$\int_0^a f(x)\, dx = \int_0^a f(a-x)\, dx;$$
$$\int_0^a f(x)\, dx = \int_0^a f(b-x)\, dx.$$

148. Find the volume of the solid of revolution obtained by rotating the area bounded by $(x-4)y^2 = x(x-3)$ and the x-axis, about the x-axis.

149. Find the Maclaurin series for $f(x) = 1/(x^4 - x^2 - 2)$.

150. If $f(x)$ is a polynomial of degree 4, and it is known that $f(1) = 1$, $f'(1) = -2$, $f''(1) = 0$, $f'''(1) = 3$, and $f''''(1) = -5$, what is $f(-2)$?

151. (a) A function $f(x)$ is said to be odd-harmonic if it satisfies
$$f(L+x) = -f(x)$$
for all x. Show that $f(x)$ must be $2L$-periodic.

(b) A function $f(x)$ is said to be odd and odd-harmonic if it satisfies
$$f(-x) = -f(x) \quad \text{and} \quad f(L-x) = f(L+x).$$
Show that $f(x)$ must be $4L$-periodic.

(c) Sketch the graph of the extension of $f(x) = x$, $0 < x < L$ to a continuous, odd, and odd-harmonic function.

(d) A function $f(x)$ is said to be even and odd-harmonic if it satisfies
$$f(-x) = f(x) \quad \text{and} \quad f(L+x) = -f(L-x).$$
Show that $f(x)$ must be $4L$-periodic.

(e) Sketch the graph of the extension of $f(x) = x$, $0 < x < L$ to an even and odd-harmonic function.

152. Find the distance from the point $(1, 2)$ to the tangent line to the curve $y = x^2 \ln x$ at the point with abscissa equal to 2.

153. Show that a function defined by the equation $xye^{xy} = 1$ cannot have a first derivative equal to 1.

154. Find all critical points of the function
$$f(x) = x^{ax}, \quad x > 0$$
where $a > 0$ is a constant.

155. Show that the equation $x^n + px + q = 0$ cannot have more than two real roots when n is an even integer, and not more than three real roots when n is an odd integer.

156. The reproduction rate of Drosophila melanogaster supposedly decreases with population. Show that this is the case if the progeny per female per day, p, when x denotes the number of flies per bottle is
$$p = 34.53 x^{-0.658} e^{-0.018x}, \quad x \geq 1.$$

157. Can a continuous function $y = f(x)$ with a first derivative satisfying
$$\frac{dy}{dx} + \sin y = 2 + x^2$$
have relative extrema?

158. For what value of k does the function $f(x) = k\cos x + 3x \sin x$ have a relative extremum at $x = \pi/3$? Is it a relative maximum or minimum?

159. Show that the angle of inclination of a curve has relative extrema at points of inflection.

160. Is the curve
$$x = t^2 - \frac{1}{t}, \quad y = t^3 + 3t, \quad t \geq 1$$
concave upward or concave downward?

161. Thurstone's learning curve is
$$y = \frac{Lx + Lc}{x + c + a}, \quad x \geq 0$$
where $L > 0$ is the limit of practice, x is the amount of formal practice, c is the equivalent previous practice, and a is the rate of learning. Sketch a graph of the learning curve.

162. One end of a ladder 20 m long slips down the side of a house at 1 m/s. What is the acceleration of the foot of the ladder when it is 15 m from the house?

163. If a point moves around the ellipse $b^2x^2 + a^2y^2 = a^2b^2$, at what points are the rates of change of its abscissa and ordinate equal?

164. If the velocity of a particle moving along the x-axis is given by $A\cos(\omega t + \phi)$, and its position at time t_1 is x_1, what is its position at time t_2?

165. If $|f(x) - g(x)| \leq \epsilon$ for $a \leq x \leq b$, what is the maximum value of
$$\left| \int_a^b [f(x) - g(x)]\, dx \right|?$$

166. Find the length of the loop of the curve $9ay^2 = x(x - 3a)^2$ where $a > 0$ is a constant.

167. A block of wood (density 500 kg/m³) is cubical with side lengths L m. If it floats in water, what work is required to lift it out of the water?

168. The lengths of the major and minor axes of an elliptic plate are $2a$ and $2b$. Let F_a denote the force on each side of the plate when it is half submerged in a fluid with major axis in the surface, and let F_b denote the force when the minor axis is in the fluid. What is the ratio F_a/F_b?

169. A thin plate with constant mass per unit area has the shape of a sector of a circle of radius r and angle α where $0 < \alpha < \pi/2$. How far from the vertex of the sector is the centre of mass?

170. Find the inverse function for
$$f(x) = \frac{e^x - e^{-x}}{e^x + e^{-x}} + 1.$$

171. Approximate
$$\int_0^{1/2} \frac{1}{1 + x^4}\, dx$$

(a) using Simpson's rule with 10 subdivisions;

(b) using the first four terms of the Maclaurin series of the integrand.

Discuss the relative merits of these techniques for this example.

172. Find the centroid of the area bounded by the x-axis and the arch of the cycloid described by
$$x = R(\theta - \sin\theta), \quad y = R(1 - \cos\theta), \quad 0 \leq \theta \leq 2\pi.$$

173. Find the length of the curve
$$r = \sin^3\left(\frac{\theta}{3}\right), \quad 0 \leq \theta \leq 3\pi.$$

174. Find the Maclaurin series for
$$f(x) = \ln[(1 + x)^{1+x}] + \ln[(1 - x)^{1-x}].$$
What is the interval of convergence of the series?

175. What are the first three terms in the Maclaurin series for the function defined by the equation
$$y + \ln(1 + xy) + x = 5.$$

176. Evaluate
$$\lim_{x \to 0} \frac{6x - x^3 - 6\sin x}{x^5}.$$

177. Find constants b_n in order that the function $y = f(x) = \sum_{n=1}^{\infty} b_n \sin nx$ will satisfy the differential equation
$$\frac{d^2y}{dx^2} - 3y = \sum_{n=1}^{\infty} \frac{1}{n^2} \sin nx.$$
Assume that the series for $f(x)$ can be differentiated term-by-term.

178. Find the distance from the point of intersection of the curves $y = 1/(x - 1)$ and $4y = x^2$ to the normal line to the curve $x^3 y + xy^2 = 6$ at the point with ordinate $y = -1$.

179. Verify that for $0 < a < b$,
$$\frac{b - a}{b} < \ln\left(\frac{b}{a}\right) < \frac{b - a}{a}.$$

180. It is known that a continuous function $y = f(x)$ always has a positive derivative which satisfies the equation
$$\frac{dy}{dx} = y^3 - 3y.$$
Show that its graph can have a point of inflection only at a point where $y = 1$ or $y = -1$.

181. The circles in Figure E.3 represent ferris wheels with radii 20 m and 10 m. Their centres are displaced horizontally by 50 m and vertically by 30 m. The larger rotates half as fast as the smaller, making one revolution every 10 seconds. How fast is the distance changing between riders at positions P and Q? Assume both ferris wheels turn counterclockwise.

FIGURE E.3

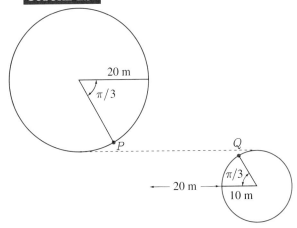

182. At time $t = 0$, a body is moving along the x-axis with velocity $v_0 > 0$. If the only force acting on it is resistance, and this force is proportional to the square of the velocity of the body, where and when does the body stop?

183. Find the length of the curve
$$y = L \ln\left(\frac{L + \sqrt{L^2 - x^2}}{x}\right) - \sqrt{L^2 - x^2}$$
from $(L, 0)$ to any other point (x, y) on the curve (see Exercise 24 in Section 9.5).

184. Find the centroid of the area bounded by the curve $|y| = x\sqrt{1 - x^2}$.

185. If the number of subdivisions in the approximation of a definite integral by Simpson's rule is doubled, does the error decrease by a factor of 16? Explain.

186. Show that a function $f(x)$ and the nth degree polynomial
$$P_n(x) = f(c) + f'(c)(x - c) + \cdots + \frac{f^{(n)}(c)}{n!}(x - c)^n$$
have the same first n derivatives at $x = c$.

187. Find a cubic polynomial approximating the function
$$f(x) = \int_0^x e^{-t^2}\, dt.$$
How accurate is it on the interval $-0.2 \leq x \leq 0.2$?

188. Show that if a continuous function $y = f(x)$, which satisfies the equation
$$\frac{dy}{dx} - y^3 = y,$$
has a point of inflection on its graph, it must be a horizontal point of inflection.

189. Find circle(s) that have centres on the line $2x + 3y = 6$ and intersect the circle $x^2 + y^2 = 1$ at only one point.

190. One hundred dollars is invested at 10% compounded quarterly. Sketch a graph of the accumulated value of the investment as a function of time t (in months) taking $t = 0$ at the time it is initially invested. Express this function in terms of the greatest integer function.

191. Evaluate
$$\lim_{y \to 0} \frac{(x + y)^{1/3} - x^{1/3}}{y}.$$

192. Evaluate
$$\lim_{x \to \infty} x\left(\sqrt{x^2 + \sqrt{x^4 + 6}} - \sqrt{2}\,x\right).$$

193. If $f(x)$ is such that $|f(x)| \leq x^2$ for all x, what is $f'(0)$?

194. Show that the angle of inclination of the tangent line to the curve $y = \text{Tan}^{-1}(2x^3)$ always lies in the interval $0 \leq \phi \leq \text{Tan}^{-1} 2$.

195. Calculate
$$\frac{d^{20}}{dx^{20}}[(x^2 + 1)\sin x].$$

196. Show that any tangent to the curve $y = (1/2)\sqrt{x - 4x^2}$ intersects the y-axis at a point equidistant from the point of tangency and the origin.

197. Let P be any point on the hyperbola $b^2 x^2 - a^2 y^2 = a^2 b^2$. If the tangent line at P cuts the asymptotes of the hyperbola at points Q and R, show that P is the midpoint of the line segment joining Q and R.

198. Show that the x-intercept of the tangent line to the curve $a/x^2 + b/y^2 = 1$ at any point is proportional to the cube of the abscissa of the point of tangency.

199. (a) Show that when the area of a circle of radius r is approximated by an inscribed polygon with n equal sides, the area of the polygon is
$$A_n = \frac{nr^2}{2}\sin\left(\frac{2\pi}{n}\right).$$
(b) Verify that $\lim_{n \to \infty} A_n = \pi r^2$.

200. Prove that the function
$$f(x) = \frac{x^2}{2} + \frac{1}{2}x\sqrt{x^2 + 1} + \ln\sqrt{x + \sqrt{x^2 + 1}}$$
satisfies the differential equation

$$x\frac{dy}{dx} + \ln\left(\frac{dy}{dx}\right) = 2y.$$

201. Verify that if a polynomial of even degree ever takes on a value opposite in sign to the coefficient of its highest-degree term, it must have at least two real zeros.

202. For what values of a and b does the function $f(x) = ax \ln x + bx^2 + x$ have relative extrema at $x = 1$ and $x = 2$? Are they relative maxima or minima?

203. A man's cottage is situated on a lake that is more or less circular with radius 2 km. Diametrically opposite is a store where he buys groceries. If the man can paddle his canoe at v km/hr and can walk along the shore at V km/hr, what combination of paddling and walking produces the quickest trip from cottage to store?

204. A fish swims upstream at a constant speed v relative to the water. The water itself has speed v_1 relative to the ground. The fish intends to reach a point at distance s upstream. The energy required is determined by the friction in water and by the time t taken. Experiments have shown that this energy is $E = cv^k t$ where $c > 0$ and $k > 2$ are constants (k depending on the shape of the fish). Given that v_1 is constant, what speed v minimizes E?

205. (a) One end of a ladder slips away from a wall at a uniform rate. Show that the speed of the other end down the wall is always increasing.

(b) Show that when the upper end falls at a constant rate, the speed of the lower end decreases.

206. An eagle is gliding horizontally 100 m above the ground at 15 m/s. It is 200 m south of a tree and is heading due north. A rabbit is 50 m east of the tree hopping along at 2 m/s toward the tree. How fast is the distance between the eagle and rabbit changing at this instant?

207. Show that
$$\lim_{x \to \infty} \frac{x - \sin x}{x + \sin x}$$
exists, but that it cannot be evaluated with L'Hôpital's rule.

208. A particle moves along the x-axis with position (in metres) given by
$$x(t) = t^3 - 9t^2 + 24t + 3, \quad t \geq 0,$$
where t is time (in seconds). Determine:

(a) the speed of the particle when it is at position $x = 39$;

(b) when the particle is speeding up;

(c) the closest it gets to the point $x = -1$ during the time interval $1 \leq t \leq 3$;

(d) its maximum speed when it is between $x = 20$ and $x = 21$.

209. A man eats a diet of 2 500 calories per day; 1 200 of them go to basic metabolism (i.e., get used up automatically). He also spends 16 calories per day for each kilogram of body mass in exercise. Assume that the storage of calories as fat is 100% efficient and that 1 kilogram of fat contains 10 000 calories. Find his mass as a function of time, and sketch this function, if initially his mass is (a) 100 kg (b) 70 kg. Is he on a diet?

210. Evaluate
$$\lim_{n \to \infty} \frac{1 - 2 + 3 - 4 + \cdots - 2n}{(3 + 8n^3)^{1/3}}.$$

211. Show that the definite integral from $x = 0$ to $x = 1$ with respect to x does not exist for the function
$$f(x) = \begin{cases} 1 & x \text{ is a rational number} \\ 0 & x \text{ is an irrational number}. \end{cases}$$

212. Without evaluating the integral
$$\int_0^{10} \frac{x}{x^3 + 16} \, dx,$$
show that it must be less than $5/6$.

213. Assuming that y is defined implicitly as a function of x by
$$\int_0^y \frac{1}{1 + t^3} \, dt + \int_0^x \cos t^2 \, dt = 5,$$
find dy/dx.

214. Evaluate
$$\int_{-a}^{a} \sin^2 x \ln\left(\frac{1 + x}{1 - x}\right) dx$$
when $0 < a < 1$.

215. (a) What is the area bounded by the line $y = 2x + 1$ and the parabola $y = x^2$?

(b) Find the line parallel to $y = 2x + 1$ which along with $y = x^2$ bounds an area equal to half that in (a).

216. Find the volume of the solid of revolution obtained by rotating the area bounded by $x^4 + y^4 = x^3$ about the x-axis.

217. A tank is in the form of a frustrum of a right-circular cone (Figure E.4). If the tank is full of oil with density ρ, what work is required to empty the tank over its top edge?

FIGURE E.4

218. The lower half of a cubical tank 2 m on each side is occupied by water, and the upper half by oil (density 0.90 grams per cubic centimetre).

    (a) What is the force on each side of the tank due to the pressure of the water and the oil?

    (b) If the oil and water are stirred to create a uniform mixture, does the force on each side change? If not, explain why not. If so, by how much does it increase or decrease?

219. Find the second moment of area of the larger loop of $y^2 = (x+1)^2(4-x^2)$ about the y-axis.

220. Find the volume cut from a right circular cylinder of radius r by a plane passing through its diameter (Figure E.5).

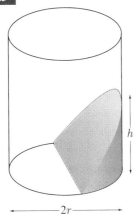

FIGURE E.5

221. A thin rod of length L has σ coulombs of positive charge uniformly distributed over each metre of length (Figure E.6). Suppose a positive charge q coulombs is placed at point P, a distance of a units along the right-bisector of the rod. According to Coulomb's law, the magnitude of the force of repulsion on q by a small amount of charge Q at point R is

$$\frac{qQ}{kr^2}$$

where k is a constant and $r = \|PR\|$. A similar charge equally distant from the other end of the rod also repels q but in a different direction. The combined repulsive force on q by these two charges is perpendicular to the rod and has magnitude

$$\frac{2qQ\cos\theta}{kr^2},$$

where θ is the angle shown.

(a) Find the total force on q.

(b) What is the force when the rod is infinitely long?

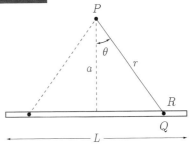

FIGURE E.6

222. (a) Your friend says that he wants to paint the area bounded by the curves

$$y = \frac{1}{x}, \quad x = 1, \quad y = 0.$$

Why would you advise him to forget it?

(b) Your friend now says that he is going to make a funnel in the shape of the surface of revolution obtained by rotating that part of $y = 1/x$ to the right of $x = 1$ about the x-axis. He intends pouring paint into the funnel until it is full. Can he do this with a finite amount of paint? If he can, what do you reply when he points out that he has now painted both sides of the area in (a)?

(c) Would you advise him to attempt to paint the outside of the funnel? How about the inside?

223. Show that the function

$$f(x) = 4 + 3\sin\left(\frac{x-2}{x+2}\right)$$

has an inverse function on the interval $x \geq 0$, and find it.

224. Prove that $\sin x$ is a transcendental function. Hint: Assume

$$P_0(x)\sin^n x + P_1(x)\sin^{n-1} x + \cdots + P_{n-1}(x)\sin x + P_n(x) = 0$$

and evaluate at $x = n\pi$.

225. Show that the function

$$f(x) = \text{Cos}^{-1}\left(\frac{a\cos x + b}{a + b\cos x}\right) - 2\text{Tan}^{-1}\left[\sqrt{\frac{a-b}{a+b}}\tan\left(\frac{x}{2}\right)\right],$$

where $0 < b \leq a$, is constant for $0 \leq x \leq \pi$. Find the value of the constant.

226. Evaluate

$$\int_0^{\pi/2} \frac{\sin x \cos x}{a^2 \cos^2 x + b^2 \sin^2 x} \, dx$$

where $ab \neq 0$.

227. Find the area bounded by $y = \ln x$ and $y = (\ln x)^2$.

228. Evaluate

$$\int_0^\pi \sqrt{1 - \cos 3x} \, dx.$$

In Exercises 229–230 evaluate the indefinite integral.

229. $\displaystyle\int \frac{1}{x^3\sqrt{x^2+5}}\,dx$ 230. $\displaystyle\int (4-x^2)^{3/2}\,dx$

231. Find the area bounded by the curve $x^4 - ax^3 + a^2 y^2 = 0$ where a is a constant.

232. Find the area bounded the the curve $(y - \text{Sin}^{-1} x)^2 = x - x^2$.

In Exercises 233–235 evaluate the indefinite integral.

233. $\displaystyle\int \frac{1}{x^2(9-x^2)^{3/2}}\,dx$ 234. $\displaystyle\int \sqrt{\frac{3+x}{3-x}}\,dx$

235. $\displaystyle\int \frac{15 + 38x - 14x^2}{15 - 14x - 11x^2 + 4x^3}\,dx$

236. In a chemical reaction, one molecule of trypsinogen yields one molecule of trypsin. In order for the reaction to take place, an initial amount of trypsin must be present. Suppose that the initial amount is y_0. Thereafter, the rate at which trypsinogen is changed into trypsin is proportional to the product of the amounts of each chemical in the reaction. Find a formula for the amount of trypsin if the initial amount of trypsinogen is A.

237. A very long, thin thread is wrapped clockwise around the circle $x^2 + y^2 = a^2$ so that one end of the string is at the point $(a, 0)$. If the thread is unwound in such a way that it is always kept taut, the path followed by the end of the string is called an involute of the circle.

(a) Find equation(s) for the involute and sketch its graph.

(b) Find a formula for the slope of the involute.

(c) What is the limit of the slope as the point $(a, 0)$ is approached along the curve?

(d) What is the slope of the curve when it crosses the positive x-axis for the first time?

238. Find the area enclosed by the astroid

$$x = a\cos^3 t, \quad y = a\sin^3 t,$$

a a constant.

239. (a) Sketch the curves

$$3(x^2 + y^2)^2 = 25(x^2 - y^2), \quad 22y = 26 - x^2.$$

How many points of intersection would you suggest for these curves.

(b) Find all points of intersection for the curves in (a) and show that at two points, the curves are tangent.

240. Find the area between the loops of $r = a\sin^3(\theta/3)$.

241. Find the area bounded by the hyperbola $b^2 x^2 - a^2 y^2 = a^2 b^2$ and the line through its focus parallel to the y-axis.

242. A model of learning proposed by Bush and Mosteller assumes that on the nth trial of a sequence of trials, the probability p_n of obtaining a certain response is related to the probability on the immediately preceding trial by

$$p_n = a + (1 - a - b)p_{n-1}$$

where a and b are constants which depend on the amounts of reward and inhibition. Prove that an explicit formula for p_n is

$$p_n = p_0(1 - a - b)^n + \frac{a}{a+b}[1 - (1 - a - b)^n]$$

where p_0 is the initial probability of response.

243. In Iraq an epidemic of methylmercury poisoning killed 459 people in 1972. Human beings became exposed to the poison when they ate homemade bread accidentally prepared from seed-wheat treated with a methylmercurial fungicide. Symptoms appeared only after weeks of exposure. Assume for simplicity that a person takes a constant daily dose d of the poison and that a certain percentage p of the accumulated poison is excreted each day. Find a formula which relates the amount of poison stored in the body to the number of days.

244. Evaluate

$$\lim_{n\to\infty} \frac{\dfrac{1}{4} + \dfrac{1}{8} + \dfrac{1}{16} + \cdots + \dfrac{1}{2^n}}{1 + \dfrac{2}{5} + \dfrac{4}{25} + \dfrac{8}{125} + \cdots + \left(\dfrac{2}{5}\right)^{n-1}}.$$

245. Show that if $\sum_{n=1}^{\infty} a_n^2$ and $\sum_{n=1}^{\infty} b_n^2$ are convergent series, then $\sum_{n=1}^{\infty} a_n b_n$ is absolutely convergent.

246. Find $f^{(6)}(0)$ if $f(x) = x^2/\sqrt{1-x^2}$.

247. Find the first three nonzero terms in the Maclaurin series for $\ln(\cos x)$.

248. Find the Maclaurin series for $\cos^4 x$.

249. Find the Maclaurin series for $x(2+3x)^{3/2}$.

250. Find $\sqrt{4.000\,000\,000\,01}$ accurate to 20 decimal places.

251. Prove that if $f(x)$ is continuous at $x = a$ and $f(a) > 0$, then there exists an open interval I containing a such that $f(x) > 0$ for all x in I.

252. (a) Show that the tangent line to the curve $y = x^3$ at any point P (other than the origin) always intersects the curve again.

(b) Verify that the slope of the curve at the second point in (a) is four times the slope at P.

253. Prove that when $f'(a)$ exists

$$\lim_{x \to a} \frac{xf(a) - af(x)}{x-a} = f(a) - af'(a).$$

254. (a) Show that if $y = (1-x)^{-a} e^{-ax}$, then

$$(1-x)\frac{dy}{dx} = axy.$$

(b) Now show that

$$(1-x)\frac{d^{n+1}y}{dx^{n+1}} - (n+ax)\frac{d^n y}{dx^n} - an\frac{d^{n-1}y}{dx^{n-1}} = 0.$$

255. Find all points on the curve $x^2 y^2 + xy = 2$ where the slope is -1.

256. Prove that the segment of the tangent line to the tractrix

$$y = \frac{a}{2}\ln\left(\frac{a+\sqrt{a^2-x^2}}{a-\sqrt{a^2-x^2}}\right) - \sqrt{a^2-x^2}$$

between the point of tangency and the y-axis always has length a.

257. Find d^2y/dx^2 if $y = |\sin x|^x$, $x > 0$.

258. A manufacturer making 144 000 units of an item during a year has a choice of different production schedules. He can make the entire lot early in the year in one production run. By doing so, he holds production costs down; but he must then carry much of the production run in inventory, and inventory costs might overbalance savings from mass production economies. Suppose that it costs $500 to make the factory ready for a production run, and $5 for each unit made after the factory has been readied. Assume that after a batch is made, the units are placed in inventory and then used up at a uniform rate such that inventory is zero when the next batch appears. If annual cost of the inventory is the cost of the average inventory, what batch size should be run to minimize annual production and inventory costs.

259. A log 20 m long has the form of a truncated cone with bases of diameters 1 m and 2 m. A beam with square cross-section is to be cut from the log so that axes of log and beam coincide. Find the dimensions of the beam with maximum volume.

260. A street light is 25 m high. A ball is thrown vertically upward from a point 10 m away from the lightpost. If the speed of the ball is 20 m/s when it is 15 m above the ground, what is the acceleration of the shadow of the ball along the ground at this instant?

261. (a) Show that the volume of the container in Figure E.7 is $abh/3$.

(b) If water is poured into the container at a rate of k cubic units per unit time, find a formula for the rate at which the depth of water is increasing when the container is half full by volume.

FIGURE E.7

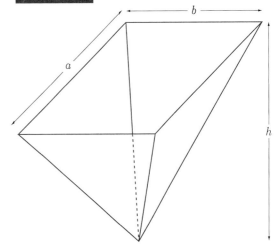

262. If a particle moves along the x-axis so that its acceleration is always proportional to its velocity, show that the percentage change in its velocity in every time interval of length T is always the same.

263. Prove that
$$\int_0^{\pi/2} f(\cos x)\, dx = \int_0^{\pi/2} f(\sin x)\, dx.$$

264. Find the volume of the barrel in Figure E.8. The cross-section is elliptic.

FIGURE E.8

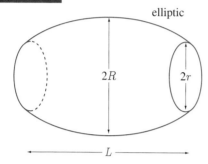

265. A piece of wood with density ρ is in the shape of a pyramid; its base is a square of side length b and its vertex is a distance a from the base. If it floats in a liquid of density $\bar{\rho}$ with its base horizontal and above the surface, find the work required to lift the block from the liquid.

266. (a) A sphere of radius r and density $\bar{\rho}$ floats partially submerged in a liquid with density ρ. Show that the depth d below the surface of the lowest point of the sphere must satisfy the equation
$$\rho(d^3 - 3rd^2) + 4r^3 \bar{\rho} = 0.$$
(b) What is d if $\bar{\rho} = \rho/10$?

267. Newton's universal law of gravitation states that the force of attraction between two point masses m and M at distance r apart has magnitude $F = GmM/r^2$, where $G = 6.67 \times 10^{-11}$ m^3/(kg·s^2). A thin uniform rod AB of length L and mass M attracts a point mass m at point C which lies at a distance d from the end closest to B on the extension of AB. What point mass placed at A has the same force of attraction on m?

268. Under what condition will the function
$$f(x) = \frac{ax+b}{cx+d}, \quad ad - bc \neq 0$$
be equal to its inverse?

269. Evaluate
$$\int_0^x \frac{1}{1+t^2}\, dt + \int_0^{1/x} \frac{1}{1+t^2}\, dt.$$

In Exercises 270–278 evaluate the indefinite integral.

270. $\displaystyle \int (\mathrm{Sin}^{-1} x)^2\, dx$ 271. $\displaystyle \int \frac{1}{x^2}\mathrm{Cos}^{-1} x\, dx$

272. $\displaystyle \int \frac{1}{\sin x \cos^2 x}\, dx$ 273. $\displaystyle \int \frac{1}{(x-4)\sqrt{x^2-16}}\, dx$

274. $\displaystyle \int x^2 \mathrm{Tan}^{-1} 2x\, dx$ 275. $\displaystyle \int e^{2x+3} \cos(3x-1)\, dx$

276. $\displaystyle \int \frac{x - \sin x}{1 - \cos x}\, dx$ 277. $\displaystyle \int \sin(2 \ln x)\, dx$

278. $\displaystyle \int \frac{x}{\cos^2 3x}\, dx$

279. The *curvature* of a curve $y = f(x)$ at a point on the curve is defined as
$$\kappa(x) = \frac{|f''(x)|}{\{1 + [f'(x)]^2\}^{3/2}}.$$

(a) Find the curvature of the parabola $y = x^2 - 2x - 3$. Where is it a maximum?

(b) Show that the only curves for which curvature is identically equal to zero are straight lines.

(c) What happens to curvature at a point of inflection on the graph of a function?

280. Show that when a curve is defined parametrically, its curvature (Exercise 279) can be expressed in the form
$$\kappa(t) = \frac{\left|\dfrac{dy}{dt}\dfrac{d^2x}{dt^2} - \dfrac{dx}{dt}\dfrac{d^2y}{dt^2}\right|}{\left[\left(\dfrac{dx}{dt}\right)^2 + \left(\dfrac{dy}{dt}\right)^2\right]^{3/2}}.$$

281. The *radius of curvature* of a curve at a point is defined as $\rho = 1/\kappa$ where κ is curvature at the point (see Exercises 279 and 280).

(a) Show that the radius of curvature of a circle is constant and equal to the radius of the circle.

(b) At what points on the ellipse $b^2 x^2 + a^2 y^2 = a^2 b^2$, $(a > b)$, is the radius of curvature a maximum and a minimum?

282. Suppose a curve $y = f(x)$ is concave upward at a point (x_0, y_0). The *circle of curvature* at (x_0, y_0) is defined as the circle above the curve with radius equal to ρ (the radius of curvature in Exercise 281),

and centre along the normal to $y = f(x)$ at (x_0, y_0) at a distance ρ from (x_0, y_0). If the curve is concave downward, the circle is drawn below the curve.

(a) Find the circle of curvature to the parabola $y = x^2 - 2x - 3$ at $(1, -4)$.

(b) Is it true that the circle of curvature at a point on a curve passes through that point? Draw a picture.

(c) Show that the circle of curvature and curve share the same tangent line at their common point.

283. Find the equation of the tangent line to the curve

$$x = 4(2\cos\theta - \cos 2\theta), \quad y = 4(2\sin\theta - \sin 2\theta),$$

$$0 \leq \theta < 2\pi,$$

at the point $(4\sqrt{3} - 2, 4 - 2\sqrt{3})$.

284. (a) The curve $r = Aq^\theta$, where $A > 0$ and $q > 1$ is called a logarithmic spiral. Sketch it.

(b) Show that the slope of the logarithmic spiral at every point at which it intersects the radial line $\theta = k$ is the same.

285. Dropping leaflets by air is a means sometimes used in advertising to reach a large group of customers. Suppose it is assumed that a constant proportion λ of the leaflets actually survive a given time period T, while the remainder are lost, destroyed, or otherwise rendered unreadable. It is further assumed that each leaflet that survives until the ith time period will reach β people, on the average, during that period. Given that N leaflets are dropped, how many people are reached by the leaflets?

286. Find the Taylor series about $x = 1$ for the function $\sqrt[3]{7 + x}$.

287. Use the Maclaurin series for $f(x) = (1+x)e^{-x} - (1-x)e^x$ to find the sum of the series $\sum_{n=1}^{\infty} n/(2n+1)!$.

288. Find point(s) on the curve $y = \ln(x^2 + 1)$ where the tangent line passes through the point $(1, -2 + \ln 2)$.

289. What point on the ellipse $x^2 + 3y^2 = 9$ is furthest from $(5, 7)$?

290. Let P be any point on the hyperbola $x^2 - y^2 = a^2$. If the tangent line at P cuts the asymptotes of the hyperbola at points Q and R, show that the area of triangle OQR is a^2.

291. Show that if $y = f(x)$ is concave upward on an interval I, then

$$\frac{f(x_1) + f(x_2)}{2} \geq f\left(\frac{x_1 + x_2}{2}\right)$$

for all x_1 and x_2 in I.

292. The rate of flow of water through a circular pipe is directly proportional to its hydraulic radius R defined as $R = A/s$. Quantity A is the cross-sectional area of the water flow in the pipe, and s is the length of the circular arc of this cross-sectional area. The degree to which the pipe is filled is measured by the angle θ from the centre of the pipe to the horizontal surface of the water. For what degree of filling is R a maximum?

293. Sand from a conveyor belt drops at the rate of 10 cubic centimetres per second. It creates a pile on a level floor in the form of a right-circular cone. The height h of the pile increases at the same rate as the diameter D of the circular area A on the floor covered by the sand. How fast is A changing when D is 0.5 metres?

294. Find the volume of the solid of revolution obtained by rotating the area bounded by the curves

$$y = x \sin x, \quad y = 0, \quad 0 \leq x \leq \pi,$$

about the x-axis.

295. A triangular plate has one vertex in the surface of a fluid. The opposite side has length a, is parallel to the surface of the fluid, and is b units below the surface. What is the force on each side of the plate?

296. If $y = f(x)$ has inverse function $x = f^{-1}(y)$, find a formula for d^2x/dy^2 in terms of derivatives of $f(x)$. Illustrate with an example.

297. Verify that

$$\int (\ln x)^n \, dx = (-1)^n n! x \sum_{r=0}^{n} \frac{(-\ln x)^r}{r!} + C.$$

298. Find the volume of the solid of revolution when the area under one arch of the cycloid of Exercise 46 in Section 10.1 is rotated about (a) the x-axis (b) the y-axis (c) the line $x = \pi R$.

299. Find the area of the surface of revolution when the lemniscate $(x^2 + y^2)^2 = a^2(x^2 - y^2)$ is rotated about the x-axis.

300. Find the area of the surface of revolution when the cardioid $r = a(1 - \cos\theta)$ is rotated about the x-axis.

301. Inscribed in a circle of radius R is a square. A circle is then inscribed in the square, a square in the circle, and so on and so on. Find the limit of the sum of the areas of all the circles and the limit of the sum of the areas of all the rectangles.

302. Are there angles at which the sine of the angle changes twice as fast as the cosine?

303. Find a cubic polynomial in x which approximates the function defined implicitly by the equation $y^5 + xy + 1 = 0$.

304. How many terms in the Maclaurin series for the integrand guarantee a value for
$$\int_1^2 \frac{\sin 2x}{x} dx$$
accurate to 10^{-3}?

305. Evaluate
$$\lim_{x \to 0} \frac{\sin(x^n)}{(\sin x)^m}$$
where n and m are positive integers.

306. Show that the segment of the tangent line to the astroid $x^{2/3} + y^{2/3} = a^{2/3}$ contained between the coordinate axes always has length a.

307. Show that the points of inflection on the curve $y = (x+1)/(x^2+1)$ are collinear.

308. At what angle(s) does the sine of the angle change half as fast as the angle itself?

309. A rectangular plate with sides of length L and l ($L > l$) is submerged in a liquid of density ρ. Its longer sides are parallel to the surface of the liquid, the plate makes an angle α with the surface, and the edge closest to the surface is at depth d. What is the force on each side of the plate?

310. The gamma function $\Gamma(\nu)$ for $\nu > 0$ is defined by
$$\Gamma(\nu) = \int_0^\infty x^{\nu-1} e^{-x} dx.$$
Show that $\Gamma(\nu + 1) = \nu \Gamma(\nu)$.

311. At what point(s) on the curve
$$x = R(1 - \cos\theta), \quad y = R(\theta - \sin\theta)$$
is the slope of the tangent line equal to 1?

312. Find a polynomial approximating the indefinite integral
$$\int \frac{1}{\sqrt[4]{1+x^4}} dx$$
to within 10^{-6} on the interval $0 \le x \le 1$.

313. Show that points of inflection of the curve $y = (\sin x)/x$ lie on the curve $y^2(4 + x^4) = 4$.

314. (a) A spherical shell with inner and outer radii 1 m and 2 m is made from styrofoam with mass 2 kg/m³. The shell is cut into two equal pieces by a plane, and the flat edge of one piece is carefully placed on the surface of a large container of water. How far will the styrofoam sink?

(b) If a small hole is drilled in the top of the floating shell, will the styrofoam sink further? If so, to what depth?

315. Two substances A and B react to form a third substance C in such a way that 2 gm of A react with 1 gm of B to produce 3 gm of C. The rate at which C is formed is equal to the product of the amounts of A and B present in the mixture. Find the amount of C present in the mixture as a function of time t when the original amounts of A and B brought together at time $t = 0$ are as follows. Sketch graphs of all three functions on the same axes.

(a) 20 gm and 10 gm (b) 20 gm and 5 gm

(c) 20 gm and 20 gm

316. Find a polynomial in x approximating $f(x) = \cos^3 x$ on $0 \le x \le 0.5$ with maximum error 10^{-6}.

317. For what values of c is the range of the function
$$f(x) = \frac{x^2 + 2x + c}{x^2 + 4x + 3c}$$
equal to $-\infty < x < \infty$?

318. Tangents at points A and B on a circle intersect at point F. The tangent at the midpoint C of the smaller arc of the circle joining A and B intersects AF and BF in D and E respectively. What is the limit of the ratio of the areas of triangles ABF and DEF as A approaches B?

319. Use the definition of a limit to prove that $\lim_{x \to 2} (x^2 + 3)/(x - 1) = 7$.

320. Find the equation of a circle with centre on the line $2y + \sqrt{3}x = 4\sqrt{3} + \sqrt{7}$ which is tangent to the ellipse $9x^2 + 4y^2 = 36$, and which satisfies the condition that the line joining the point of tangency to the centre of the circle is perpendicular to the given line.

321. (a) Sketch the curve $y^2 = x^3(a - x)$, called a piriform.

(b) Find point(s) where the tangent line to the piriform is horizontal.

(c) Where is the point of inflection on the top half of this curve?

322. The first derivative test for a relative minimum states that if $f'(x)$ changes from a negative quantity to a positive quantity as x increases through a critical point, then that critical point yields a relative minimum. Is the converse statement true? To answer, consider the function

$$f(x) = \begin{cases} \dfrac{x^2}{2}\left[1 + \sin\left(\dfrac{1}{x}\right)\right] & x \neq 0 \\ 0 & x = 0. \end{cases}$$

323. Find that point P on the ellipse $b^2 x^2 + a^2 y^2 = a^2 b^2$ so that the triangle with vertices P, $Q(0,b)$ and $R(a,0)$ has maximum possible area.

324. Two corridors, one 3 m wide and the other 6 m wide, meet at right angles. Find the length of the longest square beam, $1/3$ m by $1/3$ m, that can be transported horizontally from one corridor to the other.

325. A cylindrical tank of radius r and height H has its axis vertical. It is originally (time $t = 0$) full of water, and for $t > 0$, water leaks out through a hole in the bottom.

(a) To find the speed at which water exits through the hole consider when the depth of water is h. In a small time dt, the depth drops by an amount dh. Use the fact that during dt, the potential energy of the small disc of water with thickness $|dh|$ at height h above the bottom of the cylinder is converted into kinetic energy of the same volume exiting through the hole with speed v to show that $v = \sqrt{2gh}$ where $g = 9.81$. This formula describes an ideal situation where all potential energy is converted into kinetic energy. In less than ideal situations, it is customary to assume that $v = c\sqrt{2gh}$, where $0 < c < 1$ is a constant.

(b) Show that the rate of change of the volume of water in the cylinder is given by the two expressions

$$\dfrac{dV}{dt} = \pi r^2 \dfrac{dh}{dt} \quad \text{and} \quad \dfrac{dV}{dt} = -Av$$

where A is the cross-sectional area of the hole, and hence deduce that

$$\dfrac{dh}{dt} = -\dfrac{2\sqrt{2g}\,cA}{\pi r^2}\sqrt{h}.$$

This is sometimes called Torricelli's law.

(c) Find h as a function of t and thereby determine a formula for how long it takes for the cylinder to empty.

326. Evaluate

$$\lim_{n \to \infty} \dfrac{1}{n^2}\left[\sin\left(\dfrac{1}{n^2}\right) + 2\sin\left(\dfrac{4}{n^2}\right) + 3\sin\left(\dfrac{9}{n^2}\right) + \cdots + n\sin 1\right].$$

327. Show that when $f'(x)$ and $f''(x)$ are positive on an interval $a \leq x \leq b$, then

$$(b-a)f(a) < \int_a^b f(x)\,dx < \dfrac{1}{2}(b-a)[f(a) + f(b)].$$

328. Show that the function

$$y = \dfrac{1}{3}\int_0^x f(t)e^{t-x}\sin 3(x-t)\,dt$$

satisfies the differential equation

$$\dfrac{d^2y}{dx^2} + 2\dfrac{dy}{dx} + 10y = f(x).$$

329. Find a in order that the area in Figure E.9 be as large as possible.

FIGURE E.9

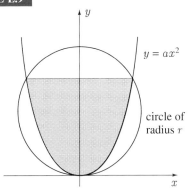

330. A square is rotated about a line which lies in its plane and passes through one of its vertices. What position of the line yields maximum volume for the resulting solid of revolution?

331. The two people in Figure E.10 each pull in the cable at 1 m/s. The curved part of the cable always has shape $y = k^{-1}\cosh kx + C$, where k and C, although independent of x, depend on the height of the minimum point above the x-axis. Find how fast this minimum point is rising when it is 5 m above the x-axis.

FIGURE E.10

332. (a) A right-circular cylinder of radius r m, length L m, and density 250 kg/m^3 floats in water with its axis horizontal. How far above the surface of the water is its axis?

(b) How much work is required to lift the cylinder out of the water?

(c) Repeat part (b) if the cylinder floats with its axis vertical.

333. A hole is drilled through a long right-circular cylinder of radius R. If the radius of the hole is $r < R$, and axes of the cylinder and hole meet at right angles, set up a definite integral for the volume removed from the cylinder by the hole.

334. For what values of k do the following integrals converge:

(a) $\int_2^\infty \dfrac{1}{x^k \ln x}\, dx$
(b) $\int_2^\infty \dfrac{1}{x(\ln x)^k}\, dx$
(c) $\int_2^\infty \dfrac{1}{(\ln x)^k}\, dx$

335. Prove that e^x is a tanscendental function. Hint: Assume

$$P_0(x)e^{nx} + P_1(x)e^{(n-1)x} + \cdots + P_{n-1}(x)e^x + P_n(x) = 0$$

and take limits as $x \to -\infty$.

336. Let P be any point on the right half of the curve $y = k\cosh(x/k)$, and Q be the foot of the perpendicular from P to the x-axis. Draw a semicircle to the left of PQ with PQ as diameter. Let R be the point on the semicircle at a distance k from Q. Show that PR is tangent to the curve.

337. Prove that when $P(x)$ is a polynomial of degree n,

$$\int e^{mx} P(x)\, dx = e^{mx} \sum_{r=0}^{n} \dfrac{(-1)^r}{m^{r+1}} P^{(r)}(x) + C$$

where $P^{(r)}(x)$ is the rth derivative of $P(x)$ and $P^{(0)}(x) = P(x)$.

338. Prove that for constant a and $n \geq 1$ an integer:

(a) $\int \sin^{2n} ax\, dx = -\dfrac{1}{a} \cos ax \sum_{r=0}^{n-1} \dfrac{(2n)!(r!)^2}{2^{2n-2r}(2r+1)!(n!)^2}$

$\sin^{2r+1} ax + \dfrac{(2n)!}{2^{2n}(n!)^2} x + C$

(b) $\int \sin^{2n+1} ax\, dx = -\dfrac{1}{a} \cos ax \sum_{r=0}^{n} \dfrac{2^{2n-2r}(n!)^2(2r)!}{(2n+1)!(r!)^2}$

$\sin^{2r} ax + C$

339. Hecht assumes that the amount S of a retinal substance which remains when exposed to light of intensity I is characterized by the equation

$$\dfrac{dS}{dt} = bI(a - S) - kS^2$$

where a denotes the amount of S prior to stimulation, and b and k are positive constants. Show that

$$S(t) = \dfrac{-bI}{2k} + \dfrac{B(1 + De^{-2kBt})}{1 - De^{-2kBt}},$$

where $B^2 = 4b^2 I^2/k^2 + abI/k$ and $D = [2k(a-B) + bI]/[2k(a+B) + bI]$.

340. Circles C_1 and C_2 in Figure E.11(a) have radius R m and circle C_1 rolls around C_2 without slipping.

FIGURE E.11

(a)

(b)

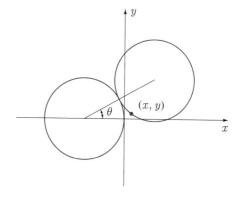

(a) Show that the curve traced out by the point $P(x,y)$ on C_1 which is initially at the origin has equations

$x = 2R\cos\theta(1 - \cos\theta),\qquad y = 2R\sin\theta(1 - \cos\theta)$

(see Figure E.11(b)).

(b) Show that if ϕ is the angle between the positive x-axis and the line joining $P(x,y)$ to the origin, then $\phi = \theta$. Hence, show that the polar equation of the curve is
$$r = 2R(1 - \cos\theta).$$
Sketch the curve.

(c) If C_1 rotates twice around C_2 every second, what is the speed of P when it is at the points with Cartesian coordinates $(-4R, 0)$ and $(0, 0)$?

341. A piece of sheet metal $1/6$ cm thick is stored as a coil. Inner and outer radii of the coil are 40 cm and 80 cm. How long is the piece of metal?

342. Find the area bounded by the curve $x^4 + y^4 = x^2 + y^2$.

343. Suppose functions $f(x)$ and $g(x)$ are positive and their graphs are concave upward on the interval $a < x < b$. Show that if they have a relative minimum at the same value of x in the interval, the graph of $f(x)g(x)$ is concave upward on $a < x < b$.

344. Show that the function
$$f(x) = \begin{cases} x^4\left[2 + \sin\left(\dfrac{1}{x}\right)\right] & x \neq 0 \\ 0 & x = 0 \end{cases}$$
has a relative minimum at $x = 0$ but this cannot be verified with the first derivative test.

345. The line $y = x + 1$ intersects the ellipse $4x^2 + 9y^2 = 36$ in two points P and Q. Find the point R on the ellipse in order that triangle PQR have maximum possible area.

346. Tangent lines to the curve $y = \sqrt{1-x}$ make triangles with the positive x- and y-axes.

(a) Find the triangle with smallest area.

(b) Find the triangle with smallest perimeter.

347. (a) Show that
$$\left[\int_a^b f(x)g(x)\,dx\right]^2 \leq \int_a^b [f(x)]^2\,dx \int_a^b [g(x)]^2\,dx.$$
Hint: Consider the following function of t,
$$\int_a^b [tf(x) + g(x)]^2\,dx.$$

(b) Verify that
$$\left[\int_a^b f(x)\,dx\right]^2 \leq (b-a)\int_a^b [f(x)]^2\,dx.$$

348. Show that if $f(x)$ is a continuous periodic function with period p, then
$$\int_a^{a+p} f(x)\,dx = \int_0^p f(x)\,dx$$
for any a.

349. Use the result of Exercise 348 to show that when $f(x)$ is an odd, p-periodic continuous function, then the function
$$\int_a^x f(t)\,dt$$
is also p-periodic for any a. Is it odd?

350. A spy satellite is in circular orbit 1 500 km above the earth's surface. The satellite's camera returns to earth a picture of every point on earth in direct "line of sight" from the satellite.

(a) At any instant, how much area does the satellite spy on?

(b) What is the total area under the satellite's surveillance in one complete revolution around the earth. Assume the earth is a perfect sphere with radius 6 370 km.

351. Find the area bounded by the curve $x^2y^2 = 4(x-1)$ and the line through its points of inflection.

352. A bead slides from rest at the origin on a frictionless wire in a vertical plane to the point (x_0, y_0) under the influence of gravity (Figure E.12). It can be shown that the time for the bead to traverse the path is
$$t = \frac{1}{\sqrt{2g}}\int_0^{x_0}\sqrt{\frac{1+(y')^2}{y}}\,dx$$
where $g = 9.81$ and $y' = dy/dx$. The problem of finding that shape $y = f(x)$ of wire that makes t as small as possible is often called the brachistochrone problem. In the area of mathematics called "calculus of variations", it is shown that $y = f(x)$ must satisfy the differential equation
$$2y\frac{d^2y}{dx^2} + \left(\frac{dy}{dx}\right)^2 + 1 = 0.$$

(a) Show that the curve defined parametrically by
$$x = a(\theta - \sin\theta), \quad y = a(1 - \cos\theta),$$
where a is a constant satisfies the differential equation.

(b) Sketch the curve in (a), called a cycloid.

(c) Show that if $2x_0 = \pi y_0$, then the value of a that makes the cycloid pass through (x_0, y_0) is $y_0/2$. Draw that part of the cycloid joining $(0,0)$ and (x_0, y_0) in this case. What is special about it?

(d) How would you find a if the condition in (c) were not satisfied?

FIGURE E.12

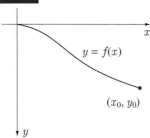

353. Find the volume of the solid of revolution when the area inside the lemniscate $(x^2 + y^2)^2 = a^2(x^2 - y^2)$ is rotated about the x-axis.

354. Find the volume of the smallest cone that can be circumscribed about a sphere of radius R.

Evaluate the integrals in exercises 355 and 356.

355. $\int (x^2 - 9)^{3/2}\, dx$ 356. $\int_{\pi/6}^{\pi/3} \ln(\tan x)\, dx$

357. Two identical rods of length L and mass M lie on the same straight line a distance d apart. Use Newton's universal law of gravitation (Exercise 267) to find the force of attraction between the rods.

358. Repeat Exercise 259 if the length of the beam is L and diameters of its ends are a and b.

359. Two military outposts are situated at points P and Q on the desert 40 km apart (Figure E.13) Parallel to PQ and distant 10 km is a road. A courier can travel 14 km/hr on the sand and 50 km/hr on the road. What path would you recommend for quickest trip from P to Q?

FIGURE E.13

360. A number x_0 is called a fixed point of a function $f(x)$ if $f(x_0) = x_0$. How many fixed points can a function have if $f'(x) < 1$ for all x?

361. Show that the function $f(x) = x^3 + px + q$ has three distinct zeros if and only if $4p^3 + 27q^2 < 0$.

362. Show that the equation

$$2 \sin \theta = \theta(1 + \cos \theta)$$

has no solution in the interval $0 < \theta < \pi$.

363. Repeat Exercise 345 with a line $y = mx + c$ which intersects the ellipse $b^2 x^2 + a^2 y^2 = a^2 b^2$ in two points.

364. A circular pasture has radius R. A cow is tied to a stake at the edge of the pasture. What length of rope permits the cow to graze on half the pasture?

365. Use the substitution $x = \sin^2 \theta$ and integration by parts to find a formula for

$$\int_0^1 x^n (1-x)^m\, dx$$

where m and n are positive integers.

366. Verify that for $n \geq 1$ an integer,

$$\int x^n \cos x\, dx = \sin x \sum_{r=0}^{\left[\frac{n}{2}\right]} \frac{(-1)^r n!}{(n-2r)!} x^{n-2r}$$

$$+ \cos x \sum_{r=0}^{\left[\frac{n-1}{2}\right]} \frac{(-1)^r n!}{(n-2r-1)!} x^{n-2r-1} + C,$$

where the square brackets indicate the greatest integer function.

367. Verify that

$$\int_0^{\pi/2} \sin^n x\, dx = \int_0^{\pi/2} \cos^n x\, dx$$

$$= \begin{cases} \frac{2 \cdot 4 \cdot 6 \cdots (n-1)}{1 \cdot 3 \cdot 5 \cdots (n)}, & n \text{ an odd integer} \\ \frac{1 \cdot 3 \cdot 5 \cdots (n-1)}{2 \cdot 4 \cdot 6 \cdots (n)} \frac{\pi}{2}, & n \text{ an even integer} \end{cases}$$

368. (a) A patrol boat at point A in Figure E.14 spots a submarine submerging at point B at a time that we call $t = 0$. The submarine, unaware of the patrol boat, follows a straight line path at constant speed v along some angle ϕ relative to BA (unknown to the patrol boat). The patrol boat heads directly toward point B at speed $V > v$ for $k/(V + v)$ units of time arriving at point C. Show that the submarine and patrol boat are equidistant from B at $t = k/(V + v)$.

(b) We set up a system of polar coordinates with B as pole and BA as polar axis. Let the distance $\|BC\|$ be denoted by r_0. Suppose that the patrol boat now follows the logarithmic spiral $r = r_0 e^{\theta/a}$ still at speed V, where $a = \sqrt{V^2/v^2 - 1}$. Show that the patrol boat must intercept the submarine.

FIGURE E.14

Path of submarine

369. Evaluate

$$\int \frac{2x^3 + 8x^2 - 3x + 5}{x^4 + 3x^3 + x^2 + 2x - 12} \, dx.$$

370. A sheet of metal a units wide and L units long is to be bent into a trough as shown in Figure E.15. End pieces are also to be attached. If edge AB must be the arc of a circle, determine the radius of the circle in order that the trough hold the biggest possible volume.

FIGURE E.15

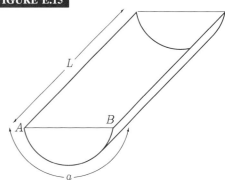

ANSWERS

4. -8
6. does not exist
8. ∞
10. does not exist
12. $4/3$
14. $1/2$
16. does not exist
18. 1
20. $x(2 - 3x^2)/(1 - 3x^2)^{3/2}$
22. $6(2x+1)^2 \ln(x+1) + (2x+1)^3/(x+1)$
24. $e^{2x}/[(e^{2x}+1)\sqrt{\ln(e^{2x}+1)}]$
26. $2x \csc^2(x^2 - 1)[1 - 2x^2 \cot(x^2 - 1)]$
28. $-1/[6\sqrt{2-x}(1+\sqrt{2-x})^{2/3}]$
30. $[\cos(x+y) - y]/[x - \cos(x+y)]$
32. $[xy + (1 + x^2 y^2) \text{Tan}^{-1}(xy)]/(1 + x^2 y^2 - x^2)$
34. $-\sec^2 x/[(1 - 2u)\sqrt{u^2 - u}]$
36. $te^{-3t}(2 - 3t)(x+1)/x$
38. $(4/3)\cot t$
40. $(3e^{-3v} + 2v)/(1 - 3e^{3v})$
42. $4/(x^2 - 4x)$
44. $16x^2/(1 + 4x^4)$
46. $1/\sqrt{x^2 + 9}$
48. $98/(x^3\sqrt{7x^2 + 49})$
50. $-15x^3\sqrt{1 - x^2}$
52. $125/[x(25 - x^2)^{3/2}]$
54. $-87x/[2(6 + 3x + 4x^2)^{3/2}]$
56. $49/(14x - x^2)^{3/2}$
58. $-5/[\sqrt{2}x\sqrt{3x^5 - 2}]$
60. $6/(4\cos^2 x + 9\sin^2 x)$
62. $x^3 \cos x$
64. $21 \cos 3x/\sqrt{1 + 49 \sin^2 3x}$
66. $(1/x^2) \text{Sec}^{-1}(2x)$
68. $10/(5 + 3e^{2x})$
70. $-1, (1, 0)$
72. $5/13, 170/2197$
76. $-2/3$
78. $x/3 - (1/6) \ln(3 + 4e^{2x}) + C$
80. $(1/3)(x-1)^3 \ln x - x^3/9 + x^2/2 - x + (1/3) \ln x + C$
82. $x/2 - (1/12) \sin 6x + C$
84. $(1/20) \cos 5x - (1/6) \cos 3x - (1/44) \cos 11x + C$
86. $\sqrt{x^2 - 3} + 5 \ln|x + \sqrt{x^2 - 3}| + C$
88. $-\sqrt{7 - x^2}/(7x) + C$
90. $\text{Sin}^{-1}(x - 1) - \sqrt{2x - x^2} + C$
92. $x + 3 \ln|x| + (2/\sqrt{3}) \text{Tan}^{-1}[(x+1)/\sqrt{3}] + C$
94. $-\cot x - (1/3) \cot^3 x + C$
96. $x/125 - (1/125) \ln(2e^x + 5) - (21/100)/(2e^x + 5) + (29/40)/(2e^x + 5)^2 + C$
98. $x/\sqrt{5 - x^2} - \text{Sin}^{-1}(x/\sqrt{5}) + C$
100. $3(x^2 - x + 5) \ln(x^2 - x + 5) - 3x^2 + 3x + C$
102. $\sec x + \cos x + C$
104. divergent
106. absolutely convergent
108. divergent
110. absolutely convergent
112. divergent
114. convergent
116. divergent
118. convergent
120. convergent
122. divergent
124. $(0.442, 0.407), (-2.442, -12.407)$
126. $\pi/4 + \alpha/2$
130. $6(\sqrt{2} + \text{Tan}^{-1}\sqrt{2})$
132. $2/\pi + (4\sqrt{2} - 5)/3$
134. $2\pi[6 \ln(3/2) - 2/3]$
136. $-1/2 + \ln 3$
138. $2/\sqrt{5}$ m, $1/\sqrt{5}$ m

ADDITIONAL EXERCISES

140. $\pi ab^3\rho/4$, $\pi a^3 b\rho/4$ **142.** $|(1-b)s| < 1$
144. $2\sqrt{3}$ m/s toward E **146.** $6x^2\cos x^2 - 4x^4\sin x^2$
148. $(15 - 8\ln 4)\pi/2$ **150.** $-187/8$
152. $4/\sqrt{5 + 16\ln 2 + 16(\ln 2)^2}$
154. $1/e$
158. $3 + \pi/\sqrt{3}$, relative maximum
160. concave upward   **162.** $-16/135$ m/s^2
164. $x_1 + (A/\omega)[\sin(\omega t_2 + \phi) - \sin(\omega t_1 + \phi)]$
166. $4\sqrt{3}a$ **168.** b/a
170. $(1/2)\ln[x/(2-x)]$ **172.** $(\pi R, 5R/6)$
174. $\sum_{n=1}^{\infty} x^{2n}/[n(2n-1)]$, $-1 \le x \le 1$
176. $-1/20$ **178.** $22/\sqrt{137}$
182. it doesn't   **184.** $(3\pi/16, 0)$
190. $100(1.025)^{[\![t/3]\!]}$ **192.** 0
202. $1/(\ln 2 - 1)$, $\ln 2/(2 - 2\ln 2)$, relative maximum at $x = 1$, relative minimum at $x = 2$
204. $kv_1/(k-1)$
206. decreasing at $62/\sqrt{21}$ m/s
208. (a) 24 m/s   (b) $2 < t < 3$, $t > 4$
(c) 19 m   (d) 7.592 m/s
210. $-1/2$ **214.** 0
216. $\pi^2/16$
218. (a) 36297 N   (b) increases by 981 N
220. $2r^2h/3$
222. (a) No   (b) Yes   (c) No, Yes
226. $(b^2 - a^2)^{-1}\ln(b/a)$ if $b^2 \ne a^2$; $1/(2a^2)$ if $b^2 = a^2$
228. $2\sqrt{2}$
230. $6\operatorname{Sin}^{-1}(x/2) + 2x\sqrt{4-x^2} + (x/4)(2-x^2)\sqrt{4-x^2} + C$
232. $\pi/4$
234. $3\operatorname{Sin}^{-1}(x/3) - \sqrt{9-x^2} + C$
236. $y_0(A + y_0)/[y_0 + Ae^{-k(A+y_0)t}]$
238. $3\pi a^2/8$ **240.** $(5\pi + 18\sqrt{3})a^2/32$
244. $3/10$ **246.** 270
248. $1 + \sum_{n=1}^{\infty}(-1)^n(2^{2n-1} + 2^{4n-3})x^{2n}/(2n)!$
250. $2.000\,000\,000\,002\,5$
258. 5367   **260.** 175.475 m/s^2
264. $\pi L(2R^2 + r^2)/3$ **266.** (b) $d = 0.3916r$
268. $f(x) = x$ or $f(x) = (ax + b)/(cx - a)$
270. $x(\operatorname{Sin}^{-1} x)^2 + 2\sqrt{1-x^2}\operatorname{Sin}^{-1} x - 2x + C$
272. $\ln|\csc x - \cot x| + \sec x + C$
274. $(x^3/3)\operatorname{Tan}^{-1}(2x) + (1/48)\ln(1 + 4x^2) - x^2/12 + C$
276. $-x\cot(x/2) + C$
278. $(x/3)\tan 3x - (1/9)\ln|\sec 3x| + C$
282. (a) $(x-1)^2 + (y + 7/2)^2 = 1/4$
286. $2 + (x-1)/12 + 2\sum_{n=2}^{\infty}(-1)^{n+1}[2 \cdot 5 \cdot 8 \cdots (3n-4)](x-1)^n/(24^n n!)$, $-7 \le x \le 9$
288. $(-1, \ln 2)$ **292.** 4.4934
294. $\pi^2(2\pi^2 - 3)/12$ **296.** $-f''(x)/[f'(x)]^3$
298. (a) $5\pi^2 R^3$ (b) $6\pi^3 R^3$ (c) $\pi R^3(9\pi^2 - 16)/6$
300. $32\pi a^2/5$

INDEX

Abscissa, 9
Absolute convergence, 542
Absolute maxima and minima, 185
Absolute value, 5
Acceleration, 198, 242
Addition of ordinates, 42
Algebraic function, 369
Allometry, 258
Alternating series, 543
    approximating sum of, 547, 584
    convergence theorem for, 544
    harmonic, 544
Amplitude, 602
Angle, between lines, 21
    between curves, 111
    radian measure of, 595
Angular displacement, 135
Antiderivative, 234
Approximation. Refer to what is being approximated
Archimedes' principle, 330
Area, of a circle, 427
    as a definite integral, 294
    of an ellipse, 428
    in polar coordinates, 484
    of a surface of revolution, 349
Arithmetic mean, 8, 170
Astroid, 472, 662, 666
Asymptote, horizontal, 68
    of a hyperbola, 34
    slanted, 183
    vertical, 65
Audible sound threshold, 610
Average, cost, 184
    price, 80
    profit, 113
    rate of change, 90
    value of a function, 285
    velocity, 89
Axes, coordinate, 9
    of an ellipse, 492
    of a hyperbola, 494

Bacteria, rate of growth, 403
Base of logarithms, 607

Bernoulli numbers, 580
Bessel functions, 558, 560, 579, 580, 589
Bifolium, 171, 484, 487
Binomial, coefficients, 116, 520
    series, 575
    theorem, 98, 520
Blood flow, 269, 351
Bouguer-Lambert law, 409
Bounded sequence, 511
Boyle's law, 208
Brachistochrone problem, 669
Brentano-Stevens law, 258, 408
Building procedure, 40
Bush and Mosteller model, 662

Calculus, Fundamental theorem of, 276, 281
Carbon dating, 407
Cardiod, 171, 477, 483, 484, 665
    area of, 486
    length of, 483, 484
Cartesian coordinates, 9
    plane, 39
Catenary, 394
Cauchy's generalized mean value theorem, 146, 471
Cell, bee's, 195
Centre of mass, 333, 334, 335
Centroid, 22, 340
Chain rule, 117
Change of variable, in definite integrals, 288
    in indefinite integrals, 248, 399
    in sigma notation, 263
Circle, 27
    arc length on, 314, 595
    area of, 427
    of curvature, 664
    equation of, 27, 28
Circumcircle, 30
Closed interval, 2
Coefficient of friction, 324, 389
Coefficients, binomial, 575
Cofunction, 133

Common ratio for geometric series, 266, 524
Comparison test for series, 534
Completeness property for reals, 560
Completing the square, 23
    for integration, 429
Composite function, 78
Compound interest, 613, 659
Concavity, 172
Conchoid of Nicomedes, 122
Conditional convergence, 543
Conic section, 488
    directrix, 499
    eccentricity, 499
    focus, 499
    in polar coordinates, 499
Continuity, of a differentiable function, 103
    of a function, 73
    on an interval, 74
    from the left, right, 74
Continuous. See Continuity
Convergence, absolute, 542
    conditional, 543
    radius of, 554
    of a sequence, 511
    of a series, 526
Coordinate axes, 9
Coordinates, Cartesian, 9
    for points on a line, 2
    polar, 472
    rectangular, 9
Corner, 103
Cosecant. See Trigonometric functions
Cosine. See Trigonometric functions
Cosine law, 597
Cotangent. See Trigonometric functions
Coulomb's law, 661
Critical point, 164
Cubic function or polynomial, 43
Curvature, circle of, 664
    of a curve, 664
    radius of, 664
Curve, curvature of, 664
    length of, 315, 469, 482

parametric definition of, 463
Cycloid, 471, 658, 665, 669
Cylindrical shell method for volumes, 309

Decay, exponential, 403
Decibel, 610
Decreasing function, 160
  sequence, 511
Definite integral, 272
Demand function, 145, 163, 374
Dependent variable, 37
Depreciation and repairs, 307, 351
Derivative(s), of composite functions, 117,
  definition of, 91
  of exponential functions, 138
  first, 91
  higher order, 113
  of hyperbolic functions, 392
  for implicitly defined functions, 122
  of an integral, 281, 283
  of inverse hyperbolic functions, 397
  of inverse trigonometric functions, 385
  of logarithm function, 137
  one-sided, 105
  for parametrically defined functions, 467
  power rule for, 98, 119
  of a power series, 567, 571
  product rule for, 107
  quotient rule for, 107
  of a sum, 98
  symbols for, 94
  of trigonometric functions, 130
Descartes, 9
Differentiable function, 105
Differential, 221
Differential equation, 238, 251
  power series solutions of, 587
Differentiation, 91. See also Derivative
  implicit, 124
  logarithmic, 142
  parametric, 467
Dipole, 484
Dirac delta function, 82, 365
Directrix, 489, 499
Disc method for volumes, 304
Discontinuity, 73
Discontinuous, 73
Diseases, 176, 440
Displacement, 198
Distance, between a point and a line, 22, 196
  between two points, 10
Divergence, of an improper integral, 264
  of a sequence, 511
  of a series, 526
Domain of a function, 37
Donut, 314, 342, 432,
Double-angle formulas, 599

, as an infinite series, 581
  as a limit, 520
Earthquake, 613
Eccentricity, 499
Elasticity of a function, 129, 145
Ellipse, 30, 490
  area of, 428
  axes of, 492
  centre of, 31, 492
  directrix of, 499
  eccentricity of, 499
  equations for in Cartesian coordinates, 31, 32, 491, 492
  equations for in polar coordinates, 499
  foci of, 490
  length of, 459, 471, 591
  parametric equations for, 471
Equation of a curve, 14
Error, function, 579, 586
  round-off, 458
  truncation, 547, 584
Escape velocity, 360
Even function, 47, 121, 402
Explicit definition of a function, 122
Exponential function, 605
  antiderivative of, 236
  derivative of, 138
Exponential growth and decay, 403
Extended power rule, 119
Extrema, absolute, 185
  relative, 165–166

Factorial, 114
Fibonacci sequence, 517
First derivative test, 167
First moment, 332, 333, 336
Fluid pressure, 326
Focal chord, 498
Focal radius, 496
Foci, of an ellipse, 490
  of a hyperbola, 493
Focus of a parabola, 489
Folium of Descartes, 130, 472
Fresnel integral, 579
Function, 37
  algebraic, 369
  average value of, 285
  composite, 78
  continuous, 73
  decreasing, 160
  derivative of (see Derivative)
  differentiable, 105
  differential of, 221
  Dirac delta, 82, 365
  discontinuity of, 73
  domain of, 37
  even, 47, 121, 402
  explicit definition of, 122
  exponential, 605

  gamma, 415
  graph of, 39
  greatest integer, 45, 82
  Heaviside unit step, 53, 82, 241
  hyperbolic, 390
  implicit definition of, 123
  increasing, 160
  inverse, 369
  inverse hyperbolic, 397
  inverse trigonometric, 374
  linear, 43
  logarithm, 607
  monotonic, 372
  odd, 47, 121, 402
  one-to-one, 369
  parametric definition of, 467
  periodic, 99
  piecewise constant, 284
  range of, 37
  rational, 44, 369
  signum, 81
  step, 449
  transcendental, 369
  trigonometric, 601
  unit pulse, 82, 365
Fundamental theorem of calculus, 276, 281

Gamma function, 415
General equation of a line, 17
General sine function, 602
Geometric mean, 8
  series, 266, 526
Gompertz growth curve, 408
Graph of a function, 39
Greatest integer function, 45, 82
Guess and fix up, 239

Half-life, 407
Hanging cable, 393, 429
Harmonic series, 525, 527
Heaviside unit step function, 53, 82, 241
Hecht's model, 668
Hooke's law, 321
Horizontal asymptote, 68
Horizontal point of inflection, 173
Hydraulic radius, 665
Hyperbola, 33, 493
  asmptotes of, 34
  axes of, 494
  centre of, 494
  directrix of, 499
  eccentricity of, 499
  equations for in Cartesian coordinates, 33, 35, 494, 495
  equations for in polar coordinates, 499
  foci of, 493
Hyperbolic functions, 390
  derivatives of, 392

Hypergeometric series, 560

Implicit differentiation, 124
    definition of a function, 123
Improper integral, 356, 459
Incentre, 30
Incircle, 30
Inclination, 20, 95
Increasing function, 160
    sequence, 511
Increment, 101, 220
Indefinite integral, 234
Independent variable, 37
Indeterminate forms, 211
Index, of refraction, 384
    of summation, 261
Induction, mathematical, 615
Inequalities, 3
Infinite limit, 65
Infinite series. See series
Infinity, limit at, 68
Inflection point, 172
Instantaneous, rate of change, 90
    velocity, 90
Integral, derivative of, 281, 283
    definite, 272
    improper, 356
    indefinite, 234
    mean value theorem for, 286
Integral test, 531
Integrand, 235
Integration, numerical, 448
    by partial fractions, 434
    by parts, 410
    of power series, 567, 571
    by reproduction, 412
    by substitutions, 248, 399
    by trigonometric substitutions, 422
Intensity, of earthquake, 613
    of sound, 610
Intercept form for a line, 16
Intercepts of a line, 16
Interest, compound, 613, 659
    continuously compounded, 613
Intermediate value theorem, 156
Interval, closed, 2
    of convergence of a power series, 552
    open, 2
Inventory, 355, 663
Inverse, function, 369
    hyperbolic functions, 397
    trigonometric functions, 374
Involute of a circle, 662

Kinetic energy, 225, 324, 591, 656
Kintsch's model, 656

Lagrange form of remainder, 476
Laplace transform, 415

Law of cosines, 597
Law of sines, 598
Left-hand, derivative, 105
    limit, 63
Legendre polynomials, 580
Lemniscate, 151, 479, 484, 665, 670
Length, of a curve, in Cartesian coordinates, 315, 469
    in polar coordinates, 482
    of a line segment, 10
L'Hôpital's rule, 211
Likelihood, 195
Limit comparison test, 535
Limit ratio test, 538
Limit root test, 539
Limiting velocity, 397
Limits, basic properties of, 59, 78
    definition of, 83
    infinite, 65
    at infinity, 68
    intuitive idea of, 57
    one-sided, 63
    of a sequence, 510, 514
Line(s), angle between, 21
    coordinate, 1
    equations of, 15-16
    intercepts of, 16
    parallel, 18
    parametric equations for, 471
    perpendicular, 18
    slope of, 15
    tangent, 91
Lissajous curve, 472
Logarithm function, definition of, 607
    derivative of, 137
Logarithmic differentiation, 142
    spiral, 665, 670
Logistic function, 408
Lower bound for a sequence, 511

Maclaurin series, 563
Magnitude of a number, 5
Marginal, cost, 102, 113
    profit, 102
    revenue, 102, 129
Mass in relativity theory, 46, 591
Mathematical induction, 615
Maximum, absolute, 185
    relative, 165
Mean, arithmetic, 8, 170
    geometric, 8
Mean value theorem, 146, 276
    Cauchy's, 146
    for integrals, 286, 300
Median of a triangle, 13, 22
Midpoint, 11
Minimum, absolute, 185
    relative, 166
Moment of inertia, 343

Moments, first, 332, 333
    second, 343
Momentum, 225
Monomolecular growth curve, 408
Monotonic, function, 372
    sequence, 511
Multiplication of ordinates, 42

Naperian logarithms, 612
Natural logarithms, 612
Newton's iterative procedure, 153, 518, 523
Newton's law of cooling, 408
Newton's universal law of gravitation, 325, 360, 664
Nondecreasing sequence, 511
Nonincreasing sequence, 511
Nonnegative series, 531
Norm of a partition, 271
Normal, line to a curve, 100
    probability density function, 184, 403
Numerical integration, 448, 585
    and improper integrals, 459

Oblate spheroid, 498
Odd function, 47, 121, 402
One-sided, derivative, 105
    limit, 63
Open interval, 2
Ordered pair, 9
Ordinate, 9
Origin, 1, 9
Orthocentre, 30
Orthogonal, curves, 111
    trajectories, 113, 122, 130
Orthonormal, 422
Ovals of Cassini, 130

-series, 533
Pappus, theorem of, 342
Parabola, 23, 489
    directrix of, 489
    eccentricity of, 499
    equation of, 23, 489
    in Cartesian coordinates, 23
    in polar coordinates, 499
    focus of, 489
    vertex of, 489
Parallel axis theorem, 347
Parallel lines, 18
Parametric equations, for a curve, 463
    for a line 471
Parametric rule for differentiation, 467
Partial fractions, 434
Partial sum of a series, 525
Pendulum, 135, 223, 383
Percentage change, 222
Period of a function, 602
Periodic function, 602

Perpendicular, bisector, 22, 30
   lines, 18
Phase shift, 602
Piecewise constant function, 284
Piriform, 666
Planck's law, 159, 219, 591
Point of inflection, 172
Point-slope form of a line, 16
Poiseuille's law, 197
Poisson distribution, 579
Polar axis, 472
Polar coordinates, 472
   area in, 484
   conics in, 499
   curves in, 475
   lengths of curves in, 482
Polar moment of inertia, 347
Pole, 472
Polynomial, 43
Power function, 142
Power rule for differentiation, 98, 119
Power series, 552
   addition of, 570
   binomial, 575
   differentiation of, 567, 571
   integration of, 567, 571
   interval of convergence of, 552
   multiplication of, 573
   radius of convergence of, 554
   sum of, 556, 580
Pressure, fluid, 326
   atmospheric, 331
Principle of mathematical induction, 616
Principal values, 375, 377
Prism, 384
Probability density function, 364, 402, 415
Product formulas for angles, 599
Product rule, 107
Profit, 102, 113
Projectile, 360, 466, 655
Prolate spheroid, 498
Pyramid, 354
Pythagorean theorem, 10

Quadrant, 9
Quadratic, equation, 24
   formula, 24
   function, 43
Quotient rule, 107

Radian measure, 595
Radioactive decay, 79, 406
Radius, of convergence, 554
   of curvature, 664
   of gyration, 347
   hydraulic, 665
Range, of a function, 37
   of a projectile, 27, 655
Rate of change, average, 90
   instantaneous, 90
   relative, 258
Rate, of flow, 269, 351
   of infection, 176
   of reproduction, 657
Rates, related, 203
Rational function, 44, 369
Rayleigh-Jeans law, 591
Real number line, 1
Rectangular, coordinates, 9
   rule, 449
Recursive sequence, 509
Reduction formula, 444
Related rates, 203
Relative, change, 222
   growth rate, 258
   maximum, 165
   miminum, 166
Relativity, mass in theory of, 46, 591
   force and acceleration in, 247
Remainder in Taylor's formula, 562
Removable discontinuity, 81
Repairs and depreciation, 351
Reproduction rate, 657
Revenue, 102
Rhombus, 56
Richter scale, 613
Riemann, integral, 272
   sum, 272
Right-handed, derivative, 105
   limit, 63
Rise, 15
Rolle's theorem, 145
Root-mean-square current, 422
Rose, 484
Round-off error, 458
Run, 15

Scale, 9
Second, derivative test, 174
   moment, 343
Sequence, 508
   bounded, 511
   explicitly defined, 509
   limit of, 510, 514
   monotonic, 511
   of partial sums, 525
   recursively defined, 509
Series, 524
   absolutely convergent, 542
   alternating (see Alternating series)
   approximating sum of, 547
   binomial, 575
   comparison test for, 534
   conditionally convergent, 543
   convergent, 526
   divergent, 526
   geometric, 266, 524
   harmonic, 525, 527
   integral test for, 531
   interval of convergence of, 552
   limit comparison test for, 535
   limit ratio test for, 538
   limit root test for, 539
   Maclaurin, 563
   nonnegative, 531
   th term test for, 528
   -, 533
   partial sum of, 525
   power (see Power series)
   radius of convergence of, 554
   sum of, 526, 580
   Taylor, 563
Shell method for volumes, 309
Sigma notation, 261
Sigmoid survival curve, 409
Sign diagram, 4
Signum function, 81, 241
Simpson's rule, 452
   error of, 455
Sine. See Trigonometric functions
Sine law, 597
Sketch, 40
Slanted Asymptote, 183
Slope, 15
   in polar coordinates, 481
   of tangent line, 91
Slope-intercept form of line, 16
Snell's law, 384
Solid of revolution, 268, 302
Speed, 200
Spheroid, 498
Spiral, logarithmic, 665, 670
Standard position of an angle, 596
Step function, 449
Strophoid, 429, 468, 487
Substitution, in definite integral, 288
   in indefinite integral, 248, 399
   in sigma notation, 263
Successive substitution method, 520, 522, 523
Sum, and difference formulas for angles, 599
   of first   cubes, 266
   of first   integers, 265
   of first   squares, 265
   partial, 525
   of power series, 556
   of series of constants, 526
Summation notation, 261
Surface of revolution, 349
Symmetry of curves, 23, 25

Table of values, 13
Tangent line, 91
Taylor series, 563, 569
Taylor's remainder formula, 561
   integral form, 569

Thurstone's learning curve, 657
Torricelli's law, 667
Tractrix, 663
Transendental function, 369
Translation of graphs, 49
Trapezoidal rule, 450
   error of, 454
Trigonometric functions, 601
   derivatives of, 130
   graphs of, 602
   integrals of, 415
   inverses of, 374
Trigonometric, identities, 597–600
   substitutions, 422
Trochoid, 471
Truncation error, 547, 584
Two-intercept formula for line, 16

Two-point formula for line, 16

Unit circle, 395
Unit hyperbola, 395
Unit pulse function, 82, 365
Units for derivatives, 101
Upper bound for sequence, 511

Variable, dependent, 37
   independent, 37
Velocity, average, 89
   instantaneous, 90, 198, 242
Vertex of a parabola, 489,
Vertical, asymptote, 65
   point of inflection, 173

Volume, of a right circular cone, 305
   of a sphere, 304
   by slicing, 348
Volume of solid of revolution, 302
   by cylindrical shell method, 309
   by disc method, 304
   intuitive idea of, 268
   by washer method, 304
Von Bertalanffy growth curve, 408

Weber-Fechner law, 408
Work, 319

Zeno's paradox, 530